Lecture Notes in Computer Science   **9981**

Commenced Publication in 1973
Founding and Former Series Editors:
Gerhard Goos, Juris Hartmanis, and Jan van Leeuwen

More information about this series at http://www.springer.com/series/7409

Paul Groth · Elena Simperl
Alasdair Gray · Marta Sabou
Markus Krötzsch · Freddy Lecue
Fabian Flöck · Yolanda Gil (Eds.)

The Semantic Web – ISWC 2016

15th International Semantic Web Conference
Kobe, Japan, October 17–21, 2016
Proceedings, Part I

 Springer

Editors

Paul Groth
Elsevier Labs
Amsterdam
The Netherlands

Elena Simperl
University of Southampton
Southampton
UK

Alasdair Gray
Heriot-Watt University
Edinburgh
UK

Marta Sabou
Vienna University of Technology
Vienna
Austria

Markus Krötzsch
Technische Universität Dresden
Dresden
Germany

Freddy Lecue
Accenture Technology Labs
Dublin
Ireland

and

Inria
Sophia Antipolis
France

Fabian Flöck
GESIS-Leibniz Institute for the Social
    Sciences
Cologne
Germany

Yolanda Gil
University of Southern California
Marina del Rey, CA
USA

ISSN 0302-9743   ISSN 1611-3349 (electronic)
Lecture Notes in Computer Science
ISBN 978-3-319-46522-7   ISBN 978-3-319-46523-4 (eBook)
DOI 10.1007/978-3-319-46523-4

Library of Congress Control Number: 2016951984

LNCS Sublibrary: SL3 – Information Systems and Applications, incl. Internet/Web, and HCI

Preface

The International Semantic Web Conference (ISWC) continues to be the premier forum for Semantic Web researchers and practitioners to gather and share exciting new findings and experiences. The community has steadily grown in size and scope over the years, covering many aspects of Semantic Web technologies that lie at the intersection of semantic technologies, data, and the Web. Basic research has renewed importance as an engine of scientific understanding and of new ideas. The broad range of applications of Semantic Web technologies in real settings help us appreciate the accomplishments of the field as well as the limitations and challenges ahead. In addition to building on well-established standards, the community is always generating shared resources and infrastructure. There is a palpable excitement as we witness the Web becoming more machine readable every day.

This volume contains the proceedings of ISWC 2016 with all the papers accepted to the main conference tracks. This year, in addition to the traditional ISWC Research Track we solicited submissions to an Applications Track and a new Resources Track. A new Journal Track was introduced to expand the scope of the conference. The main conference call for papers received 326 responses, over 60 more than the total for the 2015 conference.

The Research Track continues to be the most popular category for submissions. This year, the track solicited novel and significant research contributions addressing theoretical, analytical, empirical, and practical aspects of the Semantic Web. In addition to work building on W3C Semantic Web recommendations (e.g., RDF, OWL, SPARQL, etc.), investigations on other approaches to the intersection of semantics and the Web were encouraged. The track received 212 submissions. After a bidding process, each was reviewed by at least four anonymous members of the Program Committee of the track including one senior Program Committee member. Authors were given a chance to respond to the reviews during an author rebuttal period. The senior Program Committee member was responsible for promoting discussion among the reviewers and making a final recommendation to the program chairs. Papers were discussed in a Program Committee meeting, and the chairs made final determinations about acceptance. These proceedings include the 39 papers that were accepted for presentation at the conference.

The Applications Track solicited submissions exploring the benefits and challenges of applying semantic technologies in concrete, practical applications, in contexts ranging from industry to government and science. The track accepted submissions in three categories: (1) in-use applications providing evidence that there is actual, significant use of the proposed application or tool by the target user group, preferably outside the group that conducted the development; (2) industry applications describing a business case or motivation and demonstrating their impact in the respective industry while ideally positioning the value of the tool or system for the Semantic Web community; (3) emerging applications describing early reports on real-world projects, exposing substantial research contributions and lessons learned in terms of semantics requirements, testing of approaches or infrastructure, and evaluations of early proto-types. The track received a total of 43 submissions and accepted 12. Each submission was reviewed by at least three Program Committee members of the track. Authors had the opportunity to submit a rebuttal to the reviews to clarify questions posed by Program Committee members. The program chairs made final decisions about accep-tance: 23 submissions were emerging applications and seven of them were accepted, 15 were in-use applications and four were accepted, and five were industry applications and one was accepted.

The newly introduced Resources Track sought submissions providing a concise and clear description of a resource and its (expected) usage. Traditional resources are considered to be ontologies, vocabularies, datasets, benchmarks and replication studies, services, and software. These resources are important outputs of any scientific work. Sharing these resources with the research community does not only ensure the reproducibility of results, but also has the benefit of supporting other researchers in their own work. Although high-quality shared resources have a key role and an essential impact on the advancement of a research community, the academic acknowledgement for sharing such resources is low. Therefore, many researchers primarily focus on publishing scientific papers and lack the motivation to share their resources. An additional challenge is that resources are often shared without following best practices, for example, at non-permanent URLs that become unavailable within a few months. The Resources Track aimed to encourage resource sharing following best practices within the Semantic Web community. Besides more established types of resources, the track solicited submissions of new types of resources such as ontology design patterns, crowdsourcing task designs, workflows, methodologies, and protocols and measures. The track received 71 submissions. At least three Program Committee members for the track reviewed each paper using a structured review form that focused on best practices for publishing a resource. After an author rebuttal period and sub-sequent discussion among the reviewers, the program chairs decided on the final acceptance of 24 resource papers that are included in these proceedings and were invited to be presented at the conference.

A new Journal Track was introduced this year to invite presentations at the con-ference about recent papers in the main journals where the community publishes. This inaugural track targeted the *Journal of Web Semantics* and the *Semantic Web Journal*. Authors of papers accepted during the past year that were not previously presented at a main Semantic Web conference could self-nominate their paper. From the 49 self-nominations, the editorial boards of the respective journals chose 12. These papers are not included in these proceedings, but we list full citations of the papers that can be found in the journals.

There are 75 papers included in these proceedings for the Research, Applications, and Resources tracks. The substantial amount of papers in the Resources Track attest to the strong culture in the Semantic Web community of disseminating research products and continuing to extend the pool of shared resources, and doing so beyond ontologies and software.

The conference proceedings were meticulously assembled by Fabian Flöck as proceedings chair, who worked with the chairs to compile all the papers from the authors, produce the table of contents and the front matter, and submit everything to the publishers. Silvio Peroni and Christoph Lange served as metadata chairs, organizing structured descriptions of the contents of the proceedings so they can be made available as semantic content in linked open data format. This year we accepted paper submissions in HTML format, but only received one submission in this format.

The conference included a variety of events that are traditional at ISWC and enrich the opportunities for interaction, learning, and mentoring.

The ISWC 2016 program included invited talks from prominent researchers within and outside the field. Christian Bizer from the University of Mannheim talked about "Is the Semantic Web What We Expected? Adoption Patterns and Content-Driven Challenges." Hiroaki Kitano of from Sony Computer Science Labs, the Okinawa Institute of Science and Technology, and the Systems Biology Institute discussed "Artificial Intelligence to Win the Nobel Prize and Beyond: Creating the Engine for Scientific Discovery." Kathleen McKeown of Columbia University, gave a talk titled "At the Intersection of Data Science and Language."

The Posters and Demos session, chaired by Takahiro Kawamura and Heiko Paulheim, included 55 posters and 47 demos selected among 115 total submissions. A Lightning Talks session offered time to those who wanted to take to the stage briefly to offer late-breaking results, discussion topics, and perspectives.

Thanks to our workshop and tutorial chairs, Chiara Guidini and Heiner Stuckenschmidt, the conference started off with very successful focused and highly interactive events. Five tutorials were held on ontology design patterns, RDF-stream processing, link discovery, Semantic Web for Internet/Web of Things, and SPARQL querying benchmarks. Moreover, 15 workshops were also held to foster discussions on specific topics of interest and to catalyze emerging communities. Also prior to the conference there was a discussion to envision the future of the Semantic Web Challenge.

The doctoral consortium chairs, Philippe Cudre-Mauroux, Riichiro Mizoguchi, and Natasha Noy, reviewed submissions from students still working on their PhD, and organized an event that gave them an opportunity to share their research ideas in a critical but supportive environment, to get feedback from mentors who are senior members of the community, to explore issues related to academic and research careers, and to build relationships with other PhD students from around the world. This program was complemented by activities put together by Abraham Bernstein, Daniel Garijo, and Matthew Horridge as student coordinators, who arranged travel awards, a mentoring lunch, and other informal opportunities for students to meet other members of the Semantic Web community.

The organization of a conference goes well beyond putting together a scientific program. There were many volunteers who worked hard to support the large event that ISWC has become, with hundreds of attendees from all over the world. We are very grateful to Hideaki Takeda, who as local arrangements chair led a skilled team to support the hotel accommodations, arrange conference facilities, develop the conference website, and take care of the myriad of details involved in supporting a scientific conference. We thank all of them for making the conference a fun event and for hosting us in the beautiful city of Kobe. The city's diverse surroundings (from the modern Kobe port to the mountainous Arima hot spring) and cultural heritage (from the Ikuta shrine to Nada Sake breweries) inspired all participants to think more broadly and about the longer-term legacy of their work. We are especially thankful to Ikki Ohmukai and Kouji Kozaki as vice chairs of the local committee and Rathachai Chawuthai as the Web master.

Sponsorship is crucial to support the conference. We would like to thank our sponsorship chairs, Makoto Iwayama and Carlos Pedrinaci, for their thorough and tireless work at arranging sponsorship, and to all of our sponsors for their generous contributions. We would also like to thank Amit Sheth for submitting a proposal to the National Science Foundation that helped secure support for student travel to the conference. The continued support from the National Science Foundation is greatly appreciated.

We are also grateful to the Semantic Web Science Association (SWSA), and in particular to its chair, Natasha Noy, and its treasurer, Guus Schreiber, for their sponsorship and for maintaining all the historical records of previous conferences containing precious data and advice. We are also grateful to Steffen Staab, Ulrich Wechselberger, Jeff Heflin, and the rest of the Organizing Committee of ISWC 2015, who were always at hand to answer our questions and provide thoughtful advice.

Last but not least, we would like to thank Miel Vander Sande, our publicity chair, who took all the announcements to mailing lists, social media, and other outlets to ensure dissemination and awareness of all the conference events.

We hope that these proceedings and the events at ISWC 2016 will contribute to a lasting legacy of this conference for many years to come.

October 2016   Paul Groth & Elena Simperl
                          Program Committee Co-chairs, Research Track
                                   Alasdair Gray & Marta Sabou
                          Program Committee Co-chairs, Resources Track
                                Markus Krötzsch & Freddy Lecue
                          Program Committee Co-chairs, Applications Track
                                                  Yolanda Gil
                                                 General Chair

Conference Organization

General Chair

Yolanda Gil   University of Southern California, USA

Local Chair

Hideaki Takeda   National Institute of Informatics, Japan

Research Track Chairs

Paul Groth   Elsevier Labs, The Netherlands
Elena Simperl   University of Southampton, UK

Resources Track Chairs

Alasdair Gray   Heriot-Watt University, UK
Marta Sabou   Vienna University of Technology, Austria

Applications Track Chairs

Markus Krötzsch   TU Dresden, Germany
Freddy Lecue   Accenture Technology Labs, Ireland/Inria, France

Workshop and Tutorial Chairs

Chiara Ghidini   Fondazione Bruno Kessler, Italy
Heiner Stuckenschmidt   University of Mannheim, Germany

Posters and Demos Track Chairs

Takahiro Kawamura   Japan Science and Technology Agency, Japan
Heiko Paulheim   University of Mannheim, Germany

Journal Track Chairs

Abraham Bernstein   University of Zurich, Switzerland
Pascal Hitzler   Wright State University, USA
Guus Schreiber   VU University Amsterdam, The Netherlands

Doctoral Consortium Chairs

Philippe Cudré-Mauroux   University of Fribourg, Switzerland
Riichiro Mizoguchi   Japan Advanced Institute of Science and Technology,
                              Japan
Natasha Noy   Google, Inc., USA

Proceedings Chair

Fabian Flöck   GESIS - Leibniz Institute for the Social Sciences,
                              Germany

Sponsorship Chairs

Makoto Iwayama   Hitachi, Ltd., Japan
Carlos Pedrinaci   The Open University, UK

Metadata Chairs

Silvio Peroni   University of Bologna, Italy
Christoph Lange   University of Bonn/Fraunhofer IAIS, Germany

Publicity Chair

Miel Vander Sande   Ghent University – iMinds, Belgium

Student Coordinators

Abraham Bernstein   University of Zurich, Switzerland
Daniel Garijo   Universidad Politécnica de Madrid, Spain
Matthew Horridge   Stanford University, USA

Senior Program Committee - Research Track

Harith Alani   The Open University, UK
Philipp Cimiano   Bielefeld University, Germany
Oscar Corcho   Universidad Politécnica de Madrid, Spain
Philippe Cudré-Mauroux   University of Fribourg, Switzerland
Fabien Gandon   Inria, France
Birte Glimm   University of Ulm, Germany
Lalana Kagal   Massachusetts Institute of Technology, USA
Craig Knoblock   University of Southern California, USA
Vanessa Lopez   IBM Research, Ireland
David Martin   Nuance Communications, USA
Diana Maynard   University of Sheffield, UK

Axel-Cyrille Ngonga Ngomo	University of Leipzig, Germany
Axel Polleres	Vienna University of Economics and Business - WU Wien, Austria
Sebastian Rudolph	Technische Universität Dresden, Germany
Stefan Schlobach	Vrije Universiteit Amsterdam, The Netherlands
Mari Carmen Suárez-Figueroa	Universidad Politénica de Madrid, Spain
Maria Esther Vidal	Universidad Simon Bolivar, Venezuela

Program Committee - Research Track

Maribel Acosta	Karlsruhe Institute of Technology, Germany
Muhammad Intizar Ali	National University of Ireland, Ireland
Faisal Alkhateeb	Yarmouk University, Jordan
Pramod Anantharam	Bosch Research and Technology Center, USA
Manuel Atencia	Université Grenoble Alpes and Inria, France
Medha Atre	Unaffiliated
Nathalie Aussenac-Gilles	IRIT CNRS, France
Jie Bao	Memect, China
Payam Barnaghi	University of Surrey, UK
Christian Bizer	University of Mannheim, Germany
Eva Blomqvist	Linköping University, Sweden
Kalina Bontcheva	University of Sheffield, UK
Paolo Bouquet	University of Trento, OKKAM srl, Italy
Loris Bozzato	Fondazione Bruno Kessler, Italy
Adrian M.P. Brasoveanu	MODUL Technology GmbH, Germany
Carlos Buil Aranda	Universidad Técnica Federico Santa María, Chile
Paul Buitelaar	National University of Ireland, Ireland
Grégoire Burel	The Open University, UK
Andrea Calí	University of London, Birkbeck College, UK
Jean-Paul Calbimonte	HES-SO Valais, EPFL, France
Diego Calvanese	Free University of Bozen-Bolzano, Italy
Amparo E. Cano	Aston University, UK
Iván Cantador	Universidad Autónoma de Madrid, Spain
Irene Celino	CEFRIEL, Italy
Davide Ceolin	University Amsterdam, The Netherlands
Pierre-Antoine Champin	LIRIS, France
Gong Cheng	Nanjing University, China
Michael Compton	CSIRO, Australia
Sam Coppens	Autodesk, Inc., Ireland
Isabel Cruz	University of Illinois at Chicago, USA
Bernardo Cuenca Grau	University of Oxford, UK
Edward Curry	Digital Enterprise Research Institute, Ireland
Claudia d'Amato	University of Bari, Italy
Mathieu D'Aquin	The Open University, UK

Danica Damljanovic	Pure AI, UK
Brian Davis	National University of Ireland, Ireland
Daniele Dell'Aglio	DEIB, Politecnico di Milano, Italy
Emanuele Della Valle	DEIB, Politecnico di Milano, Italy
Elena Demidova	University of Southampton, UK
Stefan Dietze	L3S Research Center, Germany
Ying Ding	Indiana University, USA
John Domingue	The Open University, UK
Jérôme Euzenat	Inria and Université Grenoble Alpes, France
Nicola Fanizzi	Università di Bari, Italy
Anna Fensel	University of Innsbruck, Austria
Miriam Fernandez	The Open University, UK
Lorenz Fischer	Swisscom AG, Switzerland
Achille Fokoue	IBM Research, USA
Adam Funk	University of Sheffield, UK
Olaf Görlitz	Cognizant Technology Solutions, Germany
Aldo Gangemi	Université Paris 13 and CNR-ISTC, France
José María García	University of Seville, Spain
Raúl García-Castro	Universidad Politécnica de Madrid, Spain
Anna Lisa Gentile	University of Mannheim, Germany
Jose Manuel Gomez-Perez	Expert System, Spain
Tudor Groza	The Garvan Institute of Medical Research, Australia
Michael Gruninger	University of Toronto, Canada
Christophe Guéret	BBC, UK
Giancarlo Guizzardi	Federal University of Espirito Santo, Brazil
Claudio Gutierrez	Universidad de Chile, Chile
Peter Haase	metaphacts, Germany
Harry Halpin	World Wide Web Consortium, USA
Andreas Harth	Karlsruhe Institute of Technology, Germany
Olaf Hartig	Linköping University, Sweden
Tom Heath	Open Data Institute, UK
Pascal Hitzler	Wright State University, USA
Rinke Hoekstra	VU Amsterdam/University of Amsterdam, The Netherlands
Aidan Hogan	Universidad de Chile, Chile
Andreas Hotho	University of Würzburg, Germany
Geert-Jan Houben	TU Delft, The Netherlands
Wei Hu	Nanjing University, China
Eero Hyvönen	Aalto University, Finland
Luis-Daniel Ibáñez	University of Southampton, UK
Krzysztof Janowicz	University of California, Santa Barbara, USA
Ernesto Jimenez-Ruiz	University of Oxford, UK
Hanmin Jung	Korea Institute of Science and Technology Information, South Korea
Benedikt Kämpgen	FZI Forschungszentrum Informatik, Germany
Mark Kaminski	University of Oxford, UK

David Karger	Massachusetts Institute of Technology, USA
Hong-Gee Kim	Seoul National University, South Korea
Matthias Klusch	DFKI, Germany
Jacek Kopecky	University of Portsmouth, UK
Manolis Koubarakis	National and Kapodistrian University of Athens, Greece
Tobias Kuhn	Vrije Universiteit Amsterdam, The Netherlands
Werner Kuhn	University of California, Santa Barbara, USA
Steffen Lamparter	Siemens AG, Corporate Technology, Germany
Agnieszka Lawrynowicz	Poznań University of Technology, Poland
Chengkai Li	University of Texas at Arlington, USA
Juanzi Li	Tsinghua University, China
Nuno Lopes	TopQuadrant, Inc., Ireland
Chun Lu	SEPAGE S.A.S, France
Markus Luczak-Roesch	Victoria University of Wellington, New Zealand
Maria Maleshkova	Karlsruhe Institute of Technology, Germany
Erik Mannens	iMinds – Ghent University, Belgium
Robert Meusel	University of Mannheim, Germany
Peter Mika	Yahoo! Research, UK
Riichiro Mizoguchi	Japan Advanced Institute of Science and Technology, Japan
Dunja Mladenic	Jozef Stefan Institute, Slovenia
Luc Moreau	University of Southampton, UK
Boris Motik	University of Oxford, UK
Enrico Motta	The Open University, UK
Nadeschda Nikitina	Oxford University, UK
Andriy Nikolov	fluid Operations AG, Germany
Lyndon Nixon	MODUL University Vienna, Austria
Beng Chin Ooi	National University of Singapore, Singapore
Massimo Paolucci	DoCoMo Euro labs, Germany
Bijan Parsia	University of Manchester, UK
Peter Patel-Schneider	Nuance Communications, USA
Terry Payne	University of Liverpool, UK
Carlos Pedrinaci	The Open University, UK
Silvio Peroni	DASPLab, DISI, University of Bologna, Italy
Dimitris Plexousakis	Institute of Computer Science, FORTH, Greece
Guilin Qi	Southeast University, China
Ganesh Ramakrishnan	IIT Bombay, India
Maya Ramanath	IIT Delhi, India
Achim Rettinger	Karlsruhe Institute of Technology, Germany
Giuseppe Rizzo	ISMB, Italy
Dumitru Roman	SINTEF/University of Oslo, Norway
Riccardo Rosati	Sapienza Università di Roma, Italy
Marco Rospocher	Fondazione Bruno Kessler, Italy
Matthew Rowe	Lancaster University, UK

Harald Sack	Hasso Plattner Institute/University of Potsdam, Germany
Hassan Saif	The Open University, UK
Francois Scharffe	3Top, USA
Ansgar Scherp	Kiel University and ZBW – Leibniz Information Center for Economics, Germany
Marco Schorlemmer	IIIA-CSIC, Spain
Daniel Schwabe	PUC-Rio, Brazil
Juan Sequeda	Capsenta, USA
Estefania Serral	KU Leuven, Belgium
Michael Sintek	DFKI GmbH, Germany
Monika Solanki	University of Oxford, UK
Dezhao Song	Thomson Reuters, USA
Steffen Staab	University of Koblenz-Landau, Germany and University of Southampton, UK
Markus Strohmaier	University of Koblenz-Landau, Germany
Gerd Stumme	University of Kassel, Germany
Jing Sun	The University of Auckland, New Zealand
Vojtěch Svátek	University of Economics, Czech Republic
Pedro Szekely	University of Southern California, USA
Annette Ten Teije	Vrije Universiteit Amsterdam, The Netherlands
Matthias Thimm	University of Koblenz-Landau, Germany
Krishnaprasad Thirunarayan	Wright State University, USA
Nicolas Torzec	Yahoo, USA
Sebastian Tramp	eccenca, Germany
Raphaël Troncy	EURECOM, France
Anni-Yasmin Turhan	Technische Universität Dresden, Germany
Jacopo Urbani	Vrije Universiteit Amsterdam, The Netherlands
Victoria Uren	Aston University, UK
Marieke van Erp	Vrije Universiteit Amsterdam, The Netherlands
Frank van Harmelen	Vrije Universiteit Amsterdam, The Netherlands
Jacco van Ossenbruggen	CWI VU Amsterdam, The Netherlands
Ruben Verborgh	Ghent University – iMinds, Belgium
Natalia Villanueva-Rosales	University of Texas at El Paso, USA
Haofen Wang	East China University of Science and Technology, China
Kewen Wang	Griffith University, Australia
Zhichun Wang	Beijing Normal University, China
Yong Yu	Shanghai Jiao Tong University, China
Fouad Zablith	American University of Beirut, Lebanon
Sergej Zerr	University of Southampton, UK
Qingpeng Zhang	Rensselaer Polytechnic Institute, USA

Additional Reviewers - Research Track

Nitish Aggarwal
Albin Ahmeti
Saud Aljaloud
Martin Becker
Aurélien Bénel
David Berry
Georgeta Bordea
Stefano Bortoli
Stefano Braghin
Janez Brank
Michel Buffa
Elena Cabrio
Xuezhi Cao
Claudia Carapelle
David Carral
Roberto Confalonieri
Olivier Corby
Julien Corman
Minh Dao-Tran
Jérôme David
Ronald Denaux
Djellel Eddine Difallah
Anastasia Dimou
Jiwei Ding
Fajar J. Ekaputra
Basil Ell
Michael Färber
Daniel Faria
Yasmin Fathy
Jean-Philippe Fauconnier
Javier D. Fernández
Mariano Fernández-López
Giorgos Flouris
Marvin Frommhold
Irini Fundulaki
Andrés García-Silva
Camilo Garrido
Alain Giboin
Kalpa Gunaratna
Karl Hammar
Tom Hanika

Matthias Hartung
Sona Hasani
Naeemul Hassan
Daniel Hernandez
Yingjie Hu
Robert Isele
Daniel Janke
Stéphane Jean
Soufian Jebbara
Nazifa Karima
Mario Karlovcec
Yevgeny Kazakov
Felix Leif Keppmann
Robin Keskisärkkä
Christoph Kling
Magnus Knuth
Haris Kondylakis
Patrick Koopmann
Aljaz Kosmerlj
Nenad Krdzavac
Adila A. Krisnadhi
Sumant Kulkarni
Sungin Lee
Oliver Lehmberg
Tatiana Lesnikova
Mohamed Nadjib Mami
Nandana
Mihindukulasooriya
Maja Milicic Brandt
Seyed Iman Mirrezaei
Aditya Mogadala
Alexandre Monnin
Jose Mora
Raghava Mutharaju
Thomas Niebler
Charalampos Nikolaou
Nikolay Nikolov
Chifumi Nishioka
Francesco Osborne
Yaser Oulabi
Guillermo Palma

Panayiotis Papadakos
Sujan Perera
Minh Pham
Patrick Philipp
Julien Plu
María Poveda-Villalón
Freddy Priyatna
Alessandro Provetti
Lin Qiu
Kan Ren
Martin Ringsquandl
Dominique Ritze
Markus Rokicki
Marvin Schiller
Lukas Schmelzeisen
Andreas Schmidt
Xin Shuai
Hala Skaf-Molli
Cinzia Incoronata Spina
Simon Steyskal
Dina Sukhobok
Steffen Thoma
Veronika Thost
Pierpaolo Tommasi
Riccardo Tommasini
Serena Villata
Joerg Waitelonis
Simon Walk
Sebastian Walter
Wei Wang
Zhe Wang
Sanjaya Wijeratne
Ian Wood
Tianxing Wu
Yuxin Ye
Ran Yu
Veruska Zamborlini
Gensheng Zhang
Lei Zhang
Linhong Zhu

Program Committee - Resources Track

Maribel Acosta	Karlsruhe Institute of Technology, Germany
Ahmet Aker	Sheffield University, UK
Diana Bental	Heriot-Watt University, UK
Abraham Bernstein	University of Zurich, Switzerland
Eva Blomqvist	Linköping University, Sweden
Sarven Capadisli	University of Bonn, Germany
Tim Clark	Massachusetts General Hospital/Harvard Medical School, USA
Oscar Corcho	Universidad Politécnica de Madrid, Spain
Philippe Cudre-Mauroux	University of Fribourg, Switzerland
Claudia d'Amato	University of Bari, Italy
Mathieu D'Aquin	The Open University, UK
Victor de Boer	VU Amsterdam, The Netherlands
Daniele Dell'Aglio	BEIB, Politecnico di Milano, Italy
Gianluca Demartini	University of Sheffield, UK
Stefan Dietze	L3S Research Center, Germany
Ying Ding	Indiana University, USA
Michel Dumontier	Stanford University, USA
Fajar J. Ekaputra	Technische Universität Wien, Austria
Miriam Fernandez	Knowledge Media Institute, UK
Aldo Gangemi	Université Paris 13 and CNR-ISTC, France
Jose Manuel Gomez-Perez	Expert System, Spain
Alejandra Gonzalez-Beltran	University of Oxford, UK
Pascal Hitzler	Wright State University, USA
Robert Hoehndorf	King Abdullah University of Science and Technology, Saudi Arabia
Rinke Hoekstra	University of Amsterdam, The Netherlands
Matthew Horridge	Stanford University, USA
Andreas Hotho	University of Würzburg, Germany
Antoine Isaac	Europeana and Vrije Universiteit Amsterdam, The Netherlands
Simon Jupp	European Bioinformatics Institute, UK
Elmar Kiesling	TU Wien, Austria
Olga Kovalenko	TU Wien, Austria
Tobias Kuhn	VU University Amsterdam, The Netherlands
Markus Luczak-Roesch	University of Southampton, UK
Maria Maleshkova	Karlsruhe Institute of Technology, Germany
Fiona McNeill	Heriot-Watt University, UK
Chris Mungall	Lawrence Berkeley National Laboratory, USA
Vinh Nguyen	Wright State University, USA
Bijan Parsia	University of Manchester, UK
Heiko Paulheim	University of Mannheim, Germany
Silvio Peroni	University of Bologna, Italy
Valentina Presutti	STLab (ISTC-CNR), Italy

Giuseppe Rizzo	ISMB, Italy
Mariano Rodríguez Muro	IBM Research, USA
Satya Sahoo	Case Western Reserve University, USA
Cristina Sarasua	University Koblenz - Landau, Germany
Stefan Schlobach	Vrije Universiteit Amsterdam, The Netherlands
Jodi Schneider	University of Pittsburgh, USA
Stian Soiland-Reyes	University of Manchester, UK
Valentina Tamma	University of Liverpool, UK
Krishnaprasad Thirunarayan	Wright State University, USA
Ramine Tinati	University of Southampton, UK
Raphaël Troncy	EURECOM, France
Natalia Villanueva-Rosales	University of Texas at El Paso, USA
Simon Walk	Graz University of Technology, Austria
Peter Wetz	TU Wien, Austria
Marcia Zeng	Kent State University, USA
Jun Zhao	University of Oxford, UK

Additional Reviewers - Resources Track

Reihaneh Amini	Kalpa Gunaratna	Laura Rettig
Sebastian Bader	Karl Hammar	Daniel Schlör
David Carral	Robin Keskisärkkä	Saeedeh Shekarpour
Lu Chen	Adila Krisnadhi	Ruben Verborgh
Maxime Déraspe	Sarasi Lalithsena	Ran Yu
Djellel Difallah	María Poveda-Villalón	Amrapali Zaveri
Jhonatan Garcia	Sambhawa Priya	
Nuria García-Santa	José Redondo-García	

Program Committee - Applications Track

Anupriya Ankolekar	Hewlett Packard Labs, USA
Sören Auer	University of Bonn and Fraunhofer IAIS, Germany
Christian Bizer	University of Mannheim, Germany
Oscar Corcho	Universidad Politécnica de Madrid, Spain
Olivier Curé	Université Paris-Est LIGM, France
Mathieu D'Aquin	The Open University, UK
John Davies	BT, UK
Chiara Del Vescovo	British Broadcasting Corporation, UK
Mauro Dragoni	Fondazione Bruno Kessler - FBK-IRST, Italy
Fabien Gandon	Inria, France
Peter Haase	metaphacts, Germany
Aidan Hogan	Universidad de Chile, Chile
Krzysztof Janowicz	University of California, Santa Barbara, USA

Pavel Klinov	Complexible Inc., Germany
Matthias Klusch	DFKI, Germany
Craig Knoblock	University of Southern California, USA
Danh Le Phuoc	Technische Universität Berlin, Germany
Vanessa Lopez	IBM Research, Ireland
Despoina Magka	Facebook, USA
Maria Maleshkova	Karlsruhe Institute of Technology, Germany
Jeff Z. Pan	University of Aberdeen, UK
Peter Patel-Schneider	Nuance Communications, USA
Carlos Pedrinaci	The Open University, UK
Juan F. Sequeda	Capsenta Labs, USA
Dezhao Song	Thomson Reuters, USA
Kavitha Srinivas	IBM Research, USA
Federico Ulliana	Université Montpellier 2, France
Jürgen Umbrich	Vienna University of Economy and Business (WU), Austria
Jacopo Urbani	Vrije Universiteit Amsterdam, The Netherlands
Denny Vrandečić	Google, USA
Peter Yeh	Nuance Communications, USA

Additional Reviewers - Applications Track

Valerio Basile	Ali Khalili	Diego Reforgiato
Michel Buffa	Gregor Leban	Simon Scerri
Alistair Duke	Lionel Médini	Jason Slepicka
Majid Ghasemi-Gol	Michael Mrissa	Charese Smiley
Irlan Grangel	Niklas Petersen	Tobias Weller

Sponsors

IBM®

semantic software
LIBERATE YOUR DATA

ORACLE®

IOS
Press

RECRUIT
Recruit Technologies Co.,Ltd.

FUJĬTSU

g00

bla**z**egraph™   metaphacts

HITACHI
Inspire the Next

®**Rakuten**
Institute of Technology

Y**A**HOO!
JAPAN

Google

Contents – Part I

Contents – Part II

Applications

Research

Structuring Linked Data Search Results Using Probabilistic Soft Logic

Duhai Alshukaili$^{(\boxtimes)}$, Alvaro A.A. Fernandes, and Norman W. Paton

School of Computer Science, University of Manchester,
Oxford Road, Manchester M13 9PL, UK
{dhahi.alshekaili,a.fernandes,norman.paton}@manchester.ac.uk

Abstract. On-the-fly generation of integrated representations of Linked Data (LD) search results is challenging because it requires successfully automating a number of complex subtasks, such as structure inference and matching of both instances and concepts, each of which gives rise to uncertain outcomes. Such uncertainty is unavoidable given the semantically heterogeneous nature of web sources, including LD ones. This paper approaches the problem of structuring LD search results as an evidence-based one. In particular, the paper shows how one formalism (viz., probabilistic soft logic (PSL)) can be exploited to assimilate different sources of evidence in a principled way and to beneficial effect for users. The paper considers syntactic evidence derived from matching algorithms, semantic evidence derived from LD vocabularies, and user evidence, in the form of feedback. The main contributions are: sets of PSL rules that model the uniform assimilation of diverse kinds of evidence, an empirical evaluation of how the resulting PSL programs perform in terms of their ability to infer structure for integrating LD search results, and, finally, a concrete example of how populating such inferred structures for presentation to the end user is beneficial, besides enabling the collection of feedback whose assimilation further improves search result presentation.

Keywords: Linked data search · Linked data integration

1 Introduction

The idea of linked data (LD) underpins the attempt to transfer the strengths of the web of documents to data: data can be shared, searched for and browsed, building on standards for data identification, description and linking that provide low barriers to entry, facilitating new applications in areas such as data science and open government. Along with a basic model that publishing, it can be anticipated that popular types of tool can transfer successfully from the web of documents to the web of data, supporting activities such as searching and browsing. However, data and documents have important differences, and direct translations of techniques that have been successful in the web of documents can seem rather less intuitive in the web of data.

© Springer International Publishing AG 2016
P. Groth et al. (Eds.): ISWC 2016, Part I, LNCS 9981, pp. 3–19, 2016.
DOI: 10.1007/978-3-319-46523-4_1

For example, keyword search has been an important technology in the web of documents: searches give rise to ranked lists of documents, through which the user can browse. In such approaches, the user accesses the published document directly. Such an approach has been transferred to the web of data, as represented by LD search engines such as Falcons [3], Sindice [11] and Swoogle [5]. Although LD search engines are useful, their results, which take the form of collections of RDF resources, present to users the data *as published*. Thus, LD search engines tackle the question *What resources are out there that match the search?*, and not so much *What data is out there that matches the search?*. For example, searching for "Godfather actors" returns results (see Fig. 1), among others, that are about two distinct *films* in whose name the string Godfather occurs, as well as about *actors* that have appeared in those films. Assuming for the moment that the user is looking for data about actors in the US film named The Godfather (en.wikipedia.org/wiki/The_Godfather), released in 1972, filtering and structuring the results in different tables, as shown in Fig. 2, would be desirable since it distinguishes between films and actors and provides structure to the presentation of films and of actors that have appeared in them.

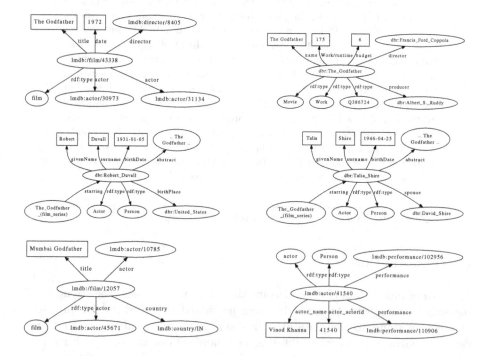

Fig. 1. Example LD search results for the term "Godfather actors"

As such, there is an opportunity to complement the work to date on search engines for LD by devising techniques to infer a structure from the returned resources. This would insulate the user from many of the heterogeneities that LD resources typically exhibit due to weak constraints on publication.

Movie				
name	date	director	budget	runtime
The Godfather	1972	Francis Ford Coppola	6	172

Actor				
givenName	surname	birthDate	birthPlace	spouse
Robert	Duvall	1931-01-05	United States	null
Talia	Shire	1946-04-25	null	David Shire

Fig. 2. Results in Fig. 1 integrated and reported as tables

The integration of search results by means of structure inference is by no means a straightforward task because there is no a *priori* target model to which the results are to be mapped, the available evidence about the relationships between different resources may be partial or misleading, and the integration must be carried out automatically, or at most with small amounts of feedback from the user. Our hypothesis is that LD search result integration can benefit from (i) combining different sources of evidence to inform the integration process, (ii) systematically managing the uncertainty associated with these sources, and (iii) making cost-effective use of feedback. To test this hypothesis, this paper investigates the use of probabilistic soft logic (PSL) [1] for combining three sources of evidence, viz., syntactic matching, domain ontologies and user feedback.

The contributions of this paper are:

1. The characterization of the LD search result integration task as one in which different sources of (partial and uncertain) evidence are brought together to inform decision making.
2. The instantiation of this characterization of the problem using PSL to combine, in a principled and uniform way, evidence from syntactic matching, from domain ontologies and from user feedback.
3. The empirical evaluation of this approach, in which it is shown how the principled, uniform use of different types of evidence improves integration quality for the end user.

2 Related Work

LD search engines such as Falcons [3] and Sindice [11] return a list of resources that match the given search terms ranked by their estimated relevance but without attempting identify their underlying structure. Our contributions build on such search engines as we infer a tabular structure over the returned results. Sig.ma [15] also builds on returned search results. Differently from us, it assumes that the search terms describe one entity (hence a singleton, never a set). Similarly to us, Sig.ma aims to build a profile (essentially a property graph, not a tabular structure as we do) that characterises the searched-for entity on the basis of the results returned by Sindice. Sig.ma uses heuristics whose only input is syntactic evidence, and accumulates information about the entity from different LD resources without using a principled evidence assimilation technique.

In contrast, we use a probabilistic framework to uniformly assimilate different kinds of evidence in a principled manner. Sig.ma does respond to feedback: the user can decide whether a resource is allowed to contribute data to the generated profile or not. However, such feedback does not affect future searches, whereas, in our approach, feedback given is assimilated as evidence and therefore influences positively the quality of future interactions.

Inferring a structure from instance-level RDF data, but not search results *per se*, has received much attention by the research community. However, this activity has focussed on finding ontology axioms such as range restriction [4,16], domain restriction [4,16], subsumption [14,16,18], and equivalence axioms [6,14,16] that apply to the data. In contrast, our approach uses PSL to populate data tables with returned search results, which are more heterogeneous, structurally and semantically, than the data the approaches cited normally target. For example, searching for a film title (e.g. The Godfather) with Falcons returns results from at least three LD resources: DBPedia, LinkedMDB and Yago. This makes the structure inference task harder, as conflicts between the vocabularies used by different datasets need to be resolved.

Given input ontologies, inductive logic programming, a statistical relational learning paradigm, is used in [6,14] to extend the ontologies with new axioms given data defined in terms of the given ontologies. In contrast, no background knowledge is needed in some other approaches. Hierarchical clustering was used in [4] to find schematic patterns, whereas association rule mining [16] and Bayesian networks [18] have also been used to infer descriptions of concepts. Our contributions also use background knowledge in the form of a PSL program resulting from a supervised learning stage in which logical rules are assigned weights. However, we view ontologies as semantic evidence, since we aim to populate a tabular structure of search results rather than extend a given ontology.

Recently, there has been a growing interest in applying statistical relational learning (SRL) techniques such Markov logic networks (MLNs) and PSL to problems where relational dependencies play a crucial role (e.g., inference over social network data) or where principled integration requires assimilation of multiple sources of evidence. In the Semantic Web context, Niepert *et al.* [10] exploited MLNs to match ontologies by modelling axioms as hard constraints that must hold for every alignment. They use similarities of concepts in the input ontologies as evidence, taking them as a seed alignment, and apply integer linear programming to maximise the probability of alignment based on the seed alignments as constrained by the ontological axioms. Their work showed that syntactic and semantic evidence can be combined using SRL techniques. Differently from them, we do not represent ontological knowledge as hard constraints since we target a tabular structure. In our approach, ontologies are used as evidence for inferring an instance of out target structure in the returned search results. This means that we treat ontological knowledge as uncertain. Similarly to us, Pujara et al. [13] used PSL to infer knowledge graphs about real-world entities from noisy extractors, and, in particular, the assimilation of lexical similarities and ontological evidence proves to be crucial in de-noising extracted graphs.

3   A Brief Introduction to PSL

In this paper, the formalism used to infer an integrated structure over returned LD search results is PSL [1], a general-purpose learning and inference framework for reasoning with uncertainty in relational domains.

In Sect. 4, we contribute a set of PSL rules and a target metamodel. The rules express probabilistic relationships between triples in search results. After the rules have been converted into a PSL program, the PSL implementation processes the returned results so as to instantiate the target metamodel, thereby yielding an integrated, structured view over the results.

A PSL program is a set of weighted first-order-logic formulae of the form $w : A \leftarrow B$, where w are non-negative weights, B is a conjunction of literals and A is a single literal. Consider the following example PSL program (adapted from [8], where rules are written as $w : B \rightarrow A$):

$$0.3 : votesFor(B, P) \overset{\sim}{\leftarrow} friend(B, A) \tilde{\wedge} votesFor(A, P) \tag{1}$$

$$0.8 : votesFor(B, P) \overset{\sim}{\leftarrow} spouse(B, A) \tilde{\wedge} votesFor(A, P) \tag{2}$$

The semantics, including the tilde-capped connectives, are briefly explained below but, intuitively, this PSL program states that, given any individuals a, b and p, instantiating (resp.) the logical variables A, B and P, a claim is made that if b is either a friend or a spouse of a and a votes for party p, then, with some likelihood, b votes for p. The respective weights assert that the influence of spouses on what party b votes for is larger than that of friends.

Softness in PSL arises from the fact that truth values are drawn from the continuous interval $[0, 1]$, i.e., if $\mathcal{A}$ is the set $\{a_1, \ldots, a_n\}$ of atoms, then an interpretation is a mapping $I : \mathcal{A} \rightarrow [0, 1]^n$, rather than only to the extreme values, i.e., either 0 (denoting falsehood) or 1 (denoting truth).

To capture the notion that different claims (expressed as rules) may have different likelihoods, a probability distribution is defined over interpretations, as a result of which rules that have more supporting instantiations are more likely. In the case of Rules (1) and (2) above, interpretations where the vote of an individual agrees with the vote of many friends, i.e., satisfies many instantiations of Rule (1) are preferred over those that do not. Moreover, where a tradeoff arises between using agreement with friends or with spouses, the latter is preferred due to the higher weight of Rule (2).

Determining the degree to which a ground rule is satisfied from its constituent atoms requires relaxing (with respect to the classical definitions) the semantics of the logical connectives for the case where terms take soft truth-values. To formalize this, PSL uses the Łukasiewicz t-norm and its corresponding co-norm, which are exact (i.e., coincide with the classical case) for the extremes and provide a consistent mapping for all other values. The relaxed connectives are notated with a capping tilde, i.e., $\tilde{\wedge}$, $\tilde{\vee}$, and $\overset{\sim}{\neg}$, with $A \overset{\sim}{\leftarrow} B \equiv \overset{\sim}{\neg} B \tilde{\vee} A$.

Atoms in a PSL rule can be user-defined. Thus, a unary predicate `IsDictWord` might be defined to have truth value 1 if the individual of which it is predicated is a dictionary word and 0 otherwise. Atoms in a PSL rule can also capture user-defined relationships between sets of individuals. Thus, the truth value of S.interests SimSetEq[] T.interests is the similarity of the respective sets of interests of the individuals S and T as computed by the user-provided definition of set similarity SimSetEq[].

In summary, the basic idea is to view logical formulas as soft constraints on their interpretations. If an interpretation does not satisfy a formula, it is taken as less likely, but not necessarily impossible. Furthermore, the more formulas an interpretation satisfies, the more likely it is. In PSL, this quantification is grounded on the relative weight assigned to each formula. The higher the weight, the greater the difference between the likelihood of an interpretation that satisfies a formula and the likelihood of one that does not.

The key tasks supported by PSL implementations are learning and inference. The rules in a PSL program can be either given (i.e., asserted) or learned from sample (or training) data. In this paper, rules (shown later) are given. Furthermore, the weight of each rule can also be given or learned from sample data. In this paper, rule weights are learned from sample data (as detailed later). The PSL weight learning process takes a PSL program (possibly with initial weights), a specification of both evidence and query predicates, and sample data. A predicate is said to be an *evidence predicate* if all its ground atoms have known truth values by observation. A predicate is said to be a *query predicate* if one or more of its ground atoms have unknown truth values. The process returns a relative non-negative weight for each rule. A positive weight denotes that a rule is supported by the sample data whereas the magnitude indicates the strength of that support. A weight of zero denotes lack of support in the sample data for that rule (but since weights are relative it does not entail impossibility).

The main purpose of a PSL program is inference. The PSL inference process takes a PSL program, evidence as data, and a query. It then computes the most probable assignment of soft truth-values to the query, i.e., the probability that the given query atom is true. A major strength of PSL is that implementations perform inference by constructing a corresponding convex optimization problem for which a solution can be efficiently computed even for large-size inputs. For detailed descriptions of the PSL learning and inference algorithms, see [1].

4   Structure Inference Over LD Search Results

Our approach can be briefly summarized as follows. Firstly, we have defined a metamodel (see Fig. 3) that characterizes the type of structure to be inferred (i.e., populated with resources returned by LD searches). Every instances of the metamodel is a tabular representation of search results that we refer to as a *report*. Thus, given the search returns, our goal is to infer that some resources are entity types and some are entities (i.e., elements in the extent of an inferred entity type), some other resources are properties of some inferred entity type and some are property values (i.e., elements in the domain of an inferred property).

Next, we have expressed the semantics of the metamodel as a set of unweighted PSL rules, which we refer to as *the baseline model*, denoted by $\mathbb{B}$

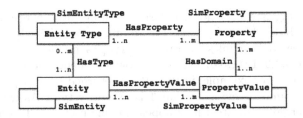

Fig. 3. The metamodel for population with LD search results.

(see Rules R1-R17 in Fig. 4). The rules in $\mathbb{B}$ build on syntactic evidence alone. In order to assimilate semantic evidence, we have added rules to $\mathbb{B}$ that build on additional ontological evidence. We refer to this as *the semantic model*, denoted by $\mathbb{S}$, where $\mathbb{S} \supset \mathbb{B}$ (see examples in Rules R18-R23 in Fig. 4). Next, using sample data from search results and from ontologies, we use a PSL implementation (github.com/linqs/psl) to learn weights for the rules in $\mathbb{S}$ and obtain the corresponding PSL program $PSL(\mathbb{S})$. This also yields $PSL(\mathbb{B})$, i.e., the subset of $PSL(\mathbb{S})$ where we only retain those rules that occur in $\mathbb{B}$. Then, given the returned results of any LD search, we can use PSL inference from $PSL(\mathbb{S})$ to instantiate the metamodel with the most probable characterization of the returned resources. In order to assimilate evidence from user feedback, we have defined a separate set of rules, which we refer to as *the feedback model*, denoted by $\mathbb{F}$ (see examples in Rules R24-R27 in Fig. 4). To obtain the weights for the rules in $\mathbb{F}$ and thus obtain the PSL program $PSL(\mathbb{F})$, we generated synthetic feedback and used the same PSL implementation. The PSL program $PSL(\mathbb{S}) \cup PSL(\mathbb{F})$ integrates LD search results given syntactic, semantic and feedback evidence.

The Baseline Model: Assimilating Syntactic Evidence. Our approach is grounded on the hypothesis that, with some likelihood, there exist relationships between RDF triples returned by search and instantiations of our metamodel that can be captured by a PSL program. Correspondingly, the first step in the construction of the baseline model $\mathbb{B}$ is to map, as a purely syntactic operation, RDF triples onto predicates that assert membership of metamodel constructs. For example, on the evidence of an RDF triple such as U `rdf:type` U', where U and U' are URIs, we may conclude that `RDFIsInstOf(U,U')`, i.e., that there is an instance-of relationship between an individual U and a type U'.

Such mappings are inherently uncertain (i.e., only hold with some likelihood) because it is impossible to capture a publisher's intentions. So, e.g., if `RDFType(Q386724)` is not a user-level type then perhaps it should not be considered to denote an entity type, and therefore not suitable to be reported as a table to the end user. Another source of uncertainty is the search itself (i.e., its associated precision and recall). For example, the RDF individual labeled `"Vinod Khana"` may be returned but should not be listed in a table describing the actors of the 1972 US-produced Godfather film. Rule uncertainty is reflected in rule weights that are learned from sample data as described in Sect. 5.

The baseline model $\mathbb{B}$ consists of Rules R1-R17 in Fig. 4. There are two subsets in $\mathbb{B}$: Rules R1-R9 infer membership for all the constructs in the metamodel in Fig. 3 except for similarity relationships, which are inferred through Rules R10-R17, comprising the second subset. The bodies of Rules R1-R9 show how RDF triples, in raw (i.e. `Triple` predicate in Fig. 4) or in mapped form (e.g. `RDFIsInstOf` in Fig. 4), provide evidence for metamodel membership predicates. For example, in Rule R1, a returned RDF triple of the form `S rdf:type T` counts as evidence that `T` is an entity type. As expected, rules heads (i.e., inferred predicates) may appear in rule bodies as further evidence. For example, in Rule R7, a returned RDF triple `(S,P,O)`, where `S` has been inferred to be an entity and `P` has been inferred to be a property, counts as evidence for inferring that `O` has domain `P`. User-defined predicates (such as `IsDictWord`) also count as evidence, as shown in Rule R2. Lexical similarity (in Rules R10-R17) also counts as evidence. For example, in Rule R8, the following count as evidence that `P` is a property of the entity type `T`: `P` appears as a predicate, and `T` as type, in the returned results, and the set of URIs of which `P` is predicated is similar to the set of URIs which are said to be of type `T`.

The bodies of Rules R10-R17 show how similarity relationships can be inferred from user-defined predicates (such as `LexSim` in, among others, Rule R10) and from user-provided definitions (possibly parametrized) of set similarity (such as `SimSetEq[]` in, among others, Rule R11). A value for `LexSim` is computed using cosine similarity for strings with length greater than fifty and Levenshtein (or edit) distance otherwise. In the case of URIs, we use the label (as given by `rdfs:label`) of the dereferenced resource the URI points to, or else the local name if no label is found. Rule R14 treats an entity as a set of property values and infers similarity between entities from the overlap of property value sets. Rule R16 uses object overlap (a metric that is commonly used to align properties [7,17]) to infer similarity of properties.

Consider again the example search results in Fig. 1. Given the baseline model, if the rules had equal weights, `EntityType(Movie)` would be inferred with a higher probability than `EntityType(Q386724)` because the former satisfies Rules R1 and R2, whereas, `Property(birthDate)` would more probable than `Property(spouse)` because the latter is a predicate of fewer resources. Instances of the `HasProperty` relationship are inferred based on their co-occurrence in RDF resources computed by set similarity predicates in PSL. Based on our running example, the probability of `HasProperty(Person, surname)` is greater than that of `HasProperty(Person, spouse)` because `surname` is a predicate of more resources than `spouse` is.

The Semantic Model: Adding Ontological Evidence. The baseline model only assimilates syntactic evidence. We can extend it with rules that assimilate evidence from ontologies to yield the model we call semantic.

Ontologies are computational artefacts that formally describe concepts and relationships in a given domain. Thus, in terms of the metamodel we target, they describe entity types and their properties and provide evidence that complements the baseline model described above. Our approach is to make use of statements in ontologies about types and properties. This is then used to ground predicates that represent ontological evidence. Table 1 shows the different kinds of evidence we extract from ontologies and how it is mapped into PSL predicates.

```
/*  Rules R1-R9: Inference of metamodel instantiations */
R1:  EntityType(T)         ← RDFSubject(S) ∧̄ RDFIsInstOf(S,T)
R2:  EntityType(T)         ← RDFSubject(S) ∧̄ RDFIsInstOf(S,T) ∧̄ IsDictWord(T)
R3:  Entity(S)             ← RDFIsInstOf(S,T) ∧̄ EntityType(T)
R4:  Property(P)           ← RDFSubjPred(S,P) ∧̄ Entity(S)
R5:  PropertyValue(0)      ← Triple(S,P,0) ∧̄ Entity(S) ∧̄ Property(P)
R6:  HasType(S,T)          ← RDFIsInstOf(S,T) ∧̄ EntityType(T)
R7:  HasDomain(0,P)        ← Triple(S,P,0) ∧̄ Entity(S) ∧̄ Property(P)
R8:  HasProperty(T,P)      ← RDFPredicate(P) ∧̄ RDFType(T) ∧̄
                             {P.RDFSubjPred(inv)} SimSetEq[URI] {T.RDFIsInstOf(inv)}
R9:  HasPropertyValue(S,0) ← Triple(S,P,0) ∧̄ Entity(S) ∧̄ Property(P)

/*  Rules R10-R17: Inference of similarity relationships */
R10: SimEntityType(T1,T2)  ← EntityType(T1) ∧̄ EntityType(T2) ∧̄
                             LexSim(T1,T2)
R11: SimEntityType(T1,T2)  ← EntityType(T1) ∧̄ EntityType(T2) ∧̄
                             {T1.HasProperty} SimSetEq[] {T2.HasProperty}
R12: SimEntity(E1,E2)      ← Entity(E1) ∧̄ Entity(E2) ∧̄ Label(E1,L1) ∧̄
                             Label(E2,L2) ∧̄ LexSim(L1,L2)
R13: SimEntity(E1,E2)      ← Entity(E1) ∧̄ Entity(E2) ∧̄ Name(E1,N1) ∧̄
                             Name(E2,N2) ∧̄ LexSim(N1,N2)
R14: SimEntity(E1,E2)      ← Entity(E1) ∧̄ Entity(E2) ∧̄
                             {E1.HasPropertyValue} SimSetEq[] {E2.HasPropertyValue}
R15: SimProperty(P1,P2)    ← Property(P1) ∧̄ Property(P2) ∧̄ LexSim(P1,P2)
R16: SimProperty(P1,P2)    ← Property(P1) ∧̄ Property(P2) ∧̄
                             {P1.HasDomain(inv)} SimSetEq[] {P2.HasDomain(inv)}
R17: SimPropertyValue(V1,V2) ← PropertyValue(V1) ∧̄ PropertyValue(V2) ∧̄
                             LexSim(V1,V2)
/*  ... */
/*  Rules R18-R20: Extending inference with ontological evidence (OE) */
R18: Entity(S) ← RDFIsInstOf(S,T) ∧̄ OntType(T)
R19: Entity(S) ← RDFIsInstOf(S,T) ∧̄ LexSim(T,OT) ∧̄ OntType(OT)
R20: Entity(S) ← RDFIsInstOf(S,T) ∧̄ OntEqType(T,OT)
/*  ... */
/*  Rules R21-R23: Extending similarity relationships with OE */
R21: SimEntityType(T1,T2) ← EntityType(T1) ∧̄ EntityType(T2) ∧̄
                             OntEqType(T1,T2)
R22: SimEntityType(T1,T2) ← EntityType(T1) ∧̄ EntityType(T2) ∧̄ OntType(OT) ∧̄
                             LexSim(T1,OT) ∧̄ LexSim(T2,OT)
R23: SimEntityType(T1,T2) ← EntityType(T1) ∧̄ EntityType(T2) ∧̄ OntType(OT) ∧̄
                             {T1.HasProperty} SimSetEq[Property] {OT.OntHasProperty} ∧̄
                             {T2.HasProperty} SimSetEq[Property] {OT.OntHasProperty}
/*  ... */
/*  Rules R24-25: Extending inference with feedback evidence */
R24: Entity(S)        ← RDFIsInstOf(S,T) ∧̄ EntityTypeFB(T,"yes")
R25: ¬Entity(S)       ← RDFIsInstOf(S,T) ∧̄ EntityTypeFB(T,"no")
R26: EntityType(S) ← ∧̄ EntityTypeFB(S,"yes")
R27: ¬EntityType(S) ← EntityTypeFB(S,"no")
/*  ... */
```

Fig. 4. PSL rules used in structure inference

The main idea here is that being defined as a concept or property in some pertinent ontology counts as additional evidence that a returned result (e.g., a resource URI) is an entity type or property, respectively, in terms of our metamodel. Thus, we use the PSL predicates in Table 1 to construct PSL rules such as R18-R20 in Fig. 4, which assimilate ontological information as evidence that a given resource S is an entity. For example, if "The Godfather" appears in a triple in the returned results as an instance of Movie, where Movie is asserted to be a concept (e.g., in Movie Ontology (www.movieontology.org/) this adds strength to the belief that "The Godfather" is an instance of Entity in our metamodel (see R18 in Fig. 4). We also use ontological statements to

Table 1. Mapping ontological statements into PSL predicates

Ontological statement	PSL predicate
T rdf:type rdfs:Class	OntType(T)
T rdf:type owl:Class	
P rdf:type rdf:Property	OntProperty(P)
P rdf:type owl:DatatypeProperty	
P rdf:type owl:ObjectProperty	
T1 owl:equivalentClass T2	OntEqType(T1,T2)
P1 owl:equivalentProperty P2	OntEqProperty(P1,P2)
T1 owl:disjointWith T2	OntDisjointType(T1,T2)
P rdfs:domain T	OntHasProperty(T,P)

supplement evidence for similarity relationships as exemplified by Rules R21-R23 in Fig. 4. Rule R21 exemplifies the direct use of ontological evidence. Rule R22 exemplifies the use of ontological evidence mediated by lexical similarity. Rule R23 exemplifies the use of set-similarity predicates where one of the sets contains elements from ontological statements. In this case, ontological statements help reconcile heterogeneity in the returned results.

Subsuming the baseline model, the semantic model $\mathbb{S} \supset \mathbb{B}$ contains Rules R1-R17 plus such rules as R18-R20 and R21-R23 that assimilate additional evidence to the predicates in the baseline model.

The Feedback Model: Assimilating User-Provided Evidence. The semantic model assimilates syntactic and semantic evidence. We can extend it with rules that assimilate user-provided evidence in the form of feedback.

There has been a growing interest in assimilating user feedback in data integration (see [2] for a general proposal, and [12] for a general methodology and recent work in the area). User feedback is even more important in environments that are characterized by large-scale heterogeneity and highly autonomous data sources as is the case in the Web of Data(WoD) [9]. Also, with recent advances in crowdsourcing, feedback for solving complex data integration challenges can now be obtained more cost-effectively. In this paper, we assume that the user knows the domain of the search term and is motivated by obtaining high-quality results from the search, which we believe to be reasonable in the context of search results personalization.

We use feedback in two ways. The first is to refine the report presented to the user. When the returned results have been structured and integrated using PSL, a report is presented to the user (as described in Sect. 6). Feedback can be provided that results in refinement of the report. For example, a prior state of the report shown in Fig. 2 could have contained another row in the Movie tables referring to the Mumbai Godfather film which the user ruled out as a false positive. In this case, feedback on inference results powers up a form of filtering, i.e., of data cleaning, generating an incentive for users to provide feedback in the first place. The second, and more significant, way we use feedback is to take advantage of accumulated, user-provided feedback to improve the inference results before a report is produced. In this case, feedback on inference results is another type of evidence that can be assimilated.

Table 2. PSL predicates used to gather user feedback (with examples)

PSL feedback predicate	Example
EntityTypeFB(uri,term)	EntityTypeFB(Q11424,no)
HasTypeFB(uri,uri,term)	HasType(dbr:The_Godfather,CreativeWork, yes)
HasPropertyFB(uri,uri,term)	HasPropertyFB(CreativeWork, author, yes)
SimEntityTypeFB(uri,uri,term)	SimEntityType(Film, Movie, yes)
SimPropertyFB(uri,uri,term)	SimPropertyFB(duration, runtime, yes)
SimEntityFB(uri,uri,term)	SimEntityFB(dbr:The_Godfather, lmdbr:film/43338, yes)

Table 2 shows the feedback predicates used in our model. The feedback is simply a comment from the user on the correctness of the inferred query predicate. Thus, we use the PSL predicates in Table 2 to construct such PSL rules as R24-R27 in Fig. 4, which assimilate user-provided feedback as evidence that a given resource S is, or is not, an entity. For example, if "War and Peace" appears in a triple in the returned results as an instance of Movie, where Movie is confirmed by feedback to be an entity type, this adds strength to the belief that "War and Peace" is an instance of Entity in our metamodel.

Weight Learning for PSL Program Generation. We generated two PSL programs $PSL(\mathbb{S})$ and $PSL(\mathbb{F})$ for inference by learning weights for the rules in $\mathbb{S}$ and $\mathbb{F}$. Firstly, we obtained search results using Sindice [11] and Falcons [3]. The search terms and vocabularies used were: for the Films domain, "The Godfather", "Godfather actors", "Casablanca" and the Movie ontology; for the Cities domain, "Berlin", "Manchester" and the GeoFeatures (www.mindswap.org/2003/owl/geo/) and GeoNames (www.geonames. org/) ontologies; for the People domain, Tim Berners-Lee, Chris Bizer and the FOAF and SWRC (ontoware.org/swrc) ontologies. We collected the top 20 results from each search engine for each search term. The total number of typed individuals in the corpus was 304 and the total number of triples was 12,160. The corpus contained, resp., 180 and 502 distinct domain types and properties.

Secondly, we annotated the corpus with the ground truth. In doing so, we took into account the relevant domain. For example, searching for "Casablanca" return results about the city and not just the film of that name. Given the sample data from search results and vocabularies, we used PSL to learn weights for the rules in the semantic model and yield the $PSL(\mathbb{S})$ program. We learn the weights discriminatively using maximum-pseudo likelihood. To reduce overfitting, we use 5-fold cross-validation and we average the weights of each rule.

To learn weights for the rules in feedback model $\mathbb{F}$ and yield the PSL program $PSL(\mathbb{F})$, we proceeded as follows. One might crowdsource sample feedback instances, but we simulated feedback acquisition to give rise to synthetic sample. The synthesis procedure we used is as follows. We take as input a sample of inferences returned by $PSL(\mathbb{S})$ and the number of users providing feedback (100, in this paper). For each worker, we randomly generate 50 feedback instances, where a feedback instance is a true/false annotation on the inference returned for query predicates. In this paper, we assume that feedback is reliable. However, introducing a per-user degree of unreliability only requires a simple change.

5   Experimental Evaluation

The goal of the experimental evaluation is to measure how well the various PSL programs, generated as described above, infer a structure from LD search results that conforms to the adopted metamodel. To provide comprehensive information about the experiments, we have made the datasets, the code, and documentation available on GitHub [1].

Exp. 1: Inference Using Syntactic and Semantic Evidence. The goal of Exp. 1 is to measure the quality of $PSL(\mathbb{B})$, where only syntactic evidence is assimilated, and then measure the quality of $PSL(\mathbb{S})$, where semantic evidence is also assimilated, thereby allowing us to measure the impact of using ontologies on the quality. We measure the quality of our program using the area under precision-recall curve (AUC) for each query predicate in our PSL model. The precision/recall curve is computed by varying the probability threshold above which a query atom is predicted to be true. This means that the measurement does not depend on setting any threshold.

We first performed PSL inference on $PSL(\mathbb{B})$ on the search results, for each of three domains in turn. We denote the measured quality as $AUC(\mathbb{B})$ with some abuse of notation. We then added semantic evidence extracted from the relevant vocabularies to the search results and performed inference on $PSL(\mathbb{S})$. We denote the measured quality by $AUC(\mathbb{S})$. We then calculate the quality impact of using semantic evidence as $\Delta = AUC(\mathbb{S}) - AUC(\mathbb{B})$. Columns 2, 3 and 4 in each subtable in Table 3 list all the measurements obtained in Exp. 1 for the corresponding domain.

Discussion. As measured in terms of the AUC, across the domains, the quality of $PSL(\mathbb{B})$ is good on average (around 0.65) if we discount the similarity relationships, which are inherently dependent on semantic evidence. This suggests that $PSL(\mathbb{B})$ uses syntactic evidence effectively but that the approach might benefit from assimilating other forms of evidence.

Assimilating ontological evidence indeed leads to improvement in the AUC, particularly w.r.t. similarity relationships, with a knock-on positive effect on the quality of the tabular representation. Thus, the corresponding average $AUC(\mathbb{S})$ increases to close to 0.7. The degree of improvement varies across

---

[1] https://github.com/duhai-alshukaili/StructuringLDSearchResults.

Table 3. AUC results for our PSL models with datasets in the test collection

Query Predicate	$AUC(\mathbb{B})$	$AUC(\mathbb{S})$	Δ	$AUC(\mathbb{S}\cup\mathbb{F})$	Δ'
Entity	.726	.736	.010	.827	.091
EntityType	.749	.812	.063	.954	.142
HasProperty	.512	.578	.057	.614	.036
HasType	.554	.604	.050	.709	.105
Property	.774	.774	.000	.779	.005
SimEntity	.083	.108	.025	.322	.214
SimEntityType	.320	.351	.030	.517	.166
SimProperty	.159	.184	.025	.621	.437

(a) Films

Query Predicate	$AUC(\mathbb{B})$	$AUC(\mathbb{S})$	Δ	$AUC(\mathbb{S}\cup\mathbb{F})$	Δ'
Entity	.645	.751	.105	.789	.039
EntityType	.828	.844	.017	.844	.000
HasProperty	.457	.511	.054	.559	.048
HasType	.557	.731	.174	.732	.000
Property	.683	.683	.000	.685	.001
SimEntity	.086	.583	.497	.612	.030
SimEntityType	.869	.891	.022	.891	.000
SimProperty	.257	.230	-.027	.537	.307

(b) People

Query Predicate	$AUC(\mathbb{B})$	$AUC(\mathbb{S})$	Δ	$AUC(\mathbb{S}\cup\mathbb{F})$	Δ'
Entity	.525	.651	.126	.882	.231
EntityType	.776	.793	.017	.793	.000
HasProperty	.470	.485	.015	.541	.056
HasType	.631	.668	.037	.820	.152
Property	.774	.780	.006	.794	.014
SimEntity	.082	.344	.261	.411	.068
SimEntityType	.648	.662	.014	.662	.000
SimProperty	.484	.386	-.098	.687	.302

(c) Cities

domains depending on the coverage of the ontologies used. For example, the $\mathbb{B}$-probability of HasProperty(dbo:Person, dbo:spouse) is 0.04, whereas its $\mathbb{S}$-probability is much higher, at 0.80, as the DBPedia ontology explicitly defines this relationship. In other cases, explicit assertion of type disjointness has a significant effect too. Thus, the $\mathbb{B}$-probability of SimEntity(dbr:Casablanca, dbr:Casablanca_(film)) is 0.55, whereas its $\mathbb{S}$-probability is much lower, at 0.01, because dbo:Work and dbo:wgs84_pos:SpatialThing are disjoint in the DBpedia ontology. In the case of Entity, the assertion of a type by an ontology acts as a reliable anchor for individuals returned in the search results. For example, the $\mathbb{B}$-probability of Entity(lmdb:film/43338) is 0.22, whereas its $\mathbb{S}$-probability is higher, at 0.38, because its type, lmdb:film, matches the type dbo:Film, in the MO. As hinted above, improvements in the inference of metatypes (e.g., Entity) has a knock-on effect on the corresponding set-similarity relationship (SimEntity in this example). In the case of SimEntity the improvement is more significant the People and Cities domain than for Films because of inherent type ambiguity. For example, searching with "Casablanca" returns films, organizations, and a city. Type ambiguity is perhaps best solved with user feedback, as the next experiment shows.

Exp. 2: Inference Using Feedback. The goal of Exp. 2 is to measure the quality of $PSL(\mathbb{S}\cup\mathbb{F})$, where feedback evidence is also assimilated, thereby allowing us to measure the impact of using feedback on the quality. We simulated the feedback evidence as being provided for the top 5% of the inference results for each feedback target. This assumes a strategy in which feedback is targeted at removing false positives. We denote the measured quality by $AUC(\mathbb{S}\cup\mathbb{F})$. We then calculate the quality impact of using semantic evidence as $\Delta' = AUC(\mathbb{S}\cup\mathbb{F}) - AUC(\mathbb{S})$. Columns 5 and 6 in each subtable in Table 3 list all the measurements obtained in Exp. 2 for the corresponding domain.

Discussion. As measured in terms of the AUC, across the domains, the quality of $PSL(\mathbb{S}\cup\mathbb{F})$ is quite good in average (around 0.7) even if we include the similarity relationships. In other words, user feedback seems to address the ambiguity issues that caused the quality of $PSL(\mathbb{S})$ to be lower for similarity relationships. Thus, although the impact of feedback is not uniformly high, it seems complementary and corrective, i.e., it improves the most where the most improvement is needed, viz., the similarity relationships, where we can observe AUC improvements that can reach 80 %, 130 %, up to almost 240 %. This, the highest improvement, was observed for SimProperty in the Films domain. The reason for this is that $PSL(\mathbb{S})$ produces many false positives for SimProperty. One possible reason is that property names (e.g., name) are often reused without qualification for very different concepts. Combining syntactic and ontological evidence seems insufficient.

6   A Deployment Case Study

Imagine a user gives "Godfather actors" as the search term. Relevant returned results come predominantly from the Linked Movie Database and the DBpedia. A few, less relevant, results come from Linked WordNet, BookMashup, and MusicBrainz. The PSL program uses the results to make inferences as to how to instantiate the target metamodel in a way that integrates the returned results into a tabular report.

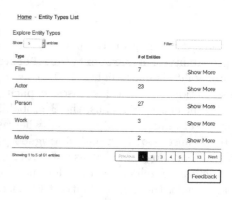

Fig. 5. Type selection

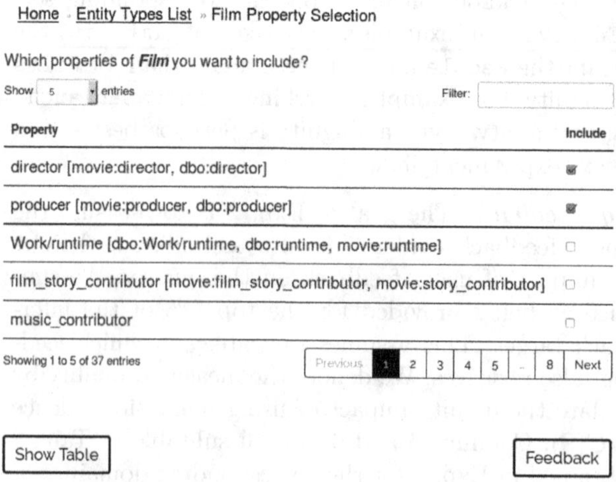

As depicted in Fig. 5, our PSL-driven user interface then provides the user with a list of postulated entity types for the given search term. The PSL query predicate behind Fig. 5 is HasType, with rows ordered by EntityType probability. At this point, the user can express an interest in one of the listed types by pressing on Show More which lists the inferred properties of the selected type. Figure 6 shows the list of inferred properties

Fig. 6. Property selection

of the postulated entity type `Film`. The PSL query predicate behind Fig. 6 is `HasProperty`, with rows ordered by `Property` probability.

At this point, the user can tick some properties to obtain the final tabular representation by clicking on `Show Table`. Figure 7 shows the table produced for entity type `Film` by selecting the properties `prequel`, `director`, and `producer`. Some properties shown (e.g. `producer`) are candidates for fusion and some entities (e.g. `The Godfather`) are candidates for deduplication. At each stage in this process, the user can intervene by clicking on the `Feedback` button to contribute feedback, which becomes evidence for use in future searches.

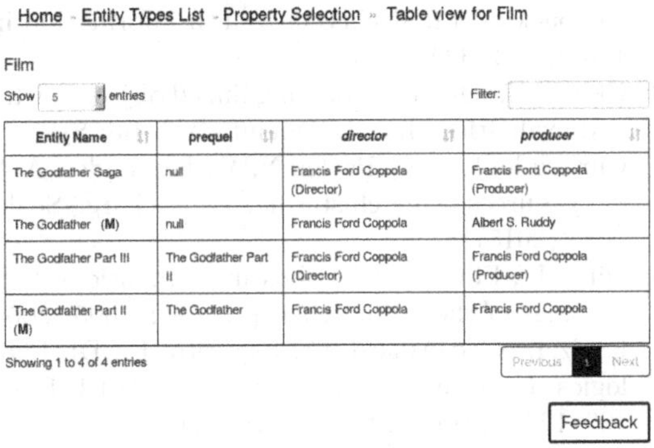

Note, therefore, that the use of a probabilistic framework allows us not only to structure and integrate the results but also to improve presentation (e.g., by ordering rows by likelihood) and to obtain targeted feedback. Note also that since the underlying PSL program models similarity relationships, the interface can make principled, uniform choices regarding deduplication (using `SimEntity` and `SimProperty`) and data fusion

Fig. 7. Data table

(using `SimProperty` and `SimPropertyValue`). In the example screenshots, some candidate property values (e.g., of `Director`) have been fused on the basis of `SimEntity`. Without this, the final table might be more heavily polluted by the natural redundancy one expects in search results.

7   Conclusions

This paper has provided empirical backing for the research hypothesis that assimilating different sources of (partial and uncertain) evidence is effective in inferring a good quality tabular structure over LD search results. The paper has described how a PSL program has been constructed with which the different sources of evidence can be assimilate in a principled and uniform way, where such sources are syntactic matching, domain ontologies and user feedback. It was shown how the PSL program can drive a user interface by mean of which the user can provide feedback that improves future quality, in a pay-as-you-go style. Moreover, the expressiveness of PSL allows the program to express similarity relationships from which, as shown, it is possible to perform immediate duplicate detection and data fusion prior to showing cleaner results to the user.

Acknowledgement. This work has been made possible by funding from the Omani National Program for Graduate Studies. Research on data integration at Manchester is supported by the VADA Programme Grant of the UK Engineering and Physical Sciences Research Council, whose support we are pleased to acknowledge.

References

1. Bach, S.H., Broecheler, M., Huang, B., Getoor, L.: Hinge-loss Markov random fields and probabilistic soft logic. CoRR abs/1505.04406 (2015)
2. Belhajjame, K., Paton, N.W., Fernandes, A.A., Hedeler, C., Embury, S.M.: User feedback as a first class citizen in information integration systems. In: CIDR. pp. 175–183 (2011)
3. Cheng, G., Qu, Y.: Searching linked objects with falcons: approach, implementation and evaluation. Int. J. Semant. Web Inf. Syst. (IJSWIS) **5**(3), 49–70 (2009)
4. Christodoulou, K., Paton, N.W., Fernandes, A.A.A.: Structure inference for linked data sources using clustering. Trans. Large-Scale Data- Knowl.-Centered Syst. **19**, 1–25 (2015)
5. Ding, L., Finin, T.W.: Boosting semantic web data access using swoogle. In: Proceedings of the 20th NCAI, pp. 1604–1605 (2005)
6. Fanizzi, N., d'Amato, C., Esposito, F.: DL-FOIL concept learning in description logics. In: Železný, F., Lavrač, N. (eds.) ILP 2008. LNCS (LNAI), vol. 5194, pp. 107–121. Springer, Heidelberg (2008)
7. Gunaratna, K., Thirunarayan, K., Jain, P., Sheth, A., Wijeratne, S.: A statistical and schema independent approach to identify equivalent properties on linked data. In: Proceedings of the 9th International Conference Semantic Systems, pp. 33–40. ACM (2013)
8. Kimmig, A., Bach, S.H., Broecheler, M., Huang, B., Getoor, L.: A short introduction to probabilistic soft logic. In: NIPS/Probabilistic Programming (2012)
9. Madhavan, J., Cohen, S., Dong, X.L., Halevy, A.Y., Jeffery, S.R., Ko, D., Yu, C.: Web-scale data integration: you can afford to pay as you go. In: Proceedings of the CIDR 2007, pp. 342–350 (2007)
10. Niepert, M., Noessner, J., Meilicke, C., Stuckenschmidt, H.: Probabilistic-logical web data integration. In: Polleres, A., d'Amato, C., Arenas, M., Handschuh, S., Kroner, P., Ossowski, S., Patel-Schneider, P. (eds.) Reasoning Web 2011. LNCS, vol. 6848, pp. 504–533. Springer, Heidelberg (2011)
11. Oren, E., Delbru, R., Catasta, M., Cyganiak, R., Stenzhorn, H., Tummarello, G.: Sindice.com: a document-oriented lookup index for open linked data. IJMSO **3**(1), 37–52 (2008)
12. Paton, N.W., Belhajjame, K., Embury, S.M., Fernandes, A.A.A., Maskat, R.: Pay-as-you-go data integration: experiences and recurring themes. In: Freivalds, R.M., Engels, G., Catania, B. (eds.) SOFSEM 2016. LNCS, vol. 9587, pp. 81–92. Springer, Heidelberg (2016). doi:10.1007/978-3-662-49192-8_7
13. Pujara, J., Miao, H., Getoor, L., Cohen, W.: Knowledge graph identification. In: Alani, H., Kagal, L., Fokoue, A., Groth, P., Biemann, C., Parreira, J.X., Aroyo, L., Noy, N., Welty, C., Janowicz, K. (eds.) ISWC 2013, Part I. LNCS, vol. 8218, pp. 542–557. Springer, Heidelberg (2013)
14. Sazonau, V., Sattler, U., Brown, G.: General terminology induction in owl. In: Arenas, M., et al. (eds.) ISWC 2015. LNCS, vol. 9366, pp. 533–550. Springer, Heidelberg (2015)

15. Tummarello, G., Cyganiak, R., Catasta, M., Danielczyk, S., Delbru, R., Decker, S.: Sig.ma: live views on the web of data. Web Semant. **8**(4), 355–364 (2010)
16. Völker, J., Niepert, M.: Statistical schema induction. In: Antoniou, G., Grobelnik, M., Simperl, E., Parsia, B., Plexousakis, D., Leenheer, P., Pan, J. (eds.) ESWC 2011, Part I. LNCS, vol. 6643, pp. 124–138. Springer, Heidelberg (2011)
17. Zapilko, B., Mathiak, B.: Object property matching utilizing the overlap between imported ontologies. In: Presutti, V., d'Amato, C., Gandon, F., d'Aquin, M., Staab, S., Tordai, A. (eds.) ESWC 2014. LNCS, vol. 8465, pp. 737–751. Springer, Heidelberg (2014)
18. Zhu, M., Gao, Z., Pan, J.Z., Zhao, Y., Xu, Y., Quan, Z.: Ontology learning from incomplete semantic web data by BelNet. In: 25th ICTAI, pp. 761–768. IEEE (2013)

The Multiset Semantics of SPARQL Patterns

Renzo Angles[1,3(✉)] and Claudio Gutierrez[2,3]

[1] Department of Computer Science, Universidad de Talca, Talca, Chile
rangles@utalca.cl
[2] Department of Computer Science, Universidad de Chile, Santiago, Chile
[3] Center for Semantic Web Research, Santiago, Chile

Abstract. The paper determines the algebraic and logic structure of the multiset semantics of the core patterns of SPARQL. We prove that the fragment formed by AND, UNION, OPTIONAL, FILTER, MINUS and SELECT corresponds precisely to both, the intuitive multiset relational algebra (projection, selection, natural join, arithmetic union and except), and the multiset non-recursive Datalog with safe negation.

1 Introduction

The incorporation of multisets (also called "duplicates" or "bags")[1] into the semantics of query languages like SQL or SPARQL is essentially due to practical concerns: duplicate elimination is expensive and duplicates might be required for some applications, e.g. for aggregation. Although this design decision in SQL may be debatable (e.g. see [6]), today multisets are an established fact in database systems [8,14].

The theory behind these query languages is relational algebra or equivalently, relational calculus, formalisms that for sets have a clean and intuitive semantics for users, developers and theoreticians [1]. The same cannot be said of their extensions to multisets, whose theory is complex (particular containment of queries) and their practical use not always clear for users and developers [8]. Worst, there exist several possible ways of extending set relational operators to multisets and one can find them in practice. As illustration, let us remind the behaviour of SQL relational operators. Consider as example the multisets $A = \{a, a, a, b, b, d\}$ and $B = \{a, b, b, c\}$. Then A UNION ALL $B = \{a, a, a, a, b, b, b, b, c, d\}$, that is, the "sum" of all the elements in both multisets (UNION DISTINCT is classical set union). A INTERSECT ALL B is $\{a, b, b\}$, i.e., the common elements in A and B, each with the minimum of the multiplicities in A and B. Regarding negation or difference, there are at least two: A EXCEPT ALL B is $\{a, a, d\}$, i.e. the arithmetical difference of the copies, and A EXCEPT B is $\{d\}$, the elements in A (with their multiplicity) after filtering out all elements occurring in B. The reader can imagine that the "rules" for combining these operators are not simple nor intuitive as they do not follow the rules of classical set operations.

---

[1] There is no agreement on terminology ([18], p. 27). In this paper we will use the word "multiset".

© Springer International Publishing AG 2016
P. Groth et al. (Eds.): ISWC 2016, Part I, LNCS 9981, pp. 20–36, 2016.
DOI: 10.1007/978-3-319-46523-4_2

Is there a rationale behind the possible extensions? Not easy to tell. Early on Dayal et al. [7] observed that there are two conceptual approaches to extend the set operators of union, intersection and negation, corresponding to the two possible interpretations of multiple copies of a tuple. The first approach treats all copies of a given tuple as being identical or indistinguishable. The second one treats all copies of a tuple as being distinct, e.g., as having an underlying identity. Each of these interpretations gives rise to a different semantics for multisets. The first one permits to extend the lattice algebra structure of sets induced by the $\subseteq$-order by defining a multiset order $\subseteq_m$ defined as $A \subseteq_m B$ if each element in A is contained in B and its multiplicity in B is bigger than in A. This order gives a lattice meet (multiset intersection) defined as the elements c present in both multisets, and with multiplicity $\min(c_A, c_B)$, where c_A, c_B are the number of copies of c in A and B respectively. This is the INTERSECT ALL operator of SQL. The lattice join of two multisets gives a union defined as the elements c present in both multisets with multiplicity $\max(c_A, c_B)$. This operator is not present in SQL. As was shown by Albert [2], there is no natural negation to add to this lattice to get a Boolean algebra structure like in sets. The second interpretation (all copies of an element are distinct) gives a poor algebraic structure. The union gives in this case an arithmetic version, where the elements in the union of the multisets A and B are the elements c present in both multisets with $c_A + c_B$ copies. This is the UNION ALL operator in SQL. Under this interpretation, the intersection loses its meaning (always gives the empty set) and the difference becomes trivial ($A - B = A$).

In order to illustrate the difficulties of having a "coherent" group of operators for multisets, let us summarize the case of SQL, that does not have a clear rationale on this point.[2] We classified the operators under those that: keep the set semantics; preserve the lattice structure of multiset order; do arithmetic with multiplicities. Let A, B be multisets, and for each element c, let c_A and c_B be their respective multiplicities in A and B.

$$
\text{union}: \begin{cases} set & \text{UNION DISTINCT (multiplicity: 1)} \\ lattice & \text{not present in SQL(*) (multiplicity: } \max(c_A, c_B)) \\ arithmetic & \text{UNION ALL (multiplicity:} c_A + c_B) \end{cases}
$$

$$
\text{intersection}: \begin{cases} set & \text{INTERSECT DISTINCT (multiplicity: 1)} \\ lattice & \text{INTERSECT ALL (multiplicity: } \min(c_A, c_B)) \\ arithmetic & \text{does not make sense} \end{cases}
$$

$$
\text{difference}: \begin{cases} set & \text{not present in SQL(**)(multiplicity: 1)} \\ lattice & \text{does not exists} \\ arithmetic & \text{EXCEPT ALL (multiplicity: } \max(0, c_A - c_B) \\ filter & \text{EXCEPT(mult:if } (c_B = 0) \text{ then } c_A \text{ else } 0 \end{cases}
$$

(*) Simulated as (A UNION ALL B) EXCEPT ALL (A INTERSECT ALL B).
(**) Simulated as SELECT DISTINCT * FROM (A EXCEPT B).

---

[2] We follow the semantics of ANSI and ISO SQL:1999 Database Language Standard.

At this point, a question arises: Are there "reasonable", "well behaved", "harmonic", groups of these operations for multisets? The answer is positive. Albert [2] proved that lattice union and lattice intersection plus a filter difference work well in certain domains. On the other hand, Dayal et al. [7] introduced the multiset versions for projection (π_X), selection (σ_C), join ($\bowtie$) and distinct (δ) and studied their interaction with Boolean operators. They showed that the lattice versions above combine well with selection ($\sigma_{P\vee Q}(r) = \sigma_P(r) \cup \sigma_Q(r)$ and similarly for intersection); that the arithmetic versions combine well with projection ($\pi_X(r \cup s) = \pi_X(r) \cup \pi_X(s)$). An important facet is the complexity introduced by the different operators. Libkin and Wong [16,17] and Grumbach et al. [9] studied the expressive power and complexity of the operations of the fragment including lattice union and intersection; arithmetic difference; and distinct.

For our purposes here, namely the study of the semantics of multisets in SPARQL, none of the above fragments help. It turns out that is a formalism coming from a logical field, the well behaved fragment of *non-recursive Datalog with safe negation* (nr-Datalog$^{\neg}$), the one that matches the semantics of multisets in SPARQL. More precisely, the natural extension of the usual (set) semantics of Datalog to multisets developed by Mumick et al. [19]. In this paper we work out the relational counterpart of this fragment, using the framework defined by Dayal et al. [7], and come up with a *Multiset Relational Algebra* (MRA) that captures precisely the multiset semantics of the core relational patterns of SPARQL. MRA is based on the operators projection (π), selection (σ), natural join ($\bowtie$), union ($\cup$) and filter difference ($\backslash$). The identification of this algebra and the proof of the correspondence with the relational core of SPARQL are the main contributions of this paper. Not less important, as a side effect, this approach gives a new relational view of SPARQL (closer to classical relational algebra and hence more intuitive for people trained in SQL); allows to make a clean translation to a logical framework (Datalog); and matches precisely the fragment of SQL corresponding to it. Table 1 shows a glimpse of these correspondences, whose details are worked in this paper.

Contributions. Summarizing, this paper advances the current understanding of the SPARQL language by determining the precise algebraic (Multiset Relational Algebra) and logical (nr-Datalog$^{\neg}$) structure of the multiset semantics of the core pattern operators in the language. This contribution is relevant for users, developers and theoreticians. For *users*, it gives an intuitive and classic view of the relational core patterns of SPARQL, allowing a good understanding of how to use and combine the basic operators of the SPARQL language when dealing with multisets. For *developers*, helps to perform optimization, design extensions of the language, and understanding the semantics of multisets allowing for example translations from SPARQL operators to the right multiset operators of SQL and vice versa. For *theoreticians*, introduces a clean framework (Multiset Datalog as defined by Mumick et al. [19]) to study from a formal point of view the multiset semantics of SPARQL patterns.

Table 1. SCHEMA OF CORRESPONDENCES OF MULTISET SPARQL PATTERNS: with SPARQL algebra operators; Relational Multiset algebra operators; Datalog rules; and SQL expressions. The operator EXCEPT in SPARQL is new (although expressible). The operator diff is a typed version of the diff SPARQL algebra operator, and \ in MRA is the multiset filter difference.

SPARQL	Multiset Relational Algebra	nr-Datalog¬	SQL
SELECT X ...	$\pi_W(...)$	$q(X) \leftarrow L_1, \ldots, L_n$	SELECT X ...
P FILTER C	$\sigma_C(r)$	$L \leftarrow L_P, C$	FROM r WHERE C
P1 . P2	$r_1 \bowtie r_2$	$L \leftarrow L_1, L_2$	r1 NATURAL JOIN r2
P1 UNION P2	$r_1 \cup r_2$	$L \leftarrow L_1$ $L \leftarrow L_2$	r1 UNION ALL r2
P1 EXCEPT P2	$r_1 \setminus r_2$	$L \leftarrow L_1, \neg L_2$	r1 EXCEPT r2

The paper is organized as follows. Section 2 presents the basic notions and notations used in the paper. Section 3 identifies a classical relational algebra view of SPARQL patterns, introducing the fragment SPARQL^R. Section 4 presents the equivalence between SPARQL^R and multiset non-recursive Datalog with safe negation, and provides explicit transformations in both directions. Section 5 introduces the Multiset Relational Algebra, a simple and intuitive fragment of relational algebra with multiset semantics, and proves that it is exactly equivalent to multiset non-recursive Datalog with safe negation. Section 6 analyzes related work and presents brief conclusions.

2   SPARQL Graph Patterns

The definition of SPARQL graph patterns will be presented by using the formalism presented in [22], but in agreement with the W3C specifications of SPARQL 1.0 [25] and SPARQL 1.1 [10].

RDF Graphs. Assume two disjoint infinite sets I and L, called IRIs and literals respectively.[3] An *RDF term* is an element in the set $T = I \cup L$. An *RDF triple* is a tuple $(v_1, v_2, v_3) \in I \times I \times T$ where v_1 is the *subject*, v_2 the *predicate* and v_3 the *object*. An *RDF Graph* (just graph from now on) is a set of RDF triples. The *union* of graphs, $G_1 \cup G_2$, is the set theoretical union of their sets of triples. Additionally, assume the existence of an infinite set V of variables disjoint from T. We will use var(α) to denote the set of variables occurring in the structure α.

A *solution mapping* (or just *mapping* from now on) is a partial function $\mu : V \to T$ where the domain of μ, dom(μ), is the subset of V where μ is defined.

---

[3] In addition to I and L, RDF and SPARQL consider a domain of anonymous resources called blank nodes. Their occurrence introduces issues that are not discussed in this paper. Based on the results in [11], we avoided blank nodes assuming that their absence does not affect the results presented in this paper.

The *empty mapping*, denoted μ_0, is the mapping satisfying that $\text{dom}(\mu_0) = \emptyset$. Given $?X \in V$ and $c \in T$, we use $\mu(?X) = c$ to denote the solution mapping variable $?X$ to term c. Similarly, $\mu_{?X \to c}$ denotes a mapping μ satisfying that $\text{dom}(\mu) = \{?X\}$ and $\mu(?X) = c$. Given a finite set of variables $W \subset V$, the restriction of a mapping μ to W, denoted $\mu_{|W}$, is a mapping μ' satisfying that $\text{dom}(\mu') = \text{dom}(\mu) \cap W$ and $\mu'(?X) = \mu(?X)$ for every $?X \in \text{dom}(\mu) \cap W$. Two mappings μ_1, μ_2 are *compatible*, denoted $\mu_1 \sim \mu_2$, when for all $?X \in \text{dom}(\mu_1) \cap \text{dom}(\mu_2)$ it satisfies that $\mu_1(?X) = \mu_2(?X)$, i.e., when $\mu_1 \cup \mu_2$ is also a mapping. Note that two mappings with disjoint domains are always compatible, and that the empty mapping μ_0 is compatible with any other mapping.

A *selection formula* is defined recursively as follows: (i) If $?X, ?Y \in V$ and $c \in I \cup L$ then $(?X = c)$, $(?X = ?Y)$ and $\text{bound}(?X)$ are atomic selection formulas; (ii) If F and F' are selection formulas then $(F \wedge F')$, $(F \vee F')$ and $\neg(F)$ are boolean selection formulas. The evaluation of a selection formula F under a mapping μ, denoted $\mu(F)$, is defined in a three-valued logic with values *true*, *false* and *error*. We say that μ satisfies F when $\mu(F) = \text{true}$. The semantics of $\mu(F)$ is defined as follows:

- If F is $?X = c$ and $?X \in \text{dom}(\mu)$, then $\mu(F) = \text{true}$ when $\mu(?X) = c$ and $\mu(F) = \text{false}$ otherwise. If $?X \notin \text{dom}(\mu)$ then $\mu(F) = \text{error}$.
- If F is $?X = ?Y$ and $?X, ?Y \in \text{dom}(\mu)$, then $\mu(F) = \text{true}$ when $\mu(?X) = \mu(?Y)$ and $\mu(F) = \text{false}$ otherwise. If either $?X \notin \text{dom}(\mu)$ or $?Y \notin \text{dom}(\mu)$ then $\mu(F) = \text{error}$.
- If F is $\text{bound}(?X)$ and $?X \in \text{dom}(\mu)$ then $\mu(F) = \text{true}$ else $\mu(F) = \text{false}$.
- If F is a Boolean combination of the previous atomic cases, then it is evaluated following a three value logic table (see [25], 17.2).

Multisets. A *multiset* is an unordered collection in which each element may occur more than once. A multiset M will be represented as a set of pairs (t, j), each pair denoting an element t and the number j of times it occurs in the multiset (called multiplicity or cardinality). When $(t, j) \in M$ we will say that t j-belongs to M (intuitively "t has j copies in M"). To uniformize the notation and capture the corner cases, we will write $(t, *) \in M$ or simply say $t \in M$ when there are ≥ 1 copies of t in M. Similarly, when there is no occurrence of t in M, we will simply say "t does not belong to M", and abusing notation write $(t, 0) \in M$, or $(t, *) \notin M$. All of them indicate that t does not occur in M.

For *multisets* of solution mappings, following the notation of SPARQL, we will also use the symbol Ω to denote a multiset and $\text{card}(\mu, \Omega)$ to denote the cardinality of the mapping μ in the multiset Ω. In this sense, we use $(\mu, n) \in \Omega$ to denote that $\text{card}(\mu, \Omega) = n$, or simply $\mu \in \Omega$ when $\text{card}(\mu, \Omega) > 0$. Similarly, $\text{card}(\mu, \Omega) = 0$ when $\mu \notin \Omega$. The domain of a multiset Ω is defined as $\text{dom}(\Omega) = \bigcup_{\mu \in \Omega} \text{dom}(\mu)$.

SPARQL Algebra. Let Ω_1, Ω_2 be multisets of mappings, W be a set of variables and F be a selection formula. The *SPARQL algebra for multisets of mappings* is composed of the operations of projection, selection, join, union, minus, difference and left-join, defined respectively as follows:

- $\pi_W(\Omega_1) = \{\mu' \mid \mu \in \Omega_1, \mu' = \mu_{|W}\}$
  where $\mathrm{card}(\mu', \pi_W(\Omega_1)) = \sum_{\mu' = \mu_{|W}} \mathrm{card}(\mu, \Omega_1)$
- $\sigma_F(\Omega_1) = \{\mu \in \Omega_1 \mid \mu(F) = true\}$
  where $\mathrm{card}(\mu, \sigma_F(\Omega_1)) = \mathrm{card}(\mu, \Omega_1)$
- $\Omega_1 \bowtie \Omega_2 = \{\mu = (\mu_1 \cup \mu_2) \mid \mu_1 \in \Omega_1, \mu_2 \in \Omega_2, \mu_1 \sim \mu_2\}$
  where $\mathrm{card}(\mu, \Omega_1 \bowtie \Omega_2) = \sum_{\mu = (\mu_1 \cup \mu_2)} \mathrm{card}(\mu_1, \Omega_1) \times \mathrm{card}(\mu_2, \Omega_2)$
- $\Omega_1 \cup \Omega_2 = \{\mu \mid \mu \in \Omega_1 \vee \mu \in \Omega_2\}$
  where $\mathrm{card}(\mu, \Omega_1 \cup \Omega_2) = \mathrm{card}(\mu, \Omega_1) + \mathrm{card}(\mu, \Omega_2)$
- $\Omega_1 - \Omega_2 = \{\mu_1 \in \Omega_1 \mid \forall \mu_2 \in \Omega_2, \mu_1 \not\sim \mu_2 \vee \mathrm{dom}(\mu_1) \cap \mathrm{dom}(\mu_2) = \emptyset\}$
  where $\mathrm{card}(\mu_1, \Omega_1 - \Omega_2) = \mathrm{card}(\mu_1, \Omega_1)$
- $\Omega_1 \setminus_F \Omega_2 = \{\mu_1 \in \Omega_1 \mid \forall \mu_2 \in \Omega_2, (\mu_1 \not\sim \mu_2) \vee (\mu_1 \sim \mu_2 \wedge (\mu_1 \cup \mu_2)(F) \neq true)\}$
  where $\mathrm{card}(\mu_1, \Omega_1 \setminus_F \Omega_2) = \mathrm{card}(\mu_1, \Omega_1)$
- $\Omega_1 \mathbin{\rlap{\bowtie}{}}_F \Omega_2 = \sigma_F(\Omega_1 \bowtie \Omega_2) \cup (\Omega_1 \setminus_F \Omega_2)$
  where $\mathrm{card}(\mu, \Omega_1 \mathbin{\rlap{\bowtie}{}}_F \Omega_2) = \mathrm{card}(\mu, \sigma_F(\Omega_1 \bowtie \Omega_2)) + \mathrm{card}(\mu, \Omega_1 \setminus_F \Omega_2)$

Syntax of Graph Patterns. A SPARQL *graph pattern* is defined recursively as follows: A tuple from $(I \cup L \cup V) \times (I \cup V) \times (I \cup L \cup V)$ is a graph pattern called a *triple pattern*.[4] If P_1 and P_2 are graph patterns then $(P_1 \text{ AND } P_2)$, $(P_1 \text{ UNION } P_2)$, $(P_1 \text{ OPT } P_2)$ and $(P_1 \text{ MINUS } P_2)$ are graph patterns. Also if C is a filter constraint (as defined below) then $(P_1 \text{ FILTER } C)$ is a graph pattern. And if W is a set of variables, $(\text{SELECT } W P_1)$ is a graph pattern.

A *filter constraint* is defined recursively as follows: (i) If $?X, ?Y \in V$ and $c \in I \cup L$ then $(?X = c)$, $(?X = ?Y)$ and $\mathrm{bound}(?X)$ are *atomic filter constraints*; (ii) If C_1 and C_2 are filter constraints then $(!C_1)$, $(C_1 \mid\mid C_2)$ and $(C_1 \;\&\&\; C_2)$ are *complex filter constraints*. Given a filter constraint C, we denote by $f(C)$ the selection formula obtained from C. Note that there exists a simple and direct translation from filter constraints to selection formulas and vice versa.

Semantics of SPARQL Graph Patterns. The evaluation of a SPARQL graph pattern P over an RDF graph G is defined as a function $[\![P]\!]_G$ (or $[\![P]\!]$ where G is clear from the context) which returns a multiset of solution mappings. Let P_1, P_2, P_3 be graph patterns and C be a filter constraint. The evaluation of a graph pattern P over a graph G is defined recursively as follows:

1. If P is a triple pattern t, then $[\![P]\!]_G = \{\mu \mid \mathrm{dom}(\mu) = \mathrm{var}(t) \wedge \mu(t) \in G\}$ where $\mu(t)$ is the triple obtained by replacing the variables in t according to μ, and each mapping μ has cardinality 1.
2. $[\![(P_1 \text{ AND } P_2)]\!]_G = [\![P_1]\!]_G \bowtie [\![P_2]\!]_G$
3. If P is $(P_1 \text{ OPT } P_2)$ then
   (a) if P_2 is $(P_3 \text{ FILTER } C)$ then $[\![P]\!]_G = [\![P_1]\!]_G \mathbin{\rlap{\bowtie}{}}_C [\![P_3]\!]_G$
   (b) else $[\![P]\!]_G = [\![P_1]\!]_G \mathbin{\rlap{\bowtie}{}}_{(true)} [\![P_2]\!]_G$
4. $[\![(P_1 \text{ MINUS } P_2)]\!]_G = [\![P_1]\!]_G - [\![P_2]\!]_G$
5. $[\![(P_1 \text{ UNION } P_2)]\!]_G = [\![P_1]\!]_G \cup [\![P_2]\!]_G$
6. $[\![(P_1 \text{ FILTER } C)]\!]_G = \sigma_{f(C)}([\![P_1]\!]_G)$
7. $[\![(\text{SELECT } W P_1)]\!]_G = \pi_W([\![P_1]\!]_G)$.

---

[4] We assume that any triple pattern contains at least one variable.

For the rest of the paper, we will call SPARQL$^{\text{W3C}}$ the fragment of graph patterns including the operators AND, UNION, OPT, FILTER, MINUS and SELECT, as defined above.

3 The Relational Fragment of SPARQL

In this section we will introduce a fragment of SPARQL which follows standard intuitions of the operators from relational algebra and SQL. We will prove that this fragment is equivalent to SPARQL$^{\text{W3C}}$. First, let us introduce the DIFF operator as an explicit way of expressing negation-by-failure[5] in SPARQL.

Definition 1 (The DIFF operator). *The weak difference of two graph patterns, P_1 and P_2, is defined as*

$$[\![(P_1 \text{ DIFF } P_2)]\!] = \{\mu_1 \in [\![P_1]\!] \mid \forall \mu_2 \in [\![P_2]\!], \mu_1 \nsim \mu_2\}$$

where $\text{card}(\mu_1, [\![(P_1 \text{ DIFF } P_2)]\!]) = \text{card}(\mu_1, [\![P_1]\!])$.

It is important to note that the DIFF operator is not defined in SPARQL 1.0 nor in SPARQL 1.1 at the syntax level. However, it can be implemented in current SPARQL engines by using the difference operator of the SPARQL$^{\text{W3C}}$ algebra. It was showed [4,13] that the operators OPT and MINUS can be simulated with the operator DIFF in combination with AND, UNION and FILTER.

In order to facilitate, and make more natural the translation from SPARQL to Relational Algebra (and Datalog), we will introduce a more intuitive notion of difference between two graph patterns. We define the domain of a pattern P, denoted $\text{dom}(P)$, as the set of variables that occur (defining the output "schema") in the multiset of solution mappings for any evaluation of P.

Definition 2 (The EXCEPT operator). *Let P_1, P_2 be graph patterns satisfying $\text{dom}(P_1) = \text{dom}(P_2)$. The except difference of P_1 and P_2 is defined as*

$$[\![(P_1 \text{ EXCEPT } P_2)]\!] = \{\mu \in [\![P_1]\!] \mid \mu \notin [\![P_2]\!]\}$$

where $\text{card}(\mu, [\![(P_1 \text{ EXCEPT } P_2)]\!]) = \text{card}(\mu, [\![P_1]\!])$.

We will denote by EXCEPT* (or outer EXCEPT) the version of this operation when the restriction on domains is not considered.[6]

Note that the restriction on the domains of P_1 and P_2 follows the philosophy of classical relational algebra. But it can be proved that EXCEPT and its outer version are simulable each other:

Lemma 1. *For each pair of graph patterns P_1, P_2 in SPARQL$^{\text{W3C}}$, and any RDF graph G, the operator EXCEPT can be simulated by EXCEPT* and vice versa.* ◆

---

[5] Recall that negation-by-failure can be expressed in SPARQL 1.0 as the combination of an optional graph pattern and a filter constraint containing the bound operator.

[6] This operation is called SetMinus in [12].

Proof. Clearly EXCEPT can be simulated by EXCEPT*. On the other direction, let us assume that $\mathrm{dom}(P_1) \neq \mathrm{dom}(P_2)$. Then $(P_1 \text{ EXCEPT}^* P_2)$ can be expressed by the pattern $(P_1' \text{ EXCEPT } P_2')$ where:

- $P_1' = (P_1 \text{ FILTER}(\neg\mathrm{bound}(?x_1)$ && ... && $\neg\mathrm{bound}(?x_n)))$ when $\mathrm{dom}(P_1) \setminus \mathrm{dom}(P_2) = \{x_1, \ldots, x_n\}$ and $P_1' = P_1$ when $\mathrm{dom}(P_1) \setminus \mathrm{dom}(P_2) = \emptyset$; and
- $P_2' = (P_2 \text{ FILTER}(\neg\mathrm{bound}(?y_1)$ && ... && $\neg\mathrm{bound}(?y_m)))$ when $\mathrm{dom}(P_2) \setminus \mathrm{dom}(P_1) = \{y_1, \ldots, y_m\}$ and $P_2' = P_2$ when $\mathrm{dom}(P_2) \setminus \mathrm{dom}(P_1) = \emptyset$.

Note that cardinalities of selected mappings are not touched.

The next lemma establishes the relationship between EXCEPT and DIFF, showing that EXCEPT can be simulated in SPARQL$^{\text{W3C}}$.

Lemma 2. *For every pair of graph patterns P_1, P_2 in SPARQL^{W3C}, and any RDF graph G, the operator EXCEPT can be simulated by DIFF and vice versa.*

Proof. The high level proof goes as follows. EXCEPT, as we saw, is equivalent to EXCEPT*. And EXCEPT* differs from DIFF only in checking compatibility of mappings (i.e. $\sim$). $[\![P_1 \text{ EXCEPT}^* P_2]\!]$ eliminates from $[\![P_1]\!]$ those mappings in $[\![P_2]\!]$ that are equal to one in $[\![P_1]\!]$; while DIFF eliminates those that are compatible with one in $[\![P_1]\!]$. That is, the difference is between the multisets $\{(\mu_1, n_1) \in \Omega_1 \mid \neg\exists\mu_2 \in \Omega_2 \wedge \mu_1 = \mu_2\}$ versus $\{(\mu_1, n_1) \in \Omega_1 \mid \neg\exists\mu_2 \in \Omega_2 \wedge \mu_1 \sim \mu_2\}$. Now, for two mappings μ_1, μ_2, equality and compatibility ($\mu_1 = \mu_2$ versus $\mu_1 \sim \mu_2$) differ only in those variables that are bound in μ_1 and unbound in μ_2 or vice versa. Thus, to simulate $=$ with $\sim$ and vice versa, it is enough to have an operator that replaces all unbound entries in mappings of Ω_1 and Ω_2 by a fresh new constant, e.g. c, call the new sets Ω_1' and Ω_2', and we will have that $\{(\mu_1, n_1) \in \Omega_1 \mid \neg\exists\mu_2 \in \Omega_2 \wedge \mu_1 \sim \mu_2\}$ is equivalent to $\{(\mu_1, n_1) \in \Omega_1' \mid \neg\exists\mu_2 \in \Omega_2' \wedge \mu_1 = \mu_2\}$. Note that cardinalities are preserved because the change between "unbound" and "c" does not change them. The rest is to express the two operations on multisets of solution mappings: the one that fills in unbound entries with a fresh constant c; and the one that changes back the values c to unbound.

Now we are ready to state the main theorem. Define SPARQL^{R} as the fragment of SPARQL$^{\text{W3C}}$ graph pattern expressions defined recursively by triple patterns plus the operators AND, UNION, FILTER and EXCEPT. Considering that DIFF is able to express OPT and MINUS (cf. [4,13]), and that the DIFF operator is expressible in SPARQL^{R} (Lemma 2), we have the following result:

Theorem 1. *SPARQL^{R} is equivalent to SPARQL^{W3C}.*

For the rest of the paper, we will concentrate our interest on SPARQL^{R}.

Note 1. An alternative proof of Theorem 1 is given as follows. (Compare [13], Lemma 12). Let θ be a function that renames variables by fresh ones.

SPARQL$^{\text{W3C}}$ contains SPARQL^{R}: The graph pattern $(P_1 \text{ EXCEPT } P_2)$ can be rewritten into an equivalent pattern $(((P_1 \text{ OPT}(\theta P_2)) \text{ FILTER } C) \text{ FILTER } C')$

where $\mathrm{dom}(P_1) = \{?x_1, \ldots, ?x_n\}$, C is $(?x_1 = \theta?x_1 \&\& \ldots \&\& ?x_n = \theta?x_n)$ and C' is $(!\,\mathrm{bound}(\theta?x_1))$.

SPARQL^R contains SPARQL^{W3C}: The graph pattern $(P_1 \, \mathrm{DIFF} \, P_2)$ can be rewritten into an equivalent graph pattern

$\quad (P_1 \, \mathrm{EXCEPT}(\mathrm{SELECT}\, W\,((P_1 \, \mathrm{AND}\, P_1') \, \mathrm{FILTER}\, C) \, \mathrm{AND}\, P_2'))$

where $W = \mathrm{dom}(P_1) = \{?x_1, \ldots, ?x_n\}$, $P_1' = \theta(P_1)$, $P_2' = \theta(P_2)$ and C is $(?x_1 = \theta?x_1 \&\& \ldots ?x_n = \theta?x_n)$.

4 SPARQL$^R \equiv$ Multiset Datalog

In this section we prove that SPARQL^R have the same expressive power of Multiset Datalog. Although the ideas of the proof are similar to those in [3] (now for SPARQL^R), we will sketch the main transformations to make the paper as self contained as possible. For notions of Datalog see Levene and Loizou [15], for the semantics of Multiset Datalog, Mumick et al. [19].

4.1 Multiset Datalog

A *term* is either a variable or a constant. A positive literal L is either a *predicate formula* $p(t_1, dots, t_n)$ where p is a predicate name and $t_1, dots, t_n$ are terms, or an *equality formula* $t_1 = t_2$ where t_1 and t_2 are terms. A negative literal $\neg L$ is the negation of a literal L. A *rule* is an expression of the form $L \leftarrow L_1 \wedge \cdots \wedge L_k \wedge \neg L_{k+1} \wedge \cdots \wedge \neg L_n$ where L is a positive literal called the *head* of the rule and the rest of literals (positive and negative) are called the *body*. A *fact* is a rule with empty body and no variables. A *Datalog program* Π is a finite set of rules and its set of facts is denoted facts(Π).

A variable x is *safe* in a rule r if it occurs in a positive predicate or in $x = c$ (c constant) or in $x = y$ where y is safe. A rule is safe it all its variables are safe. A program is *safe* if all its rules are safe. A program is non-recursive if its dependency graph is acyclic. In what follows, we only consider non-recursive and safe Datalog programs, denoted by nr-Datalog$^\neg$.

To incorporate multisets to the classical Datalog framework we will follow the approach introduced by Mumick and Shmueli [20]. The idea is rather intuitive: Each derivation tree gives rise to a substitution θ. In the standard (set) semantics, what matters is the set of the different substitutions that instantiates the distinguished literal. On the contrary, in multiset semantics the number of such instantiations also becomes relevant. As Mumick and Shmueli state [19,20], "duplicate semantics of a program is obtained by counting the number of derivation trees". Thus now we have pairs (θ, n) of substitutions θ plus the number n of derivation trees that produce θ.

A Datalog query is a pair (Π, L) where Π is a program and L is a distinguished predicate (the goal) occurring as the head of a rule. The answer to (Π, L) is the multiset of substitutions θ such that makes $\theta(L)$ true.

Normalized Datalog. Let L, L_1, L_2 be literals. We assume, without loss of generality, that any safe non-recursive Datalog program can be normalized such that it just contains rules of the following types:

- (Projection rule) $L \leftarrow L_1$ where $\text{var}(L) \subset \text{var}(L_1)$;
- (Selection rule) $L \leftarrow L_1, EQ$ where EQ is a set of equalities of the form $x_i = x_j$ such that x_i, x_j are variables or constants.
- (Join rule) $L \leftarrow L_1, L_2$ where $\text{var}(L) \subseteq \text{var}(L_1) \cup \text{var}(L_2)$; and
- (Negation rule) $L \leftarrow L_1, \neg L_2$ where $\text{var}(L_2) \subseteq \text{var}(L_1)$ and $\text{var}(L) = \text{var}(L_1)$.

4.2   From SPARQL to Datalog

The algorithm that transforms SPARQL into Datalog includes transformations of RDF graphs to Datalog facts, SPARQL queries into a Datalog queries, and SPARQL mappings into Datalog substitutions.

RDF Graphs to Datalog Facts: Let G be an RDF graph: each term t in G is encoded by a fact $iri(t)$ or $literal(t)$ when t is an IRI or a literal respectively; the set of terms in G is defined by the rules $term(X) \leftarrow iri(X)$ and $term(X) \leftarrow literal(X)$; the fact $Null(null)$ encodes the *null* value (unbounded value); each RDF triple (v_1, v_2, v_3) in G is encoded by a fact $triple(v_1, v_2, v_3)$. Recall that we are assuming that an RDF graph is a "set" of triples.

SPARQL Patterns into Datalog Rules: The transformation follows essentially the idea presented by Polleres [23]. Let P be a graph pattern and G an RDF graph. Denote by $\delta(P)_G$ the function which transforms P into a set of Datalog rules. Table 2 shows the transformation rules defined by the function $\delta(P)_G$, where the notion of compatible mappings is implemented by the rules:

$$comp(X, X, X) \leftarrow term(X), comp(X, Y, X) \leftarrow term(X) \wedge Null(Y),$$
$$comp(Y, X, X) \leftarrow Null(Y) \wedge term(X), comp(X, X, X) \leftarrow Null(X).$$

Also, an atomic filter condition C is encoded by a literal L as follows (where $?X, ?Y \in V$ and $u \in I \cup L$): if C is either $(?X = u)$ or $(?X = ?Y)$ then L is C; if C is bound$(?X)$ then L is $\neg Null(?X)$.

SPARQL Mappings to Datalog Substitutions: Let P be a graph pattern, G an RDF graph and μ a solution mapping of P in G. Then μ gets transformed into a substitution θ satisfying that for each $x \in \text{var}(P)$ there exists $x/t \in \theta$ such that $t = \mu(x)$ when $\mu(x)$ is bounded and $t = null$ otherwise.

Now, the correspondence between the multiplicities of mappings and substitutions works as follows: Each SPARQL mapping comes from an evaluation tree. A *set* of evaluation trees becomes a *multiset* of mappings. Similarly, a *set* of Datalog derivation trees becomes a *multiset* of substitutions. Thus, each occurrence of a mapping μ comes from a SPARQL evaluation tree. This tree is translated by Table 2 to a Datalog derivation tree, giving rise to an occurrence of a substitution in Datalog. Each recursive step in Table 2 carries out bottom up the correspondence between cardinalities of mappings and substitutions.

Table 2. Transforming SPARQL^R graph patterns into Datalog Rules. The function $\delta(P)_G$ takes a graph pattern P and an RDF graph G, and returns a set of Datalog rules with main predicate $p(\overline{\text{var}}(P))$, where $\overline{\text{var}}(P)$ denotes the tuple of variables obtained from a lexicographical ordering of the variables in P. If L is a Datalog literal, then $\nu_j(L)$ denotes a copy of L with its variables renamed according to a variable renaming function $\nu_j : V \to V$. *comp* is a literal encoding the notion of compatible mappings. *cond* is a literal encoding a filter condition C. $\overline{W}$ is a subset of $\overline{\text{var}}(P_1)$.

Pattern P	$\delta(P)_G$
(x_1, x_2, x_3)	$p(\overline{\text{var}}(P)) \leftarrow triple(x_1, x_2, x_3)$
$(P_1 \text{ AND } P_2)$	$p(\overline{\text{var}}(P)) \leftarrow \nu_1(p_1(\overline{\text{var}}(P_1))) \wedge \nu_2(p_2(\overline{\text{var}}(P_2)))$
	$\qquad\qquad\qquad\qquad \bigwedge_{x \in \text{var}(P_1) \cap \text{var}(P_2)} comp(\nu_1(x), \nu_2(x), x),$
	$\delta(P_1)_G, \ \delta(P_2)_G$
	$\text{dom}(\nu_1) = \text{dom}(\nu_2) = \text{var}(P_1) \cap \text{var}(P_2), \ \text{range}(\nu_1) \cap \text{range}(\nu_2) = \emptyset.$
$(P_1 \text{ UNION } P_2)$	$p(\overline{\text{var}}(P)) \leftarrow p_1(\overline{\text{var}}(P_1)) \bigwedge_{x \in \text{var}(P_2) \backslash \text{var}(P_1)} Null(x),$
	$p(\overline{\text{var}}(P)) \leftarrow p_2(\overline{\text{var}}(P_2)) \bigwedge_{x \in \text{var}(P_1) \backslash \text{var}(P_2)} Null(x),$
	$\delta(P_1)_G, \ \delta(P_2)_G$
$(P_1 \text{ EXCEPT } P_2)$	$p(\overline{\text{var}}(P_1)) \leftarrow p_1(\overline{\text{var}}(P_1)) \wedge \neg p_2(\overline{\text{var}}(P_2)),$
	$\delta(P_1)_G, \ \delta(P_2)_G$
$(\text{SELECT } W\, P_1)$	$p(\overline{W}) \leftarrow p_1(\overline{\text{var}}(P_1)),$
	$\delta(P_1)_G$
$(P_1 \text{ FILTER } C)$ and C is atomic	$p(\overline{\text{var}}(P)) \leftarrow p_1(\overline{\text{var}}(P_1)) \wedge cond$
	$\delta(P_1)_G$

Thus we have that a SPARQL query $Q = (P, G)$ where P is a graph pattern and G is an RDF graph gets transformed into the Datalog query $(\Pi, p(\overline{\text{var}}(P)))$ where Π is the Datalog program $\delta(P)_G$ plus the facts got from the transformation of the graph G, and p is the goal literal related to P.

4.3 From Datalog to SPARQL

Now we need to transform Datalog facts into RDF data, Datalog substitutions into SPARQL mappings, and Datalog queries into SPARQL queries.

Datalog Facts as an RDF Graph: Given a Datalog fact $f = p(c_1, ..., c_n)$, consider the function $\text{desc}(f)$ which returns the set of triples

$$\{(u, \text{predicate}, p), (u, \text{rdf:\_1}, c_1), \ldots, (u, \text{rdf:\_n}, c_n)\},$$

where u is a fresh IRI. Given a set of Datalog facts F, the RDF description of F will be the graph $G = \bigcup_{f \in F} \text{desc}(f)$.

Datalog Rules as SPARQL Graph Patterns: Let Π be a (normalized) Datalog program and L be a literal $p(x_1, \ldots, x_n)$ where p is a predicate in Π and each x_i is a variable. We define the function $\text{gp}(L)_\Pi$ which returns a graph pattern encoding of the program (Π, L). The translation works intuitively as follows:

(a) If predicate p is extensional, then $\mathrm{gp}(L)_\Pi$ returns the graph pattern
$((?Y, \mathrm{predicate}, p) \text{ AND}(?Y, \mathrm{rdf:\_1}, x_1) \text{ AND} \cdots \text{AND}(?Y, \mathrm{rdf\_}n, x_n))$,
where $?Y$ is a fresh variable.

(b) If predicate p is intensional and $\{r_1, \ldots, r_n\}$ is the set of all the rules in
Π where p occurs in the head, then $\mathrm{gp}(L)_\Pi$ returns the graph pattern
$(\ldots (T(r_1) \text{ UNION } T(r_2)) \ldots \text{ UNION } T(r_n))$ where $T(r_i)$ is defined as follows
(when $n = 1$ the resulting graph pattern is reduced to $T(r_1)$):

- If r_i is $L \leftarrow L_1$ then $T(r_i)$ returns SELECT $x_1, \ldots, x_n$ WHERE $\mathrm{gp}(L_1)_\Pi$.
- If r_i is $L \leftarrow L_1 \wedge EQ$, where EQ is a set of equalities of the form
  $x_i = x_j$ such that x_i, x_j are variables or constants, then $T(r_i)$ returns
  $(\mathrm{gp}(L_1)_\Pi \text{ FILTER } C)$ where C is a filter condition equivalent to EQ.
- If r_i is $L \leftarrow L_1 \wedge L_2$ then $T(r_i)$ returns $(\mathrm{gp}(L_1)_\Pi \text{ AND } \mathrm{gp}(L_2)_\Pi)$.
- If r_i is $L \leftarrow L_1 \wedge \neg L_2$ then $T(r_i)$ returns $(\mathrm{gp}(L_1)_\Pi \text{ EXCEPT}^* \mathrm{gp}(L_2)_\Pi)$.

Datalog Substitutions as SPARQL Mappings: For each substitution θ satisfy-
ing (Π, L) build a mapping μ satisfying that, if $x/t \in \theta$ then $x \in \mathrm{dom}(\mu)$ and
$\mu(x) = t$. The correspondence of multiplicities work in a similar way (via derivation tree to evaluation tree) as in the case of mappings to substitutions.

Putting together the transformation in Table 2 and the pattern obtained by
using $\mathrm{gp}(L)_\Pi$, we get the following theorem, whose proof is a long but straightforward induction on the structure of the patterns in one direction, and on the
level of Datalog in the other.

Theorem 2. *Multiset nr-Datalog* $^{\neg}$ *has the same expressive power as SPARQL^R.*

5   The Relational Version of Multiset Datalog: MRA

In this section we introduce a multiset relational algebra (called MRA), counterpart of Multiset Datalog, and prove its equivalence with the fragment of
non-recursive Datalog with safe negation.

5.1   Multiset Relational Algebra (MRA)

Multiset relational algebra is an extension of classical relation algebra having
multisets of relations instead of sets of relations. As indicated in the introduction,
there are manifold approaches and operators to extend set relational algebra with
multisets. We use the semantics of multiset operators defined by Dayal et al. [7]
for the operations of selection, projection, natural join and arithmetic union; and
add filter difference (not present there) represented by the operator "except".

Let us formalize these notions. In classical *(Set) relational algebra*, a database
schema is a set of relational schemas. A relational schema is defined as a set of
attributes. Each attribute A has a domain, denoted $\mathrm{dom}(A)$. A *relation R* over
the relational schema $S = \{A_1, \ldots, A_n\}$ is a finite *set* of tuples. An instance r of
a schema S is a relation over S. Given an instance r of a relation R with schema
S, $A_j \in S$ and $t = (a_1, \ldots, a_n) \in r$, we denote by $t[A_j]$ the tuple (a_j). Similarly
with $t[X]$ when $X \subseteq S$ and we will define $t[\emptyset] = \emptyset$.

In the *Multiset relational algebra* setting, an instance of a schema is a *multiset relation*, that is, a set of pairs (t, i), where t is a tuple over the schema S, and $i \geq 1$ is a positive integer. (For notions and notations on multisets recall Sect. 2, *Multisets*).

Definition 3 (Multiset Relational Algebra (MRA)). *Let r and r' be multiset relations over the schemas S and S' respectively. Let $A \in S$ be an attribute, $a \in \mathrm{dom}(A)$ and $I = S \cap S'$. MRA consists of the following operations:*

1. *Selection.* $\sigma_{A=a}(r) = \{(t, i) : (t, i) \in r \wedge t[A] = a\}$.
2. *Natural Join. $r \bowtie r'$ is a multiset relation over $S \cup S'$ defined as follows. Let $S'' = S' - S$. Let $t^\frown t'$ denotes concatenation of tuples. Then*

$$r \bowtie r' = \{(t^\frown(t'[S'']), i \times j) : (t, i) \in r \wedge (t', j) \in r' \wedge t[I] = t'[I]\}.$$

3. *Projection. Let $X \subseteq S$. Then:*

$$\pi_X(r) = \{(t[X], \sum_{(t_j, n_j) \in r \ s.t. \ t_j[X] = t} n_j) : (t, *) \in r\}.$$

4. *Union. Assume $S = S'$.*

$$r \cup r' = \{(t, i) : t \ i - belongs \ to \ r \ and \ t \ \notin r'\}$$
$$\cup \{(t', j) : t' \notin r \ and \ t' \ j - belongs \ to \ r'\}$$
$$\cup \{(t, i+j) : t \ i - belongs \ to \ r \ and \ t \ j - belongs \ to \ r'\}.$$

5. *Except. Assume $S = S'$.*

$$r \setminus r' = \{(t, i) \in r : (t, *) \notin r'\}.$$

As usual, we will define a query in this multiset relational algebra as an expression over an extended domain which includes, besides the original domains of the schemas, a set of variables V.

5.2   MRA $\equiv$ Multiset nr-Datalog $\neg$

This subsection is devoted to prove the following result.

Theorem 3. *Multiset relational algebra (MRA) has the same expressive power as Multiset Non-recursive Datalog with safe negation.*

From this theorem and Theorem 2 it follows:

Corollary 1. *$SPARQL^R$ is equivalent to MRA.*

Proof. The proof is based on the ideas of the proof of Theorem 3.18 in [15], extended to multisets. Let E be a relational algebra query expression over the schema R and D a database. Then it will be translated by a function $(\cdot)^{\Pi}$ to the Datalog program $\mathrm{facts}(\Pi) \cup E^{\Pi}$, where $\mathrm{facts}(\Pi)$ is the multiset of facts (over fresh predicates r^{Π} for each relation r, and having the same arity as the original schema of r):

$\text{facts}(\Pi) = \{(r^{\Pi}(t), n) : t \text{ is a tuple with multiplicity } n \text{ in schema } r \text{ in } D\}$, and E^{Π} is the translation of the expression E given by the recursive specification below. For the expression E_j, the set V_j will denote its list of attributes.

1. Base case. No operator involved. Thus the query is a member of the schema R, namely $r(x_1, \ldots, x_n)$. The corresponding Multiset Datalog query is: $out_r(x_1, \ldots, x_n) \leftarrow r^{\Pi}(x_1, \ldots, x_n)$
2. $E = \sigma_C(E_1)$, where C is a set of equalities of the form $x_i = x_j$ where x_i, x_j are variables or constants. The translation E^{Π} is the program: $out_E(x_1, \ldots, x_k) \leftarrow E_1^{\Pi}(x_1, \ldots, x_k) \wedge C$
3. $E = E_1 \bowtie E_2$. Let $V = V_2 \setminus V_1$. The translation is: $out_E(V_1, V) \leftarrow E_1^{\Pi}(V_1) \wedge E_2^{\Pi}(V_2)$
4. $E = \pi_A(E_1)$, where A is a sublist of the attributes in E_1. The translation is: $out_E(A) \leftarrow E_1^{\Pi}(V_1)$.
5. $E = E_1 \cup E_2$, where E_1 and E_2 have the same schema. The translation is: $out_E(x_1, \ldots, x_k) \leftarrow E_1^{\Pi}(x_1, \ldots, x_k) \quad out_E(x_1, \ldots, x_k) \leftarrow E_2^{\Pi}(x_1, \ldots, x_k)$
6. $E = E_1 \setminus E_2$, where E_1 and E_2 have the same schema. The translation is: $out_E(x_1, \ldots, x_k) \leftarrow E_1^{\Pi}(x_1, \ldots, x_k) \wedge \neg E_2^{\Pi}(x_1, \ldots, x_k)$

It is important to check that the resulting program is non-recursive (this is because the structure of the algebraic relational expression from where it comes is a tree). Also it is safe because in rule (6) both expressions have the same schema). Now, it needs to be shown that for each relational expression (query) E in R, $[E]_D$ and $[E^{\Pi}]$ return the same "tuples" with the same multiplicity. This is done by induction on the structure of E.

Now, let us present the transformation from Multiset Datalog to Multiset Relational Algebra. Note that we may assume a normal form for the Datalog programs as presented in Sect. 4.1. Then the recursive translation $(\cdot)^R$ from Datalog programs to MRA expressions goes as follows.

1. First translate those head predicates q occurring in ≥ 2 rules as follows. Let q be the head of rules $r_1, \ldots, r_k$, $k \geq 2$. Rename each such head q with the same set of variables V. Then the translation is $(q)^R = (q_{r_1})^R \cup \cdots \cup (q_{r_k})^R$. From now on, we can assume that, not considering these q's, all other predicates occur as head in at most one rule. Hence we will not need the subindex indicating the rule to which they belong to.
2. (Base case.) Let r be a fact $q(V)$. Then translates it as $(q_r)^R = q^R(V)$, where q^R is a fresh new schema with the corresponding arity.
3. Let r be $q(A) \leftarrow p(V)$, where A is a sublist of V. The translation is $(q_r)^R = \pi_A((p)^R)$.
4. Let r be $q(V) \leftarrow p(V) \wedge C$, where C is a set of equalities $x_i = x_j$ such that x_i, x_j are variables or constants. The translation is $(q_r)^R = \sigma_C((p)^R)$.
5. Let r be $q(X, Y, Z) \leftarrow p_1(X, Y) \wedge p_2(Y, Z)$, where X, Y, Z are disjoint lists of variables. The translation is $(q_r)^R = (p_1)^R \bowtie (p_2)^R$.
6. Let r be $q(X, Y) \leftarrow p_1(X, Y) \wedge \neg p_2(Y)$, that is the rule is safe. The translation is $(q_r)^R = (p_1)^R \setminus ((p_1)^R \bowtie (p_2)^R)$.

The arguments about multiplicity are straightforward verifications. And because the program Π is non-recursive (i.e. its dependency graph is acyclic), the recursive translation to the relational expression gives a well formed algebraic expression.

6 Related Work and Conclusions

To the best of our knowledge, the multiset semantics of SPARQL has not been systematically addressed. There are works that, when studying the expressive power of SPARQL, touched some aspects of this topic. Cyganiak [5] was among the first who gave a translation of a core fragment of SPARQL into relational algebra. Polleres [23] proved the inclusion of the fragment of SPARQL patterns with safe filters into Datalog by giving a precise and correct set of rules. Schenk [26] proposed a formal semantics for SPARQL based on Datalog, but concentrated on complexity more than expressiveness issues. Both, Polleres and Schenk do not consider multiset semantics of SPARQL in their translations. Perez et al. [21] gave the first formal treatment of multiset semantics for SPARQL. Angles and Gutierrez [3], Polleres [24] and Schmidt et al. [27] extended the set semantics to multiset semantics using this idea. Kaminski et al. [12] considered multisets in subqueries and aggregates in SPARQL. In none of these works was addressed the goal of characterizing the multiset algebraic and/or logical structure of the operators in SPARQL.

We studied the multiset semantics of the core SPARQL patterns, in order to shed light on the algebraic and logic structure of them. In this regard, the discovery that the core fragment of SPARQL patterns matches precisely the multiset semantics of Datalog as defined by Mumick et al. [19] and that this logical structure corresponds to a simple multiset algebra, namely the Multiset Relational Algebra (MRA), builds a nice parallel to that of classical set relational algebra and relational calculus. Contrary to the rather chaotic variety of multiset operators in SQL, it is interesting to observe that in SPARQL there is a coherent body of multiset operators. We think that this should be considered by designers in order to try to keep this clean design in future extensions of SPARQL.

Last, but not least, this study shows the complexities and challenges that the introduction of multisets brings to query languages, exemplified here in the case of SPARQL.

Acknowledgments. The authors have funding from Millennium Nucleus Center for Semantic Web Research under Grant NC120004. The authors thank useful feedback from O. Hartig and anonymous reviewers.

References

1. Abiteboul, S., Hull, R., Vianu, V.: Foundations of Databases. Addison-Wesley, Boston (1995)
2. Albert, J.: Algebraic properties of bag data types. In: Proceedings of the International Conference on Very Large Data Bases (VLDB), pp. 211–219 (1991)

3. Angles, R., Gutierrez, C.: The expressive power of SPARQL. In: Sheth, A.P., Staab, S., Dean, M., Paolucci, M., Maynard, D., Finin, T., Thirunarayan, K. (eds.) ISWC 2008. LNCS, vol. 5318, pp. 114–129. Springer, Heidelberg (2008)
4. Angles, R., Gutierrez, C.: Negation in SPARQL. In: Alberto Mendelzon International Workshop on Foundations of Data Management (AMW) (2016)
5. Cyganiak, R.: A relational algebra for SPARQL. Technical report HPL-2005-170, HP Labs (2005)
6. Date, C.J.: Date on Database: Writings 2000–2006. APress, New York (2006)
7. Dayal, U., Goodman, N., Katz, R.H.: An extended relational algebra with control over duplicate elimination. In: Proceedings of the Symposium on Principles of Database Systems (PODS), pp. 117–123. ACM (1982)
8. Green, T.J.: Bag semantics. In: Encyclopedia of Database Systems, pp. 201–206 (2009)
9. Grumbach, S., Libkin, L., Milo, T., Wong, L.: Query languages for bags: expressive power and complexity. SIGACT News **27**(2), 30–44 (1996)
10. Harris, S., Seaborne, A.: SPARQL 1.1 Query Language - W3C Recommendation, 21 March 2013. http://www.w3.org/TR/2013/REC-sparql11-query-20130321/
11. Hogan, A., Arenas, M., Mallea, A., Polleres, A.: Everything you always wanted to know about blank nodes. J. Web Semant. **27**(1), 42–69 (2014)
12. Kaminski, M., Kostylev, E.V., Grau, B.C.: Semantics and expressive power of subqueries and aggregates in SPARQL 1.1. In: Proceedings of the International Conference on World Wide Web (WWW), pp. 227–238. ACM (2016)
13. Kontchakov, R., Kostylev, E.V.: On expressibility of non-monotone operators in SPARQL. In: International Conference on the Principles of Knowledge Representation and Reasoning (2016)
14. Lamperti, G., Melchiori, M., Zanella, M.: On multisets in database systems. In: Calude, C.S., Pun, G., Rozenberg, G., Salomaa, A. (eds.) Multiset Processing. LNCS, vol. 2235, pp. 147–216. Springer, Heidelberg (2001)
15. Levene, M., Loizou, G.: A Guided Tour of Relational Databases and Beyond. Springer, Heidelberg (1999)
16. Libkin, L., Wong, L.: Some properties of query languages for bags. In: Proceedings of the International Workshop on Database Programming Languages (DBPL) - Object Models and Languages, pp. 97–114 (1994)
17. Libkin, L., Wong, L.: Query languages for bags and aggregate functions. J. Comput. Syst. Sci. **55**(2), 241–272 (1997)
18. Melton, J., Simon, A.R.: SQL:1999. Understanding Relational Language Components. Morgan Kaufmann Publishers, Burlington (2002)
19. Mumick, I.S., Pirahesh, H., Ramakrishnan, R.: The magic of duplicates and aggregates. In: Proceedings of the International Conference on Very Large Data Bases (VLDB), pp. 264–277 (1990)
20. Mumick, I.S., Shmueli, O.: Finiteness properties of database queries. In: Australian Database Conference, pp. 274–288 (1993)
21. Pérez, J., Arenas, M., Gutierrez, C.: Semantics of SPARQL. Technical Report TR/DCC-2006-17, Department of Computer Science, University of Chile (2006)
22. Pérez, J., Arenas, M., Gutierrez, C.: Semantics and complexity of SPARQL. ACM Trans. Database Syst. (TODS) **34**(3), 1–45 (2009)
23. Polleres, A.: From SPARQL to rules (and back). In: Proceedings of the 16th International World Wide Web Conference (WWW), pp. 787–796. ACM (2007)
24. Polleres, A.: How (well) do datalog, SPARQL and RIF interplay? In: Barceló, P., Pichler, R. (eds.) Datalog 2.0 2012. LNCS, vol. 7494, pp. 27–30. Springer, Heidelberg (2012)

25. Prud'hommeaux, E., Seaborne, A.: SPARQL query language for RDF. W3C Recommendation, 15 January 2008. http://www.w3.org/TR/2008/REC-115-sparql-query-20080115/
26. Schenk, S.: A SPARQL semantics based on datalog. In: Hertzberg, J., Beetz, M., Englert, R. (eds.) KI 2007. LNCS (LNAI), vol. 4667, pp. 160–174. Springer, Heidelberg (2007)
27. Schmidt, M., Meier, M., Lausen, G.: Foundations of SPARQL query optimization. In: Proceedings of the International Conference on Database Theory, pp. 4–33. ACM (2010)

Ontop of Geospatial Databases

Konstantina Bereta[✉] and Manolis Koubarakis

National and Kapodistrian University of Athens, Athens, Greece
{Konstantina.Bereta,koubarak}@di.uoa.gr

Abstract. We propose an OBDA approach for accessing geospatial data
stored in relational databases, using the OGC standard GeoSPARQL
and R2RML or OBDA mappings. We introduce extensions to an existing SPARQL-to-SQL translation method to support GeoSPARQL features. We describe the implementation of our approach in the system
ontop-spatial, an extension of the OBDA system Ontop for creating virtual geospatial RDF graphs on top of geospatial relational databases. We
present an experimental evaluation of our system using and extending a
state-of-the-art benchmark. To measure the performance of our system,
we compare it to a state-of-the-art geospatial RDF store and confirm its
efficiency.

1 Introduction

Currently, there is emerging interest of scientific communities from various
domains that produce and process geospatial data (e.g., earth scientists) to publish data as linked data and combine it with other data sources. Responding to
this trend, the Semantic Web community has been very active in the geospatial
domain, proposing data models, query languages, and systems for the representation and management of geospatial data. Notably, this research has led to the
development of extensions of RDF and SPARQL, such as stRDF/stSPARQL
and GeoSPARQL, that handle geospatial data. Similarly, research on geospatial
relational databases has been going on for a long time and has resulted in the
implementation of several efficient geospatial DBMS.

Despite the extensive research performed in the fields of relational databases
and the Semantic Web on the development of solutions for handling geospatial data efficiently, to the best of our knowledge, there is no OBDA system
that enables the creation of virtual, geospatial RDF graphs on top of geospatial
databases. This would be very useful for scientists that produce and process
geospatial data, as they mainly store this data in relational geospatial databases
(e.g., PostGIS) or in other geospatial data formats that are easily imported into
such databases (e.g., shapefiles). With the existing solutions in place, these scientists are forced to materialize their data as RDF in order to publish it as linked
data and/or use it in combination with other data sources. Sometimes this is
not practical and discourages users from using Semantic Web technologies. This
issue applies to the OBDA paradigm in general, but it has more impact in the
geospatial domain due to the reasons we have just described. We address these
issues by extending the OBDA paradigm with geospatial support.

© Springer International Publishing AG 2016
P. Groth et al. (Eds.): ISWC 2016, Part I, LNCS 9981, pp. 37–52, 2016.
DOI: 10.1007/978-3-319-46523-4_3

The contributions of this paper are the following:

- We introduce extensions to an existing SPARQL-to-SQL translation method in order to perform GeoSPARQL-to-SQL translation.
- We describe the implementation of our approach in the system Ontop-spatial, which is the first OBDA system for GeoSPARQL.
- We present an experimental evaluation of our system extending the benchmark Geographica [7], comparing the performance of ontop-spatial with the state-of-the-art geospatial RDF store Strabon [8]. The results show that, in most cases, ontop-spatial outperforms Strabon.

Ontop-spatial is available as free and open source software at the following link: https://github.com/ConstantB/ontop-spatial. It was developed for the Statoil use case of the EU FP7 project Optique[1], and then it was also used in the urban accountant, land management, and crisis mapping services of the EU FP7 project MELODIES[2], as well as in the maritime domain [4].

The organization of the rest of the paper is as follows. In Sect. 2 we present related work and background. In Sect. 3 we explain the GeoSPARQL-to-SQL translation. In Sect. 4 we present the system Ontop-spatial and we mention the real-world use cases in which it has been used. In Sect. 5 we present the experimental evaluation of our system. In Sect. 6 we conclude the presentation of our approach discussing its advantages and limitations, as well as its possible extensions.

2   Related Work and Background

The first area of work related to our own is research on extensions of the data model RDF and the query language SPARQL with geospatial features.

The data model stRDF and the query language stSPARQL are extensions of RDF and SPARQL 1.1 respectively, developed for the representation and querying of spatial [8] and temporal data (i.e., the valid time of triples [3]). Another framework that has been developed for the representation and querying of geospatial data on the Semantic Web is GeoSPARQL [2], which is an OGC standard. GeoSPARQL and stSPARQL were developed independently, but they have a lot of features in common, the most important of which are that they both adopt the OGC standards WKT and GML for representing geometries, and that they both support spatial analysis functions as extension functions. Their main differences derive from the fact that stSPARQL extends SPARQL 1.1, so it inherits and extends important features of SPARQL 1.1, providing support for spatial updates and spatial aggregates. Also, GeoSPARQL does not offer valid time support. Both stSPARQL and GeoSPARQL have extended SPARQL 1.1 with the topological functions defined in the OGC standard "OpenGIS Simple Feature Access for SQL" [1], and they also support the Egenhofer [6] and the RCC-8 [13] topological relation families as SPARQL 1.1 extension functions.

---

[1] http://optique-project.eu/.

[2] http://www.melodiesproject.eu/.

Since in the rest of the paper we will refer to the notation and concepts defined or followed by stSPARQL and GeoSPARQL, we briefly present them below for the convenience of the reader.

Spatial Literal. A spatial literal represents the serialization of a geometry. In stSPARQL, it is a literal of type `strdf:geometry` or its subtypes `strdf:WKT` or `strdf:GML`, as defined in [8]. Similarly, in GeoSPARQL it is a literal of type `geo:wktLiteral` or `geo:gmlLiteral`.

Spatial Term. A spatial term is either a spatial literal or a variable that can be bound to a spatial literal.

Spatial Filter. A spatial filter is a Boolean binary function $SF(t1, t2)$, where $t1, t2$ are spatial terms and SF is one of the Boolean functions of the Geometry extension of GeoSPARQL, namely `geof:sfEquals`, `geof:sfDisjoint`, `geof:stIntersects`, `geof:sfTouches`, `geof:sfCrosses`, `geof:sfWithin`, `geof:sfContains`, `geof:sfOverlaps`, and the respective *Egenhofer* and *RCC8* relation functions. These functions are defined in the Requirements 22, 23 and 24 of the GeoSPARQL standard.

Spatial Selection. A spatial selection in GeoSPARQL/stSPARQL is a SELECT query with a FILTER which is a Boolean binary function with arguments a variable and a constant.

Spatial Join. A spatial join in these languages is a query with a FILTER which is a Boolean binary function whose all arguments are variables. The definition of the spatial join in SPARQL corresponds to the definition of the spatial join in the geospatial extensions of the relational model. In the rest of this paper, spatial joins will often be denoted as $\bowtie_{sf}$, where sf is a spatial filter.

In the context of this paper, we will only consider GeoSPARQL (and, as a result, the geospatial part of stSPARQL). GeoSPARQL consists of the following six components:

- The *Core component*, which defines high level RDFS/OWL classes for spatial objects.
- The *Topology vocabulary extension*, which defines RDF properties for asserting and querying topological relations between spatial objects.
- The *Geometry extension*, which defines RDFS data types for serializing geometry data, geometry-related RDF properties, and non-topological spatial query functions for geometry objects.
- The *Geometry Topology extension*, which defines topological query functions.
- The *RDFS entailment extension*, which includes the RDF and RDFS reasoning requirements.
- The *Query Rewrite extension*, which defines rules for transforming *qualitative* spatial queries into equivalent *quantitative* queries.

The work surveyed above on extending RDF and SPARQL with geospatial functionality also gave rise to the implementation of geospatial RDF stores such as Parliament, uSeekM and Virtuoso, that implement a subset of GeoSPARQL, and Strabon [8] that implements both GeoSPARQL and stSPARQL.

There have also been systems that enable the translation of geospatial data from their native formats to RDF. GeoTriples [9] is a tool for the conversion of geospatial data from a variety of source formats (shapefiles, relational databases, XML files, etc.) to RDF using GeoSPARQL and stSPARQL vocabularies and R2RML mappings.

Another category of systems that are related to our work is SPARQL-to-SQL systems such as Ontop [5,14], Ultrawrap [15], D2RQ[3] and Morph [12]. These systems offer no geospatial functionality.

3   GeoSPARQL-to-SQL Translation

In the work described in [5,14], the authors present techniques for SPARQL-to-SQL translation using R2RML mappings. In this paper we extend their approach to support the GeoSPARQL-to-SQL translation using R2RML mappings. In this section we briefly describe how we translate the spatial extensions introduced in GeoSPARQL to Datalog and then in turn to the respective spatial extensions of SQL. A more detailed presentation of our extensions to the work described in [5,14] is omitted due to space and will appear in a longer version of this paper.

The work of [5,14] in the context of OBDA system Ontop follows the same semantics as [11] for the translation of SPARQL to Datalog. Definition 20 in [5,14] describes the valuation of filter expressions, considering only numeric binary operators in filters. We present below how to extend this definition by considering spatial filters as defined in GeoSPARQL.

Definition 1. *Evaluation of Spatial Filter Expressions.*

Let SF be a GeoSPARQL spatial filter, let v, u be variables, L_{gs} the set of literals of the datatypes `geo:wktlLiteral` *and* `geo:gmlLiteral` *and* $c \in L_{gs}$. *The valuation of SF on a substitution θ returns one of three values $\top, \bot$ and ϵ as shown below.*

$$(SF(v,c))\theta = \begin{cases} \top & if\ v \in dom(\theta)\ and\ SF(v\theta, c) = true \\ \epsilon & if\ v \notin dom(\theta)\ or\ v\theta = null \\ \bot & otherwise \end{cases}$$

$$(SF(v,u))\theta = \begin{cases} \top & if\ v, u \in dom(\theta)\ and\ SF(v\theta, u\theta) = true \\ \epsilon & if\ v\ or\ u \notin dom(\theta)\ or\ u\theta = null\ or\ v\theta = null \\ \bot & otherwise \end{cases}$$

GeoSPARQL to Datalog. In the approach described in [5,14], the SPARQL query is translated into a set of rules that comprise a Datalog program preserving the semantics of the original query. The translation algorithm is a modified version of the one presented in [11]. The intention behind this step is to optimise

---

[3] http://d2rq.org/.

the query before it gets translated into an SQL query that is eventually executed by the DBMS. The deviations of the original SPARQL-to-SQL translation algorithm of [11] proposed in [5,14] lead to a more compact encoding of rules, due to the fact that the final goal is to translate the Datalog program in SQL instead of executing it as in [11]. We follow the same approach and we extend the algorithm of [5,14] to take into account the spatial filters defined above.

We extend the algorithm by introducing a new set of distinguished predicates, namely the *distinguished spatial predicates*. We define a distinguished spatial predicate for each GeoSPARQL spatial filter [2]. Then the GeoSPARQL to Datalog translation algorithm is like the algorithm of [5,14] for SPARQL and results in Π_{QGS}, a Datalog program that corresponds to a geospatial query.

Datalog to SQL. In a similar way as in the GeoSPARQL-to-Datalog translation, we extend Definition 18 of [5,14] in order to consider distinguished spatial predicates as well: Every distinguished spatial predicate occurring in a Datalog program Π_{QGS} is translated into the equivalent geospatial SQL operator.

Mappings. In our framework we allow exactly the same mapping languages used in [5,14], namely R2RML mappings and OBDA mappings (mapping language native to Ontop).

The mapping languages offer functionalities that are useful to in our geospatial setting. For example, when geometry columns (e.g., columns storing geometries in Well-Known-Binary format) of geospatial relational tables are present in the mappings, we allow geometries to be mapped as WKT GeoSPARQL literals. Similarly, we allow the presence of geospatial SQL operators in the mappings, enabling users to manipulate their geospatial data on-the-fly (e.g., transformation of the geometries into a different Coordinate Reference System) before they are mapped to RDF.

4 Implementation

We implemented the theoretical extensions of the SPARQL-to-SQL translation framework of [5,14] discussed in Sect. 3 as an extension of the system Ontop with geospatial features focusing on *spatial selections* and *spatial joins*. We chose to extend Ontop instead of systems offering similar functionality because (i) it is open source, robust and extensible, (ii) it offers a wide range of functionalities that are useful for geospatial applications (reasoning, multiple APIs), and (iii) it implements significant SPARQL-to-SQL optimizations, producing queries that can be executed efficiently by the underlying DBMS as reported in [5,14].

Ontop-spatial supports the following components of GeoSPARQL: *Core, Topology Vocabulary extension, Geometry Topology extension, RDFS entailment extension* and the spatial filters defined in the *Geometry Extension*. It is also, to the best of our knowledge, the first GeoSPARQL implementation that supports the *Query Rewrite extension* of GeoSPARQL. The high level architecture of the system as well as an abstract overview can be seen in Figs. 1(a) and (b) respectively.

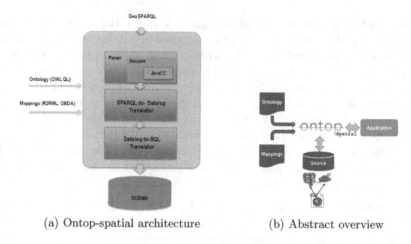

(a) Ontop-spatial architecture   (b) Abstract overview

Fig. 1. Ontop-spatial

In the following, we highlight the components of Ontop that we have extended as they are placed in the query processing workflow:

- The virtual Ontop repository takes as input an ontology and a mapping file. Mappings can be either OBDA or R2RML.
- Once Ontop-spatial receives a GeoSPARQL query, the query gets parsed. We modified the Sesame parser used by Ontop (and the javacc parser that the respective Sesame library uses), in order to extend its syntax to support geospatial operations in the filter clause of the query. Additionally, qualitative geospatial queries, (i.e., queries containing geospatial triple patterns such as `ex:feauture1 geo:overlaps ex:feature2`) are also supported as standard SPARQL triple patterns, and get transformed into their quantitative equivalents (i.e., queries with spatial filters) in the following step.
- Conventionally, the next step in Ontop is to translate the SPARQL query and the R2RML mappings into a Datalog program so that the query can be represented formally and optimized following a series of optimization steps described in detail in [5,14]. Ontop-spatial inherits these optimizations and extends the SPARQL-to-Datalog translation module. As explained in the previous section, the geospatial filters are transformed into Datalog using distinguished geospatial predicates. The same distinguished geospatial predicates are used in the case of the qualitative geospatial queries as well. As a result, both quantitative and qualitative representations of a GeoSPARQL query are transformed into the same SQL query in the following step.
- The optimized version of the query, as derived from the previous step, gets translated into SQL. Every geospatial Datalog predicate is mapped to the respective geospatial SQL operator, following the syntax of the underlying DBMS. The DBMS adapter has been extended in order to be able to identify geospatial columns in the database of the user. The PostgreSQL adapter has been modified and the Spatialite adapter has been added.

– The SQL query gets eventually executed in the underlying DBMS. Currently, the spatially-enabled databases that Ontop-spatial supports are the geospatial extensions of PostgreSQL and Sqlite, namely PostGIS and Spatialite respectively. More geospatial databases will be supported in the future.
– After the evaluation of the spatial SQL query in the DBMS, Ontop-spatial gets the results and sends them to the user. If geometries need to be projected, the SQL query that is produced returns the result as WKT. This enables Ontop-spatial to be used as a GeoSPARQL endpoint, that could serve as input endpoint for applications like linked geospatial data visualizers [10] to display the geometries that are returned as a result of a GeoSPARQL query.

Like the default version of Ontop, Ontop-spatial can be used as a web application (using Sesame workbench), as a Sesame library, as a Protege plugin, or it can be executed from the command line. The virtual geospatial graphs created by Ontop can also be materialized, creating an RDF dump, so that it can then be imported in a geospatial RDF store.

Ontop-spatial is available as free and open source software at the following link: https://github.com/ConstantB/Ontop-spatial.

Ontop-spatial in use. The motivation behind the development of Ontop-spatial was the Statoil use case of the project Optique, in order to address the issue of creating virtual RDF graphs on top of large databases that contain geometries and get frequently updated. Ontop-spatial is also being used in the urban accountant, land management and crisis mapping services of the EU FP7 project MELODIES[4]. Finally, ontop-spatial has recently be used in the Maritime security domain, in collaboration with Airbus and the University of Bolzano [4].

5 Evaluation

We conducted an empirical evaluation of our implementation based on the philosophy of Geographica[5], a benchmark for testing the performance of geospatial RDF stores [7]. Geographica consists of a *micro benchmark* and a *macro benchmark*. The *micro benchmark* is designed for testing basic geospatial operators, such as spatial selections and spatial joins. The *macro benchmark* tests the performance of the evaluated systems using queries that correspond to real application scenarios. As our aim is not to test geospatial RDF stores as done in [7], we use a modified benchmark based on the micro benchmark of Geographica as we explain later in this section.

Since there was no alternative OBDA systems that allow for posing GeoSPARQL queries over geospatial relational databases, we decided to evaluate Ontop-spatial in comparison with a geospatial RDF store. We consider that the spatiotemporal RDF store Strabon [8] is a good representative of the

---

[4] http://www.melodiesproject.eu/software-tools.
[5] http://geographica.di.uoa.gr/.

family of the geospatial RDF stores to compare with as (i) it is a state-of-the-art geospatial RDF store both in terms of functionality and performance [7,8] (ii) it supports a big subset of GeoSPARQL (apart from stSPARQL), and (iii) it uses a spatially-enabled DBMS as back-end, performing a SPARQL-to-SQL translation following a specific storage scheme as explained in [8]. This enables us to use the same DBMS (PostGIS with the same configuration and tuning) and perform a comprehensive comparison.

5.1 Datasets

Geospatial data come, in most cases, in native geospatial data formats. In a real-world scenario, a user that works with geospatial data obtains it as files in a geospatial data format (e.g., a shapefile) and stores it either in a GIS or a spatially-enabled relational database. Later on, he may convert the data into RDF and store it in a geospatial RDF store in order to combine it with other linked data.

The benchmark Geographica is based on such real-world geospatial application scenarios and for the experimental evaluation of Ontop-spatial we will also follow this approach: We will import real geospatial datasets in a spatially-enabled relational database and use it as the back-end of Ontop-spatial.

We chose to use the datasets of Geographica that are available in their original format (shapefiles). These datasets are the Corine Land Cover dataset of Greece, which is provided by the European Environment Agency (EEA), the Greek Administrative Geography (GAG), and the Hotspots dataset provided by the National Observatory of Athens. We complemented these data sources with the original raw files of OpenStreetMap data about Greece which are available as shapefiles.[6] Geographica uses the RDF versions of the same subset of the OSM datasets created by the project LinkedGeoData[7]. For the rest of this paper, we will refer to this dataset using the acronym of the resulting, RDF-ized version (LGD). We added more OSM categories to our workload (e.g., buldings, waterways, etc.), as we will exploit the fact that each one is contained in a different shapefile (so it will be imported into a different table), to stress our system as we explain later on in this section.

For the evaluation of Ontop-spatial, we imported the shapefiles in a PostGIS database using the **shp2pgsql** command as described here: https://github.com/ConstantB/Ontop-spatial/wiki/Shapefiles. In this way, each shapefile is loaded into a separate table in the database. Each one of these tables contains a column where geometries are stored in binary format (WKB) and an index has been built on that column. Then, we created the minimum set of mappings in order to pose the queries of the benchmark. We used PostgreSQL version 9.1.13 and PostGIS 2.0.3, performing the fine tuning configurations suggested here: http://geographica.di.uoa.gr.

---

[6] http://download.geofabrik.de/europe/greece.html.
[7] http://linkedgeodata.org/.

Table 3 shows information about the datasets described above, such as the disk size that each of these tables occupy, the number of tuples and the average number of points per geometry. Notice that the LGD dataset consists of 7 shapefiles/tables which is important in the OBDA setting as we will explain later on. Also, LGD-Places and LGD-Points contain only point geometries.

In order to compare the performance of our system with Strabon, we materialized the virtual geospatial RDF graphs produced by Ontop-spatial and stored them in Strabon, so that both the virtual RDF graphs produced by Ontop-spatial and the graphs stored in Strabon contain exactly the same information. The produced RDF dump consists of 5.620.482 triples and contains 855.502 geometries. The total PostGIS database size (in terms of disk usage) of Ontop-spatial is 700 MB. The respective size of the PostGIS database that was produced after loading the RDF dump to Strabon is 1665 MB, which is more than twice the disk space compared to the original database produced by importing the shapefiles directly. The reason is that in the first case the database stores the data, while in the second case the database stores the equivalent set of triples. This kind of overhead is common in RDF stores that use a relational database as back-end. Also, Strabon inherits the *per-predicate* storage scheme of the Sesame RDBMS package, so every predicate is stored in a different table and additional tables are used for dictionary encoding. According to this storage scheme, all geometries are stored in a table called *geo_values* in WKB format and the respective column is indexed using an R-tree-over-GiST index, as described in [8].

5.2 Queries

The GeoSPARQL queries that we used for the experimental evaluation of our system are a set of *spatial selections* and a set of *spatial joins*. We used some of the queries of Geographica, and some queries that are appropriate in the OBDA setting as we will explain in the rest of this section. The queries used in our evaluation are presented in Tables 1 and 2. Each query has a numeric identifier, a mnemonic label, a number that shows how many BGPs it consists of and a number that shows how many results it returns.

Both spatial selection and spatial join queries contain a spatial filter that checks if a spatial relation holds between two geometries that are given as arguments to the respective GeoSPARQL function. In the case of spatial selections, one of the arguments is a variable and the other one is a constant, which can be either a line (queries suffixed with "L" in the query label) or a polygon (using "P" suffix). In spatial join queries, both arguments of the respective spatial binary operator are variables. The first set of queries that we consider contains simple geospatial queries, i.e., queries consisting of a single triple pattern to retrieve the geometries of a dataset and a spatial filter (spatial selections 00–14 and spatial joins 00–03). Note that spatial joins require at least two triple patterns to retrieve the geometries that will be bound to the variables that are involved in the spatial filter. This kind of queries test the response time of the compared systems to perform "pure" geospatial queries (i.e., involving the least possible

```
mappingId        lgd_buildings_geometry
target           lgd:{gid} lgd:asWKT {geom}^^geo:wktLiteral .
source           select gid, geom from buildings

mappingId        lgd_landuse_geometry
target           lgd:{gid} lgd:asWKT {geom}^^geo:wktLiteral .
source           select gid, geom from landuse
```

Fig. 2. Examples of geospatial mappings for two LGD tables

```
select ?s1 ?o1 where {
       ?s1 lgd:asWKT ?o1 .
  filter(geosparql:FUNCTION(SPATIAL_CONSTANT,?o1)).}
```

Fig. 3. Template for spatial selection queries

number of triple patterns, focusing as much as possible on the evaluation of the spatial condition).

The next set of queries that we consider tackles an important issue that is crucial in OBDA systems: the generation of Union operators, deriving from the ontology and the schema of the database in the SPARQL-to-SQL translation phase. For example, the LGD dataset consists of 7 shapefiles, each one containing a column where geometries are stored. But according to the ontology, the data property that connects a spatial object with its geometry is universal for all spatial objects in the dataset. We present the mappings for two of these tables/shapefiles in Fig. 2.

Let us now consider the template for spatial selection queries in Fig. 3. The translated SQL query corresponding to a GeoSPARQL query following this template would create unions in order to fetch results deriving from all the tables it has been mapped to, that is, all seven LGD tables, and then apply the spatial selection to this union. This is the case for spatial selection queries 15–19. In order to test how our system responds by increasing/decreasing the number of unions produced in the translated query, we add an additional, thematic filter that selects a different number of LGD categories each time, thus affecting a different number of tables, and producing different number of unions, respectively. For example, consider query 19 which is shown in Listing 1.1, which contains an OR-condition in the second filter, so the respective translated query contains a union.

Listing 1.1. Query 19

```
select distinct ?s1  where {
?s1 lgd:asWKT ?o1 .
?s1 rdf:type ?type .
filter(geof:sfIntersects(GEOMETRY,?o1))
filter( ?type = lgd:Road ||
?type = lgd:Waterway ) }
```

Listing 1.2. Spatial join query 6

```
select ?s1 ?s2 where {
?s1 lgd:asWKT ?o1 .
?s2 lgd:asWKT ?o2 .
 (geo:sfIntersects(?o1,?o2))
}
```

The queries 15, 16, 17, and 18 produce 6, 4, 3, and 4 unions respectively. The presence of unions has a negative impact on the query response time, but

Table 1. Spatial selections description

No	Query	#BGPs	results
00	Equals_GADM_P	1	0
01	Contains_GADM_P	1	9
02	Contains_GADM_P	1	0
03	Equals_GADM_L	1	1
04	Overlaps_GADM_L	1	0
05	Contains_GADM_L	1	0
06	Intersects_CLC_L	1	5
07	Contains_CLC_L	1	0
08	Equals_CLC_L	1	5
09	Overlaps_CLC_L	1	0
10	Overlaps_CLC_P	1	132
11	Intersects_CLC_P	1	533
12	Contains_CLC_P	1	401
13	Equals_CLC_P	1	0
14	Intersects_LGD_P	2	2749
15	Intersects_LGD_B	2	2749
16	Intersects_LGD_PL	2	2626
17	Intersects_LGD_P	2	2522
18	Intersects_LGD_LU	2	2722
19	Intersects_LGD_ROA	2	2387
20	Intersects_LGD_bigP	1	729189
21	Intersects_LGD_P2	3	5

Table 2. Spatial joins description

No	Query	#BGPs	results
00	Within_CLC_GADM	2	34114
01	Intersects_GADM_GADM	2	1556
02	Overlaps_GADM_CLC	2	17035
03	Intersects_LGD_GADM	3	154725
04	Intersects_LGD_LGD_Mus	4	2
05	Intersects_LGD_GADM	2	819319
06	Intersects_LGD_LGD	1	3686229
07	Crosses_LGD_LGD_Roads	4	178602

Table 3. Workload characteristics

Dataset	Size	Tuples	$Avg \frac{\#points}{geometry}$
CLC	283 MB	44834	187.84
Hotspots	35 MB	37048	5
GAG	24 MB	326	3020.14
LGD-Buildings	42 MB	155474	6.5
LGD-Landuse	20 MB	40220	19.4
LGD-Places	2.4 MB	13043	1
LGD-Points	12 MB	61664	1
LGD-Railways	2 MB	4996	13.3
LGD-Roads	250 MB	514403	19
LGD-Waterways	16 MB	20565	39.84

things get even worse when unions appear in spatial joins (e.g., spatial join query 6). Since variables appear in the spatial filters that serve as the conditions of the spatial joins, all combinations of the respective tables that are involved in the corresponding mappings should be spatially joined pairwise. For example, consider the spatial join query 6 which is given in Listing 1.2. This query performs a spatial join with the condition **intersects** in *all* LGD tables that are involved in the mappings containing the predicate *lgd:asWKT*. This join is translated into the corresponding relational algebra expression as follows:

$$(L_{buildings} \cup L_{luse} \cup ... \cup L_{waterways}) \bowtie_{sf} (L_{buildings} \cup L_{luse} \cup ... \cup L_{waterways})$$

where $L_{buildings}$, L_{luse},..., $L_{waterways}$, etc. are LGD tables and sf is spatial operator corresponding to $geof : sfIntersects$ from the query. The query engine evaluates this relational algebra expression as unions of joins and all involved tables get spatially joined pairwise.

Last, in order to measure how the selectivity of the queries affect the performance of the systems, we included the spatial selection queries 20 and 21 involve the computation of the intersection of all kinds of LGD areas with a specific polygon. This polygon is large in the case of spatial selection query 20 so that many geometries will be returned, while in spatial selection query 21 this polygon is small enough so that very few LGD areas intersect with it.

5.3 Results

Experimental Set Up. The experiments were carried out on a server with the the following specifications: Intel(R) Xeon(R) CPU E5620 @ 2.40 GHz, 12 MB

L3, RAID 5, 32 GB RAM and OS: Ubuntu 12.04. All experiments were carried out with both cold and warm cache. Queries are first executed in cold cache and then in warm cache. The queries for which the system under test times out (the time out threshold is set to 40 min) are not executed in warm cache. All queries and code we used to execute the experiments in both systems, can be found in the "experiments" branch of the github repository of Ontop-spatial (folder "benchmark") at https://github.com/ConstantB/Ontop-spatial.

Query response time. The results of our experimental evaluation can be seen in Figs. 4 and 5. Response time is measured in nanoseconds and presented in logarithmic scale. A general observation is that the query response time of Ontop-spatial is better than the one of Strabon, especially when big datasets are involved, both for spatial selections and spatial joins. Strabon times out after 40 min in spatial join queries 6 and 7. In spatial selection queries 2–5, although Ontop-spatial achieves better response time than Strabon in cold cache, it gets outperformed in warm cache, as intermediate results (which are not many as the dataset involved in this query is relatively small), are more likely to be found in the cache, increasing the hit rate of the cache and decreasing I/O requests. However, such differences between executions in warm and cold cache are eliminated in larger datasets. In what follows we explain why Ontop-spatial outperforms Strabon.

Listing 1.3. Spatial join query 2

```
select ?s1 ?s2 where {
?s1 clc:asWKT ?o1 .
?s2 gag:asWKT ?o2 .
 filter(geof:sfWithin(?o1, ?o2))}
```

Listing 1.4. Spatial join query 4

```
select ?s1 ?s2 where {
?s1 lgd:asWKT ?o1 .
?s1 rdf:type lgd:Building .
?s1 lgd:type"Museum" .
?s2 lgd:asWKT ?o2 .
?s2 rdf:type lgd:Landuse .
 filter(geof:sfIntersects(?o1,?o2))}
```

Listing 1.5. Ontop-spatial SQL query

```
SELECT
 1 AS "s1QuestType", NULL AS "s1Lang",
 ('http://geo.linkedopendata.gr/clc/'
 || REPLACE(...... || '/') AS "s1",
 1 AS "s2QuestType", NULL AS "s2Lang",
 ('http://geo.linkedopendata.gr/gag/ont/'
 || REPLACE(...'/') AS "s2"
FROM
clc QVIEW1,
gag QVIEW2
WHERE
QVIEW1."gid" IS NOT NULL AND
QVIEW1."geom" IS NOT NULL AND
QVIEW2."gid" IS NOT NULL AND
QVIEW2."geometry" IS NOT NULL AND
(ST_Within(QVIEW1."geom",QVIEW2."geometry"))
```

Listing 1.6. Strabon SQL query

```
SELECT a0.subj,
u_s2.value,
a2.subj,
u_s1.value
FROM aswkt_855211 a0
INNER JOIN geo_values l_o2
ON (l_o2.id = a0.obj)
INNER JOIN geo_values l_o1 ON
 ((ST_Within(l_o1.strdfgeo,
 l_o2.strdfgeo)))
INNER JOIN aswkt_135992 a2
ON (a2.obj = l_o1.id)
LEFT JOIN uri_values u_s2
ON (u_s2.id = a0.subj)
LEFT JOIN uri_values u_s1
 ON (u_s1.id = a2.subj)
```

The queries provided in Listings 1.5 and 1.6 are the SQL translations of the GeoSPARQL spatial join query 2, which is provided in Listing 1.3. One can observe that Ontop-spatial produces the same query as one would have written by hand in a geospatial relational database. Strabon produces some extra joins, as a result of the star schema that it follows in the database (and has been

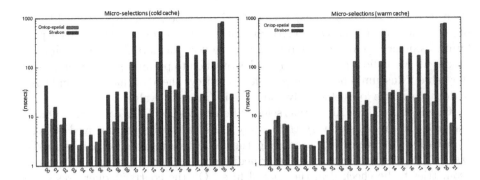

Fig. 4. Spatial Selections experiment (cold and warm cache)

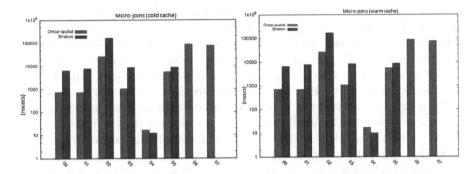

Fig. 5. Spatial Joins experiment (cold and warm cache)

inherited from the Sesame RDBMS that Strabon is built on), i.e., each predicate is stored in a different table and there are some additional tables used for dictionary encoding (tables storing URIs, one table for each different datatype, etc.). This has a negative impact on performance when many intermediate results are produced. In Strabon, geometries are stored in a single table, named *geo_values*, and are indexed on the geometry column using an R-tree-over-GiST index. On the other hand, Ontop-spatial stores each shapefile in a different table, and geometries are stored in a sepate column for each table, and a separate R-tree-over-GiST index is constructed for the geometries of each shapefile/table. As Table 3 shows, there are cases where geometries of a shapefile/table are of the same type (e.g., all contain points/linestrings/polygons), allowing Ontop-spatial to build smaller and more efficient indices.

Nevertheless, in spatial join query 4, Strabon outperforms Ontop-spatial. The query is provided in Listing 1.4. Using this query, we want to retrieve the land use of areas that intersect with Museums. This is a very selective query with respect to the thematic condition, so the PostGIS optimizer correctly chooses to perform the thematic conditions first so that only the geometries of Museums will be checked in the spatial condition that follows, and the R-tree index will be used.

Both systems execute the query very fast, with Strabon achieving nearly 4 times better performance than Ontop-spatial, as the overhead of the extra joins it performs, as described above, is reduced because very few intermediate results are produced. Also, dictionary decoding helps Strabon to perform string comparison (for value "Museum") only once, in order to retrieve the id of that value and then perform thematic joins efficiently using the id (numeric) value.

Queries 15–19 have filters that select different kinds of LGD categories. Query response time increases every time many LGD categories are involved (Query 15 asks for all categories), producing the respective number of unions in the case of Ontop-spatial and more intermediate results for Strabon, forcing more geometries to be checked in the spatial filter. On the contrary, query response time decreases when less LGD categories need to be selected.

The results of union-queries are more interesting in the case of spatial joins, shown in Fig. 5. One would expect that unions with spatial joins, as in the case of the spatial join query 6, would dramatically decrease the performance of Ontop-spatial. Indeed, query response time increases in the case of queries like query 6, but Ontop-spatial still performs better than Strabon. The explanation for this lies in the fact that each time a spatial join is performed between two different LGD tables, the optimizer chooses the one having the smaller index (and usually smaller geometries, in this case) to be nested inside the inner branch of the nested loop, where it performs an index scan. This has greater impact on the execution time of geospatial queries, as the evaluation of spatial joins is more expensive due to the cost of the evaluation of the spatial conditions.

In spatial selection query 20, the performance of the two systems is very close, while in the more selective version of the same query, i.e., spatial selection query 21, the gap in the execution times between Ontop-spatial and Strabon increases again. This happens because nearly every geometry in the workload is included in the results of the spatial selection query 20, so spatial indices are not useful in this case.

Overall, we observe that importing the shapefiles to a database and then using an OBDA approach is very efficient, as in most cases, the information that is contained in a shapefile is compact and homogeneous, as we often have one shapefile per data source. So, the SQL queries that are produced based on such a schema contain reduced amount of joins and can be executed efficiently.

6   Discussion and Conclusions

In this paper, we describe how we extended the techniques of [5,14] to develop the first geospatially-enabled OBDA system, named Ontop-spatial. By extending the OBDA system Ontop, Ontop-spatial inherits the advantages of using RDB2RDF systems in real use cases: (i) RDB-to-RDF workflow becomes less complicated, without having to use different tools for converting data into RDF and then storing it in RDF stores, (ii) no data needs to be transfered, as existing databases are used as input to the system, and (iii) mappings provide a layer of abstraction between the data manipulation/database experts and the end users.

These advantages have even greater impact when dealing with geospatial data. The domains where geospatial data are produced and used are dominated by geospatial databases and other tabular file formats that could easily be imported to a database (e.g., shapefiles). GIS practitioners use geospatial relational databases in their day-to-day tasks, either directly or as the back-end of applications to store and manipulate data (e.g., GIS have connectors for geospatial relational databases). Ontop-spatial provides a solution for combining the advantages of geospatial relational databases, for example, the wide variety of geospatial data operators and the performance achieved by the use of spatial indices, with the data modeling advantages of the RDF data model. Moreover, Ontop-spatial allows for encapsulating geospatial data manipulation functions offered by geospatial extensions to SQL (e.g., functions for transforming geometries to a different coordinate reference system) in the mappings.

On the other hand, Ontop-spatial inherits the disadvantages of the OBDA systems as well. First, in order to combine information coming from different geospatial sources, the data should be imported in databases. Second, as the database is given as input to the system, it is read-only and Ontop-spatial does not support SPARQL store or update operations; all updates should be done directly on the database level. Third, the performance of the system is heavily dependent on the ontology, the schema of the database, and the mappings, as we explained in the previous sections, which applies for OBDA approaches in general. However, our experiments showed that in many cases, our geospatially enchanced OBDA approach achieves significantly better performance than the state-of-the-art geospatial RDF store Strabon. The main reasons for this are summarized as follows:

- The database schema that is produced simply by importing the shapefiles to the database is in most cases suitable for OBDA approaches, as shapefiles contain compact and homogeneous information per dataset.
- The database produced by storing the materialized RDF dump that ontop exports in Strabon is bigger than the database that results from importing the shapefiles, even though only the RDF triples that were involved in the OBDA mappings (i.e., the *virtual* RDF triples) were exported. This happens because of (i) the normalization imposed by the RDF data model itself (i.e., triples) and (ii) the additional tables used for dictionary encoding.
- The additional joins that are created in the translated SQL queries of Strabon and the fact that geometries are stored in a single table where geospatial operators are performed increase even by more than an order of magnitude in very large workloads with many and complicated geometries, when many intermediate results are produced in queries.

In future work, we plan to continue the development of Ontop-spatial in the directions of (i) fully supporting GeoSPARQL and stSPARQL (i.e., adding also valid time support), and (ii) creating a distributed version of our extension exploiting the fact that the union-all spatial queries are parallelizable.

Acknowledgement. This work is partially supported by the EU projects Optique (318338) and MELODIES (603525). We would like to thank the Ontop development team for their support.

References

1. Open Geospatial Consortium: OpenGIS Simple Features Specification For SQL. OGC Implementation Standard (1999)
2. Open Geospatial Consortium: GeoSPARQL - A geographic query language for RDF data. OGC Candidate Implementation Standard (2012)
3. Bereta, K., Smeros, P., Koubarakis, M.: Representation and querying of valid time of triples in linked geospatial data. In: Cimiano, P., Corcho, O., Presutti, V., Hollink, L., Rudolph, S. (eds.) ESWC 2013. LNCS, vol. 7882, pp. 259–274. Springer, Heidelberg (2013)
4. Brüggemann, S., Bereta, K., Xiao, G., Koubarakis, M.: Ontology-based data access for maritime security. In: Sack, H., Blomqvist, E., d'Aquin, M., Ghidini, C., Ponzetto, S.P., Lange, C. (eds.) ESWC 2016. LNCS, vol. 9678, pp. 741–757. Springer, Heidelberg (2016). doi:10.1007/978-3-319-34129-3_45
5. Calvanese, D., Cogrel, B., Komla-Ebri, S., Kontchakov, R., Lanti, D., Rezk, M., Rodriguez-Muro, M., Xiao, G.: Ontop: answering SPARQL queries over relational databases. Semant. Web J. (2016, to appear)
6. Egenhofer, M.J.: A formal definition of binary topological relationships. In: Litwin, W., Schek, H.-J. (eds.) FODO 1989. LNCS, vol. 367, pp. 457–472. Springer, Heidelberg (1989). doi:10.1007/3-540-51295-0_148
7. Garbis, G., Kyzirakos, K., Koubarakis, M.: Geographica: a benchmark for geospatial rdf stores (long version). In: Alani, H., et al. (eds.) ISWC 2013, Part II. LNCS, vol. 8219, pp. 343–359. Springer, Heidelberg (2013)
8. Kyzirakos, K., Karpathiotakis, M., Koubarakis, M.: Strabon: a semantic geospatial DBMS. In: Cudré-Mauroux, P., et al. (eds.) ISWC 2012, Part I. LNCS, vol. 7649, pp. 295–311. Springer, Heidelberg (2012)
9. Kyzirakos, K., Vlachopoulos, I., Savva, D., Manegold, S., Koubarakis, M.: Geotriples: a tool for publishing geospatial data as RDF graphs using R2RML mappings. In: Proceedings of the ISWC 2014 Posters & Demonstrations Track, Riva del Garda, Italy, October 21, 2014, pp. 393–396 (2014)
10. Nikolaou, C., Dogani, K., Bereta, K., Garbis, G., Karpathiotakis, M., Kyzirakos, K., Koubarakis, M.: Sextant: visualizing time-evolving linked geospatial data. J. Web Sem. **35**, 35–52 (2015)
11. Polleres, A.: From SPARQL to rules (and back). In: Proceedings of the 16th International Conference on World Wide Web, WWW 2007, pp. 787–796. ACM, New York (2007)
12. Priyatna, F., Corcho, O., Sequeda, J.: Formalisation and experiences of R2RML-based SPARQL to SQL query translation using Morph. In: Proceedings of the 23rd International Conference on World Wide Web, pp. 479–490. ACM (2014)
13. Randell, D.A., Cui, Z., Cohn, A.G.: A spatial logic based on regions and connection. In: Proceedings of the 3rd International Conference on Principles of Knowledge Representation and Reasoning (KR 1992), Cambridge, MA, October 25–29, 1992, pp. 165–176 (1992)
14. Rodriguez-Muro, M., Rezk, M.: Efficient SPARQL-to-SQL with R2RML mappings. J. Web Semant. **33**(1), 141–169 (2015)
15. Sequeda, J., Miranker, D.P.: Ultrawrap: SPARQL execution on relational data. J. Web Semant. **22**, 19–39 (2013)

Expressive Multi-level Modeling
for the Semantic Web

Freddy Brasileiro[1]([✉]), João Paulo A. Almeida[1],
Victorio A. Carvalho[1,2], and Giancarlo Guizzardi[1]

[1] Ontology and Conceptual Modeling Research Group (NEMO),
Federal University of Espírito Santo (UFES), Vitória, ES, Brazil
freddybrasileiro@gmail.com, jpalmeida@ieee.org,
victorio@ifes.edu.br, gguizzardi@inf.ufes.br
[2] Research Group in Applied Informatics, Informatics Department, Federal
Institute of Espírito Santo (IFES), Colatina, ES, Brazil

Abstract. In several subject domains, classes themselves may be subject to categorization, resulting in classes of classes (or "metaclasses"). When representing these domains, one needs to capture not only entities of different classification levels, but also their (intricate) relations. We observe that this is challenging in current Semantic Web languages, as there is little support to guide the modeler in producing correct multi-level ontologies, especially because of the nuances in the constraints that apply to entities of different classification levels and their relations. In order to address these representation challenges, we propose a vocabulary that can be used as a basis for multi-level ontologies in OWL along with a number of integrity constraints to prevent the construction of inconsistent models. In this process we employ an axiomatic theory called MLT (a Multi-Level Modeling Theory).

Keywords: Multi-level modeling · Metamodeling · Semantic web · OWL

1 Introduction

The Semantic Web, or Web of Data, provides a common framework that allows data to be shared across application, enterprise, and community boundaries [1]. This is achieved by linking and publishing structured data using RDF languages, which provide a basis for producing reusable vocabularies for various domains of interest [2].

A Semantic Web vocabulary is built using the basic notion of *class*, which is present in both RDF Schema (RDFS) [3] and in the Web Ontology Language (OWL) [4]. A class (or type) is a ubiquitous notion in modern conceptual modeling approaches and is used to establish invariant features of the entities in a domain. Often, subject domains are conceptualized with entities in two levels: a level of classes, and a level of individuals which instantiate these classes. In many subject domains, however, classes themselves may also be subject to categorization, resulting in classes of classes (or metaclasses). For instance, consider the domain of biological taxonomies [5]. In this domain, a given *organism* is classified into *taxa* (such as, e.g., *Animal, Mammal, Carnivoran, Lion*), each of which is classified by a *biological taxonomic rank*

© Springer International Publishing AG 2016
P. Groth et al. (Eds.): ISWC 2016, Part I, LNCS 9981, pp. 53–69, 2016.
DOI: 10.1007/978-3-319-46523-4_4

(e.g., *Kingdom, Class, Order, Species*). Thus, to represent the knowledge underlying this domain, one needs to represent entities at different (but nonetheless related) classification levels. For example, *Cecil* (the lion killed in the Hwange National Park in Zimbabwe in 2015) is an instance of *Lion*, which is an instance of *Species*. *Species*, in its turn, is an instance of *Taxonomic Rank*. Other examples of multiple classification levels come from domains such as software development [6] and product types [7].

The need to support the representation of knowledge domains dealing with multiple classification levels has given rise to an area of investigation called *multi-level modeling* [7, 8]. A number of research initiatives have also been conducted to support multi-level modeling in the Semantic Web (e.g., [9–12]). These approaches exploit the fact that a class is itself an RDF resource and may thus be the subject or object of triples. OWL 2 explicitly adopts this strategy under the term "metamodeling", enabling the representation of facts that are stated about classes [13].

Despite these developments, the current support for the representation of *domains dealing with multiple levels of classification* in the web still lacks a number of important features. In some cases, there are no criteria or principles for the organization of vocabularies into levels, leading to problematic classification and taxonomic statements (see, e.g. [14]). Further, there has been no attention to the representation of the relations between types at different levels. For example, in the biological domain, it is key to represent that all instances of *Species* are subtypes of *Organism* (even when particular species are not represented explicitly), and that all instances of *Organism* belong to one and only one *Kingdom*.

In this paper, we address the challenges in the representation of domains with multiple levels of classification in the Semantic Web by proposing an OWL vocabulary that can be used as a basis for multi-level ontologies. By defining a taxonomy of reusable relations between types, the vocabulary enables the expression of domain rules that would otherwise not be captured. The vocabulary is based on a reference axiomatic theory called MLT [15]. The axioms and theorems of MLT are used to derive integrity constraints for multi-level vocabularies, offering guidance to prevent the construction of inconsistent vocabularies. Further, MLT rules are used to infer knowledge about the relations between types that is not explicitly stated. We focus on the support for domain metaclasses as opposed to language metaclasses, i.e., we focus on ontological instantiation instead of linguistic instantiation [16].

This paper is further structured as follows: Section 2 presents basic requirements for the representation of knowledge in domains with multiple classification levels; Sect. 3 reviews the current support for multi-level modeling in OWL as well as in related work in the literature; Sect. 4 presents briefly the MLT multi-level theory; Sect. 5 presents our approach to represent multi-level models in OWL reflecting the rules of MLT; and Sect. 5.3 presents concluding remarks.

2   Requirements for a Multi-level Approach

An essential requirement for a multi-level modeling approach is *the ability to represent entities of multiple (related) classification levels*, capturing chains of instantiation between the involved entities (requirement **R1**). To comply with this requirement, the approach must admit entities that are, simultaneously, type (class) and instance (object) [17]. This means that a multi-level approach differs from the traditional two-level scheme, in which classification (instantiation) relations can only be established between classes and individuals. As a consequence, a multi-level modeling approach *should define principles for the organization of entities into levels* (**R2**). These principles should guide the modeler on the adequate use of classification (instantiation) relations. An example of this sort of principle, which is adopted in some prominent multi-level modeling approaches, is the so-called strict metamodeling principle [17]. It assumes that each element of a level must be an instance of an element of the level above. The lack of principles to guide organization of entities into levels may lead to the construction of unsound multi-level models. For example, in [14] we assessed Wikidata and found over 22,000 violations of the strict metamodeling principle. The identified problems seem to arise from inadequate use of instantiation and subclassing and could have been prevented with guidance from the editing/modeling environment.

Another important characteristic of domains with multiple levels of classification is that there are rules that apply to the instantiation of types of different levels. This kind of rule is present in an early and important approach for multi-level modeling, named the *powertype pattern* [18, 19], which establishes a relationship between two types such that the instances of a type (the so-called "powertype" or "higher-order" type) are specializations of a lower-level type (the so-called "base type"). For example, all instances of *Dog Breed* (e.g. *Collie* and *Beagle)* specialize the base type *Dog*. In order to represent *Dog Breed,* it is thus key to establish its relation with the *Dog* type (we call this sort of relation a *structural relation,* as it governs the instantiation of types at different levels). Further, one may need to represent whether an instance of *Dog* may instantiate: (i) only one, or (ii) more than one *Dog Breed*. In biological taxonomy, another rule concerning instantiation of types at different levels is that the instances of *Biological Taxonomic Rank* obey a sort of subordination chain such that every instance of *Phylum* specializes one instance of *Kingdom*, every instance of *Class* specializes one instance of *Phylum*, and so on. Thus, an expressive multi-level approach should be able *to capture rules for the instantiation of types at different levels* (**R3**).

Finally, in various domains, there are relations which may occur between entities of different classification levels. For example, consider the following domain rules: (i) each *Car* has an *owner* (a *Person*), (ii) each *Car* is classified as instance of a *Car Model*, and (iii) each *Car Model* is designed by a *Person*. In this domain, instances of *Person* (individuals) must be related simultaneously with instances of *Car Model* (which are classes) and also with instances of *Car,* i.e., instances of instances of *Car Model.* Thus, a *multi-level modeling approach should allow the representation of domain relations between entities in different classification levels* (**R4**).

3   Related Work

In this section, we review existing approaches to support the representation of multiple levels of instantiation, with a focus on multi-level modeling in RDF languages.

3.1 RDFS(FA)

In an early effort to organize the metamodeling architecture for RDF Schema (RDFS) 1.0 [20], Pan and Horrocks proposed RDFS(FA) [9]. They observed that "RDFS uses a single primitive *rdfs:Class* to implicitly represent possibly infinite layers of classes" (as it is an instance of itself) and that this creates barriers for understanding. They show examples on how this lack of a principle of organization for levels creates a so-called "layer mistake". Inspired by the fixed UML metamodeling architecture [21], they proposed the use of four layers: Metalanguage (M), Language (L), Ontology (O) and Instance (I). The *M Layer* is responsible for defining the language, where modelling primitives of this topmost layer have no types. The *L Layer* defines a language for specifying vocabularies and each entity in this layer is an instance of an entity in the *M Layer*. Vocabularies are defined in the *O Layer* ("Person" and "Animal" are examples of classes in this layer) and each element in this layer is an instance of an element in the *L Layer*. Lastly, the *I Layer* is populated with concrete individuals, which are instances of the vocabulary defined in *O Layer*.

Figure 1 shows the result of applying this architecture to RDFS. RDFS classes are replicated in the *M* and *L Layers* with the respective prefix (*M* and *L*). In *O* layer, *Animal* and *Person* are represented as instances of *rdfsfa:LClass* (instead of *rdfs: Class*); and *John* and *Mary* in the *Instance Layer*, as an instance of *Person*.

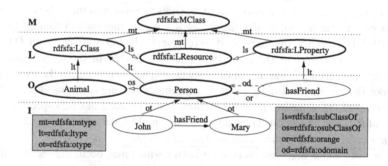

Fig. 1. Example of directed labeled graph of RDFS(FA) (from [9])

This architecture organizes the language engineering effort, but it does not aim to address the representation of domains with multiple levels of classification. In fact, it is based on the two-level scheme for the representation of domains in the *O* and *I* layers, with classes at the *O* layer, and individuals at the *I* layer, related through *rdfsfa:otype* (which represents what is known as ontological instantiation [16]). Metaclasses are only used in the *domain-independent L* layer; classes at the *O* layer are related to classes at the *L* layer through *rdfsfa:ltype* (which represents what is known as linguistic instantiation [16]). In order to represent a domain type such as *Species* one would be forced to include it in the *L* layer, specializing *rdfsfa:LClass*, which would be inadequate according to [9], as language and ontology issues would be confused. In this case, one would have to instantiate *Species* using *rdfsfa:ltype*, clearly misusing linguistic instantiation [16]. In conclusion, RDFS(FA) satisfies requirements R1 and R2 only for linguistic instantiation, but not for ontological instantiation.

3.2 OWL 2

OWL 2 [4] explicitly introduced support for *metamodeling*, enabling the representation of classes of classes. For example, in Fig. 2, two subclasses of *Eagle*, namely *Golden Eagle* and *Steppe Eagle* are defined as instances of *Species*, which means that they are member of the set of all species. In Fig. 2 (as well as in the remainder of the paper) we use a notation that is largely inspired in UML. We use UML specialization to represent the *rdfs:subClassOf* properties, and dashed arrows to represent statements, with labels to denote the names of the predicates that apply. For instance, a dashed arrow labeled *rdf:type* between *Golden Eagle* and *Species* represents that the former is an instance of the latter. Finally, we use the instance specification notation (i.e., underlining an element's name) to represent an individual (e.g. Harry).

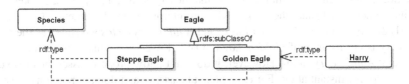

Fig. 2. OWL representation for biological taxonomic domain

OWL's multi-level modeling support is based on the notion of contextual semantics [10], often referred to as "punning", which means that a class is seen as an individual when it is an instance of another class, and that its interpretation as a class and as an individual are completely independent of each other. This "independent" interpretation means that a constraint stated to a class will not be considered when it is seen as an individual, which leads to non-intuitive interpretations [11]. For instance, consider the following statements: (i) *Harry* is an instance of *Golden Eagle*, and; (ii) *Golden Eagle* is the same as *Aquila chrysaetos*. Statement (i) treats *Golden Eagle* as a class, while statement (ii) treats *Golden Eagle* as an individual. These two aspects of *Golden Eagle* are never considered at the same time for reasoning. Thus, in this approach, it is impossible to infer that Harry is an instance of *Aquila chrysaetos*, which violates our intuition with respect to the multi-level model. We can say that while OWL 2 seems to satisfy R1 (admitting classes that are also instances), it does so only partially, given the notion of contextual semantics employed. The same can be said for the representation of relations between entities of different levels (partially satisfying R4).

OWL offers no principle of organization into levels (failing to satisfy R2). Further, punning also prevents us from correctly expressing the relation between a higher-order class and a base class in the powertype pattern, which inevitable involves considering the specializations of the base class as types and instances simultaneously (failing thus to satisfy R3). Finally, considering the open world assumption, it is also impossible to formally identify in this fragment above that Harry is an individual, as there could be unstated *rdf:type* declarations involving Harry as a class. Further, given the same assumption, it would be impossible to identify that *Species* (in isolation) is a metaclass; in other words, we cannot express when modeling *Species* (and omitting its instances) that all its instances are classes (in particular subclasses of *Organism*).

3.3   OWL FA

Later, Pan and Horrocks also proposed OWL FA [11], a metamodeling extension of OWL 1 DL, with an architecture based on RDFS(FA). They argue that OWL 1 Full supports some metamodeling by allowing users to use the built-in vocabulary without restrictions, but that leads to undecidability (as Motik pointed out [10]). They then propose a decidable extension of OWL 1 DL that can reuse existing reasoners.

While RDFS(FA) uses prefixes (M, L, O and I) to indicate the layer in which a class or axiom belongs, OWL FA intuitively introduces a layer number in its constructors and axioms, through annotations. The semantics of OWL FA [22, 23] takes into account elements that share the same URIs and interpret them dependently (in contrast to OWL 2). For instance, if *Golden Eagle* and *Aquila chrysaetos* are stated as the same and *Harry* is an instance of *Golden Eagle*, OWL FA assumes that *Harry* must be an instance of *Aquila chrysaetos*. However, it does not allow property assertions between layers except for instantiation. For example, subclassing and domain relations must be between classes at the same layer (failing thus to satisfy R4).

While RDFS(FA) allows instantiations only from *Instance Layer to Ontology Layer*, OWL FA allows the representation of multiple levels of instantiation. Thus, we understand here that identifying layers by numbers addresses the limitation of RDFS (FA) (see Sect. 3.1) thus satisfying R1 fully. Moreover, as advantages when compared to the current *multi-level modeling* support of OWL 2 (see Sect. 3.2), OWL FA: (i) interprets dependently elements that share the same URI, and; (ii) it introduces restrictions for instantiation and subclassing, providing some criteria for the organization into levels (R2). Finally, OWL FA offers no special support for the representation of constraints for the instantiation of types at different levels (not satisfying R3).

3.4   PURO

Svatek et al. [12] proposed the PURO approach which includes an OWL vocabulary that can be used as a basis for multi-level domain vocabularies. In PURO, each entity of a domain vocabulary can be annotated with a PURO term in order to clarify the entity's ontological status. The term *B-object* is used to refer to concrete individuals in the world (such as *Harry*). In contrast, the term *B-type* is used to refer to classes (such as *Eagle*). A *B-type* is analogous to an OWL class, however, *B-types* are organized into strata: instances of 1st *order B-types* are *B-objects*, instances of nth-*order B-types* are $(n - 1)^{th}$-*order B-types* (for n > 1). The OWL vocabulary supporting the PURO approach only deals with *B-objects* and first-, second- and third-order *B-types*. *B-relationship* is analogous to an object property assertion and there are variations: (i) *B-instantiation* is an assertion to indicate that an entity instantiates a *B*-type; (ii) *B-axiom* express a relationship between the extensions of two *B-types* (e.g., subclassing); and (iii) *B-fact* express information about an entity, e.g., who discovered certain species. Finally, *B-relation* is analogous to OWL Object Property.

Similarly to OWL 2 and OWL FA, PURO has the required expressivity for representing multiple levels of instantiation (R1) through the notions of *B-object* and the *B-types*. Moreover, PURO defines rules for the organization of entities along levels (R2). Finally, PURO allows modelers to express domain relations between entities of different levels (R4); an example is provided in [12] in which a musician is considered an expert in a *type* of instrument (e.g., the musician *Yo-Yo Ma* is an expert in *Violin*). However, similarly to OWL 2 and OWL FA, PURO offers no special support for the representation of constraints for the instantiation of types at different levels (not satisfying R3).

3.5 Intermediate Conclusions

Table 1 summarizes the current support provided by each of the efforts discussed here according to the requirements defined in Sect. 2. We classified this support in three categories: fully covered (+), partially covered (±) and not covered (−). Despite providing support and guidance for representing multiple levels of classification, RDFS (FA) focuses on linguistic instantiation instead of ontological instantiation, hence the partial support for R1 and R2. OWL 2 fails in the representation of relations and constraints crossing levels, due to its contextual semantics, and hence offers partial support for R1 and R4. OWL FA and PURO offer full support for R1 and R2 through annotations, and PURO also supports domain relations crossing levels (R4). Despite the efforts in all these approaches, none of them support the representation of constraints involving instantiation relations across levels (thus, not satisfying R3).

Table 1. Support for multi-level modeling in RDFS languages

Requirement	RDFS (FA)	OWL 2	OWL FA	PURO
R1 – represents entities of multiple levels of classification	±	±	+	+
R2 – offers guidance for the organization of entities into levels	±	−	+	+
R3 – represents rules for the instantiation of types at different levels	−	−	−	−
R4 – supports domain relations between entities of different levels	−	±	−	+

4   MLT: A Theory for Multi-level Modeling

Motivated by the lack of theoretical foundations for multi-level modeling, some of us have proposed a formal axiomatic theory called MLT [15] founded on the notion of (ontological) instantiation. MLT has been used successfully to analyze and improve the UML support for modeling the powertype pattern [24], to uncover problems in multi-level taxonomies on the web [14] and to provide conceptual foundations for dealing with types at different levels of classification both in core [25] and in foundational ontologies [26].

The theory is defined using first-order logic, quantifying over all possible entities (individuals and types). The *instance of* relation is represented in this formal theory by a binary predicate *iof(e,t)* that holds if an entity *e* *is instance of* an entity *t* (denoting a type). In order to accommodate the varieties of types in the multi-level setting, the notion of *type order* is used. Types having individuals as instances are *first-order types*, types whose instances are first-order types are *second-order types* and so on.

The logic constant "Individual" is used to define the conditions for entities to be considered individuals: *an entity is an instance of "Individual" iff it does not have any possible instance* (Axiom A1 in Table 2). The constant "First-Order Type" (or shortly "1stOT") *characterizes the type that applies to all entities whose instances are instances of "Individual"* (A2 in Table 2). Analogously, each *entity whose possible extension contains exclusively instances of "1stOT" is an instance of "Second-Order Type"* (or shortly "2ndOT") (A3 in Table 2). It follows from axioms A1, A2 and A3 that "Individual" is instance of "1stOT" which, in turn, is instance of "2ndOT". We call "Individual", "1stOT" and "2ndOT" the basic types of MLT. According to MLT, every possible entity must be instance of exactly one of its basic types (except the topmost type) (A4 in Table 2). We consider here only first- and second-order types. However, this scheme can be extended to consider as many orders as necessary [15].

Table 2. MLT axioms

A1	$\forall x\, iof(x, \text{Individual}) \leftrightarrow \nexists y\, iof(y,x)$
A2	$\forall t\, iof(t, 1stOT) \leftrightarrow \left(\exists y\, iof(y,t) \wedge \left(\forall x\, iof(x,t) \rightarrow iof(x, \text{Individual})\right)\right)$
A3	$\forall t\, iof(t, 2ndOT) \leftrightarrow (\exists y\, iof(y,t) \wedge (\forall t' iof(t',t) \rightarrow iof(t', 1stOT)))$
A4	$\forall x\, \left(iof(x, \text{Individual}) \vee iof(x, 1stOT) \vee iof(x, 2ndOT)\right) \vee (x = 2ndOT)$
D1	$\forall t1, t2\, specializes(t1,t2) \leftrightarrow (\exists y\, iof(y,t1) \wedge (\forall e\, iof(e,t1) \rightarrow iof(e,t2)))$
D2	$\forall\, t1, t2\, properSpecializes(t1,t2) \leftrightarrow (specializes(t1,t2) \wedge t1 \neq t2)$
D3	$\forall t1, t2\, isSubordinateTo\,(t1,t2) \leftrightarrow$ $(\exists x\, iof(x,t1) \wedge (\forall t3\, iof(t3,t1) \rightarrow (\exists t4\, iof(t4,t2) \wedge properSpecializes(t3,t4))))$
D4	$\forall t1, t2\, isPowertypeOf(t1,t2) \leftrightarrow (\exists x\, iof(x,t1) \wedge (\forall t3\, iof(t3,t1) \leftrightarrow specializes(t3,t2)))$
D5	$\forall t1, t2\, characterizes(t1,t2) \leftrightarrow (\exists x\, iof(x,t1) \wedge (\forall t3\, iof(t3,t1) \rightarrow properSpecializes(t3,t2)))$
D6	$\forall t1, t2\, completelyCharacterizes(t1,t2) \leftrightarrow$ $(characterizes(t1,t2) \wedge (\forall e\, iof(e,t2) \rightarrow \exists t3\, (iof(e,t3) \wedge iof(t3,t1))))$
D7	$\forall t1, t2\, disjointlyCharacterizes\,(t1,t2) \leftrightarrow$ $(characterizes(t1,t2) \wedge \forall e, t3, t4\, ((iof(t3,t1) \wedge iof(t4,t1) \wedge iof(e,t3) \wedge iof(e,t4)) \rightarrow t3 = t4)))$
D8	$\forall t1, t2\, partitions(t1,t2) \leftrightarrow (completelyCategorizes(t1,t2) \wedge disjointlyCategorizes(t1,t2))$

Some structural relations to support conceptual modeling are defined in MLT, starting with the ordinary specialization between types. A type t *specializes* another type t' iff *all instances of t are also instances of t'* (see definition D1 in Table 2). Since the reflexivity of the *specialization* relation may be undesired in some contexts, we define in MLT the *proper specialization* relation as follows: *t proper specializes t' iff t specializes t' and t is different from t'* (see D2 in Table 2). Additionally, MLT defines a *subordination* relation. *Subordination* between two higher-order types implies *specializations* between their instances, i.e., *t is subordinate to t'* iff every *instance of t proper specializes* an *instance of t'* (see D3 in Table 2). The definitions presented thus far guarantee that both *specializations*, *proper specializations* and *subordinations* may hold exclusively between types of the same order. We term these *intra-level relations*.

MLT also defines relations that occur between types of adjacent orders, the so-called *cross-level structural relations*. These relations are inspired on different notions of powertype in the literature. Based on the notion of *powertype* proposed by Cardelli [19] (which is founded on the notion of powerset), MLT defines a *powertype* relation between a higher-order type and a base type at a lower order: a type t *is powertype of* a base type t' iff all instances of t *specialize t'* and all possible *specializations of t' are instances of t* (see D4). Note that it follows from the axioms and definitions presented so far that "1stOT" *is powertype of* "Individual", i.e. all possible instances of "1stOT" specialize "Individual" and all possible specializations of "Individual" are instances of "1stOT". Analogously, "2ndOT" *is powertype of* "1stOT", and so on. Thus, every instance of a basic higher-order type ("1stOT" and "2ndOT") must specialize the basic type at the immediately lower level (respectively, "Individual" and "1stOT"). In other words, the notion of orders or levels in MLT can be seen as a result of the iterated application of Cardelli's notion of powertype to the basic types.

Odell [18], in turn, defined *powertype* simply as a type whose instances are subtypes of another type (the *base type*), excluding the *base type* from the set of instances of the *powertype*. Inspired on Odell's definition for powertypes, MLT defines the *characterization* relation between types at adjacent levels: *a type t characterizes a type t' iff all instances of t are proper specializations of t'* (definition D5). The *characterization relation* occurs between a higher-order type t and a base type t' when *the instances of t specialize* t' according to a specific *classification criteria*. Thus, differently from the cases involving (Cardelli's) *is powertype of* relation, there may be specializations of the base type t' that are not instances of t. For example, we may define a type named "Organism by Habitat" (with instances "Terrestrial Organism" and "Aquatic Organism") that *characterizes* "Organism", but is not a *powertype of* "Organism" since there are specializations of "Organism" that are not instances of "Organism by Habitat" (e.g. "Plant" and "Golden Eagle").

MLT defines some refinements of the cross-level relation of characterization, which are useful to capture further constraints in multi-level models. We consider that *a type t completely characterizes t' iff t characterizes t' and every instance of t' is instance of, at least, an instance of t* (D6). Moreover, *iff t characterizes t' and every instance of t' is instance of, at most, one instance of t* it is said that *t disjointly characterizes t'* (D7). Finally, a common use for the notion of powertype in the literature considers a higher-order type that, simultaneously, completely and disjointly characterizes a lower-order type. To capture this notion MLT defines the *partitions* relation. Thus, *t partitions t' iff each instance of the base type t' is an instance of exactly one instance of t* (D8). For example, considering the biological taxonomy for living beings we have that "Species" (and all other biological ranks) *partitions* "Organism".

A complete formalization of MLT in first-order logic can be found in [15], which presents proofs for all MLT theorems. Further, a formal specification in Alloy is provided in [27] and was used to verify the theorems and to simulate admissible models of the theory using the Alloy analyzer.

5 Applying MLT for Multi-level Modeling Support in OWL

Aiming to improve the OWL support for multi-level modeling, we propose (i) a vocabulary based on distinctions put forth by MLT, and (ii) a number of derivation and integrity rules reflecting axioms and theorems of MLT. The proposed vocabulary aims at providing modelers with an expressive set of constructs to support the production of multi-level ontologies in OWL. The integrity rules, in their turn, are used to verify if ontologies built using the proposed vocabulary are well-formed according to MLT rules. Finally, the derivation rules make use of MLT rules to infer information not represented explicitly by the modeler.

5.1 OWL Vocabulary Based on MLT Distinctions

The proposed vocabulary encompasses the representation of the basic types of MLT and the relations defined in the theory. The basic types of MLT are represented as instances (*rdf:type*) of *owl:Class*. The class representing the MLT *Individual* basic type is named *mlt:TokenIndividual*[1], the class representing the *First-Order Type* is named *mlt:1stOrderClass,* and the classes *mlt:2ndOrderClass* and *mlt:3rdOrderClass* represent, respectively, the *Second-order* and *Third-order basic types.* Considering that, according to MLT, instances of *Individual* are not instantiable (i.e. are not types), *mlt: TokenIndividual* does not specialize *owl:Class*. In contrast, the classes representing all other basic types have a *rdf:subClassOf* relation with *owl:Class* capturing the fact that their instances are classes (i.e. their instances are instantiable) (see Fig. 3).

Concerning the MLT relations, *instance of* relations are represented as *rdf:type* properties and *specialization* relations are represented as *rdfs:subClassOf* properties. All other intra- and cross-level relations of MLT are represented in this vocabulary in a hierarchy of instances of *owl:ObjectProperty*, including at the top: *mlt:intraLevelProperty*, which is as a super-property for all MLT intra-level relations; and *mlt:crossLevelProperty*, which is a super-property for all MLT cross-level relations. The *subordination* relation of MLT is then represented by the property *mlt:isSubordinateTo* as a sub-property of *mlt:intraLevelProperty,* while the *characterization* (*mlt:characterizes*) and the *is power type of (mlt:isPowertypeOf)* relations are represented as sub-properties of *mlt:crossLevelProperty.* Finally, each variation of

---

[1] The term "TokenIndividual" was adopted here to avoid confusion with the term "Individual" in the OWL specification. "TokenIndividual" corresponds to what we call "Individual" in [15].

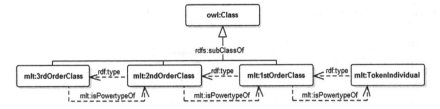

Fig. 3. Fragment of MLT vocabulary for order classes and individual.

characterization (e.g. *complete characterization, disjoint characterization* and so on) is represented as a sub-property of *mlt:characterizes*.

These properties are also used in the vocabulary definition to represent relations that occur between the basic types of MLT. To capture the fact that the basic type in one *order* is instance of the basic type in an immediately higher order, statements with *rdf:type* are defined between the classes representing the basic types (e.g., *mlt: TokenIndividual rdf:type mlt:1stOrderClass, mlt:1stOrderClass rdf:type mlt:2ndOrderClass*). Further, *mlt:isPowertypeOf* is used to represent that a basic type in an order *is the powertype of* the basic type in the immediately-lower order (Fig. 3).

The MLT vocabulary allows the representation of domain rules concerning the instantiation of types in different levels. For example, Fig. 4 illustrates a fragment of an ontology in the biological taxonomy domain applying this vocabulary. In such an ontology, *Genus* and *Species* are represented as instances of *mlt:2ndOrderClass* (and, thus, as subclasses of *mlt:1stOrderClass*) meaning that their instances (e.g. *Panthera, Panthera Onca, and so on*) must specialize *mlt:TokenIndividual*, i.e. instances of their instances are non-instantiable elements (e.g. *Cecil*, the lion, which does not possibly have instances). The domain rule that every instance of *Species* must be a subclass of an instance of *Genus* is captured by the *mlt:isSubordinateTo* property between *Species* and *Genus*. Further, the *mlt:partitions* property between *Species* and *Panthera* captures the rule that every instance of *Panthera* must be instance of exactly one instance of *Species*. Finally, *Genus mlt:partitions Organism* and *Species mlt:partitions Organism*, to capture that every organism must be instance of exactly one *Genus* and instance of exactly one instance of *Species*. Note that domain modelers only need to declare their domain classes as instances and/or specializations of the MLT basic types. (As we shall discuss later in Sect. 5.2, some of these relations can be inferred automatically, using derivation rules reflecting MLT axioms and theorems.)

Figure 5 shows an example of an ontology representing employees and their roles in a company to illustrate the use of variations of *characterization* relations to capture domain rules. To capture the rule that each *Employee* must play one or more *Business Roles* in the company, *Business Role mlt:completelyCharacterizes Employee* meaning that every instance of *Employee* must be instance of at least one instance of *Business Role*. Further, to represent that an *Employee* may play at most one *Management Role, Management Role mlt:disjointlyCharacterizes Employee*.

64   F. Brasileiro et al.

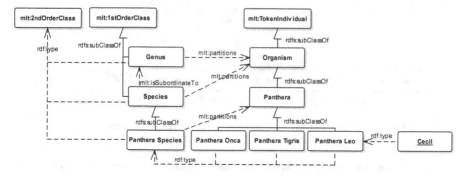

Fig. 4. Illustrating the use of *mlt:isSubordinateTo* and *mlt:partitions* properties.

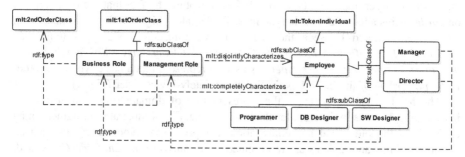

Fig. 5. Illustrating the use of *mlt:completelyCharacterizes* and *mlt:overlappinglyCharacterizes*.

5.2   Integrity Constraints and Derivation Rules Based on MLT

An important aspect of the proposed vocabulary is that it allows us to leverage rules of
the MLT formalization in order to guide modelers in producing sound models. The
rules discussed in this section ensure that the domain classes respect the stratification
into orders.

Some of these rules are expressible in pure OWL and thus were directly included in
the vocabulary. For example, a disjointness constraint (*owl:AllDisjointClasses*) is
introduced to reflect the fact that the basic types of MLT are all mutually disjoint.

The majority of the MLT rules, though, are not expressible directly in OWL, and
are represented here in SPARQL. This is the case of constraints concerning the domain
and range of MLT structural relations. For example, *mlt:isPowertypeOf*, *mlt:charac-
terizes* and all its variations must occur between classes of adjacent levels, i.e., if the
domain is a *2ndOrderClass*, then the range must be a *1stOrderClass*, if the domain is a
3rdOrderClass, then the range must be a *2ndOrderClass*, and so on.

Table 3 shows the domain/range restrictions for MLT relations.

SPARQL queries are also provided to allow the verification of rules concerning the
nature of the basic types of MLT. For example, considering that instances of Individual
must have no instances, we provide an integrity constraint to verify if there are
instances of instances of *mlt:TokenIndividual* (see Q1 in Fig. 6, which would detect
violations of this constraint).

Table 3. Domain and range restrictions for multi-level relations.

Relation name	Domain and range
rdfs:subClassOf	Classes of the same order (instances of 1st, 2nd or 3rd OrderClasses)
isSubordinateTo	Higher-order classes of the same order (2ndOrderClass or 3rdOrderClass)
rdf:type	Elements of adjacent levels.
isPowertypeOf	Classes of adjacent levels (2ndOrderClass → 1stOrderClass or
characterizes	3rdOrderClass → 2ndOrderClass)
completelyCharacterizes	
incompletelyCharacterizes	
disjointlyCharacterizes	
overlappinglyCharacterizes	

Integrity constraints are also provided to verify MLT theorems concerning characteristics of structural relations. For instance, given the definition of the *is powertype of* relation, a base class can have, at most, one higher-order class as powertype and a higher-order class may be the powertype of at most one base class. This suggests two clear integrity constraints: (i) a class can be the subject of at most one triple having *mlt: isPowertypeOf* as predicate (violations detected by Q2 in Fig. 6), and (ii) a class can be the object of at most one triple having *mlt:isPowertypeOf* as predicate. Another example is a constraint provided to allow the verification of the MLT theorem that states that if two classes *t1* and *t2* both partition the same class *t* then it is not possible for *t1* to be subclass of *t2* (Q3 in Fig. 6).

| Q1 | ```
select distinct * where{
    ?x rdf:type mlt:TokenIndividual .
    ?y rdf:type ?x .
}
``` | Q2 | ```
select distinct * where{
    ?p mlt:isPowertypeOf ?t .
    ?p mlt:isPowertypeOf ?t1 .
    FILTER (?t != ?t1) .
}
``` |
|---|---|---|---|
| Q3 | ```
select distinct * where{
    ?t1 mlt:partitions ?t .
    ?t2 mlt:partitions ?t .
    ?t1 rdfs:subClassOf ?t2 .
}
``` | Q4 | ```
select distinct * where{
    ?t2 rdfs:subClassOf+ ?t1 .
    ?t4 mlt:isPowerTypeOf ?t2 .
    ?t3 mlt:isPowerTypeOf ?t1 .
    minus{ ?p rdfs:subClassOf ?p1 . }
}
``` |
| Q5 | ```
select distinct * where{
    ?t2 mlt:isPowerTypeOf ?t1 .
    ?t3 mlt:characterizes ?t1 .
    minus{ ?t3 rdfs:subClassOf ?t2 . }
}
``` | Q6 | ```
select distinct * where{
    ?t rdf:type mlt:1stOrderClass .
    minus{
        ?t rdfs:subClassOf mlt:TokenIndividual.
    } }
``` |

Fig. 6. SPARQL queries representing MLT rules

Considering that models built using our MLT vocabulary may exhibit incomplete information, we leverage MLT axioms and theorems to allow the inference of information not represented explicitly. For example, it follows from the axioms of MLT that, if *t* is subclass of *t1* then the *powertype of t* is subclass of the *powertype of t1*. This is reflected in a query to identify cases in which the *subclass of* relation is not represented between the power types (Q4). Since, according to MLT, if $t2$ is powertype of $t1$ and $t3$ characterizes $t1$, then $t3$ is subclass of $t2$, we provide a SPARQL query to identify cases in which the *powertypeOf* and the *characterization* relations are represented but the *subclass* relations are not (Q5 in Fig. 6). Further, since every instance of a basic higher-order type must specialize the basic type at the immediately lower level, we can identify some missing relations. For example, query Q6 in Fig. 6 allows the identification of cases in which types are represented as instances of *mlt:1stOrderClass* but their subclass relations with *mlt:TokenIndividual* are not represented.

Since MLT is formalized quantifying over *all possible entities*, some MLT definitions are not expressible considering the Open World Assumption (OWA). For instance, according to MLT if $t1$ has instances such that all of them are also instances of $t2$, then we can conclude that $t1$ is a subclass of $t2$ (D1 in Table 2). This rule could not be captured in our approach since, considering the OWA, we cannot assume that all instances of an entity are represented in the knowledge base. Thus, these rules cannot be reflected in the implementation.

Finally, it is worth mentioning that, due to space limitations, we only expose here some rules to illustrate the approach. The vocabulary and the complete set of SPARQL queries is available at [27], including information on the traceability between MLT axioms and theorems and the implemented queries.

5.3 Final Considerations

Multi-level modeling addresses phenomena dealing with a number of complex notions and subtle relations that cross multiple levels of instantiation. These phenomena are ubiquitous in application domains, ranging from biology, to software engineering, from enterprise modeling to product classification [15]. Aside from the recurrence of these phenomena in practical cases, what also makes it of great importance is the fact that multi-level modeling seems to pose a significant challenge to modelers. As previously mentioned, in [14], we have empirically analyzed the presence of three anti-patterns related to multi-level modeling in Wikidata, finding over 22,000 occurrences of these anti-patterns. In fact, for one these anti-patterns, we found its manifestation in 85 % of the cases of taxonomic hierarchies spanning more than one level in Wikidata! That study clearly indicates that for complex modeling phenomena such as these, an expressive engineering support must be offered for vocabulary engineers as well as semantic web application developers. In [27], we provide a technical report showing how each of these anti-patterns found in Wikidata could be avoided by using the artifact proposed in this paper, demonstrating the relevance of MLT-OWL using real-world data.

The recognition of the importance of offering support for multi-level modeling led many researchers in the Semantic Web community to propose solutions addressing this issue. Some prominent results in that respect are reviewed in this paper, namely, RDFS (FA), metamodeling (punning) in OWL 2, OWL FA and PURO. We have shown in our analysis of these related works that all of them fail to fully support the identified modeling desiderata.

We adopted as a basis for our work a theoretically sound and well-tested formal theory (MLT) that was shown to be able to address all these multi-level modeling requirements. We then decided to offer a set of engineering tools that together would implement the modeling distinctions and axiomatization of this theory. These tools include: (i) an OWL vocabulary (capturing the formal relations put forth by this theory); (ii) a set of OWL axioms that would capture derivation and integrity rules over this vocabulary put forth by the theory; and (iii) a set of SPARQL queries that would capture those derivation and integrity rules put forth by this theory but that could not be represented in OWL directly. We strongly believe that these tools amount to an important methodological and computational contribution for guiding modelers to produce sound multi-level models in the Semantic Web.

The reason why these phenomena are recurrent in a large variety of practical application domains is because they are genuine ontological phenomena (from a philosophical point of view) [26]. As such, we advocate that truly ontological considerations cannot be eschewed from a fuller analysis of multi-level modeling. Additionally, some initiatives have demonstrated that the systematic evaluation of the ontological consistency of Semantic Web ontologies and vocabularies can greatly benefit from the use of foundational distinctions and axioms [28, 29]. In order to leverage the benefits of both a foundational ontology and a multi-level modeling theory, in [30] some of us have already combined MLT and the foundational ontology UFO [31]. A natural extension of this work is to enrich the set of engineering tools proposed here with support for the ontological distinctions and axiomatization of UFO (e.g., dealing with temporal aspects of anti-rigid concepts).

Acknowledgements. This research is funded by W3C Brasil and CNPq (grants 311313/2014-0, 485368/2013-7, 312158/2015-7 and 461777/2014-2). Freddy Brasileiro is funded by FAPES and Victorio A. Carvalho is funded by CAPES.

References

1. W3C: W3C Semantic Web Activity. https://www.w3.org/2001/sw/
2. Bizer, C., Heath, T., Berners-Lee, T.: Linked data - the story so far. Int. J. Semant. Web Inf. Syst. **5**, 1–22 (2009)
3. W3C: RDF Schema 1.1. (2014)
4. W3C: OWL 2 Web Ontology Language - Document Overview, 2nd edn. (2012)
5. Mayr, E.: The Growth of Biological Thought: Diversity, Evolution, and Inheritance. Harvard University Press, Cambridge (1982)
6. Gonzalez-Perez, C., Henderson-Sellers, B.: A powertype-based metamodelling framework. Softw. Syst. Model. **5**, 72–90 (2006)
7. Neumayr, B., Grun, K., Schrefl, M.: Multi-level domain modeling with m-objects and m-relationships. In: 6th Asia-Pacific Conference on Conceptual Modelling (2009)
8. Atkinson, C., Kühne, T.: The essence of multilevel metamodeling. In: Gogolla, M., Kobryn, C. (eds.) UML 2001. LNCS, vol. 2185, pp. 19–33. Springer, Heidelberg (2001)

9. Pan, J.Z., Horrocks, I.: Metamodeling architecture of web ontology languages. In: Proceedings of the 2001 International Semantic Web Working Symposium, pp. 131–149 (2001)
10. Motik, B.: On the properties of metamodeling in OWL. J. Log. Comput. **17**, 617–637 (2007)
11. Pan, J.Z., Horrocks, I., Schreiber, G.: OWL FA: a metamodeling extension of OWL DL. In: Proceedings of the 15th International Conference on World Wide Web, pp. 1065–1066. ACM (2006)
12. Svatek, V., Homola, M., Kluka, J., Vacura, M.: Metamodeling-based coherence checking of OWL vocabulary background models. In: Proceedings of the 10th International Workshop on OWL: Experiences and Directions (OWLED 2013) (2013)
13. W3C: OWL 2 Web Ontology Language - Structural Specification and Functional-Style Syntax, 2nd edn. (2012)
14. Brasileiro, F., Almeida, J.P.A., Carvalho, V.A., Guizzardi, G.: Applying a multi-level modeling theory to assess taxonomic hierarchies in Wikidata. In: Wiki Workshop 2016 at 25th International Conference Companion on World Wide Web, pp. 975–980 (2016)
15. Carvalho, V.A., Almeida, J.P.A.: Toward a well-founded theory for multi-level conceptual modeling. Softw. Syst. Model. (2016)
16. Atkinson, C., Kühne, T.: Model-driven development: a metamodeling foundation. IEEE Softw. **20**, 36–41 (2003)
17. Atkinson, C., Kühne, T.: Meta-level independent modelling. In: International Workshop on Model Engineering at 14th European Conference on Object-Oriented Programming, pp. 1–4 (2000)
18. Odell, J.: Power types. J. Object-Oriented Programing **7**, 8–12 (1994)
19. Cardelli, L.: Structural subtyping and the notion of power type. In: Proceedings of the 15th ACM SIGPLAN-SIGACT Symposium on Principles of Programming Languages - POPL 1988, pp. 70–79. ACM Press, New York (1988)
20. W3C: Resource Description Framework (RDF) Schema Specification 1.0 (2000)
21. OMG: Unified Modeling Language Specification v1.3 (1999)
22. Jekjantuk, N., Gröner, G., Pan, J.Z.: Modelling and reasoning in metamodelling enabled ontologies. Int. J. Softw. Inform. **4**, 277–290 (2010)
23. Gröner, G., Jekjantuk, N., Walter, T., Parreiras, F.S., Pan, J.Z.: Metamodelling and ontologies. In: Pan, J.Z., Staab, S., Aßmann, U., Ebert, J., Zhao, Y. (eds.) Ontology-Driven Software Development, pp. 151–174. Springer, Heidelberg (2013)
24. Carvalho, V.A., Almeida, J.P.A., Guizzardi, G.: Using a well-founded multi-level theory to support the analysis and representation of the powertype pattern in conceptual modeling. In: Nurcan, S., Soffer, P., Bajec, M., Eder, J. (eds.) CAiSE 2016. LNCS, vol. 9694, pp. 309–324. Springer, Heidelberg (2016). doi:10.1007/978-3-319-39696-5_19
25. Carvalho, V.A., Almeida, J.P.A.: A semantic foundation for organizational structures: a multi-level approach. In: 19th IEEE International Enterprise Distributed Object Computing Conference (EDOC 2015), pp. 50–59 (2015)
26. Carvalho, V.A., Almeida, J.P.A., Fonseca, C.M., Guizzardi, G.: Extending the foundations of ontology-based conceptual modeling with a multi-level theory. In: Johannesson, P., et al. (eds.) ER 2015. LNCS, vol. 9381, pp. 119–133. Springer, Heidelberg (2015). doi:10.1007/978-3-319-25264-3_9
27. MLT. http://nemo.inf.ufes.br/mlt
28. Fernandéz-López, M., Gómez-Pérez, A.: The integration of OntoClean in WebODE. In: EKAW 2002 Workshop on Evaluation of Ontology-Based Tools (EON2002), pp. 38–52 (2002)

29. Paulheim, H., Gangemi, A.: Serving DBpedia with DOLCE – more than just adding a cherry on top. In: Arenas, M., et al. (eds.) ISWC 2015. LNCS, vol. 9366, pp. 180–196. Springer, Heidelberg (2015). doi:10.1007/978-3-319-25007-6_11
30. Guizzardi, G., Almeida, J.P.A., Guarino, N., Carvalho, V.A.: Towards an ontological analysis of powertypes. In: International Workshop on Formal Ontologies for Artificial Intelligence (FOFAI 2015), Buenos Aires (2015)
31. Guizzardi, G.: Ontological Foundations for Structural Conceptual Models. Telematica Instituut Fundamental Research Series, Enschede (2005)

A Practical Acyclicity Notion for Query Answering Over *Horn-SRIQ* Ontologies

David Carral[1(✉)], Cristina Feier[2], and Pascal Hitzler[1]

[1] DaSe Lab, Wright State University, Dayton, USA
dcarralma@gmail.com
[2] Universität Bremen, Bremen, Germany

Abstract. Conjunctive query answering over expressive Horn Description Logic ontologies is a relevant and challenging problem which, in some cases, can be addressed by application of the chase algorithm. In this paper, we define a novel acyclicity notion which provides a sufficient condition for termination of the restricted chase over *Horn-SRIQ* TBoxes. We show that this notion generalizes most of the existing acyclicity conditions (both theoretically and empirically). Furthermore, this new acyclicity notion gives rise to a very efficient reasoning procedure. We provide evidence for this by providing a materialization based reasoner for acyclic ontologies which outperforms other state-of-the-art systems.

1 Introduction

Conjunctive query (CQ) answering over expressive Description Logic (DL) ontologies is a key reasoning task which remains unsolved for many practical purposes. Indeed, answering CQs over DL ontologies is quite intricate and often of high computational complexity [4,8,16]. Nevertheless, CQ answering over a major class of DLs, the so-called *Horn DLs*, can in some cases be addressed via application of the *chase algorithm*, a technique where all relevant consequences of an ontology are precomputed, allowing queries to be directly evaluated over the materialized set of facts. However, the chase is not guaranteed to terminate for all ontologies, and checking whether it does is not a straightforward procedure. It is thus an ongoing research endeavor to establish so-called *acyclicity conditions*; i.e., sufficient conditions which ensure termination of the chase.

The main contribution of this paper is the definition of *restricted chase acyclicity* (RCA_n), a novel acyclicity condition for *Horn-SRIQ* ontologies (the DL *Horn-SRIQ* may be informally described as the logic underpinning the deterministic fragment of OWL DL [9] minus nominals). If an ontology is proven to be RCA_n, then n-cyclic terms do not occur during the computation of the chase of such ontology and thus the chase is guaranteed to terminate.

In contrast with existing acyclicity notions [6] which deal with termination of the unrestricted, i.e. oblivious, chase of arbitrary sets of existential rules, we restrict our attention to the language *Horn-SRIQ* and seek to achieve termination of the restricted chase algorithm [3]; this is a special variant of the standard

© Springer International Publishing AG 2016
P. Groth et al. (Eds.): ISWC 2016, Part I, LNCS 9981, pp. 70–85, 2016.
DOI: 10.1007/978-3-319-46523-4_5

chase in which the inclusion of further terms to satisfy existential restrictions is avoided if such restrictions are already satisfied, and equality is dealt with via renaming. By considering such a chase algorithm we are able to devise acyclicity conditions which are more general than any other of the notions previously described.

On the theoretical side, we show that RCA_n is more general than *model-faithful acyclicity* (MFA) provided n is sufficiently large (linear in the size of ontology). As shown in [6], this is one of the most general acyclicity conditions for ontologies described to date, as it encompasses many other existing notions such as *joint acyclicity* [12], *super-weak acyclicity* [14] or the hybrid acyclicity notions presented in [2]. Furthermore, we show that deciding RCA_n membership is not harder than deciding MFA membership.

On the practical side, we empirically show that (i) RCA_n characterizes more real-world ontologies as acyclic than MFA. Furthermore, we demonstrate that (ii) the specific type of acyclicity captured by RCA_n results in a more efficient reasoning procedure. This is because acyclicity is still preserved in the case when employing renaming techniques when reasoning in the presence of equality. Thus, the use of cumbersome axiomatizations of equality such as *singularization* [14] can be avoided. Moreover, we report on an implementation of the restricted chase algorithm based on the datalog engine RDFOx [15] and show that (iii) it vastly outperforms state-of-the-art DL reasoners. To verify (i–iii), we complete an extensive evaluation with very encouraging results.

The rest of the paper is structured as follows: We start with some preliminaries in Sect. 2. Section 3 formally introduces the notions of oblivious and restricted chase, followed by an overview of MFA in Sect. 4. In Sect. 5 we introduce our new acyclicity notion RCA_n. Finally, Sects. 6 and 7 describe the evaluation of our work and list our conclusions, respectively.

An extended technical report for this paper with all the proofs and further information concerning the evaluation can be found at http://dase.cs.wright.edu/publications/acyclicity-notion-cqa-over-horn-sriq-ontologies.

2   Preliminaries

Rules. We use the standard notions of *constants*, *function symbols* and *predicates*, where $\approx$ is the equality predicate, $\top$ is universal truth, and $\bot$ is universal falsehood. *Variables, terms, atoms* and *substitutions* are defined as usual. A *fact* is a ground atom; i.e., an atom without occurrences of variables. As customary, every term t is associated with some *depth* $dep(t) \geq 0$. Furthermore, we often abbreviate a vector of terms $t_1, \ldots, t_n$ as t and identify t with the set $\{t_1, \ldots, t_n\}$. In a similar manner, we often identify a conjunction of atoms $\phi_1 \wedge \ldots \wedge \phi_n$ with the set $\{\phi_1, \ldots, \phi_n\}$. With $\phi(x)$ we stress that $x = x_1, \ldots, x_n$ are the free variables occurring in the formula ϕ.

Let t be some ground term and c some constant. Let t_c be the term obtained from t by replacing every occurrence of a constant by c, i.e., $f(d, g(e))_c = f(c, g(c))$. The notation is analogously extended to facts and sets of facts.

A term t' is a *subterm* of another term t if and only if $t' = t$, or $t = f(s)$ and t' is a subterm of some $s \in s$; if additionally $t' \neq t$, then t' is a *proper subterm* of t. A term t is *n-cyclic* if and only if there exists a sequence of terms of the form $f(s_1), \ldots, f(s_{n+1})$ such that $f(s_{n+1})$ is a subterm of t and, for every $i = 1, \ldots, n$, $f(s_i)$ is a proper subterm of $f(s_{i+1})$. We simply refer to 1-cyclic terms as *cyclic*.

A *rule* is a first-order logic (FOL) formula of one of the forms

$$\forall x \forall z [\beta(x, z) \rightarrow \exists y \eta(x, y)] \qquad \text{or} \tag{1}$$

$$\forall x [\beta(x) \rightarrow x \approx y], \tag{2}$$

where β and η are non-empty conjunctions of atoms which do not contain occurrences of constants, function symbols nor of the predicate $\approx$; x, y and z are pairwise disjoint; and $x, y \in x$. To simplify the notation, we frequently omit the universal quantifiers from rules. As customary, we refer to rules of the forms (1) and (2) as *tuple generating dependencies (TGDs)* and *equality generating dependencies (EGDs)*, respectively.

Given a set of rules $\mathcal{R}$, we define $\mathcal{R}^{\exists}$ and $\mathcal{R}^{\forall}$ as the sets of all the TGDs in $\mathcal{R}$ which do and do not contain existentially quantified variables, respectively. Moreover, let $\mathcal{R}^{\approx}$ be the set of all EGDs in $\mathcal{R}$. A *program* is a tuple $\langle \mathcal{R}, \mathcal{I} \rangle$ where $\mathcal{R}$ is a set of rules and $\mathcal{I}$ is an *instance*; i.e., a finite set of equality-free facts.

The main reasoning task we are investigating in this paper is CQ answering. Nevertheless, for the rest of the paper, we restrict our attention to the simpler task of CQ entailment of *boolean conjunctive queries* (BCQs). This is without loss of generality since CQ answering can be reduced to checking entailment of BCQs. A *BCQ*, or simply a *query*, is a formula of the form $\exists y \eta(y)$ where η is a conjunction of atoms not containing occurrences of constants, function symbols nor $\approx$.

For the remainder of the paper, we assume that $\top$ and $\bot$ are treated as ordinary unary predicates and that the semantics of $\top$ is captured explicitly in any program $\mathcal{P} = \langle \mathcal{R}, \mathcal{I} \rangle$ by including the rule $p(x_1, \ldots, x_n) \rightarrow \top(x_1) \wedge \ldots \wedge \top(x_n)$ in $\mathcal{R}$ for every predicate p with arity n occurring in $\mathcal{P}$.

We interpret programs under standard FOL semantics with true equality. As usual, a program $\mathcal{P}$ is *satisfiable* if and only if $\mathcal{P} \not\models \exists y \bot(y)$. Furthermore, given some query γ, we write $\mathcal{P} \models \gamma$ to indicate that $\mathcal{P}$ *entails* γ.

We will later employ skolemization to define the consequences of a TGD over a set of facts. The *skolemization* $sk(\rho)$ of some TGD $\rho = \beta(x, z) \rightarrow \exists y \eta(x, y)$ is the rule $\beta(x, z) \rightarrow \eta(x, y)\sigma_{sk}$ where σ_{sk} is a substitution mapping every $y \in y$ into $f_\rho^y(x)$ where f_ρ^y is a fresh function unique for every variable y and TGD ρ.

Description Logics. We next define the syntax and semantics of the ontology language *Horn-SRIQ* [13]. We assume basic familiarity with DL, and refer the reader to the literature for further details [1]. Without loss of generality, we restrict our attention to ontologies in a normal form close to the one from [13].

A *DL signature* is a tuple $\langle N_C, N_R, N_I \rangle$ where N_C, N_R and N_I are infinite countable and mutually disjoint sets of *concept names*, *role names* and

$$A_1 \sqcap \ldots \sqcap A_n \sqsubseteq B \qquad \mapsto \qquad A_1(x) \wedge \ldots \wedge A_n(x) \rightarrow B(x)$$

$$A \sqsubseteq \forall R.B \qquad \mapsto \qquad A(x) \wedge R(x,y) \rightarrow B(y)$$

$$A \sqsubseteq \, \leq 1 \, R.B \qquad \mapsto \qquad A(x) \wedge R(x,y) \wedge B(y) \wedge R(x,z) \wedge B(z) \rightarrow y \approx z$$

$$A \sqsubseteq \exists R.B \qquad \mapsto \qquad A(x) \rightarrow \exists y[R(x,y) \wedge B(y)]$$

$$S \sqsubseteq R \qquad \mapsto \qquad S(x,y) \rightarrow R(x,y)$$

$$S^- \sqsubseteq R \qquad \mapsto \qquad S(y,x) \rightarrow R(x,y)$$

$$S \circ V \sqsubseteq R \qquad \mapsto \qquad S(x,y) \wedge V(y,z) \rightarrow R(x,z)$$

Fig. 1. Mapping axioms α to rules $\Pi(\alpha)$, where $A_{(i)}, B \in N_C$, $R, S, V \in N_R$.

individuals, respectively, such that $\{\bot, \top\} \subseteq N_C$. A *role* is an element of $N_R^- = N_R \cup \{R^- \mid R \in N_R\}$. A *TBox axiom* is a formula of one of the forms given on the left hand side of the mappings in Fig. 1. TBox axioms of the form $A \sqsubseteq \exists R.B$ are also referred as *existential axioms*. An *ABox axiom* is a formula of the form $A(a)$ or $R(a,b)$ where $A \in N_C$, $R \in N_R$ and $a, b \in N_I$. An *axiom* is either a TBox or an ABox axiom. As usual, we simply refer to a set of TBox (resp. ABox) axioms as a *TBox* (resp. an *ABox*).

A *Horn-SRIQ ontology* $\mathcal{O}$ (or simply an *ontology*) is some tuple $\langle \mathcal{T}, \mathcal{A} \rangle$, where $\mathcal{T}$ and $\mathcal{A}$ are a TBox and an ABox, respectively, which satisfies the usual conditions [10].

Due to the close correspondence between ontologies and programs, we define the semantics of the former by means of a mapping into the latter. Given some TBox $\mathcal{T}$, let $\mathcal{R}_\mathcal{T} = \Pi(\mathcal{T})$. Given some ontology $\mathcal{O} = \langle \mathcal{T}, \mathcal{A} \rangle$, let $\mathcal{P}(\mathcal{O}) = (\mathcal{R}_\mathcal{T}, \mathcal{A})$ where Π is the function from Fig. 1. We say that $\mathcal{O}$ is *satisfiable* if and only if the program $\mathcal{P}(\mathcal{O})$ is satisfiable. Furthermore, $\mathcal{O}$ *entails* a query γ, written $\mathcal{O} \models \gamma$, if and only if $\mathcal{P}(\mathcal{O})$ is unsatisfiable or $\mathcal{P}(\mathcal{O})$ entails γ.

3   The Chase Algorithm

In this section we present two variants of the chase algorithm, which are somewhat similar to the oblivious and restricted chase from [3], and elaborate about how such procedures may be used to solve CQ entailment over ontologies.

Definition 1. *A fact ϕ is an* oblivious consequence *of a TGD $\rho = \beta(x, z) \rightarrow \exists y \eta(x, y)$ on a set of facts $\mathcal{F}$ if and only if there is some substitution σ with $\beta(x, z)\sigma \subseteq \mathcal{F}$ and $\phi \in sk(\eta(x, y))\sigma$ where $sk(\eta(x, y))$ is the head of the (skolemized) TGD $sk(\rho)$. A fact ϕ is a* restricted consequence *of ρ on $\mathcal{F}$ if and only if there is a substitution σ with (1) $\beta(x, z)\sigma \subseteq \mathcal{F}$ and $\phi \in sk(\eta(x, y))\sigma$, and (2) there is no substitution $\tau \supseteq \sigma$ with $\eta(x, y)\tau \subseteq \mathcal{F}$.*

The result of obliviously applying ρ to $\mathcal{F}$, *written $\rho_O(\mathcal{F})$, is the set of all oblivious consequences of ρ on $\mathcal{F}$. The result of* obliviously applying *a set of TGDs $\mathcal{R}$ to $\mathcal{F}$, written $\mathcal{R}_O(\mathcal{F})$, is the set $\bigcup_{\rho \in \mathcal{R}} \rho_O(\mathcal{F}) \cup \mathcal{F}$. The result of* restrictively applying ρ to $\mathcal{F}$ (resp., $\mathcal{R}$ to $\mathcal{F}$), *written $\rho_R(\mathcal{T})$ (resp., $\mathcal{R}_R(\mathcal{T})$), is analogously defined.*

Definition 2. *Let $\leadsto$ be some total strict order over the set of all terms such that $t \leadsto u$ only if $dep(t) \leq dep(u)$. Furthermore, we say that t is greater than u with respect to $\leadsto$ to indicate $t \leadsto u$.*

Given a set of EGDs $\mathcal{R}$ and a set of facts $\mathcal{F}$, let $\mapsto_{\mathcal{F}}^{\mathcal{R}}$ be the minimal congruence relation over terms such that $t \mapsto_{\mathcal{F}}^{\mathcal{R}} u$ if and only if there exists some $\beta(\boldsymbol{x}) \to x \approx y \in \mathcal{R}$ and some substitution σ with $\beta(\boldsymbol{x})\sigma \subseteq \mathcal{F}$, $\sigma(x) = t$ and $\sigma(y) = u$. Let $\mathcal{R}(\mathcal{F})$ be the set that is obtained from $\mathcal{F}$ by replacing all occurrences of every term t by u where u is the greatest term with respect to $\leadsto$ such that $t \mapsto_{\mathcal{F}}^{\mathcal{R}} u$.

Note that we define consequences with respect to sets of rules instead of simply (single) rules as it is customary [3]. This allows us to define the chase as a deterministic procedure (modulo $\leadsto$). Also, unlike in [3], where a lexicographic order is used to direct the replacement of terms, we employ a type of order which ensures that terms are always replaced by terms of equal or lesser depth. This effectively precludes some "deeper" terms from being introduced during the computation of the chase.

Definition 3. *Let $\mathcal{P} = \langle \mathcal{R}, \mathcal{I} \rangle$ be some program. The oblivious chase sequence of $\mathcal{P}$ is the sequence $\mathcal{F}_0, \mathcal{F}_1, \ldots$ such that $\mathcal{F}_1 = \mathcal{I}$ and, for all $i \geq 1$, $\mathcal{F}_i$ is the set of facts defined as follows.*

- *If $\mathcal{R}^{\approx}(\mathcal{F}_{i-1}) \neq \mathcal{F}_{i-1}$, then $\mathcal{F}_i = \mathcal{R}^{\approx}(\mathcal{F}_{i-1})$.*
- *If $\mathcal{F}_{i-1} = \mathcal{R}^{\approx}(\mathcal{F}_{i-1})$ and $\mathcal{F}_{i-1} \neq \mathcal{R}_O^{\forall}(\mathcal{F}_{i-1})$, then $\mathcal{F}_i = \mathcal{R}_O^{\forall}(\mathcal{F}_{i-1})$.*
- *Otherwise, $\mathcal{F}_i = \mathcal{R}_O^{\exists}(\mathcal{F}_{i-1})$.*

The restricted chase sequence of $\mathcal{P}$ is defined analogously.

For the sake of brevity, we frequently denote the oblivious (resp., restricted) chase sequence of a program $\mathcal{P}$ with $\mathcal{P}_O^1, \mathcal{P}_O^2, \ldots$ (resp., $\mathcal{P}_R^1, \mathcal{P}_R^2, \ldots$)

Definition 4. *Let $\mathcal{P}$ be some program and let $\mathcal{R}$ be some set of rules. Then, the oblivious chase of $\mathcal{P}$ is the set $OC(\mathcal{P}) = \bigcup_{i \in \mathbb{N}} \mathcal{P}_O^i$. The restricted chase of $\mathcal{P}$, written $RC(\mathcal{P})$, is defined analogously.*

The oblivious (resp., restricted) chase of $\mathcal{P}$ terminates if and only if there is some i such that, for all $j \geq i$, $\mathcal{P}_O^i = \mathcal{P}_O^j$. Furthermore, the oblivious (resp., restricted) chase of a set of rules $\mathcal{R}$ terminates if the oblivious (resp., restricted) chase of every program of the form $\langle \mathcal{R}, \mathcal{I} \rangle$ terminates.

Our definition of the chase sequence ensures that rules which do not contain existentially quantified variables are always applied with a higher priority than rules that do. Note that, by postponing the application of rules with existential variables, we may prevent them from introducing further consequences.

The (restricted or oblivious) chase of a program can be employed to solve CQ entailment [3]. I.e., a program $\mathcal{P}$ entails a query γ, written $\mathcal{P} \models \gamma$, if and only if either $OC(\mathcal{P}) \models \exists y \bot(y)$ or $OC(\mathcal{P}) \models \gamma$ (resp., $RC(\mathcal{P}) \models \exists y \bot(y)$ or $RC(\mathcal{P}) \models \gamma$). Thus, we may also use the chase to solve CQ entailment over

$$\mathcal{T} = \{\mathit{Film} \sqsubseteq \exists \mathit{isProdBy}.\mathit{Producer}, \mathit{Producer} \sqsubseteq \exists \mathit{prod}.\mathit{Film},$$
$$\mathit{isProdBy}^- \sqsubseteq \mathit{prod}, \mathit{prod}^- \sqsubseteq \mathit{isProdBy}\}$$
$$\mathcal{O} = \langle \mathcal{T}, \{\mathit{Film}(\mathit{AI})\}\rangle$$
$$\mathcal{R}_\mathcal{T} = \{\rho = \mathit{Film}(x) \to \exists y[\mathit{isProdBy}(x,y) \wedge \mathit{Producer}(y)],$$
$$v = \mathit{Producer}(x) \to \exists y[\mathit{prod}(x,y) \wedge \mathit{Film}(y)],$$
$$\mathit{isProdBy}(y,x) \to \mathit{prod}(x,y), \mathit{prod}(y,x) \to \mathit{isProdBy}(x,y)\}$$
$$\mathcal{P}(\mathcal{O}) = \langle \mathcal{R}_\mathcal{T}, \{\mathit{Film}(\mathit{AI})\}\rangle$$
$$\mathcal{P}(\mathcal{O})_R^1 = \{\mathit{Film}(\mathit{AI}), \mathit{isProdBy}(\mathit{AI}, f_\rho^y(\mathit{AI})), \mathit{Producer}(f_\rho^y(\mathit{AI}))\}$$
$$\mathcal{P}(\mathcal{O})_R^2 = \{\mathit{prod}(f_\rho^y(\mathit{AI}), \mathit{AI})\} \cup \mathcal{P}(\mathcal{O})_O^1$$
$$RC(\mathcal{P}(\mathcal{O})) = \mathcal{P}(\mathcal{O})_O^2$$
$$OC(\mathcal{P}(\mathcal{O})) = RC(\mathcal{P}(\mathcal{O})) \cup \{\mathit{prod}(f_\rho^y(\mathit{AI}), f_v^y(f_\rho^y(\mathit{AI}))), \mathit{Film}(f_v^y(f_\rho^y(\mathit{AI}))), \ldots\}$$

Fig. 2. Ontology $\mathcal{O} = \langle \mathcal{T}, \mathcal{A} \rangle$, program $\mathcal{P}(\mathcal{O})$ and the chase of $\mathcal{P}(\mathcal{O})$.

ontologies: An ontology $\mathcal{O}$ entails a query γ if and only if $OC(\mathcal{P}(\mathcal{O})) \models \exists y \perp(y)$ or $OC(\mathcal{P}(\mathcal{O})) \models \gamma$ (resp., $RC(\mathcal{P}(\mathcal{O})) \models \exists y \perp(y)$ or $RC(\mathcal{P}(\mathcal{O})) \models \gamma$).

For readability purposes, we say that the oblivious (resp. restricted) chase of some ontology $\mathcal{O}$ *terminates* if and only if the oblivious (resp. restricted) chase of $\mathcal{P}(\mathcal{O})$ terminates. The oblivious (resp. restricted) chase of some TBox $\mathcal{T}$ *terminates* if and only if if the oblivious (resp. restricted) chase of $\mathcal{R}_\mathcal{T}$ terminates.

As expected, the restricted chase has a better behavior than the oblivious chase; i.e., in some cases, the former might terminate when the latter does not:

Example 5. Let $\mathcal{O} = \langle \mathcal{T}, \mathcal{A} \rangle$ be as in Fig. 2. The figure depicts also the computation of the oblivious chase and that of the restricted chase of $\mathcal{P}(\mathcal{O})$. In this case, $RC(\mathcal{P}(\mathcal{O}))$ terminates whereas $OC(\mathcal{P}(\mathcal{O}))$ does not.

4   Model Faithful Acyclicity

In this section we briefly describe Model Faithful Acyclicity (MFA) [6], one of the most general acyclicity conditions for sets of rules. MFA guarantees the termination of the oblivious chase of a program by imposing that no cyclic term occurs in the chase. Note that, a condition such as MFA can be applied to check whether a TBox $\mathcal{T}$ is acyclic; i.e., $\mathcal{T}$ is MFA if and only if $\mathcal{R}_\mathcal{T}$ is MFA.

When one is interested in checking the termination of the oblivious chase with respect to every possible instance, it is enough to check termination with respect to a special instance, the *critical instance* [14]. The critical instance is the minimal set which contains all possible atoms that can be formed using the relational symbols which occur in TGDs and the special constant $\star$. Such a strategy is used by MFA to guarantee termination of a set of rules.

While the actual definition of MFA does not preclude the existence of EGDs, equality is assumed to be axiomatized, and thus it is treated as a regular predicate (EGDs are de facto TGDs). To reflect such treatment we will use the special predicate Eq to denote equality. However, as the following example shows, the presence of equality in a set of TGDs frequently makes the MFA membership test fail.

Example 6. Let Σ be the following set of rules and let Σ' be the set of rules that result from axiomatizing the equality predicate as usual (see Sect. 2.1 of [6]). Furthermore, let $\mathcal{I}_*(\Sigma')$ be the critical instance of Σ'.

$$\Sigma = \{A(x) \wedge B(x) \rightarrow \exists y[R(x,y) \wedge B(y)], R(z,x_1) \wedge R(z,x_2) \rightarrow Eq(x_1,x_2)\}$$
$$Eq = \{\top(x) \rightarrow Eq(x,x), Eq(x,y) \rightarrow Eq(y,x), Eq(x,z) \wedge Eq(z,y) \rightarrow Eq(x,y)\}$$
$$\Sigma' = \{A(x) \wedge Eq(x,y) \rightarrow A(y), R(x,y) \wedge Eq(x,z) \rightarrow R(z,y),$$
$$R(x,y) \wedge Eq(y,z) \rightarrow R(x,z)\} \cup \Sigma \cup Eq$$
$$\mathcal{I}_*(\Sigma') = \{A(\star), R(\star,\star), Eq(\star,\star)\}$$

The oblivious chase of $(\Sigma', \mathcal{I}_*(\Sigma'))$ does not terminate.

$$(\Sigma', \mathcal{I}_*(\Sigma'))_O^1 = \{R(\star, f(\star)), B(f(\star)), Eq(\star, f(\star))\} \cup \mathcal{I}_*(\Sigma')$$
$$(\Sigma', \mathcal{I}_*(\Sigma'))_O^2 = \{A(f(\star)), R(f(\star), f(f(\star))), B(f(f(\star))), \ldots\}$$
$$\cdots\cdots\cdots\cdots\cdots\cdots\cdots\cdots$$

To avoid this situation, the use of *singularization* [14], a somewhat "less-harmful" axiomatization of equality, is proposed in [6].

Definition 7. *A singularization of a rule ρ is the rule ρ' that results from performing the following transformation for every variable v in the body of ρ:*

- *Rename each occurrence of v using different fresh variables $v_1, \ldots, v_n$,*
- *pick some $j = 1, \ldots, n$ and add the atoms $Eq(v_1, v_j), \ldots, Eq(v_n, v_j)$ to the body of ρ and*
- *replace any occurrence of v in the head of ρ with v_j.*

Let Σ be a set of TGDs and let Eq be the set from Example 6. A singularization of Σ is a set of TGDs Σ' which contains Eq and exactly one singularization of every $\rho \in \Sigma$. Let $Sing(\Sigma)$ be the set of all possible singularizations of Σ.

Example 8. Rule $A(x) \wedge B(x) \rightarrow \exists y[R(x,y) \wedge B(y)]$ from Example 6 admits two possible singularizations: (i) $A(x_1) \wedge B(x_2) \wedge Eq(x_2, x_1) \rightarrow \exists y[R(x_1, y) \wedge B(y)]$ and (ii) $A(x_1) \wedge B(x_2) \wedge Eq(x_1, x_2) \rightarrow \exists y[R(x_2, y) \wedge B(y)]$.

Note that, for any $\Sigma' \in Sing(\Sigma)$, if Σ' is MFA, then the oblivious chase of Σ' can be used to answer queries on Σ [6]. The use of singularization along with MFA gives rise to the following acyclicity notions.

Definition 9. *For a set of TGDs Σ, if there is some $\Sigma' \in Sing(\Sigma)$ which is MFA, then Σ is said to be $MFA^{\exists}$. If every $\Sigma' \in Sing(\Sigma)$ is MFA, then Σ is $MFA^{\forall}$.*

To some extent, the use of singularization solves the problems with equality: One can check that Σ in Example 6 is MFA$^{\exists}$, but not MFA$^{\forall}$. Nevertheless, due to the high number of possible singularizations, it is frequently not feasible to check MFA$^{\exists}$ or MFA$^{\forall}$ membership. A simpler alternative is to check whether $\bigcup_{\Sigma' \in Sing(\Sigma)} \Sigma'$ is MFA. If that is the case, then Σ is said to be MFA$^{\cup}$. Note that in the case of Horn-$\mathcal{SRIQ}$ TBoxes, $|\bigcup_{\Sigma' \in Sing(\Sigma)} \Sigma'|$ is actually polynomial in $|\Sigma|$ and, as such, MFA$^{\cup}$ is more feasible to check. Thus, we will use MFA$^{\cup}$ as a baseline for the evaluation of the new acyclicity condition RCA$_n$, which is introduced in the next section.

5   Restricted Chase Acyclicity

While MFA is quite a general acyclicity condition, it has two main drawbacks:

1. It only considers the oblivious chase, which as we have seen in Example 5, might not terminate (even though the restricted chase does!), and
2. its treatment of equality via singularization is cumbersome and inefficient in practice. Not only MFA$^{\exists}$ and MFA$^{\forall}$ are difficult to check, but even after a set of TGDs are established to belong to some MFA subclass, one has to employ a singularized program for reasoning purposes.

In this section, we present RCA$_n$, an acyclicity notion with neither of these drawbacks: RCA$_n$ verifies termination of the restricted chase of a TBox and does not require the use of cumbersome axiomatizations of the equality predicate. Furthermore, unlike MFA, RCA$_n$ allows for the presence of cyclic terms in the chase up to a given depth n.

Since we are primarily interested in termination of the restricted chase of a Horn-$\mathcal{SRIQ}$ TBox, one might wonder why we do not simply check for termination of the restricted chase for such a TBox with respect to the critical instance, as it is done in the previous section with the oblivious chase. Unfortunately, this is not possible: The restricted chase of any set of existential rules always terminates with respect to the critical instance. Thus, we have to devise more sophisticated techniques to check the termination of the restricted chase. We start by introducing the notion of an overchase for a TBox.

Definition 10. *A set of facts $\mathcal{V}$ is an* overchase *for some TBox $\mathcal{T}$ if and only if, for every $\mathcal{O} = \langle \mathcal{T}, \mathcal{A} \rangle$, $RC(\mathcal{P}(\mathcal{O}))_\star \subseteq \mathcal{V}$.*

Given some TBox $\mathcal{T}$, an overchase for $\mathcal{T}$ may be intuitively regarded as an over-approximation of the restricted chase of $\mathcal{T}$.

Lemma 11. *If there exists a finite overchase for a TBox, then the restricted chase of such TBox terminates.*

Thus, to determine whether the chase of a TBox $\mathcal{T}$ terminates, we introduce a procedure to compute an overchase for $\mathcal{T}$ and a means to check its termination. We proceed with some preliminary notions and notation.

Definition 12. *Let T be some TBox and t a term. Let $\mathcal{I}(t)$ be the set of facts defined as follows: If t is of the form $f_\rho^y(s)$ where $\rho = A(x) \rightarrow \exists y[R(x,y) \wedge B(y)]$, then $\mathcal{I}(t) = \{A(s), R(s,t), B(t)\} \cup \mathcal{I}(s)$; otherwise, $\mathcal{I}(t) = \emptyset$. Furthermore, we introduce the program $\mathcal{U}(T,t) = \langle \mathcal{R}_T^\forall \cup \mathcal{R}_T^\approx, \mathcal{I}(t) \rangle$.*

Intuitively, the restricted chase of the program $\mathcal{U}(T,t)$ can be regarded as some kind of under-approximation of the facts that must occur in the chase of every program of the form $\mathcal{P}(\langle T, \mathcal{A} \rangle)$ where t occurs. I.e., if t occurs in the restricted chase sequence of any program $\mathcal{P}(\langle T, \mathcal{A} \rangle)$, then the facts in the restricted chase of $\mathcal{U}(T,t)$ must also occur (up to renaming) in the chase sequence of such program. Furthermore, due to the special priority of application of the rules during the computation of the chase, the facts in the restricted chase of $\mathcal{U}(T,t)$ must occur in the restricted chase sequence of every program of the form $\mathcal{P}(\langle T, \mathcal{A} \rangle)$ before any successors of t are introduced.

Example 13. Let $\mathcal{O}$, ρ and v be the ontology and rules from Example 5. Then, by Definition 12:

$$\mathcal{I}(f_\rho^y(AI)) = \{Film(AI), isProdBy(AI, f_\rho^y(AI)), Producer(f_\rho^y(AI))\} \text{ and}$$
$$RC(\mathcal{U}(T, f_\rho^y(AI))) = \{prod(f_\rho^y(AI), AI)\} \cup \mathcal{I}(f_\rho^y(AI)).$$

All the facts in the restricted chase of $\mathcal{U}(T,t)$ occur in the restricted chase sequence of $\mathcal{P}(\mathcal{O})$ before any successors of term $f_\rho^y(AI)$ are introduced. This is because the rule $isProdBy(y,x) \rightarrow prod(x,y)$ is applied with a higher priority than the rule $v = Producer(x) \rightarrow \exists y[prod(x,y) \wedge Film(y)]$.

Given a TBox T and some term of the form $f_\rho^y(t)$, we can in some cases conclude that such a term may never occur during the computation of the restricted chase of every program of the form $\mathcal{P}(\langle T, \mathcal{A} \rangle)$ by carefully inspecting the facts in the set $\mathcal{U}(T,t)$.

Definition 14. *Let T be a TBox and t a term of the form $f_\rho^y(s)$ where $\rho = A(x) \rightarrow \exists y[R(x,y) \wedge B(y)]$. We say that a term t is* restricted *with respect to T if and only if there is some term u with $\{R([s], u), B(u)\} \subseteq RC(\mathcal{U}(T,s))$ where $[s] = [v]$, if s is replaced by v during the computation of the restricted chase sequence; and $[s] = s$, otherwise.*

We often simply say that a term is "restricted", instead of "restricted with respect to T," if the TBox T is clear from the context.

Lemma 15. *Let T be a TBox and t a restricted term. Then, for every possible $\mathcal{O} = \langle T, \mathcal{A} \rangle$, $t \notin RC(\mathcal{P}(\mathcal{O}))$.*

Proof (Sketch). Let t be a term of the form $f_\rho^y(s)$ where $\rho = A(x) \rightarrow \exists y(R(x,y) \wedge B(y))$. We can verify that, if t occurs during the computation of the chase sequence, then every fact $RC(\mathcal{U}(T,s))$ will also be included in such chase sequence before any new terms are introduced. Thus, if t is indeed restricted, there must be some u with $R([s], u)$ and $B(u)$ occurring in the chase sequence. Therefore, by the definition of the chase, the term t may never be derived.

∀-rule	if	there is some TGD of the form $\rho = \beta(\boldsymbol{x}, \boldsymbol{y}) \to \eta(\boldsymbol{x}) \in \mathcal{R}_{\mathcal{T}}$
	then	$\mathcal{V}_{\mathcal{T}} \to \rho_R(\mathcal{V}_{\mathcal{T}}) \cup \mathcal{V}_{\mathcal{T}}$
∃-rule	if	there is some TGD of the form $\rho = A(x) \to \exists y[R(x,y) \wedge B(y)] \in \mathcal{R}_{\mathcal{T}}$ and there exists some substitution σ such that (i) $A(x)\sigma \subseteq \mathcal{V}_{\mathcal{T}}$ and (ii) $f_\rho^y(x)\sigma$ is not restricted with respect to $\mathcal{T}$
	then	$\mathcal{V}_{\mathcal{T}} \to \{R(x, f_\rho^y(x)), B(f_\rho^y(x))\}\sigma \cup \mathcal{V}_{\mathcal{T}}$
≈-rule	if	there is some EGD $\beta(\boldsymbol{x}, \boldsymbol{y}) \to x \approx y \in \mathcal{R}_{\mathcal{T}}$ and there exists some substitution σ such that $\beta(\boldsymbol{x}, \boldsymbol{y})\sigma \subseteq \mathcal{V}_{\mathcal{T}}$
	then	$\mathcal{V}_{\mathcal{T}} \to \{Eq(x,y), Eq(y,x)\}\sigma \cup \mathcal{V}_{\mathcal{T}}$
Eq-rule	if	there are some terms t, u and u_i where $i = 1, \ldots, n$ and some predicate p such that (i) $p \neq Eq$, (ii) $\{Eq(t,u), p(u_1, \ldots, u_n)\} \subseteq \mathcal{V}_{\mathcal{T}}$, (iii) $dep(t) \leq dep(u)$ and (iv) $u = u_j$ for some $j = 1, \ldots, n$
	then	$\mathcal{V}_{\mathcal{T}} \to \{p(u_1, \ldots, u_n)\}[u/t] \cup \mathcal{V}_{\mathcal{T}}$

Fig. 3. Expansion rules for the construction of $\mathcal{V}_{\mathcal{T}}$.

Example 16. Let $\mathcal{T}$, ρ and v be the TBox and rules from Example 5. We proceed to show that the term $f_\rho^y(f_v^y(AI))$ is restricted. First, we compute the restricted chase of $\mathcal{U}(\mathcal{T}, f_v^y(AI))$.

$$RC(\mathcal{U}(\mathcal{T}, f_v^y(AI))) = \{Producer(AI), prod(AI, f_v^y(AI)),$$
$$Film(f_v^y(AI)), isProdBy(f_v^y(AI), AI)\}$$

Note that $\{isProdBy(f_v^y(AI), AI), Producer(AI)\} \subseteq RC(\mathcal{U}(\mathcal{T}, f_v^y(AI)))$. Thus, $f_\rho^y(f_v^y(AI))$ is restricted with respect to $\mathcal{T}$ and, by Lemma 15, it may not occur in the restricted chase of a program of the form $\mathcal{P}(\langle \mathcal{T}, \mathcal{A} \rangle)$. Furthermore, by Definition 14, if $f_\rho^y(f_v^y(AI))$ is restricted, then every term of the form $f_\rho^y(f_v^y(c))$, where c is a constant, is also restricted.

With Definition 14 and Lemma 15 in place, we proceed with the definition of a procedure to construct an overchase for some given TBox $\mathcal{T}$.

Definition 17. *Let $\mathcal{T}$ be a TBox. We define $\mathcal{V}_{\mathcal{T}}$ as the set initially containing every fact in $\mathcal{I}_*(\mathcal{R}_{\mathcal{T}})$ which is then expanded by repeatedly applying the rules in Fig. 3 (in non-deterministic order).*

Lemma 18. *The set $\mathcal{V}_{\mathcal{T}}$ is an overchase of the TBox $\mathcal{T}$.*

Proof (Sketch). The lemma can be proven via induction on chase sequence of any ontology of the form $\mathcal{O} = \langle \mathcal{T}, \mathcal{A} \rangle$. Note that, $\mathcal{O}_R^0 \subseteq \mathcal{V}_{\mathcal{T}}$ by the definition of $\mathcal{V}_{\mathcal{T}}$. It can be verified that, for every possible derivation of a set of facts during the computation of the chase of $\mathcal{O}$, such facts will always be contained in $\mathcal{V}_{\mathcal{T}}$.

Corollary 19. *The restricted chase of some TBox $\mathcal{T}$ terminates if $\mathcal{V}_{\mathcal{T}}$ is finite.*

Example 20. Let $\mathcal{T}$ be the TBox from Example 5. Then $\mathcal{V}_{\mathcal{T}}$ is as follows.

$$\mathcal{V}_{\mathcal{T}} = \{Film(\star), isProdBy(\star), Producer(\star), prod(\star, \star),$$
$$isProdBy(\star, f_\rho^y(\star)), Producer(f_\rho^y(\star)), prod(\star, f_v^y(\star)), Producer(f_v^y(\star))\}$$

Note that terms $f_\rho^y(f_v^y(\star))$ and $f_v^y(f_\rho^y(\star))$ are restricted and thus, they are not included in $\mathcal{V}_\mathcal{T}$. Since $\mathcal{V}_\mathcal{T}$ is finite, we can conclude termination of the restricted chase of the TBox $\mathcal{T}$.

In the previous example, we were able to ascertain termination of the restricted chase of $\mathcal{T}$ after verifying that the set $\mathcal{V}_\mathcal{T}$ is finite. A sufficient condition for finiteness of $\mathcal{V}_\mathcal{T}$ is to only allow cyclic terms up to a certain depth in this set. We use such condition to formally define RCA_n.

Definition 21. *A TBox $\mathcal{T}$ is RCA_n if and only if there are no n-cyclic terms in $\mathcal{V}_\mathcal{T}$. An ontology $\langle \mathcal{T}, \mathcal{A} \rangle$ is RCA_n if and only if $\mathcal{T}$ is RCA_n.*

Theorem 22. *If a TBox $\mathcal{T}$ is RCA_n then the restricted chase of $\mathcal{T}$ terminates.*

We proceed with several results regarding the complexity of deciding RCA_n membership and reasoning over RCA_n ontologies.

Theorem 23. *Deciding whether some TBox $\mathcal{T}$ is RCA_n is in* EXPTIME.

Theorem 24. *Let $\mathcal{O} = \langle \mathcal{T}, \mathcal{A} \rangle$ be some RCA_n ontology and γ a query. Then, checking whether $\mathcal{O} \models \gamma$ is* EXPTIME-*complete.*

To close the section, we present several results in which we theoretically compare the generality of RCA_n to $MFA^\cup$.

Theorem 25. *$MFA^\cup$ does not cover RCA_1.*

Proof. The TBox $\mathcal{T}$ from Example 5 is RCA_1 but not $MFA^\cup$.

Theorem 26. *If $\mathcal{T}$ is $MFA^\cup$ then $\mathcal{T}$ is RCA_n for every $n > |\mathcal{T}^\exists|$ where $\mathcal{T}^\exists$ is the set of all existential axioms in $\mathcal{T}$.*

6   Evaluation

6.1   An Empirical Comparison of RCA_n and $MFA^\cup$

In this section we include an empirical comparison of the generality of RCA_n and $MFA^\cup$. For our experiments, we use the TBoxes of the ontologies in the OWL Reasoner Evaluation workshop (ORE, https://www.w3.org/community/owled/ore-2015-workshop/) and Ontology Design Patterns (ODP, http://www.ontologydesignpatterns.org) datasets. The former is a large repository used in the ORE competition containing a large corpus of ontologies. The latter contains a wide range of smaller ontologies that capture design patterns commonly used in ontology modeling. The ORE dataset is rather large, and thus we restrict our experiments to the 294 ontologies with the smallest number of existential axioms, while skipping the 77 ontologies with the largest number of existential axioms. The number of such axioms contained in an ontology is a useful metric to predict the "hardness" of acyclicity membership tests; i.e. running these experiments would be very time-intensive, while our results, reported below, already indicate

ORE

∃-Axioms	Avg. Size	Count	MFA$^{\cup}$	RCA$_1$	RCA$_2$	RCA$_3$
1-5	175	70	70.0	87.1	92.9	92.9
6-10	219	48	58.3	83.3	83.3	83.3
11-25	916	54	83.3	85.2	91	91
26-100	521	42	54.8	59.5	61.9	61.9
101-500	1290	42	26.2	26.2	28.6	28.6
501-1922	5052	38	60.5	60.5	60.5	60.5
1-1922	1362	294	60.9	70.1	73.1	73.1

ODP

∃-Axioms	Size	Total	MFA$^{\cup}$	RCA$_1$	RCA$_2$	RCA$_3$
1-12	39	18	73.7	100.0	100.0	100.0

Fig. 4. Results for the ORE and ODP Repositories.

that for such very hard TBoxes MFA$^{\cup}$ and RCA$_n$ will likely not differ much (while they differ significantly for ontologies with a lower count of existential axioms).

Only *Horn-SRIQ* TBoxes which cannot be expressed in any of the OWL 2 profiles were considered in our experiments. This is because all OWL 2 RL TBoxes are acyclic (with respect to every applicable acyclicity notion known to us), and there already exist effective algorithms and efficient implementations that solve CQ answering over OWL 2 EL and OWL 2 QL ontologies [11,17,18] (albeit, if these do not include complex roles).

The results from our experiments are summarized in Fig. 4. The evaluated TBoxes are sorted into brackets depending on the number of existential axioms they contain. For each bracket we provide the average number of axioms in the ontologies ("Avg. Size"), the number of ontologies ("Count"), and, for every condition "X" considered, the percentage of "X acyclic" ontologies

RCA$_2$ and RCA$_3$ turned out to be indistinguishable with respect to the TBoxes considered and thus, we limit our evaluation to RCA$_n$ with $n \leq 3$. Our tests reveal that RCA$_2$ is significantly more general than MFA$^{\cup}$, particularly when it comes to TBoxes with a low count of existential axioms. However note that reasoning over ontologies with few (existential) axioms is in general not trivial: All of the ontologies considered in our materialization tests (see Fig. 5) contain less than 20 existential axioms. For TBoxes containing from 1 to 10 existential axioms in the ORE dataset, more than half of the ontologies which are not MFA$^{\cup}$ are RCA$_2$. Furthermore, the 4 ontologies in the ODP dataset which are not MFA$^{\cup}$ are RCA$_2$. Interestingly, in both repositories we could not find any ontology that is MFA$^{\cup}$ but not RCA$_1$. Thus, with respect to the TBoxes in our corpus, RCA$_1$ already proves to be more general than MFA$^{\cup}$.

In total, we looked at 312 ontologies, 62 % and 75 % of which are MFA$^{\cup}$ and RCA$_2$, respectively. To gauge the significance of this improvement, we roughly compare these numbers with the results presented in [6]. In that paper, the authors consider a total of 336 ontologies, of which 49 %, 58 % and 68 % are

Triples Count	Restricted		Oblivious		PAGOdA		Konc.
	C	Q1-Q4	C	Q1-Q4	P	Q1-Q4	R
2.8M	10	0 0 0 0	45	0 0 TO 0	89	OM 4 1 0	75
5.1M	21	0 0 0 0	138	0 0 TO 3	147	OM 1 2 0	214
6.7M	28	0 0 0 0	1029	2 0 TO 0	203	OM 2 3 1	506
8.1M	36	37 0 0 0	TO	- - - -	263	OM 2 2 6	1347
9.0M	37	0 0 0 0	OM	- - - -	113	1 1 1 1	198
17.8M	72	0 0 0 0	OM	- - - -	232	2 2 3 3	987
26.2M	107	0 0 0 0	OM	- - - -	378	4 10 12 5	3491
33.9M	141	0 1 0 0	OM	- - - -	521	6 21 21 12	TO
2.8M	8	0 0 0 1	70	0 0 0 74	51	OM 0 0 0	51
5.7M	16	0 0 0 2	158	1 1 1 154	99	OM 1 1 0	118
8.4M	26	0 0 0 3	242	1 1 2 186	142	OM 2 1 1	220
11.4M	37	1 0 0 5	341	2 2 3 311	197	OM 3 1 1	315
2.2M	11	0 0 0 0	56	0 0 0 1	61	28 0 TO 1	53
4.5M	27	2 0 0 0	133	0 0 1 2	121	60 0 TO 2	125
6.6M	42	3 1 1 0	216	1 1 2 3	186	TO 0 TO 5	292
8.9M	58	5 1 2 1	310	1 2 4 6	260	TO 0 TO 5	644

Fig. 5. Results for Reactome, Uniprot, LUBM and UOBM (sorted from top to bottom in the above table).

weakly acyclic [7], *jointly acyclic* [12] and MFA$^\cup$, respectively. Even though the comparison is not over the same TBoxes, we verify that the improvement in generality of our notion is in line with previous iterations of related work.

6.2   A Materialization Based Reasoner

We now report on an implementation of the restricted chase as defined in Sect. 3. Moreover, we also present an implementation of the oblivious chase with singularization, i.e., the chase as it must be used if we employ MFA$^\cup$ (see Sect. 4). We use the datalog engine RDFOx [15] in both implementations.

We evaluate the performance of our chase based implementations against Konclude [19], a very efficient OWL DL reasoner, and PAGOdA [20], a hybrid approach to query answering over ontologies. PAGOdA combines a datalog reasoner with a fully-fledged OWL 2 reasoner in order to provide scalable 'pay-as-you-go' performance and is, to the best of our knowledge, the only other implementation that may solve CQ answering over *Horn-SRIQ* ontologies with completeness guarantees, albeit only in some cases. Nevertheless, PAGOdA was able to solve all the queries (that is, all of which for which it did not time-out or run out of memory) in this evaluation in a sound and complete manner.

We consider two real-world ontologies in our experiments, Reactome and Uniprot, and two standard benchmarks, LUBM and UOBM, all of which contain a large amount of ABox axioms. Axioms in these ontologies which are not expressible in *Horn-SRIQ* were pruned. Furthermore, one extra axiom had to be removed from Uniprot for it to be both MFA$^\cup$ and RCA$_1$ acyclic.

The results from our experiments are summarized in Fig. 5. For each ontology, we consider four samples of the original ABox. The number of triples contained in each one of these is indicated at the beginning of each row, under the column "Triples Count". As previously mentioned, we consider four different implementations: These include the two aforementioned variants of the chase ("Restricted" and "Oblivious"), PAGOdA ("PAGOdA") and Konclude ("Konc."). For both chase based implementations, we check the time it takes to compute the chase ("C") and then the time to solve each of the four queries crafted for each ontology ("Q1–Q4"). In a similar manner, we list the time PAGOdA takes to preprocess each ontology ("P") plus the time it takes to answer the queries ("Q1–Q4"). Finally, we list the time Konclude takes to solve realization; i.e., the task of computing every fact of the form $A(a)$ entailed by an ontology (note that Konclude cannot solve arbitrary CQ answering). Time-outs, indicated with "TO," were set at 1 h for materialization and 5 min for queries. We make use of the acronym "OM" to indicate that an out-of-memory error occurred. Sometimes, a time-out or an out of memory error prevents us from answering the queries: Such a situation is indicated with "-." All experiments were performed on a MacBook Pro with 8 GB of RAM and a 2.4 GHz Intel Core i5 processor.

For each ontology, we consider four different queries which are listed in the App. Section B included in the extended technical report. A summarized description of these queries, in which we ignore unary predicates, can be found in Fig. 6. For every ontology, the query Q1 is of the form $\exists x, y, z R(x, y) \wedge R(z, y)$ where R is an existentially quantified role occurring in the TBox. It appears that PAGOdA has trouble with this kind of query, whereas the chase based implementations efficiently solve it in all but one case. This is probably due to the design of the hybrid reasoner which considers under and over approximations to provide complete answers to CQ: It appears that queries as the one previously considered find a large number of matches in the upper bound which slows down the performance of this reasoner. Queries Q2, and Q3 and Q4 are acyclic and cyclic, respectively (a query is acyclic if the shape of its body is acyclic). Even though it is well-known that answering acyclic CQs can be reduced to satisfiability [5], we included such a type of query in our evaluation in an attempt to verify whether solving acyclic queries is simpler than cyclic queries (this is indeed the case theoretically). Nevertheless, our experiments do not reveal any significant differences.

First, note that computing the restricted chase employing renaming techniques to deal with equality is way more efficient than computing the oblivious chase with singularization. We conjecture that this is because the efficient built-in capabilities of RDFOx to deal with equality and the fact that the rules that result from the application of singularization are rather cumbersome. Second, see that our proposed algorithm is also superior to PAGOdA when it comes to CQ answering. Third, the implementation of the restricted chase outperforms the DL reasoner Konclude by an order of magnitude when it comes to solve materialization of the larger samples considered (note that, by computing the chase of a program we already solve materialization). It is clear that our implementation also scales much better than the OWL DL reasoner.

$q_1(w,y):$ $\mathsf{pE}(w,z),\mathsf{pE}(y,z)$

$q_2(x,z):$ $\mathsf{mPE}(z,w),\mathsf{mPE}(z,w),\mathsf{p}(y,z),\mathsf{pC}(x,y)$

$q_3(x,z):$ $\mathsf{fL}(x,w),\mathsf{fL}(x,y),\mathsf{sIB}(w,z),\mathsf{sIB}(y,z)$

$q_4(x,z):$ $\mathsf{p}(w,z),\mathsf{p}(y,z),\mathsf{pC}(x,w),\mathsf{pC}(x,y)$

$q_1(x,y):$ $\mathsf{cC}(x,z),\mathsf{cC}(y,z)$

$q_2(x):$ $\mathsf{tF}(w,x),\mathsf{IO}(x,y),\mathsf{d}(x,z)$

$q_3(x):$ $\mathsf{tF}(w,y),\mathsf{tF}(w,x),\mathsf{d}(y,z),\mathsf{d}(x,z)$

$q_4(x):$ $\mathsf{II}(x,w),\mathsf{cC}(w,z),\mathsf{II}(x,y),\mathsf{cC}(y,z)$

$q_1(x,z):$ $\mathsf{wF}(x,y),\mathsf{wF}(z,y),\mathsf{pA}(x,z)$

$q_2(x):$ $\mathsf{a}(x,y),\mathsf{tO}(y,z),\mathsf{mO}(y,w)$

$q_3(x,z):$ $\mathsf{tO}(y,z),\mathsf{a}(x,y),\mathsf{tC}(x,z)$

$q_4(x):$ $\mathsf{pA}(x,z),\mathsf{pA}(x,y),\mathsf{a}(z,y),$ $\mathsf{mO}(z,w),\mathsf{mO}(y,w)$

$q_1(x,y):$ $\mathsf{tC}(x,z),\mathsf{tC}(y,z)$

$q_2(x):$ $\mathsf{tAO}(x,y),\mathsf{pA}(z,x),\mathsf{tC}(w,y),\mathsf{wF}(x,v)\}$

$q_3(x,y):$ $\mathsf{iFO}(x,y),\mathsf{I}(x,z)$

$q_4(x,y):$ $\mathsf{hDDF}(x,z),\mathsf{hDDF}(y,z),\mathsf{hMDF}(x,w),$ $\mathsf{hMDF}(y,w),\mathsf{wF}(x,v),\mathsf{wF}(y,v)$

Fig. 6. Summarized queries for Reactome (top left), Uniprot (top right), LUBM (bottom left) and UOBM (bottom right).

7   Conclusions and Future Work

We introduce a novel acyclicity notion for *Horn-$\mathcal{SRIQ}$* TBoxes and prove it to be, theoretically and empirically, more general than previously existing conditions [6]. To the best our knowledge, this is the first acyclicity notion (for ontologies or rules) which considers termination of the restricted chase algorithm. Moreover, our contribution is also relevant in practice: Based on our ideas, we produce an implementation which vastly outperforms state-of-the-art reasoners.

As future work, we plan to lift our acyclicity condition to the case of general rules; i.e., not only those resulting from the translation of *Horn-$\mathcal{SRIQ}$* TBoxes. We also intend to work on further optimizing our implementation of the RCA$_n$ membership check and our restricted chase based algorithm.

Acknowledgements. We wish to thank Bernardo Cuenca Grau for extensive discussions on the subject and valuable feedback. This work was supported by the National Science Foundation under awards 1017255 *III: Small: TROn – Tractable Reasoning with Ontologies* and 1440202 *EarthCube Building Blocks: Collaborative Proposal: GeoLink – Leveraging Semantics and Linked Data for Data Sharing and Discovery in the Geosciences*; the *ERC grant 647289* and the *European Research Council grant CODA 647289*.

References

1. Baader, F., Calvanese, D., McGuinness, D., Nardi, D., Patel-Schneider, P. (eds.): The Description Logic Handbook: Theory, Implementation, and Applications, 2nd edn. Cambridge University Press, New York (2007)
2. Baget, J., Garreau, F., Mugnier, M., Rocher, S.: Extending acyclicity notions for existential rules. In: ECAI 2014, Frontiers in Artificial Intelligence and Applications, vol. 263, pp. 39–44. IOS Press (2014)
3. Calì, A., Gottlob, G., Kifer, M.: Taming the infinite chase. In: Brewka, G., Lang, J. (eds.) Proceedings of 11th International Conference on Principles of Knowledge Representation and Reasoning (KR 2008), pp. 70–80. AAAI Press (2008)

4. Calvanese, D., Eiter, T., Ortiz, M.: Answering regular path queries in expressive dls via alternating tree-automata. Inf. Comput. **237**, 12–55 (2014)
5. Carral Martínez, D., Hitzler, P.: Extending description logic rules. In: Simperl, E., Cimiano, P., Polleres, A., Corcho, O., Presutti, V. (eds.) ESWC 2012. LNCS, vol. 7295, pp. 345–359. Springer, Heidelberg (2012)
6. Grau, B.C., Horrocks, I., Krötzsch, M., Kupke, C., Magka, D., Motik, B., Wang, Z.: Acyclicity notions for existential rules and their application to query answering in ontologies. JAIR **47**, 741–808 (2013)
7. Fagin, R., Kolaitis, P.G., Miller, R.J., Popa, L.: Data exchange: semantics and query answering. Theor. Comput. Sci. **336**(1), 89–124 (2005)
8. Glimm, B., Lutz, C., Horrocks, I., Sattler, U.: Conjunctive query answering for the description logic SHIQ. J. Artif. Intell. Res. (JAIR) **31**, 157–204 (2008)
9. Hitzler, P., Krötzsch, M., Parsia, B., Patel-Schneider, P.F., Rudolph, S. (eds.): OWL 2 Web Ontology Language: Primer. W3C Recommendation (27 October 2009). http://www.w3.org/TR/owl2-primer/
10. Horrocks, I., Kutz, O., Sattler, U.: The even more irresistible SROIQ. In: Proceedings, Tenth International Conference on Principles of Knowledge Representation and Reasoning, United Kingdom, June 2–5, 2006, pp. 57–67. AAAI Press (2006)
11. Kontchakov, R., Lutz, C., Toman, D., Wolter, F., Zakharyaschev, M.: The combined approach to query answering in dl-lite. In: Lin, F., Sattler, U., Truszczynski, M. (eds.) KR 2010. AAAI Press (2010)
12. Krötzsch, M., Rudolph, S.: Extending decidable existential rules by joining acyclicity and guardedness. In: IJCAI, pp. 963–968 (2011)
13. Krötzsch, M., Rudolph, S., Hitzler, P.: Complexities of horn description logics. ACM Trans. Comp. Log. **14**(1), 2:1–2:36 (2013)
14. Marnette, B.: Generalized schema-mappings: from termination to tractability. In: PODS, pp. 13–22 (2009)
15. Nenov, Y., Piro, R., Motik, B., Horrocks, I., Wu, Z., Banerjee, J.: RDFox: a highly-scalable RDF store. In: Arenas, M., et al. (eds.) ISWC 2015. LNCS, vol. 9367, pp. 3–20. Springer, Heidelberg (2015). doi:10.1007/978-3-319-25010-6_1
16. Rudolph, S., Glimm, B.: Nominals, inverses, counting, and conjunctive queries or: Why infinity is your friend! CoRR abs/1401.3849 (2014)
17. Stefanoni, G., Motik, B., Horrocks, I.: Introducing nominals to the combined query answering approaches for $\mathcal{EL}$. In: AAAI (2013)
18. Stefanoni, G., Motik, B., Krötzsch, M., Rudolph, S.: The complexity of answering conjunctive and navigational queries over OWL 2 EL knowledge bases. J. Artif. Intell. Res. (JAIR) **51**, 645–705 (2014)
19. Steigmiller, A., Liebig, T., Glimm, B.: Konclude: system description. J. Web Sem. **27**, 78–85 (2014)
20. Zhou, Y., Grau, B.C., Nenov, Y., Kaminski, M., Horrocks, I.: Pagoda: pay-as-you-go ontology query answering with a datalog reasoner. J. Artif. Intell. Res. (JAIR) **54**, 309–367 (2015)

Containment of Expressive SPARQL Navigational Queries

Melisachew Wudage Chekol[1(✉)] and Giuseppe Pirrò[2]

[1] Data and Web Science Group, University of Mannheim, Mannheim, Germany
mel@informatik.uni-mannheim.de
[2] Institute for High Performance Computing and Networking,
ICAR-CNR, Rende, Italy
pirro@icar.cnr.it

Abstract. Query containment is one of the building block of query optimization techniques. In the relational world, query containment is a well-
studied problem. At the same time it is well-understood that relational
queries are not enough to cope with graph-structured data, where one
is interested in expressing queries that capture *navigation* in the graph.
This paper contributes a study on the problem of query containment
for an expressive class of navigational queries called Extended Property Paths (EPPs). EPPs are more expressive than previous navigational
extension of SPARQL (e.g., nested regular expressions) as they allow to
express path conjunction and path negation, among others. We attack
the problem of EPPs containment and provide complexity bounds.

1 Introduction

Research in graph query languages has emerged as a consequence of the intrin-
sic limitations of relational query languages when it comes to the possibility to
express recursion and navigation. This lead to the design of languages like Regular Path Queries (RPQs) and their extensions, including 2RPQs [2] that include
the possibility to traverse paths backwards, Extended Regular Path Queries
(ERPQs) [1] that offer higher expressive power by allowing to express queries that
also capture graphs [11,13]. Another well-studied class of queries is that of Nested
Regular Expressions (NREs) [26], that were originally proposed as the navigational core of SPARQL. The current SPARQL 1.1 standard introduced Property
Paths (PPs) as navigational core; PPs offer very limited expressive power due to
the lack of features to express any type of test within a path. To cope with these
issues and design a language as close as possible to the current W3C standard,
the language of Extended Property Paths (EPPs) has been recently proposed [12].
EPPs are more expressive than NREs, PPs and other navigational extensions of
SPARQL thanks to the possibility to express path conjunction, negation and more
powerful types of tests (e.g., checking the values of nodes reached via a nested
expression) while keeping query evaluation tractable.

Related Work. The problem of query containment is one of the main pillars of static
analysis and query optimization. In what follows we will not consider containment
under constraints as we tackle containment of EPPs without constraints.

© Springer International Publishing AG 2016
P. Groth et al. (Eds.): ISWC 2016, Part I, LNCS 9981, pp. 86–101, 2016.
DOI: 10.1007/978-3-319-46523-4_6

In the relational world, query containment for conjunctive queries (CQs) is now well-understood; it is NP-complete for basic CQs [5] and union of CQs [29] while it is Π_2^p when considering arithmetic comparison and also bag semantics [6,19].

For graph queries, containment of 2RPQs has been shown to be PSPACE-complete [1,2]; the complexity jumps to EXPTIME-hard under the presence of functionality constraints and to 2EXPTIME when considering expressive description logics constraints [4]. The problem becomes exponential if the query on the right hand side has a tree structure (cf. for example, [3]); for *extended* RPQs containment is undecidable for Boolean queries over a fixed alphabet [1].

In the Semantic Web, the containment of PSPARQL has been studied in Chekol et al. [8], this work provides lower bounds for upper bound complexity results reported in that paper; for NRE, Reutter [28] has shown a PSPACE upper bound; for SPARQL PPs, Kostylev et al. [20] show that the containment ranges from EXPSPACE-complete for OPT-free queries to PSPACE-complete if the right-hand side query is a pattern without projection. The study in [27] provides complexity analysis for several fragments of SPARQL: the results range from NP-completeness for AND-UNION queries to undecidability for the full SPARQL.

Another related language is XPath; the problem of XPath 2.0 query containment has been studied in ten Cate and Lutz [30]. They showed that the introduction of path intersection alone (i.e., the language CoreXPath ($\cap$) leads to 2EXPTIME-completeness. Finally, Kostylev et al. [22] studied the problem of containment of navigational XPath queries (i.e., the GXpath langauge), with results ranging from undecidability (when negation is considered) to EXP-TIME completeness for the positive fragment.

Contributions. We study query containment for EPPs, a significant extension of property paths (PPs), the current navigational core of SPARQL, and NREs for which containment has been already studied. We resort on two main ingredients: (i) an encoding of EPPs into the μ-calculus; (ii) the notion of (RDF) transition systems to check the validity test of μ-calculus formulae and provide an *upper bound* on the containment of EPPs and SPARQL with EPPs. For *lower bounds* we make connections between EPPs, PDL and XPath 2.0.

Automata theoretic notions and a reduction into validity test in a logic have been widely used to address the problem of query answering and containment [2,3,10,15,20,24]. Contrary to the automata techniques, the logic based approaches are fairly implementable. In this respect, it has been shown [16] that logical combinators can provide an exponential gain in succinctness in terms of the size of a logical formula thus allowing to study containment for expressive query languages in exponential-time, even though their direct formulation into the underlying logic results in an exponential blow up of the formula size.

Outline. The remainder of the paper is organized as follows. We provide some background in Sect. 2. The language of EPPs is introduced in Sect. 3. Section 4 describes an encoding of EPPs into the μ-calculus. Section 5 discusses the containment of EPPs. We conclude in Sect. 6.

2   Preliminaries

This section provides some background about the machineries used in this paper. We start with RDF and SPARQL and then give a brief overview about μ-calculus that will be used to tackle the containment of EPPs.

RDF and SPARQL. An RDF triple[1] is a tuple of the form $\langle s, p, o \rangle \in \mathbf{I} \times \mathbf{I} \times \mathbf{I} \cup \mathbf{L}$, where $\mathbf{I}$ (IRIs) and $\mathbf{L}$ (literals) are countably infinite sets. The set of terms of an RDF graph will be $terms(G)$ while $nodes(G)$ will be the set of terms used as a subject or object of a triple. An RDF graph G is a set of triples. To query RDF data, a standard query language, called SPARQL, has been defined. The semantics of a SPARQL query [25] is defined in terms of solution mappings. A (solution) mapping m is a partial function $m : \mathcal{V} \to \mathbf{I} \cup \mathbf{L}$. The SPARQL semantics uses a function $[\![Q]\!]_G$ that evaluates a query Q on a graph G and gives a multiset (bag) of mappings in the general case. However, when considering SPARQL with patterns using recursive PPs (i.e., using *, +) the standard introduces auxiliary functions (called ALP) that return sets of mapping.

μ **-calculus.** The μ-calculus ($\mathcal{L}_\mu$) is a logic obtained by adding fixpoint operators to ordinary modal logic [23]. For the purpose of this paper we will make usage of the μ-calculus with nominals and converse programs [31]. The syntax of the μ-calculus includes countable sets of *atomic propositions* AP, a set of *variables* Var, a set of *programs and their respective converses* $Prog = \{s, p, o, \bar{s}, \bar{p}, \bar{o}\}$ used to allow navigation in a graph. A μ-calculus formula φ is defined inductively as:

$$\varphi :: = \top \mid q \mid X \mid \neg\varphi \mid \varphi \wedge \psi \mid \varphi \vee \psi \mid \langle a \rangle \varphi \mid [a]\varphi \mid \mu X \varphi \mid \nu X \varphi$$

where $q \in AP, X \in Var$ and $a \in Prog$ is a transition program or its converse $\bar{a}$. The greatest fixpoint ν and least fixpoint operator μ introduce general and finite recursion in graphs, respectively. The semantics of the μ-calculus is given over a transition system (aka Kripke structure) $K = (S, R, L)$, where S is a non-empty set of nodes, $R : Prog \to 2^{S \times S}$ is the transition function, and $L : AP \to 2^S$ assigns a set of nodes to each atomic proposition or nominal where it holds, such that $L(p)$ is a *singleton* for each nominal p. For converse programs, R can be extended as $R(\bar{a}) = \{(s', s) \mid (s, s') \in R(a)\}$ where $s, s' \in S$.

Definition 1 *(Model of a formula). For a sentence φ and a transition system $K = (S, R, L)$, K is model of φ, denoted $K \models \varphi$, if there exists $s \in S$ such that $K, s \models \varphi$ if and only if $s \in [\![\varphi]\!]^K$ – s is an element of the answer to the evaluation of φ over K. If a sentence has a model, then it is called satisfiable.*

If a μ-calculus formula ψ appears under the scope of a least μ or greatest ν fixed point operator over all the programs $\{s, p, o, \bar{s}, \bar{p}, \bar{o}\}$ as, $\mu X.\psi \vee \langle s \rangle X \vee \langle p \rangle X \vee \cdots$ or $\nu X.\psi \wedge \langle s \rangle X \wedge \langle p \rangle X \wedge \cdots$, then, for legibility, we denote the formulae by $lfp(X, \psi)$ and $gfp(X, \psi)$, respectively.

---

[1] We do not consider bnodes.

RDF Transition System. An RDF transition system [9] is a labeled transition system, $K_G = (S = S' \cup S'', R, L)$, representation of an RDF graph G where two sets of nodes, S' and S'', are introduced: one set S'' for each triple (called triple node) and the other set S' for each subject, predicate, and object of each triple. A triple node (the black node in the figure below) is connected to its subject, predicate, and object nodes. For instance, the RDF graph $\boxed{x} \xrightarrow{z} \boxed{y}$ can be turned into the RDF transition

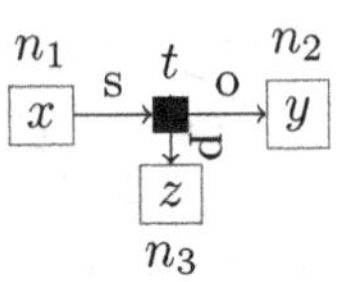

Fig. 1. Encoding of an RDF triple into a transition system.

system in Fig. 1, where $S' = \{n_1, n_2, n_3\}$, $S'' = \{t\}$, $R(s) = \{(n_1, t)\}$, $R(p) = \{(t, n_3)\}$, $R(o) = \{(t, n_2)\}$, and $L(x) = \{n_1\}, L(y) = \{n_2\}, L(z) = \{n_3\}$. Navigation from one node to another, in RDF transition systems, is done by using a set of transition programs $\{s, p, o\}$ and their converses. Since the μ-calculus with converse lacks functionality for number restrictions one cannot impose that each triple node is connected to exactly one node for each subject, predicate, and object node. However, one can impose a lighter restriction to achieve this by taking advantage of the technique introduced in [14] and adopted in [7]. Since it is not possible to ensure that there is only one successor, then we restrict all the successors to bear the same constraints; thus, they become interchangeable. This is achieved by rewriting the formulas using a function func such that all occurrences of $\langle a \rangle \varphi$ (existential formulas) are replaced by $\langle a \rangle \top \wedge [a]\varphi$; in Definition 2, func is defined inductively on the structure of a μ-calculus formula. When checking for query containment, we assume that the formulas are rewritten using the function func.

Definition 2. func *is inductively defined on the structure of a μ-calculus formula as follows:*

$$\mathsf{func}(\top) = \top \qquad\qquad \mathsf{func}(\bot) = \bot$$
$$\mathsf{func}(q) = q \quad q \in AP \cup Nom \qquad \mathsf{func}(X) = X \quad X \in Var$$
$$\mathsf{func}(\neg\varphi) = \neg\mathsf{func}(\varphi) \qquad \mathsf{func}(\varphi \wedge \psi) = \mathsf{func}(\varphi) \wedge \mathsf{func}(\psi)$$
$$\mathsf{func}(\varphi \vee \psi) = \mathsf{func}(\varphi) \vee \mathsf{func}(\psi) \qquad \mathsf{func}(\langle a \rangle \varphi) = \langle a \rangle \top \wedge [a]\mathsf{func}(\varphi) \quad a \in Prog$$
$$\mathsf{func}(\langle a \rangle \varphi) = \langle a \rangle \mathsf{func}(\varphi) \qquad \mathsf{func}([a]\varphi) = [a]\mathsf{func}(\varphi)$$
$$\mathsf{func}(\mu X.\varphi) = \mu X.\mathsf{func}(\varphi) \qquad \mathsf{func}(\nu X.\varphi) = \nu X.\mathsf{func}(\varphi)$$

Figure 2 shows an RDF graph G and an excerpt of its corresponding RDF transition system K_G. Each RDF triple requires the introduction of a transition node (i.e., black nodes in the figure), where: the subject of the RDF triple has an incoming edge to the transition node labeled as s and the predicate and object have outgoing edges labeled as p and o, respectively. At this point, we shall define the model of formula ψ in terms of RDF transition systems.

Definition 3 *(RDF transition system model). An RDF transition system K_G is considered as a model of formula ψ if there exists a node s in the transition system where ψ holds, i.e., $K_G, s \models \psi$. If a formula has a model, then it is called satisfiable.*

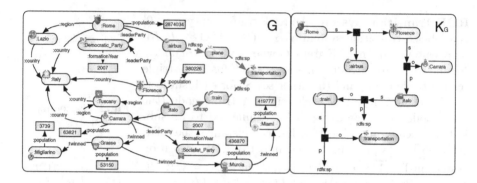

Fig. 2. An graph G and an excerpt of its corresponding RDF transition system K_G

The following lemma links RDF transition systems with transition systems when encoding in μ-calculus a query expressed in some RDF query language.

Lemma 1 (Chekol et al. [7]). *Let φ be a μ-calculus encoding of an EPPs query, φ is satisfied by some RDF transition system K_G if and only if* $\mathrm{func}(\varphi)$ *is satisfied by some transition system K.*

The above lemma serves as a basis for the encoding of EPPs that will be encoded in μ-calculus formulas and then can be interpreted over RDF transition systems.

3   Extended Property Paths

We now introduce our navigational extension of SPARQL called Extended Property Paths (EPPs) [12].

Syntax. EPPs extend PPs and NREs-like languages with path conjunction/negation, repetitions, and more types of tests. The syntax of EPPs is given below:

$$e ::= \text{`}^{\wedge}\text{'} \, e \mid e \text{ `+'} \mid e \text{ `?'} \mid e \text{`*'} \mid e \text{ `/'} \, e \mid e \text{ `|'} \, e \mid$$
$$\text{`(' } e \text{ `)'} \mid [\mathrm{P}]^1 \text{ test } [\mathrm{P}]^2 \mid e \text{ `\&' } e \mid e \text{ `\sim' } e \mid e\text{`\{'}l, h\text{`\}'}$$
$$\text{test} ::= \text{`!' test} \mid \text{ test `\&\&' test} \mid \text{ test `||' test} \mid \text{`(' test `)'} \mid \mathbf{base}$$
$$\mathbf{base} ::= \mathbf{iri} \mid \text{`}^{\wedge}\text{'}\mathbf{iri} \mid \text{`TP('}\mathrm{P} \text{ `,' } e \text{ `)'} \mid \text{`T('EExp')'}$$
$$\mathrm{P} ::= \text{`\_s'} \mid \text{`\_p'} \mid \text{`\_o'}$$

[1]Default is _s; [2]Default is _o. We refer to EPPs without path repetition, path complement as **cEPPs**.

EPPs introduce the following new features: path conjunction ($e_1 \& e_2$), path negation ($e_1 \sim e_2$), path repetitions between l and h times ($e\{l, h\}$), and different types of tests (test) *within* a path. EPPs allow to specify the *starting* and *ending* position (P) of a test; it is possible to test from each of the subject, predicate and object positions in RDF triples, mapped in the EPPs syntax to the position symbols $\_s$, $\_p$ and $\_o$, respectively. Positions do not need to be always specified; by default a test starts from the subject ($\_s$) and ends on the object ($\_o$) of the triple being traversed. There are different types of tests (test); a test can be a simple check for the existence of a IRI in forward/reverse direction, a *nested* EPP, i.e., TP(P, e), which corresponds to the evaluation of the expression e starting from a position P (of the last triple traversed) and returns true iff there exists at least one node that can be reached via e. A test can also be of type T; here, **EExp** (not reported here for sake of space) extends the production [110] in the SPARQL grammar[2] which enables to use in EPPs tests available in SPARQL as built-in conditions. Tests can be combined via the logical operators AND (&&), OR (||) and NOT (!).

Positions and Tests. EPPs tests can be coupled with positions. To formally explain the reasoning behind tests and positions, we make usage of a function $\Pi(P,t)$, which projects the element in position P of a triple t. If we have $t=\langle u_1,p_1,u_2\rangle$ and the test T($\_p=p_1$) then $\Pi(\_p,\langle u_1,p_1,u_2\rangle)=p_1$ that checks $p_1=p_1$, and, in this case, returns true; however, it returns false for T($\_o=p_3$).

Example 1 (**Path Conjunction, Negation and Tests**). Find pairs of cities located in the same country but not in the same region; such cities must be governed by the same political party, which has been founded before 2010.

NREs-based languages (and PPs) *cannot* express such request due to the lack of path conjunction/negation. With EPPs it can be expressed as shown in Fig. 3.

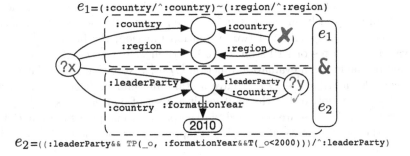

Fig. 3. EPP expression for Example 1.

The expression is the conjunction (&) of the two sub-expressions e_1 and e_2. In the sub-expression e_1, the symbol ^ denotes backward navigation; from the object to the subject of a triple. Path negation ∼ enables to discard from the set of cities in the same country (i.e., :country/^:country) those that are in the same region (i.e., :region/^:region). Path conjunction & enables to keep

[2] http://www.w3.org/TR/sparql11-query/#rExpression This is not considered when dealing with containment of EPPs.

from the set of nodes satisfying the first subexpression (e_1) those that also satisfy the second (e_2), i.e. governed by the same party, which has been founded before 2010. Note that PPs and NREs-based languages lack tests like TP, to check the existence of a path (via :formationYear) to nodes having value <2010.

Example 2 (**Positions and Tests**). Consider again the expression in Example 1. The expression including both default positions[3] and using positions to traverse backward edges is shown in Fig. 4.

As an example, the test TP (in Fig. 4) starts from the position _o, that is, the object of the last triple traversed.

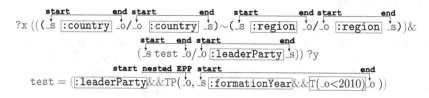

Fig. 4. An EPP expression with positions.

Definition 4. *An EPP query q has the form q:: = $(\mathcal{V} \cup \mathbf{I}, e, \mathcal{V} \cup \mathbf{I} \cup \mathbf{L})$.*

Semantics. For the purpose of this paper we will focus our attention on set semantics. This is in line with recent related work that studied the containment of PPs in SPARQL [20] where PPs were given a set-based semantics. Besides, the containment of conjunctive queries in the relational world when considering bag semantics is still an open issue[4]. Table 1 shows the semantics of EPPs. The semantics makes usage of two evaluation function; the first $\mathbf{E}[\![\cdot]\!]^G$ is used to evaluate all forms of EPPs on an RDF graph G but tests; these latter are handled by the boolean evaluation function $\mathbf{E}_T[\![\cdot]\!]^G_t$. Note that in this latter case tests consider a triple $t \in G$ and have a start and end position. By observing rule R9 one may notice that the final result is a set of pair of nodes in the graph G; these pairs of nodes are obtained via the position mapping function Π that projects the two elements of the triple t appearing in position P_1 and P_2 if t satisfies the test.

4   Translating cEPPs into $\mathcal{L}_\mu$

The goal of this section is to provide the first building block toward tackling the containment of EPP expressions; this concerns the encoding of **cEPPs** into $\mathcal{L}_\mu$. We shall start by introducing the notion of path pattern translation.

---

[3] These are automatically added during the parsing phase.

[4] It is claimed that for basic CQs the complexity of query containment under bag semantics is in Π_2^p [6].

Table 1. Set-based semantics for EPPs.

R1	$\mathbf{E}[\![\hat{}\, e]\!]^G := \{(u,v) : (v,u) \in \mathbf{E}[\![e]\!]^G\}$
R2	$\mathbf{E}[\![e_1/e_2]\!]^G := \{(u,v) : \exists w\ (u,w) \in \mathbf{E}[\![e_1]\!]^G \wedge (w,v) \in \mathbf{E}[\![e_2]\!]^G\}$
R3	$\mathbf{E}[\![(e)^*]\!]^G := \{(u,u) \mid u \in nodes(G)\} \cup \bigcup_{i=1}^{\infty} \mathbf{E}[\![e_i]\!]^G \mid$
	$e_1 = e \wedge e_i = e_{i-1}/e$
R4	$\mathbf{E}[\![(e)^+]\!]^G := \bigcup_{i=1}^{\infty} \mathbf{E}[\![e_i]\!]^G \mid e_1 = e \wedge e_i = e_{i-1}/e$
R5	$\mathbf{E}[\![(e)?]\!]^G := \{(u,u) \mid u \in nodes(G)\} \cup \mathbf{E}[\![e]\!]^G$
R6	$\mathbf{E}[\![(e_1\|e_2)]\!]^G := \{(u,v) : (u,v) \in \mathbf{E}[\![e_1]\!]^G \vee (u,v) \in \mathbf{E}[\![e_2]\!]^G\}$
R7	$\mathbf{E}[\![(e_1\&e_2)]\!]^G := \{(u,v) : (u,v) \in \mathbf{E}[\![e_1]\!]^G \wedge (u,v) \in \mathbf{E}[\![e_2]\!]^G\}$
R8	$\mathbf{E}[\![(e_1 \sim e_2)]\!]^G := \{(u,v) : (u,v) \in \mathbf{E}[\![e_1]\!]^G \wedge (u,v) \notin \mathbf{E}[\![e_2]\!]^G\}$
R9	$\mathbf{E}[\![P_1\ test\ P_2]\!]^G := \{(\Pi(P_1,t), \Pi(P_2,t))) \mid triple\ t \in G \wedge \mathbf{E}_T[\![test]\!]_t^G\}$
R10	$\mathbf{E}_T[\![u]\!]_t^G := \Pi(\_p, t) = u$
R11	$\mathbf{E}_T[\![T(EExp)]\!]_t^G := \texttt{EvalSPARQLBuilt-in(EExp, t)}$
R12	$\mathbf{E}_T[\![TP(P,e)]\!]_t^G := \exists v : (\Pi(P,t), v) \in \mathbf{E}[\![e]\!]^G$
R13	$\mathbf{E}_T[\![test_1\&\&test_2]\!]_t^G := \mathbf{E}_T[\![test_1]\!]_t^G \wedge \mathbf{E}_T[\![test_2]\!]_t^G$
R14	$\mathbf{E}_T[\![test_1\|\|test_2]\!]_t^G := \mathbf{E}_T[\![test_1]\!]_t^G \vee \mathbf{E}_T[\![test_2]\!]_t^G$
R15	$\mathbf{E}_T[\![!test]\!]_t^G := \neg \mathbf{E}_T[\![test]\!]_t^G$

Definition 5 *(Path pattern translation). Given a **cEPPs** query $q :: = (\alpha, e, \beta)$ where $\alpha \in \mathcal{V} \cup \boldsymbol{I}$ and $\beta \in \mathcal{V} \cup \boldsymbol{I} \cup \boldsymbol{L}$, its translation into $\mathcal{L}_\mu$ is given by the following (sub)-formula:* $\mathsf{E}_\mu[q] = (\langle \bar{s} \rangle \alpha \wedge \langle p \rangle \mathsf{RE}_\mu(e, \beta))$, where $\alpha, \beta \in AP \cup \top$.

The formula states the existence of the encoding of the path pattern (via $\mathsf{E}_\mu[e]$) *somewhere* in a transition system and thus it is quantified by μ (least fixed point) so as to propagate the sub-formula to the entire transition system. μ encodes a reflexive transitive closure over all the programs and is denoted by $lfp(X, \mathsf{E}_\mu[q])$ [9]. The variables α and β are encoding as nominals whereas the IRIs in path expressions are encoded as atomic propositions. The encoding, shown in Table 2, makes usage of three functions

- RE_μ: takes an **cEPPs** expression e and a $\mathcal{L}_\mu$ formula φ and builds inductively an encoding based on the structure of the path expression.
- T_μ: takes a path test expression and inductively builds an $\mathcal{L}_\mu$ formula. The test positions are translated into transition programs in $\mathcal{L}_\mu$, i.e., _s becomes the transition program $\bar{s}$ whereas _p and _o become p and o programs, respectively.
- Γ: takes two (complex) **cEPPs** expressions and on the basis of the end position of the first and start position of the second returns transition programs expressed in $\mathcal{L}_\mu$ for the RDF transition system.

We now prove the correctness of the translation of **cEPPs** into $\mathcal{L}_\mu$ formulas as shown in the following Lemma.

Lemma 2. *Let q be a **cEPPs** query, for every RDF transition system K_G whose associated RDF graph is G, it holds that* $\mathbf{E}[\![q]\!]^G \neq \varnothing$ *iff* $[\![\mathsf{E}_\mu(q)]\!]^{K_G} \neq \varnothing$.

Table 2. Encoding of EPPs into $\mathcal{L}_\mu$. The function Γ takes two (complex) EPP expressions and on the basis of the end position of the first and start position of the second returns transition programs expressed in $\mathcal{L}_\mu$ for the RDF transition system. Note that if $\mathrm{P} = \_\mathrm{s}$ in line (14), (resp., $\mathrm{P_1} = \_\mathrm{s}$, $\mathrm{P_2} = \_\mathrm{s}$ in line (9)) then the translation into $\mathcal{L}_\mu$ is the transition program $\bar{s}$.

$$
\begin{aligned}
&(1) \quad \mathsf{RE}_\mu(iri, \varphi) &&= \begin{cases} \langle p \rangle iri \wedge \langle o \rangle \varphi & \text{if } \varphi \in AP \\ \langle p \rangle iri \wedge \varphi & \text{otherwise} \end{cases} \\
&(2) \quad \mathsf{RE}_\mu(\hat{\ }\mathrm{e}, \varphi) &&= \mathsf{RE}_\mu(\mathrm{e}, \langle \bar{s} \rangle \varphi) \\
&(3) \quad \mathsf{RE}_\mu(\mathrm{e_1 \mid e_2}, \varphi) &&= \mathsf{RE}_\mu(\mathrm{e_1}, \varphi) \vee \mathsf{RE}_\mu(\mathrm{e_2}, \varphi)) \\
&(4) \quad \mathsf{RE}_\mu(\mathrm{e_1/e_2}, \varphi) &&= \mathsf{RE}_\mu(\mathrm{e_1}, \Gamma(\mathrm{e_1}, \mathrm{e_2})\mathsf{RE}_\mu(\mathrm{e_2}, \varphi)) \\
&(5) \quad \mathsf{RE}_\mu((\mathrm{e_1 \& e_2}), \varphi) &&= (\mathsf{RE}_\mu(\mathrm{e_1}, \varphi) \wedge \mathsf{RE}_\mu(\mathrm{e_2}, \varphi)) \\
&(6) \quad \mathsf{RE}_\mu(\mathrm{e?}, \varphi) &&= \mathsf{RE}_\mu(\mathrm{e}, \varphi) \vee \langle \bar{s} \rangle \varphi \\
&(7) \quad \mathsf{RE}_\mu(\mathrm{e^+}, \varphi) &&= \mu X.\mathsf{RE}_\mu(\mathrm{e}, \varphi) \vee \mathsf{RE}_\mu(\mathrm{e}, \Gamma(\mathrm{e}, \mathrm{e})X) \\
&(8) \quad \mathsf{RE}_\mu(\mathrm{e^*}, \varphi) &&= \mathsf{RE}_\mu(\mathrm{e^+}, \varphi) \vee \langle \bar{s} \rangle \varphi \\
&(9) \quad \mathsf{RE}(\mathrm{P_1 \; test \; P_2}, \varphi) &&= \langle \mathrm{P_1} \rangle \top \wedge \mathsf{T}_\mu(\mathrm{test}) \wedge \langle \mathrm{P_2} \rangle \top \wedge \varphi \\
&(10) \quad \mathsf{T}_\mu(iri) &&= \langle p \rangle iri \\
&(11) \quad \mathsf{T}_\mu(\mathrm{test_1 \; \&\& \; test_2}) &&= \mathsf{T}_\mu(\mathrm{test_1}) \wedge \mathsf{T}_\mu(\mathrm{test_2}) \\
&(12) \quad \mathsf{T}_\mu(\mathrm{test_1 \; \| \; test_2}) &&= \mathsf{T}_\mu(\mathrm{test_1}) \vee \mathsf{T}_\mu(\mathrm{test_2}) \\
&(13) \quad \mathsf{T}_\mu(\mathrm{!test}) &&= \neg \mathsf{T}_\mu(\mathrm{test}) \\
&(14) \quad \mathsf{T}_\mu(\mathrm{TP}(\mathrm{P}, \mathrm{e})) &&= \langle \mathrm{P} \rangle \mathsf{RE}_\mu(\mathrm{e}, \top) \\
&(15) \quad \Gamma(\mathrm{e_1}, \mathrm{e_2}) &&= \langle o \rangle \langle s \rangle \\
&(16) \quad \Gamma(\mathrm{e_1}, \hat{\ }\mathrm{e_2}) &&= \langle o \rangle \langle \bar{o} \rangle \\
&(17) \quad \Gamma(\hat{\ }\mathrm{e_1}, \mathrm{e_2}) &&= \langle o \rangle \langle s \rangle \\
&(18) \quad \Gamma(\hat{\ }\mathrm{e_1}, \hat{\ }\mathrm{e_2}) &&= \langle o \rangle \langle \bar{o} \rangle
\end{aligned}
$$

Proof ($\Rightarrow$). Assume that there exists an RDF G such that the evaluation of q over G is nonempty, i.e., $\mathbf{E} [\![q]\!] G \neq \varnothing$. We can build such graph from the canonical instance of q and it can be produced using a function θ as shown below:

- if $(\alpha, iri, \beta) \in q$, then $\theta((\alpha, iri, \beta)) = (\alpha, y, \beta) \in G$,
- if $(\alpha, \mathrm{e}, \beta) \in q$, then $\theta((\alpha, \mathrm{e}, \beta)) \in G$,
- if $(\alpha, \hat{\ }\mathrm{e}, \beta) \in q$, then $\theta((\alpha, \hat{\ }\mathrm{e}, \beta)) = (\beta, \mathrm{e}, \alpha) \in G$,
- if $(\alpha, \mathrm{e_1/e_2}, \beta) \in q$, then $\theta((\alpha, \mathrm{e_1}, y)) \in G$ and $\theta((y, \mathrm{e_2}, \beta)) \in G$,
- if $(\alpha, \mathrm{e_1 | e_2}, \beta) \in q$, then $\theta((\alpha, \mathrm{e_1}, \beta)) \in G$ or $\theta((\alpha, \mathrm{e_2}, \beta)) \in G$,
- if $(\alpha, \mathrm{e_1 \& e_2}, \beta) \in q$, then $\theta((\alpha, \mathrm{e_1}, \beta)) \in G$ and $\theta((\alpha, \mathrm{e_2}, \beta)) \in G$,
- if $(\alpha, \mathrm{e^+}, \beta) \in q$, then $\bigcup_{i=1}^n \theta((\alpha, \mathrm{e}^i, \beta)) \in G$, where e^i denotes the composition of e i times and $n \in \mathbb{N}$,
- if $(\alpha, \mathrm{e^*}, \beta) \in q$, then $(\alpha, u, \alpha) \in G$ or $(\beta, u, \beta) \in G$ or $\theta((\alpha, \mathrm{e^+}, \beta)) \in G$, where u is a fresh IRI,
- if $(\alpha, \mathrm{P_1} \; iri \; \mathrm{P_2}, \beta) \in q$, then $(\alpha, iri, \beta) \in G$,
- if $(\alpha, \mathrm{P_1} \; \mathrm{test} \; \mathrm{P_2}, \beta) \in q$, then $\theta((\alpha, \mathrm{P_1} \; \mathrm{test} \; \mathrm{P_2}, \beta)) \in G$,

- if $(\alpha, P_1 \text{ test}_1 P_2 \ \&\& \ P_3 \text{ test}_2 P_4, \beta) \in q$, then
  $\theta((\alpha, P_1 \text{ test}_1 P_2, \beta)) \in G$ and $\theta((\alpha, P_3 \text{ test}_2 P_4, \beta)) \in G$,
- if $(\alpha, P_1 \text{ test}_1 P_2 \ || \ P_3 \text{ test}_2 P_4, \beta) \in q$, then
  $\theta((\alpha, P_1 \text{ test}_1 P_2, \beta)) \in G$ or $\theta((\alpha, P_3 \text{ test}_2 P_4, \beta)) \in G$,
- if $(\alpha, P_1 \text{ !test } P_2, \beta) \in q$, then we introduce a set of fresh IRI's iri_i that do not appear in test such that $(\alpha, iri_i, \beta) \in G$ and $\theta(\alpha, \text{test}, \beta) \notin G$,
- if $(\alpha, P_1 \text{ TP}(P, e) P_2, \beta) \in q$, then $(\alpha, y, \beta) \in G$ and $\theta(y, e, r) \in G$ or $\theta(\alpha, e, r) \in G$ or $\theta(\beta, e, r) \in G$, where y and r are fresh IRIs.

Since G is an instance of q, the evaluation of q over G is not empty. Now, we construct an RDF transition system $K_G = (S, R, L)$ in the same way as it is done in Definition 9 of [7]. It is possible to verify K_G is a model of $\mathsf{E}_\mu(q)$ by working inductively on the construction of $\mathsf{E}_\mu(q)$. This is because atomic propositions encoding the constants and distinguished variables are true in K_G as they exist in G; therefore, $\mathsf{E}_\mu(q)$ is satisfiable in K_G. To elaborate, if l is a distinguished variable (i.e., either x or z) or a constant e in q, then

- for l a distinguished variable, l is satisfiable in K_G since $[\![l]\!]^{K_G}$ is non empty,
- for $l = e$ an EPP, its encoding $\mathsf{RE}_\mu(e, \top)$ is satisfiable in K_G if e is satisfiable in K_G, this can be proved inductively on the structure of the encoding.

Thus, since K_G is an RDF transition system, we obtain that $[\![\mathsf{E}_\mu(q)]\!]^{K_G} \neq \varnothing$.

$(\Leftarrow)$ Assume that $[\![\mathsf{E}_\mu(q)]\!]^{K_G} \neq \varnothing$. In order to test if the evaluation of q over an RDF graph obtained from K_G is non empty, we produce an RDF graph G from K_G as done in Lemma 4 of [7]. Thus, since G is a technical construction obtained from an RDF transition system associated to q, it holds that $\mathbf{E}[\![q]\!]^G \neq \varnothing$.

Example 3. The μ-calculus encoding of the EPPs in Example 1 is given below. $\Gamma(et1, et2)$ is a shorthand for $\Gamma(\text{:leaderParty}\&\& \ \text{TP}(\_o, \text{:formationYear})$, ^:leaderParty).

$$\mathsf{E}_\mu(\mathsf{e1} \ \& \ \mathsf{e2}) = \textit{lfp}\big(X, \langle \bar{s} \rangle \top \wedge \mathsf{RE}_\mu(\mathsf{e1} \ \& \ \mathsf{e2}, \top)\big)$$

$$\mathsf{RE}_\mu(\mathsf{e1} \ \& \ \mathsf{e2}, \top) = \mathsf{RE}_\mu(\mathsf{e1}, \top) \wedge \mathsf{RE}_\mu(\mathsf{e2}, \top)$$

$$\mathsf{RE}_\mu(\mathsf{e1}, \top) = \mathsf{RE}_\mu((\text{:country}/\text{^:country}) \sim (\text{:region}/\text{^:region}), \top)$$

$$= \mathsf{RE}_\mu((\text{:country}/\text{^:country}), \top) \wedge \neg\mathsf{RE}_\mu((\text{:region}/\text{^:region}), \top)$$

$$= \text{blue}(\mathsf{RE}_\mu(\text{:country}, \Gamma(\text{:country}, \text{^:country})\mathsf{RE}_\mu(\text{^:country}, \top))\text{blue}) \wedge$$
$$\neg\text{blue}(\mathsf{RE}_\mu(\text{:region}, \Gamma(\text{:region}, \text{^:region})\mathsf{RE}_\mu(\text{^:region}, \top))\text{blue})$$

$$= \text{blue}(\langle p \rangle\text{:country} \wedge \langle o \rangle\langle \bar{o} \rangle(\langle p \rangle\text{:country} \wedge \top)\text{blue}) \wedge \neg\text{blue}(\langle p \rangle\text{:region} \wedge \langle o \rangle\langle \bar{o} \rangle(\langle p \rangle\text{:region} \wedge \langle \bar{s} \rangle\top)\text{blue})$$

$$\mathsf{RE}_\mu(\mathsf{e2}, \top) = \mathsf{RE}_\mu(\text{blue}(\text{:leaderParty} \ \&\& \ \text{TP}(\_o, \text{:formationYear})\text{blue})/\text{^:leaderParty}, \top)$$

$$= \mathsf{RE}_\mu(\text{:leaderParty} \ \&\& \ \text{TP}(\_o, \text{:formationYear}), \Gamma(et1, et2)\mathsf{RE}_\mu(\text{^:leaderParty}, \top))$$

$$= \mathsf{RE}_\mu(\text{:leaderParty}, \Gamma(et1, et2)\mathsf{RE}_\mu(\text{^:leaderParty}, \top)) \wedge$$
$$\mathsf{RE}_\mu(\text{TP}(\_o, \text{:formationYear}), \Gamma(et1, et2)\mathsf{RE}_\mu(\text{^:leaderParty}, \top))$$

$$= \text{blue}(\langle p \rangle\text{:leaderParty} \wedge \langle o \rangle\langle \bar{o} \rangle(\langle p \rangle\text{:leaderParty} \wedge \top)\text{blue}) \wedge$$
$$\mathsf{RE}_\mu(\text{TP}(\_o, \text{:formationYear}), \langle o \rangle\langle \bar{o} \rangle(\langle p \rangle\text{:leaderParty} \wedge \top))$$

$$= \text{blue}(\langle p \rangle\text{:leaderParty} \wedge \langle o \rangle\langle \bar{o} \rangle(\langle p \rangle\text{:leaderParty} \wedge \top)\text{blue}) \wedge$$
$$(\langle \bar{s} \rangle\top \wedge \mathsf{T}_\mu(\text{TP}(\_o, \text{:formationYear})) \wedge \langle o \rangle\top \wedge \langle o \rangle\langle \bar{o} \rangle(\langle p \rangle\text{:leaderParty} \wedge \top))$$

$$= \text{blue}(\langle p \rangle\text{:leaderParty} \wedge \langle o \rangle\langle \bar{o} \rangle(\langle p \rangle\text{:leaderParty} \wedge \top)\text{blue}) \wedge$$
$$(\langle \bar{s} \rangle\top \wedge \langle o \rangle\text{:formationYear} \wedge \langle p \rangle\top \wedge \langle o \rangle\top \wedge \langle o \rangle\langle \bar{o} \rangle(\langle p \rangle\text{:leaderParty} \wedge \top))$$

So far, we have shown that our encoding of **cEPPs** queries into the μ-calculus is correct. Next, we formally present the reduction of the containment test of **cEPPs** into unsatisfiability test in the μ-calculus.

5   Containment Results

In this section we state our complexity results. We start with **cEPPs** and then move to SPARQL with **cEPPs**. We start with the containment of **cEPPs** which can be stated as follows:

Problem: cEPPs Containment
Input: **cEPPs** q_1 and q_2
Output: Is $q_1 \sqsubseteq q_2$?

Given two **cEPPs** queries q_1 and q_2, $q_1 \sqsubseteq q_2$ if and only if for any RDF graph G, the answers of q_1 are included in the answers of q_2, i.e., $\mathbf{E}[\![\ q_1\ ]\!]^G \subseteq \mathbf{E}[\![\ q_2\ ]\!]^G$.

In the previous section, we have presented various functions to produce $\mathcal{L}_\mu$ formulas corresponding to the encodings of **cEPPs**. Hence, the problem of containment can be reduced to formula unsatisfiability test in $\mathcal{L}_\mu$ as:

Problem: cEPPs Containment via $\mathcal{L}_\mu$
Input: $\mathcal{L}_\mu$ encodings $\mathsf{E}_\mu(q_1)$ and $\mathsf{E}_\mu(q_2)$
Output: Is $\mathsf{E}_\mu(q_1) \wedge \neg \mathsf{E}_\mu(q_2)$ unsatisfiable?

Therefore, we obtain that $q_1 \sqsubseteq q_2 \Leftrightarrow \mathsf{E}_\mu(q_1) \wedge \neg \mathsf{E}_\mu(q_2)$ is unsatisfiable. We now show that our decision procedure is sound and complete.

Theorem 1 (Soundness). *Given two **cEPPs** queries q_1 and q_2 if $\mathsf{E}_\mu(q_1) \wedge \neg \mathsf{E}_\mu(q_2)$ is unsatisfiable, then $q_1 \sqsubseteq q_2$.*

Proof. We show the contrapositive. If $q_1 \not\sqsubseteq q_2$, then $\mathsf{E}_\mu(q_1) \wedge \neg \mathsf{E}_\mu(q_2)$ is satisfiable. One can verify that every RDF graph G in which there is at least one tuple satisfying q_1 but not q_2 can be turned into a RDF transition system model for $\mathsf{E}_\mu(q_1) \wedge \neg \mathsf{E}_\mu(q_2)$. To do so, consider an RDF graph G and assume that there is exists a mapping $m \in \mathbf{E}[\![\ q_1\ ]\!]^G$ and $m \notin \mathbf{E}[\![\ q_2\ ]\!]^G$. Let us construct an RDF transition system K_G from G. In fact, we produce G which contains at least a canonical instantiation of q_1, i.e., by replacing the variables in q_1 with their mappings in m. It has been shown that an RDF graph can be turned into an RDF transition system (cf. Lemma 1). At this point, it remains to verify that $[\![\mathsf{E}_\mu(q_1)]\!]^{K_G} \neq \varnothing$ and $[\![\mathsf{E}_\mu(q_2)]\!]^{K_G} = \varnothing$.

Let us construct the formulas $\mathsf{E}_\mu(q_1)$ and $\mathsf{E}_\mu(q_1)$ by first skolemizing the distinguished variables using their mappings in m. Consequently, from Lemma 2 one obtains, $[\![\mathsf{E}_\mu(q_1)]\!]^{K_G} \neq \varnothing$. However, $[\![\mathsf{E}_\mu(q_2)]\!]^{K_G} = \varnothing$, this is because the atomic propositions in the formula corresponding to the constants and variables are not satisfied in K. This implies that $[\![\neg \mathsf{E}_\mu(q_2)]\!]^{K_G} \neq \varnothing$. This is justified by the fact that if a formula φ is satisfiable in an RDF transition system, then $[\![\varphi]\!]^{K_G} = S$ thus $[\![\neg\varphi]\!]^{K_G} = \varnothing$. So far we have: $[\![\mathsf{E}_\mu(q_1)]\!]^{K_G} \neq \varnothing$ and $[\![\neg \mathsf{E}_\mu(q_2)]\!]^{K_G} \neq \varnothing$. Without loss of generality, $[\![\mathsf{E}_\mu(q_1) \wedge \neg \mathsf{E}_\mu(q_2)]\!]^{K_G} \neq \varnothing$. Therefore, $\mathsf{E}_\mu(q_1) \wedge \neg \mathsf{E}_\mu(q_2)$ is satisfiable.

Theorem 2 (Completeness). *Given two **cEPPs** queries q_1 and q_2, if $\mathsf{E}_\mu(q_1) \wedge \neg \mathsf{E}_\mu(q_2)$ is satisfiable, then $q_1 \not\sqsubseteq q_2$.*

Proof. $\mathsf{E}_\mu(q_1) \wedge \neg\mathsf{E}_\mu(q_2)$ is satisfiable $\Rightarrow \exists K_G.[\![\mathsf{E}_\mu(q_1) \wedge \neg\mathsf{E}_\mu(q_2)]\!]^{K_G} \neq \varnothing$. Consequently, K_G is an RDF transition system as shown in Proposition 1 of [7].

Using $K_G = (S' \cup S'', R, L)$, we construct an RDF graph G such that we can show that $\mathbf{E}[\![\ q_1\ ]\!]^G \not\subseteq \mathbf{E}[\![\ q_2\ ]\!]^G$ and hence $q_1 \not\sqsubseteq q_2$ holds:

- for each `iri` or constant u in q_1 and q_2, we create an RDF triple (α, u, β) such that $\{(\alpha, \beta) \mid \exists n \in [\![u]\!]^{K_G} \wedge n' \in S'' \wedge (\alpha, n') \in R(s) \wedge (n', n) \in R(p) \wedge (n', \beta) \in R(o)\}$.

Thus, it remains to show that $\mathbf{E}[\![\ q_1\ ]\!]^G \not\subseteq \mathbf{E}[\![\ q_2\ ]\!]^G$. From our assumption, we obtain the following:

$$[\![\mathsf{E}_\mu(q_1) \wedge \neg\mathsf{E}_\mu(q_2)]\!]^{K_G} \neq \varnothing \Rightarrow [\![\mathsf{E}_\mu(q_1)]\!]^{K_G} \neq \varnothing \text{ and } [\![\neg\mathsf{E}_\mu(q_2)]\!]^{K_G} \neq \varnothing.$$
$$\Rightarrow [\![\mathsf{E}_\mu(q_1)]\!]^{K_G} \neq \varnothing \text{ and } [\![\mathsf{E}_\mu(q_2)]\!]^{K_G} = \varnothing.$$

Note here that, if a formula φ is satisfiable in an RDF transition system K, then $[\![\varphi]\!]^{K_G} = S$ as shown in [7]. Consequently, using Lemma 2 and G, we obtain $\mathbf{E}[\![\ q_1\ ]\!]^G \neq \varnothing$ and $\mathbf{E}[\![\ q_2\ ]\!]^G = \varnothing$ because G contains all those triples that satisfy q_1 and not q_2. Therefore, we get $\mathbf{E}[\![\ q_1\ ]\!]^G \not\subseteq \mathbf{E}[\![\ q_2\ ]\!]^G$ and thus $q_1 \not\sqsubseteq q_2$.

From Theorems 1 and 2, we obtain the following result.

Proposition 1 (Complexity of *cEPPs* Containment). *The containment of **cEPPs** queries can be determined in a double exponential amount time.*

Note that the size of the encoding n is exponential. Hence, we obtain a 2EXPTIME upper bound for containment. If we remove path conjunction from the **cEPPs** we obtain an EXPTIME upper bound; this is due to the complexity of satisfiability in the μ-calculus which is $2^{\mathcal{O}(n^2 \log n)}$ where n is the size of the encoding (which is exponential for **cEPPs**). To provide a lower bound to the problem of containment in **cEPPs**, we utilize the close connection between **cEPPs** and PDL (Propositional Dynamic Logic) with path intersection and converse [17]. Like $\mathcal{L}_\mu$, PDL formulas are interpreted over transition systems (aka. Kripke structures). **cEPPs** without tests are exactly the same as that of PDL navigational programs (which are regular expressions). PDL formulas can be interpreted over RDF transitions systems by using a fixed set of programs $\{s, p, o, \bar{s}, \bar{p}, \bar{o}\}$ as we have done for $\mathcal{L}_\mu$. Likewise, **cEPPs** expressions can be evaluated over RDF transition systems by a simple rewriting. Given a **cEPPs** expression e that can be expressed over a standard RDF graph, it can be turned into an expression e' that can be expressed over an RDF transition system by a bijective rewriting function ϑ. ϑ is defined inductively:

$$\vartheta(iri) = p/\mathtt{T}(\_o = iri)/\hat{}p/o \qquad \vartheta(\hat{}e) = p/\mathtt{T}(\_o = iri)/\hat{}p/\hat{}s$$
$$\vartheta(\mathsf{e}_1/\mathsf{e}_2) = \vartheta(\mathsf{e}_1)/pos(\mathsf{e}_1, \mathsf{e}_2)/\vartheta(\mathsf{e}_2) \qquad \vartheta((\mathsf{e})^*) = (\vartheta(\mathsf{e}))^*$$
$$\vartheta((\mathsf{e})^+) = (\vartheta(\mathsf{e}))^+ \qquad \vartheta(\mathsf{e}_1 \mid \mathsf{e}_2) = \vartheta(\mathsf{e}_1) \mid \vartheta(\mathsf{e}_2)$$
$$\vartheta(\mathsf{e}_1 \& \mathsf{e}_2) = \vartheta(\mathsf{e}_1) \ \& \ \vartheta(\mathsf{e}_2) \qquad \vartheta(\mathrm{P}_1 \text{ test } \mathrm{P}_2) = \vartheta_\mathsf{T}(\mathrm{P}_1)/\vartheta_\mathsf{T}(\text{test})/\vartheta_\mathsf{T}(\mathrm{P}_2)$$
$$\vartheta_\mathsf{T}(iri) = p/\mathtt{T}(\_o = iri)/\hat{}p \qquad \vartheta_\mathsf{T}(\text{test}_1 \ \&\& \ \text{test}_2) = \vartheta_\mathsf{T}(\text{test}_1) \ \&\& \ \vartheta_\mathsf{T}(\text{test}_2)$$
$$\vartheta_\mathsf{T}(\text{test}_1 \mid\mid \text{test}_2) = \vartheta_\mathsf{T}(\text{test}_1) \mid\mid \vartheta_\mathsf{T}(\text{test}_2) \qquad \vartheta_\mathsf{T}(\mathrm{TP}(\mathrm{P}, \mathsf{e})) = \vartheta_\mathsf{T}(\mathrm{P})/\vartheta(\mathsf{e})$$
$$\vartheta_\mathsf{T}(!\text{test}) = !\vartheta_\mathsf{T}(\text{test}) \qquad \vartheta_\mathsf{T}(\_\mathsf{s}) = s \quad \vartheta_\mathsf{T}(\_\mathsf{p}) = p \quad \vartheta_\mathsf{T}(\_\mathsf{o}) = o$$
$$pos(\mathsf{e}_1, \mathsf{e}_2) = s \qquad pos(\mathsf{e}_1, \hat{}\mathsf{e}_2) = \hat{}o \qquad pos(\hat{}\mathsf{e}_1, \mathsf{e}_2) = s \qquad pos(\hat{}\mathsf{e}_1, \hat{}\mathsf{e}_2) = \hat{}o$$

where s, p and o are navigational programs of an RDF transition system. In order to show that **cEPPs** tests are just a syntactic variant of PDL node formulas (aka XPath node expressions), we prove the following.

Lemma 3. *Satisfiability of* **cEPPs** *queries on RDF transition systems can be reduced in polynomial time to satisfiability of* **cEPPs** *queries on RDF graphs.*

Proof. We can turn an RDF transition system into a standard RDF graph by using a bijective mapping function similar to the one in [7]. It is also possible to transform a **cEPPs** *expression* e *over an RDF transition system* K_G *into a* **cEPPs** *expression* e' *that can be expressed over a standard RDF graph* G *via the function* ϑ. *Thus, it remains to show that,* e *is satisfiable in* K_G *iff* e' *is satisfiable in* G. *This can be done inductively on the construction of the expressions.*

Theorem 3. *Testing containment of* **cEPPs** *queries is 2EXPTIME-hard.*

The above result is obtained by examining the connection between **cEPPs** and Propositional Dynamic Logic with path intersection (ICPDL). **cEPPs** can be as seen as a notational variant of ICPDL with the fixed set of atomic programs $\{s, p, o, \bar{s}, \bar{p}, \bar{o}\}$. For example, the **cEPPs** expression $\vartheta(a/\mathrm{TP}(\_s, b^+)/c) = s/a/...$ corresponds to the ICPDL formula $a \circ \langle s \rangle \langle b^+ \rangle \circ \langle c \rangle$.

The double exponential upper bound is unavoidable as it has already been shown that path conjunction makes the containment problem very difficult for XPath; it has been proved that the complexity of the containment of CoreXPath(∗, ∩) expressions is 2EXPTIME-complete [30]. Consider now the language of **cEPPs** without path intersection (**cEPPs**$_{nn}$). This language is closely related to PDL and CoreXPath [30].

Theorem 4. *The containment problem of* **cEPPs**$_{nn}$ *is EXPTIME-complete.*

The following result follows from the containment problem for GXPath$_{reg}$ (Graph XPath) [21] and the satisfiability problem for PDL with negation on paths [18].

Theorem 5. *The containment problem of EPPs is undecidable.*

5.1   Containment of SPARQL with EPPs

In this section we briefly discuss the containment problem for SPARQL queries *with EPPs* (from here onwards, we refer to it as just SPARQL). We restrict ourselves to the union of conjunctive fragment of SPARQL, i.e., *AND-UNION fragment of SPARQL, without projection.*

Definition 6. *A SPARQL query* q *is defined inductively as:* $q:: = (\mathcal{V} \cup \boldsymbol{I}, e, \mathcal{V} \cup \boldsymbol{I} \cup \boldsymbol{L}) \mid q$ **AND** $q' \mid q$ **UNION** q'.

The problem of the containment of SPARQL queries can be reduced into unsatisfiability test in $\mathcal{L}_\mu$ as given below:

> **Problem: SPARQL Containment via $\mathcal{L}_\mu$**
> Input: $\mathcal{L}_\mu$ encodings $E(q_1)$ and $E(q_2)$
> Output: Is $E(q_1) \wedge \neg E(q_2)$ unsatisfiable?

In the following, we extend the EPPs encoding function E_μ to translate the containment test of q_1 in q_2 into the μ-calculus.

Definition 7 (SPARQL containment). *The encoding of a SPARQL query q is obtained inductively as follows:*

$$E_\mu((\alpha, e, \beta)) = lfp(X, \langle \bar{s} \rangle \alpha \wedge \langle p \rangle RE_\mu(e, \beta))$$
$$E_\mu(q \textbf{ AND } q') = E_\mu(q) \wedge E_\mu(q')$$
$$E_\mu(q \textbf{ UNION } q') = E_\mu(q) \vee E_\mu(q')$$

The variables on the left hand side query are encoded into nominals and the IRIs in the path expressions are encoded into atomic propositions; to encode variables that appear on the right hand side query, we follow two steps: (1) if a variable appears uniquely in the query, it is encoded into $\top$, and (2) if a variable appears multiple times, it is encoded by using the IRIs in the path expression (the triple pattern in which it appears). This technique has been already used in [9]. For instance, if the query is $q = (x, e_1, y) \textbf{ AND } (y, e_2, z)$, then the encoding of x (resp. z) is $\top$ and $y \rightarrow \langle \bar{o} \rangle \langle p \rangle e_1$ or $y \rightarrow \langle s \rangle \langle p \rangle e_2$. Thus, the encoding becomes:

$$E_\mu(qp) = E_\mu((\top, e_1, \langle \bar{o} \rangle \langle p \rangle e_1) \textbf{ AND } (\langle \bar{o} \rangle \langle p \rangle e_1, e_2, \top)) \vee$$

$$E_\mu((\top, e_1, \langle s \rangle \langle p \rangle e_2) \textbf{ AND } (\langle s \rangle \langle p \rangle e_2, e_2, \top))$$

The size of the encoding for the containment problem is polynomial in the number of variables that appear more than once. To be more precise, for a given query, the size of the encoding is: $\Pi_{x \in multiVar} |x|$ where *multiVar* is the set of variables occurring more than once, and $|x|$ is the number of times variable x appears in the query. Note that the size of the encoding is linear if all the variables on the right-hand side query appear uniquely.

Theorem 6 (Soundness and Completeness). *Given two SPARQL queries q_1 and q_2, $E_\mu(q_1) \wedge \neg E_\mu(q_2)$ is unsatisfiable if and only if $q_1 \sqsubseteq q_2$.*

From the above theorem, we obtain the following result:

Proposition 2. *Given two SPARQL queries q_1 and q_2, the containment test can be solved in a double exponential amount of time in the size of the encoding $|E_\mu(q_1)| + |E_\mu(q_2)|$.*

This result is not surprising, as we have shown in Proposition 1 that the complexity of the containment of EPPs is 2EXPTIME in the worst case. Furthermore, if projection was part of the SPARQL fragment we consider here, then there is a further jump in the complexity of containment, i.e., the complexity increases by one exponential. Thereby, we obtain 3EXPTIME upper bound for the containment of SPARQL queries with projection. This is due to an exponential blow up in the size of the encoding as one needs to take care of multiple occurring non-distinguished variables.

6   Conclusions and Future Work

We have discussed a preliminary study of the problem of query containment for an expressive class of navigational queries captured by the EPPs language. Our study leverages μ-calculus to encode EPPs and then use this encoding to get a 2EXPTIME complexity upper bound. This bound remains the same when considering EPPs in SPARQL without projection. However, if one considers projection, there is an exponential increase in the complexity which results in a 3EXPTIME bound. While the results obtained are of theoretical interest, from a practical point of view an implementation is available (http://sparql-qc-bench. inrialpes.fr/) which can be extended for EPPs. Furthermore, an additional benefit of using μ-calculus is that by exploiting *logical combinators* the size of the encoding can be reduced by upto exponentiation. Thus, the complexity bounds that we obtained are not prohibitive in terms of a practical implementation.

A natural line of future research is to provide a tighter complexity bound for the problem of containment of EPPs. Moreover, the investigation of how the inclusion of constraints in EPPs affects the complexity of query containment is in our research agenda.

References

1. Barcelo, P., Libkin, L., Lin, A.W., Wood, P.T.: Expressive languages for path queries over graph-structured data. ACM Trans. Database Syst. (TODS) **37**, 31 (2012)
2. Calvanese, D., De Giacomo, G., Lenzerini, M., Vardi, M.Y.: Containment of conjunctive regular path queries with inverse. In: Proceedings of 7th International Conference on the Principles of Knowledge Representation and Reasoning (KR 2000), pp. 176–185 (2000)
3. Calvanese, D., Giacomo, G.D., Lenzerini, M.: Conjunctive query containment and answering under description logic constraints. ACM Trans. Comput. Logic (TOCL) **9**(3), 22 (2008)
4. Calvanese, D., Ortiz, M., Simkus, M.: Containment of regular path queries under description logic constraints. In: IJCAI Proceedings-International Joint Conference on Artificial Intelligence, vol. 22, p. 805 (2011)
5. Chandra, A.K., Merlin, P.M.: Optimal implementation of conjunctive queries in relational data bases. In: Proceedings of the Ninth Annual ACM Symposium on Theory of Computing, STOC 1977, pp. 77–90. ACM, New York (1977)
6. Chaudhuri,S., Vardi, M.Y.: Optimization of real conjunctive queries. In: Proceedings of the Twelfth ACM SIGACT-SIGMOD-SIGART Symposium on Principles of Database Systems, PODS 1993, pp. 59–70. ACM, New York (1993)
7. Chekol, M.W.: Static analysis of semantic web queries. Ph.D. thesis, Université de Grenoble (2012)
8. Chekol, M.W., Euzenat, J., Genevès, P., Layaïda, N.: PSPARQL query containment. In: Proceedings of DBPL (2011)
9. Chekol, M.W., Euzenat, J., Genevès, P., Layaïda, N.: SPARQL query containment under RDFS entailment regime. In: Gramlich, B., Miller, D., Sattler, U. (eds.) IJCAR 2012. LNCS, vol. 7364, pp. 134–148. Springer, Heidelberg (2012)

10. Chekol, M.W., Euzenat, J., Genevès, P., Layaïda, N.: SPARQL query containment under SHI axioms. In: Proceedings of AAAI, pp. 10–16 (2012)
11. Fionda, V., Pirrò, G.: Querying graphs with preferences. In: 22nd ACM International Conference on Information and Knowledge Management, CIKM 2013, San Francisco, CA, USA, October 27– November 1, pp. 929–938 (2013)
12. Fionda, V., Pirrò, G., Consens, M.P.: Extended property paths: writing more SPARQL queries in a succinct way. In: Proceedings of the Twenty-Ninth AAAI Conference on Artificial Intelligence, 25–30 January, Austin, Texas, USA, pp. 102–108 (2015)
13. Fionda, V., Pirrò, G., Gutierrez, C.: NautiLOD: a formal language for the web of data graph. Trans. Web 9(1), 5: 1–5: 43 (2015)
14. Genevès, P., Layaïda, N.: A system for the static analysis of XPath. ACM Trans. Inf. Syst. 24(4), 475–502 (2006)
15. Genevès, P., Layaïda, N., Schmitt, A.: Efficient static analysis of XML paths and types. In: Proceedings of PLDI, pp. 342–351. ACM (2007)
16. Genevès, P., Schmitt, A. : Expressive logical combinators for free. In: IJCAI, pp. 311–317 (2015)
17. Göller, S., Lohrey, M., Lutz, C.: PDL with intersection and converse: satisfiability and infinite-state model checking. J. Symb. Logic 74(01), 279–314 (2009)
18. Harel, D., Kozen, D., Tiuryn, J.: Dynamic Logic. MIT Press, Cambridge (2000)
19. Ioannidis, Y.E., Ramakrishnan, R.: Containment of conjunctive queries: beyond relations as sets. ACM Trans. Database Syst. 20(3), 288–324 (1995)
20. Kostylev, E.V., Reutter, J.L., Romero, M., Vrgoč, D.: SPARQL with property paths. In: Arenas, M., et al. (eds.) ISWC 2015. LNCS, vol. 9366, pp. 3–18. Springer, Heidelberg (2015)
21. Kostylev, E.V., Reutter, J.L., Vrgoc, D.: Containment of data graph queries. In: ICDT, pp. 131–142 (2014)
22. Kostylev, E.V., Reutter, J.L., Vrgoč, D.: Static analysis of navigational XPath over graph databases. Inf. Process. Lett. 116(7), 467–474 (2016)
23. Kozen, D.: Results on the propositional μ-calculus. Theoret. Comput. Sci. 27(3), 333–354 (1983)
24. Krötzsch, M., Rudolph, S.: Conjunctive queries for EL with composition of roles. In: Proceedings of the 2007 International Workshop on Description Logics (DL 2007), Brixen-Bressanone, near Bozen-Bolzano, 8–10 Italy 2007, June 2007
25. Pérez, J., Arenas, M., Gutierrez, C.: Semantics and complexity of SPARQL. ACM Trans. Database Syst. (TODS) 34(3), 16 (2009)
26. Pérez, J., Arenas, M., Gutierrez, C.: nSPARQL: a navigational language for RDF. Web Semant. Sci. Serv. Agents WWW 8(4), 255–270 (2010)
27. Pichler, R., Skritek, S.: Containment and Equivalence of Well-designed SPARQL, pp. 39–50 (2014)
28. Reutter, J.L.: Containment of nested regular expressions. CoRR, abs/1304.2637 (2013)
29. Sagiv, Y., Yannakakis, M.: Equivalences among relational expressions with the union and difference operators. J. ACM 27(4), 633–655 (1980)
30. ten Cate, B., Lutz, C.: The complexity of query containment in expressive fragments of XPath 2.0. J. ACM (JACM) 56(6), 31 (2009)
31. Vardi, M.Y., Murano, A., Lutz, C., Bonatti, P.A.: The complexity of enriched mu-calculi. Log. Methods Comput. Sci. 4, 1–27 (2008)

WebBrain: Joint Neural Learning of Large-Scale Commonsense Knowledge

Jiaqiang Chen[1], Niket Tandon[2], Charles Darwis Hariman[3], and Gerard de Melo[4(✉)]

[1] IIIS, Tsinghua University, Beijing, China
[2] Allen Institute for Artificial Intelligence, Seattle, WA, USA
[3] Max Planck Institute for Informatics, Saarbrücken, Germany
[4] Rutgers University, Piscataway, NJ, USA
gdm@demelo.org
http://gerard.demelo.org

Abstract. Despite the emergence and growth of numerous large knowledge graphs, many basic and important facts about our everyday world are not readily available on the Web. To address this, we present WebBrain, a new approach for harvesting commonsense knowledge that relies on joint learning from Web-scale data to fill gaps in the knowledge acquisition. We train a neural network model to learn relations based on large numbers of textual patterns found on the Web. At the same time, the model learns vector representations of general word semantics. This joint approach allows us to generalize beyond the explicitly extracted information. Experiments show that we can obtain representations of words that reflect their semantics, yet also allow us to capture conceptual relationships and commonsense knowledge.

1 Introduction

Motivation. In the past decade, massive amounts of machine-readable knowledge have become available, both in large knowledge graphs such as DBpedia, YAGO, and GeoNames, as well as through the widespread adoption of standards such as schema.org for Web pages. Additionally, information extraction techniques allow us to mine further knowledge from natural language text. To date, such data has mainly been used for improved information interchange and integration, e.g. for better Web search results on entity-focused queries or for novel kinds of visualizations that combine information from different sources.

However, while there are numerous bots and services that scour the Web to exploit a particular (often hard-coded) kind of information, we still lack intelligent systems that more flexibly draw advanced conclusions from information found on the Web. Among the missing ingredients, the lack of required world knowledge stands out as particularly relevant. This includes knowledge that is

---

This work was supported in part by China 973 Program Grants 2011CBA00300, 2011CBA00301, NSFC Grants 61033001, 61361136003, 61550110504, and the Samsung R&D Institute of China.

P. Groth et al. (Eds.): ISWC 2016, Part I, LNCS 9981, pp. 102–118, 2016.
DOI: 10.1007/978-3-319-46523-4_7

less of the factual, encyclopedic kind found in DBpedia, but more related to a general understanding of everyday objects and concepts in the world. In some cases, such commonsense knowledge can be expressed as subject-predicate-object triples, similar to those used for factual knowledge. Relevant predicates include `causes` (e.g., fire causes heat), `hasProperty` (e.g., ice has the property of being cold and ice cream has the property of being sweet), `partOf` (e.g., legs as parts of humans), and `usedFor` (e.g., that keys can be used to open doors).

Goal. We ultimately aim at a system capable of guessing the truth of commonsense facts (e.g. whether a dog can fly), based on knowledge seen on the Web. However, procuring this sort of knowledge from the Web is non-trivial because shared assumptions about the world are often taken for granted such that it would be rare if not strikingly odd for someone to write that tractors are inedible or that radiologists are capable of breathing. Thus, information extraction alone is insufficient for equipping computational systems with commonsense knowledge.

Overview and Contribution. In this paper, we propose a joint learning approach to acquire commonsense knowledge both from explicit and implicit textual information. explicit triples and on large-scale word co-occurrence information. We optimize matrix representations of relations explicitly mined from large amounts Web data using a custom information extraction approach designed to minimize noise when applied to Web-scale data. At the same time, concepts are modeled as vectors trained on large-scale text following the word2vec CBOW approach to capture generic semantics [22].

As a result, our approach jointly learns representations of words and relations to better reflect our natural understanding of them. In particular, we are able to exploit general Web-scale semantics when learning commonsense relationships, e.g. inferring from eagles being capable of flying that hawks are likely also capable of the same. Our experiments show that we can obtain representations that simultaneously capture conceptual relationships as well as word meanings.

2   Background and Related Work

Commonsense knowledge acquisition has been studied for many years now. Traditionally, such knowledge was modeled by human experts, an approach best exemplified by the Cyc project [18], a decades-long commercial effort at creating a large axiomatic rule base. The SUMO ontology [24] shares this goal, but relies on open source principles and more collaborative development processes. However, in both cases, contributing requires significant expertise and effort in knowledge modeling. Although feasible for specific domains, it is difficult to obtain extensive amounts of commonsense knowledge in this way.

For large-scale commonsense knowledge acquisition, there are two more promising directions. The well-known ConceptNet project [12] relies on crowdsourcing, aiming at much simpler commonsense knowledge propositions. Another approach is to turn to large-scale data mining. Many information extraction papers follow the bootstrapping method proposed by Hearst [13], who used linguistic patterns to mine `isA` relationships. However, pattern-based approaches tend to extract only few facts and suffer from significant problems with noise. Several approaches have been proposed to improve bootstrapping in general [25,31]. Another route is to develop improved algorithms catering to particular kinds of information, e.g. temporal knowledge [9], properties and attributes [30], or activity knowledge [32].

Irrespective of whether one relies on crowdsourcing or data mining, however, it is necessary to generalize and expand beyond what has explicitly been acquired. For instance, we may have obtained that Samoyed dogs have fur but we may not have explicitly found this to be the case for shiba inus as well.

This leads us to the task of knowledge base completion [26]. When relying on machine learning, this typically becomes a relation prediction task. Given a a training set of true example instances of relations, i.e. triples, the goal is to learn a model that can then be used to predict whether a new, previously unseen triple is true or false. While the relation itself will have occurred in the training set, the specific triple will be new.

One approach is to consider this a tensor or matrix completion problem. For instance, if we view a relation as a matrix between subjects and objects storing their truth values, then relation prediction boils down to filling in the missing values to complete the matrix. Previous work in this area includes AnalogySpace [28], which relied on singular value decomposition applied to ConceptNet extractions. Nickel et al.'s RESCAL [23] uses tensor factorization to model relationships, targeting collective classification and entity resolution. Sutskever et al. [29] propose Bayesian clustered tensor factorization to model relational data. Jiang et al. [16] proposed a generative probabilistic model for relation prediction based not only on existing triples but also on information extraction.

A more recent line of work uses neural networks for relation prediction. Bordes et al. [4–6,15] proposed several neural network architectures to capture relation triples, the most well-known of these being the TransE approach, which models the relation as a translation from a vector for the subject to a vector for the object. Numerous variations have been proposed that modify how the relation is modeled. For instance, Socher et al. [27] propose neural tensor networks (NTN), in which each relation is represented as a tensor. TransH [35] models relations as translations on hyperplanes. TransR [20] adds extra projections of entity vectors for each specific relation, or, in the CTransR variant, for each cluster of relations. PTransE [19] attempts to consider inference via property paths to improve the prediction of a triple (for example, x bornInCity y, y cityInState z helps us predict x bornInState z).

Our approach differs from all such previous efforts by learning to generalize not just based on the existing triples, as done by matrix and tensor methods as well as the TransE-related neural models, but by additionally exploiting semantic information derived from large-scale text statistics. As we show in our experiments, pure relational modeling does not result in semantically satisfactory word vector representations. Our joint model alleviates this problem by enabling the choice of word representations to benefit from large amounts of raw text, similar to the way humans draw on general semantic associations as well as more explicit information. Since the word vectors are constrained to reflect semantic similarity, we use more flexible matrix representations of relations rather than simple translations as in the TransE model. Compared to the NTN model, in contrast, we model relations in a less flexible way, so as to ensure a mutual influence between commonsense relations and word representations. Compared to Jenatton et al. [15], which in turn is related to the NTN model, we do not use any 1-gram or 2-gram features, but directly use the product as a scoring function. We also forgo requiring that the matrix be the sum of rank-1 matrices. This enables our approach to scale to much larger data sets such as DBpedia.

Our joint model learns word representations that allow us to better transfer knowledge between related concepts. We rely on Mikolov et al.'s word2vec CBOW approach [22], who simplified previous neural language models [1] for significantly greater scalability. They also introduced the Skip-Gram model as an alternative, but in our approach, we build on the CBOW variant, as it is faster to optimize. There have been other proposals to adapt the word2vec models. Several approaches aim at improving word vectors using additional knowledge of semantic similarity [8,37]. These are based on generic semantic similarity rather than capturing specific kinds of relations. Hill and Korhonen [14] presented a model for multi-modal representations, training on large amounts of image labels and text, with a minor addition of 638 abstract concept descriptions.

Bollegala et al. [3] proposed a method for obtaining improved word vectors by exploiting information about the lexical patterns they occur in. This approach is aimed at obtaining vectors that better reflect word analogies but does not address our model's goal of relation prediction. Xie et al. [36] exploit entity description glosses but do not use large-scale text. Wang et al. [34] proposed the *probabilistic TransE* model, capturing Freebase triples following the TransE model, but also viewing the co-occurrences of two phrases as a relationship that should likewise be modeled as a translation. Their model uses two vectors per phrase and an alignment component to connect entities to phrases. Our model uses just a single vector per word, so mutual dependencies between word vectors are exploited to a greater degree, while the relation modeling is less constrained due to the use of matrices, so a greater divergence from the word relationships is enabled. Toutanova et al. [33] also attempt to model relationships between two entities found in text, but use syntactic dependency trees and apply a convolutional neural network over them to obtain relation representations. In contrast, we exploit any occurrence of a word, not just explicit relationships between two entities in a sentence.

3   Web-Scale Knowledge Bootstrapping

Pattern-Based Information Extraction. For knowledge acquisition, it is well-known that one can attempt to induce patterns based on matches of seed facts, and then use pattern matches to mine new knowledge [13]. Unfortunately, this bootstrapping approach suffers from significant noise when applied to large

Web-scale data [31], which appears necessary in order to mine adequate amounts of training data. Specifically, we rely on Google's large Web N-Gram dataset (see Sect. 5). This problem is exacerbated by the fact that we are aiming at commonsense knowledge, which is not typically expressed using explicit relational phrases. We thus devise a custom bootstrapping approach designed to minimize noise when applied to Web-scale data.

We assume that we have a set of relations $\mathcal{R}$, a set of seed facts $S(r)$ for a given $r \in \mathcal{R}$, as well as a domain(r) and range(r), specified manually to provide the domain and range of a given r as its type signature. For pattern induction, we look for co-occurrences of words in the seed facts within the n-grams data (for $n = 3,4,5$). Any match is converted into a pattern based on the words between the two occurrences, e.g. *that apple is red* would become $\langle x \rangle$ *is* $\langle y \rangle$.

Pattern Scoring. The acquired patterns are still rather noisy. To score the reliability of patterns, we rely on a ranking function that rewards patterns with high distinct seed support but also discounts patterns that occur across multiple dissimilar relations [31]. The intuition is that a good pattern should match many of the seed facts, but should not be overly generic so as to apply to many relations (as, e.g., $\langle x \rangle$ *or* $\langle y \rangle$). A pattern with many matches for both hasLocation and partOf is less likely to be a reliable one.

Still, a pattern that matches isa or hasLocation may also match a relation such as conceptuallyRelatedTo. To allow for this, we first define a relatedness score between relations. We can either provide these scores manually, or consider Jaccard overlap statistics computed directly from the seed assertion data. Let p be a candidate pattern and $r \in \mathcal{R}$ be the relation under consideration. We define $|S(r,p)|$ as the number of distinct seeds $s \in S(r)$ under the relation r that p matches. We then define the score of the pattern p for relation r as:

$$\phi(r,p) = \sum_{r' \in \mathcal{R}, r' \neq r} \frac{|S(r,p)|}{|S(r)|} - (1 - \text{sim}(r,r'))\frac{|S(r',p)|}{|S(r')|}$$

where $\text{sim}(r,r')$ is the similarity between relations r and r'. At the end, we choose the top-k ranked patterns as the relevant patterns for the extraction phase.

Assertion Extraction. We apply the chosen patterns to find new occurrences in our (Google Web N-grams) data. For instance, $\langle x \rangle$ *is* $\langle y \rangle$ could match *the sun is bright*, yielding (*sun, bright*) as an assertion for the hasProperty relation. To filter out noise from these candidate assertions, we check if the extracted words match the required domain and range specification for the relation, using WordNet's hypernym taxonomy. Finally, we rank the candidate assertions analogously to the candidate patterns, but treating the patterns as seeds.

4   Representation Learning and Prediction

Figure 1 provides an overview of our approach. Having extracted triples from text, the next step is to train a model for learning commonsense word and

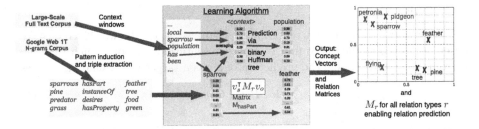

Fig. 1. Schematic overview of our approach.

relation representations. Our model takes the extracted knowledge and learns to generalize it by simultaneously drawing not only on the extractions but also on large amounts of generic text. For instance, we may have observed that pigeons fly but not that sparrows do. Our model aims to infer the latter based on both the extracted facts as well as general semantic relatedness between pigeons and sparrows. For the latter, we use word co-occurrences observed in large amounts of text. While word co-occurrences provide only weak signals of semantic related-ness, we can benefit from their large quantities. Thus, their overall contribution may make up for some of the sparsity of the explicitly extracted knowledge.

At the same time, word representations can also benefit from the joint learn-ing setup. We learn word meanings not only from general context information, but also from the extracted relationships. For instance, we may have extracted that roses tend to have the property of being red, or that fire causes heat. From a cognitive perspective, commonsense knowledge about concepts is salient and should also guide the meaning representation of words.Distributional semantics has a long history, which has often been linked to J.R. Firth's famous quote that one shall "know a word by the company that it keeps". Still, while it is indis-putable that regular contexts play a vital role in meaning acquisition, words and concepts are often acquired by other means than just from general con-texts. Depending on the situation, humans may pay special attention to certain cognitively salient features and relationships of an object (e.g., appearance and function). In our approach, we thus train concept representations jointly with relationship representations, exploiting both the general contextual information and the mined relationship data.

This also touches on the long-standing dispute about whether mental rep-resentations of concepts are best modeled using discrete symbolic methods or in connectionist models based on numerical information processing. Whilst this has sometimes been regarded as an irreconcilable dichotomy, models like the one we propose here learn neural representations of words but also capture explicit symbolic relationships between them.

4.1 Objective

For the general word co-occurrence information, we adopt the word2vec CBOW objective [22]. The idea is to find vector representations of words such that the surrounding words enable the prediction of a given target word. One maximizes

$$\sum_w \log P(w \mid C(w)), \tag{1}$$

where w denotes a word token in a large corpus and $C(w)$ denotes the context words. In the CBOW model, w is represented by a dense, real-valued word vector v_w and the context $C(w)$ is represented by the average of the vectors for the surrounding words. The resulting context vector is used to predict v_w.

Simultaneously, we jointly optimize for modeling the explicit relationships mined earlier. We assume that we have extracted labeled relationships between words. These can be viewed as (s, r, o) tuples consisting of a left word s (the subject), a predicate (relation) r, and a right word o (the object). We wish to use matrices and vectors to capture the information that the extracted relational data provides. We still assume every word is mapped to a vector, but additionally map each relation type to a specific matrix. To learn these representations, we seek to maximize a scoring function over all relation triples. We define

$$f(s, r, o) = v_s^{\mathsf{T}} M_r v_o,$$

where v_s is the word vector for the subject s of the relation triple and v_o is the word vector for the object o, while M_r is a matrix for relation r. If M_r is the identity matrix, then this function will compute a simple dot product measuring the similarity of the two vectors. If M_r is some other form of diagonal matrix, f would compute a weighted vector similarity. Other forms of M_r can capture transformations of the two vectors. The word vectors, described by v_s and v_o here, are jointly modified by both the CBOW and the relation modeling components, while the relation matrices M_r are only modified by the latter. For this relation modeling, we rely on the following loss function to quantify the error:

$$L_{s,r,o,l} = -l \, \log(\sigma(f(s, r, o))) - (1 - l) \, \log(1 - \sigma(f(s, r, o))), \tag{2}$$

where $\sigma(\cdot)$ is the sigmoid function $\sigma(x) = \frac{1}{1+e^{-x}}$ and l is the label of the training triple s,r,o (1 for positive training examples and 0 for negative ones).

Finally, we train our model to learn representations both from the relations and using the word2vec CBOW objective, to exploit the contextual statistics from large raw text corpora, thus making our representations more meaningful, as we will show later on in the experiments. Our overall loss function is as follows (with a parameter β to control the relative contributions):

$$\sum_{(s,r,o,l)} -l \, \log(\sigma(f(s, r, o))) - (1 - l) \, \log(1 - \sigma(f(s, r, o)))$$

$$+ \beta \sum_w - \log P(w \mid C(w)) \tag{3}$$

4.2   Training

During training, we seek to minimize this loss by simultaneously optimizing both parts of the objective. For the CBOW part, we follow the well-known negative sampling procedure [22]. For the relation tuples, the training procedure is as follows. Given a training tuple, we generate k random negative examples by replacing the subject or object with a random word. Every triple is mapped to the corresponding word vectors and relation matrices. For each triple, both negative and positive ones, we optimize the loss function mentioned above.

For this optimization, we rely on stochastic gradient descent, which involves repeatedly picking random training examples, evaluating the current model on them, and making small updates if the model gives an incorrect answer. The direction of an update step is given by the gradient of the objective function, while the learning rate is a small factor that determines how much we move the model parameters (in our case, the values in the word vectors and relation matrices) in this direction. The gradients with respect to the second sum in Eq. 3 are as for the standard word2vec CBOW model, while for the first sum they are as follows:

$$\frac{\partial L_{s,r,o,l}}{\partial v_s} = (1-l)\, M_r v_o \quad \frac{\partial L_{s,r,o,l}}{\partial v_o} = (1-l)\, M_r^{\mathsf{T}} v_s \quad \frac{\partial L_{s,r,o,l}}{\partial M_r} = (1-l)\, v_s v_o^{\mathsf{T}}$$

Although individual updates with respect to the two parts of the objective function may pull the model in different directions, stochastic gradient descent normally finds stable solutions in the long run. In our case, this is expected because objects with similar extracted properties are also likely to be similar from a word semantics perspective. Our experimental results confirmed this.

5   Experiments

5.1   Data and Extraction

General Corpus. For our experiments, we rely on two datasets. The first is a frequently used dump of the English Wikipedia[1] that serves as our general corpus for word representation learning. We normalize the text to lower case and remove special characters, obtaining 1,205,009,210 tokens after this preprocessing.

Extraction Corpus. In order to extract relations, we turn to a Web-scale resource based on much larger quantities of text, the Google Web 1T N-gram dataset[2]. Although this data is limited to short 5-grams, it is well-suited for the kind of general commonsense knowledge relationships between words that we are targeting.

Seed Data. In order to bootstrap the extraction process, we rely on seed facts taken from the ConceptNet dataset [21] to induce patterns for each commonsense relation in ConceptNet. Examples of these relations include atLocation,

---

[1] http://nlp.stanford.edu/data/WestburyLab.wikicorp.201004.txt.bz2.
[2] https://catalog.ldc.upenn.edu/LDC2006T13.

causes, hasProperty, motivatedByGoal, partOf, and usedFor. This data is
rather noisy, mostly due to incorrect natural language analysis of crowdsourced
statements. We further find that more than 90 % are named entities, since ConceptNet also imports from other existing knowledge sources. We remove these
and also generally filter out all concepts not in WordNet, a lexicon containing
mostly general words. At this point, we obtain a size of nearly 192 K assertion
triples. We then applied a score threshold of 5 (i.e., we enforce that at least
five crowdsourcing annotators agree) to filter out further noise. Additionally, we
required that the subject and object of each triple match our domain(r) and
range(r) for the involved relation r. This is checked using the WordNet taxonomy [10]. For instance, for the hasProperty relation, we accept (*apple*, *red*),
because *apple* is classified as a physical noun in WordNet and *red* as an adjective. A manual annotation of two random samples of size 200 revealed that the
raw ConceptNet facts had an accuracy of only 53 %, while the filtered seed facts
had an accuracy of 99 %.

5.2   Extraction

We then follow the extraction approach described in Sect. 3. Applying the seeds
to our Web-scale n-grams, we obtain large numbers of patterns. Table 1 shows
the top patterns for a few example relations.

These patterns then give rise to vast quantities of commonsense relation
triples, each consisting of word pairs as well as the relation between them. We
extract triples for 24 different relation types.

After filtering for noise we are left with a total of 1,160,136 extractions,
e.g. (*abbey*, *church*) for the isA relation, or (*telephone*, *notice*) for the usedFor
relation[3]

Before training, we filter out triples that contain words appearing less than
100 times in Wikipedia and obtain 1,158,141 triple instances. We split the data
and use 118,826 triples each for validation and testing, and the remaining ones
for training. We also obtain a human-created gold dataset as ground truth, by
taking a human-verified subset of ConceptNet with over 20,000 triples.

Table 1. Top-k patterns for some relations

AtLocation	IsA	UsedAs	MadeOf	HasProperty
X across Y	X was only Y	X used to Y	X made of Y	X is very Y
X inside Y	X except Y	X is used to Y	X repair Y	X can be Y
X outside Y	X called Y	X designed to Y	X is made of Y	X is too Y
X near Y	Y is X	X was used to Y	X made from Y	X may be Y
X under Y	X means any Y	X to help Y	X cast Y	X is as Y

---

[3] See http://gerard.demelo.org/webbrain/.

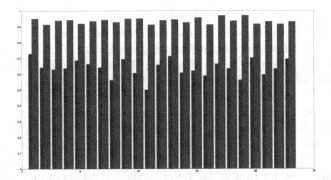

Fig. 2. Accuracy for every relation type comparing the fixed vector baseline (dark blue) and the joint training of WebBrain (brighter red). (Color figure online)

Our goal is to use this knowledge in order to train a neural network so as to learn word vectors that reflect semantic and commonsense knowledge, and to be able to generalize this to new commonsense not directly observed in the data. This should enable the representations to capture cognitively salient information inherent or associated with word meanings, e.g. that a *tyre* is part of a *car*. We can train the word vector representations by jointly optimizing for the relations and optimizing for the word2vec CBOW model. Raw text like that from Wikipedia provides regular contexts, while the triples describe common sense relationships, thus contributing different kinds of information to the representations.

5.3 Training

We consider several experimental setups. In the first setup, we pre-initialize the word representations using vectors from the word2vec CBOW model, utilizing the information from the Wikipedia text corpus. To test if these representations alone can successfully be used to reflect the relations, in this first setup, we fix the embeddings during the training and just modify the relation matrices. The training proceeds for 10 iterations.

In the second setup, we pre-initialize the vectors in the same way but allow both the vectors and the relation matrices to be modified during the 10 training iterations. In the third setup, instead of pre-initializing the vectors, we train the relational data jointly with the word2vec CBOW model, optimizing both simultaneously. In all setups, we use a vector dimensionality of 50 in order to reduce the runtime. In the relation-only setups, following the literature, we normalize the word vectors during the training and use a standard learning rate of 0.01. For each training triple, we generate 5 negative examples by randomly replacing its left or right word with a random word in the vocabulary. The vocabulary is created with words appearing at least 100 times in the Wikipedia.

When we train the triples jointly with the word2vec objective, we optimize both objectives simultaneously until the CBOW architecture has completed 3 epochs. This is done in parallel in several threads that can run on multiple cores. Alongside with several threads for the word2vec model, we create additional threads for our relational objective function from Sect. 4. This time, we do not normalize the word vectors in the relational thread, as we are training jointly and this would not make good use of the raw text. Instead of varying both β and the learning rates, we simply factor the variation of different possible choices of β into the choice of learning rate for the relational component. We fixed the CBOW learning rate at 0.05, while tuning the relational one on the validation data but also checking the WS-353 word similarity dataset. While there is no separate held-out dataset available for these word similarities, the much larger MEN dataset was not used for tuning. We describe these datasets in more detail later on. Ultimately, we arrived at the a much lower rate of just 0.002 for the relational data, which avoids distorting the vectors too much. With higher learning rates, we obtain almost the same results in terms of relation prediction, but the word vectors become overly biased towards those predictions and the word similarities correlate less strongly with human judgements.

For comparison, we also experiment with the TransE model as a representative example of methods that only use the existing relation triples without relying on information from large-scale text. Following the literature, we set the starting learning rate to 0.001 and require that the margin between positive triple and its corresponding negative samples be at least 1. We pre-initialize the model with the word2vec vectors and train it for 500 iterations, which suffices for convergence.

The final vectors and matrices successfully separate the positive training triples and randomly generated negative ones, as indicated in Fig. 3. The y axis here reports the $v_s^\mathsf{T} M_r v_o$ scores. We can see that if we fix the word vectors and just optimize the matrix, the scores of positive and negative examples mix. If we allow the word vectors to change, the scores of the positive and negative examples are better separated.

5.4   Evaluation and Analysis

We attempt to discriminate between positive triples (from the test and gold sets) and random triples to assess whether our model can successfully capture the relations and classify unseen triples. In our model, the positive $v_s^\mathsf{T} M_r v_r$ scores are usually bigger than random ones, while in the TransE model, positive examples usually have smaller scores. We use the validation data set to choose the threshold. For our model, test examples with scores below the threshold are classified as negative and those with larger scores are classified as positive. For the TransE model, the opposite applies.

The classification results are presented in Table 2. Results on the test split reflect how well our model learns to predict the extractions, while results on the gold data set reflect to what extent the predicted relationships really hold true from a ground truth perspective. If we fix the vectors to be the ones from word2vec ("relations only, fixed vectors"), the accuracy is rather low, suggesting

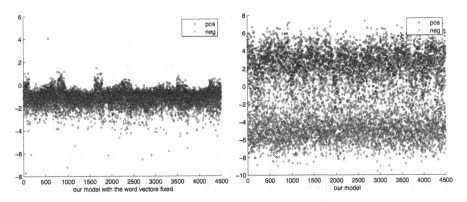

Fig. 3. Comparing positive examples in the validation set (dark blue circles) with negative ones (light red diamonds). Left: $v_s M_r v_o$ scores for WebBrain when vectors are fixed to the original vectors from word2vec. Right: scores for regular WebBrain. (Color figure online)

Table 2. Relationship modeling results

Approach	Test	Gold
TransE	0.976	0.892
WebBrain: relations only, fixed vectors	0.761	0.700
WebBrain: relations only	0.983	0.897
WebBrain joint text+relation model	0.969	0.935

that the word2vec vectors are not well-suited for capturing relational similarities based on commonsense knowledge. However, when we allow our algorithm to modify the vectors ("relations only"), the resulting model achieves a good accuracy. We obtain 0.983 on the test data set and 0.897 for the gold, which is comparable with the TransE model despite our scalable training procedure. This setting corresponds to using only the first term of Eq. 3.

If we train our relational model jointly with the word2vec CBOW model, i.e. using the full objective given by Eq. 3, we see a slight reduction in accuracy on the test set, i.e. in predicting the original extractions. This is understandable, given that we are no longer optimizing for the goal of predicting relations exclusively. However, we obtain a significant improvement on the gold data set, showing that the model better reflects the real properties of concepts. This suggests that the joint training with general word semantics gives WebBrain better generalization capabilities than relation prediction models only considering the training triples.

In Fig. 2, we see that the advantage of joint training is consistent across relation types. For each relation, we plot the fixed vector baseline (left, dark) and our joint training method (right, lighter). The 24 different relations are plotted along the x axis, while the y axis corresponds to the accuracy in $[0, 1]$.

Table 3. Spearman's ρ for word similarity data

Approach	WS-353	MEN
WebBrain: text only	0.621	0.668
WebBrain: relations only	0.316	0.302
WebBrain joint text+relation model	0.632	0.679

Word Representations. We also evaluate the word representations directly, using two semantic relatedness datasets to assess the semantic similarities reflected in the word vectors. One is the well-known WS-353 [11] dataset, while the second is a significantly larger one called MEN [7]. Both contain English word pairs with similarity judgements elicited from human assessors. We calculate the cosine distance of word vectors for the word pairs in these datasets and compare them to the scores from the human annotations using Spearman's ρ.

Table 3 shows how the resulting word representations fare on the WS-353 and MEN datasets. For the vectors trained just on the relational data, e.g. with the TransE model, the result is significantly worse than for the text-only model. This means that, after training, the vectors are optimized for the relations and fail to reflect much of the information that the raw corpus data provides. However, if we train jointly, we observe better correlations, indicating that the vectors are able to maintain the semantic information from the raw corpus contexts.This shows that our model can modify the word vectors to reflect commonsense relations without degrading the quality of general word similarities.

5.5   Additional Experiment on DBpedia

Data. We additionally evaluate the model's performance on DBpedia [2]. We focus on extractions from the English Wikipedia, using the "mapping-based types" and "mapping-based properties (cleaned)" data. We consider only URIs from within DBpedia (starting with "http://dbpedia.org/resource/"), since others are not part of DBpedia itself and thus the data only contains very sparse, incomplete information about them. After preprocessing, we obtain a total of 15,109,444 triples describing 4,222,635 entities and 675 distinct relation types. We split this data by reserving 15,110 triples as validation data for tuning and 15,110 as a test data set for evaluation. All remaining triples are used as training data.

Training and Evaluation. To determine the optimal parameter settings, we rely on the validation data and choose the vector size for the entities from {30, 50, 100}. We run the experiment for {20, 40, 60, 80, 100} iterations. Based on these options, we select the best-performing set of parameters in terms of their accuracy on the validation data, as explained below. Following the literature, we normalize the vectors after every stochastic gradient descent step. For every

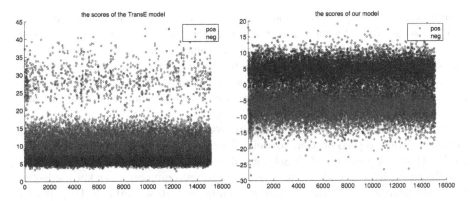

Fig. 4. Scores on the DBpedia validation data, computed as $\|v_s + v_r - v_o\|$ for the TransE model (left) and $v_s M_r v_o$ for ours (right), with legend as in Fig. 3.

iteration, we first shuffle the training triples. We decrease the learning rate linearly until it reaches 0.0001 of the starting learning rate. At that point, we hold it constant at that value, i.e. 0.0001 of the starting learning rate.

For comparison, we again also consider the TransE model on this data. We set the parameters for the TransE model as described in the original paper. The initial learning rate is set to 0.001, and the optimization proceeds for 1000 iterations. The vector size is also chosen from $\{30, 50, 100\}$.

We test the model's performance by discriminating between true and random triples. After the training, their classification scores should be different. True triples should have larger scores than random negative scores and we indeed observe this result. Figure 4 plots the scores for the validation data set. We can see that the TransE model already achieves reasonably good results. Most positive triples have smaller scores. For our model, although we do not use a max-margin approach, the scores of positive and negative triples separate naturally as a result of the training objective. True triples usually have positive scores, while randomly generated ones have negative scores.

We use the validation data set to choose the best threshold that separates positive from negative triples and then test the model's ability to discriminate these on the test data set. For the TransE model, the best result is obtained with 50 dimensions and a threshold of 7.0, reaching an accuracy of 0.8245. For our model, setting the dimensionality to 50 and running for 60 iterations, the threshold is -2.26 and the accuracy is 0.8831. This shows that our model can successfully predict relationships even for the rich set of entities in DBpedia.

6   Conclusion

We have proposed WebBrain, a novel approach for knowledge acquisition and modeling, that makes a further step towards the goal of equipping computers with commonsense knowledge to enable more intelligent systems. Our model combines multiple objectives to model word meanings and relationships. We rely on large-scale Web information extraction and on general corpus co-occurrences to train our model. While we remain far from genuine commonsense understanding and intelligence, our approach is able to learn vector representations of words and relations that reflect both their general semantics and basic commonsense facts about the world, giving accurate answers even for knowledge that has not been observed in text.

In future work, our joint training approach could also be evaluated with further relation representation models. This seems particularly promising for models that incorporate additional reasoning capabilities [19] or constraints [17]. Finally, we wish to extend our approach to combine common-sense knowledge as extracted from text with the kinds of encyclopedic facts available in DBpedia so as to obtain a more complete model of world knowledge. We believe that models of this sort that combine heterogeneous kinds of inputs will enable us to put semantic resources to use in advanced intelligent systems.

References

1. Bengio, Y., Ducharme, R., Vincent, P., Janvin, C.: A neural probabilistic language model. J. Mach. Learn. Res. **3**, 1137–1155 (2003)
2. Bizer, C., Lehmann, J., Kobilarov, G., Auer, S., Becker, C., Cyganiak, R., Hellmann, S.: DBpedia - a crystallization point for the web of data. Web Semant. **7**(3), 154–165 (2009)
3. Bollegala, D., Maehara, T., Kawarabayashi, K.: Embedding semantic relations into word representations. In: Proceedings of IJCAI, pp. 1222–1228 (2015)
4. Bordes, A., Weston, J., Collobert, R., Bengio, Y.: Learning structured embeddings of knowledge bases. In: Proceedings of AAAI (2011)
5. Bordes, A., Glorot, X., Weston, J., Bengio, Y.: Joint learning of words and meaning representations for open-text semantic parsing. In: Proceedings of AISTATS (2012)
6. Bordes, A., Usunier, N., Garcia-Duran, A., Weston, J., Yakhnenko, O.: Translating embeddings for modeling multi-relational data. Adv. Neural Inf. Process. Syst. **26**, 2787–2795 (2013)
7. Bruni, E., Tran, N.K., Baroni, M.: Multimodal distributional semantics. J. Artif. Int. Res. **49**(1), 1–47 (2014)
8. Chen, J., Tandon, N., de Melo, G.: Neural word representations from large-scale commonsense knowledge. In: Proceedings of WI/IAT 2015 (2015)
9. Espinosa, J.A., Lieberman, H.: EventNet: inferring temporal relations between commonsense events. In: Gelbukh, A., de Albornoz, Á., Terashima-Marín, H. (eds.) MICAI 2005. LNCS (LNAI), vol. 3789, pp. 61–69. Springer, Heidelberg (2005)
10. Fellbaum, C.: WordNet: An Electronic Lexical Database. The MIT Press, Cambridge (1998)
11. Finkelstein, L., Evgenly, G., Yossi, M., Ehud, R., Zach, S., Gadi, W., Eytan, R.: Placing search in context: the concept revisited. In: Proceedings of WWW (2001)
12. Havasi, C., Speer, R., Alonso, J.: ConceptNet 3: a flexible, multilingual semantic network for common sense knowledge. In: Proceedings of RANLP (2007)
13. Hearst, M.A.: Automatic acquisition of hyponyms from large text corpora. In: Proceedings of COLING, pp. 539–545 (1992)

14. Hill, F., Korhonen, A.: Learning abstract concept embeddings from multi-modal data: since you probably can't see what I mean. In: Proceedings of EMNLP, pp. 255–265 (2014)
15. Jenatton, R., Roux, N.L., Bordes, A., Obozinski, G.R.: A latent factor model for highly multi-relational data. Adv. Neural Inf. Process. Syst. **25**, 3167–3175 (2012)
16. Jiang, X., Huang, Y., Nickel, M., Tresp, V.: Combining information extraction, deductive reasoning and machine learning for relation prediction. In: Simperl, E., Cimiano, P., Polleres, A., Corcho, O., Presutti, V. (eds.) ESWC 2012. LNCS, vol. 7295, pp. 164–178. Springer, Heidelberg (2012)
17. Krompaß, D., Baier, S., Tresp, V.: Type-constrained representation learning in knowledge graphs. In: Proceedings of ISWC (2015)
18. Lenat, D.B.: CYC: a large-scale investment in knowledge infrastructure. Commun. ACM **38**, 33–38 (1995)
19. Lin, Y., Liu, Z., Luan, H., Sun, M., Rao, S., Liu, S.: Modeling relation paths for representation learning of knowledge bases. In: Proceedings of EMNLP (2015)
20. Lin, Y., Liu, Z., Sun, M., Liu, Y., Zhu, X.: Learning entity and relation embeddings for knowledge graph completion. In: Proceedings AAAI 2015, AAAI Press (2015)
21. Liu, H., Singh, P.: ConceptNet– a practical commonsense reasoning tool-kit. BT Technol. J. **22**(4), 211–226 (2004)
22. Mikolov, T., Sutskever, I., Chen, K., Corrado, G.S., Dean, J.: Distributed representations of words and phrases and their compositionality. In: Advances in Neural Information Processing Systems, vol. 26, pp. 3111–3119 (2013)
23. Nickel, M., Tresp, V., Kriegel, H.P.: A three-way model for collective learning on multi-relational data. In: Proceedings of ICML (2011)
24. Niles, I., Pease, A.: Towards a standard upper ontology. In: Proceedings of FOIS (2001)
25. Pantel, P., Pennacchiotti, M.: Espresso: leveraging generic patterns for automatically harvesting semantic relations. In: Proceedings of COLING/ACL 2006 (2006)
26. Paulheim, H.: Knowledge graph refinement: a survey of approaches and evaluation methods. Semantic Web (2016)
27. Socher, R., Chen, D., Manning, C.D., Ng, A.: Reasoning with neural tensor networks for knowledge base completion. Adv. Neural Inf. Process. Syst. **26**, 926–934 (2013)
28. Speer, R., Havasi, C., Lieberman, H.: Analogyspace: reducing the dimensionality of common sense knowledge. In: Proceedings of AAAI, AAAI Press (2008)
29. Sutskever, I., Tenenbaum, J.B., Salakhutdinov, R.R.: Modelling relational data using Bayesian clustered tensor factorization. Adv. Neural Inf. Process. Syst. **22**, 1821–1828 (2009)
30. Tandon, N., de Melo, G., Suchanek, F.M., Weikum, G.: WebChild: harvesting and organizing commonsense knowledge from the web. In: Proceedings of WSDM (2014)
31. Tandon, N., de Melo, G., Weikum, G.: Deriving a Web-scale common sense fact database. In: Proceedings of AAAI 2011, AAAI Press, Palo Alto, CA, USA (2011)
32. Tandon, N., Weikum, G., de Melo, G., De, A.: Lights, camera, action: Knowledge extraction from movie scripts. In: Proceedings of WWW (2015)
33. Toutanova, K., Chen, D., Pantel, P., Poon, H., Choudhury, P., Gamon, M.: Representing text for joint embedding of text and knowledge bases. In: Proceedings of EMNLP, ACL (2015)
34. Wang, Z., Zhang, J., Feng, J., Chen, Z.: Knowledge graph and text jointly embedding. In: Proceedings of EMNLP, pp. 1591–1601 (2014)

35. Wang, Z., Zhang, J., Feng, J., Chen, Z.: Knowledge graph embedding by translating on hyperplanes. In: Proceedings of AAAI 2014, pp. 1112–1119 (2014)
36. Xie, R., Liu, Z., Jia, J., Luan, H., Sun, M.: Representation learning of knowledge graphs with entity descriptions. In: Proceedings of AAAI (2016)
37. Yu, M., Dredze, M.: Improving lexical embeddings with semantic knowledge. In: Proceedings of ACL 2014 (2014)

Efficient Algorithms for Association Finding and Frequent Association Pattern Mining

Gong Cheng$^{(\boxtimes)}$, Daxin Liu, and Yuzhong Qu

National Key Laboratory for Novel Software Technology,
Nanjing University, Nanjing, China
{gcheng,yzqu}@nju.edu.cn, dxliu@smail.nju.edu.cn

Abstract. Finding associations between entities is a common information need in many areas. It has been facilitated by the increasing amount of graph-structured data on the Web describing relations between entities. In this paper, we define an association connecting multiple entities in a graph as a minimal connected subgraph containing all of them. We propose an efficient graph search algorithm for finding associations, which prunes the search space by exploiting distances between entities computed based on a distance oracle. Having found a possibly large group of associations, we propose to mine frequent association patterns as a conceptual abstract summarizing notable subgroups to be explored, and present an efficient mining algorithm based on canonical codes and partitions. Extensive experiments on large, real RDF datasets demonstrate the efficiency of the proposed algorithms.

Keywords: Association finding · Canonical code · Distance oracle · Frequent association pattern mining · Graph search

1 Introduction

Finding associations between entities has found applications in many areas. For instance, social networking services suggest friends based on known associations between people. Security agents are interested in associations between suspected terrorists. In recent years, the increasing amount of graph-structured data on the Web, like RDF data, has made association finding easier than extracting from Web text [14]. In such a graph describing relations between entities, associations between entities are reflected by paths or subgraphs connecting them. Finding such connections is also an essential component of some semantic search and question answering systems [18].

Existing research efforts mainly focus on finding, ranking, and filtering associations between two entities [2,3,5–8,10,15,20], which are usually defined as paths connecting them in a graph. Given multiple (i.e., two or more) entities, a more general notion of association naturally builds on the paths between all pairs of entities, but requires a more concise structure [4,12,17]. In this work, we define an association connecting multiple entities in a graph as a minimal connected subgraph that contains all of them. Then two challenges arise:

© Springer International Publishing AG 2016
P. Groth et al. (Eds.): ISWC 2016, Part I, LNCS 9981, pp. 119–134, 2016.
DOI: 10.1007/978-3-319-46523-4_8

(a) how to efficiently find associations in a possibly very large graph, and (b) how to help users explore a possibly large set of associations that have been found. Both challenges are addressed in this paper. Our contribution is threefold.

- We propose an efficient algorithm for finding associations based on graph search and path merging. To prune the search space, distances between entities are exploited, and a distance oracle is used to achieve a trade-off between time for computing and space for materializing distances.
- To help users explore a large group of associations, complementary to the existing ranking approaches [4,12,17], we propose to identify its notable subgroup(s) that match a common conceptual structure called a frequent association pattern, which provides a high-level abstract of major results. Our efficient algorithm for mining frequent association patterns calculates frequency based on canonical codes of association patterns, and reduces calculations using partitions of associations.
- We carry out extensive experiments based on large, real RDF datasets. The results demonstrate the efficiency of the proposed algorithms.

In this paper, we focus on the efficiency of algorithms for finding associations and mining frequent association patterns. The effectiveness of using frequent association patterns for exploring associations between two entities has been demonstrated in [6]. The effectiveness in a multiple-entity setting will be empirically tested in future work.

The remainder of this paper is structured as follows. Section 2 provides preliminaries. Sections 3 and 4 introduce our algorithms for finding associations and mining frequent association patterns, respectively. Section 5 presents experiments. Section 6 discusses related work. Section 7 concludes the paper with future work.

2   Preliminaries

We deal with a directed unweighted *entity-relation graph* $G = \langle E, A, R, l \rangle$ characterizing binary relations over entities, where

- E is a set of entities as vertices,
- A is a set of arcs, each arc $a \in A$ directed from its tail vertex $t(a) \in E$ to its head vertex $h(a) \in E$,
- R is a set of binary relations on entities, and
- $l : A \mapsto R$ labels each arc $a \in A$ with a relation $l(a) \in R$.

Let C be the set of all classes. For each entity $e \in E$, let $T(e) \subseteq C$ be e's types, and we assume that each entity has at least one type, i.e., $T(e) \neq \emptyset$. Figure 1 shows an entity-relation graph to be used as a running example in this paper. An RDF graph (i.e., a set of RDF triples) can be regarded as an entity-relation graph if considering only the triples connecting two entities; T is given by the rdf:type property. In this paper, we will stick to the above graph notation rather than RDF because our approach is not specific to RDF but also applies to other kinds of graph-structured data.

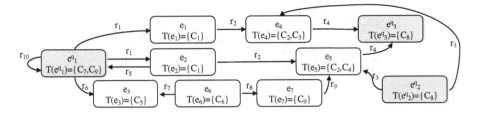

Fig. 1. An example entity-relation graph, with three query entities in grey.

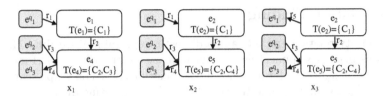

Fig. 2. Three associations connecting e_1^q, e_2^q, and e_3^q.

Given a set of n *query entities* $e_1^q, \ldots, e_n^q \in E$, an *association* x connecting $e_1^q, \ldots, e_n^q$ is a minimal subgraph of G that contains all of them and is connected; no proper subgraph of it also has these properties. Therefore, the underlying graph of x is a tree (i.e., having no parallel edges, loops, or cycles), and the leaves only come from query entities; otherwise x would not be minimal. For consistency, e_1^q is always designated as the root of x. Figure 2 illustrates three associations connecting the three query entities in the running example.

Note that in this paper, the arcs in a path or in a rooted tree are not required to all go the same direction, since an arc a directed from $t(a)$ to $h(a)$ labeled with a relation $l(a) = r$ can be equivalently treated as an arc directed from $h(a)$ to $t(a)$ labeled with a relation $\widehat{r}$ that represents the inverse of r. For the same reason, later in our algorithms, every arc can be traversed in both directions in graph search.

The *diameter* of an association x, denoted by $diam(x)$, is the greatest distance between any pair of entities in x. Given a *diameter constraint* λ, a *valid association* has a diameter of λ or less. For instance, given $\lambda = 3$, Fig. 2 shows all the valid associations connecting the three query entities in the running example; all of them have a diameter of 3. An *invalid association* has a diameter larger than λ. We will focus on valid associations because such shorter-distance associations usually represent stronger connections between entities and thus are more attractive to users.

An *association pattern* matched by an association x is a directed graph obtained by replacing each non-query entity in x with one of its types. For instance, x_1 and x_2 in Fig. 2 match z_1 in Fig. 3; x_1 also matches z_2. Since an association is tree-structured and the leaves only come from query entities, an association pattern also has these properties, and e_1^q is designated as its root for consistency.

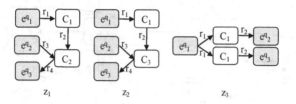

Fig. 3. Three association patterns.

3   Association Finding

Given an entity-relation graph G and a diameter constraint λ, we aim to find all the valid associations connecting a set of n query entities $e_1^q, \ldots, e_n^q$. Firstly, we present a basic algorithm for finding valid associations based on graph search and path merging. Then, we prune the search space by exploiting distances between vertices. Finally, to achieve a trade-off between time for computing and space for materializing distances, we discuss the use of distance oracles.

3.1   A Basic Algorithm

Our basic algorithm is inspired by the following theorem.

Theorem 1. *An association x connecting a set of query entities can be decomposed into a set of (possibly overlapping) paths of length $\left\lfloor \frac{diam(x)+1}{2} \right\rfloor$ or less that have query entities as their start vertices and have a common end vertex.*

For instance, x_1 in Fig. 2, with $diam(x_1) = 3$, can be decomposed into three paths of length $\left\lfloor \frac{3+1}{2} \right\rfloor = 2$ or less: $e_1^q r_1 e_1$, $e_2^q r_3 e_4 \widehat{r_2} e_1$, and $e_3^q \widehat{r_4} e_4 \widehat{r_2} e_1$; all of them start from query entities and have e_1 as a common end vertex.

Proof. Let p be a longest path in x, having a length of $diam(x)$. Let e' be an entity in the middle of p, i.e., the two paths p_1 and p_2 connecting the start and end vertex of p to e' have a length of $\left\lfloor \frac{diam(x)+1}{2} \right\rfloor$ or $\left\lfloor \frac{diam(x)+1}{2} \right\rfloor - 1$. Then for every leaf e of x, the path connecting e to e' must have a length of $\left\lfloor \frac{diam(x)+1}{2} \right\rfloor$ or less; otherwise we can merge such a path with p_1 or p_2 to obtain a path longer than $diam(x)$, which contradicts that the diameter of x is $diam(x)$. Therefore, x can be decomposed into a set of paths of length $\left\lfloor \frac{diam(x)+1}{2} \right\rfloor$ or less, each connecting a leaf of x (which is a query entity) to e'.

Following this theorem, we develop Algorithm 1 for finding all the valid associations by searching for and merging paths. Specifically, all the paths of length $\left\lfloor \frac{\lambda+1}{2} \right\rfloor$ or less starting from each query entity are found by searching G in a breadth-first manner (line 3–4). For instance, when $\lambda = 3$, starting from e_1^q in Fig. 1, four paths of length 1 and four paths of length 2 are found:

$$P_1 = \{e_1^q r_1 e_1, \; e_1^q r_1 e_2, \; e_1^q \widehat{r_5} e_2, \; e_1^q r_6 e_3, \\ e_1^q r_1 e_1 r_2 e_4, \; e_1^q r_1 e_2 r_2 e_5, \; e_1^q \widehat{r_5} e_2 r_2 e_5, \; e_1^q r_6 e_3 \widehat{r_7} e_6 \}. \tag{1}$$

Algorithm 1. A Basic Algorithm for Association Finding

Data: An entity-relation graph G, a set of n query entities $e_1^q, \ldots, e_n^q$, and a diameter constraint λ.

Result: A set of valid associations connecting $e_1^q, \ldots, e_n^q$.

1 $X = \emptyset$;   /* a set of associations */
2 $codes = \emptyset$;   /* a set of canonical codes for associations */
3 **for** $i = 1$ **to** n **do**
4       $P_i =$ the set of all paths of length $\lfloor \frac{\lambda+1}{2} \rfloor$ or less starting from e_i^q found by searching G in a breadth-first manner;
5 **foreach** $\langle p_1, \ldots, p_n \rangle \in (P_1 \times \cdots \times P_n)$ **do**
6   **if** $p_1, \ldots, p_n$ *have a common end vertex* **then**
7   Merge $p_1, \ldots, p_n$ to form a connected subgraph x of G;
8   **if** x *is minimal* **then**
             /* x is minimal if its underlying graph is a tree, and the leaves only come from $e_1^q, \ldots, e_n^q$.   */
9   **if** $diam(x) \leq \lambda$ **then**
10                  $code(x) =$ the canonical code of x;
11   **if** $code(x) \notin codes$ **then**
12   Add $code(x)$ to $codes$;
13   Add x to X;
14 **return** X

Then, all possible combinations of such paths are examined (line 5–13); each combination consists of one path starting from each query entity, i.e., one from P_1, one from P_2, ..., one from P_n. If all the paths in a combination have a common end vertex (e.g., $e_1^q r_1 e_1$, $e_2^q r_3 e_4 \widehat{r_2} e_1$, and $e_3^q \widehat{r_4} e_4 \widehat{r_2} e_1$ in Fig. 1), they will be merged into a subgraph x of G (e.g., x_1 in Fig. 2) that is potentially a valid association to be found (line 6–7). However, before adding x to the results X (line 13), it has to satisfy three requirements.

Firstly, x should be minimal; that is, its underlying graph is a tree, and the leaves only come from query entities (line 8). These tests can be carried out within a single depth-first search of x.

Secondly, x should be valid, i.e., $diam(x) \leq \lambda$ (line 9). This test is needed because when λ is odd, it is possible that x is formed by merging paths of length $\lfloor \frac{\lambda+1}{2} \rfloor = \frac{\lambda+1}{2}$ so that $diam(x) = \lambda + 1 > \lambda$.

Thirdly, the same association should not be added to X multiple times. For instance, x_1 in Fig. 2 can be formed twice by merging the paths in two different combinations: one with e_1 as a common end vertex and the other with e_4. To avoid such duplicates, we generate a *canonical code* for x (line 10), denoted by $code(x)$, so that two associations will have the same canonical code if and only if they are *isomorphic* to each other, i.e., they have the same set of entities as vertices and there is a bijection between their arcs that preserves adjacency and arc labels. If it is the first time $code(x)$ is seen, x will be added to X (line 11–13).

There have been various ways of defining and generating canonical codes for trees [11], assuming a total order ($\preceq$) on each set of sibling vertices. We adopt the following recursive definition, and implement $\preceq$ by the alphabetical order of entity identifiers (e.g., URIs).

- For a tree T with a single vertex e, we define

$$code(T) = e\$\,, \tag{2}$$

  where $\$$ is a special symbol not in the alphabet for naming entities and relations.
- For a tree T with more than one vertex, assuming its root is e and the arcs connecting e to its children $e_1, \ldots, e_k$ (subject to $e_1 \preceq \cdots \preceq e_k$) are labeled with relations $r_1, \ldots, r_k$, respectively, we define

$$code(T) = e r_1 code(T_1) \cdots r_k code(T_k)\$\,, \tag{3}$$

  where $T_1, \ldots, T_k$ are the subtrees rooted at $e_1, \ldots, e_k$, respectively.

Such a code can be generated for x via a depth-first search of x. For instance, for x_1 in Fig. 2 with e_1^q always designated as its root, assuming $e_2^q \preceq e_3^q$, we have

$$code(x_1) = e_1^q r_1 e_1 r_2 e_4 \widehat{r_3} e_2^q \$ r_4 e_3^q \$\$\$. \tag{4}$$

Let Δ be the maximum of the degrees of vertices in G. In the algorithm, the number of paths that can be found from a query entity is bounded by $O(\Delta^{\lfloor \frac{\lambda+1}{2} \rfloor})$. Given n query entities, there are $O(\Delta^{\lfloor \frac{\lambda+1}{2} \rfloor n})$ combinations of paths to examine; in practice we can index paths by their end vertices to significantly improve the performance. The time for checking one combination of paths for the three requirements of a valid association is linear with its size, which is bounded by $O(n\lambda)$. Overall, the algorithm takes $O(\Delta^{\lfloor \frac{\lambda+1}{2} \rfloor n} n\lambda)$ time, but n and λ are both very small in practice.

3.2 Distance-Based Search Space Pruning

To improve the performance of Algorithm 1, we notice that some paths found in graph search will not be merged into any valid association. For instance, when $\lambda = 3$, among the eight paths in P_1 as shown in Eq. (1), $e_1^q r_6 e_3$ and $e_1^q r_6 e_3 \widehat{r_7} e_6$ eventually do not take part in any valid association in Fig. 2. If we can prune the search space to exclude such paths, graph search will end earlier (line 4) and there will be much fewer combinations of paths to be examined (line 5–13), so that the performance of the algorithm can be improved.

We prune the search space by exploiting distances between entities in the entity-relation graph G. Let $dist$ return the distance between two entities in G. For instance, in Fig. 1, we have $dist(e_1^q, e_3) = 1$ and $dist(e_2^q, e_3) = 4$. When searching G for the set of paths P_i starting from a query entity e_i^q and arriving at an entity e via a path $p_{e_i^q e}$ from e_i^q to e, the search space may then be pruned depending on the distances between e and other query entities, i.e., $dist(e_j^q, e)$ for $j \neq i$.

Specifically, if $dist(e_j^q, e) > \lfloor \frac{\lambda+1}{2} \rfloor$ for any other query entity e_j^q ($j \neq i$), $p_{e_i^q e}$ can be excluded from P_i *safely* (i.e., not affecting the final results X) because it will not take part in any valid association since P_j is not likely to contain a path from e_j^q to e of length $\lfloor \frac{\lambda+1}{2} \rfloor$ or less. For instance, given $\lambda = 3$, when searching the graph in Fig. 1 starting from e_1^q and arriving at e_3 via the path $e_1^q r_6 e_3$, this path will be excluded from P_1 because $dist(e_2^q, e_3) = 4 > 2 = \lfloor \frac{3+1}{2} \rfloor$.

Further, let $ln(p)$ be the length of a path p. If $ln(p_{e_i^q e}) + dist(e_j^q, e) > 2 \lfloor \frac{\lambda+1}{2} \rfloor$ for any other query entity e_j^q ($j \neq i$), which implies $dist(e_j^q, e) > \lfloor \frac{\lambda+1}{2} \rfloor$ since $ln(p_{e_i^q e}) \leq \lfloor \frac{\lambda+1}{2} \rfloor$, we can safely exclude from P_i not only $p_{e_i^q e}$ but also all the paths that extend $p_{e_i^q e}$ (i.e., having $p_{e_i^q e}$ as a prefix); in other words, the entire branch of search stemming from $p_{e_i^q e}$ can be pruned. For instance, given $\lambda = 3$, when searching the graph in Fig. 1 starting from e_1^q and arriving at e_3 via the path $e_1^q r_6 e_3$, we will not only exclude this path from P_1 but also prune the branch of search stemming from it because $ln(e_1^q r_6 e_3) + dist(e_2^q, e_3) = 1 + 4 = 5 > 2 \lfloor \frac{3+1}{2} \rfloor$; as a result, the path $e_1^q r_6 e_3 \widehat{r_7} e_6$ will be implicitly excluded from P_1. We prove the safeness by showing that any path $p_{e_i^q e'}$ from e_i^q to an entity e' that extends $p_{e_i^q e}$ will not take part in any valid association. Specifically, $p_{e_i^q e'}$ is composed of $p_{e_i^q e}$ from e_i^q to e and $p_{ee'}$ from e to e'. If it can be merged with some path $p_{e_j^q e'} \in P_j$ ($j \neq i$) from e_j^q to e' into a valid association, we will have $ln(p_{e_j^q e'}) \leq \lfloor \frac{\lambda+1}{2} \rfloor$ and

$$
\begin{aligned}
2 \left\lfloor \frac{\lambda+1}{2} \right\rfloor &= \left\lfloor \frac{\lambda+1}{2} \right\rfloor + \left\lfloor \frac{\lambda+1}{2} \right\rfloor \\
&\geq ln(p_{e_i^q e'}) + ln(p_{e_j^q e'}) \\
&= ln(p_{e_i^q e}) + ln(p_{ee'}) + ln(p_{e_j^q e'}) \\
&\geq ln(p_{e_i^q e}) + dist(e_j^q, e),
\end{aligned}
\tag{5}
$$

which contradicts $ln(p_{e_i^q e}) + dist(e_j^q, e) > 2 \lfloor \frac{\lambda+1}{2} \rfloor$.

3.3   Distance Computation

The above pruning strategy requires knowing distances between entities. When the entity-relation graph is large, e.g., consisting of millions of vertices and billions of arcs, obtaining distances will be nontrivial. On the one hand, online computing distances would be time-consuming and lead to unacceptable latency. On the other hand, materializing offline computed distances between all pairs of entities would be a challenge. To achieve a trade-off between time for computing and space for materializing distances, we turn to distance oracles [16].

A *distance oracle* is a data structure that, after preprocessing a graph, allows for fast distance computation. Specifically, the graph is offline processed to compute certain information (e.g., distances between each vertex and some landmark vertices) to be materialized in a distance oracle; its size is usually much smaller than the size of materializing distances between all pairs of vertices. By using a distance oracle, computing the distance between two vertices can be reasonably fast, though not as fast as looking up a materialized distance.

There are two types of distance oracles: exact and approximate. Given two vertices between which the distance is d, an *exact distance oracle* will return d, whereas an *approximate distance oracle* will return a value that is in the range of $[d, \alpha d + \beta]$ where $\alpha \geq 1$ and $\beta \geq 0$, which is said to have stretch (α, β). Different approximate distance oracles have different trade-offs between stretch, size, and time. Practical approximate distance oracles usually have stretch $\alpha = 2$ or $\alpha = 3$. However, such a distance oracle is not particularly useful for small-world graphs in which distances between vertices are typically very small [16]. As we will see in Sect. 5.1, some widely used entity-relation graphs are exactly small-world graphs. Therefore, we choose to implement a state-of-the-art exact distance oracle [1], to be used in distance-based pruning.

4   Frequent Association Pattern Mining

Having found a possibly large group of associations, we aim to identify its notable subgroup(s) that match a common conceptual structure, i.e., a frequent association pattern, to provide a high-level abstract of major results. Specifically, given a group of associations X, the *frequency* of an association pattern z, denoted by $f_X(z)$, is the number of associations in X that match z. Given a threshold $\tau \in [0, 1]$, we aim to find all the *frequent association patterns* z for which $\frac{f_X(z)}{|X|} \geq \tau$. Note that existing solutions to frequent tree pattern mining [11] do not apply here because their resulting subtrees may not contain all the query entities. In the following, we firstly present a basic algorithm. Then we improve its performance by partitioning X.

4.1   A Basic Algorithm

The idea is to firstly, for each association in X, enumerate all the association patterns it matches; for instance, x_1 in Fig. 2 matches z_1 and z_2 in Fig. 3. Then we calculate the frequency of each association pattern and identify frequent ones; to this end, the main problem is to judge whether two association patterns enumerated for different associations are isomorphic to each other. Since an association pattern is tree-structured, we intend to generate a canonical code for each enumerated association pattern by reusing the way of defining and generating canonical codes presented in Sect. 3.1, and then count the occurrence of each canonical code as the frequency of the corresponding association pattern.

Recall that in Sect. 3.1, the definition of canonical code relies on a predefined total order ($\preceq$) on each set of sibling vertices; there, we implement $\preceq$ by the alphabetical order of entity identifiers, considering that sibling vertices in an association are always different entities with different identifiers. However, if sibling entities in an association have a common type, the corresponding sibling vertices in an association pattern will represent the same class; for instance, in Fig. 3, the two children of e_1^q in z_3 both represent C_1. Hence, the alphabetical order of entity and class identifiers fails to give a total order on such a set of sibling vertices. If we still use this order and break ties arbitrarily, different canonical codes may be generated for isomorphic association patterns, leading to incorrect calculation of frequency. For instance, the canonical code for z_3 in Fig. 3 could be

$$e_1^q r_1 C_1 r_2 e_2^q \$\$ r_1 C_1 r_2 e_3^q \$\$\$$$
$$\text{or } e_1^q r_1 C_1 r_2 e_3^q \$\$ r_1 C_1 r_2 e_2^q \$\$\$, \tag{6}$$

depending on how to order the two children of e_1^q.

To obtain a unique canonical code, a less efficient solution is to generate codes in all possible orders and choose the lexicographically smallest one [11]. Differently, we propose a more efficient solution that directly generates a unique code by implementing $\preceq$ in a different way that exploits query entities. Specifically, instead of directly ordering sibling vertices by their identifiers (which may represent the same class), for each sibling vertex v that is not a query entity, we choose a query entity as its *proxy* to be ordered by entity identifiers, which is the one with the alphabetically smallest entity identifier in the subtree rooted at v. Since subtrees rooted at sibling vertices contain different sets of query entities, the proxies chosen are different. This successfully gives a total order on each set of sibling vertices, and thus ensures a unique canonical code for isomorphic association patterns. For instance, assuming e_2^q alphabetically precedes e_3^q, the unique canonical code for z_3 in Fig. 3 will be

$$e_1^q r_1 C_1 r_2 e_2^q \$\$ r_1 C_1 r_2 e_3^q \$\$\$$$
$$\text{but not } e_1^q r_1 C_1 r_2 e_3^q \$\$ r_1 C_1 r_2 e_2^q \$\$\$, \tag{7}$$

because the proxy for the upper child of e_1^q in Fig. 3 is e_2^q, which alphabetically precedes e_3^q, the proxy for the lower child of e_1^q. Proxies for all the vertices in an association pattern z can be found within a single depth-first search of z.

The size of an association is bounded by $O(n\lambda)$. Let γ be the maximum number of types that an entity can have. An association can match $O(\gamma^{n\lambda})$ association patterns, and thus $O(|X|\gamma^{n\lambda})$ canonical codes will be generated. Generating one canonical code takes $O(n\lambda)$ time, plus $O(n\lambda)$ time for finding proxies. Overall, the algorithm takes $O(|X|\gamma^{n\lambda}n\lambda)$ time to generate all the canonical codes to be counted, but γ, n, λ are all very small in practice.

4.2   Partitioning-Based Performance Improvement

Enumerating association patterns and generating canonical codes for them can be time-consuming. To improve the performance, we aim to divide X into mutually disjoint partitions, and ensure that only the associations in the same partition can match a common association pattern. Then, when mining frequent association patterns, we can ignore partitions containing fewer than $\tau|X|$ associations, without spending time processing association patterns they match.

We observe that two associations can match a common association pattern only if they: (a) consist of the same number of vertices, and (b) have the same set of arc labels (i.e., relations). We divide X based on a combination of these two metrics. For instance, x_1 and x_2 in Fig. 2 will be put in the same partition because both of them consist of five vertices and their arc labels are both $\{r_1, r_2, r_3, r_4\}$, whereas x_3 is in a different partition because its arc labels are $\{r_2, r_3, r_4, r_5\}$.

5   Experiments

We tested the performance of the proposed algorithms on an E3-1226 v3 with 24GB memory for JVM. Entity-relation graphs and entities' types were stored in memory. Distance oracles were stored in a MySQL database on disk.

5.1   Datasets and Test Queries

Datasets. Experiments were conducted on two widely used RDF datasets.

- LinkedMDB[1] provided RDF data about movies and related entities like actors and directors. After filtering out RDF triples involving literals or `rdf:type`, an entity-relation graph was obtained, consisting of 1,327,069 entities as vertices and 2,132,796 arcs. Entities' types were derived from RDF triples involving `rdf:type`.
- DBpedia[2] provided encyclopedic RDF data extracted from Wikipedia. After filtering out RDF triples involving literals, an entity-relation graph was obtained from the Mapping-based Properties dataset, consisting of 4,337,485 entities as vertices and 15,007,564 arcs. Entities' types were derived from the Mapping-based Types dataset.

For entities having no type information, `owl:Thing` was added to be their type.
     To characterize the two entity-relation graphs, we randomly selected 10,000 pairs of entities from each graph, and tested whether they were connected by paths and if so, calculated the distance between them. As shown in Table 1, in LinkedMDB, most pairs of entities (77.20 %) were connected, and their average and median distances were 6.61 and 7, respectively, showing the small-world effect, which was even more pronounced on DBpedia. The results revealed two findings.

Table 1. Distance between entities

	% of connected entities	Distance between connected entities	
		Average	Median
LinkedMDB	77.20 %	6.61	7
DBpedia	96.41 %	5.06	5

---

[1] http://www.cs.toronto.edu/~oktie/linkedmdb/linkedmdb-latest-dump.zip.
[2] http://wiki.dbpedia.org/Downloads2015-04.

- An exact (not approximate) distance oracle was needed for effective distance-based pruning on such small-world graphs as discussed in Sect. 3.3.
- The diameter constraint had to be set to a small value (≤ 4), because larger values would require searching almost the entire entity-relation graph, and could find too many paths and associations to fit in memory.

Test Queries. Test queries were constructed under different settings of diameter constraint (λ) and number of query entities (n). For each combination of $\lambda \in \{2, 4\}$ and $n \in \{2, 3, 4, 5\}$, we randomly selected 1,000 sets of n query entities from each of the two entity-relation graphs as test queries.

5.2 Association Finding

Algorithms. Three algorithms for association finding were tested:

- BSC: the basic algorithm described in Sect. 3.1, which can be regarded as an extension of the existing bi-directional BFS algorithm for finding paths between two entities [10],
- PRN: the improved algorithm using distance-based search space pruning described in Sects. 3.2 and 3.3, and
- PRN-1: a variant of PRN that would not try to prune the search space at the last level of search, and thus might exclude fewer paths than PRN but could reduce the number of distance computations, achieving a different trade-off.

In PRN and PRN-1, the distance between two vertices would be cached in memory after being computed for the first time. However, to avoid distorting the results of performance tests, the cache would be cleared after every single run of an algorithm on a test query.

Results. We ran each algorithm five times on each test query, and took the median running time. Then we calculated the average running time per query used by each algorithm on all the test queries under each setting of λ and n.

As shown in Fig. 4 on a logarithmic scale, when $\lambda = 2$, all the three algorithms were very fast on both datasets, using not more than 4ms per query. PRN and

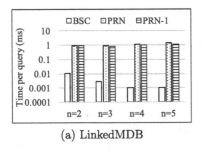
(a) LinkedMDB

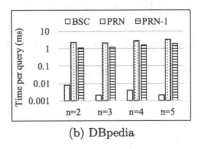
(b) DBpedia

Fig. 4. Running time of association finding under $\lambda = 2$.

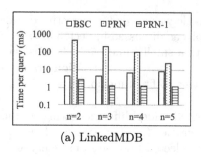 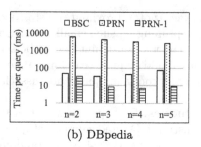

(a) LinkedMDB   (b) DBpedia

Fig. 5. Running time of association finding under $\lambda = 4$.

Table 2. Number of distance computations

Dataset	Algorithm	$\lambda = 2$				$\lambda = 4$			
		$n = 2$	$n = 3$	$n = 4$	$n = 5$	$n = 2$	$n = 3$	$n = 4$	$n = 5$
LinkedMDB	PRN	2.0	3.0	4.0	5.0	3,055.3	1,525.6	836.0	144.5
	PRN-1	2.0	3.0	4.0	5.0	2.9	4.0	4.8	5.7
DBpedia	PRN	2.2	3.0	4.0	5.0	32,530.2	24,061.5	19,057.1	15,346.5
	PRN-1	2.0	3.0	4.0	5.0	5.7	8.9	9.1	13.0

PRN-1 were relatively slow because the search space was very small when $\lambda = 2$, so that distance computation for pruning took more time than it saved.

Distance-based pruning proved to be effective when the search space became large. As shown in Fig. 5 on a logarithmic scale, when $\lambda = 4$, PRN-1 used not more than 34ms per query, being 55%–548% faster than BSC on Linked-MDB, and 40%–712% faster on DBpedia. The difference rose when increasing n because given a larger number of query entities (i.e., n), more distances could be exploited in graph search and the search space would be more likely to be pruned.

PRN was slower than BSC and PRN-1 because, compared with PRN-1, it also tried to prune the search space at the last level of search, which required computing distances between much more pairs of entities, as shown in Table 2. However, each of those computations could exclude at most one path, as opposed to a possibly large branch of search stemming from a path when pruning at earlier levels of search, thereby being cost-ineffective.

5.3 Frequent Association Pattern Mining

Approaches. Two algorithms for frequent association pattern mining were tested:

- BSC: the basic algorithm described in Sect. 4.1, whose running time was independent of the relative frequency threshold (τ), and
- PRT: the improved algorithm using partitions described in Sect. 4.2, with $\tau = 5\%$ or $\tau = 25\%$.

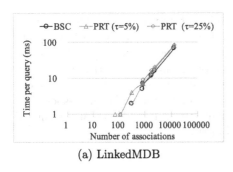

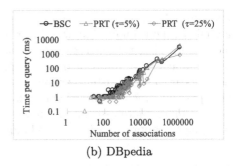

Fig. 6. Running time of frequent association pattern mining under $\lambda = 4$.

Results. We ran each algorithm five times on each test query that resulted in at least two associations when $\lambda = 4$, and took the median running time. Then we calculated the average running time per query used by each algorithm on all the test queries resulting in the same number of associations.

As shown in Fig. 6 on a log-log scale, all the algorithms were reasonably fast on both datasets for 10,000 or fewer associations, using not more than 21ms and 65ms per query on LinkedMDB and DBpedia, respectively. For larger sets of associations on DBpedia, hundreds or thousands of milliseconds was used. Actually, the reported running time had the potential to be reduced by easily parallelizing the algorithms, e.g., enumerating association patterns for different associations in parallel, and generating canonical codes for different association patterns in parallel.

When the number of associations was small, the difference between BSC and PRT was not significant. On most queries resulting in 5,000 or more associations on DBpedia, PRT was 13 %–722 % faster than BSC when $\tau = 25$ %, showing the effectiveness of using partitions. However, PRT was slower than BSC on some queries particularly when $\tau = 5$ % because only very small partitions could be occasionally ignored so that computing partitions took more time than it saved.

5.4   Discussion

In the experiments, we found two limitations of our approach.

Firstly, to find associations, although PRN-1 was very fast when $\lambda \in \{2, 4\}$, using not more than 34ms per query on two fairly large datasets, it frequently used the memory up when we tried to increase λ to 6. That was due to the small-world effect; there were indeed quite many associations to find when $\lambda = 6$. If some of such long-distance associations were believed to be useful according to a certain ranking criterion, graph search could leverage the criterion to prune the search space and return not all but top-ranked associations. However, that would be a different research problem having its own applications [13].

Secondly, to mine frequent association patterns, associations were partitioned so that it was possible to avoid enumerating association patterns for some associations and generating canonical codes for them. However, to put an association into the right partition according to the number of its vertices and the set of its arc labels, the running time was linear with its size, being asymptotically equivalent to the time for generating its canonical code. Therefore, partitioning could not fundamentally improve the performance of the mining algorithm, and did not appear to be consistently superior to the basic algorithm in the experiment. One possibly essential improvement would be to integrate frequent association pattern mining into association finding. For instance, it would be interesting to combine our approach with the techniques in [19].

6   Related Work

Numerous research efforts have been made to find associations between *two* entities, and they define association in different ways [3,7,15]. In a seminal work [3], four types of associations are discussed. Among others, an association between two entities can be a *path* in an entity-relation graph that connects the two entities. Although recent attempts propose to merge certain paths to better explain relatedness between two entities [7,15], the path-based straightforward definition is adopted by most of the subsequent researches, which mainly focus on two problems: how to efficiently find all the paths of a limited length between two entities [10], and how to help users explore such a possibly very large set of paths [2,5,6,8,20]. Concerning the latter problem, one line of work studies the ranking of paths to show users more important paths earlier [2,5]. Complementary to that, other solutions allow users to filter paths by specifying keywords appearing on the paths [20], relations and classes of entities contained in the paths [8], or frequent patterns of the paths [6].

Different from the above efforts, in this work we aim to find associations between *multiple* (i.e., two or more) entities in an entity-relation graph. It goes beyond simply finding paths between all pairs of entities [9], but requires consolidating those paths into concise structures. For instance, in [4,12,17], their goal is to find an optimal association between multiple entities that is a subgraph connecting those entities via a limited number of other entities and maximizing a "goodness" function. In [13], the goal is to find top-k minimum-cost Steiner trees connecting those entities. Differently, we deal with unweighted graphs because we aim to find not top-ranked associations but all the associations having a limited diameter, and then identify their frequent patterns to provide a conceptual abstract of them. This extends our previous work on mining frequent patterns of paths connecting two entities [6], and complements the existing approaches to ranking associations between multiple entities [4,12,13,17].

Compared with a recent work on mining frequent patterns of associations connecting multiple entities in an entity-relation graph [19], our work has made two technical advances. Firstly, in [19], associations are efficiently found by merging paths of a limited length that are materialized in an index. However, it has two limitations: (a) the size of that index increases exponentially with the length of path, and may not be affordable for large datasets and long paths, and (b) when a larger diameter constraint is given, the index may have to be rebuilt to include longer paths. By comparison, to achieve a trade-off between time for computing and space for materializing, we materialize not paths but only a distance oracle which has a fixed, affordable size; using that, paths not taking part in any valid association can be efficiently pruned. Besides, once a distance oracle is built, it can work with arbitrarily large diameter constraints. Secondly, in [19], an association pattern (which is tree-structured) is formed by merging path patterns. That may result in structurally isomorphic association patterns that trivially differ in the designation of root. We eliminate such duplicates by defining and generating a canonical code for each pattern.

7 Conclusion

We have presented efficient algorithms for finding associations connecting a set of query entities in graph-structured data, and mining their frequent association patterns to summarize major results for exploration. Experiment results show that our algorithms are reasonably fast on large, real datasets. They can find applications in many areas where finding associations is a common information need. The novel idea of using a distance oracle to compute distances for pruning the search space may also benefit the study of other research problems such as semantic search and query processing over graph-structured data.

As discussed at the end of the experiments, to further improve the performance of our algorithms, one promising direction is to incorporate ranking criteria (if any) into graph search, and to embed frequent association pattern mining in association finding. This will be our future work. Besides, we have found that sometimes a large number of frequent association patterns can be found, some of which have overlapping meanings and some are not so meaningful to users. It inspires us to consider selecting appropriate ones from all the frequent association patterns, to help users effectively explore associations.

Acknowledgments. This work was supported by the 863 Program under Grant 2015AA015406, the NSFC under Grant 61572247 and 61223003, and the Fundamental Research Funds for the Central Universities.

References

1. Akiba, T., Iwata, Y., Yoshida, Y.: Fast exact shortest-path distance queries on large networks by pruned landmark labeling. In: 2013 ACM SIGMOD International Conference on Management of Data, pp. 349–360 (2013)
2. Anyanwu, K., Maduko, A., Sheth, A.: SemRank: ranking complex relationship search results on the semantic web. In: 14th International Conference on World Wide Web, pp. 117–127 (2005)
3. Anyanwu, K., Sheth, A.: ρ-queries: enabling querying for semantic associations on the semantic web. In: 12th International Conference on World Wide Web, pp. 690–699 (2003)
4. Chen, C., Wang, G., Liu, H., Xin, J., Yuan, Y.: SISP: a new framework for searching the informative subgraph based on PSO. In: 20th ACM International Conference on Information and Knowledge Management, pp. 453–462 (2011)

5. Chen, N., Prasanna, V.K.: Learning to rank complex semantic relationships. Int. J. Semant. Web Inf. Syst. **8**(4), 1–19 (2012)
6. Cheng, G., Zhang, Y., Qu, Y.: Explass: exploring associations between entities via top-K ontological patterns and facets. In: Mika, P., et al. (eds.) ISWC 2014, Part II. LNCS, vol. 8797, pp. 422–437. Springer, Heidelberg (2014)
7. Fang, L., Das Sarma, A., Yu, C., Bohannon, P.: REX: explaining relationships between entity pairs. Proc. VLDB Endow. **5**(3), 241–252 (2011)
8. Heim, P., Hellmann, S., Lehmann, J., Lohmann, S., Stegemann, T.: RelFinder: revealing relationships in RDF knowledge bases. In: Chua, T.-S., Kompatsiaris, Y., Mérialdo, B., Haas, W., Thallinger, G., Bailer, W. (eds.) SAMT 2009. LNCS, vol. 5887, pp. 182–187. Springer, Heidelberg (2009)
9. Heim, P., Lohmann, S., Stegemann, T.: Interactive relationship discovery via the semantic web. In: Aroyo, L., Antoniou, G., Hyvönen, E., ten Teije, A., Stuckenschmidt, H., Cabral, L., Tudorache, T. (eds.) ESWC 2010, Part I. LNCS, vol. 6088, pp. 303–317. Springer, Heidelberg (2010)
10. Janik, M., Kochut, K.J.: BRAHMS: a workbench RDF store and high performance memory system for semantic association discovery. In: Gil, Y., Motta, E., Benjamins, V.R., Musen, M.A. (eds.) ISWC 2005. LNCS, vol. 3729, pp. 431–445. Springer, Heidelberg (2005)
11. Jiménez, A., Berzal, F., Cubero, J.-C.: Frequent tree pattern mining: a survey. Intell. Data Anal. **14**(6), 603–622 (2010)
12. Kasneci, G., Elbassuoni, S., Weikum, G.: MING: mining informative entity relationship subgraphs. In: 18th ACM Conference on Information and Knowledge Management, pp. 1653–1656 (2009)
13. Kasneci, G., Ramanath, M., Sozio, M., Suchanek, F.M., Weikum, G.: STAR: steiner-tree approximation in relationship graphs. In: IEEE 25th International Conference on Data Engineering, pp. 868–879 (2009)
14. Luo, G., Tang, C., Tian, Y.-L.: Answering relationship queries on the web. In: 16th International Conference on World Wide Web, pp. 561–570 (2007)
15. Pirró, G.: Explaining and suggesting relatedness in knowledge graphs. In: Arenas, M., et al. (eds.) ISWC 2015. LNCS, vol. 9366, pp. 622–639. Springer, Berlin (2015)
16. Sommer, C.: Shortest-path queries in static networks. ACM Comput. Surv. **46**(4), 45 (2014)
17. Tong, H., Faloutsos, C.: Center-piece subgraphs: problem definition and fast solutions. In: 12th ACM SIGKDD International Conference on Knowledge Discovery and Data Mining, pp. 404–413 (2006)
18. Tran, T., Cimiano, P., Rudolph, S., Studer, R.: Ontology-Based interpretation of keywords for semantic search. In: Aberer, K., et al. (eds.) ASWC 2007 and ISWC 2007. LNCS, vol. 4825, pp. 523–536. Springer, Heidelberg (2007)
19. Yang, M., Ding, B., Chaudhuri, S., Chakrabarti, K.: Finding patterns in a knowledge base using keywords to compose table answers. Proc. VLDB Endow. **7**(14), 1809–1820 (2014)
20. Zhou, M., Pan, Y., Wu, Y.: Conkar: constraint keyword-based association discovery. In: 20th ACM International Conference on Information and Knowledge Management, pp. 2553–2556 (2011)

A Reuse-Based Annotation Approach for Medical Documents

Victor Christen[✉], Anika Groß, and Erhard Rahm

Department of Computer Science, University of Leipzig, Leipzig, Germany
{christen,gross,rahm}@informatik.uni-leipzig.de

Abstract. Annotations are useful to semantically enrich documents and other datasets with concepts of standardized vocabularies and ontologies. In the medical domain, many documents are not annotated at all and manual annotation is a difficult process making automatic annotation methods highly desirable to support human annotators. We propose a reuse-based annotation approach that utilizes previous annotations to annotate similar medical documents. The approach clusters items in documents such as medical forms according to previous ontology-based annotations and uses these clusters to determine candidate annotations for new items. The final annotations are selected according to a new context-based strategy that considers the co-occurrence and semantic relatedness of annotating concepts. The evaluation based on previous UMLS annotations of medical forms shows that the new approaches outperform a baseline approach as well as the use of the MetaMap tool for finding UMLS concepts in medical documents.

Keywords: Semantic annotation · Medical documents · Ontology · UMLS

1 Introduction

The annotation of data with concepts of standardized vocabularies and ontologies has gained increasing significance due to the huge number and size of available datasets as well as the need to deal with the resulting data heterogeneity. In the biomedical domain, gene or protein functions are thus often described by concepts of the Gene Ontology(GO) [2], scientific publications can be annotated with Medical Subject Headings (MESH) [14], and electronic health records (EHRs) can be semantically classified by concepts of SNOMED CT [7]. Annotations of medical documents such as EHRs can also support advanced analyses, e.g. significant co-occurrences between the use of certain drugs and negative side effects in terms of occurring diseases [12]. Still many medical documents are not annotated at all, impeding data analysis and data integration. For instance, more than 200.000 trials are registered on http://clinicaltrials.gov and every study requires a set of so-called case report forms (CRFs), e.g. to ask for the medical history of probands. For every new clinical trial, CRFs are usually built from scratch, although previous forms might already cover similar topics.

© Springer International Publishing AG 2016
P. Groth et al. (Eds.): ISWC 2016, Part I, LNCS 9981, pp. 135–150, 2016.
DOI: 10.1007/978-3-319-46523-4_9

Question			Annotations
Confirmed(1) diagnosis(2) of AML(3) according to the WHO definition (except(4) for acute promyelocytic leukaemia, APL(5))	1	C0750484	label:confirmation synonyms: confirmatory, confirm
	2	C0011900	label: diagnosis (observable entity) synonyms: diagnostic, diagnosis (DX) ; DX ;...
	3	C0023467	label: AML - acute myeloid leukaemia synonyms: acute myeloid leukaemia ; acute granulocytic leukaemia ;ANLL; ...
	4	C1554961	label: exception
○ yes ○ no	5	C0023487	label: acute promyelocytic Leukemia synonyms: APL; acute myeloid leukaemia, PML/RAR-alpha;...

Fig. 1. Example medical form item and associated annotations to UMLS concepts.

CRF annotations are helpful to search for existing form collections, e.g., in the MDM repository of medical data models [4].

To improve the value of medical documents for analysis, reuse and data integration it is thus crucial to annotate them with concepts of ontologies. Since the number, size and complexity of medical documents and ontologies can be very large, a manual annotation process is time-consuming or even infeasible. Hence, automatic annotation methods become necessary to support human annotators with recommendations for manual verification. Figure 1 shows an exemplary annotation for one item in a medical form (CRF) on eligibility criteria for a clinical trial on acute myeloid leukaemia (AML). Such an item comprises a question as well as a response field or a list of answer options. The shown question has been manually annotated based on a reference mapping with five concepts of the Unified Medical Language System (UMLS) [3], a comprehensive knowledge base integrating many biomedical ontologies. The associated UMLS concepts relate to different terms of the item text (italicized) as indicated by the numbers (1) to (5).

The automatic annotation of medical documents is challenging for several reasons. In particular, it is difficult to correctly identify relevant terms and medical concepts within natural language sentences such as the items (questions) occurring in medical forms. This is because concepts typically have several synonyms that may occur in sentences in different variations. Furthermore, concepts are often described by labels or synonyms consisting of several words, e.g., *AML-Acute myeloid leukaemia (C0023467)*, that can match many irrelevant terms in the items to be annotated. We might further need to identify complex many-to-many mappings between items and ontology concepts without knowing a priori how many medical concepts should be associated per item. Moreover, UMLS is very large (2.8 mio. concepts) making it difficult to identify the best fitting concepts for annotation.

We recently proposed already an initial approach to annotate medical forms with UMLS concepts by extracting terms from items and matching these terms to UMLS concepts based on linguistic ontology matching techniques [5]. The study revealed the mentioned challenges and showed the difficulty of automatically achieving high quality annotations especially for long natural language sentences. Moreover, we observed frequent errors due to the high number of available concept synonyms and misleading terms in synonyms. In this study we aim at improving the quality of annotations and reducing the manual annotation effort by reusing already determined and manually verified annotations. This assumes that there are similar questions in different medical forms of a domain of interest so that previous annotations can be reapplied. For this purpose, we propose and evaluate a new reuse-based annotation approach for annotating medical forms and documents.

Specifically, we make the following contributions:

- To enable annotation reuse, we propose to cluster all previously annotated items that are annotated with the same medical concept. For such annotation clusters, we identify representative features that are more compact than the large set of terms in concept labels and synonyms. We use these clusters and their features to find likely annotations for new items that are similar to already annotated ones.
- We propose a new context-based strategy to select the most promising annotations from a set of previously determined candidates. The strategy considers both the semantic relatedness of the annotating concepts as well as their co-occurrence in previously annotated items.
- We evaluate the proposed approaches based on reference mappings between a set of medical forms and UMLS and compare them with a baseline annotation approach as well as with using the MetaMap tool [1] to identify UMLS concepts within medical documents.

The remainder of this paper is organized as follows. We first provide a more formal problem definition and introduce a base workflow for determining annotations (Sect. 2). We then propose our new reuse-based annotation approach and the context-based selection strategy (Sect. 3). Section 4 presents evaluation results for the new approaches. Finally, we discuss related work in Sect. 5 and conclude in Sect. 6.

2 Preliminaries

We first present the formal definition of the annotation problem we address. Next we present a base workflow to determine annotation mappings for medical forms. This workflow has already been proposed in [5] and serves as a basis for our new approach that can reuse previous annotations (Sect. 3).

2.1 Problem Definition

We are given a set of medical forms $\mathcal{F}$ and an ontology $\mathcal{O}$. Each form $F \in \mathcal{F}$ consists of a set of items $\{i_1, i_2, \ldots i_k\}$ where each item has a question q and a response part. The response may be provided as free text or by selecting an answer from a list of possible values (as in Fig. 1). While the list of possible answers may include valuable information for the annotation of items, in this work we concentrate on using the question parts for finding suitable annotations. An ontology $\mathcal{O}$ consists of a set of concepts $C_{\mathcal{O}} = \{c_1, c_2, \ldots, c_l\}$ and a set of relations $R_{\mathcal{O}} = \{(c_1, c_2, rel\_type_1), \ldots (c_i, c_j, rel\_type_k)\}$ interrelating the ontology concepts by certain relationship types, e.g. $is-a$, $part-of$ or domain-specific relationships such as $is-located-in$. The concepts in $\mathcal{O}$ are typically described by an id, a label and several synonyms as shown on the right side of Fig. 1. The goal is to annotate each question (item) with one or several concepts from the given ontology $\mathcal{O}$. More specifically, we aim to determine an annotation mapping $\mathcal{M}_{F,\mathcal{O}} = \{(q, c, sim)|q \in F, c \in \mathcal{O}, sim[0, 1]\}$ for each form F. An annotation (q, c, sim) in these mappings indicates that question q is semantically described by concept c; the similarity value sim indicates the strength of the association according to the underlying method to compute the annotations.

Note that a question may be annotated by several concepts and that a concept may describe several questions. The challenge is to develop automatic methods that can determine annotation mappings of good quality (recall, precision). Ideally, all questions are correctly annotated, i.e. they are annotated with the ontology concepts that provide the best semantic description for the questions. A secondary goal is to efficiently determine the annotation mappings in a short time, even for large form collections and large ontologies.

2.2   Base Workflow

In our previous work [5] we used the basic workflow shown in Algorithm 1 to determine annotation mappings for medical forms. The input of the workflow is a set of forms $\mathcal{F}$, an ontology $\mathcal{O}$, and a similarity threshold δ. First, we normalize the label and synonyms of ontology concepts by removing stop words, transforming all string values to lower case and removing delimiters. The same preprocessing steps are applied for each form F_i. We identify an intermediate annotation mapping $\mathcal{M}'_{F_i,\mathcal{O}}$ by lexicographically comparing each question with the label and synonyms of ontology concepts. For this purpose we apply three string similarity measures, namely trigram, TF/IDF as well as a longest common sequence string similarity approach. We keep an annotation (q, c, sim), if the maximal similarity of the three string similarity approaches exceeds the threshold δ. Finally, we select annotations from the intermediate result by not only choosing the concepts with the highest similarity but also by considering the

---

Algorithm 1. annotation method $\mathcal{A}$

Input: Set of forms $\mathcal{F}$, ontology $\mathcal{O}= (C_\mathcal{O}, R_\mathcal{O})$, threshold δ
Output: Annotation mapping $\mathcal{M}_{\mathcal{F},\mathcal{O}}$

1 $\mathcal{O} \leftarrow$ preprocess $(\mathcal{O})$;
2 $\mathcal{M}_{\mathcal{F},\mathcal{O}} \leftarrow \emptyset$;
3 **foreach** $F_i \in \mathcal{F}$ **do**
4       $F_i \leftarrow$ preprocess (F_i);
5       $\mathcal{M}'_{F_i,\mathcal{O}} \leftarrow$ identifyCandidates $(F_i, C_\mathcal{O}, \delta)$;
6       $\mathcal{M}_{F_i,\mathcal{O}} \leftarrow$ selectAnnotations $(\mathcal{M}'_{F_i,\mathcal{O}})$;
7       $\mathcal{M}_{\mathcal{F},\mathcal{O}} \leftarrow \mathcal{M}_{\mathcal{F},\mathcal{O}} \cup \mathcal{M}_{F_i,\mathcal{O}}$;
8 **return** $\mathcal{M}_{\mathcal{F},\mathcal{O}}$;

---

similarity among the concepts. For this purpose, we group the concepts associated with a question based on their mutual similarity and only choose the concept with the highest similarity per group in order to avoid the redundant selection of highly similar concepts. This group-based selection proved to be quite effective in [5] albeit it only considers the string-based (linguistic) similarity between questions and concepts, and among concepts.

3 Reuse-Based Annotation Approach

In this section we outline an extended workflow to determine annotation mappings that reuses previously found annotations for similar questions. The goal is to reduce the complexity of the annotation problem by avoiding to search a very large ontology for finding concepts that describe or match terms of a question to annotate. By reusing verified annotations we also hope to achieve a good annotation quality since the previous annotations may include concepts that are difficult to find by common match techniques based on linguistic similarity. The reuse approach is also motivated by the existence of a high number of related forms in a specific domain, e.g. dealing with a specific disease. It would thus be desirable to reuse the annotations of a subset of these forms to more quickly and effectively annotate the remaining ones. The proposed approach is not limited to the annotation of medical forms but could be generalized for other medical documents such as electronic health records (EHRs) where we would associate medical concepts from an ontology to specific sentences or sections of the document rather than to questions.

We will first outline the new workflow for reuse-based annotation and then provide more details about its main steps, i.e., the generation of so-called annotation clusters (Sect. 3.2), determination of candidate annotations (Sect. 3.3) and a context-based strategy for selecting the final annotations (Sect. 3.4).

3.1 Workflow for Reuse-Based Annotation

The workflow for the reuse-based annotation approach is shown in Algorithm 2. Its input includes a set of verified annotation mappings containing the annotations for reuse. The result is a set of annotation mappings $\mathcal{M}_{\mathcal{F},\mathcal{O}}$ for the input forms $\mathcal{F}$ w.r.t. ontology $\mathcal{O}$. In the first step, we use the verified annotations to determine a set of *annotation clusters* $\mathcal{AC} = \{ac_{c_1}, ac_{c_2}, \ldots, ac_{c_m}\}$. For each concept c_i used in the verified annotations, we have an annotation cluster ac_{c_i} containing all questions that are associated to this concept. To calculate the similarity between an unannotated question and the questions of an annotation cluster we determine for each cluster a *representative* (feature set) $ac_{c_i}^{fs}$ consisting of relevant term groups in this cluster. These term groups are identified based on common terms between the questions $q \in ac_{c_i}$ and the description (label and synonyms) of the corresponding concept of ac_i.

After these initial steps we determine the annotation mapping for each unannotated input form F_i (lines 3–7 in Algorithm 2). We first preprocess a form as

---

Algorithm 2. Extended annotation method $\mathcal{A}^{reuse}$

Input: Set of unknown forms $\mathcal{F}$, ontology $\mathcal{O} = (C_{\mathcal{O}}, R_{\mathcal{O}})$, set of verified annotation mappings $\mathcal{M}_{\mathcal{F},\mathcal{O}}^{verified}$, threshold δ
Output: Annotation mapping $\mathcal{M}_{\mathcal{F},\mathcal{O}}$

1   $\mathcal{AC} \leftarrow$ determineAnnotationCluster $(\mathcal{M}_{\mathcal{F},\mathcal{O}}^{verified})$;
2   $\mathcal{AC} \leftarrow$ determineFeatureSets $(\mathcal{AC}, \mathcal{O})$;
3   $\mathcal{O} \leftarrow$ preprocess $(\mathcal{O})$;
4   **foreach** $F_i \in \mathcal{F}$ **do**
5 | $F_i \leftarrow$ preprocess (F_i);
6 | $\mathcal{M}_{F_i,\mathcal{O}}^{Reuse} \leftarrow$ identifyCandidatesByReuse $(F_i, \mathcal{AC}, \delta)$;
7 | $F_i' \leftarrow$ findUnannotatedQuestions $(F_i, \mathcal{M}_{F_i,\mathcal{O}}^{Reuse})$;
8 | $\mathcal{M}_{F_i',\mathcal{O}}^{reduced} \leftarrow$ identifyCandidates $(F_i', \mathcal{O}, \delta)$;
9 | $\mathcal{M}_{F_i,\mathcal{O}}' \leftarrow \mathcal{M}_{F_i',\mathcal{O}}^{reduced} \cup \mathcal{M}_{F_i,\mathcal{O}}^{Reuse}$;
10 | $\mathcal{M}_{F_i,\mathcal{O}} \leftarrow$ selectAnnotationsByContext $(\mathcal{M}_{F_i,\mathcal{O}}')$;
11 | $\mathcal{M}_{\mathcal{F},\mathcal{O}} \leftarrow \mathcal{M}_{\mathcal{F},\mathcal{O}} \cup \mathcal{M}_{F_i,\mathcal{O}}$;
12 **return** $\mathcal{M}_{\mathcal{F},\mathcal{O}}$;

---

in the base approach of Algorithm 1. Then we determine an annotation mapping $\mathcal{M}_{F_i,\mathcal{O}}^{Reuse}$ for the form based on the annotation clusters. Depending on the degree of reusable annotations the determined mapping is likely to be incomplete. We thus identify all questions that are not yet covered by the first mapping. For these questions we apply the base algorithm to match them to the whole ontology and obtain a second annotation mapping (line 7). We then take the union of the two partial mappings to obtain the intermediate mapping $\mathcal{M}_{F_i,\mathcal{O}}'$. Finally, we apply a new strategy to select the annotations for the final mapping $\mathcal{M}_{\mathcal{F},\mathcal{O}}$. This selection strategy considers the context of concepts, their linguistic similarity as well as their co-occurrences in previous annotations.

3.2   Generation of Annotation Clusters and Representatives

We build annotation clusters from verified annotation mappings by creating a cluster for each applied ontology concept c_k and associating to it all input questions that are assigned to this concept. Formally, an annotation cluster ac_{c_k} is represented as triple:

$$ac_{c_k} := (c_k, Q_{c_k}, ac_{c_k}^{fs}).$$

It includes the concept c_k, the set of questions Q_{c_k} annotated with c_k, as well as a cluster representative or feature set $ac_{c_k}^{fs}$. The purpose of the cluster representative is to provide a compact cluster description that is more suitable for finding further annotations than the free text questions or the label and synonym terms of the ontology concept.

A feature set is formed by terms or groups of terms that frequently co-occur in the questions of the cluster and that are similar to the synonym description

C0023467	$Q_{C0023467}$	$ac^{fs}_{C0023467}$
ANLL, AML, Acute myelocytic leukaemia, AML - Acute myeloid leukaemia, acute myelogenous leukemia (AML) ⋮	1. Previous induction-type chemotherapy for MDS or AML 2. Relapsed or treatment refractory AML 3. Patients with relapsed AML 4. Patients older than 60 years with acute myeloid leukemia according to FAB (>30 % bone marrow blasts) not qualifying for, or not consenting to, standard induction chemotherapy or immediate allografting	AML, acute myeloid leukemia, acute promyelocytic leukemia, acute myelodysplastic leukaemia ⋮
32 synonyms	25 questions	9 term groups

Fig. 2. Sample annotation cluster $ac_{C0023467}$ for UMLS concept *C0023467* with its set of associated questions $Q_{C0023467}$ and feature set $ac^{fs}_{C0023467}$.

of the corresponding concept. To identify frequently co-occurring terms, we use a frequent itemset mining algorithm where the frequency of term groups has to exceed a given *min_support*. Moreover, we only keep term groups that maximize the overlap between the terms of a question and the synonyms or the label of a concept, i.e., we do not use term groups that build a subset of another frequently occurring term group. The resulting feature sets build representatives for the annotation clusters that will be used to identify new annotations by matching unannotated forms to cluster representatives.

As an example, Fig. 2 shows the resulting annotation cluster $ac_{C0023467}$ for UMLS concept *C0023467* about the disease *Acute Myeloid Leukaemia*. In the UMLS ontology, this concept is described by a set of 32 synonyms (Fig. 2 left). The annotation cluster also contains 25 questions associated to this concept in the verified annotation mappings. Most questions only relate to some of the synonym terms of the concept while other synonyms remain unused. So the abbreviation 'AML' that is a part of some synonyms is often used but the abbreviation 'ANLL' does not occur in the medical forms used to build the annotation clusters. For this example, we generate only 9 relevant term groups, i.e., the representative feature set of the cluster is much more compact than the free text questions and large synonym set.

3.3 Identification of Annotation Candidates

To reuse the confirmed annotations for unannotated forms we have to determine the annotation clusters (and thus their concepts) that match best the new questions to be annotated. One difficulty is that we need to find several annotations per question, i.e., we aim at identifying several annotation clusters. Since we may find too many related annotation clusters it is also important to select the most promising ones from the set of candidates.

We first describe how we determine the set of candidate annotation clusters. The example in Fig. 1 showed that annotating concepts typically refer to some portion, i.e., succeeding terms, of the question text. Our approach to find matching annotation clusters thus uses a sliding window with a specified size *wnd_size* that partitions a given question into smaller portions according to the order of words in the question. Every text portion is compared with the feature

set of every existing annotation cluster using a linguistic similarity measure. For
this linguistic comparison we apply a soft TF/IDF string similarity function.
TF/IDF weights the different terms based on their significance in all considered
documents. A soft variant of TF/IDF is more robust than TF/IDF w.r.t. different word forms. An annotation cluster and thus its concept is an annotation
candidate for a given question, if the linguistic similarity exceeds a threshold δ
for one portion of the question.

In the final selection of annotations, we want to avoid choosing similar annotations referring to the same medical concept. We therefore group the annotation
candidates per question that relate to the same tokens and text portions of a
question. For selecting the best matching concept per candidate group we apply
the context-based selection strategy to be described next.

3.4   Context-Based Selection of Annotations

The input for the final selection of annotations is a set of grouped candidate
concepts for each question in the medical forms $\mathcal{F}$. To determine the final annotations per question, we rank the candidate concepts within each group based on
a combination of both linguistic and context-based similarity among the candidate concepts. For this purpose, we calculate an aggregated similarity ($aggSim$)
for each question and candidate concept based on weighted linguistic ($lsim$) and
context ($csim$) similarity scores:

$$aggSim_{q,Candidates}(c_k) = \omega_{lsim} \cdot lsim(q, c_k) + \omega_{csim} \cdot csim(c_k, Candidates)$$

The linguistic similarity between candidate concepts is determined by the
linguistic similarity of their concept descriptions, similarly as in the selection
strategy of the base approach (Sect. 2.2). The calculation of the context-based
similarity is more involved and will be described below. For each question in the
set of input forms, we select the concepts with the highest $aggSim$ value per
candidate group to obtain the final set of annotations.

For the context-based similarity between candidate concepts we consider two
criteria: first, the degree to which concepts co-occurred in the annotations for the
same question within the verified annotation mapping, and second, the degree of
semantic (contextual) relatedness of the concepts w.r.t. the ontological structure.
The goal is to give a high contextual similarity (and thus a high chance of being
selected) to frequently co-occurring concepts and to semantically close concepts.
These concepts are more likely to fit the context of a question which is typically
about one subject, e.g. different medical aspects such as medications for a specific
disease.

For the context-based selection of candidate concepts, we construct a *context
graph* $G_q = (V_q, E_q)$ for each question q. The vertices V_q represent candidate
concepts that are interconnected by two kinds of edges in E_q to express that
concepts have co-occurred in previous annotations or that concepts are semantically related within the ontology. In both cases we assign distance scores to
the edges that will be used to calculate the context similarity between concepts.

Figure 3a shows the sample input for annotation selection consisting of a question and the set of grouped candidate concepts. In the context graph of the question (Fig. 3b), green edges interconnect concepts that have co-occurred before and red edges interconnect semantically related concepts.

To determine the co-occurrence score between concepts c_1 and c_2 we count how often the two concepts have been annotated to the same question and compute the following normalized overlap of their annotation clusters:

$$cooccDist(c_1, c_2) = 1 - \frac{|ac_{c_1} \cap ac_{c_2}|}{|ac_{c_1}|}.$$

Concepts that often co-occur thus have a small distance score.

We further assign a semantic distance between concepts in the context graph based on the shortest path between two considered concepts in the ontological structure (see Fig. 3c), similarly to common techniques [18]. The ontological structure consists of the $is-a$, $part-of$ relationships and further domain specific relationships. We determine the semantic distance between two candidate concepts by summarizing the weighted distances of each relationship within the shortest path. We currently use the same distance 1 for each relationship type. Hence the semantic distance between two concepts corresponds to the path length, e.g., distance 4 for the concept pair in the example of Fig. 3c.

Based on the context graph and its distance scores we compute a context-based similarity for each concept by computing the distance to all other concepts in the candidate set of a question. Thereby, we favor concepts that often co-occur and those with a close semantic relatedness for our selection, i.e. selected concepts should have a small distance to other annotated concepts. We use the closeness centrality measure cc that computes the reciprocal of the sum of all distances d between a vertex v and all other vertices w in the graph G:

$$cc(v) = \frac{1}{\sum_{w \in G} d(v, w)}$$

We adopt a modified version of the closeness centrality to compute the context-based similarity score as follows. In our graph concepts can be isolated

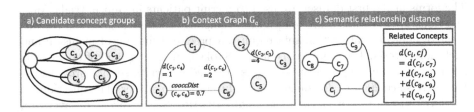

Fig. 3. Context-based similarity computation. (a) Candidate concept groups for one question; (b) context graph with different edges for concept co-occurrence (green edges) and semantic relatedness (red edges); (c) computation of semantic relatedness between concepts with related concepts from UMLS. (Color figure online)

in case they do not co-occur with any other concepts and have a very different semantic context (e.g., concept c_5 in the context graph of Fig. 3b). Such isolated concepts should get a lower similarity score than concepts in a larger subgraph of G_q. However, isolated concepts have infinite distances d to all other vertices such that $cc(v)$ would often converge to zero. To compute a normalized context-based similarity score $csim(c_i) \in [0,1]$ for each concept c_i in the set of vertices V of the context-graph G_q, we sum up single reciprocal values of distances and normalize it by the number of concepts in the context-graph:

$$csim(c_i, V) = \frac{\sum_{c_j \in V \setminus \{c_i\}} \frac{1}{d(c_i, c_j)}}{|V| - 1}$$

Concepts with a small distance to every other concept in the graph have high $csim$ values meaning they are highly related to the other candidate concepts due to annotation co-occurrences and relationships from UMLS.

For instance, the context similarity for the concept c_4 is computed by the semantic distance $d(c_4, c_1) = 1$ and the co-occurrence distance $cooccDist(c_4, c_6) = 0.7$. The distances to the other concepts in the context graph are infinite. Therefore, we get the following context-similarity $csim(c_4) = \frac{\frac{1}{1} + \frac{1}{0.7} + \frac{1}{\infty} + \frac{1}{\infty} + \frac{1}{\infty}}{6-1} \approx 0.49$.

4   Evaluation

We now evaluate the proposed reuse-based annotation approach for medical forms and compare it with the baseline approach and the MetaMap tool. In the next subsection we introduce the used datasets and workflow configurations. We then evaluate the annotation quality compared to the baseline approach (Sect. 4.2) and analyze the effectiveness of the context-based selection strategy (Sect. 4.3). Finally, we provide the comparison with MetaMap (Sect. 4.4).

4.1   Evaluation Setting

Our evaluation uses medical forms about eligibility criteria EC and about quality assurance QA w.r.t cardiovascular procedures from the MDM platform [4]. The forms in the first dataset are used to recruit patients in clinical trials. Most questions in this dataset are long natural language sentences since the recruitment of clinical trial participants requires a precise definition of inclusion and exclusion criteria. The sentences contain ∼8 tokens on average and often mention several medical concepts. The QA forms are used by health service providers in Germany since 2000 to document the quality of their services. The questions of the QA forms are shorter than the eligibility criteria (∼3 tokens on average), therefore a question is probably annotated with only one concept. The forms will be annotated with concepts of a reduced version of UMLS [3] covering all UMLS concepts that possess at least one preferred label or synonym (∼1 Mio. concepts with ∼7 Mio. labels/synonyms). Moreover, we do not consider general concepts (∼12000 concepts) that are associated with one of the following

Table 1. Statistics on the reuse and evaluation datasets for EC and QA

Dataset	EC_{RD1}	EC_{RD2}	EC_{eval}	QA_{RD1}	QA_{RD2}	QA_{Eval}
#forms	100	200	25	16	32	23
#items	1638	3125	310	453	795	609
#annotations	6911	13027	578	694	1054	668

semantic types: *Qualitative Concept, Quantitative Concept, Functional Concept, Conceptual Entity.*

To evaluate the quality of automatically generated annotations, we use manually created reference mappings from the MDM portal [4]. These reference mappings might not be perfect ("a silver standard") since the huge size of UMLS makes it hard to manually identify the most suitable concepts for each item. We divide the set of input forms into disjoint reuse and evaluation datasets. For both use cases, EC and QA, we consider two reuse datasets of different sizes to study the impact of the amount of reusable annotations. Table 1 shows the number of forms, items and verified annotations for the reuse and evaluation datasets. To analyze the quality of the resulting annotation mappings, we compute precision, recall and F-measure using the union of all annotated form items in the evaluation dataset.

For our reuse-based annotation workflow, we set a fixed window size *wnd_size* of five tokens for the *Candidate Identification* and fixed weights ω_lsim/ω_csim to 0.5 for the *Context-based Selection*. In our experiments, we observe that these parameters only slightly affected the results for the considered datasets. We evaluate different thresholds $\delta = \{0.5, 0.6, 0.7\}$ to present the recall and precision trends. For the selection strategy we consider both the previously proposed group-based strategy [5] as well as the new context-based strategy. Note that we can use the group-based strategy not only for the base workflow but also in the reuse-based approach by setting the weight ω_{csim} for the context similarity to 0.

4.2   Reuse-Based Annotation

Figure 4 shows evaluation results w.r.t. the mapping quality (precision, recall, F-measure) for the baseline approach and the different configurations of the reuse-based approach for the two datasets. For the baseline approach we only show the results for the best threshold of $\delta = 0.7$ for both datasets. The reuse-based approaches uses the context-based selection strategy. We observe that the reuse-based approach can significantly improve the annotation quality and that the improvement grows with the amount of annotations that we can reuse. Compared to the baseline approach, the reuse of existing annotations increases the F-measure from 39.1 % to 50.7 % for the EC dataset and from 57.5 % to 59 % for the QA dataset for the best threshold setting of $\delta = 0.6$. Using more existing annotations (EC_{RD2} and QA_{RD2}) improves the mapping quality - and especially recall - compared to the smaller reuse datasets (EC_{RD1} and QA_{RD1}) since annotation clusters and their feature sets become more accurate and are thus

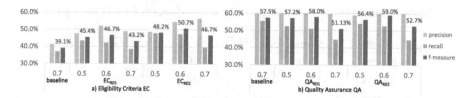

Fig. 4. Results on the quality of annotation results for the baseline and reuse-based annotation using the *EC* dataset and the *QA* dataset with both configurations.

more valuable to match to unannotated questions. The reuse-based approach is especially effective for the EC dataset where we could apply more annotations (Table 1) to build the annotation clusters compared to the QA dataset. The results confirm that matching questions to the feature sets of annotation clusters (*reuse-based*) helps to identify more correct annotations than trying to find the best matches in the UMLS (*baseline*). At the same time, the reuse-based approaches with the context-based selection strategy usually improve precision compared to the baseline approach.

An added benefit is that the execution time of the reuse-based approaches is lower than for the baseline approaches since matching questions with the compact annotation clusters is much faster than matching with the large UMLS ontology. Overall, runtimes could be reduced by half for our experiments compared to the baseline. Moreover, the execution time depends on the number of reused forms and the coverage of reused annotation clusters.

4.3   Context-Based Selection

To analyze the effectiveness of the proposed context-based selection strategy (CS), we now compare its use with the group-based selection strategy (GS) that was used in the baseline approach but can also be applied for the reuse-based approaches. Table 2 shows the resulting mapping quality for the two selection strategies for the different EC and QA reuse configurations and threshold 0.6 that led to the best mapping quality for the reuse-based approach. The results show that the context-based selection strategy improves F-measure in all cases (up to 2.2%) compared to the simpler group-based approach. While recall is generally reduced this is more than outweighed by an increase in precision by up to almost ~7% (EC_{RD2}). This indicates that considering the context eliminates many false candidates.

Table 2. Results on the quality of annotation results for the group-based (GS) and context-based (CS) selection strategies for both datasets

Dataset$_{configuration}$	EC_{RD1}		EC_{RD2}		QA_{RD1}		QA_{RD2}	
Selection-strategy	gs	cs	gs	cs	gs	cs	gs	cs
Precision	45.9%	**52.1%**	47.9%	**54.5%**	61.9%	**67.0%**	60.4%	**66.9%**
Recall	**43.6%**	42.2%	**49.2%**	47.3%	51.0%	**51.2%**	**54.6%**	52.8%
f-measure	44.7%	**46.7%**	48.5%	**50.7%**	55.9%	**58.0%**	57.4%	**59.0%**

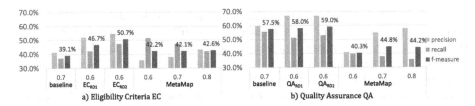

Fig. 5. Comparison of the quality for the resulting annotation mappings from the baseline approach, reuse-based approach and MetaMap.

4.4 Comparing Reuse-Based Annotation Approach with MetaMap

We finally compare our reuse-based annotation method with the MetaMap tool that is commonly used for annotating medical documents (see Sect. 5). We generate the annotations with a local installation of a MetaMap server and the MetaMapAPI and use the provided word sense disambiguation service and the configuration considering several variants for a concept. We select annotations based on the generated MetaMap score. This score ranges from 0 to 1000 and is computed by applying several ranking functions for each identified term. If MetaMap generates more than one annotation per question, we select the annotations with an aggregated score above a threshold. We normalize the scores by dividing by 1000 for comparing with our approach and evaluate different thresholds $\delta = \{0.6, 0.7, 0.8\}$ for selecting the candidates.

Figure 5 shows the results for the two datasets and different configurations. Our reuse-based approach outperforms Meta-Map in terms of mapping quality for each dataset. For the EC dataset, F-Measure is improved by $\sim$4 % (EC_{RD1}) and $\sim$8.6 % (EC_{RD2}) indicating that the computed annotation clusters allow a more effective identification of annotations than with the original concept definition. In addition, our approach benefits from using the ontological relationships for selecting annotations resulting in a much better precision than using MetaMap (54.5 % for EC_{RD2} than compared to 43.1 %). While MetaMap achieved a better F-Measure than the baseline approach for the EC dataset it performed poorly for the QA dataset where its best F-Measure of 44.8 % was much lower for the baseline approach and reuse-based approaches (57.5 and 59 %), mainly because of a very low recall for Metamap.

A positive side of MetaMap is its high performance due to the use of an indexed database for finding annotations. Its runtimes were up to 13 times faster than for the baseline approach and it was also faster than the reuse-based approach. In future work we will study whether the use of MetaMap in combination with the reuse approach, either as an alternative or in addition to the baseline approach, can further improve the annotation quality.

5   Related Work

The automatic annotation of medical forms and documents with concepts of standardized vocabularies is related to the well-studied fields of ontology matching [8,20] and entity linking [22]. Both research domains provide useful generic methods to identify concepts or names in full-text documents and match them to concepts or entities of a knowledge base or standardized vocabulary. These techniques can also be applied to the medical domain. In fact, our base workflow proposed in [5] uses linguistic ontology matching techniques to map terms of medical forms to the concepts and their synonyms of the UMLS ontology. Entity linking approaches focus on the identification of named entities in text documents and their linking to corresponding entities of a knowledge base for enrichment. Many approaches (e.g. [6,16,24]) use a dictionary-based strategy to identify entity occurrences by searching the whole knowledge base.

Moreover, there are many approaches to select the correct entities from a set of candidates (e.g. [6,9,11]). For instance, in [9] co-occurrences of entities in Wikipedia articles are transformed into a graph model to consider the global interdependence between different candidate entities in a document.

There is also some research focusing on the manual or automatic annotation of certain kinds of medical documents. The MetaMap tool [1] considered in our evaluation applies information retrieval methods such as tokenization, lexical lookup and term-based ranking methods to retrieve UMLS concepts within medical documents. There is evidence in the literature that MetaMap results are not fine-grained enough [15], contain many spurious annotations [19] and do not cover mappings to longer medical terms [21]. These observations confirm that a correct annotation of medical documents with UMLS concepts is challenging. Our reuse-based approach could significantly outperform MetaMap due to its use of annotation clusters derived from verified annotations and due to its context-based approach to select and disambiguate concept candidates.

In the medical domain, the standardization of eligibility criteria has become an active field of research and datasets from this subdomain are often used for method evaluation (e.g. [10,13,17,23]). For instance, the study in [23] identified the most frequent ECs in clinical trial forms and performed a manual annotation of eligibility criteria top terms. In [10], similar clinical trials have been clustered by performing nearest neighbor search using annotated eligibility criteria, and the application of a dictionary-based pre-annotation method [13] showed to improve the speed of manual annotation for clinical trial announcements. In [17], a set of eligibility criteria in the context of clinical trials on breast cancer is formalized by defining eligibility criteria specific patterns in order to improve their comparability.

In contrast to previous research we propose a novel reuse-based annotation approach for medical documents. Our method is especially valuable to annotate documents from different biomedical domains with ontology concepts, i.e. it is not restricted to a specific medical subdomain. The proposed use of annotation clusters and their feature sets has not been explored before. Furthermore, we apply a novel context-based selection of annotations considering both, the co-occurrences of verified annotations as well as the semantic relatedness of concepts. Our comparative evaluation showed that the new approaches outperform previous annotation schemes including tools like MetaMap.

6   Conclusion

We proposed and evaluated a new reuse-based approach to semantically annotate medical documents such as CRFs with concepts of an ontology. The approach utilizes already found and verified annotations for similar CRFs. It builds so-called annotation clusters combining all previously annotated questions related to the same medical concept. Clusters are represented by features covering meaningful term groups from the annotated questions and concept description. New questions are matched with these cluster representatives to find candidates for annotating concepts. We further presented a context-based selection strategy to identify the most promising annotations based on the semantic relatedness of concept candidates and well as known co-occurrences from previous annotations. In a real-world evaluation, our methods showed to be effective and we could generate valuable recommendations to reduce the manual annotation effort. Moreover, reusing annotation clusters is more efficient than searching a large knowledge base such as UMLS for suitable annotation candidates.

For future work, we plan to evaluate further annotation approaches, in particular the combined use of several reuse-based and other techniques. For example, the MetaMap tool alone was inferior to the reuse-based scheme but it could be used in a combined scheme to find further annotation candidates. We also plan to build a reuse repository covering annotation clusters and their feature sets for different medical subdomains. Such a repository can be used to efficiently and effectively identify annotations for new medical documents. It further enables a semantic search for existing medical document annotations. This can be useful to define new medical forms by finding and reusing suitable annotated items instead of creating new forms from scratch.

Acknowledgment. This work is funded by the German Research Foundation (DFG) (grant RA 497/22-1, "ELISA - Evolution of Semantic Annotations").

References

1. Aronson, A.R., Lang, F.-M.: An overview of MetaMap: historical perspective and recent advances. J. Am. Med. Inform. Assoc. **17**(3), 229–236 (2010)
2. Ashburner, M., Ball, C.A., Blake, J.A., Botstein, D., Butler, H., et al.: Gene ontology: tool for the unification of biology. Nat. Genet. **25**(1), 25–29 (2000)
3. Bodenreider, O.: The Unified Medical Language System (UMLS): integrating biomedical terminology. Nucleic Acids Res. **32**(suppl. 1), D267–D270 (2004)
4. Breil, B., Kenneweg, J., Fritz, F., et al.: Multilingual medical data models in ODM format-a novel form-based approach to semantic interoperability between routine health-care and clinical research. Appl. Clin. Inf. **3**, 276–289 (2012)
5. Christen, V., Groß, A., Varghese, J., Dugas, M., Rahm, E.: Annotating medical forms using UMLS. In: Ashish, N., Ambite, J.-L. (eds.) DILS 2015. LNCS, vol. 9162, pp. 55–69. Springer, Heidelberg (2015)
6. Cucerzan, S.: Large-scale named entity disambiguation based on wikipedia data. In: Proceedings of Joint Conference on Empirical Methods in Natural Language Processing and Computational Natural Language Learning (EMNLP-CoNLL), pp. 708–716 (2007)

7. Donnelly, K.: SNOMED-CT: the advanced terminology and coding system for eHealth. Stud. Health Technol. Inform.-Med. Care Compunetics **3**(121), 279–290 (2006)

8. Euzenat, J., Shvaiko, P.: Ontology Matching. Springer, Heidelberg (2007)

9. Han, X., Sun, L., Zhao, J.: Collective entity linking in web text: a graph-based method. In: Proceedings of the 34th International ACM SIGIR Conference, pp. 765–774 (2011)

10. Hao, T., Rusanov, A., Boland, M.R., et al.: Clustering clinical trials with similar eligibility criteria features. J. Biomed. Inform. **52**, 112–120 (2014)

11. Kulkarni, S., Singh, A., Ramakrishnan, G., Chakrabarti, S.: Collective annotation of Wikipedia entities in web text. In: Proceedings of the 15th ACM SIGKDD Conference, pp. 457–466 (2009)

12. LePendu, P., Iyer, S., Fairon, C., Shah, N.H., et al.: Annotation analysis for testing drug safety signals using unstructured clinical notes. J. Biomed. Semant. **3**(S–1), S5 (2012)

13. Lingren, T., Deleger, L., Molnar, K., et al.: Evaluating the impact of pre-annotation on annotation speed and potential bias: natural language processing gold standard development for clinical named entity recognition in clinical trial announcements. J. Am. Med. Inform. Assoc. **21**(3), 406–413 (2014)

14. Lowe, H.J., Barnett, G.O.: Understanding and using the medical subject headings (MeSH) vocabulary to perform literature searches. J. Am. Med. Assoc. (JAMA) **271**(14), 1103–1108 (1994)

15. Luo, Z., Duffy, R., Johnson, S., Weng, C.: Corpus-based approach to creating a semantic lexicon for clinical research eligibility criteria from UMLS. AMIA Summits Transl. Sci. Proc. **2010**, 26 (2010)

16. Mihalcea, R., Csomai, A.: Wikify! linking documents to encyclopedic knowledge. In: Proceedings of the 16th ACM CIKM, pp. 233–242 (2007)

17. Milian, K., Hoekstra, R., Bucur, A., ten Teije, A., van Harmelen, F., Paulissen, J.: Enhancing reuse of structured eligibility criteria and supporting their relaxation. J. Biomed. Inform. **56**, 205–219 (2015)

18. Pesquita, C., Faria, D., Falcao, A.O., Lord, P., Couto, F.M.: Semantic similarity in biomedical ontologies. PLoS Comput. Biol. **5**(7), e1000443 (2009)

19. Ogren, P., Savova, G., Chute, C.: Constructing evaluation corpora for automated clinical named entity recognition. In: Proceedings of the (LREC) Conference, pp. 3143–3150 (2008)

20. Rahm, E.: Towards large-scale schema and ontology matching. In: Bellahsene, Z., Bonifati, A., Rahm, E. (eds.) Schema Matching and Mapping. Data-Centric Systems and Applications, pp. 3–27. Springer, Heidelberg (2011)

21. Ren, K., Lai, A.M., Mukhopadhyay, A., et al.: Effectively processing medical term queries on the UMLS Metathesaurus by layered dynamic programming. BMC Med. Genomics **7**(Suppl. 1), 1–12 (2014)

22. Shen, W., Wang, J., Han, J.: Entity linking with a knowledge base: issues, techniques, and solutions. IEEE Trans. Knowl. Data Eng. **27**(2), 443–460 (2015)

23. Varghese, J., Dugas, M., et al.: Frequency analysis of medical concepts in clinical trials and their coverage in MeSH and SNOMED-CT. Meth. Inf. Med. **54**(1), 83–92 (2015)

24. Zhang, W., Tan, C.L., Sim, Y.C., Su, J.: NUS-I2R: learning a combined system for entity linking. In: Proceedings of the 3rd Text Analysis Conference (TAC), NIST (2010)

Knowledge Representation on the Web Revisited: The Case for Prototypes

Michael Cochez[1,2,4]($\boxtimes$), Stefan Decker[1,2], and Eric Prud'hommeaux[3]

[1] Fraunhofer Institute for Applied Information Technology FIT,
53754 Sankt Augustin, Germany
{michael.cochez,stefan.decker}@fit.fraunhofer.de
[2] Informatik 5, RWTH Aachen University, 52056 Aachen, Germany
[3] World Wide Web Consortium (W3C), Stata Center, MIT, Cambridge, USA
eric@w3.org
[4] Department of Mathematical Information Technology, University of Jyvaskyla,
40014 University of Jyvaskyla, Finland

Abstract. Recently, RDF and OWL have become the most common knowledge representation languages in use on the Web, propelled by the recommendation of the W3C. In this paper we examine an alternative way to represent knowledge based on Prototypes. This Prototype-based representation has different properties, which we argue to be more suitable for data sharing and reuse on the Web. Prototypes avoid the distinction between classes and instances and provide a means for object-based data sharing and reuse.

In this paper we discuss the requirements and design principles for Knowledge Representation based on Prototypes on the Web, after which we propose a formal syntax and semantics. We further show how to embed knowledge representation based on Prototypes in the current Semantic Web stack and report on an implementation and practical evaluation of the system.

Keywords: Linked data · Knowledge representation · Prototypes

1 Introduction and Motivation

In earlier days of Knowledge Representation, *Frames* [19,20] and Semantic Networks [23] were accepted methods of representing static knowledge. These had no formal semantics but subsequent works (e.g., KL-ONE [2]) introduced reasoning with concepts, roles, and inheritance, culminating in Hayes's 1979 [10] formalization of Frames. This formalization included instances formalized as elements of a domain (individuals) and classes (or concepts) as sets in a domain (unary predicates). This formalization was subsequently used as a basis for Description Logics (DL) and the investigation of expressiveness vs. tractability [16], which lead to Description Logic systems and reasoners such as SHIQ [11] and FaCT [12]. Finally, the Semantic Web effort led to the combination of Description Logics with Web Technologies such as RDF [6], which subsequently

© Springer International Publishing AG 2016
P. Groth et al. (Eds.): ISWC 2016, Part I, LNCS 9981, pp. 151–166, 2016.
DOI: 10.1007/978-3-319-46523-4_10

evolved into the Web Ontology Language OWL [13]. However, the formalization of Frames only covered some modeling primitives which were in use at the time. Specifically Prototype-based systems, which do not make a distinction between instances and classes, did not get much attention for knowledge representation (cf. Karp [15]). Exceptions exists, for instance, THEO [21], which is a Frame Knowledge Representation System deviating from the — now common — instance–class distinction by using only one type of frame, with the authors arguing that the distinction between instances and classes is not always well defined. Also several programming Languages based on prototypes were successfully developed (SELF [27], JavaScript [8] and others), but the notion of Prototypes as a Knowledge Representation mechanism was not formalized and remained unused in further developments. As noted in [24], these knowledge representation mechanisms may now be again relevant for applications. In this paper we develop a syntax and formal semantics for a language based on prototypes for the purpose of enabling knowledge representation and knowledge sharing on the Web. We argue that such a system has distinctive advantages compared to other representation languages.

This paper is augmented by a separate technical report in which we detail the software which we wrote to support prototype knowledge representation. [3] The report also includes experiments which show how the system performs in a web environment.

2   A Linked Prototype Layer on the Web

2.1   Idea and Vision

Tim Berners-Lee stated the motivation for creating the Web as:

> The dream behind the Web is of a common information space in which we communicate by sharing information.[1]

We aim to optimize the sharing and reuse of structured data. Currently, on the Semantic Web, this sharing is typically achieved by either querying a SPARQL endpoint or downloading a graph or an ontology. We call this *vertical sharing*: top-down sharing where a central authority or institution shares an ontology or graphs. We would like to enable *horizontal sharing*: sharing between peers where individual pieces of instance data can be used and reused. Note that this mode of sharing appears much closer to the intended spirit of the Web. Languages like OWL evolved driven by the AI goal of intelligent behavior and sound logical reasoning [14]. They don't emphasize or enable horizontal sharing - the sharing and reuse of individual objects in a distributed environment. Rather, their goal is to represent axioms and enable machines to reason. Imagine a prototype, for example, an `Oil Painting` with properties and values for those properties, that lives at a particular addressable location on the Web. This prototype `Oil Painting` can be reused in a number of different ways (see Fig. 1):

---

[1] https://www.w3.org/People/Berners-Lee/ShortHistory.html.

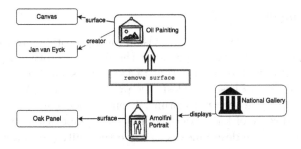

Fig. 1. The figure shows three prototypes and different relations between them. The Arnolfini Portrait is a specialization of the Oil Painting but also displayed at the National Gallery, London.

- First, by specializing the Oil Painting prototype(i.e., using it as a template by linking to it), and either specializing or changing its properties. For example, whereas the Oil Painting has a value Canvas for its surface property, the Arnolfini Portrait prototype has the value Oak Panel. However, the value for the creator property (Jan van Eyck) remains the same. To accomplish this, current Semantic Web infrastructure would require one to copy the initial object to a new object before changing its properties. Note, however, that a this also means that the newly created object looses its heritage, meaning that it will not receive any updates which are made to object in the inheritance chain later on.
- Second, by either directly or indirectly referring to it as a value of a property. For instance, in Fig. 1 the prototype National Gallery has a property displays, which links to the prototype Arnolfini Portrait, which is based on the Oil Painting prototype. This usage of entities is currently also possible using RDF. (But, see also the discussion in Sect. 4.3.)

These two ways to reusing objects on the Web create a distributed network of interlinked objects, requiring horizontal as well as vertical sharing:

- Vertical sharing is enabled by specializing an object or prototype. The prototype that is being specialized defines the vocabulary and structure for the new object, realizing the task of ontologies. For example, a museum can publish a collection of prototypes that describe the types of artifacts on display (e.g., Oil Painting), which can then be used to describe more specific objects.
- Horizontal sharing is enabled by reusing prototypes and only changing specific attributes or linking to other prototypes as attribute values. For example, a specific oil painting by painter Jan van Eyck can be used as a template by describing how other oil paintings differ from it, or a specific oil painting can be the attribute value for the National Gallery prototype. This creates a network of prototypes across the Web.

2.2 Requirements

In the previous section, we presented a vision for a prototype layer on the Web. In this section we discuss requirements for the linked prototype layer. Some of these requirements are based on actual tasks that user communities want to perform while others are based on desirable principles of the World Wide Web.

The linked prototype layer must primarily enable *sharing and reuse of knowledge*. Sharing and reuse of knowledge requires an explicit distributed network of entities. In particular we desire means to share vertically (i.e., provide a central vocabulary or ontology that many can refer to) and share horizontally (i.e., provide concrete reusable entities). Further, it must be possible for the knowledge to evolve over time and anyone should be able to define parts of the network. This implies that central authority should be avoided as much as possible. Preferably, the realization of the prototype layer should be achieved using facilities which the Semantic Web already provides, such as RDF and IRIs, in order to leverage existing data resources. Finally, the designed system should still retain a certain level of familiarity.

2.3 Design Principles

While designing the prototype-based system, we were inspired by design principles, such as the KISS Principle (as defined in [28]), and *worse-is-better* (as coined by R.P. Gabriel [9]). On the intersection between these principles lies the idea of simplicity. The worse-is-better approach encourages dropping parts of the design that would cause complexity or inconsistency.

Our goal was explicitly not to enable sophisticated reasoning, but rather provide a simple object or prototype layer for the Web.

We use the idea of prototypes as suggested in early Frame Systems [15] as well as in current programming languages such as Javascript [8]. Prototypes fulfill the requirements to support the reusabilty and horizontal shareability since it is possible to just refer to an existing prototype that exists elsewhere on the Web, ensuring horizontal shareability. Furthermore a collection of prototypes published by an authority can still serve the function of a central ontology, ensuring vertical shareability.

3 Prototypes

In this section we introduce our approach for knowledge representation on the web, based on prototypes. First, we provide an informal overview of the approach, illustrating the main concepts. Then we introduce a formal syntax and semantics.

3.1 Informal Presentation

To illustrate the prototype system we use an example about two Early Netherlandish painters, the brothers *van Eyck*. First, we look at a simple

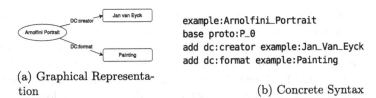

```
example:Arnolfini_Portrait
base proto:P_0
add dc:creator example:Jan_Van_Eyck
add dc:format example:Painting
```

(a) Graphical Representation

(b) Concrete Syntax

Fig. 2. The prototype representation of the Arnolfini Portrait

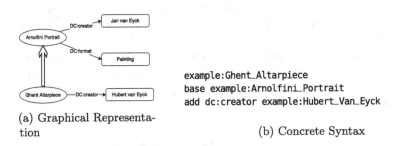

```
example:Ghent_Altarpiece
base example:Arnolfini_Portrait
add dc:creator example:Hubert_Van_Eyck
```

(a) Graphical Representation

(b) Concrete Syntax

Fig. 3. Deriving the prototype representation of the Ghent Altarpiece from the Arnolfini Portrait

representation of the *Arnolfini Portrait* in Fig. 2.[2] This figure contains the prototype of the portrait which is derived from the empty prototype (P$_\emptyset$, see Sect. 3.2) and has two properties. The first property is `dc:creator` and has value Jan van Eyck[3]. The second property describes the format of the artwork. We also display the example using a concrete syntax.

Next we will start making use of the prototype nature of the representation. Starting from the Arnolfini Portrait, we derive the *Ghent Alterpiece*. This painting was created by the same painter, but also his brother *Hubert van Eyck* was involved in the creation of the work. Figure 3 illustrates how this inheritance works in practice; we create a prototype for the second work and indicate that its base is the first one (using the big open arrow). Then, we add a property asserting that the other brother is also a creator of the work. The resulting prototype has the properties we defined directly as well as those inherited from its base.

---

[2] In the illustrations, we loosely write identifiers like **Arnolfini Portrait** for prototypes, properties and their values. However, the proposed systems requires the use of IRIs for identifiers, just like RDF. The concrete syntax examples reflect this. Note that our syntax does not support prefixes as supported by RDF Turtle syntax. If we write `dc:creator` we mean an IRI with scheme `dc`.

[3] For illustrative purposes we use different graphical shapes for the prototypes under consideration and the values of their properties. However, as will become clear in the sections below, all values are themselves prototypes.

Often, there will be a case where the base prototype has properties which are not correct for the derived prototype. In the example shown in Fig. 4 we added the `example:location` property to the Arnolfini Portrait with the value `National Gallery, London`. The Ghent Altarpiece is, however, located in the `Saint Bavo Cathedral, Ghent`. Hence, we first remove the `example:location` property from the Arnolfini Portrait before we add the correct location to the second painting. In effect, the resulting prototype inherits the properties of its base, can remove unneeded ones, and add its own properties as needed.

Another way to arrive at the same final state would be to derive from a base without any properties and add all the properties needed. The predefined empty prototype (`proto:P_0`) has no properties. All other prototypes derive from an already existing prototype; circular derivation is not permitted. Now, we will let the prototype which we are creating derive directly from the empty prototype and add properties. This flattening of inherited properties produces the prototype's *fixpoint*. The fixpoint of the prototype created in Fig. 4 can be found in Fig. 5.

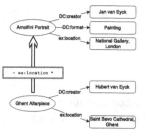

```
example:Arnolfini_Portrait
base proto:P_0
add dc:creator example:Jan_Van_Eyck
add dc:format example:Painting
add example:location example:National_Gallery

example:Ghent_Altarpiece
base example:Arnolfini_Portrait
rem example:location *
add dc:creator example:Hubert_Van_Eyck
add example:location example:Saint_Bavo
```

(a) Graphical Representation

(b) Concrete Syntax

Fig. 4. Removing properties while deriving the Ghent Altarpiece from the Arnolfini Portrait

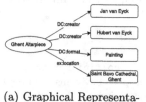

```
example:Ghent_Altarpiece
base proto:P_0
add dc:creator example:Jan_Van_Eyck
add dc:format example:Painting
add dc:creator example:Hubert_Van_Eyck
add example:location example:Saint_Bavo
```

(a) Graphical Representation

(b) Concrete Syntax

Fig. 5. The result of removing properties while deriving the Ghent Altarpiece from the Arnolfini Portrait.

In the proposed system we apply the closed world and the unique name assumptions. If the system used the open world assumption and one would ask whether the Arnolfini Portrait is located in Beijing, the system would only be able to answer that it does not know. In a closed world setting, the system will answer that the painting is not in Beijing. This conclusion is not based on the fact that the system sees that the painting is located in England, but because of the fact that there is no indication that it would be in Beijing. Under the *non*-unique name assumption, the system would not be able to answer how many paintings it knows about. Instead, it would only be able to tell that there are one or more. Without the unique name assumption, the resource names `Ghent Altarpiece` and `Arnolfini Portrait` may refer to the same real-world instance.

3.2 Formal Presentation

The goal of this section is to give a formal presentation of the concepts discussed in the previous section. We separate the formal definition into two parts. First, we define the syntax of our prototype language. Then, we present the semantic interpretation and a couple of definitions which we used informally above.

Prototype Syntax. In this section we define the formal syntax of prototype-based knowledge bases. We define a set of syntactic material first, before we define the language.

Definition 1 (Prototype Expressions). *Let ID be a set of absolute IRIs according to RFC 3987 [7] without the IRI* `proto:P_0`*. The IRI* `proto:P_0` *is the empty prototype and will be denoted as* $P_\emptyset$*. We define expressions as follows:*

- *Let $p \in ID$ and $r_1, \ldots, r_m \in ID$ with $1 \leq m$. An expression $(p, \{r_1, \ldots, r_m\})$ or $(p, *)$ is called a* simple change expression*. p is called the simple change expression ID, or its property. The set $\{r_1, \ldots, r_m\}$ or $*$ are called the* values *of the simple change expression.*
- *Let $id \in ID$ and $base \in ID \cup P_\emptyset$ and add and $remove$ be two sets of simple change expressions (called change expressions) such that each simple change expression ID occurs at most once in each of the add and remove sets and $*$ does not occur in the add set. An expression $(id, (base, add, remove))$ is called a* prototype expression*. id is called the prototype expression ID.*

Let `PROTO` *be the set of all prototype expressions. The tuple $PL = (P_\emptyset, ID, \text{PROTO})$ is called the* Prototype Language*.*

Informally, a prototype expression contains the parts of a prototype which we introduced in the previous subsection. It has an id, a base (a reference to the prototype it derives from), and a description of the properties which are added and removed.

As an example, we could write down the example of Fig. 4 using this syntax. The prototype expression of the Arnolfini Portrait would look like this:

```
(example:Arnolfini_Portrait,(proto:P_0,
{(dc:creator,{example:Jan_Van_Eyck}),
(dc:format,{example:Painting}),
(example:location,{example:National_Gallery})},
∅))
```

The prototype for the Altarpiece would be written down as follows:

```
(example:Ghent_Altarpiece,(example:Arnolfini_Portrait,
{(dc:creator,{example:Hubert_Van_Eyck}),
(example:location,{example:Saint_Bavo})},
{(example:location,*)}))
```

This syntax is trivially transformable into the concrete syntax which we used in Fig. 4b and the other examples in the previous subsection.

Definition 2 (*dom*). *The domain of a finite subset* $S \subseteq$ PROTO, *i.e.,* $dom(S)$ *is the set of the prototype expression IDs of all prototype expressions in* S.

Definition 3 (Grounded). *Let* $PL = (\text{P}_\emptyset, ID, \text{PROTO})$ *be the Prototype Language. Let* $S \subseteq$ PROTO *be a finite subset of* PROTO. *The set* $\mathcal{G}$ *is defined as:*

1. $\text{P}_\emptyset \in \mathcal{G}$
2. If there is a prototype $(id, (base, add, remove)) \in S$ *and* $base \in \mathcal{G}$ *then* $id \in \mathcal{G}$.
3. $\mathcal{G}$ *is the smallest set satisfying (1) and (2).*

S *is called grounded iff* $\mathcal{G} = dom(S) \cup \{\text{P}_\emptyset\}$. *This condition ensures that all prototypes derive (recursively) from* $\text{P}_\emptyset$ *and hence ensures that no cycles occur.*

To illustrate how cycles are avoided by this definition, imagine that $S = \{(A, (\text{P}_\emptyset, \emptyset, \emptyset)), (B, (C, \emptyset, \emptyset)), (C, (B, \emptyset, \emptyset)), \}$. What we see is that there is a cycle between B and C. If we now construct the set $\mathcal{G}$, we get $\mathcal{G} = \{\text{P}_\emptyset, A\}$ while $dom(S) \cup \{\text{P}_\emptyset\} = \{A, B, C, \text{P}_\emptyset\}$, and hence the condition for being grounded is not fulfilled.

Definition 4 (Prototype Knowledge Base). *Let* $PL = (\text{P}_\emptyset, ID, \text{PROTO})$ *be the Prototype Language. Let* $KB \subseteq$ PROTO *be a finite subset of* PROTO. KB *is called a* Prototype Knowledge Base *iff 1)* KB *is grounded, 2) no two prototype expressions in* KB *have the same prototype expression ID, and 3) for each prototype expression* $(id, (base, add, remove)) \in KB$, *each of the values of the simple change expressions in add are also in* $dom(KB)$.

Definition 5 (R). *Let* KB *be a prototype knowledge base and* $id \in ID$. *Then, the resolve function* R *is defined as:* $R(KB, id) =$ *the prototype expression in* KB *which has prototype expression ID equal to id.*

Prototype Semantics

Definition 6 (Prototype-Structure). *Let SID be a set of identifiers. A tuple $pv = (p, \{v_1, \ldots, v_n\})$ with $p, v_i \in SID$ is called a Value-Space for the ID-Space SID. A tuple $o = (id, \{pv_1, \ldots, pv_m\})$ with $id \in SID$ and Value-Spaces $pv_i, 1 \leq i \leq m$ for the ID-Space SID is called a Prototype for the ID-Space SID. A Prototype-Structure $O = (SID, OB, I)$ for a Prototype Language PL consists of an ID-Space SID, a Prototype-Space OB consisting of all Prototypes for the ID-Space SID and an interpretation function I, which maps IDs from PL to elements of SID.*

Definition 7 (Herbrand-Interpretation).
Let $O = (SID, OB, I_h)$ be a Prototype-Structure for the prototype language $PL = (\text{P}_\emptyset, ID, PROTO)$. I_h is called a Herbrand-Interpretation if I_h maps every element of ID to exactly one distinct element of SID.

As per the usual convention used for Herbrand-Interpretations, we assume that ID and SID are identical.

Next, we define the meaning of the constituents of a prototype. We start with the interpretation functions I_s and I_c which give the semantic meaning of the syntax symbols related to change expressions. These functions (and some of the following ones) are parametrized (one might say contextualized) by the knowledge base. This is needed to link the prototypes together.

Definition 8 (I_s). *Interpretation for the values of a simple change expression Let KB be a prototype knowledge base and v the values of a simple change expression. Then, the interpretation for the values of the simple change expression $I_s(KB, v)$ is a subset of SID defined as follows:*

$$SID, if v = *$$
$$\{I_h(r_1), I_h(r_2), \ldots, I_h(r_n)\}, \text{ if } v = \{r_1, \ldots, r_n\}$$

Definition 9 (I_c). *Interpretation of a change expression. Let KB be a prototype knowledge base and a function $ce = \{(p_1, vs_1), (p_2, vs_2), \ldots\}$ be a change expression with $p_1, p_2, \cdots \in ID$ and the vs_i be values of the simple change expressions. Let $W = ID \setminus \{p_1, p_2, \ldots\}$. Then, the interpretation of the change expression $I_c(KB, ce)$ is a function defined as follows (We will refer to this interpretation as a change set, note that this set defines a function):*

$$\{(I_h(p_1), I_s(KB, vs_1)), (I_h(p_2), I_s(KB, vs_2)), \ldots\} \cup \bigcup_{w \in W} \{(I_h(w), \emptyset)\}$$

Next, we define J which defines what it means for a prototype to have a property.

Definition 10 (J). *The value for a property of a prototype. Let KB be a prototype knowledge base and $id, p \in ID$. Let $R(KB, id) = (id, (b, r, a))$ (the resolve* *function applied to id). Then the value for the property p of the prototype id, i.e., $J(KB, id, p)$ is:*

$$I_c(KB, a)(I_h(p)), \quad if \ b = \mathbf{P}_\emptyset$$
$$(J(KB, b, p) \setminus I_c(KB, r)(I_h(p))) \cup I_c(KB, a)(I_h(p)), otherwise$$

Informally, this function maps a prototype and a property to (1) the set of values defined for this property in the base of the prototype (2) minus what is in the remove set (3) plus what is in the add set.

As an example, let us try to find out what the value for the creator of the Ghent Altarpiece described in the example of the previous subsection would evaluate to assuming that these prototypes were part of a Prototype Knowledge Base KB. For brevity we will write `example:Ghent_Altarpiece` as GA, `example:Arnolfini_Portrait` as AP, `dc:creator` as creator, `example:Jan_Van_Eyck` as JVE, and `example:Hubert_Van_Eyck` as HVE.

Concretely, we have to evaluate $J(KB, GA, creator)$ = $(J(KB, AP, creator) \setminus I_c(KB, \emptyset)(creator)) \cup I_c(KB, add)(creator)$ where add is the add change set of the GA prototype expression. First we compute the recursive part, $J(KB, AP, creator)$ = $I_c(KB, add_{ap})(creator)$ = $\{(creator, \{JVE\}), \dots\}(creator)$ = $\{JVE\}$. Where add_{ap} is the add change set of the AP prototype expression. The second part (what is removed) becomes $I_c(KB, \emptyset)(creator) = \emptyset$. The final part (what this prototype is adding) becomes $I_c(KB, add)(creator)$ = $\{(creator, \{HVE\}), \dots\}(creator)$ = $\{HVE\}$. Hence, the original expression becomes $(\{JVE\} \setminus \emptyset) \cup \{HVE\}$ = $\{JVE, HVE\}$ as expected.

Definition 11 (FP). *The interpretation of a prototype expression is also called its fixpoint. Let pe = $(id, (base, add, remove)) \in KB$ be a prototype expression. Then the interpretation of the prototype expression in context of the prototype knowledge base KB is defined as $FP(KB, pe)$ = $(I_h(id), \{(I_h(p), J(KB, id, p)) | p \in ID, J(KB, id, p)) \neq \emptyset\})$, which is a Prototype.*

Definition 12 (I_{KB}: Interpretation of Knowledge Base). *Let O = (SID, OB, I_h) be a Prototype-Structure for the Prototype Language PL = $(\mathbf{P}_\emptyset, ID, PROTO)$ with I_h being a Herbrand-Interpretation. Let KB be a Prototype-Knowledge Base. An interpretation I_{KB} for KB is a function that maps elements of KB to elements of OB as follows: $I_{KB}(KB, pe)$ = $FP(KB, pe)$*

This concludes the definition of the syntactic structures and semantics of prototypes and prototype knowledge bases. For the semantics, we have adopted Herbrand-Interpretations, which are compatible with the way RDF is handled in SPARQL.

4   Inheritance

Our discussion of inheritance is based on the work by Lieberman [17], Cook et al. [5], de la Rocque Rodriguez [25], and Taivalsaari [26]. The combination of these works provides a wide overview of different forms of inheritance. Despite the fact that the focus of these works is on object oriented programming (OOP) we chose them because prototype-based systems are much more developed in OOP than in knowledge representation. Many of the OOP concepts and concerns also apply to how inheritance mechanisms can be applied in Knowledge Representation.

Broadly speaking, inheritance means that an entity receives properties from another one because of a relation between the two. Two types of inheritance are common: class-based and prototype-based. In class-based systems there is a distinction between objects and classes. An object is an instantiation of a class or, as some say, a class is a blueprint for an object. A new class can be inherited from another one and will typically inherit all properties and methods from the base or parent class. The values associated with these properties are typically defined in the context of the instances. Prototype-based systems on the other hand only have one type of things: prototypes. A new prototype can be made by *cloning* an existing prototype (i.e., the base). The freshly created object now inherits from the earlier defined one and the values are defined directly on the prototypes. As we argued above, we chose the prototype-based inheritance to allow for both horizontal and vertical sharing. In the next sections we will describe the consequences of the choice of prototype-based inheritance.

4.1   Prototype Dynamics

There are essentially two ways to achieve prototype-based inheritance. The first one, *concatenation*, would copy all the content from the original object to the newly created one and apply the needed changes to the copy. The second one, *delegation*, keeps a reference to the original object and only stores the changes needed in the newly created object. We decided to follow the second option (for now) because it more closely resembles what one would expect from a system on the web. Instead of centralizing all information into one place, one links to information made available by others. This type of inheritance makes it possible to automatically make use of enhancements made in the base prototypes. Furthermore, the option of making a copy of the object one extends from is still available; we will discuss this further in Sect. 4.3. Note that this is also a space-time trade-off. Copying will occupy more space, but make look-up faster while delegation will be slower, but only the parts which have been changed have to be stored. Another option is to get parts of both worlds by caching frequently used prototypes for a set amount of time. In this case, one may retrieve outdated values. In our technical report [3], we describe a possible approach towards caching using existing HTTP mechanisms.

When parts of a knowledge base are not in the control of the knowledge engineer who is adding new information, it might be tempting to recreate certain prototypes to make sure that the prototypes one is referring to do not change over time, rendering the newly added information invalid.

4.2 A Prototype Is-not-a Class

In class-based object oriented languages, deriving a class A from a base class B usually implies that an instance of A can be used wherever an object of type B is expected. In other words, the objects instantiated from the classes A and B follow the Liskov substitution principle [18]. Since class-based object-orientation is currently most common in popular programming languages, one might be tempted to emulate classes in a prototype-based language. Imagine, for instance, that we want to create a prototype *employee* to represent an employee of a company. One might be tempted to give this *employee* a property *name*, with some default value since all employees will have a name in the end. However, this is not necessary, or even desired, when working with prototype-based systems. Instead, the *employee* should only have properties with values which all or most employees have in common, like for example the company they work for. Any more specific properties should instead be put on the employees themselves. Moreover, the fact that a prototype derives from the created *employee* does not have any implication beyond the inherited properties. Put another way, there is no *is-a* relation between a concrete employee and the *employee* prototype from which it was derived. This is also clearly visible from the fact that a derived prototype has the ability to remove properties from the base. Moreover, any other prototype with the properties needed to qualify for being an employee can be seen as an employee; independently from whether it derives from the *employee* prototype or not. Next, we will discuss what it means to be 'seen' as an employee.

4.3 Object Boundaries

Applications usually need to work with data with predictable properties. For instance, the employees from the example in the previous section need to have a name, gender, birthday, department, and social security number in order for the application to work. Hence, there is a need to specify the properties a prototype needs to have in order to be used for a specific application. This idea is not new and has also been identified in other knowledge representation research. Named Graphs are often used for this purpose, but they don't capture shared ownership or inheritance. Further, resource shapes[4] and shape expressions [22] have the core idea of determining whether a given RDF graph complies with a specification. The main goal of these is checking some form of constraints, but they could as well be used to identify instances in a dataset.

This need has been identified in many places in OOP literature. An object oriented programming language which allows variables to contain any object which fulfills a given interface definition is said to have a structural type system. Recent examples of programming languages with such type system include OCaml and Go, but to our knowledge the first programming language to use it was Emerald [1] and later School [25]. In these languages, if objects have a given

---

[4] https://www.w3.org/Submission/2014/SUBM-shapes-20140211/.

set of operations (according to what they called an *abstract type* in Emerald, *type* in School, or *interface* in Go), they would be treated a being an instance of, or assignable to, a variable of that type.

One of the arguments against structural type systems is that it might happen that an object has the properties (or methods) of the type by accident. We can envision this happening in OOP because the names of methods have little semantic meaning connected to them (does the write() method write something to the disk or to the printer?). However, in a Semantic Web setting, the property names are themselves IRIs and chosen carefully not to clash with existing names (a http://xmlns.com/foaf/0.1/workplaceHomepage will always be 'The workplaceHomepage of a person is a document that is the homepage of a organization that they work for.'[5]). In other words the property names in the system under consideration in this paper do in principle not suffer from this problem.

5   Future Work

Since most past work in the research community has been focused on class-based knowledge representation, there are still many areas unexplored related to prototype-based knowledge representation on the web.

5.1   Relation to RDF and OWL

In this paper, we are suggesting a knowledge sharing language based on prototypes. Future work will need to investigate how to layer the prototype language on top of RDF. While most of the conversion and layering should be straightforward (e.g., the IRI of a prototype expression would also be the IRI of the RDF resource), some challenges remain. For example, one would need to define a protocol working on RDF graphs in order to locate and interact with a prototype. However, we believe that these challenges can be overcome.

5.2   A Hint of Class?

In this paper, we presented prototypes as a possible alternative to class-based systems such as OWL for Knowledge representation on the Web - at least for the purpose of scalable Knowledge Sharing. However, both ways - prototypes and class-based representations, have different use cases and reasons to exist: OWL is focusing on enabling reasoning whereas prototypes are focusing on enabling Knowledge Sharing. Exploring the exact boundaries of their respective use cases still remains a topic for future work.

Another interesting future research path would be the discovery of 'hidden' classes in the knowledge base. A hidden class would be formed by a group of objects with similar characteristics. These classes would be automatically discovered, perhaps with techniques like Formal Concept Analysis (FCA) [29], by

---

[5] definition of `foaf:workplaceHomepage` from http://xmlns.com/foaf/spec/.

collecting a large number of prototypes from the Web. Another approach to this would be to perform a hierarchical clustering of the prototypes with a scalable technique as proposed in [4]. After this clustering, it might be possible to extract a class hierarchy from the generated dendrogram.

5.3   Variations, Evaluations and Large Scale Benchmarks

The prototype system introduced in this paper is only an initial exploration. There are numerous variations possible by making different choices for the inheritance model (e.g. concatenation, multiple inheritance, etc.), the allowed values (intervals, literals, etc.), and solutions for resolving the values for non-local prototypes. These choices will have different implications for implementations and good evaluation metrics and large scale benchmarks should be designed to compare them. We presented initial work in this direction in a technical report [3] and publicly available software https://github.com/miselico/ knowledgebase (LGPLv3). We benchmarked the system using several synthetic data sets and observed that the theoretical model presented offers the scalability needed for use in production environment in a typical distributed web architecture.

6   Conclusions

During the last decade, Knowledge Representation (KR) research has been dominated by W3C standards whose development was influenced by the state of the mind that researchers in the involved research communities had at the time of creation. Several choices which were made which have far reaching consequences on the way knowledge representation is done on the Web today.

In this paper we tried to take a step back and investigate another option for KR which, in our opinion, has properties more suitable to deliver on the goals of horizontal and vertical sharing. Concretely, we introduced a system in which everything is represented by what we call prototypes and the relations between them. Prototypes enable both vertical sharing by the inheritance mechanism and horizontal sharing by direct reference to any prototype. We provided a possible syntax and semantics for the Prototype system and performed experiments with an implementation. The experiments showed that the proposed system easily scales up to millions of prototypes. However, many question still remain to be answered. First and foremost, this kind of Knowledge Representation needs to get traction on the Web, which is a considerable challenge - but one we believe can be achieved based on early feedback we obtained. Furthermore, a larger deployment of this kind of system would need a clear mechanism for resolving non-local prototypes. We did some experiments in this direction in a technical report using existing web technologies like HTTP for this, but still there are many options to investigate. We would like to see what kind of options others come up with to introduce useful parts of class-based systems into the prototype world. Finally, we hinted towards finding 'hidden' classes in the Prototype system. This would not only be an academic exercise, but would be very useful to be able to compress knowledge base representations and reduce communication costs. We hope that this paper contributes constructively to the field of Knowledge Representation on the Web and that in the future, more researchers will explore different directions to see how far we can reach.

Acknowledgments. Stefan Decker would like to thank Pat Hayes, Eric Neumann, and Hong-Gee Kim for discussions about Prototypes and Knowledge Representation in general.

Michael Cochez performed parts of this research at the Industrial Ontologies Group of the University of Jyväskylä, Finland, and at the Insight Centre for Data Analytics in Galway, Ireland.

References

1. Black, A.P., Hutchinson, N.C., Jul, E., Levy, H.M.: The development of the emerald programming language. In: Proceedings of the Third ACM SIGPLAN Conference on History of Programming Languages, HOPL III, pp. 11-1-11-51. ACM, New York (2007). http://doi.acm.org/10.1145/1238844.1238855
2. Brachman, R.J.: A structural paradigm for representing knowledge. Technical report, BBN Report 3605, Bolt, Beraneck and Newman Inc., Cambridge, MA (1978)
3. Cochez, M., Decker, S., Prud'hommeaux, E.G.: Knowledge representation on the web revisited: tools for prototype based ontologies. In: arXiv (2016). https://arxiv.org/abs/1607.04809, arXiv:1607.04809 [cs.AI]
4. Cochez, M., Mou, H.: Twister tries: Approximate hierarchical agglomerative clustering for average distance in linear time. In: Proceedings of the 2015 ACM SIGMOD International Conference on Management of Data, pp. 505–517. ACM (2015)
5. Cook, W.R., Hill, W., Canning, P.S.: Inheritance is not subtyping. In: Proceedings of the 17th ACM SIGPLAN-SIGACT Symposium on Principles of Programming Languages, POPL 1990, pp. 125–135. ACM, New York (1990). http://doi.acm.org/10.1145/96709.96721
6. Decker, S., Fensel, D., van Harmelen, F., Horrocks, I., Melnik, S., Klein, M., Broekstra, J.: Knowledge representation on the web. In: Proceedings of the 2000 Description Logic Workshop (DL 2000), CEUR, vol. 33, pp. 89–98 (2000). (http://ceur-ws.org/)
7. Duerst, M., Suignard, M.: Internationalized resource identifiers (IRIS). RFC 3987, RFC Editor, January 2005. http://www.rfc-editor.org/rfc/rfc3987.txt
8. European Computer Manufacturers Association and others: Standard ecma-262 ecmascrippt 2015 language specification, June 2015
9. Gabriel, R.: The rise of "worse is better". In: LISP: Good News, Bad News, How to Win Big 2, 5 (1991)
10. Hayes, P.J.: The logic of frames. In: Metzing, D. (ed.) Frame Conceptions and Text Understanding, pp. 46–61. Walter de Gruyter and Co., Berlin (1979)
11. Horrocks, I., Sattler, U., Tobies, S.: Practical reasoning for expressive description logics. In: Ganzinger, H., McAllester, D., Voronkov, A. (eds.) Proceedings of the 6th International Conference on Logic for Programming and Automated Reasoning (LPAR 1999). LNAI, vol. 1705, pp. 161–180. Springer, Heidelberg (1999)

12. Horrocks, I.: Using an expressive description logic: FaCT or fiction? In: Proceedings of the 6th International Conference on Principles of Knowledge Representation and Reasoning (KR 1998), pp. 636–647 (1998)
13. Horrocks, I., Patel-Schneider, P.F., Harmelen, F.: From SHIQ and RDF to OWL: the making of a web ontology language. J. Web Semant. **1**(1), 7–26 (2003)
14. Israel, D.J., Brachman, R.J.: Some remarks on the semantics of representation languages. In: Brodie, M.L., Mylopoulos, J., Schmidt, J.W. (eds.) On Conceptual Modelling: Perspectives from Artificial Intelligence, Databases, and Programming Languages, pp. 119–142. Springer, New York (1984)
15. Karp, P.D.: The design space of frame knowledge representation systems. Technical report, SRI International Artificial Intelligence (1993)
16. Levesque, H.J., Brachman, R.J.: A fundamental tradeoff in knowledge representation and reasoning (revised version). In: Brachman, R.J., Levesque, H.J. (eds.) Readings in Knowledge Representation, pp. 41–70. Kaufmann, Los Altos (1985)
17. Lieberman, H.: Using prototypical objects to implement shared behavior in object-oriented systems. In: Conference Proceedings on Object-oriented Programming Systems, Languages and Applications, OOPLSA 1986, pp. 214–223. ACM, New York (1986). http://doi.acm.org/10.1145/28697.28718
18. Liskov, B.: Keynote address - data abstraction and hierarchy. SIGPLAN Not. **23**(5), 17–34 (1987). http://doi.acm.org/10.1145/62139.62141
19. Minsky, M.: A framework for representing knowledge. Technical report, Massachusetts Institute of Technology, Cambridge, MA, USA (1974)
20. Minsky, M.: A framework for representing knowledge. In: Haugeland, J. (ed.) Mind Design: Philosophy, Psychology, Artificial Intelligence, pp. 95–128. MIT Press, Cambridge (1981)
21. Mitchell, T.M., Allen, J., Chalasani, P., Cheng, J., Etzioni, O., Ringuette, M., Schlimmer, J.C.: Theo: a framework for self-improving systems. In: Architectures for Intelligence, pp. 323–356 (1991)
22. Prud'hommeaux, E., Labra Gayo, J.E., Solbrig, H.: Shape expressions: an RDF validation and transformation language. In: Proceedings of the 10th International Conference on Semantic Systems, pp. 32–40. ACM (2014)
23. Quillian, M.R.: Semantic memory. Technical report, DTIC Document (1966)
24. Rector, A.L.: Defaults, context, and knowledge: alternatives for owl-indexed knowledge bases. In: Altman, R.B., Dunker, A.K., Hunter, L., Jung, T.A., Klein, T.E. (eds.) Pacific Symposium on Biocomputing, pp. 226–237. World Scientific, Singapore (2004). http://dblp.uni-trier.de/db/conf/psb/psb2004.html#Rector04
25. Rodriguez, N.D.L.R., Ierusalimschy, R., Rangel, J.L.: Types in school. SIGPLAN Not **28**(8), 81–89 (1993). http://doi.acm.org/10.1145/163114.163125
26. Taivalsaari, A.: On the notion of inheritance. ACM Comput. Surv. (CSUR) **28**(3), 438–479 (1996)
27. Ungar, D., Smith, R.B.: Self. In: Ryder, B.G., Hailpern, B. (eds.) Proceedings of the Third ACM SIGPLAN History of Programming Languages Conference (HOPL-III), San Diego, California, USA, 9–10 June 2007, pp. 1–50. ACM (2007). http://doi.acm.org/10.1145/1238844.1238853
28. Victor, T., Dalzell, T.: The Concise New Partridge Dictionary of Slang and Unconventional English. Routledge, London (2007)
29. Wille, R.: Formal concept analysis as mathematical theory of concepts and concept hierarchies. In: Ganter, B., Stumme, G., Wille, R. (eds.) Formal Concept Analysis. LNCS (LNAI), vol. 3626, pp. 1–33. Springer, Heidelberg (2005)

Updating DL-Lite Ontologies Through First-Order Queries

Giuseppe De Giacomo[1], Xavier Oriol[2(✉)], Riccardo Rosati[1],
and Domenico Fabio Savo[1]

[1] Sapienza Università di Roma, Rome, Italy
{giacomo,rosati,savo}@dis.uniroma1.it
[2] Universitat Politècnica de Catalunya, Barcelona, Spain
xoriol@essi.upc.edu

Abstract. In this paper we study instance-level update in $DL\text{-}Lite_A$, the description logic underlying the OWL 2 QL standard. In particular we focus on formula-based approaches to ABox insertion and deletion. We show that $DL\text{-}Lite_A$, which is well-known for enjoying first-order rewritability of query answering, enjoys a first-order rewritability property also for updates. That is, every update can be reformulated into a set of insertion and deletion instructions computable through a non-recursive datalog program. Such a program is readily translatable into a first-order query over the ABox considered as a database, and hence into SQL. By exploiting this result, we implement an update component for $DL\text{-}Lite_A$-based systems and perform some experiments showing that the approach works in practice.

1 Introduction

In this paper we study effective techniques to perform updates over $DL\text{-}Lite$ ontologies. In particular, we focus on $DL\text{-}Lite_A$, which is the most expressive member of the $DL\text{-}Lite$ family of Description Logics (DLs) [4,5]. $DL\text{-}Lite_A$ includes virtually all constructs of the OWL 2 QL profile of the W3C OWL 2 standard. In addition, it includes the most typical cardinality restrictions on the participation in roles of UML class diagrams, i.e., any combination of mandatory participation and functional participation.

The crucial characteristic of $DL\text{-}Lite_A$ ontologies is that they enable the so-called *ontology-based data access* by virtue of first-order rewritability of query answering, that is, every (union of) conjunctive query over a $DL\text{-}Lite_A$ ontology can be rewritten into a first-order query to be evaluated over the ABox only (i.e., the individual data) considered as a database. This property, on the one hand, gives us a very low worst-case computational complexity bound w.r.t. data, namely AC^0 data complexity. On the other hand, it gives us a very effective practical technique to deal with ontologies that include very large ABoxes (i.e., a lot of individual data): perform the rewriting; transform the first-order query into SQL, or SPARQL, depending on how data are stored; and perform the resulting query exploiting a data management engine to take advantage of all optimizations available for these standard languages.

© Springer International Publishing AG 2016
P. Groth et al. (Eds.): ISWC 2016, Part I, LNCS 9981, pp. 167–183, 2016.
DOI: 10.1007/978-3-319-46523-4_11

When we come to updates over ontologies, several approaches are available in the literature [7,10,17,20]. In particular, in this paper we are interested in the so-called *instance-level* update: we add and delete (or erase) facts about individuals only. Namely, we change the ABox, while we keep the TBox unchanged. This is the most common form of update in practice, since it is essentially concerned with keeping the intensional part of the ontology fixed, while changing freely the individual data (indeed, the ABox changes are typically frequent whereas the TBox typically evolves slowly). Even in this specific kind of updates, there are sophisticated semantic issues to consider in general. One crucial issue is that, in practice, we need the result of the update to be still in the same language as the original ontology, in order to keep using the same system [20]. The most promising approaches that enjoy this property are the so-called *formula-based* approaches [9,13,14,23], in which the update is seen as a change of the ontology axioms. Again, several forms of *formula-based* instance-level updates have been considered [6,18,19,22]. Interestingly, however, for the DLs in the *DL-Lite* family, virtually all proposals in the literature reduce to two main approaches: the one in which we simply act on the ABox assertions explicitly stated in the ontology, and another one in which we act also on the ABox assertions that are not present but logically entailed through the use of the TBox. Notice that, while the first approach is syntax-dependent (i.e., updating logically equivalent ontologies that are stated through different assertions may give rise to logically different resulting ABoxes), the second one is not. In both cases, the semantics have been clarified, their computational tractability established, and ad-hoc algorithms are available. Though, for both approaches, there are essentially no implemented tools yet.

In this paper we look again at the problem of instance-level formula-based update in $DL\text{-}Lite_A$, and we establish a result that may turn out to be crucial to generate efficient implementations: like query answering, updating an ontology is *first-order rewritable*. That is, given an update specification, we can rewrite it into a set of addition and deletion instructions over the ABox which can be characterized as the result of a first-order query. This means that (i) updating a $DL\text{-}Lite_A$ ontology is AC^0 in data complexity, and, (ii) updates can be processed by widely used data management engines, e.g., based on SQL or SPARQL. We proof this by showing that every update can be reformulated into a datalog program that generates the set of insertion and deletion instructions to change the ABox while preserving its consistency w.r.t. the TBox. Since the obtained datalog program is non-recursive, it can be further translated as first-order queries over the ABox considered as a database. Exploiting this result, we implement an update component for $DL\text{-}Lite_A$-based systems and perform some experiments over (a $DL\text{-}Lite_A$ version of) the LUBM ontology [15] with increasing ABox sizes, showing that the approach works in practice.

As far as we know, this is the first time that the first-order rewritability property for $DL\text{-}Lite_A$ ontology updating is defined, proved, and empirically evaluated. It is important to mention here that some previous work has been done in the context of RDF triplestores [2,3], but only for the more restricted case of RDFS (with class disjunctions), which is a proper subset of the expressiveness of *DL-Lite$_A$*, the language we deal with in this paper.

2   Preliminaries

In this section, we first present the notion of Description Logic (DL) ontology, then we provide the definition of the specific DL considered in this work, and finally we summarize some datalog basic concepts and notation.

2.1   Description Logic Ontologies

Let $\mathcal{S}$ be a signature of symbols for individual (object and value) constants, and atomic elements, i.e., concepts, value-domains, attributes, and roles. If $\mathcal{L}$ is a DL, then an $\mathcal{L}$-ontology $\mathcal{O}$ over $\mathcal{S}$ is a pair $\langle \mathcal{T}, \mathcal{A} \rangle$, where $\mathcal{T}$, called *TBox*, is a finite set of intensional assertions over $\mathcal{S}$ expressed in $\mathcal{L}$, and $\mathcal{A}$, called *ABox*, is a finite set of instance assertions, i.e., assertions on individuals, over $\mathcal{S}$ expressed in $\mathcal{L}$. Different DLs allow for different kinds of concept, attribute, and role expressions, and different kinds of TBox and ABox assertions over such expressions. In this paper we assume that ABox assertions are always *atomic*, i.e., they correspond to ground atoms, and therefore we omit to refer to $\mathcal{L}$ when we talk about ABox assertions.

The semantics of a DL ontology is given in terms of interpretations. An interpretation is a *model* of an ontology $\mathcal{O} = \langle \mathcal{T}, \mathcal{A} \rangle$ if it satisfies all assertions in $\mathcal{T} \cup \mathcal{A}$, where the notion of satisfaction depends on the constructs allowed by the specific DL in which $\mathcal{O}$ is expressed. We denote the set of models of $\mathcal{O}$ with $Mod(\mathcal{O})$.

Let $\mathcal{T}$ be a TBox in $\mathcal{L}$, and let $\mathcal{A}$ be an ABox. We say that $\mathcal{A}$ is $\mathcal{T}$-*consistent* if $\langle \mathcal{T}, \mathcal{A} \rangle$ is satisfiable, i.e., if $Mod(\langle \mathcal{T}, \mathcal{A} \rangle) \neq \emptyset$, $\mathcal{T}$-inconsistent otherwise. The $\mathcal{T}$-*closure* of $\mathcal{A}$ with respect to $\mathcal{T}$, denoted $\mathsf{cl}_{\mathcal{T}}(\mathcal{A})$, is the set of all atomic ABox assertions that are formed with individuals in $\mathcal{A}$, and are logically implied by $\langle \mathcal{T}, \mathcal{A} \rangle$. Note that if $\langle \mathcal{T}, \mathcal{A} \rangle$ is an $\mathcal{L}$-ontology, then $\langle \mathcal{T}, \mathsf{cl}_{\mathcal{T}}(\mathcal{A}) \rangle$ is an $\mathcal{L}$-ontology as well, and is logically equivalent to $\langle \mathcal{T}, \mathcal{A} \rangle$, i.e., $Mod(\langle \mathcal{T}, \mathcal{A} \rangle) = Mod(\langle \mathcal{T}, \mathsf{cl}_{\mathcal{T}}(\mathcal{A}) \rangle)$. $\mathcal{A}$ is said to be $\mathcal{T}$-*closed* if $\mathsf{cl}_{\mathcal{T}}(\mathcal{A}) = \mathcal{A}$.

2.2   The Description Logic *DL-Lite$_A$*

The *DL-Lite* family [4] is a family of low-complexity DLs particularly suited for dealing with ontologies with very large ABoxes. It constitutes the basis of OWL 2 QL, a tractable profile of OWL 2, the official ontology specification language of the World Wide Web Consortium (W3C)[1].

We now present the DL *DL-Lite$_A$*, which is one of the most expressive logics in the family. *DL-Lite$_A$* distinguishes concepts from *value-domains*, which denote sets of (data) values, and roles from *attributes*, which denote binary relations

---

[1] http://www.w3.org/TR/2008/WD-owl2-profiles-20081008/.

between objects and values. Concepts, roles, attributes, and value-domains in this DL are formed according to the following syntax:

$$
\begin{aligned}
B &\longrightarrow A \mid \exists Q \mid \delta(U) & E &\longrightarrow \rho(U) \\
C &\longrightarrow B \mid \neg B & T &\longrightarrow \top_D \mid T_1 \mid \cdots \mid T_n \\
Q &\longrightarrow P \mid P^- & R &\longrightarrow Q \mid \neg Q \\
V &\longrightarrow U \mid \neg U
\end{aligned}
$$

where A, P, and U are symbols in $\mathcal{S}$ denoting respectively an atomic concept name, an atomic role name and an attribute name, $T_1, \ldots, T_n$ are n pairwise disjoint unbounded value-domains, $\top_D$ denotes the union of all domain values. Furthermore, P^- denotes the inverse of P, $\exists Q$ denotes the objects related to by the role Q, $\neg$ denotes negation, $\delta(U)$ denotes the *domain* of U, i.e., the set of objects that U relates to values, and $\rho(U)$ denotes the *range* of U, i.e., the set of values related to objects by U.

A *DL-Lite$_A$* TBox $\mathcal{T}$ contains intensional assertions of the form:

$B \sqsubseteq C$	*(concept inclusion)*	$E \sqsubseteq T$	*(value-domain inclusion)*
$Q \sqsubseteq R$	*(role inclusion)*	$U \sqsubseteq V$	*(attribute inclusion)*
(funct Q)	*(role functionality)*	(funct U)	*(attribute functionality)*

A concept inclusion assertion expresses that a (basic) concept B is subsumed by a (general) concept C. Analogously for the other types of inclusion assertions. Inclusion assertions that do not contain (resp. contain) the symbols '$\neg$' in the right-hand side are called *positive inclusions* (resp. *negative inclusions*). Role and attribute functionality assertions are used to impose that roles and attributes are actually functions respectively from objects to objects and from objects to domain values.

Finally, a *DL-Lite* TBox $\mathcal{T}$ satisfies the following condition: each role (resp., attribute) that occurs (in either direct or inverse direction) in a functional assertion, is not specialized in $\mathcal{T}$, i.e., it does not appear in the right-hand side of assertions of the form $Q \sqsubseteq Q'$ (resp., $U \sqsubseteq U'$).

A *DL-Lite$_A$* ABox $\mathcal{A}$ is a finite set of assertions of the form $A(a)$, $P(a,b)$, and $U(a,v)$, where A, P, and U are as above, a and b are object constants in $\mathcal{S}$, and v is a value constant in $\mathcal{S}$.

We refer to [21] for the semantics of a *DL-Lite$_A$* ontology. Here, we present an example of one such ontology.

Example 1. We consider a slightly modified version of the LUBM ontology [15] about the university domain. We know that a Person can be either a Professor or a Student, where every Student takes (takesCourse role) at least one Course, and every Professor can be either a FullProfessor or an AssociateProfessor. Finally, we know that john is a FullProfessor and that bob is a Student. The corresponding ontology $\mathcal{O}$ is:

$T = \{$ Student $\sqsubseteq$ Person   Professor $\sqsubseteq$ Person
     FullProfessor $\sqsubseteq$ Professor   AssociateProfessor $\sqsubseteq$ Professor
     Student $\sqsubseteq \neg$Professor   FullProfessor $\sqsubseteq \neg$AssociateProfessor
     Student $\sqsubseteq \exists$takesCourse $\exists$takesCourse$^-$ $\sqsubseteq$ Course $\}$

$\mathcal{A} = \{$ FullProfessor(john), Student(bob) $\}$

$\square$

A notable characteristic of $DL\text{-}Lite_A$ is that both satisfiability checking and conjunctive query answering are First-Order (FO) rewritable. Intuitively, FO-rewritability of satisfiability (resp., query answering) captures the property that we can reduce satisfiability checking (resp., query answering) to evaluating a FO query over the ABox $\mathcal{A}$ considered as a relational database. We remark that FO-rewritability of a reasoning problem that involves the ABox of an ontology (such as satisfiability or query answering) is tightly related to low data complexity of the problem. Indeed, since the evaluation of a First-Order Logic query (i.e., an SQL query without aggregation) over an ABox is in AC^0 in data complexity [1], the FO-rewritability of a problem has as the immediate consequence that the problem is in AC^0 in data complexity.

2.3 Datalog Concepts and Notation

A *term* T is either a *variable* or a *constant*. An *atom* is formed by a *n*-ary *predicate* p together with n terms, i.e., $p(T_1, \ldots, T_n)$. We may write $p(\overline{T})$ for short. If all the terms $\overline{T}$ of an atom are constants, we call the atom to be *ground*. A *literal* is either an atom $p(\overline{T})$, a negated atom $\neg p(\overline{T})$, or an inequality $T_i \neq T_j$.

A predicate p is said to be *derived* (or *intensional*) if the evaluation of an atom $p(\overline{T})$ depends on some derivation rules, otherwise, it is said to be *base* (or *extensional*). A *derivation rule* is a rule of the form $p(\overline{T_p}) \leftarrow \phi(\overline{T})$, where $p(\overline{T_p})$ is an atom called the *head* of the rule, and $\phi(\overline{T})$ is a conjunction of literals called the *body*. All derivation rules must be *safe*, i.e., every variable appearing in the head or in a negated or inequality literal of the body should also appear in a positive literal of the body. Additionally, all the predicates must be *stratified*, i.e., it should be possible to partition the set of predicates P into several pairwise disjoint *strata* $P_1 \cup \ldots \cup P_m$ s.t. for each predicate $p \in P_i$, each predicate appearing in the derivation rules of p should belong to a stratum P_j with $j < i$, if it appears in a negated literal, or, $j \leq i$, if it only appears in positive literals.

Finally, a *datalog program* is a set of derivation rules together with a set of *facts*, where a fact is a ground atom of a non-derived predicate.

3   Formula-Based Approach for Updating DL Ontologies

In the following, we first present the intuitions on ontology update, then we define two distinct formula-based update semantics, and we argue that, for the case of $DL\text{-}Lite_A$, these two semantics capture virtually all other formula-based update semantics proposed so far. Then, we show that the *careful semantics*, a different formula-based update semantics proposed in the literature, is not uniquely defined in the case of *DL-Lite$_A$*, contradicting a result stated in [6], which makes this update semantics inappropriate in our approach due to its inherent nondeterminism.

3.1   Update Semantics for *DL-Lite$_A$*

In the formula-based approaches to the update, the objects of change are sets of formulae. That is, the result of the change is explicitly defined in terms of a formula, by resorting to some minimality criterion with respect to the formula expressing the original ontology.

Thus, an *update* is a set $\mathcal{U}$ of operations of two types: insertion operations, denoted by i(α), and deletion operations denoted by d(α), where α is an ABox assertion. Intuitively, updating a consistent ontology with an insertion operation i($A(o)$), where $A(o)$ is a concept ABox assertion, means changing the extensional level of the ontology in such a way that the ontology resulting from the update is still consistent and entails the fact $A(o)$. Conversely, updating a consistent ontology with a deletion operation d($A(o)$), means changing the extensional level of the ontology in such a way that the ontology resulting from the update is still consistent and does not entail the fact $A(o)$.

After adding new facts into an ontology, one may find that the revised ontology becomes inconsistent. A strategy to overcome such a situation is to remove part of the original ABox to the aim of preserving consistency. Similarly, if the goal is to update the ontology by deleting a fact, we might need to retract several facts from the original ABox that entailed it. When applying these modifications to the original ABox, one should respect the *minimal change principle*, a widely accepted principle of the knowledge base evolution literature [8,11,16]. This principle states that the ontology resulting from the update should be as *close* as possible to the original one. In updating an ontology at the instance level following the formula-based approach, the goal becomes the preservation of the facts contained in the original ABox. In what follows we formalize this idea.

Given an ontology $\mathcal{O} = \langle \mathcal{T}, \mathcal{A} \rangle$, an update $\mathcal{U}$, and an ABox $\mathcal{A}'$, we say that $\mathcal{A}'$ *accomplishes the update* of $\mathcal{O}$ with $\mathcal{U}$ if it satisfies all the insertions/deletions in $\mathcal{U}$ minimally. To formalize this notion, we first need to introduce the set $\mathcal{A}_{\mathcal{U}}^+$, which denotes the set of ABox assertions appearing in $\mathcal{U}$ in insertion operations, and the set $\mathcal{A}_{\mathcal{U}}^-$, which denotes the set of ABox assertions appearing in $\mathcal{U}$ in deletion operations.

Definition 1. Let $\mathcal{O} = \langle \mathcal{T}, \mathcal{A} \rangle$ be an ontology, $\mathcal{U}$ an update, and $\mathcal{A}'$ be an ABox. $\mathcal{A}'$ *accomplishes the update* of $\mathcal{O}$ with $\mathcal{U}$ if $\mathcal{A}' = \mathcal{A}'' \cup \mathcal{A}_{\mathcal{U}}^+$ for some maximal subset $\mathcal{A}''$ of $\mathcal{A}$ s.t. $\mathcal{A}'' \cup \mathcal{A}_{\mathcal{U}}^+$ is $\mathcal{T}$-consistent and $\langle \mathcal{T}, \mathcal{A}' \rangle \not\models \beta$ for each $\beta \in \mathcal{A}_{\mathcal{U}}^-$.

It easy to see that, by definition, if such ABox $\mathcal{A}'$ exists, it also satisfies $\langle \mathcal{T}, \mathcal{A}' \rangle \models \alpha$ for each $\alpha \in \mathcal{A}_{\mathcal{U}}^+$ since $\mathcal{A}_{\mathcal{U}}^+ \subseteq \mathcal{A}'$. In order to ensure its existence, note that $\mathcal{U}$ has to respect both of the following conditions:

(i) $Mod(\langle \mathcal{T}, \mathcal{A}_{\mathcal{U}}^+ \rangle) \neq \emptyset$, which means that the set of facts we are adding is consistent with the TBox of the ontology.

(ii) $\mathcal{A}_{\mathcal{U}}^{-} \cap \mathsf{cl}_{\mathcal{T}}(\mathcal{A}_{\mathcal{U}}^{+}) = \emptyset$, which means that the update is not asking for deleting and inserting the same knowledge at the same time.

Given a TBox $\mathcal{T}$ and an update $\mathcal{U}$, we say that $\mathcal{U}$ is *coherent* with $\mathcal{T}$ if $\mathcal{U}$ respects both the above conditions with respect to a TBox $\mathcal{T}$.

Given a consistent ontology $\mathcal{O} = \langle \mathcal{T}, \mathcal{A} \rangle$ and an update $\mathcal{U}$ coherent with $\mathcal{T}$, there might be more than one ABox accomplishing the update of $\mathcal{O}$ with $\mathcal{U}$. This fact leads to different update semantics, each one addressing this issue by means of a different criterium, like the *Cross Product Approach* [9], the *When In Doubt Throw It Out* principle [14,18,19,23], allowing the user to choose the update [22], or even nondeterminism [6]. Fortunately, when the TBox of the ontology is expressed in *DL-Lite$_A$*, the ABox accomplishing the update is uniquely defined [6]. Hence, the application of all the above approaches leads to the same result, which can be defined as follows:

Definition 2. Let $\mathcal{O} = \langle \mathcal{T}, \mathcal{A} \rangle$ be a consistent *DL-Lite$_A$* ontology and $\mathcal{U}$ be an update coherent with $\mathcal{T}$. The result of updating $\mathcal{O}$ with $\mathcal{U}$, denoted by $\mathcal{O} \circ \mathcal{U}$, is the ontology $\langle \mathcal{T}, \mathcal{A}' \rangle$, where $\mathcal{A}'$ is the ABox accomplishing the update of $\mathcal{O}$ with $\mathcal{U}$.

When dealing with ontology updating, there is a fundamental philosophical aspect that has to be considered: one has to decide if the formulae explicitly given in our ontology provide a justification for our knowledge (foundational semantics) or if they are just used as a finite representation of our knowledge (coherence semantics) [11,12]. Depending on this point of view, one may or may not need to preserve a fact that is entailed in the ontology despite not being explicitly asserted. The choice depends on the particular application and personal preferences (we refer to [12] for more details).

Clearly, the update semantics given in Definition 2 embraces the foundational theory. Depending on the specific scenario, and the particular application at hand, this semantics might be considered inappropriate. This motivates the definition of the following update semantics [6,18] for *DL-Lite$_A$* ontologies based on the coherence theory, in which the objects of the update is not the original ABox, but its deductive closure with respect to the TBox.

Definition 3. Let $\mathcal{O} = \langle \mathcal{T}, \mathcal{A} \rangle$ be a consistent *DL-Lite$_A$* ontology and let $\mathcal{U}$ be an update coherent with $\mathcal{T}$. The result of updating $\mathcal{O}$ with $\mathcal{U}$ according to the coherence semantics, denoted by $\mathcal{O} \bullet \mathcal{U}$, is the ontology $\langle \mathcal{T}, \mathcal{A}' \rangle$, where $\mathcal{A}'$ is the ABox accomplishing the update of $\langle \mathcal{T}, \mathsf{cl}_{\mathcal{T}}(\mathcal{A}) \rangle$ with $\mathcal{U}$.

3.2 Careful Semantics in *DL-Lite$_A$*

An alternative formula-based update semantics based on the coherence theory is the *Careful semantics* [6] which was proposed with the aim of preventing *unexpected information*. Formally, an ontology updated according to the careful semantics should not entail a role constraint ϕ (i.e., a rule of the form $\exists x(R(o,x)) \wedge (x \neq c_1) \wedge \cdots \wedge (x \neq c_n))$, unless ϕ is entailed by the original

ABox, or the update itself. In practice, the careful update semantics encompasses deleting more ABox assertions so that the final ontology does not entail any new role constraint ϕ. However, although the careful update semantics was thought to be uniquely defined [6, Theorem 16], it can bring to several solutions as we show in the following example.

Example 2. Consider the *DL-Lite$_A$* ontology $\mathcal{O} = \langle \mathcal{T}, \mathcal{A} \rangle$ where:

$$\mathcal{T} = \{ \ A \sqsubseteq \exists R_A, \quad R_A \sqsubseteq R, \quad \exists R_A^- \sqsubseteq \neg \exists R_B^-,$$
$$B \sqsubseteq \exists R_B, \quad R_B \sqsubseteq R, \quad \exists R_A^- \sqsubseteq \neg \exists R_C^-,$$
$$C \sqsubseteq \exists R_C, \quad R_C \sqsubseteq R, \quad \exists R_B^- \sqsubseteq \neg \exists R_C^-,$$
$$D \sqsubseteq \exists R_D, \quad R_D \sqsubseteq R, \quad \exists R_C^- \sqsubseteq \neg \exists R_D^- \}$$
$$\mathcal{A} = \{ \ A(o), B(o) \ \}$$

and the update $\mathcal{U} = \{\mathsf{i}(C(o)), \mathsf{i}(D(o))\}$. It is easy to see that the ABox $\mathcal{A}' = \mathcal{A} \cup \mathcal{A}_\mathcal{U}^+$ is $\mathcal{T}$-consistent and that it accomplishes the update of $\mathcal{O}$ with $\mathcal{U}$. Moreover, $\langle \mathcal{T}, \mathcal{A}' \rangle \models \varphi$, where $\varphi = \exists x (R(o, x)) \land (x \neq c_1 \land (x \neq c_2)))$ (since the negative inclusions in $\mathcal{T}$ imply that in every model $\mathcal{I}$ of $\langle \mathcal{T}, \mathcal{A}' \rangle$ there are three distinct individuals d_a, d_b, d_c such that $\langle o, d_a \rangle \in R_A^{\mathcal{I}}$, $\langle o, d_b \rangle \in R_B^{\mathcal{I}}$, $\langle o, d_c \rangle \in R_C^{\mathcal{I}}$). However, since neither $\langle \mathcal{T}, \mathcal{A} \rangle \models \varphi$ nor $\langle \mathcal{T}, \mathcal{A}_\mathcal{U}^+ \rangle \models \varphi$, we have that $\mathcal{A}'$ does not accomplish the update of $\mathcal{O}$ with $\mathcal{U}$ carefully. Conversely, both the ABoxes $\{A(o)\} \cup \mathcal{A}_\mathcal{U}^+$ and $\{B(o)\} \cup \mathcal{A}_\mathcal{U}^+$ accomplish the update of $\mathcal{O}$ with $\mathcal{U}$ carefully. This is because the only role-constraining formula $\exists x (R(o, x)) \land (x \neq c_1))$ that both entail with $\mathcal{T}$, is also entailed by $\langle \mathcal{T}, \mathcal{A}_\mathcal{U}^+ \rangle$. Hence, we have more than one ABox that accomplishes the update of $\mathcal{O}$ with $\mathcal{U}$ carefully. $\square$

4   Foundational-Semantic Updates Through Datalog

Now, our intention is, given a *DL-Lite$_A$* ontology $\langle \mathcal{T}, \mathcal{A} \rangle$, and some update $\mathcal{U}$, to define a datalog program $\mathcal{D}$ that permits querying whether $\mathcal{U}$ is coherent with $\mathcal{T}$ and, in such a case, allows for generating a set of insertion/deletion instructions that should be applied to $\mathcal{A}$ to accomplish $\mathcal{U}$ according to Definition 2 (foundational-semantic updates).

For ease of presentation, from now on we assume that the TBox $\mathcal{T}$ does not contain inclusions involving attributes and value-domains. However, all the results presented in the next two sections can be easily extended to TBoxes containing such kinds of axioms.

Formally, the datalog program $\mathcal{D}$ contains a derived predicate *incoherent_update*, together with a pair of derived predicates *ins_a/del_a* for each concept/role A such that:

– *incoherent_update()* is true iff $\mathcal{U}$ is not coherent with $\mathcal{T}$.

and, in case *incoherent_update()* is false,

– *ins_a($\bar{o}$)* is true iff the assertion $A(\bar{o})$ was not in $\mathcal{A}$, but $A(\bar{o}) \in \langle \mathcal{T}, \mathcal{A} \rangle \circ \mathcal{U}$.
   That is, *ins_a* captures the assertions of $\mathcal{A}$ that should be inserted into $\mathcal{A}$ to accomplish the (foundational-semantic) update $\mathcal{U}$.

– $del\_a(\overline{o})$ is true iff the assertion $A(\overline{o})$ was in $\mathcal{A}$, but $A(\overline{o}) \notin \langle \mathcal{T}, \mathcal{A} \rangle \circ \mathcal{U}$. That is, del_a captures the assertions of $\mathcal{A}$ that should be deleted from $\mathcal{A}$ to accomplish the (foundational-semantic) update $\mathcal{U}$.

Briefly, the main idea of the translation is to map each ABox assertion in $\mathcal{A}$, and each operation in $\mathcal{U}$ into different datalog facts. Then, we map each assertion in the closure of $\mathcal{T}$ into several datalog derivation rules that define the *incoherent_update*, *ins_a*$(\overline{X})$, *del_a*$(\overline{X})$ predicates. In the following, we formally describe how to obtain such a datalog program $\mathcal{D}$. Then, we prove that the set of instructions generated in $\mathcal{D}$ are sound and complete to obtain $\langle \mathcal{T}, \mathcal{A} \rangle \circ \mathcal{U}$.

4.1 Translation Rules

Translation of $\mathcal{A}$ and $\mathcal{U}$. All the assertions in $\mathcal{A}$ and operations in $\mathcal{U}$ are translated as different facts in $\mathcal{D}$. In particular:

Each assertion $A(\overline{o}) \in \mathcal{A}$ is translated as the fact $a(\overline{o})$.
Each operation $i(A(\overline{o})) \in \mathcal{U}$ is translated as the fact *ins_a_request*$(\overline{o})$.
Each operation $d(A(\overline{o})) \in \mathcal{U}$ is translated as the fact *del_a_request*$(\overline{o})$.

Intuitively, *ins_a_request*$(\overline{o})$/*del_a_request*$(\overline{o})$ means that the ontology has received the request to insert/delete the ABox assertion $A(\overline{o})$. Since according to the Definition 2 all the insertions/deletions requested should be applied, we define the datalog rules:

```
ins_a(X) :- ins_a_request(X), not a(X).
del_a(X) :- del_a_request(X), a(X).
incoherent_update() :- ins_a_request(X), del_a_request(X).
```

for each atomic concept A. Note that *incoherent_update* becomes true in case we request for the insertion and deletion of the same axiom. Similarly, we define the rules *ins_p(X, Y)*/*del_p(X,Y)* for each atomic role P.

Translation of $cl(\mathcal{T})$. We translate positive and negative/functional axioms in the closure of $\mathcal{T}$ differently. In particular, for each positive inclusion axiom $B \sqsubseteq A$ in the closure of $\mathcal{T}$, where A is an atomic concept, we define the rules:

```
del_b(X) :- b(X), del_a_request(X).
incoherent_update() :- ins_b_request(X), del_a_request(X).
```

Intuitively, when we request for deleting $A(o)$, we have to delete any other ABox assertion $B(o)$ that entails $A(o)$. Note that it cannot be accomplished if there is a request for inserting $B(o)$, so, this case makes *incoherent_update* true. We define similar rules when the left-hand side of the axiom is of the form $\exists P$, and also for role inclusion axioms.

Note that we translate the closure of $\mathcal{T}$, instead of $\mathcal{T}$ itself, to be able to capture deletions that are propagated along the concept/role hierarchy. E.g. if in our example we have $\mathcal{U} = d(\mathsf{Person}(\mathsf{john}))$, the translated datalog program

$\mathcal{D}$ generates the deletion of FullProfessor(john) because of the translation of the assertion FullProfessor $\sqsubseteq$ Person appearing in $cl(\mathcal{T})$:

```
del_fullprof(X) :- fullprof(X), del_person_request(X).
```

Differently, for each negative inclusion axiom $B \sqsubseteq \neg A$ in $cl(\mathcal{T})$, we define the rules:

```
del_b(X) :- b(X), ins_a_request(X).
del_a(X) :- ins_b_request(X), a(X).
incoherent_update() :- ins_a_request(X), ins_b_request(X).
```

Intuitively, if we insert $A(o)$ when we have $B(o)$ in the ABox, we have to delete $B(o)$. In the case where the requested update tries to insert both things, we reach a contradiction and thus, *incoherent_update* becomes true. We define similar rules for role negative inclusions, negative inclusions involving the $\exists$ constructor, and functional axioms. In this last case, we require using the inequality built-in predicate to check whether the requested role assertion insertion is going to violate the functional axiom. E.g., given a functional axiom defined over R, we define:

```
del_r(X,Y) :- r(X,Y), ins_r_request(X,Z), Y<>Z.
incoherent_update() :- ins_r_request(X,Y),ins_r_request(X,Z),
    Y<>Z.
```

Again, note that since we translate the closure of $\mathcal{T}$, the rules are able to capture deletions due to inconsistencies generated by propagation. E.g. if in our previous example we have the update $\mathcal{U} = $ i(AssociateProfessor(bob)), $\mathcal{D}$ generates the deletion of Student(bob) because of the first rule obtained when translating the assertion Student $\sqsubseteq \neg$AssociateProfessor appearing in $cl(\mathcal{T})$:

```
del_student(X) :- student(X), ins_assocprof_request(X).
del_assocprof(X) :- assocprof(X), ins_student_request(X).
```

4.2   Datalog Program Soundness and Completeness

The update generated by the datalog program $\mathcal{D}$ is sound in the sense that, for every axiom $A(\bar{o})$ that should be inserted/deleted according to $\mathcal{D}$, $A(\bar{o})$ should be truly inserted/deleted according to the foundational-semantic update. Formally:

Theorem 1. *Given a consistent ontology $\langle \mathcal{T}, \mathcal{A} \rangle$, and an update $\mathcal{U}$, the datalog program $\mathcal{D}$ obtained through the translation defined in Sect. 4.1, satisfies that: if incoherent_update() is true in $\mathcal{D}$, $\mathcal{U}$ is incoherent with $\mathcal{T}$, otherwise, for each concept/role A, if ins_a($\bar{o}$) is true in $\mathcal{D}$, then, $A(\bar{o}) \in \langle \mathcal{T}, \mathcal{A} \rangle \circ \mathcal{U} \setminus \mathcal{A}$, and if del_a($\bar{o}$) is true in $\mathcal{D}$, then, $A(\bar{o}) \in \mathcal{A} \setminus \langle \mathcal{T}, \mathcal{A} \rangle \circ \mathcal{U}$.*

Proof (Sketch). If *incoherent_update()* is true, it can only be because of a rule generated when translating the update $\mathcal{U}$, the positive axioms of $cl(\mathcal{T})$, or the negative/functional axioms of $cl(\mathcal{T})$. The rules generated in the first two cases are true only if $\mathcal{A}_{\mathcal{U}}^{-} \cap \mathcal{A}_{\mathcal{U}}^{+} \neq \emptyset$ and $\mathcal{A}_{\mathcal{U}}^{-} \cap cl_{\mathcal{T}}(\mathcal{A}_{\mathcal{U}}^{+}) \neq \emptyset$, respectively. The rules of the third case are true only if $Mod(\langle \mathcal{T}, \mathcal{A}_{\mathcal{U}}^+ \rangle) = \emptyset$. Thus, if $incoherent\_update()$ is true, $\mathcal{U}$ is incoherent with $\mathcal{T}$.

If $ins\_a(\bar{o})$ is true, it is because of a rule generated when translating $\mathcal{U}$, which can only be true if $A(\bar{o}) \notin \mathcal{A}$, and $A(\bar{o}) \in \mathcal{A}_{\mathcal{U}}^+$, thus $A(\bar{o}) \in \langle \mathcal{T}, \mathcal{A} \rangle \circ \mathcal{U} \setminus \mathcal{A}$.

If $del\_a(\bar{o})$ is true, it can only be because of (1) a rule generated when translating $\mathcal{U}$, where in such case we have $A(\bar{o}) \in \mathcal{A}$, and $A(\bar{o}) \in \mathcal{A}_{\mathcal{U}}^-$, thus $A(\bar{o}) \in \mathcal{A} \setminus \langle \mathcal{T}, \mathcal{A} \rangle \circ \mathcal{U}$; or (2) a rule generated when translating a positive axiom in $\mathcal{T}$, where in such case we have that $A(\bar{o}) \in \mathcal{A}$ and that for some $B(\bar{o}) \in \mathcal{A}_{\mathcal{U}}^-$, $A(\bar{o}) \models_{\mathcal{T}} B(\bar{o})$, thus, $A(\bar{o}) \in \mathcal{A} \setminus \langle \mathcal{T}, \mathcal{A} \rangle \circ \mathcal{U}$; or (3) a rule generated when translating a negative/functional axiom in $cl(\mathcal{T})$ where in such case we have $A(\bar{o}) \in \mathcal{A}$ and $Mod(\langle \mathcal{T}, \mathcal{A}_{\mathcal{U}}^+ \cup \{A(\bar{o})\} \rangle) = \emptyset$, and thus, $A(\bar{o}) \in \mathcal{A} \setminus \langle \mathcal{T}, \mathcal{A} \rangle \circ \mathcal{U}$.   □

Conversely, $\mathcal{D}$ is also complete in the sense that any axiom insertion/deletion of $A(\bar{o})$ that should be applied according to the foundational-semantic update is also generated in $\mathcal{D}$. Formally:

Theorem 2. *Given a consistent ontology $\langle \mathcal{T}, \mathcal{A} \rangle$, and an update $\mathcal{U}$, the datalog program $\mathcal{D}$ obtained through the translation defined in Sect. 4.1, satisfies that: if $\mathcal{U}$ is incoherent with $\mathcal{T}$, then, incoherent_update() is true in $\mathcal{D}$, otherwise, for each concept/role A, if $A(\bar{o}) \in \langle \mathcal{T}, \mathcal{A} \rangle \circ \mathcal{U} \setminus \mathcal{A}$, then, ins_a($\bar{o}$) is true in $\mathcal{D}$, and if $A(\bar{o}) \in \mathcal{A} \setminus \langle \mathcal{T}, \mathcal{A} \rangle \circ \mathcal{U}$, then, del_a($\bar{o}$) is true in $\mathcal{D}$.*

Proof (Sketch). First, if $\mathcal{U}$ is incoherent with $\mathcal{T}$, it is immediate to verify that then, $incoherent\_update()$ is true in $\mathcal{D}$. So, from now on we assume that $\mathcal{U}$ is coherent with $\mathcal{T}$. Moreover, since $\mathcal{U}$ is coherent with $\mathcal{T}$, $\langle \mathcal{T}, \mathcal{A} \rangle \circ \mathcal{U} \setminus \mathcal{A} = \mathcal{A}_{\mathcal{U}}^+ \setminus \mathcal{A}$, and by definition of $\mathcal{D}$, it easily follows that, for each concept/role A, if $A(\bar{o}) \in \mathcal{A}_{\mathcal{U}}^+ \setminus \mathcal{A}$, $ins\_a(\bar{o})$ is true in $\mathcal{D}$. Finally, we prove that for every assertion deleted from $\mathcal{A}$ there is a corresponding deletion instruction in $\mathcal{D}$. To this aim, we define the following algorithm:

Algorithm ComputeDeletedAssertions($\mathcal{T}, \mathcal{A}, \mathcal{U}$)
Input: *DL-Lite$_A$* TBox $\mathcal{T}$, ABox $\mathcal{A}$, update $\mathcal{U}$ coherent with $\mathcal{T}$
Output: ABox $\mathcal{A}_d = \mathcal{A} \setminus \langle \mathcal{T}, \mathcal{A} \rangle \circ \mathcal{U}$
begin
    $\mathcal{A}_d = \emptyset$;
    for each $C(a) \in \mathcal{A}_{\mathcal{U}}^+$ **do begin**
        for each $D(a) \in \mathcal{A}$ such that $\mathcal{T} \models C \sqsubseteq \neg D$ do $\mathcal{A}_d = \mathcal{A}_d \cup \{D(a)\}$;
        for each $R(a,x) \in \mathcal{A}$ such that $\mathcal{T} \models C \sqsubseteq \neg \exists R$ do $\mathcal{A}_d = \mathcal{A}_d \cup \{R(a,x)\}$;
        for each $R(x,a) \in \mathcal{A}$ such that $\mathcal{T} \models C \sqsubseteq \neg \exists R^-$ do $\mathcal{A}_d = \mathcal{A}_d \cup \{R(x,a)\}$
    end;
    for each $R(a,b) \in \mathcal{A}_{\mathcal{U}}^+$ **do begin**
        for each $S(a,b) \in \mathcal{A}$ such that $\mathcal{T} \models R \sqsubseteq \neg S$ do $\mathcal{A}_d = \mathcal{A}_d \cup \{S(a,b)\}$;
        for each $S(b,a) \in \mathcal{A}$ such that $\mathcal{T} \models R \sqsubseteq \neg S^-$ do $\mathcal{A}_d = \mathcal{A}_d \cup \{S(b,a)\}$;
        for each $C(a) \in \mathcal{A}$ such that $\mathcal{T} \models \exists R \sqsubseteq \neg C$ do $\mathcal{A}_d = \mathcal{A}_d \cup \{C(a)\}$;
        for each $C(b) \in \mathcal{A}$ such that $\mathcal{T} \models \exists R^- \sqsubseteq \neg C$ do $\mathcal{A}_d = \mathcal{A}_d \cup \{C(b)\}$;
        for each $S(a,x) \in \mathcal{A}$ such that $\mathcal{T} \models \exists R \sqsubseteq \neg \exists S$ do $\mathcal{A}_d = \mathcal{A}_d \cup \{S(a,x)\}$;
        for each $S(x,a) \in \mathcal{A}$ such that $\mathcal{T} \models \exists R \sqsubseteq \neg \exists S^-$ do $\mathcal{A}_d = \mathcal{A}_d \cup \{S(x,a)\}$;
        for each $S(b,x) \in \mathcal{A}$ such that $\mathcal{T} \models \exists R^- \sqsubseteq \neg \exists S$ do $\mathcal{A}_d = \mathcal{A}_d \cup \{S(b,x)\}$;

for each $S(x, b) \in \mathcal{A}$ such that $\mathcal{T} \models \exists R^- \sqsubseteq \neg \exists S^-$ do $\mathcal{A}_d = \mathcal{A}_d \cup \{S(x, b)\}$
end;
for each $C(a) \in \mathcal{A}_{\mathcal{U}}^-$ do begin
    for each $D(a) \in \mathcal{A}$ such that $\mathcal{T} \models D \sqsubseteq C$ do $\mathcal{A}_d = \mathcal{A}_d \cup \{D(a)\}$;
    for each $R(a, x) \in \mathcal{A}$ such that $\mathcal{T} \models \exists R \sqsubseteq C$ do $\mathcal{A}_d = \mathcal{A}_d \cup \{R(a, x)\}$;
    for each $R(x, a) \in \mathcal{A}$ such that $\mathcal{T} \models \exists R^- \sqsubseteq C$ do $\mathcal{A}_d = \mathcal{A}_d \cup \{R(x, a)\}$
end;
for each $R(a, b) \in \mathcal{A}_{\mathcal{U}}^-$ do begin
    for each $S(a, b) \in \mathcal{A}$ such that $\mathcal{T} \models S \sqsubseteq R$ do $\mathcal{A}_d = \mathcal{A}_d \cup \{S(a, b)\}$;
    for each $S(b, a) \in \mathcal{A}$ such that $\mathcal{T} \models S \sqsubseteq R^-$ do $\mathcal{A}_d = \mathcal{A}_d \cup \{S(b, a)\}$
end;
return $\mathcal{A}_d$
end

It can easily be shown that the ABox returned by such an algorithm is equal to $\mathcal{A} \setminus \langle \mathcal{T}, \mathcal{A} \rangle \circ \mathcal{U}$. Moreover, it is easy to see that, for each concept/role $\mathcal{A}$, if $A(\bar{o})$ belongs to the ABox returned by ComputeDeletedAssertions$(\mathcal{T}, \mathcal{A}, \mathcal{U})$, then $del\_a(\bar{o})$ is true in $\mathcal{D}$. □

5   Coherent-Semantic Updates Through Datalog

The previous datalog program $\mathcal{D}$ generates the set of insertions/deletions that should be applied to an ABox $\mathcal{A}$ to accomplish an update $\mathcal{U}$ according to the foundational-semantics. Now, our purpose is to modify this datalog program to deal with the coherent-semantics as described in Definition 3.

Briefly, to accomplish the coherent-semantics, we need to generate more insertion instructions in $\mathcal{D}$. This is because in the coherent-semantics we need to keep the updated ABox as *close* as possible to the $\mathcal{T}$-closure of the original ABox, instead of the ABox itself. For instance, if in our previous example we apply the update $\mathcal{U} = \{d(\mathsf{Student}(\mathsf{bob}))\}$ with coherent-semantics, besides deleting the assertion $\mathsf{Student}(\mathsf{bob})$, we also need to apply the insertion $\mathsf{Person}(\mathsf{bob})$ since $\mathsf{Person}(\mathsf{bob})$ appears in $cl_{\mathcal{T}}(\mathcal{A})$.

Thus, in practice, we only need to extend our datalog program $\mathcal{D}$ to (1) additionally capture those assertions $A(\bar{o})$ entailed by assertions $B(\bar{o})$ that are requested for deletion, and (2) derive their insertion in case they do not get in conflict with the assertions in $\mathcal{A}_{\mathcal{U}}^+$. Intuitively, we do (1) by considering an additional derived predicate $ins\_a\_closure$ for each concept/role A; then, we use this new predicate to define new derivation rules for ins_a in case they do not get in conflict with any axiom in $\mathcal{A}_{\mathcal{U}}^+$, thus accomplishing (2).

In the following, we first define how we obtain these new derivation rules, and then we prove that the insertion/deletion instructions generated by this extended datalog program $\mathcal{D}$ are sound and complete with respect to the coherent-semantics.

5.1 Translation Rules

Capturing Closure Insertions Due to Deletions. For each positive inclusion axiom $B \sqsubseteq A$ in the closure of T, where A is an atomic concept, let $A_1, ldots, A_m$ be all the atomic concepts having a positive inclusion axiom of the form $A \sqsubseteq A_i$ in the TBox closure of T, then we define the rules:

```
ins_a_closure(X) :- del_b(X), not a(X), not ins_a_request(X),
    not
del_a_request(X), not del_a1_request(X),\ldots, not
    del_am_request(X).
```

For example, for the assertion FullProfessor $\sqsubseteq$ Professor, we define the rules:

```
ins_prof_closure(X) :- del_fullprof(X), not prof(X), not
    ins_prof_request(X), not del_prof_request(X), not
    del_person_request(X).
```

Intuitively, when we delete a FullProfessor(o), we might need to insert Professor(o) because of the closure of the semantics. However, such *closure insertion* is not necessary if Professor(o) is already in the ABox, or if there is a request for its insertion, or if it is requested for deletion (either Professor(o) itself or its parent concepts Person(o)). We define similar rules for role positive inclusion axioms and positive inclusion axioms in which the left-hand side uses the $\exists$ constructor.

Defining New Insertions Due to Closure Insertions. Once we have defined the predicates *ins_a_closure*, we use them for defining new insertions in case they do not get in conflict with the assertions in $\mathcal{A}_{\mathcal{U}}^{+}$. To do so, for each atomic concept A, let $B_1, \ldots, B_n$ be all the concepts having a negative inclusion axiom with A in the TBox closure of T, then we define the rules:

```
ins_a(X) :- ins_a_closure(X), not ins_b1_request(X)\ldots not
    ins_bn_request(X).
```

Following the previous example, we would define:

```
ins_prof(X):-ins_prof_closure(X), not ins_student_request(X).
```

Intuitively, any derived *closure insertion* of Professor(o) should be applied only if it does not get in conflict with any negative inclusion axiom. Such a conflict might arise if there is a request to insert some Student(o) because of the negative inclusion assertion Student $\sqsubseteq \neg$Professor. Similarly, we define the rules for roles.

5.2 Datalog Program Soundness and Completeness

We finally state that the generated insertion/deletions instructions generated by the datalog program $\mathcal{D}$ is sound and complete with respect to the coherent-semantics (the proof of the following theorem can be obtained by easily extending the proofs of Theorems 1 and 2).

Theorem 3. *Given a consistent ontology* $\langle \mathcal{T}, \mathcal{A} \rangle$, *and an update* $\mathcal{U}$, *the datalog program* $\mathcal{D}$ *obtained through the translation defined in Sects. 4.1 and 5.1, satisfies that:*

(i) incoherent_update() is true in $\mathcal{D}$ *iff* $\mathcal{U}$ *is incoherent with* $\mathcal{T}$;
(ii) if $\mathcal{U}$ *is coherent with* $\mathcal{T}$, *then for each concept/role* A, *ins_a($\bar{o}$) is true in* $\mathcal{D}$ *iff* $A(\bar{o}) \in \langle \mathcal{T}, \mathcal{A} \rangle \bullet \mathcal{U} \setminus \mathcal{A}$, *and del_a($\bar{o}$) is true in* $\mathcal{D}$ *iff* $A(\bar{o}) \in \mathcal{A} \setminus \langle \mathcal{T}, \mathcal{A} \rangle \bullet \mathcal{U}$.

6   Implementation and Experiments

To show the feasibility and scalability of our technique, we have developed a Java program that, given a closed $DL\text{-}Lite_A$ TBox, builds the datalog program that generates the insertion/deletion instructions for applying a coherent-semantic update. Furthermore, the program translates this datalog into standard SQL queries. Since these queries depend only on the TBox, but not on the ABox nor the requested update, all of them are created in compilation time and stored in the database as SQL views. Thus, on runtime, the user can generate the instructions by means of inserting the operations s/he wants to perform in the *ins_a_request/del_a_request* tables of the database and querying these views.

We have run the experiments using a $DL\text{-}Lite_A$ approximation of the LUBM benchmark, an ontology describing university concepts (e.g., teachers, departments, etc.) with 75 basic concept/roles and 243 assertions. For our purposes, we have removed those axioms not expressible in $DL\text{-}Lite_A$, and added 20 disjointness/functional assertions to increase the complexity of the updates. Thus, our final ontology consisted of 195 axioms.

Regarding the data, we have created different ABoxes of increasing size (from 10^5 to $3.5 * 10^7$ assertions). To do so, we have modified the UBA Data Generator to create a single university, but with an increasing number of connected departments, teachers, etc. Due to this increasing number of connected objects, the updates became more complex when increasing the data size. Then, we have defined an update request by means of selecting 3 tuples to delete, and 3 tuples to insert. Such tuples were selected in a way to ensure several interactions with the TBox assertions, thus, generating several insertions/deletions.

In Fig. 1 we summarize the results we have obtained using the MySQL 5.7 DBMS, running on a Windows 8.1 over an Intel Core i7-4710HQ, with 8GB of RAM [2]. In particular, we show the times to generate the instructions (x points in the first diagram), the time to generate and execute the instructions (+ points in the first diagram), and the number of instructions generated (x points in the second diagram). We also depict the different trend lines in the diagrams.

As it can be seen, our method has generated from 139 insertion/deletion instructions in 12 s for the smallest ABox, to 479 instructions in 16 s for the largest. Thus, although there is a constant time penalty of about 12 s to generate the instructions, the time increment in function of the ABox size is small. Adding this time to the time to execute the instructions, we got a total cost

---

[2] More experiment details and results at www.essi.upc.edu/~xoriol/dllitea/.

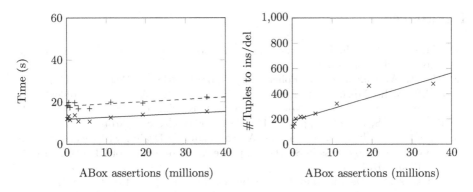

Fig. 1. Experimental results

near to 20s. We argue that this low time increment behavior is due to the fact that, in *DL-Lite$_A$*, an update request only causes updates *locally*, i.e., the unique tuples to insert/delete are a subset of those that are *connected* to the requested insertions/deletions. Thus, since ABoxes tends to increase its size by considering more objects, rather than infinitely augmenting the connectivity between them, increasing the ABox size barely increases the generated instructions, as can be seen in the second diagram. Hence, we argue that our approach can be effectively used in practice with large ABoxes.

7 Conclusions

In this paper we have shown that the *DL-Lite* family, in particular *DL-Lite$_A$*, enjoys the first-order rewritability of instance level updates. Apart from the theoretical interest, this result gives us a practical and effective technique to perform updates over *DL-Lite* ontologies.

Although we have not considered any specific syntax to express the update, what we proposed here is fully compatible with SPARQL update operators studied in [2]. There, the set of insertions and deletions are defined through unions of conjunctive queries over the current ontology. We can immediately extend our approach in the same way, producing update operators that are equivalent to the ones defined in [2] in the case of RDFS, but that deal with the more expressive *DL-Lite$_A$* and OWL 2 QL languages.

There are several directions for future work, but maybe the most compelling one, encouraged by the practical applicability of our results, is to extend our datalog-based approach blurring the distinction between TBox and ABox assertions, in line with the use of SPARQL over OWL 2 QL ontologies.

Acknowledgments. This research has been partially supported by the EU under FP7 project Optique (grant no. FP7-318338), by the Ministerio de Economía y Competitividad (project TIN2014-52938-C2-2-R), and by the Sapienza project "Immersive Cognitive Environments".

References

1. Abiteboul, S., Hull, R., Vianu, V.: Foundations of Databases. Addison Wesley Publ. Co., Boston (1995)
2. Ahmeti, A., Calvanese, D., Polleres, A.: Updating RDFS ABoxes and TBoxes in SPARQL. In: Mika, P., Tudorache, T., Bernstein, A., Welty, C., Knoblock, C., Vrandečić, D., Groth, P., Noy, N., Janowicz, K., Goble, C. (eds.) ISWC 2014, Part I. LNCS, vol. 8796, pp. 441–456. Springer, Heidelberg (2014)
3. Ahmeti, A., Calvanese, D., Polleres, A., Savenkov, V.: Dealing with inconsistencies due to class disjointness in SPARQL update. In: Proceedings of DL 2015, CEUR, vol. 1350 (2015)
4. Calvanese, D., De Giacomo, G., Lembo, D., Lenzerini, M., Rosati, R.: Tractable reasoning and efficient query answering in description logics: the DL-Lite family. J. Autom. Reason. **39**(3), 385–429 (2007)
5. Calvanese, D., De Giacomo, G., Lembo, D., Lenzerini, M., Rosati, R.: Data complexity of query answering in description logics. Artif. Intell. **195**, 335–360 (2013)
6. Calvanese, D., Kharlamov, E., Nutt, W., Zheleznyakov, D.: Evolution of *DL-Lite* knowledge bases. In: Patel-Schneider, P.F., Pan, Y., Hitzler, P., Mika, P., Zhang, L., Pan, J.Z., Horrocks, I., Glimm, B. (eds.) ISWC 2010, Part I. LNCS, vol. 6496, pp. 112–128. Springer, Heidelberg (2010)
7. Giacomo, G., Lenzerini, M., Poggi, A., Rosati, R.: On instance-level update erasure in description logic ontologies. J. Log. Comput. Spec. Issue Ontol. Dyn. **19**(5), 745–770 (2009)
8. Eiter, T., Gottlob, G.: On the complexity of propositional knowledge base revision, updates and counterfactuals. Artif. Intell. **57**, 227–270 (1992)
9. Fagin, R., Ullman, J.D., Vardi, M.Y.: On the semantics of updates in databases. In: Proceedings of PODS 1983, pp. 352–365 (1983)
10. Flouris, G., Manakanatas, D., Kondylakis, H., Plexousakis, D., Antoniou, G.: Ontology change: classification and survey. Knowl. Eng. Rev. **23**(2), 117–152 (2008)
11. Flouris, G., Plexousakis, D.: Handling ontology change: survey and proposal for a future research direction. Technical report TR-362 FORTH-ICS, Institute of Computer Science, Forth, Greece (2005)
12. Gärdenfors, P.: Propositional logic based on the dynamics of belief. J. Symb. Log. **50**(2), 390–394 (1985)
13. Ginsberg, M.L.: Counterfactuals. Artif. Intell. **30**(1), 35–79 (1986)
14. Ginsberg, M.L., Smith, D.E.: Reasoning about action I: a possible worlds approach. Technical report KSL-86-65, Knowledge Systems, AI Laboratory (1987)
15. Guo, Y., Pan, Z., Heflin, J.: LUBM: a benchmark for OWL knowledge base systems. J. Web Semant. **3**(2–3), 158–182 (2005)
16. Katsuno, H., Mendelzon, A.: On the difference between updating a knowledge base and revising it. In: Proceedings of KR 1991, pp. 387–394 (1991)
17. Kharlamov, E., Zheleznyakov, D., Calvanese, D.: Capturing model-based ontology evolution at the instance level: the case of DL-Lite. J. Comput. Syst. Sci. **79**(6), 835–872 (2013)
18. Lenzerini, M., Savo, D.F.: On the evolution of the instance level of DL-Lite knowledge bases. In: Proceedings of DL 2011, CEUR, vol. 745 (2011). http://www.ceur-ws.org
19. Lenzerini, M., Savo, D.F.: Updating inconsistent description logic knowledge bases. In: Proceedings of ECAI 2012 (2012)

20. Liu, H., Lutz, C., Milicic, M., Wolter, F.: Updating description logic ABoxes. In: Proceedings of KR 2006, pp. 46–56 (2006)
21. Poggi, A., Lembo, D., Calvanese, D., De Giacomo, G., Lenzerini, M., Rosati, R.: Linking data to ontologies. In: Spaccapietra, S. (ed.) Journal on Data Semantics X. LNCS, vol. 4900, pp. 133–173. Springer, Heidelberg (2008)
22. Stojanovic, L., Maedche, A., Motik, B., Stojanovic, N.: User-driven ontology evolution management. In: Proceedings of the 13th International Conference on Knowledge Engineering and Knowledge Management, pp. 133–140 (2002)
23. Winslett, M.: Updating Logical Databases. Cambridge University Press, Cambridge (1990)

Are Names Meaningful? Quantifying Social Meaning on the Semantic Web

Steven de Rooij, Wouter Beek$^{(\boxtimes)}$, Peter Bloem, Frank van Harmelen, and Stefan Schlobach

Department of Computer Science, VU University Amsterdam,
Amsterdam, Netherlands
{s.rooij,w.g.j.beek,p.bloem,frank.van.harmelen,k.s.schlobach}@vu.nl

Abstract. According to its model-theoretic semantics, Semantic Web IRIs are individual constants or predicate letters whose names are chosen arbitrarily and carry no formal meaning. At the same time it is a well-known aspect of Semantic Web *pragmatics* that IRIs are often constructed mnemonically, in order to be meaningful to a human interpreter. The latter has traditionally been termed 'social meaning', a concept that has been discussed but not yet quantitatively studied by the Semantic Web community. In this paper we use measures of mutual information content and methods from statistical model learning to quantify the meaning that is (at least) encoded in Semantic Web names. We implement the approach and evaluate it over hundreds of thousands of datasets in order to illustrate its efficacy. Our experiments confirm that many Semantic Web names are indeed meaningful and, more interestingly, we provide a quantitative lower bound on how much meaning is encoded in names on a per-dataset basis. To our knowledge, this is the first paper about the interaction between social and formal meaning, as well as the first paper that uses statistical model learning as a method to quantify meaning in the Semantic Web context. These insights are useful for the design of a new generation of Semantic Web tools that take such social meaning into account.

1 Introduction

The Semantic Web constitutes the largest logical database in history. Today it consists of at least tens of billions of atomic ground facts formatted in its basic assertion language RDF. While the meaning of Semantic Web statements is formally specified in community Web standards, there are other aspects of meaning that go beyond the Semantic Web's model-theoretic or formal meaning [12].

Model theory states that the particular IRI chosen to identify a resource has no semantic interpretation and can be viewed as a black box: "urirefs are treated as logical constants."[1] However, in practice IRIs are not chosen randomly, and similarities between IRIs are often used to facilitate various tasks on RDF data, with ontology alignment being the most notable, but certainly not the only one.

---

[1] See https://www.w3.org/TR/2002/WD-rdf-mt-20020429/#urisandlit.

© Springer International Publishing AG 2016
P. Groth et al. (Eds.): ISWC 2016, Part I, LNCS 9981, pp. 184–199, 2016.
DOI: 10.1007/978-3-319-46523-4_12

Our aim is to evaluate (a lower bound on) the amount of information the IRIs carry about the structure of the RDF graph.

A simple example: Taking RDF graphs G (Listing 1.1) and H (Listing 1.2) as an example, it is easy to see that these graphs are structurally isomorphic up to renaming of their IRIs. This implies that, under the assumption that IRIs refer to objects in the world and to concepts, graphs G and H denote the same models.[2]

Listing 1.1. Serialization of graph G.

```
abox:item1024  rdf:type       tbox:Tent    .
abox:item1024  tbox:soldAt    abox:shop72  .
abox:shop72    rdf:type       tbox:Store   .
```

Listing 1.2. Serialization of graph H.

```
fy:jufn1024  pe:ko9sap_     fyufnt:Ufou   .
fy:jufn1024  fyufnt:tmffqt  fy:aHup       .
fy:aHup      pe:ko9sap_     fyufnt:70342  .
```

Even though graphs G and H have the same formal meaning, an intelligent agent – be it human or not – may be able to glean more information from one graph than from the other. For instance, even a human agent that is unaware of RDF semantics may be inclined to think that the object described in graph G is a tent that is sold in a shop. Whether or not the constant symbols `abox:item1024` and `fy:jufn1024` denote a tent is something that cannot be glanced from the formal meaning of either graph. In this sense, graph G may be said to purposefully mislead a human agent in case it is not about a tent sold in a shop but about a dinosaur trodding through a shallow lake. Traditionally, this additional non-formal meaning has been called *social meaning* [11].

While social meaning is a multifarious notion, this paper will only be concerned with a specific aspect of it: *naming*. Naming is the practice of employing sequences of symbols to denote concepts. Examples of names in model theory are individual constants that denote objects and predicate letters that denote relations. *The claim we want to substantiate in this paper is that in most cases names on the Semantic Web are meaningful.* This claim cannot be proven by using the traditional model-theoretic approach, according to which constant symbols and predicate letters are arbitrarily chosen. Although this claim is widely recognized among Semantic Web practitioners, and can be verified after a first glance at pretty much any Semantic Web dataset, there have until now been no attempts to quantify the *amount* of social meaning that is captured by current naming practices. We will use mutual information content as our quantitative measure

---

[2] Notice that the official semantics of RDF [13] is defined in terms of a Herbrand Universe, i.e., the IRI `dbr:London` does not refer to the city of London but to the syntactic term `dbr:London`. Under the official semantics graphs G and H are therefore *not* isomorphic and they do *not* denote the same models. The authors believe that RDF names refer to objects and concepts in the real world and not (solely) to syntactic constructs in a Herbrand Universe.

of meaning, and will use statistical model learning as our approach to determine this measure across a large collection of datasets of varying size.

In this paper we make the following contributions:

1. We prove that Semantic Web names are meaningful.
2. We quantify *how much* meaning is (at least) contained in names on a per-dataset level.
3. We provide a method that scales comfortably to datasets with hundreds of thousands of statements.
4. The resulting approach is implemented and evaluated on a large number of real-world datasets. These experiments do indeed reveal substantial amounts of social meaning being encoded in IRIs.

To our knowledge, this is the first paper about the interaction between social and formal meaning, as well as the first paper that uses statistical model learning as a method to quantify meaning in the Semantic Web context. These insights are useful for the design of a new generation of Semantic Web tools that take such social meaning into account.

2   Method

RDF Graphs and RDF Names. An RDF graph G is a set of atomic ground expressions of the form $p(s, o)$ called triples and often written as $\langle s, p, o \rangle$, where s, p and o are called the subject, predicate and object term respectively. Object terms o are either IRIs or RDF literals, while subject and predicate terms are always IRIs. In this paper we are specifically concerned with the social meaning of RDF names that occur in the subject position of RDF statements. This implies that we will not consider unnamed or blank nodes, nor RDF literals which only appear in the object position of RDF statements [5].

IRI Meaning Proxies. What IRIs on the Semantic Web mean is still an open question, and in [11] multiple meaning theories are applied to IRI names. However, none of these different theories of meaning depend on the IRI trees, neither their structure nor their string-labels. Thus, whatever theory of IRIs is discussed in the literature, it is always independent of the string (the name) that makes up the IRI. The goal of this paper is to determine if there are some forms of meaning for an IRI that correlate with the choice of their name (as defined by the IRI trees above).

For this purpose, we will use the same two "proxies" for the meaning of an IRI that were used in [10]. The first proxy for the meaning of an IRI s the *type-set* of x: the set of classes $Y^C(x)$ to which an IRI x belongs. The second proxy for the meaning of an IRI x is the *property-set* of x: the set of properties $Y^P(x)$ that are applied to IRI x. Using the standard intension (Int) and extension (Ext) functions for RDF semantics [13] we define these proxies in the following way:

$$\textbf{Type-set: } Y^C(x) := \{c \,|\, \langle Int(x), Int(c) \rangle \in Ext(Int(\texttt{rdf:type}))\}$$
$$\textbf{Property-set: } Y^P(x) := \{p \,|\, \exists o. \, \langle Int(x), Int(o) \rangle \in Ext(Int(p))\}$$

Notice that every subject term has a non-empty property-set (every subject term must appear in at least one triple) but some subject terms may have an empty type-set (in case they do not appear as the subject of a triple with the rdf:type predicate). We will simply use Y in places where both Y^C and Y^P apply. Since we are interested in relating names to their meanings we will use X to denote an arbitrary IRI name and will write $\langle X, Y \rangle$ for a pair consisting of an arbitrary IRI name and either of its meaning proxies.

Mutual Information. Two random variables X and Y are *independent iff* $P(X, Y) = P(X) \cdot P(Y)$ for all possible values of X and Y. Mutual information $I(X; Y)$ is a measure of the *dependence* between X and Y, in other words a measure of the discrepancy between the joint distribution $P(X, Y)$ and the product distribution $P(X) \cdot P(Y)$:

$$I(X; Y) = E[\log P(X, Y) - \log P(X) \cdot P(Y)],$$

where E is the expectation under $P(X, Y)$. In particular, there is no mutual information between X and Y (i.e. $I(X; Y) = 0$) when X and Y are independent, in which case the value of X carries no information about the value of Y or vice versa.

Information and Codes. While the whole paper can be read strictly in terms of probability distributions, it may be instructive to take an information theoretical perspective, since information theory inspired many of the techniques we use. Very briefly: it can be shown that for any probability distribution $P(X)$, there exists a prefix-free encoding of the values of X such that the codeword for a value x has length $-\log P(x)$ bits (all logarithms in this paper are base-2). "Prefix-free means" that no codeword is the prefix of another, and we allow non-integer codelengths for convenience. The inverse is also true: for every prefix free encoding (or "code") for the values of X, there exists a probability distribution $P(X)$, so that if element x is encoded in $L(x)$ bits, it has probability $P(x) = 2^{-L(x)}$ [4, Theorem 5.2.1].

Mutual information can thus be understood as the expected number of bits we waste if we encode an element drawn from $P(X, Y)$ with the code corresponding to $P(X)P(Y)$, instead of the optimal choice, the code corresponding to $P(X, Y)$.

Problem Statement and Approach. We can now define the central question of this paper more precisely. Let o be an IRI. Let $n(o)$, $c(o)$ and $p(o)$ be its name (a Unicode string), its type-set and its predicate-set respectively. Let O be a random element so that $P(O)$ is a uniform distribution over all IRIs in the domain. Let $X = n(O)$, $Y^C = c(O)$ and $Y^P(O)$. As explained, we use Y^C and Y^P as *meaning proxies*, if the value of X can be reliably used to predict the value of Y^C or Y^P, then we take X to contain information about its meaning. The treatment is the same for both proxies so we will use Y as a symbol for a meaning proxy in general to report results for both.

We take the IRIs from an RDF dataset and consider them to be a sequence of randomly chosen IRIs from the dataset's domain with names $X_{1:n}$ and corresponding meanings $Y_{1:n}$. Our method can now be stated as follows:

> If we can show that there is *significant* mutual information between the name X of an IRI and its meaning Y, then we have shown that the IRIs in this domain carry information about their meaning.

This implies a *best-effort* principle: if we can predict the value of Y from the value of X we have shown that X carries meaning. However, if we did not manage this prediction, there may yet be smarter methods to do so and we have not proved anything. For instance, an IRI that seems to be a randomly generated string could always be an encrypted version of a meaningful one. Only by cracking the encryption could we prove the connection. Thus, we can prove conclusively that IRIs carry meaning, but not prove conclusively that they do not.

Of course, even randomly generated IRIs might, through chance, provide *some* information about their meaning. We use a *hypothesis test* to quantify the amount of evidence we have. We begin with the following null hypothesis:

> H_0: There is no mutual information between the IRIs $X_{1:n}$ and their meanings $Y_{1:n}$.

There are two issues when calculating the mutual information between names and meaning proxies for real-world data:

1. **Computational cost:** The straightforward method for testing independence between random variables is the use of a χ^2-test. Unfortunately, this results in a computational complexity that is impractical for all but the smallest datasets.
2. **Data sparsity:** For many names there are too few occurrences in the data in order for a statistical model to be able to learn its meaning proxies. In these cases we must learn predict the meaning from attributes shared by different IRIs with the same meaning (clustering "similar" IRIs together).

To reduce computational costs, we develop a less straightforward *likelihood ratio test* that does have acceptable computational properties. To combat data-sparsity, we exploit the hierarchical nature of IRIs to group together IRIs that share initial segments. Where we do not have sufficient occurrences of the full IRI to make a useful prediction, we can look at other IRIs that share some prefix, and make a prediction based on that.

Hypothesis Testing. The approach we will use is a basic statistical hypothesis test: we formulate a null hypothesis (that the IRIs and their meanings have no mutual information) and then show that under the null hypothesis, the structure we observed in the data is very unlikely.

Let $X_{1:n}, Y_{1:n}$ denote the data of interest and let P_0 denote the true distribution of the data under the null hypothesis that X and Y are independent:

$$P_0(Y_{1:n}|X_{1:n}) = P_0(Y_{1:n}).$$

We will develop a likelihood ratio test to disprove the null hypothesis. The likelihood ratio Λ is the ratio between the probability of the data if the null hypothesis is true, divided by the probability of the data under an *alternative model* P_1, which in this case attempts to exploit any dependencies between names and semantics of terms. We are free to design the alternative model as we like: the better our efforts, the more likely we are to disprove P_0, if it can be disproven. We can never be sure that we will capture all possible ways in which a meaning can be predicted from its proxy, but, as we will see in Sect. 4, a relatively straightforward approach suffices for most datasets.

Likelihood Ratio. The likelihood ratio Λ is a test statistic contrasting the probability of the data under P_0 to the probability under an alternative model P_1:

$$\Lambda = \frac{P_0(Y_{1:n}|X_{1:n})}{P_1(Y_{1:n}|X_{1:n})} = \frac{P_0(Y_{1:n})}{P_1(Y_{1:n}|X_{1:n})}$$

If the data is sampled from P_0 (as the null hypothesis states) it is extremely improbable that this alternative model will give much higher probability to the data than P_0. Specifically:

$$P_0(\Lambda \leq \lambda) \leq \lambda \qquad (1)$$

This inequality gives us a *conservative* hypothesis test: it may underestimate the statistical significance, but it will never *over*estimate it. For instance, if we observe data such that $\Lambda \geq 0.01$, the probability of this event under the null hypothesis is less than 0.01 and we can reject H_0 with significance level 0.01. The true significance level may be even lower, but to show that, a more expensive method may be required. To provide an intuition for what (1) means, we can take an information theoretic perspective. We rewrite:

$$P_0(-\log \Lambda \geq k) \leq 2^{-k} \qquad \text{with } k = -\log \lambda$$
$$-\log \Lambda = (-\log P_0(Y_{1:n} \mid X_{1:n})) - (-\log P_1(Y_{1:n} \mid X_{1:n}))$$

That is, if we observe a likelihood ratio of Λ, we know that the code corresponding to P_1 is $-\log \Lambda$ bits more efficient than P_0. Under P_0, the probability of this event is less than 2^{-k} (i.e. less than one in a billion for as few as 30 bits). Both codes are provided with $X_{1:n}$, but the first ignores this information while the second attempts to exploit it to encode $Y_{1:n}$ more efficiently. Finally, note that H_0 does not actually specify P_0, only that it is independent of $X_{1:n}$, so that we cannot actually compute Λ. We solve this by using

$$\hat{P}(Y = y) = \frac{|\{i \mid Y_i = y\}|}{n}$$

in place of P_0. $\hat{P}$ is guaranteed to upper-bound any P_0 (note that it "cheats" by using information from the dataset).[3] This means that by replacing the unknown P_0 with $\hat{P}$ we increase Λ, making the hypothesis test *more* conservative.

3   The Alternative Model

As described in the previous section, we must design an alternative model that gives higher probability to datasets where there is mutual information between IRIs and their meanings.[4] *Any* alternative model yields a valid test, but the better our design, the more likely it is we will be to be able reject the null-hypothesis, and the more strongly we will be able to reject it.

As discussed in the previous section, for many IRIs, we may only have one occurrence. From a single occurrence of an IRI we cannot make any meaningful predictions about its predicate-set, or its type-set. To make meaningful predictions, we cluster IRIs together. We exploit the hierarchical nature of IRIs by storing them together in a *prefix-tree* (also known as a *trie*). This is a tree with labeled edges where the root node represents the empty string and each leaf node represents exactly one IRI. The tree branches at every internal node into subtrees that represent (at least) two distinct IRIs that have a common prefix. The edge labels are chosen so that their concatenation along a path starting at the root node and ending in some node n always results in the common prefix of the IRIs that are reachable from n. In other words: leaf nodes represent full IRIs and non-leaf nodes represent IRI prefixes. Since one IRI may be a strict prefix of another IRI, some non-leaf nodes may represent full IRIs as well.

For each IRI in the prefix tree, we choose a node to represent it: instead of using the full IRI, we represent the IRI by the prefix corresponding to the node, and use the set of all IRIs sharing that prefix to predict the meaning. Thus, we are faced with a trade-off: if we choose a node too far down, we will have too few examples to make a good prediction. If we choose a node too far up, the prefix will not contain any information about the meaning of the IRI we are currently dealing with.

Once the tree has been constructed we will make the choice once for all IRIs by constructing a *boundary*. A boundary B is a set of tree nodes such that every path from the root node to a leaf node contains exactly one node in B. Once the boundary has been selected we can use it to map each IRI X to a node n_X in B. Multiple IRIs can be mapped onto the same boundary node. Let X^B denote the node in the prefix tree for IRI X and boundary B. We use $\mathcal{B}$ to denote the set of all boundaries for a given IRI tree.

For now, we will take the boundary as a given, a parameter of the model. Once we have described our model $P_1(Y_{1:n} \mid X_{1:n}, B)$ with B as a parameter, we will describe how to deal with this choice.

---

[3] A detailed proof for this, and for (1) is shared as an external resource at http://wouterbeek.github.io/iswc2016_appendix.pdf.

[4] Or, equivalently, we must design a code which exploits the information that IRIs carry about their meaning to store the dataset efficiently.

We can now describe our model P_1. The most natural way to describe it, is as a sampling process. Note that we do not actually implement this process, it is simply a construction. We only compute the probability $P_1(Y_{1:n} \mid X_{1:n}, B)$ that a given set of meanings emerges from this process. Since we will use an IRI's boundary node boundary in place of the full IRI, we can rewrite

$$P_1(Y_{1:n} \mid X_{1:n}, B) = P_1(Y_{1:n} \mid X_{1:n}^B).$$

When viewed as a sampling process, the task of P_1 is to label a given sequence of IRIs with randomly chosen meanings. Note that when we view P_0 this way, it will label the IRIs independently of any information about the IRI, since $P_0(Y_{1:n} \mid X_{1:n}) = P_0(Y_{1:n})$. For P_1 to assign datasets with meaningful IRIs a higher probability than P_0, P_1 must assign the same meaning to the same boundary node more often than it would by chance.

We will use a Pitman-Yor process [16] as the basic structure of P_1.

We assign meanings to the nodes X_i^B in order. At each node, we decide whether to sample its meaning from the global set of possible meanings $\mathcal{Y}$ or from the meanings that we have previously assigned to this node.

Let $\mathcal{Y}_i$ be the set of meanings that have previously been assigned to node X_i^B: $\mathcal{Y}_i = \{y_j \mid j \leq i \wedge X_j^B = X_{i+1}^B\}$.

With probability $\frac{(|\mathcal{Y}_i|+1)/2}{i+\frac{1}{2}}$, we choose a meaning for X_i^B that has not been assigned to it before (i.e. $y \in \mathcal{Y} - \mathcal{Y}_i$). We then choose meaning y with probability $\frac{|\{j \leq i : Y_j = y\}| + \frac{1}{2}}{i + |\mathcal{Y}|\frac{1}{2}}$ [5]. Note that both probabilities have a self-reinforcing effect: every time we choose to sample a new meaning, we are more likely to do so in the future, and every time this results in a particular meaning y, we are more likely to choose y in the future.

If we do not choose to sample a new meaning, we draw y from the set of meanings previously assigned to X_i^B. Specifically:

$$P(Y_i = y \mid X_i^B) = \frac{|\{j \leq i \mid X_j^B = X_{i+1}^B, Y_j = y\}| - \frac{1}{2}}{i + \frac{1}{2}}.$$

Note that, again, the meanings that have been assigned often in the past are assigned more often in the future. These "the rich-get richer"-effects mean that the Pitman-Yor process tends to produce power-law distributions.

Note that this sampling process makes no attempt to map the "correct" meanings to IRIs: it simply assigns random ones. It is unlikely to produce a dataset that actually looks natural us. Nevertheless, a natural dataset with mutual information between IRIs and meanings still has a much higher probability under P_1 than under P_0, which is all we need to reject the null hypothesis.

While it may seem from this construction that the order in which we choose meanings has a strong influence on the probability of the sequence, it can in fact

---

[5] The Pitman-Yor process itself does not specify which new meaning we should choose, only that a new meaning should be chosen. This distribution on meanings in $\mathcal{Y}$ is inspired by the Dirichlet-Multinomial model.

be shown that every permutation of any particular sequence of meanings has the same probability (the model is *exchangeable*). This is a desirable property, since the order in which IRIs occur in a dataset is usually not meaningful.

To compute the probability of $Y_{1:n}$ for a given set of nodes $X_{1:n}$ we use

$$P_1(Y_{1:n} \mid X_{1:n}^B) = \prod_{i=0}^{n-1} P_1(Y_{i+1} \mid Y_{1:i}, X_{1:n}^B) \qquad \text{with}$$

$$P_1(Y_{i+1} = y \mid Y_{1:i}, X_{1:n}^B)$$

$$= \begin{cases} \dfrac{(|\mathcal{Y}_i| + 1)\frac{1}{2}}{i + \frac{1}{2}} \cdot \dfrac{|\{j \leq i : Y_j = y\}| + \frac{1}{2}}{i + |\mathcal{Y}|\frac{1}{2}} & \text{if } y \notin \mathcal{Y}_i, \\[2ex] \dfrac{|\{1 \leq j \leq i \mid X_j^B = X_{i+1}^B, Y_j = y\}| - \frac{1}{2}}{i + \frac{1}{2}} & \text{otherwise.} \end{cases}$$

Choosing the IRI Boundary. We did not yet specify which boundary results in clusters that are of the right size, i.e., which boundary choice of boundary gives us the highest probability for the data under P_1, and thus the best chance of rejecting the null hypothesis.

Unfortunately, which boundary B is best for predicting the meanings Y cannot be determined a priori. To get from $P_1(Y \mid X, B)$ to $P_1(Y \mid X)$, i.e. to get rid of the boundary parameter, we take a Bayesian approach: we define a prior distribution $W(B)$ on all boundaries, and compute the marginal distribution on $Y_{1:n}$:

$$P_1(Y_{1:n} \mid X_{1:n}) = \sum_{B \in \mathcal{B}} W(B) P_1(Y_{1:n} \mid X_{1:n}, B) \qquad (2)$$

This is our complete alternative model.

To define $W(B)$, remember that a boundary consists of IRI prefixes that are nodes in an IRI tree (see above). Let $\mathrm{lcp}(x_1, x_2)$ denote the longest common prefix of the IRIs denoted by tree nodes x_1 and x_2. We then define the following distribution on boundaries:

$$W(B) := 2^{-|\{\mathrm{lcp}(x_1, x_2) \mid x_1, x_2 \in B\}|}$$

Here, the set of prefixes in the exponent corresponds to the nodes that are in between the root and some boundary node, including the boundary nodes themselves. Therefore, the size of this set is equal to the number of nodes in the boundary plus all internal nodes that are closer to the root. Each such node divides the probability in half, which means that W can be interpreted as the following generative process: starting from the root, a coin is flipped to decide for each node whether it is included in the boundary (in which case its descendants are not) or not included in the boundary (in which case we need to recursively flip coins to decide whether its children are).

The number of possible boundaries $\mathcal{B}$ is often very large, in which case computing 2 takes a long time. We therefore use a heuristic (Algorithm 1) to lower-bound (2), by using only those terms that contribute the most to the total.

Algorithm 1. Heuristic calculation for the IRI boundary.

1: **procedure** MARGINALPROBABILITY($X_{1:n}$, $Y_{1:n}$, IRI tree with root r)
2:     $B \leftarrow \{r\}$ ▷ The boundary in the sum in (2)
3:     $Q \leftarrow \{r\}$ ▷ Queue of boundary states to be expanded
4:     $best\_term \leftarrow W(B)P_1(Y_{1:n} \mid X_{1:n}, B)$ ▷ Largest term found
5:     $acc \leftarrow best\_term$ ▷ Accumulated probability
6:     **while** $Q \neq \emptyset$ **do**
7:         $n \leftarrow shift(Q)$
8:         $B' \leftarrow B \setminus \{n\} \cup children(n)$
9:         $term \leftarrow W(B)P_1(Y_{1:n} \mid X_{1:n}, B')$
10:        $acc \leftarrow acc + term$
11:        **if** $term \geq best\_term$ **then**
12:            $(B, best\_term) \leftarrow (B', term)$
13:            $add(Q, children(n))$
        return acc ▷ Approx. $P_1(Y_{1:n} \mid X_{1:n})$ from below

---

Starting with the single-node boundary containing only the root node, we recursively expand the boundary. We compute P_1 for all possible expansion of each boundary we encountered, but we recurse only for the one which provides the largest contribution.

Note that this only weakens the alternative model: the probability under the heuristic version of P_1 is always lower than it would be under the full version, so that the resulting hypothesis tests results in a higher p-value. In short, this approximation may result in fewer rejections of the null hypothesis, but when we do reject it, we know that we would also have rejected it if we had computed P_1 over all possible boundaries. If we cannot reject, there may be other alternative models that would lead to a rejection, but that is true for the complete P_1 in (2) as well. Algorithm 1 calculates the probability of the data under the alternative model, requiring only a single pass over the data for every boundary that is tested.

4   Evaluation

In the previous section we have developed a likelihood ratio test which allows us to verify the null hypothesis that names are statistically independent from the two meaning proxies. Moreover, the alternative model P_1, provides a way of quantifying how much meaning is (at least) shared between IRI names X and meaning proxies Y.

Since we calculate P_1 on a per-dataset basis our evaluation needs to scale in terms of the number of datasets. This is particularly important since we are dealing with Semantic Web data, whose open data model results in a very heterogeneous collection of real-world datasets. For example, results that are obtained over a relatively simple taxonomy may not translate to a more complicated ontology. Moreover, since we want to show that our approach and its corresponding implementation scale, the datasets have to be of varying size and some of them have to be relatively big.

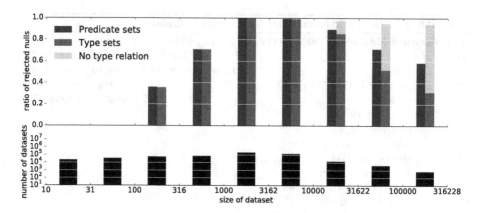

Fig. 1. The fraction of datasets for which we obtain a significant result at significance level $\alpha = 0.01$. Note that we group the datasets in logarithmic bins (i.e., the bin edges $\{e_i\}$ are chosen so that the values $\{\log e_i\}$ are linearly spaced. As explained in Sect. 2, all datasets have predicate-sets but not all datasets have type-sets. The fraction of datasets with no type-set is marked in gray.

For this experiment we use the LOD Laundromat data collection [1], a snapshot of the LOD Cloud that is collected by the LOD Laundromat scraping, cleaning and republishing framework. LOD datasets are scraped from open data portals like Datahub[6] and are automatically cleaned and converted to a standards-compliant format. The data cleaning process includes removing 'stains' from the data such as syntax errors, duplicate statements, blank nodes and more.

We processed $544,504$ datasets from the LOD Laundromat data collection, ranging from 1 to $129,870$ triples. For all datasets we calculate the Λ-value for the two meaning proxies Y^C and Y^P, noting that if $\Lambda < \alpha$, then $p < \alpha$ also, and we can reject the null-hypothesis with significance level at least α. We choose $\alpha = 0.01$ for all experiments.

Figure 1 shows the frequency with which the null hypothesis was rejected for datasets in different size ranges.

The figure shows that for datasets with at least hundreds of statements our method is usually able to reliably refute the null hypothesis at a very strong significance level of $\alpha = 0.01$. $6,351$ datasets had no instance/class-assertions (i.e., `rdf:type`-statements) whatsoever (shown in gray in Fig. 1). For these datasets it was therefore not possible to obtain results for Y^C.

Note that we may *not* conclude that no datasets with less than 100 statements contain meaningful IRIs. We had too little data to show meaning in the IRIs with our method, but other, more expensive methods may yet be successful.

In Fig. 2 we explore the correlation between the results for type-sets Y^C and property-sets Y^P. As it turns out, in cases where we do find evidence for social meaning the evidence is often overwhelming, with a $p\Lambda$-value exponentially small in terms of the number of statements. It is therefore instructive to consider not

---

[6] See http://datahub.io.

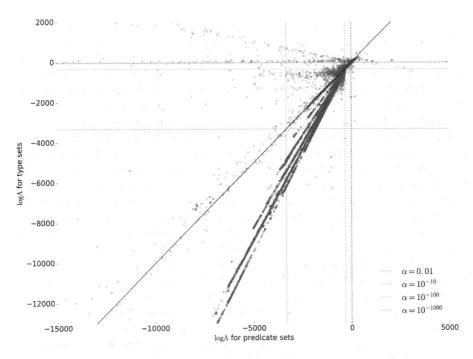

Fig. 2. This figure shows $\log \Lambda$ for both meaning proxies, for each dataset. Datasets that appear below a horizontal line provide sufficient evidence (at that α) to refute the claim that Semantic Web names do not encode type-sets Y^C. Datasets that appear to the left of a vertical line provide sufficient evidence (at that α) to refute the claim that Semantic Web names do not encode property-sets Y^P. Datasets containing no instance/class- or rdf:type-relations are not included.

the Λ-value itself but its binary logarithm. A further reason for studying $\log \Lambda$ is that $-\log \Lambda$ can be seen not only as a measure of evidence against the null hypothesis that Y and X are independent, but also as a conservative estimate of the mutual information $I(X{:}Y)$: predicting the meanings from the IRIs instead of assuming independence allows us to encode the data more efficiently by at least $-\log \Lambda$ bits.

In Fig. 2, the two axes correspond to the two meaning proxies, with Y^P on the horizontal and Y^C on the vertical axis. To show the astronomical level of significance achieved for some datasets, we have indicated several significance thresholds with dotted lines in the figure. The figure shows results for $544,504$ datasets[7] and as Fig. 2 shows, the overwhelming majority of these indicate very

---

[7] Datasets with fewer than $1,000$ statements are not included in order to get a clear picture of what happens in case we have sufficient data to refute the null, as indicated by our observations from Fig. 1. A zoomed out version of Fig. 2, scaling to $\log(p)$ values of $-300,000$ is available at https://goo.gl/r3uxpA, but is not included in this paper because its scale is no longer suitable for print.

strong support for the encoding of meaning in IRIs, measured both via mutual information content with type-sets and with property-sets. Recall that $-\log\Lambda$ is a lower bound for the amount of information the IRIs contain about their meaning. For datasets that appear to the top-left of the diagonal property-sets Y^P provide more evidence than type-sets Y^C. For points to the bottom-right of the diagonal, type-sets Y^C provide more evidence than property-sets Y^P.

Only very few datasets appear in the upper-right quadrant. Manual inspection has shown that these are indeed datasets that use 'meaningless' IRIs. There are some datasets where the $\log\Lambda$ for property-sets is substantially higher than zero; this probably occurs when there are very many property-sets so that the alternative model has many parameters to fit, whereas the null model is a maximum likelihood estimate so it does not have to pay for parameter information.

Datasets that cluster around the diagonal are ones that yield comparable results for Y^C and Y^P. There is also a substantial grouping around the horizontal axis: these are the datasets with poor `rdf:type` specifications. There is some additional clustering visible, reflecting that there is structure not only within individual Semantic Web datasets but also between them. This may be due to a single data creator releasing multiple datasets that share a common structure. These structures may be investigated further in future research.

The results reported on until now have been about the *amount of evidence against the null hypothesis*. In our final figure we report about the *amount of information that is encoded in Semantic Web names*. For this we ask ourselves the information theoretic question: how many bits of the schema information

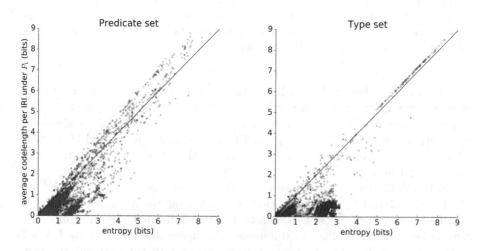

Fig. 3. Measuring the amount of information that is encoded in Semantic Web names. The horizontal axis shows the entropy of the empirical distribution of Y for a given dataset, a lower-bound for the information contained in the meaning of the average IRI. The vertical axis shows the number of bits used to encode the average meaning by the code corresponding to P_1. This is an upper bound, since P_1 may not be the optimal model. Datasets containing no type relations are not included in the right-hand figure.

in Y can be compressed by taking into account the name X? Again we make a conservative estimate: the average number of bits required to describe Y is underestimated by the empirical entropy, whereas the average number of bits we need to encode Y with our alternative model, given by $-\log(P_1(Y_{1:n}|X_{1:n})/n$, is an overestimate (because P_1 is an ad-hoc model rather than the true distribution). Again, we only consider datasets with more than $1,000$ statements.

The results in Fig. 3 show that for many datasets more than half of the information in Y, and sometimes almost all of it, *can* in fact be predicted by looking at the IRI. On the other hand, for datasets of high entropy the alternative model P_1 tends not to compress a lot. Pending further investigation, it is unclear whether this later result is due to inefficiency in the alternative model or because the IRIs in those datasets are just less informative.

5   Related Work

Statistical Observations. Little is known about the information theoretic properties of real-world RDF data. Structural properties of RDF data have been observed to follow a power-law distribution. These structural properties include the size of documents [6] and frequency of term and schema occurrence [6,15,19]. Such observations have been used as heuristics in the implementation of triple stores and data compressors.

The two meaning proxies we have used were defined by [10] who report the empirical entropy and the mutual information of both Y^C and Y^P for various datasets. However, we note that the distribution underlying Y^C and Y^P, as well as the joint distribution on pairs $\langle Y^P, Y^C \rangle$, is unknown and has to be *estimated* from the observed frequencies of occurrence in the data. This induces a bias in the reported mutual information. Specifically, the mutual information may be substantial even though the variables Y^C and Y^P are in fact *independent*. Our approach in Sect. 2 avoids this bias.

Social Meaning. The concept of social meaning on the Semantic Web was actively discussed on W3C mailing lists during the formation of the original RDF standard in 2003–2004. social meaning is similar to what has been termed the "human-meaningful" approach to semantics by [9]. While social meaning has been extensively studied from a philosophical point of view by [11], to the best of our knowledge there are no earlier investigations into its *empirical* properties.

Perhaps most closely related is again the work in [10]. They study the same two meaning proxies (which we have adopted from their work), and report on empirical entropy and mutual information of *between* two quantities. That is essentially different from our work, where we study the entropy and mutual information content not between these two quantities, but between each of them and the IRIs whose formal meaning they capture. Thus, [10] tells us whether type-sets are predictive of predicate-sets, whereas our work tells us whether IRIs are predictive of their type- and predicate-sets.

Naming RDF Resources. Human readability and memorization are explicit design requirements for URIs and IRIs. [3,8,20] At the same time, best practices have been described that advise against putting "too much meaning" into IRIs [20]. This mainly concerns aspects that can easily change over time and that would, therefore, conflict with the permanence property of so-called 'Cool URIs' [2]. Examples of violations of best practices include indicators of the status of the IRI-denoted resource ('old', 'draft'), its access level restrictions ('private', 'public') and implementation details of the underlying system ('/html/', '.cgi').

Several guidelines exist for minting IRIs with the specific purpose of naming RDF resources. [17] promotes the use of the aforementioned Cool URIs due to the improved referential permanence they bring and also prefers IRIs to be mnemonic and short. In cases in which vocabularies have evolved over time the date at which an IRI has been issued or minted has sometimes been included as part of that IRI for versioning purposes.

6   Conclusion and Future Work

In this paper we have shown that Semantic Web data contains social meaning. Specifically, we have quantitatively shown that the social meaning encoded in IRI names significantly coincides with the formal meaning of IRI-denoted resources.

We believe that such quantitative knowledge about encoded social meaning in Semantic Web names is important for the design of future tools and methods. For instance, ontology alignment tools already use string similarity metrics between class and property names in order to establish concept alignments [18]. The Ontology Alignment Evaluation Initiative (OAEI) contains specific cases in which concept names are (consistently) altered [7]. The analytical techniques provided in this paper can be used to predict a priori whether or not such techniques will be effective on a given dataset. Specifically, datasets in the upper right quadrant of Fig. 2 are unlikely to yield to those techniques.

Similarly, we claim that social meaning should be taken into account when designing reasoners. [14] already showed how the names of IRIs could be used effectively as a measure for semantic distance in order to find coherent subsets of information. This is a clear case where social meaning is used to support reasoning with formal meaning. Our analysis in the current paper has shown that such a combination of social meaning and formal meaning is a fruitful avenue to pursue.

References

1. Beek, W., Rietveld, L., Bazoobandi, H.R., Wielemaker, J., Schlobach, S.: LOD laundromat: a uniform way of publishing other people's dirty data. In: Mika, P., et al. (eds.) ISWC 2014, Part I. LNCS, vol. 8796, pp. 213–228. Springer, Heidelberg (2014)
2. Berners-Lee, T.: Cool URIs don't change (1998). http://www.w3.org/Provider/Style/URI.html.en
3. Berners-Lee, T., Fielding, R., Masinter, L.: Uniform resource identifier: generic syntax, January 2005. http://www.rfc-editor.org/info/rfc3986

4. Cover, T.M., Thomas, J.A.: Elements of Information Theory. Wiley-Interscience, Hoboken (2006)
5. Cyganiak, R., Wood, D., Lanthaler, M.: RDF 1.1 concepts and abstract syntax (2014)
6. Ding, L., Finin, T.W.: Characterizing the semantic web on the web. In: Cruz, I., Decker, S., Allemang, D., Preist, C., Schwabe, D., Mika, P., Uschold, M., Aroyo, L.M. (eds.) ISWC 2006. LNCS, vol. 4273, pp. 242–257. Springer, Heidelberg (2006)
7. Dragisic, Z., Eckert, K., Euzenat, J., Faria, D., Ferrara, A., Granada, R., Ivanova, V., Jiménez-Ruiz, E., Kempf, A.O., Lambrix, P., et al.: Results of the ontology alignment evaluation initiative 2014. In: Proceedings of the 9th International Conference on Ontology Matching, vol. 1317, pp. 61–104 (2014)
8. Duerst, M., Suignard, M.: Internationalized resource identifiers, January 2005. http://www.rfc-editor.org/info/rfc3987
9. Farrugia, J.: Model-theoretic semantics for the web. In: Proceedings of the 12th Internaional Conference on WWW, pp. 29–38. ACM (2003)
10. Gottron, T., Knauf, M., Scheglmann, S., Scherp, A.: A systematic investigation of explicit and implicit schema information on the linked open data cloud. In: Cimiano, P., Corcho, O., Presutti, V., Hollink, L., Rudolph, S. (eds.) ESWC 2013. LNCS, vol. 7882, pp. 228–242. Springer, Heidelberg (2013)
11. Halpin, H.: Social Semantics: The Search for Meaning on the Web. Semantic Web and Beyond, 13th edn. Springer, Heidelberg (2013)
12. Halpin, H., Thompson, H.: Social meaning on the web: from Wittgenstein to search engines. IEEE Intell. Syst. **24**(6), 27–31 (2009)
13. Hayes, P.J., Patel-Schneider, P.F.: RDF 1.1 semantics (2014)
14. Huang, Z., van Harmelen, F.: Using semantic distances for reasoning with inconsistent ontologies. In: Sheth, A.P., Staab, S., Dean, M., Paolucci, M., Maynard, D., Finin, T., Thirunarayan, K. (eds.) ISWC 2008. LNCS, vol. 5318, pp. 178–194. Springer, Heidelberg (2008)
15. Oren, E., Delbru, R., Catasta, M., Cyganiak, R., Stenzhorn, H., Tummarello, G.: Sindice.com: a document-oriented lookup index for Open Linked Data. Int. J. Metadata Semant. Ontol. **3**(1), 37–52 (2008)
16. Pitman, J., Yor, M.: The two-parameter Poisson-Dirichlet distribution derived from a stable subordinator. Ann. Probab. **25**(2), 855–900 (1997)
17. Sauermann, L., Cyganiak, R.: Cool URIs for the semantic web (2006)
18. Stoilos, G., Stamou, G., Kollias, S.D.: A string metric for ontology alignment. In: Gil, Y., Motta, E., Benjamins, V.R., Musen, M.A. (eds.) ISWC 2005. LNCS, vol. 3729, pp. 624–637. Springer, Heidelberg (2005)
19. Theoharis, Y., Tzitzikas, Y., Kotzinos, D., Christophides, V.: On graph features of semantic web schemas. IEEE Trans. Knowl. Data Eng. **20**(5), 692–702 (2008)
20. Théraux, O.: Common HTTP implementation problems, January 2003. http://www.w3.org/TR/chips/

User Validation in Ontology Alignment

Zlatan Dragisic[1]([✉]), Valentina Ivanova[1], Patrick Lambrix[1]([✉]),
Daniel Faria[2], Ernesto Jiménez-Ruiz[3], and Catia Pesquita[4]

[1] Linköping University and the Swedish e-Science Research Centre,
Linköping, Sweden
{zlatan.dragisic,patrick.lambrix}@liu.se
[2] Gulbenkian Science Institute, Oeiras, Portugal
[3] University of Oxford, Oxford, UK
[4] LaSIGE, Faculdade de Ciências, Universidade de Lisboa, Lisboa, Portugal

Abstract. User validation is one of the challenges facing the ontology alignment community, as there are limits to the quality of automated alignment algorithms. In this paper we present a broad study on user validation of ontology alignments that encompasses three distinct but interrelated aspects: the profile of the user, the services of the alignment system, and its user interface. We discuss key issues pertaining to the alignment validation process under each of these aspects, and provide an overview of how current systems address them. Finally, we use experiments from the Interactive Matching track of the Ontology Alignment Evaluation Initiative (OAEI) 2015 to assess the impact of errors in alignment validation, and how systems cope with them as function of their services.

1 Introduction

The growth of the ontology alignment area in the past years has led to the development of many ontology alignment systems. In most cases, these systems apply fully automated approaches where an alignment is generated for a given pair of input ontologies without any human intervention. However, after many editions of the Ontology Alignment Evaluation Initiative (OAEI), it is becoming clear to the community that there are limits to the performance (in terms of precision and recall of the alignments) of automated systems, as adopting more advanced alignment techniques has brought diminishing returns [21,40]. This is likely due to the complexity and intricacy of the ontology alignment process, with each task having its particularities, dictated by both the domain and the design of the ontologies. Thus, automatic generation of mappings should be viewed only as a first step towards a final alignment, with validation by one or more users being essential to ensure alignment quality [12].

Having users validate an alignment enables the detection and removal of erroneous mappings, and potentially the addition of alternative mappings, or altogether new ones, not detected by the alignment system. Additionally, if user validation is done during the alignment process, it enables the adjustment of

© Springer International Publishing AG 2016
P. Groth et al. (Eds.): ISWC 2016, Part I, LNCS 9981, pp. 200–217, 2016.
DOI: 10.1007/978-3-319-46523-4_13

system settings, the selection of the most suitable alignment algorithms, and the incorporation of user knowledge in them [40]. While users can make mistakes, experiments have shown that user validation is still beneficial up to an error rate of 20 % [26], although the exact error threshold will depend on the alignment system and how it makes use of the user input.

The relevance of user involvement in ontology alignment is evidenced by the fact that nearly half of the challenges facing the community identified in [46] are directly related to it. These include *explanation of matching results* to users, fostering *user involvement* in the matching process, and *social and collaborative matching*. Moreover, the lack of evaluation of the quality and effectiveness of user interventions was identified as one of the general issues after six years of experience in the OAEI [12], leading to the introduction of the Interactive Matching track in the OAEI 2013 campaign [40] where user validation was simulated using an oracle. This track was extended in 2015 [4] to also take into account erroneous user feedback to the systems as well as additional use cases.

There have been earlier studies addressing user involvement in ontology alignment and identifying and evaluating the requirements and techniques involved therein [14,17,21,28]. More recently, the requirements for fostering user support for large-scale ontology alignment were identified and current systems were evaluated [24]. However, these studies focused mostly on the user interface of alignment systems. While that is a critical aspect for user involvement, there are other important aspects which have been largely unaddressed, such as how systems cope with erroneous user input or how they maximize the value of limited input.

In this paper we present a broader study of user validation in ontology alignment. In Sect. 2, we identify the key issues regarding user validation of ontology alignments by reviewing existing systems and literature related to ontology alignment, as well as drawing from our experience in the field. These issues pertain to three categories: the user profile, the alignment systems' services and their user interfaces. In Sect. 3, we first assess how current systems deal with the identified issues in a qualitative evaluation (Subsect. 3.1), then use the experiments from the Interactive Matching track of the OAEI 2015 campaign to show how some of these issues impact alignment quality (Subsect. 3.2). While the experiments from the OAEI Interactive track considered the erroneous input as a function solely of user knowledge, here we discuss them in light of different aspects of user expertise.

2   Issues Regarding User Alignment Validation

Alignment validation requires users to first become familiar with the ontologies and their formal representations, and to grasp the view of the ontology modelers, before being able to understand and decide on the mappings provided by an alignment system or creating mappings by themselves [15]. Thus, it is a cognitively demanding task that involves a high memory load and complex decision making, and is inherently error-prone because of different levels of expertise, differences in interpretation or perception, and human biases [17].

There are three categories of issues that affect alignment validation: the profile of the user, i.e., his domain and technical expertise, and his expertise with the alignment system (Subsect. 2.1); the system services, concerning how systems formulate user interactions and how they capitalize on user input (Subsect. 2.2); and the user interfaces, including the impact of visualization and interaction strategies on the alignment validation process (Subsect. 2.3).

2.1 User Profile

The **domain expertise of the user** concerns his knowledge about the domain of the aligned ontologies, and therefore his ability to assess the correctness of a mapping conceptually (e.g., whether two ontology classes mapped as equivalent actually represent the same concept in the domain). Thus, domain expertise is critical for alignment quality, and the lack thereof is likely to be the main source of erroneous input from a user, particularly in specialized domains with complex terminology such as the life sciences.

The **technical expertise of the user** pertains to his knowledge about ontologies themselves, and his experience in knowledge engineering and modeling, and therefore his ability to assess the correctness of a mapping formally (i.e., whether a mapping is logically sound given the constraints of the two ontologies). While domain knowledge is critical for alignment validation, domain experts are often not familiar with knowledge engineering concepts and formal representations [5], and may have difficulty grasping the consequences of a mapping in the context of the ontologies, or even in perceiving subtle differences in modeling that make that mapping incorrect.

While alignment system users will usually fall under the categories of domain expert or knowledge engineer, it should be noted that domain and technical expertise are not disjoint. Indeed, the development of tools like Protégé has allowed domain experts to delve into knowledge engineering [20]. Nevertheless, the differences between these two user types are important for the design of every knowledge-based system, and should be addressed both when designing the system and when building support for it. For instance, in order to assist users with limited technical expertise, alignment systems should provide information about the structure of the ontologies and the entailments of a mapping in a manner that is intuitive to understand. Likewise, in order to assist users with limited domain expertise, systems should provide detailed contextual and conceptual information about the mapping. Indeed, a recent study showed that, given enough contextual help, the quality of the validation of non-domain experts can approximate that of domain experts [36] – although this is likely to depend on the domain in question.

The final aspect of user expertise is **expertise with the alignment system**, which concerns the user's familiarity with the functionality of the system and its visual representations. Novice users can face comprehension difficulties and make erroneous decisions, not for lack of domain or technical expertise, but because they cannot fully acquire the information made available about a mapping or its entailments. It is up to the alignment system to be as intuitive as possible in both functionality and visual representations so that novice users can focus on the alignment process and are not limited by their lack of expertise with the system [35]. In this context, it is important to consider that different visual representations are suited for conveying different types of information, as we will detail in Subsect. 2.3. Systems should also provide support to expert users in the form of shortcuts or customizations, so that they can speed up their work.

Users can be expected to make mistakes in alignment validation [5,22], be that due to lack of domain expertise, technical expertise, or expertise with the alignment system. However, the possibility of user errors is often disregarded in existing alignment systems. On the one hand, it is true that users are generally expected to make less errors than automated systems, and experiments have shown that up to an error rate of 20 %, user input is still beneficial [26]. On the other hand, there are risks to taking user input for granted, particularly when that input is given during the alignment process, and inferences are drawn from it, leading to the potential propagation of errors. An example of this is given in [26], where user validated relations during an alignment repair step are fixed, meaning that they cannot be removed during subsequent steps, and other potentially correct relations may have to be removed instead.

User errors can be prevented to some extent by warning the user when contradicting validations are made [23] or by preemptively removing mappings that lead to logical conflicts. In a multi-user setting (e.g. [7,43]), errors may be diluted through a voting strategy, where the mapping confidence is proportional to the consensus on the mapping, by accepting the decision made by a majority of the users [43], or by adopting a more skeptical approach where full agreement between the users is required [7]. However, given the limited availability of users for alignment validation, systems cannot rely on having multiple users to prevent user errors.

One way of assessing the impact of the user profile on the alignment quality is by simulating the user input by an oracle with different error rates [33], which is the strategy we have adopted in our evaluation (Subsect. 3.2).

2.2   System Services

Alignment validation is an extensive task, particularly when large ontologies are involved, as alignments can include several thousand mappings. Since users capable of performing alignment validation are a scarce and valuable resource, alignment systems cannot expect them to be able to validate a whole alignment. Rather, they must limit their demand for user intervention and exploit that intervention to maximize its value, wherein lies one of the main challenges of alignment validation [26,38].

With regard to demand for user intervention, several strategies have been implemented by alignment systems for limiting the number of mapping suggestions to be validated by the user (**suggestions selection** - {Sys.e}). The simplest and most common of these consists of employing threshold values for different alignment algorithms. Other, more sophisticated filtering approaches include filtering with respect to principles (e.g., consistency, locality, and conservativity) [25] or quality checks [3], selecting only "problematic" mappings where different alignment algorithms disagree [8], and using a similarity propagation graph to select the most informative questions to ask the user [44].

One common strategy that both reduces demand for user intervention and exploits that intervention is to automatically reject alternative mappings for a concept when the user validates one of that concept's mappings [26,31].

With regard to exploiting user interventions, systems can adopt different strategies depending on the **stage of involvement** of the user in the alignment process: *before* ({Sys.a}), *after* ({Sys.b}), *during* ({Sys.c}), or *iterative* ({Sys.d}).

When the validation happens *before* the matching process, the user provides an initial partial alignment which is then used by the system to guide the matching process. The partial alignment can be used in the preprocessing phase to reduce the search space [30], as input for the alignment algorithms [11,30], or to select and configure the algorithms to use [29,40,42,49].

When the validation is performed *after* the automatic alignment process, the input of the user cannot be exploited for aligning the ontologies. However, many systems still filter out mapping suggestions which are in conflict with user validations before proceeding to a final reasoning and diagnosis phase [23,26,31,37].

When the validation is done *during* the alignment process, input from the user can be extrapolated through the use of **feedback propagation** ({Sys.f}) techniques to fully exploit it. When the validation is *iterative*, the user is asked for feedback on several iterations of the alignment process, where in each iteration the alignment from the previous iteration is improved [29].

Feedback propagation techniques usually consist of propagating mapping confidence from validated mappings to those in their neighborhood, be that neighborhood defined from the structure of the ontologies [31,37,44] or from the pattern of similarity scores from the various alignment algorithms [8,30]. They usually require that the validation is done during the alignment process or is iterative, but one form of feedback propagation that systems can implement regardless of when the validation takes place in the alignment process, is conflict detection ({Sys.g}) [18,26]. This consists of testing user validated relations against the ontologies, report on the violation of logical constraints (e.g., unsatisfiable classes), and possibly ask for revalidations of certain relations to resolve the conflict.

Demand for user involvement in the matching process can be evaluated by measuring the number of questions (mapping suggestions) the system asks the user, and comparing it to the actual size of the alignment produced by the system. The effectiveness with which systems exploit user involvement can be evaluated by measuring their improvement in performance (in terms of precision and recall) over the fully automated process, and relating it with the number of questions asked.

2.3 User Interface

A graphical user interface (UI) is an indispensable part of every interactive system, as the visual system is humans' most powerful perception channel. Alignment validation is a cognitively demanding task that involves a high memory load – ontologies are complex knowledge-bases, and validating each mapping requires considering the structure and constraints of two ontologies while also keeping in mind other mappings and their logical consequences – and thus is all but impossible without visual support.

Given the complexity of ontologies and alignments, a critical aspect of visualizing them is not overwhelming the user. Humans apprehend things by using their working memory, which is limited in capacity (it can typically hold 3 ± 1 items) and thus can be easily overwhelmed when too much information is presented [48]. However, this limitation can be expanded by grouping similar things, a process called "chunking", which can be exploited by visualization designers to facilitate cognition and reduce memory load [39]. For instance, encoding properties of entities and mappings with different graphical primitives facilitates their identification and enables their chunking.

Another critical aspect of ontology alignment visualization is providing the user with sufficient information to be able to decide on the validity of each mapping, which includes lexical and structural information in the ontologies, and potentially other related mappings. This naturally competes with the need not to overwhelm the user with information, and a balance between the two must be struck. As we discussed in Subsect. 2.1, different user types are likely to have different information requirements, and alignment systems must cater to all.

The **Visual Information Seeking Mantra**, {UI.a}, [45] defines seven low-level tasks to be supported by information visualization interfaces in order to enable enhanced data exploration and retrieval: overview, zoom, filter, details-on-demand, relate, history, and extract. The former six of these were further refined for the purpose of ontology visualization [27], and all are relevant in the context of striking a balance between providing information and avoiding memory overload.

Providing enhanced information while addressing the working memory limits is also the goal of the field of **visual analytics**, {UI.b}, which combines data mining and interactive visualization techniques to aid analytic reasoning and obtain insights into (large) data sets. The application of visual analytics to ontology alignments facilitates their exploration and can provide quick answers to questions of interest from the users [2,6,8,32,34].

Another technique at the disposal of alignment systems is that of providing **alternative views** {UI.c} [6,17,29,34]. Different views may be more suitable for performing different tasks – for instance, graphs are better for information perception, whereas indented lists are better for searching [19] – and by providing alternate views, systems need not condense all relevant information into a single view, and thus avoid overwhelming the user. Also relevant in this context are maintaining the user focus in one area of the ontology [37], and preserving the user's mental map (e.g., by ensuring that the layout of the ontology remains constant).

Two strategies that facilitate chunking are grouping mappings together by different criteria to help identify patterns {UI.d}, and distinguishing between different types of mappings and their provenance {UI.e} – particularly between validated and candidate mappings [17]. Color-coding is a common and effective technique for implementing both strategies.

With regard to facilitating the decision making process, showing context and definitions of terms {UI.f} is essential, and providing recommendations and/or ranking ({UI.g}) facilitates the process by allowing the user to focus on a specific set of mappings. Also important is the **explanation of mapping suggestions** by presenting the provenance of, or justification for, a mapping ({UI.h}). Likewise, the user should be provided with feedback about the consequences of his decision {UI.i} about a mapping with regard to the alignment and ontologies, possibly through a trial execution [17].

Justifications have been identified as one of the future challenges of ontology alignment, given that many alignment systems merely present confidence values for mappings as a form of justification [38]. They require particular attention to the user type: domain experts will require detailed contextual information and a clear explanation of how a mapping suggestion was inferred, whereas for knowledge engineers summarized provenance information might suffice.

Three distinct justification approaches have been identified [13]: proof presentation, strategic flow, and argumentation. In the proof presentation approach, the explanation for why a mapping suggestion was created is given in the form of a proof, which can be a formal proof, a natural language explanation (e.g., [17, 47]), or a visualization (e.g., [29]). In the strategic flow approach the explanation is in the form of a decision flow which describes the provenance of the acquired mapping suggestion (e.g., [10, 15]). Finally, in the argumentation approach, the system gives arguments for or against certain mapping suggestions, which can be used to achieving consensus in multi-user environments.

In addition to providing visual information to support the decision process, alignment systems need to provide functionalities for the user to interact with the alignment in order to validate it. The most basic level of interaction is to allow the user to either accept or reject mapping suggestions {UI.j}. Additionally, the functionality of adding a mapping manually or refining a mapping suggestion {UI.k} is also important, since the system may not have captured a mapping that is required according to the user, or may not have correctly identified the mapping relationship [1, 6, 15, 16, 29].

An important functionality is searching and filtering {UI.l}, which contributes to minimize the user's cognitive load [1, 6, 17, 34]. It is relevant to enable searching/filtering both of the ontologies (e.g., to analyze the structural context of a mapping suggestion, or look for a concept to map manually) [17, 34] and of the mapping suggestions themselves [6, 17, 29].

Given the extension of the validation process, allowing the user to add metadata in the form of annotations {UI.m} [17, 29], and accommodating interruptions or sessions {UI.n} are important functionalities. However, while many systems enable interruptions through saving and loading the ontologies and alignment, this often does not preserve the provenance information.

Finally, allowing the creation of temporary mappings {UI.o} in order to test decisions is a relevant functionality for supporting the decision process [34], as is enabling trial execution to help the user understand the consequences of his decisions {UI.i} [17].

3   Evaluation

In this section, we assess how state of the art ontology alignment systems take into account the three aspects discussed above: user profile, system services, and user interface. We start by surveying systems and evaluating them qualitatively with regard to key features of their interfaces and services for processing user input in Subsect. 3.1. Then, we assess the impact of erroneous user input to the system through a series of experiments from the Interactive Matching track of OAEI 2015 in Subsect. 3.2. In these experiments, we simulate user input with varying error rates, which can reflect lack of user of expertise, but also limitations of the system's user interface. How these errors impact the alignment is dependent on the system services.

3.1   Qualitative Evaluation

We performed a qualitative evaluation of state of the art systems that incorporate user validation in the alignment process and have a mature user interface: AgreementMaker [6,8,9], AIViz [34], AML [18,41], CogZ/Prompt [16,17,37], COMA [1], LogMap [26], RepOSE [23], and SAMBO [29,31]. The results of this evaluation are summarized in Table 1.

Regarding system services, the majority of the systems ask for validations after running the matching algorithms (AgreementMaker, AIViz, AML, LogMap, and RepOSE), CogZ/Prompt involves the user during the alignment process, while SAMBO and COMA allow validations both before and after the alignment process. AgreementMaker, COMA, SAMBO, and RepOSE also allow multiple iterations of the alignment process, and allow for user involvement in multiple validation sessions.

The majority of the systems use some form of thresholds to select mapping suggestions to present to the user for validation. AML and AgreementMaker use a more refined strategy for identifying "problem" mappings to present to the user, which relies on the variance of the similarity scores of their various alignment algorithms. LogMap presents mapping suggestions that cause the violation of alignment principles such as consistency, locality, and conservativity.

With respect to feedback propagation, most systems implement at least a conflict detection mechanism, such as checking if the validated mapping contradicts previously validated mappings or results in an incoherent or inconsistent integrated ontology (AML, CogZ/Prompt, LogMap, SAMBO, RepOSE). AIViz does not implement such mechanisms and accepts user's feedback without any additional steps. AgreementMaker employs a blocking propagation strategy where the

Table 1. Issues regarding user interaction in ontology alignments addressed by state-of-the-art systems

	Issue	Agreement maker {Sys.b+d}	AlViz {Sys.b}	AML {Sys.b}	CogZ Prompt {Sys.c}	COMA {Sys.a,b+d}	LogMap {Sys.b}	SAMBO {Sys.a,b+d}	RepOSE {Sys.b+d}
System services — Stage of involvement	{Sys.a}-before, {Sys.b}-after, {Sys.c}-during, {Sys.d}-iterative	✓	✓ -	✓	-	✓ -	✓	✓ -	✓ -
Suggestions selection	{Sys.e} threshold/advanced filtering	✓	-	✓	✓	✓	✓	✓	✓
Feedback	{Sys.f} recomputation	✓	-	-	✓	✓	-	✓	-
propagation	{Sys.g} conflict detection/ blocking/revalidation	✓ -(*)	-	✓ -	✓ - -	-	✓ - -	✓ - -	✓ - -
User interface — Alignment presentation	{UI.a} 7 visual info-seeking tasks	✓	✓	✓	✓	✓ - -	-	✓ - -	✓ - -
	{UI.b} visual analytics	✓	✓ -	✓	-	-	-	-	-
	{UI.c} alternative views	✓	✓	✓	✓	-	-	✓	-
	{UI.d} grouping	✓	-	✓	✓	-	-	✓	✓
	{UI.e} validated/candidate mappings	✓ -	✓ -(**)	✓	✓	✓ - -(**)	✓ -	✓ -	✓ -
	{UI.f} metadata & context	✓	-	✓	✓	✓	✓ -	✓ - -	✓
	{UI.g} ranking/recommendations	-	✓ - -	✓ - -	✓ - -	-	✓ - -	-	✓ - -
Mapping explanation	{UI.h} provenance & justification	✓ - -	✓ - -	✓ -	✓ -	✓ - -	✓ - -	-	✓ - -
	{UI.i} impact of decisions/ consequences of actions	✓ -	✓ -	-	✓ - -	-	✓ -	-	✓ - -
Alignment interaction	{UI.j} accept/reject mapping	✓	✓ -	✓	✓	✓ -	✓	✓	✓
	{UI.k} create/refine mapping	✓	✓	✓	✓	✓	-	✓	✓ -
	{UI.l} search	-	✓	✓	✓	✓	-	✓	-
	{UI.m} user annotation	-	-	-	✓ -	-	-	✓	-
	{UI.n} session	✓ -	✓ -	✓ -	✓ -	✓	✓	✓	✓ -
	{UI.o} create temporary mapping	-	✓ - -	✓ -	- - -	-	-	-	-

In the table ✓ marks that all of the listed items are supported by the system while - marks that the issue is not covered by the system. Combinations such as ✓- and ✓- - mark that one or two of the listed items are not supported. (*) in a multi-user environment; (**) candidate and validate mappings cannot be distinguished in the user interface user can control to how many similar instances the validation is propagated. Revalidation is supported by AML and RepOSE as a part of the conflict resolution phase. AgreementMaker, CogZ/Prompt, COMA, RepOSE and SAMBO employ some form of recomputation, where the user's input is used to guide the matching process. For example, AgreementMaker propagates the user's decision to similar mappings thus increasing/decreasing the similarity value.

Regarding the representation of the ontologies and the alignment, systems typically represent ontologies as trees or graphs. Graphs are usually used as an additional representation (AlViz, CogZ) and rarely as a main representation (AML, RepOSE). Mappings are typically represented as links between corresponding nodes, or sometimes as a list/table of pairs (AML, SAMBO, CogZ, COMA, LogMap). The list/table view is used to support different interactions by systems. About half of the systems support more than one view of the alignments and ontologies, often a tree and a graph view which are more suitable for different alignment tasks [19]. Most of the systems employ strategies to group the mappings together: SAMBO presents all mappings for a particular concept together, CogZ, AML, LogMap, and RepOSE show the local neighborhoods of a mapping up to different number of levels. AgreementMaker and AlViz combine the different views with clustering algorithms and interaction techniques to support the comparison of the similarity values calculated by the different matchers (AgreementMaker) or clustering nodes of the ontologies according to a selected relationship (AlViz).

Most systems also provide detailed information for mappings individually, such as the context of the mapping and its state (e.g., whether it is accepted). However few systems provide interface support for features regarding explaining the mappings, such as why the system has suggested the mapping or how the current validation would affect other candidate or validated mappings. Most systems provide only a similarity value or employ color coding as a form of explanation for the mapping, which is insufficient for users to make informed decisions (one exception is CogZ which shows a short natural language explanation for the mapping). Thus our evaluation survey confirms findings from [24] that explanations for mappings suggestions are not well supported by the user interfaces of alignment systems, and continues to be a challenge for the alignment community [46]. Ranking and recommendation functionalities are also rarely provided by systems.

Interactions for accepting, rejecting and creating mappings manually are supported by most of the systems but the different systems do not always present this information to the user – rejected mappings for instance are rarely shown. AlViz and COMA do not distinguish between validated and candidate mappings, thus the user cannot keep track of already visited mappings. Creating temporary mappings is supported by AlViz, AML, and CogZ. Interactions to support the 7 information visualization seeking tasks are provided to a different extent by the different systems with overview usually supported and filter, history and relate rarely supported. Search is often supported but a previous survey of some of these systems found serious limitations [24]. Two systems (CogZ and SAMBO) allow the user to annotate mappings during the validation process. Sessions are directly (COMA, LogMap, SAMBO) or indirectly (by saving and loading files) supported by all systems.

3.2   Experiments in the OAEI Campaign

The Interactive Matching track of OAEI was extended in 2015 to take into account erroneous user validations and assess varying error rates and their impact of the performance of alignment systems [4]. Systems were evaluated according to their performance in terms of precision and recall versus the reference alignments, as well as in terms of number of interactions required and time between interactions.

Although the original purpose of introducing error rates in the Interactive Matching track was to simulate users with different expertise levels, the results can be interpreted and discussed in a broader light with regard to how system services are affected by and cope with errors, irrespective of their cause.

3.2.1   Setup and Systems

The OAEI SEALS client[1] allows interactive systems to pose questions regarding the correctness of a mapping to an oracle, which will simulate a user by checking the reference alignment from the respective OAEI task, and answering with a predefined error rate. In this experiment the error rates considered were 0 % (perfect oracle), 10 %, 20 % and 30 %. Systems were evaluated on three datasets from the OAEI: Conference (16 small ontologies), Anatomy (2 medium-sized ontologies) and LargeBio (3 large ontologies). For the sake of brevity, we will present only results for the Anatomy track, as the other results are similar (they can be found in [4]).

The systems that participated in the OAEI 2015 Interactive track were AML, JarvisOM, LogMap and ServOMBI. We note that not all of these systems have user interfaces, but they implement an interface to communicate with the oracle, so we can automatically evaluate the impact of the user input to the resulting alignment. We could not evaluate other systems with this experimental setup, as it requires compliance with the OAEI's SEALS client.

Apart from JarvisOM, which involves the user during the computation of the alignment, the systems all make use of user interactions exclusively in post-alignment steps. Both LogMap and AML request feedback on selected mapping suggestions and filter mapping suggestions based on the user validations. The former interacts with the user to decide on mapping suggestions which are not clear-cut cases, whereas the latter employs a query limit and other strategies to minimize user interactions. ServOMBI asks the user to validate all of its mapping suggestions and uses the validations and a stable marriage algorithm to decide on the final alignment. JarvisOM is based on an active learning strategy known as

---

[1] The SEALS client is the infrastructure used in the OAEI to automate the evaluation of ontology matching systems http://oaei.ontologymatching.org/2016/seals-eval.html.

query-by-committee: at every iteration JarvisOM asks the user for pairs of entities that have the highest disagreement between committee members and lower average euclidean distance, and at the last iteration, the classifiers committee is used to generate the alignment.

3.2.2 Results and Discussion

The evaluation results for the Anatomy track are shown in Table 2. As expected, the performance of all systems improves when they have access to an all-knowing oracle (Or^0 in the table) in comparison with their non-interactive performance

Table 2. Interactive anatomy alignment evaluation

Oracle	System	P/F/R	P/F/R Or	TReq	DReq	TP	TN	FP	FN	Size
N/A	AML	.96/.94/.93	-	–	–	–	–	–	–	1477
	JarvisOM	.36/.17/.11	-	–	–	–	–	–	–	458
	LogMap	.92/.88/.85	-	–	–	–	–	–	–	1397
	ServOMBI	.96/.75/.62	-	–	–	–	–	–	–	971
Or^0	AML	.97/.96/.95	.97/.96/.95	312	312	73	239	0	0	1491
	JarvisOM	.86/.75/.67	.86/.75/.67	7	7	4	3	0	0	1173
	LogMap	.98/.91/.85	.98/.91/.85	590	590	287	303	0	0	1306
	ServOMBI	1/.76/.62	1/.76/.62	2136	1128	955	173	0	0	935
Or^{10}	AML	.96/.95/.95	.97/.96/.95	317.3	317.3	66.3	218	23	10	1502
	JarvisOM	.76/.68/.67	.76/.68/.67	7	7	3.3	3	0.3	0.3	1475
	LogMap	.96/.89/.83	.96/.89/.83	609	609	261.3	288.3	33.7	25.7	1302
	ServOMBI	1/.71/.55	1/.74/.59	2198.7	1128	857.3	156.3	16.7	97.7	843
Or^{20}	AML	.94/.94/.94	.97/.96/.95	321.7	321.7	66.3	186.7	52.3	16.3	1525
	JarvisOM	.53/.60/.71	.53/.60/.71	8	8	4.7	1	1.3	1	2055
	LogMap	.95/.88/.82	.95/.88/.81	630	630	233	274	69	54	1321
	ServOMBI	.99/.66/.49	1/.71/.55	2257	1128	767.3	131.3	41.7	187.7	758
Or^{30}	AML	.93/.93/.94	.97/.96/.95	306	306	54	168.7	61.3	22	1526
	JarvisOM	.51/.49/.53	.51/.49/.53	7.3	7.3	4	1.7	1	0.7	1509
	LogMap	.94/.87/.82	.92/.86/.80	663	663	200.7	270.7	105.3	86.3	1334
	ServOMBI	.99/.60/.43	1/.68/.52	2329.7	1128.3	663.3	129	44.3	291.7	659

Systems were evaluated with user interactions simulated by an oracle with different error rates (Or^x corresponds to an error rate of x%) and without user interactions (N/A). The "P/F/R" column shows the Precision, F-measure and Recall obtained in the task; the "P/F/R Or" column shows the same parameters with respect to oracle, i.e., as if the errors made by the oracle were instead correct; "TReq" and "DReq" correspond respectively to the total number of requests and the number of distinct requests made by the system to the oracle; "TP", "TN", "FP" and "FN" are respectively the number of True Positive, True Negative, False Positive and False Negative answers given by the oracle; and "Size" indicates the number of mappings in the alignment produced by the system. All values in interactive settings with non-zero error rate are averages over 3 runs, to dilute the variance of the oracle errors.

(N/A in the table). Also as expected, when we increase the oracle's error rate, we observe that the performance of all systems deteriorates. However, it takes an error rate of 30 % for the user interaction not to be beneficial to most systems, which corroborates the observations in [26]. The way in which the systems exploit user interactions, how they benefit from them, and how they are affected by errors are very different.

AML is the only system that improves more in terms of recall than in terms of precision with user interactions, because it exploits them in part to test mappings with lower similarity scores than it accepts in non-interactive mode. This is why it is the system that asks the most negative questions from the oracle, proportionally. As a result, when the error rate increases, AML's precision drops below the non-interactive precision (at 20 %), but its recall remains higher than the non-interactive recall. AML is also the only system that is affected linearly by the errors, as evidenced by the fact that its performance as measured against the oracle (i.e., assuming the oracle errors are instead correct) remains constant at all error rates. This means that, unlike the other three systems, AML does not extrapolate from the user feedback about a mapping to decide on the classification of multiple mapping candidates. While extrapolation (be it through active learning, feedback propagation, or other techniques) is an effective strategy for reducing user demand, it also implies that the system will be more heavily impacted by user errors.

JarvisOM is the system that most depends on user interactions, as evidenced by the very poor quality of its non-interactive alignment. Thus, it is the system that most improves with user interactions, and the only one that improves substantially in both precision and recall. It is also the one that makes the least requests from the oracle – only 7–8 requests per alignment – as it uses these requests in an active learning approach rather than to validate a final alignment. This means it is the system that extrapolates the most from the user feedback, which as expected, makes it the one that is most affected by user errors – its F-measure drops by 26 % between 0 and 30 % errors. However, it depends so heavily on user interaction, that even at 30 % errors, its results are still better than the non-interactive ones. JarvisOM is also the system where the impact of the errors most deviates from linearity, precisely because it extrapolates from so few mappings. Another curious consequence of this is that its alignment size fluctuates considerably, increasing to almost double between 0 and 20 % errors, but then decreasing again at 30 % errors. It should be noted that JarvisOM behaves very differently in the Conference track [4], showing a linear impact of the errors, as in that case less inferences are drawn from its 7–8 oracle requests because they represent ~50 % of the Conference alignments (whereas in Anatomy they represent 0.5 %).

LogMap improves only with regard to precision with user interactions, which is curious considering it is the most balanced system regarding positive versus negative oracle answers. This means that, in this particular task, the positive questions LogMap asks the oracle all correspond to mappings it would also accept in its non-interactive setting, whereas the negative questions allow it to exclude some mappings that it would also (erroneously) accept. Due to the balance between its questions, when presented with user errors, LogMap is affected with regard to both precision and recall in approximately equal measure. However, since its precision increased substantially with user interactions, it remains higher than the non-interactive precision at all error rates, unlike the recall. Another interesting observation about LogMap is that the number of requests it makes increases slightly but steadily with the error rate, whereas other systems show stable rates. This increase is tied to the fact that user errors can lead to more complex decision trees when interaction is used in filtering steps and inferences are drawn from the user feedback. For instance, during alignment repair, if the user indicates that a mapping that would be removed by the system to solve a conflict is correct, the system may have to ask the user about one or more alternative mappings to solve that conflict, thus increasing the number of requests. In this context, the present query-based evaluation does not accurately reflect an interface-based alignment validation, where the user could be shown all the mappings that cause a conflict simultaneously.

ServOMBI is the system that improves the least with user interaction, showing an increase of only 1 % F-measure, and like LogMap improves only with regard to precision. It is also the system that makes the most oracle requests, as it asks the oracle about every mapping candidate it finds, and the only system that makes redundant questions (its total number of requests is almost double that of the distinct ones). Interestingly, it is also the only system that produces alignments that do not contain all the mappings identified as positive by the user, as some are apparently discarded by its stable marriage algorithm. Because it makes so many oracle requests, ServOMBI is strongly affected by user errors, so much so that at only 10 % errors, user interaction is no longer beneficial in terms of F-measure. In fact, since 85 % of the questions ServOMBI asks the oracle are positive, the system would have a better performance (72 % F-measure) by simply accepting all its mapping candidates than it does at 10 % errors. Because of its strong bias towards positive questions, ServOMBI feels the impact of the errors mostly in terms of recall and alignment size, whereas precision is hardly affected. However, given the number of false positive questions returned by the oracle at 30 % errors, we would expect a drop in precision as well, but it remains almost constant as the errors increase. This attests to the ability of this system's stable marriage algorithm to filter out user errors. Interestingly, the number of total oracle requests made by ServOMBI increased with the error rate, even though the number of distinct requests remains constant – as it should, considering the system already asks the user about all mapping candidates it identifies. This means that ServOMBI is making more redundant questions.

4   Conclusions

Despite the advances in automated ontology alignment techniques, user validation remains critical to ensure alignment quality, due to the complexity and diversity of ontologies and their domains. In this broad study of user validation in ontology alignment, we encompassed three distinct but interrelated aspects: the profile of the user; the ontology alignment systems services; and their user interfaces. We assessed the services and user interfaces of state of the art systems in a qualitative evaluation, and investigated the impact of errors in alignment validation through a series of experiments that revealed how systems cope with it, depending on their services.

The profile of the user is a key factor to take into account in alignment validation, as systems will not be able to rely exclusively on domain experts for validation, and even domain experts require extensive support for deciding on the validity of mappings – particularly if they have little technical expertise regarding ontologies and knowledge engineering. Thus, it is up to alignment systems' user interfaces to provide rich contextual information on each mapping. However, they have to balance that need with the need not to overwhelm the users with too much information, as humans have limited working memory. To that end, systems must ensure that their user interfaces convey information in an intuitive manner, and that while all required information is ready on-click, it is not all shown simultaneously. A strategy that many systems have implemented to achieve this is to provide different views of the alignment and/or each mapping.

In order to support user decisions, alignment systems' user interfaces should provide detailed explanations about mappings, and allow users to interact with the alignment in multiple ways, so as to make clear the consequences of accepting or rejecting a mapping. Allowing users to manually annotate mappings, and enabling validation over multiple sessions are also important features, due to the complexity and extensiveness of the validation task. However, these are all aspects where most current alignment systems have room for improvement.

Given the limited availability of users for alignment validation, systems should be able to prioritize the mapping suggestions they present to the users, by focusing on mappings about which they are unsure and/or those which cause conflicts. Systems can further exploit user input by extrapolating on it through feedback propagation techniques. However, as our experiments have shown, extrapolating will increase the impact of user errors, so systems should consider the profile of the user when deciding whether or not to employ feedback propagation. One possible strategy for that would be to ask the user how confident he is about each mapping, and only extrapolating on his decision when his confidence is high.

Our study should serve as a starting point towards establishing guidelines and best practices for good user interface design in the context of ontology alignment, which our evaluation of state of the art systems has shown to be necessary. Furthermore, we expect our study to help guide the development of alignment systems with regard to exploiting user interactions and coping with user errors.

For future work, we will aim to extend our evaluation by making usability assays with real users having varying degrees of expertise. We will also refine our experimental setup to better mirror the manual validation process, namely by considering the scenario where the user chooses between different conflicting mappings, rather than evaluating them independently, and by having the user provide a confidence value rather than a binary classification.

Acknowledgments. This work has been supported by SeRC, CUGS, the EU projects VALCRI (FP7-IP-608142) and Optique (FP7-ICT-318338), the EPSRC projects ED3 and DBOnto, and the Fundação para a Ciência e Tecnologia through the funding of the LaSIGE research unit (UID/CEC/00408/2013) and project PTDC/EEI-ESS/4633/2014.

References

1. Aumüller, D., Do, H.H., Maßmann, S., Rahm, E.: Schema and ontology matching with COMA++. In: SIGMOD, pp. 906–908 (2005)
2. Aurisano, J., Nanavaty, A., Cruz, I.: Visual analytics for ontology matching using multi-linked views. In: VOILA, pp. 25–36 (2015)
3. Beisswanger, E., Hahn, U.: Towards valid, reusable reference alignments-ten basic quality checks for ontology alignments, their application to three different reference data sets. J. Biomed. Semant. **3**(S-1), S4 (2012)
4. Cheatham, M., et al.: Results of the ontology alignment evaluation initiative 2015. In: OM, pp. 60–115 (2015)
5. Conroy, C., Brennan, R., Sullivan, D.O., Lewis, D.: User evaluation study of a tagging approach to semantic mapping. In: Aroyo, L., Traverso, P., Ciravegna, F., Cimiano, P., Heath, T., Hyvönen, E., Mizoguchi, R., Oren, E., Sabou, M., Simperl, E. (eds.) ESWC 2009. LNCS, vol. 5554, pp. 623–637. Springer, Heidelberg (2009)
6. Cruz, I., Antonelli, F., Stroe, C.: Agreementmaker: efficient matching for large real-world schemas and ontologies. Proc. VLDB Endowment **2**(2), 1586–1589 (2009)
7. Cruz, I., Loprete, F., Palmonari, M., Stroe, C., Taheri, A.: Quality-based model for effective and robust multi-user pay-as-you-go ontology matching. Semant. Web J. **7**(4), 463–479 (2016)
8. Cruz, I., Stroe, C., Palmonari, M.: Interactive user feedback in ontology matching using signature vectors. In: ICDE, pp. 1321–1324 (2012)
9. Cruz, I., Sunna, W., Makar, N., Bathala, S.: A visual tool for ontology alignment to enable geospatial interoperability. J. Vis. Lang. Comput. **18**(3), 230–254 (2007)
10. Dhamankar, R., Lee, Y., Doan, A., Halevy, A., Domingos, P.: iMAP: discovering complex semantic matches between database schemas. In SIGMOD, pp. 383–394 (2004)
11. Duan, S., Fokoue, A., Srinivas, K.: One size does not fit all: customizing ontology alignment using user feedback. In: Patel-Schneider, P.F., Pan, Y., Hitzler, P., Mika, P., Zhang, L., Pan, J.Z., Horrocks, I., Glimm, B. (eds.) ISWC 2010, Part I. LNCS, vol. 6496, pp. 177–192. Springer, Heidelberg (2010)
12. Euzenat, J., Meilicke, C., Stuckenschmidt, H., Shvaiko, P., Trojahn, C.: Ontology alignment evaluation initiative: six years of experience. In: Spaccapietra, S. (ed.) Journal on Data Semantics XV. LNCS, vol. 6720, pp. 158–192. Springer, Heidelberg (2011)
13. Euzenat, J., Shvaiko, P.: User involvement. In: Euzenat, J., Shvaiko, P. (eds.) Ontology Matching, pp. 353–375. Springer, Heidelberg (2013)
14. Falconer, S., Noy, N.: Interactive techniques to support ontology matching. In: Bellahsene, Z., Bonifati, A., Rahm, E. (eds.) Schema Matching and Mapping, pp. 29–51. Springer, Heidelberg (2011)
15. Falconer, S., Noy, N., Storey, M.-A.: Towards understanding the needs of cognitive support for ontology mapping. In: OM (2006)
16. Falconer, S., Noy, N., Storey, M.-A.: Ontology mapping - a user survey. In: OM, pp. 49–60 (2007)

17. Falconer, S.M., Storey, M.-A.D.: A cognitive support framework for ontology mapping. In: Aberer, K., Choi, K.-S., Noy, N., Allemang, D., Lee, K.-I., Nixon, L.J.B., Golbeck, J., Mika, P., Maynard, D., Mizoguchi, R., Schreiber, G., Cudré-Mauroux, P. (eds.) ASWC 2007 and ISWC 2007. LNCS, vol. 4825, pp. 114–127. Springer, Heidelberg (2007)

18. Faria, D., Martins, C., Nanavaty, A., Oliveira, D., Sowkarthiga, B., Taheri, A., Pesquita, C., Couto, F.M., Cruz, I.F.: AML results for OAEI 2015. In: OM (2015)

19. Fu, B., Noy, N., Storey, M.-A.: Eye tracking the user experience-an evaluation of ontology visualization techniques. Semant. Web J. (2014)

20. Gennari, J.H., Musen, M.A., Fergerson, R.W., Grosso, W.E., Crubzy, M., Eriksson, H., Noy, N.F., Tu, S.W.: The evolution of Protégé: an environment for knowledge-based systems development. Int. J. Hum.-Comput. Stud. **58**(1), 89–123 (2003)

21. Granitzer, M., Sabol, V., Onn, K.W., Luckose, D., Tochtermann, K.: Ontology alignment–a survey with focus on visually supported semi-automatic techniques. In: Future Internet, pp. 238–258 (2010)

22. Ivanova, V., Bergman, J.L., Hammerling, U., Lambrix, P.: Debugging taxonomies, their alignments: the ToxOntology-MeSH use case. In: WoDOOM, pp. 25–36 (2012)

23. Ivanova, V., Lambrix, P.: A unified approach for aligning taxonomies and debugging taxonomies and their alignments. In: Cimiano, P., Corcho, O., Presutti, V., Hollink, L., Rudolph, S. (eds.) ESWC 2013. LNCS, vol. 7882, pp. 1–15. Springer, Heidelberg (2013)

24. Ivanova, V., Lambrix, P., Åberg, J.: Requirements for and evaluation of user support for large-scale ontology alignment. In: Gandon, F., Sabou, M., Sack, H., d'Amato, C., Cudré-Mauroux, P., Zimmermann, A. (eds.) ESWC 2015. LNCS, vol. 9088, pp. 3–20. Springer, Heidelberg (2015)

25. Jiménez-Ruiz, E., CuencaGrau, B., Horrocks, I., Berlanga, R.: Logic-based assessment of the compatibility of UMLS ontology sources. J. Biomed. Semant. **2**(S–1), S2 (2011)

26. Jiménez-Ruiz, E., Cuenca Grau, B., Zhou, Y., Horrocks, I.: Matching, large-scale interactive ontology : algorithms and implementation. In: ECAI, pp. 444–449 (2012)

27. Katifori, A., Halatsis, C., Lepouras, G., Vassilakis, C., Giannopoulou, E.G.: Ontology visualization methods - a survey. ACM Comput. Surv. **39**(4), 10 (2007)

28. Lambrix, P., Edberg, A.: Evaluation of ontology merging tools in bioinformatics. In: Pacific Symposium on Biocomputing, pp. 589–600 (2003)

29. Lambrix, P., Kaliyaperumal, R.: A session-based approach for aligning large ontologies. In: Cimiano, P., Corcho, O., Presutti, V., Hollink, L., Rudolph, S. (eds.) ESWC 2013. LNCS, vol. 7882, pp. 46–60. Springer, Heidelberg (2013)

30. Lambrix, P., Liu, Q.: Using partial reference alignments to align ontologies. In: Aroyo, L., Traverso, P., Ciravegna, F., Cimiano, P., Heath, T., Hyvönen, E., Mizoguchi, R., Oren, E., Sabou, M., Simperl, E. (eds.) ESWC 2009. LNCS, vol. 5554, pp. 188–202. Springer, Heidelberg (2009)

31. Lambrix, P., Tan, H.: SAMBO - a system for aligning and merging biomedical ontologies. J. Web Semant. **4**(3), 196–206 (2006)

32. Lambrix, P., Tan, H.: A tool for evaluating ontology alignment strategies. In: Spaccapietra, S., Atzeni, P., Fages, F., Hacid, M.-S., Kifer, M., Mylopoulos, J., Pernici, B., Shvaiko, P., Trujillo, J., Zaihrayeu, I. (eds.) Journal on Data Semantics VIII. LNCS, vol. 4380, pp. 182–202. Springer, Heidelberg (2007)

33. Lambrix, P., Wei-Kleiner, F., Dragisic, Z., Ivanova, V.: Repairing missing is-a structure in ontologies is an abductive reasoning problem. In: WoDOOM, pp. 33–44 (2013)

34. Lanzenberger, M., Sampson, J., Rester, M., Naudet, Y., Latour, T.: Visual ontology alignment for knowledge sharing and reuse. J. Knowl. Manag. **12**(6), 102–120 (2008)
35. Nielsen, J.: Usability Engineering (1993)
36. Noy, N., Mortensen, J., Alexander, P., Musen, M.: Mechanical turk as an ontology engineer? In: ACM Web Science, pp. 262–271 (2013)
37. Noy, N., Musen, M.: Algorithm and tool for automated ontology merging and alignment. In: AAAI, pp. 450–455 (2000)
38. Otero-Cerdeira, L., Rodríguez-Martínez, F.J., Gómez-Rodríguez, A.: Ontology matching: a literature review. Expert Syst. Appl. **42**(2), 949–971 (2015)
39. Patterson, R.E., Blaha, L.M., Grinstein, G.G., Liggett, K.K., Kaveney, D.E., Sheldon, K.C., Havig, P.R., Moore, J.A.: A human cognition framework for information visualization. Comput. Graph. **42**, 42–58 (2014)
40. Paulheim, H., Hertling, S., Ritze, D.: Towards evaluating interactive ontology matching tools. In: Cimiano, P., Corcho, O., Presutti, V., Hollink, L., Rudolph, S. (eds.) ESWC 2013. LNCS, vol. 7882, pp. 31–45. Springer, Heidelberg (2013)
41. Pesquita, C., Faria, D., Santos, E., Neefs, J.-M., Couto, F.M.: Towards visualizing the alignment of large biomedical ontologies. In: Galhardas, H., Rahm, E. (eds.) DILS 2014. LNCS, vol. 8574, pp. 104–111. Springer, Heidelberg (2014)
42. Ritze, D., Paulheim, H.: Towards an automatic parameterization of ontology matching tools based on example mappings. In: OM, pp. 37–48 (2011)
43. Sarasua, C., Simperl, E., Noy, N.: CrowdMap: Crowdsourcing ontology alignment with microtasks. In: ISWC, pp. 525–541 (2012)
44. Shi, F., Li, J., Tang, J., Xie, G., Li, H.: Actively learning ontology matching via user interaction. In: Bernstein, A., Karger, D.R., Heath, T., Feigenbaum, L., Maynard, D., Motta, E., Thirunarayan, K. (eds.) ISWC 2009. LNCS, vol. 5823, pp. 585–600. Springer, Heidelberg (2009)
45. Shneiderman, B.: The eyes have it: a task by data type taxonomy for information visualizations. In: IEEE Symposium on Visual Languages, pp. 336–343 (1996)
46. Shvaiko, P., Euzenat, J.: Ontology matching: state of the art and future challenges. Knowl. Data Eng. **25**(1), 158–176 (2013)
47. Shvaiko, P., Giunchiglia, F., da Silva, P.P., McGuinness, D.L.: Web explanations for semantic heterogeneity discovery. In: Gómez-Pérez, A., Euzenat, J. (eds.) ESWC 2005. LNCS, vol. 3532, pp. 303–317. Springer, Heidelberg (2005)
48. Smith, E., Kosslyn, S., Psychology, C.: Mind and Brain (2013)
49. Tan, H., Lambrix, P.: A method for recommending ontology alignment strategies. In: Aberer, K., Choi, K.-S., Noy, N., Allemang, D., Lee, K.-I., Nixon, L.J.B., Golbeck, J., Mika, P., Maynard, D., Mizoguchi, R., Schreiber, G., Cudré-Mauroux, P. (eds.) ASWC 2007 and ISWC 2007. LNCS, vol. 4825, pp. 494–507. Springer, Heidelberg (2007)

Seed, an End-User Text Composition Tool for the Semantic Web

Bahaa Eldesouky[1(✉)], Menna Bakry[2], Heiko Maus[1], and Andreas Dengel[1]

[1] Knowledge Management Department, German Research Center for Artificial
Intelligence (DFKI), Kaiserslautern, Germany
{bahaa.eldesouky,heiko.maus,andreas.dengel}@dfki.de
[2] German University in Cairo (GUC), Cairo, Egypt
menna.nabil@guc.edu.eg

Abstract. Despite developments of Semantic Web-enabling technologies, the gap between non-expert end-users and the Semantic Web still exists. In the field of semantic content authoring, tools for interacting with semantic content remain directed at highly trained individuals. This adds to the challenges of bringing user-generated content into the Semantic Web. In this paper, we present *Seed*, short for Semantic Editor, an extensible knowledge-supported natural language text composition tool for non-experienced end-users. It enables automatic as well as semi-automatic creation of standards based semantically annotated textual content with focus on the task of text composition. We point out the structure of *Seed*, compare it with related work and explain how it excels at utilizing Linked Open Data and state of the art Natural Language Processing to realize user-friendly generation of textual content for the Semantic Web. We also present experimental evaluation results involving a diverse group of 120 participants, which showed that *Seed* helped end-users easily create and interact with semantic content with nearly no prerequisite knowledge.

Keywords: Semantic web · Semantic content authoring · Semantic text composition · Microdata · LOD · NLP · HCI

1 Introduction

Since the advent of the Semantic Web vision [12], the web has gradually evolved from a structure of interlinked documents to that of interlinked data. This vision drove rapid developments in technologies essential for its realization. Developments in Semantic Web enabling technologies can be seen in the field of modeling and structuring data, where crowd-sourced knowledge repositories such as DBPedia [13] and Freebase [14] have grown into huge graphs of entities containing millions of interrelated concepts, which comprise a web of LOD (Linked Open Data) [8]. In addition to public knowledge repositories, there are private ones, which focus on individual or group knowledge [29] (e.g. corporate knowledge repositories). Also, research on NLP (Natural Language Processing)

© Springer International Publishing AG 2016
P. Groth et al. (Eds.): ISWC 2016, Part I, LNCS 9981, pp. 218–233, 2016.
DOI: 10.1007/978-3-319-46523-4_14

techniques underwent great developments in both its syntactical and semantic variations [17]. Formats for embedding semantic content in web pages (e.g. RDFa [5], Microformats [4] and Microdata [6]) have also seen growth in their number and adoption rate. However, despite those developments, a gap between non-expert end-users and the Semantic Web still exists. This so-called semantic gap [31] is more evident in the process of creating structured information on the Web, where tools remain directed almost entirely at highly trained individuals [11].

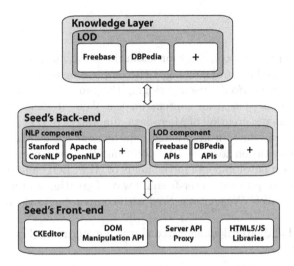

Fig. 1. Architecture of *Seed*

1.1   Related Work

Research about technologies that allow for user-friendly consumption and interaction with the existing web of data is gaining traction. SemCards [33] provides an intermediate ontological representational level that allows end-users to create rich semantic networks for their information sphere. OntoAnnotate [32] is an ontology-based annotation environment for web pages based on RDF [25] and RDFschema [15]. RDFauthor [34] bases on making arbitrary XHTML views with integrated RDFa annotations editable. OntosFeeder [23] is a WYSIWYG tool for annotating text for the news/journalism domain. In [7], authors can use Epiphany to get RDFa enhanced versions of their articles that link to Linked Data models.

In [21], the authors surveyed 31 recent primary studies, that dealt with *Semantic Content Authoring* (SCA) of textual content. Special focus was made on 4 of them, namely OntoWiki [9,28], SAHA 3 [24], Loomp also known as One Click Annotator [26] and RDFaCE [22]. The authors defined SCA as the tool-supported manual composition process aiming at the creation of documents which have one of two types:

(a) fully semantic (i.e. their original data model uses a semantic knowledge representation formalism).
(b) based on a non-semantic representation enriched with semantic representations during the authoring process. [21, p. 2]

Among the variety of SCA tools previously mentioned, the One Click Annotator [19] and RDFaCE have the most similar goals to those of *Seed*. This is why we will closely compare them in a later section.

1.2   Contribution

In this paper, we present *Seed*, short for semantic editor, an extensible knowledge-supported Web-based natural language text composition tool, which targets non-experienced end-users. *Seed* aims at bridging the gap between normal users and semantically annotated textual content on the Web. It enables automatic as well as semi-automatic creation of Microdata-annotated [6] HTML-based textual content without any domain knowledge requirements regarding the underlying technology or annotation formats. We point out the structure of *Seed*, explain how it builds upon developments in the fields of NLP, LOD and other Semantic Web technologies to provide a user-friendly way of creating and interacting with knowledge on the Web. We contrast *Seed* with comparable works and show what distinguishes it:
As discussed and later demonstrated through experimental evaluation in this paper, we show that:

- *Seed*'s focus on non-expert end-users makes semantics completely transparent to authors by focusing on the process of text composition, the actual interest of end-users, rather than semantic annotation or the underlying semantic analysis.
- It realizes more aspects of end-user SCA systems mentioned in [21] such as:
  - Real-time annotation during composition, which encourages users to review and interact with annotations making them more reliable. In that regard, comparable systems, are better described as a posteriori annotation tools.
  - Inline annotations behave like normal text reacting to inserting, deleting or updating characters all while preserving correct clean semantic markup.
  - Going beyond annotation to enable interaction and exploration of knowledge
- It is rigorously evaluated in terms of scale of the evaluation and evaluated aspects diversity, a much needed practice in Semantic Web research. Our experimental user study involved 120 participants from various backgrounds, age-groups and nationalities, we evaluated the usability of *Seed*, the quality of content it produces, and the subjective opinion of participants about the value of using *Seed* to explore, modify, and create semantic content.

This paper proceeds as follows: Sect. 2 explains the architecture and implementation of *Seed* showing how it builds on open Web technologies and standards. Section 3 assesses *Seed* in comparison with two related works and points out what distinguishes it as a SCA tool. In Sect. 4, we discuss in detail the setup and results of an experimental evaluation study. Finally, we wrap up and present examples of future work in the conclusion.

2 Seed Architecture and Implementation

As shown in Fig. 1, *Seed* consists of 3 loosely-coupled main components: (a) Knowledge layer, (b) Back-end and (c) Web front-end. Mutual communication between these components uses standard Web APIs (e.g. REST -based [18] Web services) to promote interoperability.

2.1 Knowledge Layer

This is a logical component of *Seed*, which represents the collective body of structured information available on the Semantic Web. Possible sources of knowledge integrable in this layer include not only public LOD sources such as DBPedia, but also any ontology-based knowledge repository. The current implementation of *Seed* integrates two LOD sources in its knowledge layer, namely DBPedia and Freebase. Other knowledge sources can be integrated as shown in Fig. 1.

2.2 Back-End

The second logical component of *Seed* is subdivided into two sub-components:

NLP Component. This sub-component utilizes state-of-the-art NLP toolkits to perform tasks such as part of speech tagging (POS), named entity recognition (NER), coreference resolution, ... etc. The implementation is carried out in a modular way that eases integrating or swapping various NLP toolkits as implied in Fig. 1. The current implementation of *Seed* specifically builds upon Stanford CoreNLP [27] and Apache OpenNLP [1] to provide a server API capable of real-time analysis of the text being authored. It also extracts named entity candidates, which are then processed to discover knowledge. This component currently supports English and German.

LOD Component. Together with the NLP component, this component communicates in real-time with LOD sources to extract information about potential entities extracted from the text. It is responsible for performing entity disambiguation and providing contextual information about discovered entities. The LOD component does not enforce a specific vocabulary or domain on the front-end. This has the following benefits for end-user oriented use-cases:

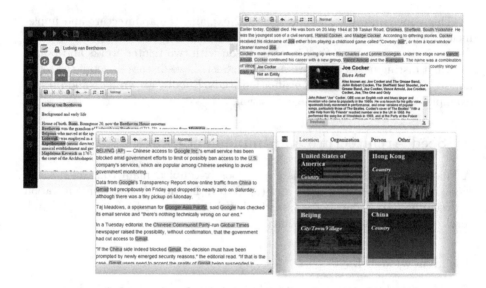

Fig. 2. Screenshots of multiple configurations of *Seed*'s front-end

- freeing the creator of the user interface (UI) in which *Seed* is to be embedded, from being restricted by the back-end's choice and
- elimination of the mental overload on the end-user incurred in understanding the vocabulary and learning to use it.

However, it is also possible to extend the front-end to enforce a specific vocabulary if the application domain or the use-case at hand dictates it.

2.3 Web Front-End

The current prototype of *Seed*'s front-end is meant to run in the browser (see Fig. 2 for various configurations of *Seed* in the browser). It is implemented as a pluggable component suitable for any Web-based UI. Therefore, it is written completely in HTML5 and JavaScript (JS). Nonetheless, it is also possible to embed it in non-Web GUIs. The only prerequisite is the availability of an HTML capable UI element. This flexibility in integrating *Seed* makes it highly portable and facilitates reaching end-users dealing with different UI types or different types of devices. The front-end component consists of the following logical sub-components.

CKEditor. At the core of the front-end, *Seed* builds upon CKEditor [2], the open source WYSIWYG HTML editor. We have extended CKEditor with the following components:

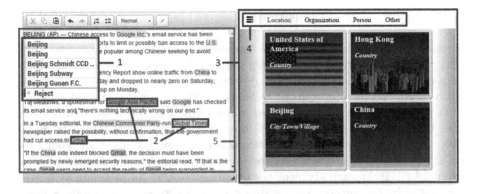

Fig. 3. Main parts of *Seed*'s front-end: 1- Dropdown menu for viewing, selecting or rejecting an annotation, 2- Suggested as well as confirmed inline annotations, 3- Entity side pane for viewing more information about entities in the text, 4- Controls for Faceted viewing/browsing, 5- Text composition area

DOM Manipulation API. In order to implement inline editing of HTML content including semantic markup in a reliable usable way, we have built upon JavaScript (JS) native HTML Document Object Model (DOM) manipulation constructs, jQuery [3] as well as CKEditor's own JS APIs to implement a basic API for monitoring and interacting with the HTML DOM for text editing purposes.

Server API Proxy. This is JS code that handles communication with the server in near real-time. It consumes standard RESTful APIs provided by the NLP and LOD components of *Seed*'s back-end to update the semantic representation of the text as it changes.

Semantic Annotator. The semantic annotator is a JS/HTML extension code responsible for:

- building upon the DOM manipulation API of *Seed* to add, remove or update Microdata annotations during editing. Annotations are applied in the form of HTML Microdata markup,
- maintaining a client side representation of the knowledge in the text in the form of entities and their metadata,
- binding between entities and their arbitrary UI manifestations (labels, highlights, information panes, images, ... etc.).

HTML5/JS Libraries. For the creation of the various elements of the front-end, we have used the following main third-party JS/HTML5/CSS libraries: (jQuery, jQuery Mobile, Mutation Summary and Bootstrap)

As shown in Fig. 3, various parts of the *Seed*'s UI allow authors to interact with the textual and knowledge content of the text being composed.

3   Comparative Assessment

Despite obviously increasing research interest in the field of SCA, our literature survey of related work pointed to only a handful of approaches to SCA, which focus on semantic text composition of natural language text. Fewer are works, which target end-users in contrast to Semantic Web computer professionals.

In this section, we will assess *Seed* in comparison with two other SCA tools (One Click Annotator and RDFaCE) due to their conceptual similarity. Both tools also follow a bottom up semantic authoring approach like *Seed*. They target, at least in part, end-users. Our assessment will start by a tabular comparison inspired by [21]. We will augment the quality attributes for assessing SCA systems listed by the authors with additional metrics we suggest to form a basis for the comparison. According to [21], those quality attributes adopt the point of view of SCA users (i.e. end-users in our case). For each quality attribute, concrete UI features should realize the respective quality attribute [21, p. 7].

After the tabular comparison, we will discuss selected comparative aspects in detail to point out the significance of *Seed*.

A review of the condensed comparison in Table 1 reveals many advantages of *Seed* over One Click Annotator and RDFaCE. For the sake of brevity, we will elaborate on some of those advantages and suffice to the table entries for the rest.

No prerequisite knowledge: *Seed* requires no prerequisite technical knowledge about formats of embedding semantic content, underlying knowledge representation models, vocabularies or even the basic terminology of the Semantic Web. For example, annotating text with information about its semantics, modifying existing annotations and exploring knowledge about entities beyond the textual content being authored take place through familiar user interaction scenarios. In contrast, other similar works vary from requiring knowledge of triple representations to learning specific vocabularies.

Real-time annotation: One Click Annotator and RDFaCE require text authors to explicitly request the annotation of content. This reduces the productivity of the text composition task and increases the mental load on authors, which in turn discourages end-users from reviewing and possibly correcting annotations of the content. *Seed*, on the other hand, continuously analyzes authored text and proactively annotates it with semantic information. The reduced effort required from authors, is expected to encourage them to review and interact with annotations, thus producing more reliable semantic annotations.

Native annotations: In *Seed*, annotations behave like normal text. They react to inserting, deleting or updating characters intuitively and consistently, thus producing correct semantic markup. In comparable systems like RDFaCE, once annotations are created, attempts to modify them break the semantic markup.

Focus on knowledge: In addition to generating semantic content, *Seed* focuses on enabling users to consume the underlying knowledge in a high level fashion and through different views. As shown in Figs. 2 and 3, it provides multiple views

Table 1. SCA quality attributes assessment of *Seed*. Table layout adapted from [21, p. 12]

	One click annotator	RDFaCE	Our system (Seed)
Online demo	N/A	rdface.aksw.org/	tiny.cc/seed-demo
Experimental evaluation	12 non-expert participants, paper prototype user study	16 experienced participants	**120 non-expert & expert subjects, online experiment**
Usability			
Single entry point	Yes	Yes	Yes
Faceted browsing	No	No	**Yes**
Faceted viewing	Yes	Yes	Yes
Inline editing	No	Yes	Yes
View editing	No	No	**Yes**
Native annotations	No	No	**Yes**
Automation			
Automatic annotation	No	External NLP APIs	Own server API
Generalizability			
Multiple ontologies support	Yes	Yes	Yes
Ontology modification support	No	No	No
Heterogeneous content formats	no	No	No (only standard HTML5)
Collaboration			
Access control	No	No	**Yes**
Standard formats support	RDF & RDFa	RDFa & Microdata	Microdata
UIs for social collaboration	No	No	(Depends on Web UI)
Customizability			
Living UIs	No	No	No
Providing different semantic views	No	Yes	Yes
Evolvability			
Resource consistency	No	No	**Yes**
Document & annotation consistency	Yes	Yes	Yes
Versioning	No	No	**Yes**
Proactivity			
Resource suggestion	Yes	Yes	Yes
Real-time semantic tagging	No	No	**Yes**
Concept reuse	Yes	No	Yes
Real-time validation	No	No	**Yes**
Portability			
Cross-browser compatibility	Yes	Yes	Yes
Mobile UI support	No	No	No
Accessibility			
Accessible UIs	No	No	No
Interoperability			
Standard formats	RDF & RDFa	RDFa & Microdata	Microdata
Semantic syndication	No	No	No
Scalability			
Caching support	No	No	No
Storage strategy	Server-side triple store	On the fly client-side triple store	**Triple store or live LOD + client-side in-memory storage**

to the underlying semantics on annotations. By means of a faceted view, authors can have a high level view of the entities mentioned in the text and are able to explore information about the entities derived from LOD sources.

Rigorous evaluation: To our knowledge, no other comparable system has been as rigorously evaluated as *Seed*, neither in terms of the scale of the evaluation, the target test-subject audience or the diversity of evaluated aspects. RDFaCE for example, has been evaluated by 16 participants from a computer science background participating in a LOD workshop [22].

4   Experimental Evaluation

A comparative experimental evaluation with other works such as RDFaCE and One Click Annotator, which involved 120 participants was not feasible in the scope of our study. Reasons include the lack of publicly accessible functioning prototypes/demos of other works. Besides, the scale of the experiment and the practical time limits for an online evaluation made it impossible for us to evaluate other works using the same procedure without substantially shrinking the population. So, we focused instead on evaluating *Seed* while providing enough information for reproducing the evaluation by others [1].

4.1   Goals

As previously mentioned, an important yet missing aspect of research on SCA is user studies of sizable scale involving ordinary non-expert users. Most of the studies in the field propose conceptual ideas, which are seldom put to reasonable evaluation. Therefore, we have set out to target a large group of Web-users with the following goals in mind:

1. Show that Seed is a highly usable and easy-to-learn semantic text composition tool, which hides the complexity of the underlying technology, thus enabling Semantic Web end-users to focus on the process of textual content generation.
2. Enable end-users with no prerequisite knowledge to produce standards based semantically annotated textual content.
3. Proactively help end-users to explore, and interact with knowledge from the Semantic Web (LOD in our case) while composing textual content.

4.2   Design

The evaluation was designed as a within-subjects repeated measures experiment. All participants were exposed to the same conditions. The independent variables of the study were:

1. The number of text passages
2. The length of each passage
3. The number of entities in each passage

---

[1] Evaluation data available at http://tiny.cc/seed-iswc2016-data.

4.3   Procedure

We have set up an evaluation website at http://tiny.cc/seed-demo and prepared the experiment, which consisted of the following stages:

- User registration, where participants were asked to provide information about themselves for demographic profiling and validation purposes.
- Once registered, participants watched a 3 min. video[1] that explained the concept of *Seed* in a non-technical way. We refrained from detailed descriptions of technical aspects of the system in order to properly measure its learnability by non-experts.
- Participants were then asked to review and annotate 3 text passages using *Seed*. Every participant started with a pre-loaded text. The user then reviewed automatic annotations by *Seed* as well as annotation suggestions that (s)he could confirm, modify, reject or augment.
- Afterwards, participants were asked to type in a predefined text passage into *Seed*, which gets annotated in real time and reviewed during writing. Then, participants are asked questions to test their understanding of the text and validate their attention to the experiment. Answers helped us later pre-process data and eliminate non-serious participants.
- Finally, participants were asked to fill in a standardized usability questionnaire to assess participants' satisfaction with the perceived usability of the system, then answer additional questions about *Seed*.

4.4   Participants

The evaluation received 256 registrations, of which 120 completed the experiment. Table 2 shows demographic information about participants.

Table 2. Demographics of the participants population

Characteristics	Percentages
Age	15–25 years (47 %), 25–35 years (42 %), 35–45 years (9 %), 55–65 years (2 %)
Gender	Males (61 %), Females (39 %)
Profession	Undergraduate students (31 %), graduate students (23 %), computer professionals (20 %), non computer professionals (19 %), researchers (7 %)
Nationality	Egypt (59 %), Germany (14 %), India (8 %), Jordan (3 %), Pakistan (3 %), Palestine (2 %), others (11 %)

---

[1] Seed, the Semantic Editor - http://tiny.cc/seed-video.

4.5 Measures

For the assessment of semantic annotations, we calculated precision, recall and F-1 scores of annotations done by participants for all texts. For assessing learnability, we measured the time required for reviewing and annotating texts in the repeated measures part of the experiment. For assessing the perceived usability of the system, we measured the System Usability Scale score (SUS) for *Seed*

4.6 Semantic Annotations Assessment

For the choice of texts to be authored and annotated in the experiment, we provided all participants with a set of text passages to achieve as much consistency as possible in regard to the length of text, the number of entities mentioned, their types and the subject domain. To compensate for the small size of dataset we targeted a large participants population. The outcome of the experiment was then assessed against a ground truth version of the set of texts.

The texts used in the experiment were produced as follows. We selected 3 representative text passages from different subject domains (news articles, wiki articles, and blog posts) to control the subject domain familiarity variable. The fourth passage was arbitrarily selected to be from the wiki articles domain.

To create a ground truth for assessing annotations in the text passages, 3 different human annotators separately annotated named entities of type person, location or organization by hand. We restricted types of annotations in the ground truth to the three mentioned types to parallel the most widely used 3-class model in state of the art NLP tools. Only annotations agreed upon by 2 or more annotators were added.

In order to evaluate *Seed*'s ability to produce correct annotations during text composition, we calculated Precision, Recall and F-1 scores for annotations in all passages submitted by participants as shown in Table 3.

Table 3. Annotation performance measures assessment

	Avg. recall	Avg. precision	Avg. F1
Passage 1	0.80	0.79	0.79
Passage 2	0.89	0.97	0.93
Passage 3	0.76	0.88	0.80
Passage 4	0.89	0.93	0.91

For the calculation of the performance measure values, we considered an entity annotation correct if it was:

- correctly recognized (i.e. there is an entity and its token delimiters were identified by the author)
- disambiguated and correctly mapped to a LOD entity from DBPedia or Freebase

The values in Table 3 show that *Seed* helped automatically annotate the majority of the entities in the text already at the text authoring stage.

The following interesting observations resulted from the performance measures assessment for all texts:

On average 38.2 % percent of participants annotated more entities in total than existed in the ground truth. This is a valuable remark because it shows that *Seed*'s support for annotation during text composition goes beyond the limitations of state-of-the-art NLP models. On average 13.5 % percent of participants submitted more correct annotations than the total number of annotations in the ground truth. This shows that the annotations submitted by participants are not only more but also correct.

4.7 Usability Evaluation

System Usability Scale Score. At the end of the experiment, we prompted users to fill in a questionnaire which consisted of a standard SUS form in addition to two questions we added. As defined by [16], scores of individual items in a SUS are not meaningful on their own. So, we calculated the overall SUS score for *Seed* across the population of participants, which resulted in an overall SUS score with mean: 73.56, median: 75, standard deviation of 13.71. According to [16], this means *Seed* has above average usability. In order to assess the statistical significance of the SUS results, we performed a one sample Z-test on the SUS scores of the participants.

Following Sauro's notion in [30], we defined our hypotheses as follows:

- Null Hypothesis, H_0: It's predicated that Seed's SUS score is at most around average ($\mu \leq 70$).
- Alternate hypothesis, H_a: Seed's SUS score above average ($\mu > 70$).

The results of the Z-test showed that SUS scores for Seed in our experiment ($\mu = 73.96, \sigma = 13.94$) are significantly higher than the predicated SUS score of 70 (z = 2.71, p = 0.0034). According to [10], we can confidently say that Seed's SUS score is between good and excellent.

Interactive Inline Annotations. In order to assess the effect of interactive inline annotations in authored text in *Seed* on its overall usability, we asked participants the following two alternating tone questions with answers varying on a 5-degree scale (0 to 4), from "strongly disagree" to "strongly agree" respectively:

- The annotated entities helped me to understand the written content
- Entities annotated in the text distracted me from reading the content

A clear majority of users found the annotations not distracting. The answers to the negatively formulated question had a median= 1 and a mean= 1.42. They also found them helpful in understanding the content of the text they were annotating. Answers to the positively formulated question had a median= 3 and a mean= 2.85.

Real-Time Annotation. In order to assess the value of real-time annotation, we asked participants the following question:

"Which of the following options would you prefer more?"

(a) Annotating entities as you write them
(b) Annotating once after you finish typing the text
(c) No preference

According to responses, $55.5555\% \simeq 55.6\%$ of the participants chose (a), $38.8888\% \simeq 38.9\%$ chose (b) while $5.5555\% \simeq 5.6\%$ expressed no preference. It is worth mentioning that many of the users who chose (b) justified their choice by the relative simplicity of the topic of the 3 passages or by the irrelevance to a personal context of theirs.

Faceted Viewing and Knowledge Discovery. These important features of *Seed*'s UI aim at enabling end-users to easily explore and consume knowledge about content of the text being authored. In order to evaluate these features, we asked users questions whose answers are not contained in a passage, but are available through the entity summary side pane of *Seed* as well as through the interactive annotation information pane. The results of users answers were as follows:

- For the questions, whose answers required looking into the information in the entity side pane or in the interactive annotation info pane, 94.9 % of the participants managed to find the correct answer.
- For the question, whose answer is most easily accessible by faceted browsing, 51.5 % of the participants managed to find the answer.

To check whether participants had looked up the answer elsewhere, we asked them how they found it. For those who correctly answered at least one question, 93.9 % did so using *Seed*'s features. This showed that *Seed* successfully helped participants discover knowledge about the content. The results hint at the need for further inspection of the design of the faceted browsing feature (Fig. 3).

Learnability. To assess how fast participants learned to use *Seed*, we carried out the following:

- We measured the time required for annotating each of the first 3 text passages for all of participants.
- Outliers were eliminated using a two-sided Iglewicz and Hoaglin's robust test for multiple outliers [20].
- To account for varying length of the texts, we calculated the time per word in each passage.
- In order to check for statistically significant differences in mean times per word required for annotating texts, a repeated measures ANOVA with a Greenhouse-Geisser correction determined that mean time per word differed statistically significantly between passages ($F(1.89, 177.696) = 17.09$, $P < .0005$). Post hoc tests using the Bonferroni correction revealed that

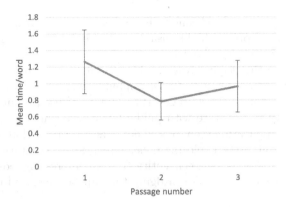

Fig. 4. Mean time per word required to annotate text passages. An overall decreasing trend is seen, which signals the speed of learning of users

time/word decreased from passage 1 to passage 2 ($1.26 \pm 0.77s$ vs $0.79 \pm 0.45s$, respectively), which was statistically significant ($p < .0005$). Also, time/word decreased from passage 1 to passage 3 ($1.26 \pm 0.77s$ vs $0.96 \pm 0.62s$, respectively), which was also statistically significant ($p = 0.005$). However, time/word slightly increased from passage 2 to passage 3 ($0.79 \pm 0.45s$ vs $0.96 \pm 0.62s$, respectively), which was not statistically significant ($p = 0.055$). Therefore, we can conclude that an overall decreasing trend exists from passage 1 on one hand and passage 2 or 3 on the other hand.

To explain the slight increase in passage 3 time, we further inspected annotation data and qualitative feedback collected in the experiment. Deteriorating performance measures for passage 3 combined with re-occurring comments regarding passage 3 about the inability to annotate overlapping entities such as "Old Town Hall" and "New Town Hall" in a sentence containing the text "Old and New Town Hall" provide an explanation for the apparent increase. It also highlights a technical limitation in dealing with overlapping entities in HTML based semantic annotations. Attempts to annotate pairs of overlapping entities is not easily doable in HTML markup due to its hierarchical nature (Fig. 4).

5   Conclusion

In previous sections, we highlighted the importance of bridging the gap between end-users and the Semantic Web. We presented *Seed*, a user-friendly semantic text composition tool, which brings technical non-experts closer to the Semantic Web. It allows them to benefit from, interact with, and create semantic content in the form of semantically annotated HTML-based text. We showed how it realizes real-time annotation during authoring, thus encouraging end-users to review, possibly add annotations as they write. Using rigorous experimental evaluation involving a sizable, diverse population of 120 participants, we assessed our hypotheses about *Seed*. Results showed that it enabled users to produce semantically annotated textual content in a reliable way. By means of a standard SUS evaluation, *Seed* proved highly usable, not only in annotating content, but also in exploring knowledge about it.

The outcome of this paper gives insight into future work research questions. The loosely-coupled architecture of *Seed* combined with the fact that it supports German as well as English, encourages us to explore its use for multilingual content. *Seed*'s ability to integrate with public knowledge sources motivates us to explore its use in application scenarios where personal rather than public knowledge is more relevant. Also, exploring richer semantic representations embedded in the text (e.g. relations between entities) is an interesting possibility. This in turn will further contribute to bridging the gap between end-users and the Semantic Web.

References

1. Apache OpenNLP. http://opennlp.apache.org/index.html. Accessed 18 July 2016
2. CKEditor. http://ckeditor.com. Accessed 18 July 2016
3. jQuery. http://jquery.com/. Accessed 18 July 2016
4. Microformats. http://microformats.org/. Accessed 18 July 2016
5. RDFa Core 1.1 - 2nd edn., 22 August 2013. http://www.w3.org/TR/rdfa-syntax/. Accessed 18 July 2016
6. HTML Microdata, 29 October 2013. http://www.w3.org/TR/microdata/. Accessed 18 July 2016
7. Adrian, B., Hees, J., Herman, I., Sintek, M., Dengel, A.: Epiphany: adaptable RDFa generation linking the web of documents to the web of data. In: Cimiano, P., Pinto, H.S. (eds.) EKAW 2010. LNCS, vol. 6317, pp. 178–192. Springer, Heidelberg (2010)
8. Auer, S., Bryl, V., Tramp, S.: Linked Open Data-Creating Knowledge Out of Interlinked Data: Results of the LOD2 Project, vol. 8661. Springer, Heidelberg (2014)
9. Auer, S., Dietzold, S., Riechert, T.: OntoWiki – a tool for social, semantic collaboration. In: Cruz, I., Decker, S., Allemang, D., Preist, C., Schwabe, D., Mika, P., Uschold, M., Aroyo, L.M. (eds.) ISWC 2006. LNCS, vol. 4273, pp. 736–749. Springer, Heidelberg (2006)
10. Bangor, A., Kortum, P., Miller, J.: Determining what individual SUS scores mean: adding an adjective rating scale. J. Usability Stud. **4**(3), 114–123 (2009)
11. Benson, E., Karger, D.R.: End-users publishing structured information on the web: an observational study of what, why, and how. In: Proceedings of the 32nd Annual ACM Conference on Human Factors in Computing Systems, pp. 1265–1274. ACM (2014)
12. Berners-Lee, T., Hendler, J., Lassila, O., et al.: The semantic web. Sci. Am. **284**(5), 28–37 (2001)
13. Bizer, C., Lehmann, J., Kobilarov, G., Auer, S., Becker, C., Cyganiak, R., Hellmann, S.: DBpedia - a crystallization point for the web of data. Web Semant.: Sci. Serv. Agents World Wide Web **7**(3), 154–165 (2009)
14. Bollacker, K., Evans, C., Paritosh, P., Sturge, T., Taylor, J.: Freebase: a collaboratively created graph database for structuring human knowledge. In: Proceedings of the 2008 ACM SIGMOD International Conference on Management of data, pp. 1247–1250. ACM (2008)
15. Brickley, D., Guha, R.V.: Resource Description Framework (RDF) Schema Specification 1.0: W3C Candidate Recommendation, 27 March 2000

16. Brooke, J.: SUS - a quick and dirty usability scale. In: Jordhan, P., Thomas, B., Weerdmeester, B.A., McClelland, I. (eds.) Usability Evaluation in Industry, vol. 189. Taylor and Francis, London (1996)
17. Cambria, E., White, B.: Jumping NLP curves: a review of natural language processing research. IEEE Comput. Intell. Mag. 9(2), 48–57 (2014)
18. Fielding, R.T.: Architectural styles and the design of network-based software architectures. Ph.D. thesis, University of California, Irvine (2000)
19. Heese, R., Luczak-Rösch, M., Oldakowski, R., Streibel, O., Paschke, A.: One click annotation. In: CEUR Workshop Proceedings, vol. 514 (2009)
20. Iglewicz, B., Hoaglin, D.C.: How to Detect and Handle Outliers, vol. 16. ASQ Press, Milwaukee (1993)
21. Khalili, A., Auer, S.: User interfaces for semantic authoring of textual content: a systematic literature review. Web Semant.: Sci. Serv. Agents World Wide Web 22, 1–18 (2013)
22. Khalili, A., Auer, S., Hladky, D.: The RDFa content editor - from WYSIWYG to WYSIWYM. In: 2012 IEEE 36th Annual Computer Software and Applications Conference (COMPSAC), pp. 531–540. IEEE (2012)
23. Klebeck, A., Hellmann, S., Ehrlich, C., Auer, S.: OntosFeeder – a versatile semantic context provider for web content authoring. In: Antoniou, G., Grobelnik, M., Simperl, E., Parsia, B., Plexousakis, D., De Leenheer, P., Pan, J. (eds.) ESWC 2011, Part II. LNCS, vol. 6644, pp. 456–460. Springer, Heidelberg (2011)
24. Kurki, J., Hyvönen, E.: Collaborative metadata editor integrated with ontology services and faceted portals. In: ORES-2010 Ontology Repositories and Editors for the Semantic Web, p. 7 (2010)
25. Lassila, O., Swick, R.R., et al.: Resource Description Framework (RDF) Model and Syntax Specification (1998)
26. Luczak-Rösch, M., Heese, R.: Linked data authoring for non-experts. In: LDOW (2009)
27. Manning, C.D., Surdeanu, M., Bauer, J., Finkel, J., Bethard, S.J., McClosky, D.: The Stanford coreNLP natural language processing toolkit. In: Proceedings of 52nd Annual Meeting of the Association for Computational Linguistics: System Demonstrations, pp. 55–60 (2014)
28. Salas, P.E.R., Martin, M., Da Mota, F.M., Auer, S., Breitman, K.K., Casanova, M.A.: OLAP2DataCube: an ontowiki plug-in for statistical data publishing. In: TOPI 2012, pp. 79–83 (2012)
29. Sauermann, L., Elst, L., Dengel, A.: PIMO - a framework for representing personal information models. In: I-SEMANTICS Conference, 5–7 September 2007, Graz, Austria, pp. 270–277. J.UCS, Know-Center, Austria (2007)
30. Sauro, J.: A Practical Guide to the System Usability Scale: Background, Benchmarks and Best Practices. Measuring Usability LLC (2011)
31. Siorpaes, K., Simperl, E.: Human intelligence in the process of semantic content creation. World Wide Web 13(1–2), 33–59 (2010)
32. Staab, S., Maedche, A., Handschuh, S.: Creating Metadata for the Semantic Web - An Annotation Environment and the Human Factor. Institute AIFB (2000)
33. Thórisson, K.R., Spivack, N., Wissner, J.M.: SemCards: a new representation for realizing the semantic web. In: Nguyen, N.T., Kowalczyk, R., Chen, S.-M. (eds.) ICCCI 2009. LNCS, vol. 5796, pp. 425–436. Springer, Heidelberg (2009)
34. Tramp, S., Heino, N., Auer, S., Frischmuth, P.: RDFauthor: employing RDFa for collaborative knowledge engineering. In: Cimiano, P., Pinto, H.S. (eds.) EKAW 2010. LNCS, vol. 6317, pp. 90–104. Springer, Heidelberg (2010)

Exception-Enriched Rule Learning from Knowledge Graphs

Mohamed H. Gad-Elrab[1]([✉]), Daria Stepanova[1], Jacopo Urbani[2], and Gerhard Weikum[1]

[1] Max Planck Institute of Informatics, Saarbrücken, Germany
{gadelrab,dstepano,weikum}@mpi-inf.mpg.de
[2] VU University Amsterdam, Amsterdam, The Netherlands
jacopo@cs.vu.nl

Abstract. Advances in information extraction have enabled the automatic construction of large knowledge graphs (KGs) like DBpedia, Freebase, YAGO and Wikidata. These KGs are inevitably bound to be incomplete. To fill in the gaps, data correlations in the KG can be analyzed to infer Horn rules and to predict new facts. However, Horn rules do not take into account possible exceptions, so that predicting facts via such rules introduces errors. To overcome this problem, we present a method for effective revision of learned Horn rules by adding exceptions (i.e., negated atoms) into their bodies. This way errors are largely reduced. We apply our method to discover rules with exceptions from real-world KGs. Our experimental results demonstrate the effectiveness of the developed method and the improvements in accuracy for KG completion by rule-based fact prediction.

1 Introduction

Motivation and Problem. Recent advances in information extraction have led to huge graph-structured knowledge bases (KBs) also known as knowledge graphs (KGs) such as NELL [4], DBpedia [2], YAGO [22] and Wikidata [8]. These KGs contain millions or billions of relational facts in the form of subject-predicate-object (SPO) triples.

As such KGs are automatically constructed, they are incomplete and contain errors. To complete and curate a KG, inductive logic programming and data mining techniques (e.g., [5,11,29]) have been used to identify prominent patterns, such as *"Married people live in the same place"*, and cast them in the form of Horn rules, such as: $r_1 : livesIn(Y, Z) \leftarrow isMarriedTo(X, Y), livesIn(X, Z)$.

This has twofold benefits. First, since KGs operate under the Open World Assumption (OWA) (i.e., absent facts are treated as unknown rather than false), the rules can be used to derive additional facts. For example, applying the rule r_1 mined from the graph in Fig. 1a, the missing living place of Dave can be deduced based on the data about his wife Clara. Second, rules can be used to eliminate erroneous facts in the KG. For example, assuming that *livesIn* is a functional

© Springer International Publishing AG 2016
P. Groth et al. (Eds.): ISWC 2016, Part I, LNCS 9981, pp. 234–251, 2016.
DOI: 10.1007/978-3-319-46523-4_15

relation, Amsterdam as a living place of Alice could be questioned as it differs from her husband's.

State of the Art and its Limitations. Methods for learning rules from KGs are typically based on inductive logic programming or association rule mining (see [11] and references given there). However, these methods are limited to Horn rules where all predicates in the rule body are positive. This is insufficient to capture rules that have exceptions, such as *"Married people live in the same place unless one is a researcher"*: $r_2 : livesIn(Y, Z) \leftarrow isMarriedTo(X, Y), livesIn(X, Z), not\ researcher(Y)$. This additional knowledge could be an explanation for Alice living in an unexpected place. If r_2 often holds, then one can no longer complete the missing living place for Dave by assuming that he lives with his wife Clara. Thus, understanding exceptions is crucial for KG completion and curation.

Our goal is to learn rules with exceptions, also known as *nonmonotonic rules*. Learning nonmonotonic rules under the Closed World Assumption (CWA) is a well-studied problem that lies at the intersection of inductive and abductive logic programming (e.g., [26,27]). However, these methods cannot be applied to KGs treated under the OWA.

Approach and Contribution. We present a novel method that takes a KG and a set of Horn rules as input and yields a set of exception-enriched rules as output. The output rules are no longer necessarily Horn clauses (e.g., rule r_2 above could be in our output). So we essentially we tackle a variant of a *theory revision* problem [30] under OWA.

Our method proceeds in four steps. First, we compute what we call "exception witnesses": predicates that are potentially involved in explaining exceptions (e.g., *researcher* in our example). Second, we generate nonmonotonic rule candidates that we could possibly add to our KG rules. Third, we devise quality measures for nonmonotonic rules to quantify their strength w.r.t the KG. In contrast to prior work, we do not merely give measures for individual rules in isolation, but also consider their cross-talk through a new technique that we call "partial materialization". Fourth and last, we rank the nonmonotonic rules by their strengths and choose a cut-off point such that the obtained rules describe the KG's content as well as possible with awareness of exceptions.

The salient contributions of our paper are:

- A framework for nonmonotonic rule mining as a knowledge revision task, to capture exceptions from Horn rules and overcome the limitations of prior work on KG rule mining.
- An algorithm for computing exception candidates, measuring their quality, and ranking them based on a novel technique that considers *partial materialization* of judiciously selected rules.
- Experiments with the YAGO3 and IMDB KGs where we show the gains of our method for rule quality as well as fact quality when performing KG completion.

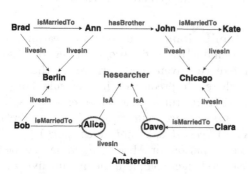

	bornInUS	livesInUS	stateless	immigrant	singer	poet	hasUSPass
p1	✓	✓		✓			✓
p2	✓	✓					✓
p3	✓	✓			✓	✓	✓
p4	✓	✓					✓
p5	✓	✓	✓				
p6	✓			✓			
p7	✓			✓			
p8	✓		✓	✓			
p9	✓			✓		✓	
p10	✓			✓	✓	✓	✓
p11	✓				✓	✓	✓

(a) Rule mining for KG completion and KG cleaning   (b) US inhabitants KG

Fig. 1. Examples of knowledge graphs.

The rest of the paper is structured as follows. Section 2 introduces necessary notation and definitions. Section 3 presents our approach to nonmonotonic rule mining. Section 4 gives details on computing exceptions and revision candidates. Section 5 describes how we measure the quality of rules. Section 6 presents an experimental evaluation of how exception-enriched rules can improve the KG quality. Section 7 discusses related work.

2   Preliminaries

Knowledge Graphs. On the Web, knowledge graphs (KG) are often encoded using the RDF data model [16], which represents the content of the graph with a set of triples of the form $\langle subject\ predicate\ object \rangle$. These triples encode positive facts about the world, and they are naturally treated under the OWA.

In this work, we focus on KGs without blank nodes or schema (i.e. TBox). For simplicity, we represent the triples using unary and binary predicates. The unary predicates are the objects of the RDF *isA* predicate while the binary ones correspond to all other RDF predicates. We call this the factual representation $\mathcal{A}_\mathcal{G}$ (the subscript $\mathcal{G}$ is omitted when clear from context) of the KG $\mathcal{G}$ defined over the signature $\Sigma_{\mathcal{A}_\mathcal{G}} = \langle \mathbf{C}, \mathbf{R}, \mathcal{C} \rangle$, where $\mathbf{C}$, $\mathbf{R}$ and $\mathcal{C}$ are resp. sets of unary predicates, binary predicates and constants.

Example 1. The factual representation of the graph $\mathcal{G}$ from Fig. 1a among others contains the following facts: $ism(brad, ann); ism(bob, alice); li(brad, berlin); r(alice); r(dave); hb(ann, john); li(alice, amsterdam); li(bob, berlin); li(clara, chicago),$ where *ism,li,hb,r* stand for *isMarriedTo, livesIn, hasBrother,* and *researcher* respectively. The signature of $\mathcal{A}_\mathcal{G}$ is $\Sigma_{\mathcal{A}_\mathcal{G}} = \langle \mathbf{C}, \mathbf{R}, \mathcal{C} \rangle$, where $\mathbf{C} = \{r\}$, $\mathbf{R} = \{ism, li, hb\}$ and $\mathcal{C} = \{john, ann, brad, dave, clara, alice, kate, bob, chicago, berlin, amsterdam\}$. $\square$

In this work, we focus primarily on mining rules over unary predicates. Binary relations can be translated into multiple unary ones by concatenating the binary predicate and one of its arguments, e.g. the binary predicate *livesIn* of Fig. 1a can be translated into three unary ones *livesInAmsterdam, livesInBerlin, livesInChicago*. We apply this conversion to a KG in the input so that it consists of a collection of unary facts.

Nonmonotonic Logic Programs. Nonmonotonic Logic Programs We define a logic program in the usual way [21]. In short, a *(nonmonotonic) logic program* P is a set of *rules* of the form

$$H \leftarrow B, not\ E \tag{1}$$

where H is a standard first-order atom of the form $a(X)$ known as the rule head and denoted as $Head(r)$, B is a conjunction of positive atoms of the form $b_1(Y_1), \ldots, b_k(Y_k)$ to which we refer as $Body^+(r)$ and $not\ E$, with slight abuse of notation, denotes the conjunction of atoms $not\ b_{k+1}(Y_{k+1}), \ldots, not\ b_n(Y_n)$. Here, not is the so-called *negation as failure (NAF)* or *default negation*. The negated part of the body is denoted as $Body^-(r)$. The rule r is *positive* or *Horn* if $Body^-(r) = \emptyset$. $X, Y_1, \ldots, Y_n$ are tuples of either constants or variables whose length corresponds to the arity of the predicates $a, b_1, \ldots, b_n$ respectively. The signature of P is given as $\Sigma_P = \langle \mathbf{P}, \mathcal{C} \rangle$, where $\mathbf{P}$ and $\mathcal{C}$ are resp. sets of predicates and constants occurring in P.

A logic program P is *ground* if it consists of only ground rules, i.e. rules without variables. Ground instantiation $Gr(P)$ of a nonground program P is obtained by substituting variables with constants in all possible ways. The *Herbrand universe* $HU(P)$ (resp. *Herbrand base* $HB(P)$) of P, is the set of all constants occurring in P, i.e. $HU(P) = \mathcal{C}$ (resp. the set of all possible ground atoms that can be formed with predicates in $\mathbf{P}$ and constants in $\mathcal{C}$). We refer to any subset of $HB(P)$ as a *Herbrand interpretation*. By $MM(P)$ we denote the set-inclusion minimal Herbrand interpretation of a ground positive program P.

An interpretation I of P is an *answer set* (or *stable model*) of P iff $I \in MM(P^I)$, where P^I is the *Gelfond-Lifschitz (GL) reduct* [12] of P, obtained from $Gr(P)$ by removing (i) each rule r such that $Body^-(r) \cap I \neq \emptyset$, and (ii) all the negative atoms from the remaining rules. The set of answer sets of a program P is denoted by $AS(P)$.

Example 2. Consider the program

$$P = \left\{ \begin{array}{l} (1)\ bornInUS(alex);\ (2)\ bornInUS(mat);\ (3)\ immigrant(mat); \\ (4)\ livesInUS(X) \leftarrow bornInUS(X), not\ immigrant(X) \end{array} \right\}$$

The ground instantiation $Gr(P)$ of P is obtained by substituting X with *mat* and *alex*. For $I=\{bornInUS(alex), bornInUS(mat), immigrant(mat), livesInUS(alex)\}$, the GL-reduct P^I of P contains the rule $livesInUS(alex) \leftarrow bornInUS(alex)$ and the facts (1)-(3). As I is a minimal model of P^I, it holds that I is an answer set of P. □

The answer set semantics for nonmonotonic logic programs is based on the CWA, under which whatever can not be derived from a program is assumed to be false. Nonmonotonic logic programs are widely applied for formalizing common sense reasoning from incomplete information.

Definition 1 (Rule-based KG completion). *Let $\mathcal{G}$ be a KG and $\mathcal{A}$ its factual representation over the signature $\Sigma_\mathcal{A} = \langle \mathbf{C}, \mathbf{R}, \mathcal{C} \rangle$. Let, moreover, $\mathcal{R}$ be a set of rules mined from $\mathcal{G}$, i.e. rules over the signature $\Sigma_\mathcal{R} = \langle \mathbf{C} \cup \mathbf{R}, \mathcal{C} \rangle$. Then completion of $\mathcal{G}$ (resp. $\mathcal{A}$) w.r.t. $\mathcal{R}$ is a graph $\mathcal{G}_\mathcal{R}$ constructed from any answer set $\mathcal{A}_\mathcal{R} \in AS(\mathcal{R} \cup \mathcal{A})$.*

Example 3. Consider a factual representation $\mathcal{A}$ of a KG $\mathcal{G}$ given in a tabular form in Fig. 1b, where a tick appears in an intersection of a row s and a column o, if $o(s) \in \mathcal{A}$ (resp. $\langle s \quad isA \quad o \rangle \in \mathcal{G}$). Suppose we are given a set of rules $\mathcal{R} = \{r_1, r_2\}$, where

$$r_1 : livesInUS(X) \leftarrow bornInUS(X), not \; immigrant(X);$$
$$r_2 : livesInUS(X) \leftarrow hasUSPass(X).$$

The program $\mathcal{A} \cup \mathcal{R}$ has a single answer set $\mathcal{A}_\mathcal{R} = \mathcal{A} \cup \{livesInUS(p_i) \mid i=6, 7, 11\}$, from which the completion $\mathcal{G}_\mathcal{R}$ of $\mathcal{G}$ can be reconstructed. □

3 Learning Exception-Enriched Rules

Horn Rule Revision. Before we formally define our problem, we introduce the notion of an incomplete data source following [7].

Definition 2 (Incomplete data source). *An incomplete data source is a pair $G = (\mathcal{G}^a, \mathcal{G}^i)$ of two KGs, where $\mathcal{G}^a \subseteq \mathcal{G}^i$ and $\Sigma_{\mathcal{A}_{\mathcal{G}^a}} = \Sigma_{\mathcal{A}_{\mathcal{G}^i}}$. We call $\mathcal{G}^a$ the available graph and $\mathcal{G}^i$ the ideal graph.*

The graph $\mathcal{G}^a$ is the graph that we have available as input. The ideal graph $\mathcal{G}^i$ is the perfect completion of $\mathcal{G}^a$, which is supposed to contain all correct facts with entities and relations from $\Sigma_{\mathcal{A}_{\mathcal{G}^a}}$ that hold in the current state of the world.

Given a potentially incomplete graph $\mathcal{G}^a$ and a set of Horn rules $\mathcal{R}_H$ mined from $\mathcal{G}^a$, our goal is to add default negated atoms (exceptions) to the rules in $\mathcal{R}_H$ and obtain a revised ruleset $\mathcal{R}_{NM}$ such that the set difference between $\mathcal{G}^a_{\mathcal{R}_{NM}}$ and $\mathcal{G}^i$ is smaller than between $\mathcal{G}^a_{\mathcal{R}_H}$ and $\mathcal{G}^i$. If in addition the set difference between $\mathcal{G}^a_{\mathcal{R}_{NM}}$ and $\mathcal{G}^i$ is the smallest among the ones produced by other revisions $\mathcal{R}'_{NM}$ of $\mathcal{R}_H$, then we call $\mathcal{R}_{NM}$ an *ideal nonmonotonic* revision. For single rules such revision is defined as follows:

Definition 3 (Ideal nonmonotonic revision). *Let $G = (\mathcal{G}^a, \mathcal{G}^i)$ be an incomplete data source. Moreover, let $r : a \leftarrow b_1, \ldots, b_k$ be a Horn rule mined from $\mathcal{G}^a$. An ideal nonmonotonic revision of r w.r.t. $\mathcal{G}$ is any rule*

$$r' : \quad a \leftarrow b_1, \ldots, b_k, not \; b_{k+1}, not \; b_n, \tag{2}$$

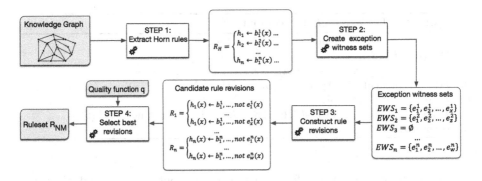

Fig. 2. Exception-enriched rule learning: general overview

such that $\mathcal{G}^i \triangle \mathcal{G}^a_{r'} \subset \mathcal{G}^i \triangle \mathcal{G}^a_r$ [1], i.e. the completion of $\mathcal{G}^a$ based on r' is closer to $\mathcal{G}^i$ than the completion of $\mathcal{G}^a$ based on r, and $\mathcal{G}^a_{r''} \triangle \mathcal{G}^i \subset \mathcal{G}^a_{r'} \triangle \mathcal{G}^i$ for no other nonmonotonic revision $r'' \neq r'$ of r. If $k=n$, then the revision coincides with the original rule.

In our work, we assume that the ideal graph $\mathcal{G}^i$ is not available (otherwise nothing would need to be learnt). Therefore, we cannot verify whether a revision is ideal for $\mathcal{R}_H$. What we can do, however, is to estimate using some quality functions whether a given revision produces an approximation of $\mathcal{G}^i$ that is better than the approximation produced by the original Horn ruleset. For this purpose, we introduce a generic quality function q which receives as input a revision $\mathcal{R}_{NM}$ of the ruleset $\mathcal{R}_H$ and a graph $\mathcal{G}$, and returns a real value that reflects the quality of the revised set $\mathcal{R}_{NM}$. We can now formally define our problem:

Problem: quality-based Horn rule revision

Given: KG $\mathcal{G}$, set of nonground Horn rules $\mathcal{R}_H$ mined from $\mathcal{G}$, quality function q

Find: set of rules $\mathcal{R}_{NM}$ obtained by adding default negated atoms to $Body^-(r)$ for some $r \in \mathcal{R}_H$, such that $q(\mathcal{R}_{NM}, \mathcal{G})$ is maximal.

Note that so far we did not specify the details of the quality function q. In our approach, we estimate the quality of a ruleset by exploiting well-established measures proposed in the field of data mining [3]. Even though none of these measures can offer any sort of guarantee, our hypothesis is that they still indicate to some extent the percentage of correctly predicted facts obtained as a result of completing a KG based on a given ruleset. We discuss in Sect. 5 in more details how q can be defined.

Approach Overview. Figure 2 illustrates the main phases of our approach. In Step 1, we launch an off-the-shelf algorithm to mine Horn rules from the input KG. We use FPGrowth [13], but any other, e.g., [5,11] can be likewise applied, i.e., our overall revision approach is independent of the concrete technique used

---

[1] $\mathcal{G}_1 \triangle \mathcal{G}_2 = (\mathcal{G}_1 \backslash \mathcal{G}_2) \cup (\mathcal{G}_2 \backslash \mathcal{G}_1)$.

Algorithm 1: *ComputeEWS*: compute $EWS(r, \mathcal{A})$

Input: KB $\mathcal{A}$, rule $r : a(X) \leftarrow b_1(X), \ldots, b_k(X)$
Output: $EWS(r, \mathcal{A})$
(a) $N \leftarrow NS(r, \mathcal{A})$; $A \leftarrow ABS(r, \mathcal{A})$
(b) $E^+ \leftarrow \{not\_a(c) \mid c \in A\}$; $E^- \leftarrow \{not\_a(c) \mid c \in N\}$
(c) $\mathcal{R}_e \leftarrow Learn(E^+, E^-, \mathcal{A})$
(d) $EWS \leftarrow \{\text{predicate } p \text{ in } Body^+(r') \mid r' \in \mathcal{R}_e, \text{ s.t. }, p \text{ is not in } Body^+(r)\}$
(e) **return** EWS

---

for Horn rule mining. Then, for each rule we compute *normal* and *abnormal* instance sets, defined as:

Definition 4 (r-(ab)normal instance set). *Let $\mathcal{A}$ be the factual representation of a KG $\mathcal{G}$ and $r : a(X) \leftarrow b_1(X), \ldots, b_k(X)$ a Horn rule mined from $\mathcal{G}$. Then,*

- $NS(r, \mathcal{A}) = \{c \mid b_1(c), \ldots, b_k(c), a(c) \in \mathcal{A}\}$ *is an r-normal instance set;*
- $ABS(r, \mathcal{A}) = \{c \mid b_1(c), \ldots, b_k(c) \in \mathcal{A}, a(c) \notin \mathcal{A}\}$ *is an r-abnormal instance set.*

Example 4. For $\mathcal{A}$ from Fig. 1b and the rule $r : livesInUS(X) \leftarrow bornInUS(X)$, r-normal and r-abnormal instance sets are given as $NS(r, \mathcal{A}) = \{p1, \ldots, p5\}$ and $ABS(r, \mathcal{A}) = \{p6, \ldots, p11\}$ respectively. □

Intuitively, if the given data was complete, then the r-normal and r-abnormal instance sets would exactly correspond to instances for which the rule r holds (resp. does not hold) in the real world. Since the KG is potentially incomplete, this is no longer the case and some r-abnormal instances might in fact be classified as such due to data incompleteness. In order to distinguish between the "wrongly" and "correctly" classified instances in the r-abnormal set, in Step 2 we construct *exception witness sets* (EWS), which are defined as follows:

Definition 5 (Exception witness set (EWS)). *Let $\mathcal{A}$ be the factual representation of a KG $\mathcal{G}$ and let r be a Horn rule mined from $\mathcal{G}$. An r-exception witness set $EWS(r, \mathcal{A}) = \{e_1, \ldots, e_l\}$ is a maximal set of predicates, such that*

(i) $e_i(c') \in \mathcal{A}$ for some $c' \in ABS(r, \mathcal{A})$, $1 \leq i \leq m$ and
(ii) $e_1(c), \ldots, e_m(c) \notin \mathcal{A}$ for all $c \in NS(r, \mathcal{A})$.

Example 5. For $\mathcal{A}$ and r from Example 4 $EWS(r, \mathcal{A}) = \{immigrant\}$ is an r-exception witness set. For $\mathcal{A}' = \mathcal{A} \setminus \{p5\}$ it holds that $EWS(r, \mathcal{A}') = \{immigrant, stateless\}$. □

After EWSs are computed for all rules in $\mathcal{R}_H$, we use them to create potential revisions in Step 3. Then, we rank the newly created revisions and select the best ones using different criteria (Step 4). These selected rules will constitute the new $\mathcal{R}_{NM}$.

4   Computing Exception Witnesses and Potential Rule Revisions

In this section we describe how we calculate the exception witness sets for Horn rules (Fig. 2, Step 2) and how we create potential rule revisions (Fig. 2, Step 3).

Computing Exception Witness Sets. For constructing exception witness sets we use the algorithm *ComputeEWS* (Algorithm 1), which given a factual representation $\mathcal{A}$ of a KG and a rule $r \in \mathcal{R}_H$ as input, outputs the set $EWS(r, \mathcal{A})$.

The algorithm works as follows: First in (a) r-normal $NS(r, \mathcal{A})$ and r-abnormal $ABS(r, \mathcal{A})$ instance sets are found and stored resp. in N and A. Then in (b) the fresh predicate not_a the facts $not\_a(c)$ are added to E^+ for all $c \in ABS(r, \mathcal{A})$. In the same step the facts $not\_a(c)$ for $c \in N$ are stored in E^-. In (c), a variant of a classical inductive learning procedure $Learn(E^+, E^-, \mathcal{A})$, e.g., [23] is employed to induce a set of hypothesis $\mathcal{R}_e$ in the form of Horn rules with unary atoms, s.t. $\mathcal{A} \cup \mathcal{R}_e \models e$ for as many as possible $e \in E^+$, and $\mathcal{A} \cup \mathcal{R}_e \not\models e'$ for all $e' \in E^-$. Finally, in (d) the bodies of rules in $\mathcal{R}_e$ not containing predicates from $Body^+(r)$ are put in EWS, which is output in (e).

The correctness of *ComputeEWS* follows from the correctness of the procedure *Learn*. Indeed, by (d) for $p \in EWS$, a rule r' with p occurring in $Body(r')$ exists in $\mathcal{R}_e$. Since $r' \cup \mathcal{A} \not\models not\_a(c)$ for $not\_a(c) \in E^-$, we have that $p(c) \notin \mathcal{A}$ for r-normal c due to (a) and (b). Moreover, $p(c') \in \mathcal{A}$ for some r-abnormal c', as otherwise $r' \notin \mathcal{R}_e$. Hence, (i) and (ii) of Definition 5 hold, i.e. EWS is an exception witness set for r w.r.t. $\mathcal{A}$.

Constructing Candidate Rule Revisions. After all EWSs are calculated for Horn rules in $\mathcal{R}_H$, we construct a search space of potential revisions by adding to rule bodies exceptions in the form of default negated atoms. More specifically, for every $r_i : a(X) \leftarrow b_1(X), \ldots, b_k(X)$ in $\mathcal{R}_H$ we create $m = |EWS(r_i, \mathcal{A})|$ revision candidates, i.e. rules $r_i^{e_j}$, s.t. $Head(r_i^{e_j}) = Head(r_i)$, $Body^+(r_i^{e_j}) = Body(r_i)$, $Body^-(r_i) = e_j(X)$, where $e_j \in EWS(r_i, \mathcal{A})$. We denote with $\mathcal{R}_i$ the set of all $r_i^{e_j}$.

Example 6. For $EWS(r, \mathcal{A}') = \{immigrant, stateless\}$ from Example 5 in Step 3 revision candidates $r^{im} : livesInUS(X) \leftarrow bornInUS(X), not\ immigrant(X)$ and $r^{st} : livesInUS(X) \leftarrow bornInUS(X), not\ stateless(X)$ are created.   □

5   Rules Quality Assessment

Given a potential R_{NM}, the function q should approximate the closeness between the completion $\mathcal{G}^a_{\mathcal{R}_{NM}}$ of the input KG $\mathcal{G}^a$ and the ideal KG $\mathcal{G}^i$. In this work, we follow usual practice in data mining and adapt standard association rule measures to our needs. Let rm be a generic rule measure, e.g. one defined in Table 1[2]. Then, naively generalizing rm for rulesets by taking the average of rm

---

[2] Table 1 reports the definition of confidence, lift and Jaccard coefficient – three commonly-used rule measures [1]. Here, $n(B)$ (resp. $n(H)$) denotes the number of

Table 1. Rule evaluation measures for a rule r w.r.t. $\mathcal{A}$

Rule measure	Formula for $r : H \leftarrow B$
Confidence	$conf(r, \mathcal{A}) = \dfrac{n(HB)}{n(B)}$
Lift	$lift(r, \mathcal{A}) = \dfrac{n(HB)}{n(H) * n(B)}$
Jaccard coef	$jc(r, \mathcal{A}) = \dfrac{n(HB)}{n(H) + n(B) - n(HB)}$

values for all rules in a given set we obtain

$$q_{rm}(\mathcal{R}_{NM}, \mathcal{A}) = \frac{\sum_{r \in \mathcal{R}_{NM}} rm(r, \mathcal{A})}{|\mathcal{R}_{NM}|} \tag{3}$$

In our case, q_{rm} alone is not sufficiently representative for being the target quality function q for two reasons: (1) it does not penalize rules with noisy exceptions[3]; (2) it does not measure how many contradicting beliefs our revisions reflect.

Example 7.

(1) For $r : livesInUS(X) \leftarrow hasUSPass(X), not\ poet(X)$ and $\mathcal{A}$ (from Fig. 1b) we have $conf(r, \mathcal{A}) = 1$, as all 3 non-poets with US passports live in the US, i.e., r gets the highest individual score based on confidence. However, $poet$ is a noisy exception due to p_3, who is a poet possessing a US passport and living in the US.

(2) Let $\mathcal{R}_{NM} = \{r_1 : lu(X) \leftarrow hu(X), st(X), r_2 : lu(X) \leftarrow bu(X), not\ im(X), r_3 : im(X) \leftarrow st(X)\}$, where lu, hu, bu, st, im stand for $livesInUS, hasUSPass, bornInUS, stateless$ and $immigrant$. Although im in r_2 may perfectly fit as exception w.r.t. some (unspecified here) original KG; once the KG is completed based on r_1 and r_3, im might become noisy for r_2. Indeed, r_1 can easily bring new instances c in lu, while r_3 can predict facts $im(c)$. If this is the case, i.e., $r_2 \in \mathcal{R}_{NM}$ becomes noisy after other rules in $\mathcal{R}_{NM}$ are applied, then intuitively rules in $\mathcal{R}_{NM}$ do not agree on the beliefs about $\mathcal{G}^i$ they express.

□

To resolve the above issues we introduce an additional quality function $q_{conflict}$, next to q_{rm}, whose purpose is to evaluate the ruleset w.r.t (1) and (2). To measure $q_{conflict}$ for $\mathcal{R}_{NM}$, we create an extended set of rules $\mathcal{R}^{aux}$, which contains every revised rule $r : a(X) \leftarrow b(X), not\ e(X)$ in $\mathcal{R}_{NM}$ and its auxiliary version $r_{aux} : not\_a(X) \leftarrow b(X), e(X)$, where not_a is a fresh predicate collecting instances that are not in a. Notice that r_{aux} is meaningless, and thus

---

instances for which the body (resp. head) of a rule $H \leftarrow B$ is satisfied in $\mathcal{A}$ or in data mining terminology the number of transactions in $\mathcal{A}$ with items from B (resp. H).

[3] e is a *noisy* exception for r if $e(c) \in \mathcal{A}$ for some r-normal c.

void in $\mathcal{R}^{aux}$, for rules r with positive bodies. Formally, we define $q_{conflict}$ as follows

$$q_{conflict}(\mathcal{R}_{NM}, \mathcal{A}) = \sum_{p \in pred(\mathcal{R}^{aux})} \frac{|\{c \mid p(c), not\_p(c) \in \mathcal{A}_{\mathcal{R}^{aux}}\}|}{|\{c \mid not\_p(c) \in \mathcal{A}_{\mathcal{R}^{aux}}\}|} \quad (4)$$

where $pred(\mathcal{R}^{aux})$ is the set of predicates appearing in $\mathcal{R}^{aux}$.

Intuitively, $\mathcal{A}_{\mathcal{R}^{aux}}$ contains both positive predictions of the form $p(c)$ and negative ones $not\_p(c)$ produced by the rules in $\mathcal{R}^{aux}$. The function $q_{conflict}$ computes the ratio of "contradicting" pairs $\{p(c), not\_p(c)\}$ over the number of $not\_p(c)^4$ in $\mathcal{A}_{\mathcal{R}^{aux}}$, which reflects how much the rules in $\mathcal{R}_{NM}$ disagree with each other on beliefs about the ideal KG $\mathcal{G}^i$ they express. The smaller $q_{conflict}$, the better is the ruleset $\mathcal{R}_{NM}$.

Revision Based on Partial Materialization. Our goal in Step 4 is to find a set of revisions $\mathcal{R}_{NM}$, for which $q_{rm}(\mathcal{R}_{NM}, \mathcal{A})$ is maximal and $q_{conflict}(\mathcal{R}_{NM}, \mathcal{A})$ is minimal.

To determine such globally best set $\mathcal{R}_{NM}$ many candidate rule combinations have to be checked, which is unfortunately not feasible because of the large size of our $\mathcal{A}$ and EWS. Therefore, we propose an approach where we incrementally build $\mathcal{R}_{NM}$ by considering every $r_i \in \mathcal{R}_H$ and choose the best revision $r_i^j \in \mathcal{R}_i$ for it. In order to select the best r_i^j, we use a special ranking function, which estimates how well a rule r at hand describes the data and how noisy its exceptions are. In the remaining of this section, we will propose four different ranking functions, starting from the simplest to the most sophisticated one.

Naive-**ranker.** The first implementation, which we call *rank_naive*, calculates the average value of the rm scores of r and r_{aux} and uses it to rank the rules. Formally, the average is computed by the following function:

$$est_{rm}(r, \mathcal{A}) = \frac{\overbrace{rm(H \leftarrow B, not\ E, \mathcal{A})}^{r} + \overbrace{rm(not\_H \leftarrow B, E, \mathcal{A})}^{r_{aux}}}{2} \quad (5)$$

where rm is one of the measures in Table 1. E.g., plugging in *conf* instead of *rm*, gives

$$est_{conf}(r, \mathcal{A}) = \frac{1}{2}\left(\frac{n(BH) - n(BHE)}{n(B) - n(BE)} + \frac{n(BE) - n(BHE)}{n(BE)}\right) \quad (6)$$

where $n(X)$ is the number of transactions with items from X.

Example 8. For r and $\mathcal{A}$ from Example 7 (1) $est_{conf}(r, \mathcal{A}) = 0.75$, i.e., due to noisiness of *poet* the value of est_{conf} decreased. □

PM-**ranker.** The main problem of *rank_naive* is that it does not exploit any knowledge about the properties that a final revision $\mathcal{R}_{NM}$ might have. In other

---

[4] Ratio over the number of $p(c)$ instead of $not\_p(c)$ is possible, but then $q_{conflict}$ is smaller and less representative.

244   M.H. Gad-Elrab et al.

words, ranking of revisions of a rule at hand is completely independent from ranking of revisions for other rules. To address this issue, we propose a second implementation called *revision based on partial materialization* (denoted as *rank_pm*). Here, the idea is to apply est_{rm} for a rule r not on $\mathcal{A}$ but on completion of $\mathcal{A}$ based on other rules, which according to our estimates constitute some approximation of $\mathcal{R}_{NM}$.

Example 9. Consider a rule $r_1{:}lu(X)\leftarrow bu(X), not\;\; im(X)$, and suppose there is only a single other rule $r_2{:}lu(X)\leftarrow hu(X)$ given, for which $EWS(r_2,\mathcal{A}) = \emptyset$ for $\mathcal{A}$ from Fig. 1b. This knowledge can be exploited when ranking r_1. We have $est_{conf}(r_1,\mathcal{A}) = 0.8$, while $est_{conf}(r_1,\mathcal{A}_{r_2}) = 0.875$ due to the materialized fact $livesInUS(p11)$. This increase gives us an indication that r_1 agrees with r_2 on predictions it makes.

On the contrary, for $r_3\;:\;lu(X)\;\leftarrow\;hu(X), not\;\;\;pt(X)\}$, where pt stands for *poet* and $r_4\;:\;lu(X)\leftarrow bu(X)$ we have $est_{conf}(r_3,\mathcal{A})=0.75$, but $est_{conf}(r_3,\mathcal{A}_{r_4})=0.5$, which witnesses that beliefs of r_3 and r_4 contradict. □

The function *rank_pm* first constructs the temporary rule set $\mathcal{R}^t$, which contains, for every rule $r_i \in \mathcal{R}_H$, a rule r_i^t with all exceptions from $EWS(r_i,\mathcal{A})$ incorporated, i.e., $\mathcal{R}^t$ predicts the smallest number of facts, which are also predicted by any possible revision $\mathcal{R}_{NM}$. Then, for each $r_i \in \mathcal{R}_H$, we compute the est_{rm} value for all revision candidates r_i^j based on $\mathcal{A}_{\mathcal{R}^t\backslash r_i^t}$. Formally,

$$rank\_pm(r_i^j,\mathcal{A}) = est_{rm}(r_i^j,\mathcal{A}_{\mathcal{R}^t\backslash r_i^t}) \tag{7}$$

Once the scores for all revision candidates r_i^j for r_i are computed, we pick the revision with the highest score, add it to the current snapshot of $\mathcal{R}_{NM}$ and move to r_{i+1}.

OPM-**ranker**. With *rank_pm*, facts inferred by rules of low quality might have a significant impact on more promising rules. To handle this issue, we propose a variation of *rank_pm* called *revision with ordered partial materialization* (abbr. *rank_opm*), which proceeds as follows. First we rank Horn rules based on some rm' (possibly same as rm) and obtain an ordered list $os_{\mathcal{R}_H}$. Then we go through $os_{\mathcal{R}_H}$ and for every rule r_i we compute a snapshot $\mathcal{A}_i$ of $\mathcal{A}$ by materializing only those rules $r_k^t \in \mathcal{R}^t$, for which r_k is ordered higher in the list $os_{\mathcal{R}_H}$ than r_i. More formally,

$$rank\_opm_{rm}(r_i,\mathcal{A}) = est_{rm}(r_i,\mathcal{A}_i) \tag{8}$$

where $\mathcal{A}_i = \mathcal{A}_{\mathcal{R}^t\backslash\{r_k^t\,|\,os_{\mathcal{R}_H}[k]=r_k;\,i\geq k\}}$.

OWPM-**ranker**. With *rank_opm* as we have defined it, the facts inferred by rules count the same as the true facts in $\mathcal{A}$. Since the predicted facts are inferred based on statistically-supported assumptions, it is natural to distinguish them from the facts that are explicitly present in $\mathcal{A}$. To achieve this, we propose one last ranking function that exploits weights assigned to facts. Here, there is a clear distinction between facts from $\mathcal{A}$ (which get maximal weight) and the predicted facts (which inherit weights from rules that inferred them). We call this method *revision with ordered weighted partial materialization* (abbr. *rank_owpm*).

The method *rank_owpm* differs from *rank_opm* in that weights are used to estimate the revisions' scores. It is convenient (and a common practice) to assign weights of probabilistic nature between 0 and 1 (e.g., *confidence* can be exploited). There are several ways to produce weighted partial materialization; for example, using probabilistic logic programming systems, e.g., Problog [9] or PrASP [25].

However, normally, in such systems facts predicted by some rules in a ruleset at hand are used as input to other rules, i.e., uncertainty is propagated through rule chains, which might be undesired in our setting. To avoid such propagation, when computing weighted partial materialization of $\mathcal{A}$ we keep predicted facts (i.e., derived using rules) separately from the explicit facts (i.e., those in $\mathcal{A}$), and infer new facts using only $\mathcal{A}$.

The method *rank_owpm* works as follows. Initially, we sort the rules in $\mathcal{R}_H$ and create the $\mathcal{A}_i$s with the same procedure as described for *rank_opm*. The only difference is that here every inferred fact in $\mathcal{A}_{\mathcal{R}^t}$ receives a specific weight that corresponds to $rm(r', \mathcal{A})$, where r' is the positive version of the rule that inferred the fact[5]. If the same fact is derived by multiple rules, we keep the highest weight.

The weights play a role when we evaluate a rule w.r.t. the partially materialized KG. To this end, we slightly change the rm function so that it considers weighted facts (we denote such function as rm^w). E.g., $conf^w(r, \mathcal{A})$ calculates a weighted sum of the instances for which the head (resp. body) of r is satisfied w.r.t. $\mathcal{A}$ (instead of a normal sum used in $conf$). Formally, *rank_owpm* computes a score for a revision r_i^j as follows:

$$rank\_owpm_{rm^w}(r_i^j, \mathcal{A}) = est_{rm^w}(r_i^j, \mathcal{A}_i^w) \tag{9}$$

where $\mathcal{A}_i^w$ is the weighted version of $\mathcal{A}_i$ from Eq. 8. In the following section, we will analyze the performance of these four functions on some realistic KGs.

6 Evaluation

Experimental Setup. We considered two knowledge graphs: a slice of almost 10M facts from YAGO3 [22], a general purpose KG, and an RDF version of IMDB[6] data with 2M facts, a well known domain-specific KG of movies and artists. We chose these two KGs in order to evaluate our method's performance on both general-purpose and domain-specific KGs. Our experiments were performed on a machine with 40 cores and 400GB RAM. The used datasets and the experimental code are publicly available[7].

Outline. First we evaluate different configurations of our method using the quality functions q_{rm} and $q_{conflict}$, defined in Sect. 5. Then, we report the results of a

---

[5] We cannot consider the entire rule (i.e. with all exceptions attached), since standard measures like *confidence* will return values very close to 1 for such rules.

[6] http://imdb.com.

[7] http://people.mpi-inf.mpg.de/~gadelrab/rules_iswc.

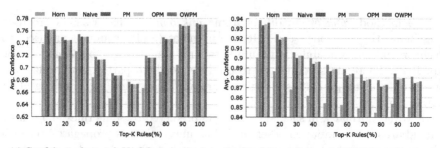

(a) Confidence for top-k YAGO revised rules   (b) Confidence for top-k IMDB revised rules

Fig. 3. Average rules' confidence on YAGO and IMDB (higher is better).

manual assessment that we performed to evaluate the quality of the predictions, reporting good and bad examples produced by our method.

6.1 Evaluation of the Revision Steps

Step 1. Initially, we considered the Horn rules produced by AMIE [11]. However, they mainly focus on unsupported binary predicates and the only unary rules are restricted to the *isA* predicate, which was too limiting for us. Therefore, we first propositionalized the original KG, and then mined the Horn rules using the association rule mining implementation based on standard FPGrowth [13] offered by SPMF Library[8]. In order to avoid over-fitting rules as well as to reduce the computation, we limited the extraction to rules with maximum four body atoms, a single head atom, a minimum support of $0.0001 \times \#$ *entities* and a minimum confidence of 0.25 for YAGO. Since IMDB is smaller and more connected, we set a higher minimum support of $0.005 \times \#$ *entities* and confidence of 0.6. On our machine, this process took approx. 10 seconds on YAGO and 2.5 second on IMDB, and it generated about 10 K and 25K rules respectively.

Steps 2 and 3. We implemented a simple inductive learning procedure, which performs manipulations on the set of facts instantiating the rule and its body to get the EWS. The generation of EWSs with minimum support of 0.05 took about 50 seconds for YAGO and 30 seconds for IMDB. The execution time is significantly affected by the size and distribution of the predicates in the KG. We could find EWSs for about 6 K rules mined from YAGO, and 22 K rules mined from IMDB. On average, the EWSs for the YAGO's rules contained 3 exceptions, and 28 exceptions on IMDB.

Step 4. We evaluated the quality of our rule selection procedure w.r.t. two dimensions, which reflect the two q proposed in Sect. 5: *average of the rules' confidence* (q_{conf}), and the *number of conflicts* ($q_{conflict}$). The average confidence shows how well the revised rules adhere to the input. The number of conflicts indicates how consistent the revised rules set is w.r.t the final predictions

---

[8] http://www.philippe-fournier-viger.com/spmf/.

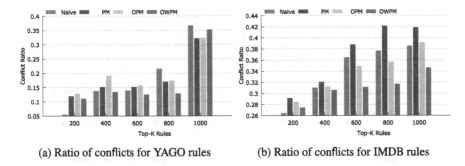

(a) Ratio of conflicts for YAGO rules   (b) Ratio of conflicts for IMDB rules

Fig. 4. Ratio of conflicts on YAGO and IMDB (lower is better).

it makes. Due to space constraints, we report the results using only confidence as rule evaluation function (Eq. 6) and lift as rule ordering criterion, as we found this combination to be a good representative.

6.2 Exception-Enriched Rules vs. Horn Rules

Figure 3 reports the obtained average rules' confidence using the four ranking functions to select the best revisions. *Horn* reports the average confidence of the original Horn rules; while *Naive*, *PM*, *OPM* and *OWPM* are our ranking methods described in Sect. 5. For both inputs, we show the results on the top $10, \ldots, 100\%$ rules ranked by lift.

We make three observations. (i) In general enriching Horn rules with exceptions increases the average confidence (approx. 11 % for YAGO, 3.5 % for IMDB). This indicates that our method is useful to mine rules that reflect the data more precisely. It is also worth mentioning that along with the increase in confidence, the average coverage of the revised rules dropped only by 13 % for YAGO and 4 % for IMDB (i.e. the rules do not become too specific). (ii) The comparison between the four ranking methods shows that the highest confidence is achieved by the non-materialized (*Naive*) function followed by the weighted one (*OWPM*). (iii) Since we used lift for ordering the rules, and it is not neccessarily correlated with confidence, one can see that the confidence drops for around top 60 % of the YAGO rules, and then slightly increases again. For IMDB a smooth confidence decrease is observed with the addition of lower-ranked rules.

The higher value of *Naive* was expected, since this procedure is designed to maximize the confidence. However, confidence alone is not a good indicator to determine the overall rule's quality, as we explained in Sect. 5. Figure 4 shows the number of conflicts (for YAGO and IMDB) that were obtained by executing the revised rules and their corresponding auxiliary versions (r_{aux}) using the DLV system [19]. Unfortunately, DLV was unable to scale to the entire ruleset; hence, we used up to 1000 rules. In our experiment, a conflict occurs when we derive both $p(c)$ and $not\_p(c)$. The graphs report the ratio between the number of conflicts and negated derived facts. From them, we observe that both *OPM* and

$Y_1 : isMountain(X) \leftarrow isLocatedInAustria(X), isLocatedInItaly(X),$
$\quad\quad\quad \textbf{not}[isRiver(X)|isLocatedInRussia(X)]$
$Y_2 : bornInUSA(X) \leftarrow actedInMovie(\overline{X}), createdMovie(X), isPerson(X),$
$\quad\quad\quad \textbf{not}[\overline{wonFilmfareAwards(X)}|bornInNewYork(X)]$
$Y_3 : isPoliticianOfUSA(X) \leftarrow \overline{bornInUSA(X)}, isGovernor(X),$
$\quad\quad\quad \textbf{not}[isPoliticianOfPuertoRico(X)|isPoliticianOfHawaii(X)]$

$I_1 : hasLanguageEnglish(X) \leftarrow hasGenreDrama(X), hasGenreTriller(X), hasGenreCrime(X),$
$\quad\quad\quad \textbf{not}[\underline{producedInIndia(X)}|createdByNovelist(X)]$
$I_2 : hasGenreAnimation(X) \leftarrow \overline{directedByActor(X)}, hasLanguageEnglish(X), producedInUSA(X),$
$\quad\quad\quad hasGenreFamily(X), \textbf{not}[\underline{hasGenreDrama(X)}|producedIn1984(X)]$

Fig. 5. Anecdotal example rules (Y = YAGO, I = IMDB) with good and bad exceptions

OWPM produce less conflicts than the *Naive* function in most of the cases. By comparing the *OPM* and *OWPM* functions, we find that the weighted version is better, especially on the IMDB dataset when we can reduce the conflicts from 775 to 685 on a base of about 2000 negated facts.

We executed the top-1000 revised rules using DLV and counted the number of derivations that our exceptions prevented. For YAGO with the original Horn rules, the reasoner inferred 924591 new triples. Our exception-enriched ruleset decreased the number of inferred triples to 888215 (*Naive*), 892707 (*PM*), 892399 (*OPM*), and 891007 (*OWPM*). For IMDB we observed a smaller reduction. With the Horn rules the reasoner derived 38609 triples, while with the revised rules the inference set decreased to 36069 (*Naive*), 36355 (*PM*), 36021 (*OPM*), and 36028 (*OWPM*) triples.

Unfortunately, there is no automatic way available to assess whether the removed inference consists of genuine errors. Therefore, we selected the revised ruleset produced by the *OWPM* function and sampled 259 random facts from YAGO (we selected three facts for each binary predicate to avoid skewness). Then, we manually consulted online resources like Wikipedia to determine whether these triples were indeed incorrect. We found that 74.3 % of these triples consisted of factual mistakes. This number provides a first empirical evidence that our method is indeed capable of detecting good exceptions and hence can improve the general quality of the Horn rules.

We conclude reporting some anecdotal examples of rules on YAGO and IMDB in Fig. 5. Between the brackets we show examples of both good (underlined) and bad exceptions. In some cases, the rules have high quality exceptions such as rule *Y*1. In others, we found that the highest ranked exceptions mainly refer to disjoint classes of the head. The complete list of mined rules with the scores given to the determined exceptions is available in our repository.

7 Related Work

The problem of automatically learning patterns from KGs and exploiting them for predicting new facts has gained a lot of attention in the recent years. Approaches for predicting unseen data in KGs can be roughly divided into two groups: statistics-based, and logic-based. The firsts apply well-known techniques like tensor factorization, or neural-embedding-based models (see [24] for overview). The second group focuses more on logical rule learning (e.g., [11,29]). The most relevant works for us are in the last group. These, however, typically focus on learning Horn rules, rather than nonmonotonic (i.e., exception-enriched) as we do.

In the association rule mining community, some works concenrated on finding (interesting) exception rules (e.g. [28]), which are defined as rules with low support (rare) and high confidence. Our work differs from this line of research because we do not necessarily look for rare interesting rules, but care about the quality of their predictions. Another relevant stream of research is concerned with learning Description Logic TBoxes or schema (e.g., [18]). However, these techniques focus on learning concept definitions rather than nonmonotonic rules.

In the context of inductive and abductive logic, learning nonmonotonic rules from complete datasets [10] was studied in several works ([6,15,17,26,26,27]. These methods rely on CWA and focus on describing a dataset at hand exploiting negative example, which are explicitly given unlike in our setting.

Learning Horn rules in presence of incompleteness was studied in hybrid settings in [14,20]. There a background theory or a hypothesis can be represented as a combination of a DL ontology and Horn rules. While the focus of this work is on the complex interaction between reasoning components and the learned rules are positive, we are concerned with techniques for deriving nonmonotonic rules with high predictive quality from huge KGs.

8 Conclusions and Future Work

We have presented a method for mining nonmonotonic rules from KGs: first learning a set of Horn rules, and then revising them by adding negated atoms into their bodies with the goal of improving the quality of a rule set for data prediction. To select the best revision from potential candidates we devised rule-set ranking measures, based on data mining measures and the novel concept of partial materialization. We evaluated our method with various configurations on both general-purpose and domain-specific KGs and observed significant improvements over a baseline Horn rule mining.

There are various directions for future work. First, we look into extracting evidence for or against exceptions from text and web corpora. Second, our framework can be enhanced by partial completeness assumptions for certain predicates (e.g., all countries are available in KG) or constants (e.g., knowledge about Barack Obama is complete). Finally, an overriding future direction is to extend our work to more complex nonmonotonic rules with higher-arity predicates, aggregates and disjunctions in rule heads.

Acknowledgments. We thank Thomas Eiter, Francesca A. Lisi and the anonymous reviewers for their constructive feedback about this work. The research was partially funded by the NWO VENI project 639.021.335.

References

1. Agrawal, R., Carey, M.J., Livny, M.: Concurrency control performance modeling: alternatives and implications. In: Performance of Concurrency Control Mechanisms in Centralized Database Systems, pp. 58–105 (1996)
2. Auer, S., Bizer, C., Kobilarov, G., Lehmann, J., Cyganiak, R., Ives, Z.G.: DBpedia: a nucleus for a web of open data. In: Aberer, K., et al. (eds.) ASWC 2007 and ISWC 2007. LNCS, vol. 4825, pp. 722–735. Springer, Heidelberg (2007)
3. Azevedo, P.J., Jorge, A.M.: Comparing rule measures for predictive association rules. In: Kok, J.N., Koronacki, J., Lopez de Mantaras, R., Matwin, S., Mladenič, D., Skowron, A. (eds.) ECML 2007. LNCS (LNAI), vol. 4701, pp. 510–517. Springer, Heidelberg (2007)
4. Carlson, A., Betteridge, J., Kisiel, B., Settles, B., Hruschka Jr., E.R., Mitchell, T.M.: Toward an architecture for never-ending language learning. In: Proceedings of AAAI (2010)
5. Chen, Y., Goldberg, S., Wang, D.Z., Johri, S.S.: Ontological pathfinding: mining first-order knowledge from large knowledge bases. In: Proceedings of SIGMOD 2016, pp. 835–846 (2016)
6. Corapi, D., Russo, A., Lupu, E.: Inductive logic programming as abductive search. In: Proceedings of ICLP, pp. 54–63 (2010)
7. Darari, F., Nutt, W., Pirrò, G., Razniewski, S.: Completeness statements about RDF data sources and their use for query answering. In: Alani, H., et al. (eds.) ISWC 2013, Part I. LNCS, vol. 8218, pp. 66–83. Springer, Heidelberg (2013)
8. Erxleben, F., Günther, M., Krötzsch, M., Mendez, J., Vrandečić, D.: Introducing wikidata to the linked data web. In: Mika, P., et al. (eds.) ISWC 2014, Part I. LNCS, vol. 8796, pp. 50–65. Springer, Heidelberg (2014)
9. Fierens, D., den Broeck, G.V., Renkens, J., Shterionov, D.S., Gutmann, B., Thon, I., Janssens, G., Raedt, L.D.: Inference and learning in probabilistic logic programs using weighted boolean formulas. TPLP **15**(3), 358–401 (2015)
10. Flach, P.A., Kakas, A.C.: Abduction and Induction: Essays on Their Relation and Integration. Applied Logic Series, vol. 18. Springer, Heidelberg (2000)
11. Galárraga, L., Teflioudi, C., Hose, K., Suchanek, F.M.: Fast rule mining in ontological knowledge bases with AMIE+. VLDB J. **24**, 707–730 (2015)
12. Gelfond, M., Lifschitz, V.: The stable model semantics for logic programming. In: Proceedings of ICLP/SLP, pp. 1070–1080 (1988)
13. Han, J., Pei, J., Yin, Y., Mao, R.: Mining frequent patterns without candidate generation: a frequent-pattern tree approach. Data Min. Knowl. Discov. **8**(1), 53–87 (2004)
14. Józefowska, J., Lawrynowicz, A., Lukaszewski, T.: The role of semantics in mining frequent patterns from knowledge bases in description logics with rules. TPLP **10**(3), 251–289 (2010)
15. Katzouris, N., Artikis, A., Paliouras, G.: Incremental learning of event definitions with inductive logic programming. Mach. Learn. **100**(2–3), 555–585 (2015)
16. Lassila, O., Swick, R.R.: Resource description framework (RDF) model and syntax specification (1999)
17. Law, M., Russo, A., Broda, K.: Inductive learning of answer set programs. In: Fermé, E., Leite, J. (eds.) JELIA 2014. LNCS, vol. 8761, pp. 311–325. Springer, Heidelberg (2014)
18. Lehmann, J., Auer, S., Bühmann, L., Tramp, S.: Class expression learning for ontology engineering. J. Web Sem. **9**(1), 71–81 (2011)

19. Leone, N., Pfeifer, G., Faber, W., Eiter, T., Gottlob, G., Perri, S., Scarcello, F.: The DLV system for knowledge representation and reasoning. ACM TOCL **7**(3), 499–562 (2006)
20. Lisi, F.A.: Inductive logic programming in databases: from datalog to DL+log. TPLP **10**(3), 331–359 (2010)
21. Lloyd, J.W.: Foundations of Logic Programming, 2nd edn. Springer, Heidelberg (1987)
22. Mahdisoltani, F., Biega, J., Suchanek, F.M.: YAGO3: a knowledge base from multilingual Wikipedias. In: Procedings of CIDR (2015)
23. Muggleton, S., Feng, C.: Efficient induction of logic programs. In: ALT, pp. 368–381 (1990)
24. Nickel, M., Murphy, K., Tresp, V., Gabrilovich, E.: A review of relational machine learning for knowledge graphs. Proc. IEEE **104**(1), 11–33 (2016)
25. Nickles, M., Mileo, A.: A hybrid approach to inference in probabilistic nonmonotonic logic programming. In: Proceedings of ICLP, pp. 57–68 (2015)
26. Ray, O.: Nonmonotonic abductive inductive learning. J. Appl. Logic **3**(7), 329–340 (2008)
27. Sakama, C.: Induction from answer sets in nonmonotonic logic programs. ACM Trans. Comput. Log. **6**(2), 203–231 (2005)
28. Taniar, D., Rahayu, W., Lee, V., Daly, O.: Exception rules in association rule mining. Appl. Math. Comput. **205**(2), 735–750 (2008)
29. Wang, Z., Li, J.: RDF2Rules: learning rules from RDF knowledge bases by mining frequent predicate cycles. CoRR abs/1512.07734 (2015)
30. Wrobel, S.: First order theory refinement. In: Raedt, L.D. (ed.) Advances in Inductive Logic Programming, pp. 14–33. IOS Press, Amsterdam (1996)

Planning Ahead: Stream-Driven Linked-Data Access Under Update-Budget Constraints

Shen Gao[1(✉)], Daniele Dell'Aglio[1,2], Soheila Dehghanzadeh[3],
Abraham Bernstein[1], Emanuele Della Valle[2], and Alessandra Mileo[3]

[1] Department of Informatics, University of Zurich, Zurich, Switzerland
shengao@ifi.uzh.ch
[2] DEIB, Politecnico di Milano, Milano, Italy
daniele.dellaglio@polimi.it
[3] INSIGHT Research Center, NUI Galway, Galway, Ireland

Abstract. Data stream applications are becoming increasingly popular on the web. In these applications, one query pattern is especially prominent: a join between a continuous data stream and some background data (BGD). Oftentimes, the target BGD is large, maintained externally, changing slowly, and costly to query (both in terms of time and money). Hence, practical applications usually maintain a local (cached) view of the relevant BGD. Given that these caches are not updated as the original BGD, they should be refreshed under realistic budget constraints (in terms of latency, computation time, and possibly financial cost) to avoid stale data leading to wrong answers. This paper proposes to model the join between streams and the BGD as a bipartite graph. By exploiting the graph structure, we keep the quality of results good enough without refreshing the entire cache for each evaluation. We also introduce two extensions to this method: first, we consider a continuous join between recent portions of a data stream and some BGD to focus on updates that have the longest effect. Second, we consider the future impact of a query to the BGD by proposing to delay some updates to provide fresher answers in future. By extending an existing stream processor with the proposed policies, we empirically show that we can improve result freshness by 93 % over baseline algorithms such as Random Selection or Least Recently Updated.

1 Introduction

Real-time processing of massive, dynamically generated stream-data has become increasingly popular on the Web [18]. In stream processing, one common task is to enrich the streams with external background data (BGD). This kind of tasks has to deal with two **V**'s of "Big Data" at the same time: *Velocity*, the rapidly changing nature of the stream data; *Variety*, integrating data from different sources[1]. RDF Stream Processing (RSP) has provided necessary languages to declare this task. Current RSP languages, such as C-SPARQL [3],

---

[1] http://www.ibmbigdatahub.com/infographic/four-vs-big-data.

© Springer International Publishing AG 2016
P. Groth et al. (Eds.): ISWC 2016, Part I, LNCS 9981, pp. 252–270, 2016.
DOI: 10.1007/978-3-319-46523-4_16

SPARQL$_{stream}$ [4], and CQELS-QL [16], support complex queries that involve both streams and remote BGD. However, these RSP engines are not optimized for remote BGD access. Usually, they continuously fetch BGD to match newly arrived stream data ignoring the communication and potential financial cost of such operations. To improve BGD access, RSP engines may adopt local views (or caches), as done in database systems [9]. However, the remote BGD is not always static. Indeed, even in the mostly static linked-data realm, information changes [13]. Hence, the *freshness* of local views in the RSP engine degrades over time as updates in BGD do not propagate to the local view. To address this problem, RSP engines have to *maintain* the local view, by identifying the out-of-date (or *stale*) data items and replacing them with the up-to-date (or *fresh*) values retrieved from the remote. Examples of such updating behavior include the identification of opinion makers in social media based on a stream of posts and (slowly-changing) contact-networks as BGD or traffic prediction based on position data fetched from mobile phones.

Maintaining a local view can take time. Given that a federated query evaluation can spend up to 95 % of its time on accessing remote data [19], query evaluation under response time constraints becomes a major challenge. To ensure a certain response time, only a limited number of remote accesses can be allowed. Additionally, BGD providers may impose constraints such as API rate limits, e.g., Twitter[2]. Lastly, other communication and financial constraints may have to be considered, since accessing BGD can cost money, computation power or energy (in both the RSP engine and the remote service). Returning to the above examples, computing updated network metrics for opinion makers is computationally expensive, and fetching location updates from cell phones burdens scarce battery power. In this paper, we consider these constraints as a limited *budget* that restricts the number of BGD accesses. We study the problem of how to utilize the limited budget so that it can provide fresher response to the query.

To optimally manage BGD accesses under realistic budget constraints, this paper proposes to allocate budget only to carefully selected "important" data that could lead to more *fresh* join results. Our algorithms exploit characteristics of the join between the stream and the BGD to improve the response freshness. Specifically, our contribution is threefold. First, we propose an algorithm that employs a bipartite graph to model the join selectivity between stream and BGD. It favors the update of data items with a higher selectivity within a budget constraint. This problem decomposes to two scenarios: one can be tackled with a local optimal approach; a second is NP-hard requiring a greedy heuristic approach. This encodes Hypothesis **H1: A maintenance processes exploiting join selectivity improves response freshness**. Second, we extend the above model to favor data items that have a longer impact on the response freshness, which leads to hypothesis **H2: Leveraging the definition of the sliding window and BGD change frequencies can improve response freshness**. Third, we explore the trade-off between the current and future importance of data elements. We present an algorithm that exploits the change frequencies, join

---

[2] https://dev.twitter.com/rest/public/rate-limiting.

selectivity, and the sliding window all together to delay some current refreshes in favor of future, more important ones. It encodes hypothesis **H3: Considering both current and future evaluations for budget allocation can further improve response freshness.**

Outline: Section 2 introduces some background of RSP and BGD access. Section 3 reviews related work. Section 4 formalizes the problem. Our solutions and their optimization are in Sect. 5. Section 6 provides evaluation results of our hypotheses on both real and synthetic data sets.

2   Background

An **RDF stream** S is a potentially unbounded sequence of timestamped informative units (d_i, t_i) ordered by the temporal dimension, where t_i is the timestamp (as in [3,4,16], we consider the time as discrete) and d_i is a set of RDF statements. An RDF statement is a triple $(s, p, o) \in (I \cup B) \times I \times (I \cup B \cup L)$, where I, B, and L identify the sets of IRIs, blank nodes and literals, respectively. An **RDF term** is an element of the set $T = I \cup B \cup L$.

RSP Query Languages [3,4,7,16] extend SPARQL[3] with operators to cope with streams. They enable the registration of queries over RDF streams. RSP queries are evaluated in a continuous fashion, i.e., results are computed at different time instances as the data flows in the streams. Given a query q, the answer $Ans(q)$ is a stream, to which the results of the evaluations are appended. This work focuses on the RSP query languages that support the **time-based sliding window** operator $\mathbb{W}$, which is defined through the parameter ω, the width, and β, the slide, and generates a sequences of fixed windows, i.e., portions of S in a time interval $(o, c]$ [3,4,16]. Given a time-based sliding window and two generated consecutive windows W_i and W_{i+1}, defined in $(o_i, c_i]$ and $(o_{i+1}, c_{i+1}]$, two constraints hold: $c_i - o_i = c_{i+1} - o_{i+1} = \omega$ and $o_{i+1} - o_i = \beta$.

Let V be a set of variables (disjoint with I, B and L), graph patterns are expressions defined recursively as: (1) a basic graph pattern, i.e., a set of triple patterns $(ts, tp, to) \in (I \cup B \cup V) \times (I \cup V) \times (I \cup B \cup L \cup V)$, is a graph pattern; (2) let P_1 and P_2 be graph patterns, P_1 *JOIN* P_2 or P_1 *UNION* P_2 is a graph pattern; (3) let P be a graph patterns and $u \in I \cup V$, *SERVICE* u P or $WINDOW$ u P is also a graph pattern. Other graph pattern expressions are possible (e.g. OPTIONAL, FILTER) but are not presented for the sake of space[3].

Like SPARQL, the evaluation semantics of RSP Query Languages rely on the notion of **solution mapping**, i.e., a partial function that maps variables to RDF terms, i.e., $\mu : V \to T$. A full formalization of RSP Query Languages is in [7]. We briefly describe the semantics of WINDOW, SERVICE, and JOIN in RSP Query Languages. Evaluating a WINDOW clause results the content of a sliding window, similarly to what GRAPH does in SPARQL, which refers to the content of a named graph in the data set. The SERVICE retrieves mappings from SPARQL

---

[3] Cf. https://www.w3.org/TR/sparql11-query/ for additional reference.

endpoints by submitting a graph pattern [2]. JOIN can be formally defined as: let $dom(\mu) \subset V$ be the set of variables mapped by μ, two mappings μ_1 and μ_2 are **compatible** (denoted with $\mu_1 \sim \mu_2$) if they assign the same values to the common variables, i.e., $\forall v \in dom(\mu_1) \cap dom(\mu_2), \mu_1(v) = \mu_2(v)$. We name **joining variables** the elements in $dom(\mu_1) \cap dom(\mu_2)$.

As explained, this paper focuses on queries containing the graph pattern:

$$(WINDOW \ u^W \ P^W) \ JOIN \ (SERVICE \ u^S \ P^S),$$

where P^W and P^S are two graph patterns that share one or more variables, u^S is the address of a service BGD in remote and u^W is an IRI denoting a sliding window operator $\mathbb{W}$ defined through ω, β and applied to a stream S.

Local View. Existing RDF stream engines leverage a nested loop join strategy to fetch data from BGD. It follows that evaluating the above graph pattern can be expensive: each request to BGD has a latency, computational and, possibly, financial cost. In the SPARQL endpoint of our experiments (see Sect. 6), each invocation takes 4.6 ms. Hence, during one second, it can only accommodate up to 200 requests. In real scenarios, SPARQL endpoints are exposed over Internet, and each quest can take more than 500 ms [19].

For this reason, we previously proposed to use a local view $\mathcal{R}$ to store the result of P^S in the RSP engine [5]. $\mathcal{R}$ stores the results of the SERVICE clause so that the engine computes the results of the query without invoking the SPARQL endpoint of BGD at each evaluation. However, given that the content of BGD changes over time, the mappings in $\mathcal{R}$ become outdated, and the evaluation of the SERVICE clause produces different solution mappings can leading to wrong results. We consider these outdated results invalid. Therefore, each mapping $\mu^{\mathcal{R}} \in \mathcal{R}$ can be classified as *fresh* or *stale*: $\mu^{\mathcal{R}}$ is *fresh* at time t, if it is contained in the result set by evaluating the SERVICE clause over BGD at t; it is *stale* otherwise (i.e., if BGD changes, it produces different results when evaluating of the SERVICE clause over $\mu^{\mathcal{R}}$ and the remote BGD). In the following, we assume that mappings in BGD change with fixed intervals. This happens, e.g., in data warehouses, where updates are scheduled, or in data generated by sensors or automatic processes, where data is updated with fixed interval. As in [6], we define the freshness of an answer $Ans(q)$ as $\frac{|fresh(Ans(q))|}{|Ans(q)|}$.

Maintenance Process. To ensure the freshness of the local view over time, we introduce a maintenance process MP that refreshes a portion of $\mathcal{R}$. MP selects a set of mappings $\mathcal{E} \subseteq \mathcal{R}$ to refresh within each evaluation of the queries over BGD. The design of MP is the key to the freshness of $Ans(q)$: if the process correctly identifies the stale mappings and puts them in $\mathcal{E}$, then both the freshness of $\mathcal{R}$ and $Ans(q)$ increase. Note, however, that if the number of refresh queries sent to BGD is too high, the presence of $\mathcal{R}$ does not bring any advantage. In practice, MP has to consider (i) Quality of Service requirements associated to the query, e.g., responsiveness; (ii) system reactiveness, e.g., each evaluation should terminate before the next one starts; (iii) constraints imposed by the BGD providers on the number of requests during a time interval. We capture these aspects by introducing a notion of **refresh budget** value Γ, defined as the number of refresh queries that can be sent to BGD in a given time period under the above constraints. In our Hypotheses 1 and 2, we assume that Γ is evenly distributed over n evaluations, when the stream rate is stable. In Hypothesis 3, in order to deal with unstable stream rate, we relax such assumption by allowing to move budget between evaluations. We use $\gamma = \lfloor \Gamma/n \rfloor$ to denote the maximum refresh budget available in one evaluation.

3   Related Work

Traditional databases usually materialize remote BGD locally. Sophisticated optimizations of retrieving remote data on-demand have been introduced to improve availability, scalability and query processing performance [8,9,14]. The drawback of materialization is that local data becomes stale when the remote data changes. Those works are neither in stream processing context, nor considering budget constraints on remote access.

In Complex Event Processing (CEP), the incoming events not only need to be matched with specified event patterns, but also need to be enriched [10,22]. During enrichment, it usually needs to access remote BGD through APIs defined by service providers [11]. These API providers usually apply constraints on the number of accesses to restrict the massive loads of requests, as the computation and communication costs involved are shown to be intensive. Given the repetitive nature of the access to BGD [17], caching techniques can improve on response latency. However, when a cache becomes outdated, refreshing it raises the trade-off between latency and freshness [1]. More remote accesses could provide fresher response, but take longer time. Authors in [14] addresses this trade-off in a web setting, where updates of the remote BGD are pushed into the system [9]. However, this work does not consider the constraints of service providers or the view maintenance without updates being pushed into the system.

In RDF processing, SPARQL 1.1 standardizes the access to remote BGD by introducing the federated extension [2] and the SERVICE clause. Broadly, there are two ways of accessing BGD: either one pulls the whole data into the query processor [15] or one 'federates' query-execution and transfers the data for individual operations over the network [12], defining new join strategies that can efficiently process both local and remote data [15]. Extending static RDF processing, RSP technologies deal with data of different velocity and variety. C-SPARQL [3] performs query matching on subsets of the information flow defined by windows. CQELS [16] implements its native query operators, which can be adaptively optimized to improve performance. MorphStream [4] allows querying relational data streams over a set of stream-to-ontology mappings. INSTANS [20] is a semantic event processing platform, which compiles a query into a Rete-like structure. All those systems are optimized for processing streams. They support the SERVICE clause as described above but do not consider budget-constrained updates in the local view. Hence, our solution is orthogonal to these and other RSP engines.

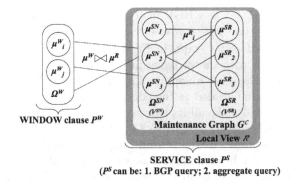

Fig. 1. WINDOW/SERVICE clauses and the maintenance graph

Our previous work [5] studied the maintenance process of local view for queries where each mapping in the WINDOW clause joins with exactly one mapping in the SERVICE one. In this paper, we tackle a more general join relationship between WINDOW and SERVICE clauses, i.e., we extend the 1:1 join relationship to M:N and propose a flexible budget allocation method that further improves the maintenance process.

4   Problem Definition

Given the graph pattern expression P^S in the SERVICE clause, we define two sets of variables: first, $V^{SR} \subset var(P^S)$ contains the variables in $var(P^S)$ that are related to the changing part in BGD. In other words, V^{SR} captures the dynamicity of BGD and contains the information needed to construct the refresh queries that are sent to remote BGD. Second, V^{SN} are the common variables that join the P^S and P^W clauses, i.e., $V^{SN} = var(P^S) \setminus V^{SR}$. We model the relationship between V^{SR} and V^{SN} as a bipartite graph. The maintenance process MP exploits the graph to identify the candidate set $\mathcal{E}$ for refreshing. The MP builds a bipartite graph (*maintenance graph*, Fig. 1) out of $\mathcal{C}$, which is a subset of $\mathcal{R}$. Mappings in $\mathcal{C}$ are (1) stale and (2) belong to the candidate set of the current window (i.e., they have compatible mappings in the result set Ω^W of the WINDOW clause). The maintenance graph has signature $G^{\mathcal{C}} = (\Omega^{SN}, \Omega^{SR}, E)$, where Ω^{SN} (Ω^{SR}) is the set of mappings with domain V^{SN} (V^{SR}), and E are the mappings $\mu^{\mathcal{R}}$ in $\mathcal{C}$, modeled as edges connecting elements of Ω^{SN} and Ω^{SR}.

Different subqueries in P^S have different optimization goals. In this work, we consider: (1) P^S is a Basic Graph Pattern (BGP) query; (2) P^S is an aggregate query[4].

Case 1: P^S is a BGP query. By differentiating V^{SR} and V^{SN}, we split $\mu^{\mathcal{R}}$ into two mappings $\mu = \mu^{SR} \cup \mu^{SN}$ such that $dom(\mu^{SR}) \subseteq V^{SR}$ and

---

[4] We assume that the aggregation is performed locally in the query processor and not in the remote BGD. It happens, e.g., when BGD is not SPARQL 1.1 compliant.

$dom(\mu^{SN}) \subseteq V^{SN}$. As P^S is a BGP query, each mapping $\mu_k^{\mathcal{R}}$ consists a μ_i^{SN} and a μ_j^{SR}. Updating one μ_j^{SR} can ensure all its corresponding $\mu_k^{\mathcal{R}}$ are fresh. As an example, consider the graph in Fig. 1, where $\mathcal{C} = \{\mu_1^{\mathcal{R}}, \ldots, \mu_6^{\mathcal{R}}\}$. Ω^{SR} contains the mappings with the variables in V^{SR}, i.e., $\{\mu_1^{SR}, \mu_2^{SR}, \mu_3^{SR}\}$ (on the right); Ω^{SN} contains the other mappings, i.e., $\{\mu_1^{SN}, \mu_2^{SN}, \mu_3^{SN}\}$ (in the middle). The mappings in $\mathcal{R}$ are encoded as the edges in E (e.g., (μ_1^{SN}, μ_1^{SR}) represents $\mu_1^{\mathcal{R}}$). Updating μ_1^{SR} will make all its three corresponding mappings to be fresh: (μ_1^{SN}, μ_1^{SR}), (μ_2^{SN}, μ_1^{SR}), and (μ_3^{SN}, μ_1^{SR}). Given Ω^W (on the left) as the solution of the WINDOW clause and γ as the refresh budget at the current iteration, the maintenance process can be summarized as: what is the subset of Ω^{SR} to refresh can maximize the number of fresh join results between μ^W and $\mu^{\mathcal{R}}$? Formally, it can be modeled as the following optimization problem:

$$\text{Sub. } u_j^{SR} = 0 \text{ or } 1 \quad \forall j = [1, |\Omega^{SR}|] \tag{1}$$

$$\sum_{j=1}^{|\Omega^{SR}|} u_j^{SR} \leq \gamma \tag{2}$$

$$f_i^{SN} = \sum \mu_j^{SR} \quad \forall \mu_j^{SR} : (\mu_i^{SN}, \mu_j^{SR}) \in E \quad \forall i = [1, |\Omega^{SN}|] \tag{3}$$

$$c_i^{SN} = |\{\mu^W : \mu^W \in \Omega_W \wedge \mu^W \text{comp. with}(\mu_i^{SN}, \mu_j^{SR})\}| \; \forall i = [1, |\Omega^{SN}|] \tag{4}$$

$$\text{Max. } \sum_{i=1}^{|\Omega^{SN}|} f_i^{SN} * c_i^{SN} \tag{5}$$

The optimization is subject to: in Formula (1), the value of u_j^{SR} shows whether the j-th stale mapping is updated ($u_j^{SR} = 1$) or not ($u_j^{SR} = 0$). The total number of updates is limited by γ, as in Formula (2). Formula (3) defines f_i^{SN} as the number of fresh mappings μ_i^{SN} will have. Each μ_i^{SN} may have several related μ_j^{SR}. By summing all its refreshed μ_j^{SR}, we have the total number of fresh mappings for μ_i^{SN}. As discussed above, this is because each updated μ_i^{SR} produces one fresh $\mu_k^{\mathcal{R}}$ (μ_i^{SN}, μ_j^{SR}) . Overall, Formula (1) to (3) give the total number of fresh $\mu^{\mathcal{R}}$ in the SERVICE clause. Since each $\mu^{\mathcal{R}}$ may have several compatible mappings in the WINDOW clause, Formula (4) introduce c_i^{SN} to represent the number of compatible mappings of $\mu_k^{\mathcal{R}}$ in the window. Finally, our optimization goal is to maximize the total number of join results between WINDOW and SERVICE clauses, which could be defined as the product of c_i^{SN} and f_i^{SN}, as shown in Formula (5).

Case 2. P^S is an aggregate query. In this case, the maintenance graph $G^{\mathcal{C}}$ is constructed as the previous case: Ω^{SN} contains mappings with variables used for join, and Ω^{SR} contains mappings with dynamic values. However, Ω^{SR} in this case does not directly participate in the join, but are needed for aggregation.

Consider the example in Fig. 1: $\mathcal{C} = \{\mu_1^{\mathcal{R}}, \mu_2^{\mathcal{R}}, \mu_3^{\mathcal{R}}\}$: $\mu_1^{\mathcal{R}}$ contains the value of the aggregate variables by using the data stored in μ_1^{SR}; $\mu_2^{\mathcal{R}}$ has an aggregate computed from μ_1^{SR} and μ_3^{SR}; $\mu_3^{\mathcal{R}}$ is computed from μ_1^{SR}, μ_2^{SR} and μ_3^{SR}. The edges in this case represent the mappings required to compute the aggregates, e.g., (μ_2^{SN}, μ_1^{SR}) and (μ_2^{SN}, μ_3^{SR}) indicate that the mapping $\mu_2^{\mathcal{R}}$ should be computed by using both the fresh values of μ_1^{SR} and μ_3^{SR}. The maintenance problem is still to choose a subset of μ^{SR} to maximize the fresh join results. However, in this case, updating one μ_j^{SR} cannot ensure its corresponding μ_k^R is fresh. To have a fresh μ_k^R, we need all its related μ^{SR} to be fresh. Therefore, the problem can be modeled as:

$$\text{Sub. } u_j^{SR} \leq 1 \quad \forall j = [1, |\Omega^{SR}|] \tag{6}$$

$$\sum_{j=1}^{|\Omega^{SR}|} u_j^{SR} \leq \gamma \tag{7}$$

$$f_i^{SN} = \prod \mu_j^{SR} \quad \forall \mu_j^{SR} : (\mu_i^{SN}, \mu_j^{SR}) \in E \quad \forall i = [1, |\Omega^{SN}|] \tag{8}$$

$$c_i^{SN} = |\{\mu^W \colon \mu^W \in \Omega_W \land \mu^W \text{ comp. with} (\mu_i^{SN}, \mu_j^{SR})\}| \; \forall i = [1, |\Omega^{SN}|] \tag{9}$$

$$\text{Max. } \sum_{i=1}^{|\Omega^{SN}|} f_i^{SN} * c_i^{SN} \tag{10}$$

The constraints in Formula (6) and (7) are same with Case 1. Formula (8) uses f_i^{SN} to model the fact that the i-th mapping μ_i^{SN} is fresh ($f_i^{SN} = 1$) iff all its related μ^{SR} are refreshed. For example, to have a fresh result of μ_2^{SN}, both μ_1^{SR} and μ_3^{SR} have to be 1; otherwise, $f_i^{SN} = 0$. Formula (9) is same with Case 1. Finally, the objective function in Formula (10) maximizes the number of fresh mappings produced by the join.

Overall, both Case 1 and 2 can be treated as binary integer programming problems. However, Case 2 can be seen as an extension of the knapsack problem, which is NP-hard, e.g., packing a μ^{SN} has a cost (the number of its μ^{SR}). We can only afford a certain number of μ^{SR}, but need to maximize the number of μ^{SN}. Furthermore, after choosing a μ^{SN} and its related μ^{SR} to pack, those μ^{SR} might contribute to other μ^{SN}. Therefore, choosing different μ^{SR} will have different influence on the following decisions. Currently, there is no optimal way to find the best subset of μ^{SR}.

5   Maintenance Algorithms

In this section, we propose a set of budget allocation algorithms. Section 5.1 proposes two greedy algorithms, SBM_{BGP} and SBM_{Agg}, for the problems in Case 1 and 2, respectively. They aim at maximizing the freshness of the current slide evaluation. Because the sliding window operator supplies information about future evaluations (i.e., elements stay in the window for different periods), Sect. 5.2 shows how to exploit this information to improve the maintenance process. Section 5.3 discusses how to flexibly manage the budget to optimize the overall response freshness. The basic idea is to uniformly allocate Γ to n evaluations (i.e., $\gamma = \lfloor \Gamma/n \rfloor$). When it is worthwhile, the solution trades the current remote accesses for the future ones.

5.1   Selectivity-Based Maintenance (SBM)

To maximize the number of fresh join results, we propose the SBM_{BGP} algorithm for Case 1, where P^S is a BGP query; and the SBM_{Agg} for Case 2, where P^S

is an aggregate query. In both cases, we start from the maintenance graph G^C defined above.

SBM$_{BGP}$. The objective function of Case 1 (Formula (5)) aims at maximizing the number of fresh mappings produced by the join. Based on G^C, SBM$_{BGP}$ first computes a score for each mapping $\mu^{SR} \in \Omega^{SR}$, which represent the total number of the fresh join mappings that would be generated if μ^{SR} is updated:

$$score_{SBM}(\mu^{SR}) = \sum_{\mu_i^{SN}:(\mu_i^{SN},\mu^{SR})\in E} c_i \qquad (11)$$

Based on the selectivity of μ^{SR}, the number of results it will have equals to the sum of each its connected μ^{SN} times μ^{SN}'s compatible mappings in the window. Then, SBM$_{BGP}$ picks μ^{SR} with the highest scores under the budget γ to refresh. If there are more than γ data with the same highest score, our algorithm chooses among them randomly.

SBM$_{Agg}$. This case aims to maximize the number of fresh aggregate results. A mapping μ^{SN} produces a fresh aggregate result only if all its connected μ^{SR} are fresh. As discussed, fining the optimal set of μ^{SN} is a NP-hard problem. We propose a heuristic algorithm: SBM$_{Agg}$. It tries to utilize the budget on those "cheap" μ^{SN}, which connects to less stale μ^{SR}. Specifically, SBM$_{Agg}$ picks the mapping $\mu^{\bar{S}N}$ with the smallest amount of connected μ^{SR} and puts those μ^{SR} in $\mathcal{E}$. Then, $\mu^{\bar{S}N}$ and the mappings in $\mathcal{E}$ are removed from the maintenance graph G^C, and a new iteration starts again. It ends when γ elements have been moved into $\mathcal{E}$. If the budget left $\gamma\prime$ is less than $\mu^{\bar{S}N}$, we will randomly choose $\gamma\prime$ amount of stale μ^{SR}.

5.2   The Impact-Based Maintenance (IBM)

The two SBM algorithms maximize the freshness of the current evaluation, but do not consider future evaluations. As shown in [5], a maintenance process MP can take into account the sliding window and the changing frequency of the background data to have a prediction on what will be stale in future. We combine this idea with SBM to improve the performance of MP.

Before presenting the solution, we first introduce the concept of ranking data by a score based two properties from [5], which quantify the impact of a mapping in future window evaluations. Consider a set of solution mappings Ω^W resulted from the evaluation of a WINDOW clause and a local view $\mathcal{R}$, where each mapping in Ω^W can have *only one* compatible mapping in $\mathcal{R}$.

The first property is the *remaining lifetime*, denoted with L. Let $\mu^{\mathcal{R}}$ be a mapping in $\mathcal{R}$, and let μ^W be its only compatible mapping in Ω^W computed at time t in a sliding window $\mathbb{W} = (S, \omega, \beta)$. The L value of μ^S at time t^{now} is computed as $\lceil(t+\omega-t^{now})/\beta\rceil$. It represents the number of evaluations, in which μ^S will be involved. For example, given a sliding window $\mathbb{W} = (S, \omega = 150, \beta = 30)$ and a mapping μ^W with timestamp $t = 100$, the L value of the compatible mapping $\mu^{\mathcal{R}}$ at time 100 is $L(\mu^{\mathcal{R}}, 100) = \lceil(100 + 150 - 100)/30\rceil = 5$; at time 160, it is $L(\mu^{\mathcal{R}}, 100) = \lceil(100 + 150 - 160)/30\rceil = 3$. The second property is the *number of evaluations before the next expiration,* denoted with B. Given a stale mapping $\mu^{\mathcal{R}}$, B represents the number of evaluations that $\mu^{\mathcal{R}}$ would be fresh, if refreshed now. B is computed as $B(\mu^{\mathcal{R}}, t^{now}) = \lceil (t^{exp} - t^{now})/\beta \rceil$, where t^{exp} is the next time on which $\mu^{\mathcal{R}}$ would become stale. t^{exp} is processed by exploiting the change rate interval information of $\mu^{\mathcal{R}}$. At time $t^{now} = 100$, the value of B is $B(\mu^{\mathcal{R}}, 100) = 3$, i.e., if $\mu^{\mathcal{R}}$ is refreshed now, it would remain fresh for the next three evaluations (evaluations at 100, 130, and 160; at 190, $\mu^{\mathcal{R}}$ will be stale).

Now, L and B can be combined to assign a *score* to the elements in $\mathcal{C}$ (i.e., the stale mappings in the local view currently involved). Intuitively, the *score* of the mapping $\mu^{\mathcal{R}}$ represents *how many future correct results are attainable if $\mu^{\mathcal{R}}$ is refreshed now*. The *score* of $\mu^{\mathcal{R}}$ at time t^{now} is computed as $score(\mu^{\mathcal{R}}, t^{now}) = \min\{L(\mu^{\mathcal{R}}, t^{now}), B(\mu^{\mathcal{R}}, t^{now})\}$. If $B(\mu^{\mathcal{R}}, t^{now}) < L(\mu^{\mathcal{R}}, t^{now})$ $\mu^{\mathcal{R}}$, it can generate at most $B(\mu^{\mathcal{R}}, t^{now})$ fresh join mappings, before it becomes stale while remaining in the window; otherwise, it generates $L(\mu^{\mathcal{R}}, t)$ fresh join results and will leave the window before it becomes stale. Based on this *score*, we extend the two SBM algorithms so that they also consider the future impact of a refresh. Given the maintenance graph $G^{\mathcal{C}} = (\Omega^{SN}, \Omega^{SR}, E)$ as defined in Sect. 4 (M:N bipartite graph), the extensions, namely IBM_{BGP} and IBM_{Agg}, can cope with the stale mappings μ^{SR} appearing in different mappings $\mu^{\mathcal{R}}$ of the local view.

IBM_{BGP}. We assign a score for the stale mappings in Ω^{SR}, as with SBM_{BGP}. The formula proposed above for B is still valid for the elements in Ω^{SR}. However, L cannot be directly associated with mappings in Ω^{SR} because they are related to the mappings computed by the WINDOW clause Ω^{W} through Ω^{SN}.

$$L(\mu^{\mathcal{R}}, \mu^{W}, t^{now}) = \lceil (t_{\mu^W} + \omega - t^{now})/\beta \rceil \tag{12}$$

$$score(\mu^{\mathcal{R}}, \mu^{W}, t^{now}) = \min\{L(\mu^{\mathcal{R}}, \mu^{W}, t^{now}), B(\mu^{SR}, t^{now})\} \tag{13}$$

$$score(\mu^{\mathcal{R}}, t^{now}) = \sum_{\mu^W c.w. \mu^{\mathcal{R}}} score(\mu^{\mathcal{R}}, \mu^{W}, t^{now}) \tag{14}$$

$$score_{IBM_{bgp}}(\mu^{SR}, t^{now}) = \sum_{\mu^{\mathcal{R}} = (\mu^{SN}, \mu^{SR}) \in E} score(\mu^{\mathcal{R}}, t^{now}) \tag{15}$$

IBM_{BGP} associates the remaining lifetime L to the pair of compatible mappings $(\mu^{\mathcal{R}}, \mu^{W})$ as defined in Formula (12): the function considers the arriving time t_{μ^W} of μ^{W} as well, in order to cope with the fact that there are multiple compatible mappings for a μ^{SR}. This extension allows defining a score for each pair $(\mu^{\mathcal{R}}, \mu^{W})$, as in Formula (13). It represents the number of fresh mappings that are potentially generated by joining $\mu^{\mathcal{R}}$ and μ^{W} in the current and the following evaluations, if a μ^{SR} is refreshed. Formula (14) computes the score of a mapping $\mu^{\mathcal{R}}$ in the local view, which sums the scores of $\mu^{\mathcal{R}}$ with compatible mappings in Ω^{W}. Finally, IBM_{BGP} assigns the score to the mappings in Ω^{SR} by Formula (15): it represents the total number of fresh join mappings that will be generated, if μ^{SR} is refreshed. We select γ mappings of μ^{SR} with the highest scores to refresh. Section 5.4 discusses why IBM_{BGP} is a local optimal solution.

IBM_{Agg}. As discuss in Sect. 4, budget allocation in this case is a NP-hard problem. When future evaluations of the current data are considered, the complexity increases further, due to the additional level of combinatorial optimization.

Therefore, IBM$_{Agg}$ exploits a *score* function to improve the basic SBM$_{Agg}$ algorithm. An aggregate value for a $\mu^{SN} \in \Omega^{SN}$ is fresh only when all the required mappings $\mu^{SR} \in \Omega^{SR}$ are fresh.

$$L\left(\mu^{SN}, t^{now}\right) = \left\lceil \left(\max_{t:\mu^W \in \Omega^W \wedge \mu^W \text{ c.w. } \mu^{SN}} \{t\} + \omega - t^{now} \right) / \beta \right\rceil \qquad (16)$$

$$score_{IBM_{agg}}(\mu^{SN}, t^{now}) = \min \left\{ L(\mu^{SN}, t^{now}), \min_{\mu^{SR}:(\mu^{SN}, \mu^{SR}) \in E} \{B(\mu^{SR}, t^{now})\} \right\} \qquad (17)$$

IBM$_{Agg}$ computes the score of the mappings in Ω^{SN} when two or more of them have the same lowest amount of connected μ^{SR}. Specifically, Formula (16) computes the remaining lifetime of μ^{SN}, which takes the most recent timestamp of the compatible mappings of a μ^{SN} in Ω^W. Formula (17) reports the function to compute the score, which considers two factors: (1) μ^{SN} will continue to generate fresh mappings as long as all its related mappings μ^{SR} are fresh; (2) their compatible mappings of μ^{SN} still remain in the window.

5.3 Flexible Budget Allocation (FBA)

Above solutions only consider the fixed amount of refresh budget γ assigned in the current evaluation. However, fixing γ may be inefficient as the number of refresh requests changes over time. Saving current budget for future updates may improve result freshness, if the future ones can generate more results.[5] The semantics of the sliding window allow inferring how long each element in the current window will be involved in future joins. We propose FBA to allocate the refresh budget by considering both current and future evaluations. Specifically, FBA iterates from the current to the future ω/β slides (window length/slide length). At each iteration, it identifies the maintenance graph G_i^C and the stale data Ω_i^{SR}. It calculates the number of future fresh results for each μ^{SR} in every Ω_i^{SR} at their corresponding evaluation time and orders μ^{SR} by their scores. FBA allocates total $n \times \gamma$ budgets to the Top-$(n \times \gamma)$ μ^{SR} with the largest scores. Note that this set contains both current and future stale μ^{SR}. If the number of μ^{SR} in the current evaluation is less than γ, it means FBA delays the budgets of current μ^{SR} to some future ones.

5.4 Discussion

SBM$_{BGP}$ and IBM$_{BGP}$ are optimal for Case 1. For a BGP query, choosing the top-γ data in Ω^{SR} based on deg^{SR}, which is the number of μ^{SR}'s associated elements in Ω^{SN}. It gives the local optimal solution at the current time without considering the future impact of Ω^{SR}. This is because the top-γ of Ω^{SR} is the

---

[5] We acknowledge that not all types of budget can be saved for future (e.g., a fixed amount of bandwidth cannot be saved). Other types of budgets, such as a supplier charges per request, a limited data plan, or limited power can be saved.

set with the largest sum of deg^{SR}, since the sum of deg^{SR} exactly equals the number of fresh results. Therefore, SBM_{BGP} gives the local optimal solution. The same reason applies to IBM_{BGP}, where γ mappings with the largest *score* also gives the most results, as *score* accurately reflects the number of future results. Note that since the future elements in stream are not predictable (with certainty), there is no global optimal solution for BGP query.

Complexity. Both the SBM and IBM only consider data in the current evaluation. $\text{SBM}_{BGP}/\text{IBM}_{BGP}$ visits each μ^W and μ^R to count the number of mappings and calculate scores for μ^{SR}, which both take linear time of $\mathcal{O}(|\Omega^W| + |\Omega^R| + |\Omega^{SR}|)$. Then, choosing the Top-γ mapping take $\mathcal{O}(|\Omega^{SR}| \log |\Omega^{SR}|)$ time. SBM_{Agg} takes $\mathcal{O}(|\Omega^{SN}|^2 \log |\Omega^{SN}|)$ time, as whenever updating a μ^{SN}, we have to update all its related μ^{SR}. IBM_{Agg}, as an extension of SBM_{Agg}, has the same complexity. FBA has the same time complexity as IBM, since they have the same way of ranking and choosing data to refresh, except that IBM chooses data only in the current slide; FBA does this for a fixed number of future slides.

6 Experiments

Experiment environment. We implemented the maintenance process in a real RSP system: C-SPARQL [3]. The system registers continuous federated queries with WINDOW and SERVICE clauses (as in Sect. 2) and continuously evaluates the query per window on the incoming stream. Each evaluation joins the content of the current WINDOW with the results of the SERVICE clause. For evaluating the SERVICE clause, we have implemented a local view in C-SPARQL to cache remote BGD data (as in Sect. 4). Before executing the SERVICE clause, different maintenance algorithms will select a candidate set $\mathcal{E}$ from Ω^{SR} to refresh. For each data in $\mathcal{E}$, the SERVICE clause will request its fresh value from the remote server. We used Fuseki 2.0.0 as the remote BGD server and ran it with the C-SRAPQL engine on the same machine. The delay of each remote access under this setting is much smaller than querying an actual remote server.

Experiment data sets. We employ a *real* data set and several *synthetic* data sets to investigate the performance of our solutions. The real data set was recorded from Twitter. The synthetic ones were constructed by resembling the real one, but using a generator that can alter its characteristics. Each data set broadly contains three kinds of data: the remote BGD, the local view $\mathcal{R}$, and the input stream. We discuss different parameters of our data sets below and report their values in each experiment.

The remote BGD. In BGD, data change according to each one's own change interval ChR. In $real_{twitter}$ data, we use the number of a user's followers as the BGD. When a tweet mentions several users, the SERVICE clause will provide the number of followers for each mentioned user in Case 1 (BGP query); it will provide the sum of the followers for all mentioned users in Case 2 (aggregate query). We collected the follower number of 100 selected users every minute for four hours by using the Twitter search API [5]. We noticed that the distribution of ChR is highly skewed: only few users have a very dynamic changing number of followers, while others are stable. Roughly, it resembles a Beta distribution with $\alpha = 50$ and $\beta = 1$. In *synthetic* data, our data generator outputs data with different ChR-distributions. The generator has two parameters: the skewness of the distributions and the correlation with the selectivity of data elements in the local view. The latter controls whether data that changes more frequently can have either a higher or a lower selectivity.

The local view $\mathcal{R}$. In $\mathcal{R}$, we model the relationship between Ω^{SN} and Ω^{SR} as a bipartite graph. Therefore, we are mainly interested in the deg^{SR} of Ω^{SR}. After analyzing the $real_{twitter}$ data set, we observed a skewed distribution of the deg^{SR} that can be modeled as a Zipf distribution with a skewness parameter of 0.2. In the *synthetic* data, we can tune two aspects of deg^{SR}: the skewness and the correlation with the stream/remote data.

The stream and the sliding window. By using the Twitter stream API, the *real* data set collects a stream of tweets that contains the mentions of the monitored Twitter users described above. The *synthetic* data set generates the streaming data through a Poisson process [21]. To verify Hypothesis 3, the stream is generated with a non-homogeneous Poisson process, where the data arrival rate changes over time, e.g., $\lambda_i = 0.95\lambda_0 \cdot (i \bmod 2) + \lambda_0 \cdot ((i+1) \bmod 2)$, where λ_0 is the initial expected arrival interval and i is incremented along the time. The input query has a sliding window length of $\omega = 4$ seconds and slides every $\beta = 1$ second. Each experiment has 50 evaluations, and the first 10 % is used as a warm-up period.

We first use synthetic data sets to verify our hypotheses and study the performance of our algorithms. The performance and the computational overhead on the real data set are reported as well. The average response freshness is used as the Key Performance Indicator (KPI). As discussed in Sect. 2, it is the ratio of fresh results to total number of results, within each evaluation. The number of fresh results is acquired by comparing the current result set to the corresponding set acquired by the original C-SPARQL engine[6], where all results are fresh, since it queries BGD without budget constraints.

The baseline algorithms. We choose two baseline algorithms: (1) Least Recently Update (LRU), which selects the least recently updated *stale* data from $\mathcal{R}$; (2) Random (RAND), which randomly chooses *stale* data from $\mathcal{R}$. All algorithms pick at most γ (the refresh budget) candidates to refresh.

Resulting synthetic data sets. The default settings of the synthetic data set are: Ω^{SN} and Ω^{SR} in $\mathcal{R}$ contains 50 data elements each. There are 1000 edges μ^R between Ω^{SN} and Ω^{SR}. Each μ^R randomly connects a pair of μ^{SN} and μ^{SR}. Every μ^{SR} has a change interval ChR randomly chosen from $[100, 3000]$ ms. A stream trace generated from a Poisson distribution decides the arrival time of each μ^{SN}. For the Poisson distribution, each μ^{SN} chooses its λ (the expected arrival interval) randomly from $[1000, 2000]$ ms. The default budget γ is 10.

---

[6] http://streamreasoning.org/larkc/csparql/CSPARQL-ReadyToGoPack-0.9.zip.

6.1   Verifying Hypotheses H1 and H2

H1 and H2 are tested together by comparing the response freshness among RAND, LRU, SBM and IBM in both subquery cases:

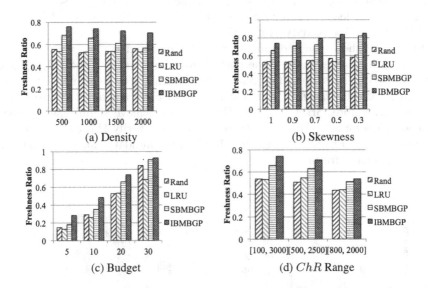

Fig. 2. SBM$_{BGP}$ and IBM$_{BGP}$ outperform baselines under different settings.

Case 1. In the four settings of Fig. 2, both SBM$_{BGP}$ and IBM$_{BGP}$ greatly improve the response freshness of the baselines by $22\%/43\%/93\%$ (Min/Average/Max). These different settings show how the performance improvement generalizes.

In Fig. 2(a), we show the performance of using $\mathcal{R}$ with different densities, i.e., the number of edges $|\mu^R|$ in $\mathcal{R}$ is set to be 500, 1000, 1500, and 2000. Note that 2500 ($|\Omega^{SN}| \times |\Omega^{SR}|$) edges will form a fully connected $\mathcal{R}$. First, we observe that the performance of RAND and LRU remain roughly stable over different densities. The reason is that they select the refresh candidates $\mathcal{E}$ "blindly" without considering deg^{SR}—the selectivity of μ^{SR}. Therefore, the percentage of edges being updated remains the same for different densities. On the other hand, in higher densities, the performance improvement of SBM$_{BGP}$ and IBM$_{BGP}$ decreases a bit. The reason is that in a denser graph the difference of deg^{SR} among μ^{SR} becomes less significant. SBM$_{BGP}$ and IBM$_{BGP}$ always choose the μ^{SR} with the highest deg^{SR}, however, the percentage of the chosen μ^{SR} to the total number of μ^{SR} in G^C becomes smaller. Hence, SBM$_{BGP}$ and IBM$_{BGP}$ favor sparse graphs.

Figure 2(b) plots the performance on graphs with different distributions of μ^{SR}'s selectivity. We set the selectivity to follow different Zipf's distributions, with skewness parameter s to be 1 (uniform), 0.8 (slightly skewed), 0.5 (skewed), and 0.3 (highly skewed). Figure 2(b) shows that the performance improvement of SBM_{BGP} and IBM_{BGP} is more significant in skewed graphs. The reason is same with the second observation above: SBM_{BGP} and IBM_{BGP} refresh the μ^{SR} with the highest deg^{SR}. In a skewed graph, the percentage of the selected μ^{SR} increases, which leads to more fresh results. Therefore, SBM_{BGP} and IBM_{BGP} favor skewed μ^{SR} selectivity distribution.

Figure 2(c) shows the performance with different budgets i.e., $\gamma = 5$, 10, 20, and 30. With a larger budget, the performance improvement of SBM_{BGP} and IBM_{BGP} becomes less. In an extreme case of having a large enough budget to cover most of the *stale* μ^{SR}, different subsets of $\mathcal{E}$ do not affect the freshness anymore. Therefore, SBM_{BGP} and IBM_{BGP} can achieve significant improvement with less budget. The above three experiments verify **H1**: considering the selectivity deg^{SR} of μ^{SR} enables choosing better candidates for refreshing and improves response freshness.

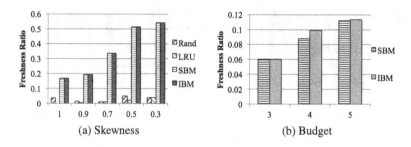

(a) Skewness   (b) Budget

Fig. 3. SBM_{Agg} and IBM_{Agg} outperform baselines in subquery Case 2.

Regarding H2, Fig. 2(d) shows the performance results of BGD change intervals that are randomly chosen from different ranges: $[100, 3000]$, $[500, 2000]$, and $[800, 1200]$ ms. We can make these comparisons: first, IBM_{BGP} always has a higher freshness than SBM_{BGP}. Second, having a wider range for ChR leads to better improvement in IBM_{BGP}. The reason is that IBM_{BGP} chooses μ^{SR} with larger "impact", i.e., larger score, since the score indicates that μ^{SR} makes more results in the current and future slides. Therefore, this experiment verifies **H2** and shows that IBM_{BGP} favors larger ranges of change intervals.

Case 2. When P^S is an aggregate query, in all of the above cases, we observed similar performance improvements of SBM_{Agg} over RAND and LRU. To save space, we just show the results with different skewnesses to demonstrate the performance in Fig. 3(a). Besides the freshness improvement, we notice that in most cases IBM_{Agg} performs similarly with SBM_{Agg}. This is because IBM_{Agg} is designed to be at least as good as SBM_{Agg}. Only when several μ^{SN} have the same amount of connected μ^{SR}, SBM_{Agg} will choose the one with the lowest score. Furthermore, for different μ^{SN}, when the overlapping between their associated

μ^{SR} is small, the chance of μ^{SN}s have different scores is larger and the effect of SBM_{Agg} is, therefore, more significant. Figure 3(b) investigates this by plotting the results of a special case: a very sparse graph (100 edges) and a tiny budget, e.g., 3 to 5. In these cases, IBM_{Agg} outperforms SBM_{Agg} by up to 12.5 %.

6.2 Verifying Hypothesis H3

We compare the performance of FBA with IBM in Case 1 with three refresh budgets, $\gamma = 5$, 10, 15 in Fig. 4(a). They track the accumulated number of stale results over time. By increasing the budget, the gap between FBA and IBM becomes more significant. Furthermore, when $\gamma = 15$, after the first 15 iterations, FBA makes the accumulated stale result increases very slowly, i.e., the freshness ratio of the answer is almost 100 %, while IBM still keep producing stale results. To explain the improvement, Fig. 4(b) plots the actual amounts of budget that are consumed over time. For IBM, the consumption of budget will always be a vertical line for different budgets. For FBA, when $\gamma = 5$, the line fluctuates a bit. With larger budgets, the lines fluctuate more. It shows that FBA moves budgets between different slides to improve freshness.

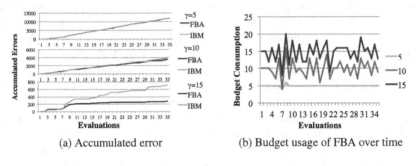

(a) Accumulated error   (b) Budget usage of FBA over time

Fig. 4. The performance of FBA under different budgets.

Results on a real data set. Figure 5 plots the results on a real data set with different budgets for both cases. We can observe that IBM always achieves the best freshness and SBM also outperforms the two baseline algorithms. When we decrease the budget, the performance improvement of IBM and SBM increases. These results confirm our findings in Fig. 2(c).

Computational overhead. We finally report the computational overhead and the average remote access delay. Under the default setting $\gamma = 20$, the total latency of a slide evaluation is about 94.4 ms. The delay of querying the BGD server accounts for 92 ms on average (4.6 ms per request); the computational overhead is only about 2.3 ms (2.5 % of the overall latency). Note that, the current setting has the remote BGD server running locally. When requests are sent over Internet, the computational overhead will become even more negligible while the performance gain will become more substantial.

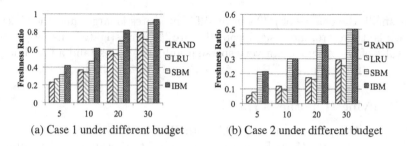

Fig. 5. SBM and IBM outperform baselines in a real data set.

7   Conclusions and Future Work

In this paper, we studied the problem of accessing remote background data (BGD) from an RDF Stream Processing (RSP) context. When BGD is large, stored remotely, and/or changing over time, accessing it can be expensive, waste resources, and deteriorate the response time. Hence, a local view is often used to speed up the BGD accesses, but maintaining it is often subject to refresh budget constraints. This paper proposes to efficiently allocate the budget for refreshing the local view. Specifically, our solution relies on a bipartite graph to model the join between stream data and BGD. It exploits the graph structure to improve response freshness for two kinds of SERVICE subqueries: a BGP query (Case 1) and an aggregate query (Case 2). Our solution, SBM, exploits a set of basic algorithms that leverage the selectivity of the join between the stream and the background data. Experiments show that it can significantly improve the response freshness up to 25 % compared to baseline algorithms (i.e., RAND and LRU). An also introduce an improved approach, IBM, that takes the future impact of refreshes into account and improves the performance up to 55.6 % over the SBM. Finally, we propose the FBA optimization that flexibly allocates budget considering not only the current but also future data. As a result FBA significantly improves over all other solutions and maintains a freshness of close to 100 % even in the light of limited update budget.

Our findings have the following limitations: first, we propose a greedy heuristic algorithm for Case 2. We hope to investigate a more advanced approximate approach in the future. Second, the current approach focuses on BGP. Some SPARQL operators (e.g., OPTIONAL) can introduce new challenges and require non-trivial extensions of our model. Third, we currently focus on stream querying. In future, we plan to extend our current optimization problem to reasoning over both stream and BGD. For example, we plan to investigate how to ensure stream consistency over a background knowledge base under a given budget.

Even in the light of these limitations we believe that this paper highlights an important problem in RSP—the joint evaluation of stream and BGD under budget constraints—and provides solutions for different subqueries. As such it paves the way for truly scalable RSP systems in real-world environments, where the integration of stream and BGD is ubiquitous.

Acknowledgments. This research has been partially funded by Science Foundation Ireland (SFI) grant No. SFI/12/RC/2289, EU FP7 CityPulse Project grant No. 603095 and the IBM Ph.D. Fellowship Award 2014 granted to Dell' Aglio.

References

1. Abadi, D.J.: Consistency tradeoffs in modern distributed database system design: cap is only part of the story. Computer **2**, 37–42 (2012)
2. Aranda, C.B., Arenas, M., Corcho, Ó., Polleres, A.: Federating queries in SPARQL 1.1: syntax, semantics and evaluation. J. Web Semant. **18**(1), 1–17 (2013)
3. Barbieri, D.F., Braga, D., Ceri, S., Della Valle, E., Grossniklaus, M.: Querying RDF streams with C-SPARQL. SIGMOD Rec. **39**(1), 20–26 (2010)
4. Calbimonte, J., Jeung, H., Corcho, Ó., Aberer, K.: Enabling query technologies for the semantic sensor web. Int. J. Semant. Web Inf. Syst. **8**(1), 43–63 (2012)
5. Dehghanzadeh, S., Dell'Aglio, D., Gao, S., Della Valle, E., Mileo, A., Bernstein, A.: Approximate continuous query answering over streams and dynamic linked data sets. In: Cimiano, P., Frasincar, F., Houben, G.-J., Schwabe, D. (eds.) ICWE 2015. LNCS, vol. 9114, pp. 307–325. Springer, Heidelberg (2015)
6. Dehghanzadeh, S., Parreira, J.X., Karnstedt, M., Umbrich, J., Hauswirth, M., Decker, S.: Optimizing SPARQL query processing on dynamic and static data based on query time/freshness requirements using materialization. In: Supnithi, T., Yamaguchi, T., Pan, J.Z., Wuwongse, V., Buranarach, M. (eds.) JIST 2014. LNCS, vol. 8943, pp. 257–270. Springer, Heidelberg (2015)
7. Dell'Aglio, D., Della Valle, E., Calbimonte, J., Corcho, Ó.: RSP-QL semantics: a unifying query model to explain heterogeneity of RDF stream processing systems. Int. J. Semant. Web Inf. Syst. **10**(4), 17–44 (2014)
8. Gançarski, S., Naacke, H., Pacitti, E., Valduriez, P.: The leganet system: freshness-aware transaction routing in a database cluster. Inf. Syst. **32**, 320–343 (2007)
9. Guo, H., Larson, P.-Å., Ramakrishnan, R.: Caching with good enough currency, consistency, and completeness. In: VLDB, pp. 457–468. VLDB Endowment (2005)
10. Hasan, S., O'Riain, S., Curry, E.: Towards unified and native enrichment in event processing systems. In: DEBS, pp. 171–182. ACM (2013)
11. Hinze, A., Sachs, K., Buchmann, A.: Event-based applications and enabling technologies. In: DEBS, p. 1. ACM (2009)
12. Ji, Y., Jerzak, Z., Nica, A., Hackenbroich, G., Fetzer, C.: Optimization of continuous queries in federated database and stream processing systems. In: BTW 2015. LNI, vol. 241, pp. 403–422. GI (2015)
13. Käfer, T., Umbrich, J., Hogan, A., Polleres, A.: Towards a dynamic linked data observatory. LDOW at WWW (2012)
14. Labrinidis, A., Roussopoulos, N.: Exploring the tradeoff between performance and data freshness in database-driven web servers. VLDB J. **13**(3), 240–255 (2004)
15. Ladwig, G., Tran, T.: SIHJoin: querying remote and local linked data. In: Antoniou, G., Grobelnik, M., Simperl, E., Parsia, B., Plexousakis, D., De Leenheer, P., Pan, J. (eds.) ESWC 2011, Part I. LNCS, vol. 6643, pp. 139–153. Springer, Heidelberg (2011)
16. Le-Phuoc, D., Dao-Tran, M., Xavier Parreira, J., Hauswirth, M.: A native and adaptive approach for unified processing of linked streams and linked data. In: Aroyo, L., Welty, C., Alani, H., Taylor, J., Bernstein, A., Kagal, L., Noy, N., Blomqvist, E. (eds.) ISWC 2011, Part I. LNCS, vol. 7031, pp. 370–388. Springer, Heidelberg (2011)

17. Lee, R., Xu, Z.: Exploiting stream request locality to improve query throughput of a data integration system. IEEE Trans. Comput. **58**(10), 1356–1368 (2009)
18. Margara, A., Urbani, J., van Harmelen, F., Bal, H.: Streaming the web: reasoning over dynamic data. J. Web Semant. **25**, 24–44 (2014)
19. Montoya, G., Vidal, M.-E., Corcho, O., Ruckhaus, E., Buil-Aranda, C.: Benchmarking federated SPARQL query engines: are existing testbeds enough? In: Cudré-Mauroux, P., et al. (eds.) ISWC 2012, Part II. LNCS, vol. 7650, pp. 313–324. Springer, Heidelberg (2012)
20. Rinne, M., Solanki, M., Nuutila, E.: RFID-based logistics monitoring with semantics-driven event processing. In: DEBS, pp. 238–245 (2016)
21. Sharaf, M., Chrysanthis, P., Labrinidis, A.: Preemptive rate-based operator scheduling in a data stream management system. In: AICCSA, pp. 46–59 (2005)
22. Teymourian, K., Paschke, A.: Plan-based semantic enrichment of event streams. In: Presutti, V., d'Amato, C., Gandon, F., d'Aquin, M., Staab, S., Tordai, A. (eds.) ESWC 2014. LNCS, vol. 8465, pp. 21–35. Springer, Heidelberg (2014)

Explicit Query Interpretation and Diversification for Context-Driven Concept Search Across Ontologies

Chetana Gavankar[1,2,3(✉)], Yuan-Fang Li[3], and Ganesh Ramakrishnan[2]

[1] IITB-Monash Research Academy, Mumbai, India
chetanagavankar@gmail.com
[2] IIT Bombay, Mumbai, India
[3] Monash University, Melbourne, Australia

Abstract. Finding relevant concepts from a corpus of ontologies is useful in many scenarios, such as document classification, web page annotation, and automatic ontology population. Many millions of concepts are contained in a large number of ontologies across diverse domains. A SPARQL-based query demands the knowledge of the structure of ontologies and the query language, whereas user-friendlier and, simpler keyword-based approaches suffer from false positives. This is because concept descriptions in ontologies may be ambiguous and may overlap. In this paper, we propose a keyword-based concept search framework, which (1) exploits the structure and semantics in ontologies, by constructing *contexts* for each concept; (2) generates the interpretations of a query; and (3) balances the relevance and diversity of search results. A comprehensive evaluation against the domain-specific BioPortal and the general-purpose Falcons on widely-used performance metrics demonstrates that our system outperforms both.

Keywords: Ontology concept search · Query interpretation · Diversification

1 Introduction

The current breed of Semantic Web search engines can be broadly grouped into three categories: (1) those that search for ontologies [12,15], (2) those that search for individual resources [15,20], and (3) those that search for concepts that represent a group of individuals [17,27].[1] Searching concepts across ontologies represents an ideal granularity middle ground and has applicability in ontology mapping, ontology merging, bootstrapping ontology population, entity annotation, web page classification, and link prediction, all real world applications. With structured content (e.g., knowledge graph) increasing on the web, searching concepts across these is a challenge. In certain domains such as life sciences,

---

[1] Throughout this paper we use the terms *concept* and *class* interchangeably.

© Springer International Publishing AG 2016
P. Groth et al. (Eds.): ISWC 2016, Part I, LNCS 9981, pp. 271–288, 2016.
DOI: 10.1007/978-3-319-46523-4_17

there are many overlapping domain ontologies that contain concepts and properties that describe and link concepts. In such a scenario, concept search in itself is a very important task.

To the best of our knowledge, existing concept search approaches can be divided into two types on basis of the nature of the input queries: (1) *SPARQL queries* [23], which as precise input queries, lead to exact results. However, it requires knowledge of writing SPARQL queries and knowledge of the structure of the ontologies that are to be queried. In reality, learning SPARQL may be an additional burden, and often the structure might not be known to the user. (2) *Keyword-based approaches* [15,27], typically use the standard information retrieval techniques such as tf-idf-based and PageRank-inspired algorithms. However, these approaches do not make use of the structure and semantics in ontologies to capture the *intents* of queries with multiple keywords. In our preliminary work [17] we proposed a concept search framework that only considers relevance. Extending it, in this work, we incorporate diversification of search results and propose a context-based diversification framework that automatically captures fine-grained query intents in the top-k results. We incorporated inferred knowledge using reasoners and refined context further to include annotation properties of widely used vocabularies such as SKOS.

In this paper, we propose a novel keyword-based concept search framework that optimizes both the *relevance* and *diversity* of search results. In order to improve search relevance, our framework interprets a query by constructing *contexts* for concepts from ontology axioms. We exploit the rich and inherent structure and semantics of ontologies and adopt an *explicit* query interpretation approach [14] in our concept search problem. A keyword query can be ambiguous with multiple intents. Our framework returns the subset of relevant results that contain the most relevant as well as the most diverse results that cover these intents. Our diversification approach achieves the goal of capturing *fine-grained* intents in the top-k results by using the structure of the ontology.

The technical contributions of our concept search framework are three-fold: (1) the proposal and design of *contexts* of concepts for their retrieval, (2) explicit, context-based query interpretation based on co-occurrences among keywords in a query, and, (3) explicit, context-based diversification of top-k results using fine-grained search intents.

We have conducted extensive experiments that compare our framework against two concept search systems: the domain-specific BioPortal and the general-purpose Falcons. Our evaluation shows that our framework outperforms both systems by a large margin for both relevance and diversity.

2   Related Work

We relate our work with the broad areas of search approaches and systems.

Semantic Search Approaches. Semantic search engines such as Sindice [32], Swoogle [12,15], and Watson [11], enable keyword-based search for the ontologies and entities within them. Sindice [32] provides a search interface by using keywords, URI's and inverse functional properties. Swoogle [12,15] has developed algorithms to rank the importance of documents, individuals and RDF graphs. The existing semantic search approaches do not leverage the structure and semantics in ontologies to capture the *intents* of queries with multiple keywords. BioPortal [26] and Falcons [27] are state-of-the-art concept search engines. The Falcons system retrieves concepts, the textual descriptions of which match the keyword query. The system then ranks the results according to the relevance and popularity of the concepts. The BioPortal system provides multiple search functions across ontologies, individuals, and concepts. The BioPortal concept search system is based on the precise or partial matching of the preferred name with the search string. BioPortal use ontology popularity to rank concept search results. We differ in our approach from both these concept search systems in the aspect of searching by using query interpretation and search result diversification techniques.

Indexing and Ranking. SchemEX [24] is an indexing approach for search across the linked open data (LOD) using structured queries. SchemEX consists of three schema layers of RDF classes, RDF types, and equivalence classes with each layer supporting different types of structured queries. However, our framework supports keyword queries and makes use of contexts based on a richer set of ontology constructs. Blanco et al. [5] propose *r-vertical* index (reduced version of their vertical index) for the RDF entity search problem. The *r-vertical* index is built by manually categorizing RDF properties in three fields (important, unimportant and neutral). In comparison, our index is built using context information of concepts in the ontologies suitable for our concept search problem. Recent work in the area of Semantic Web resources ranking has largely focused on adapting and modifying the PageRank algorithm. ReConRank [19] is PageRank-inspired [22] algorithm for Semantic Web data. It uses node degree to rank Semantic Web resources in a manner analogous to the PageRank algorithm. ReConRank combines ranks from the RDF graph data sources and their linkage. AKTiveRank [3] ranks ontologies on the basis of how well they cover the specified search terms. The Linked open vocabularies (LOV) [4] search system ranks results on the basis of the popularity of the term in the LOD datasets and in the LOV ecosystem. Butt et al. [6,7], use offline ranking with the popularity of the concept within the ontology and the popularity of the ontology that contains the concept as the ranking features. Blanco et al. [5] propose instance/entity search using BM25F ranking function. Their ranking function does not exploit proximity information or term dependencies. The existing approaches do not directly exploit the structure in ontologies for indexing and ranking. Dali et al. [9], propose the *learning to rank* (LTR) [25] approach by using query-independent frequency-based features to rank the results of structured queries. We build the context-based inverted index to interpret the queries, and rank the results of keyword queries using query-based features in the LTR algorithm.

Query Processing and Interpretation. There has been work on structured query processing over LOD and related ontologies. The work on Top-k exploration of query candidates on the (RDF) graph data [31] proposes an intermediate step of converting keyword queries to structured queries. The user needs to select the correct SPARQL query interpretation to retrieve search results. However, our method internally interprets the keyword query without needing to explicitly generate the candidate SPARQL query. Fu and Anyanwu [16] generate query interpretations using the query history as contextual information. However, the queries may not always be iterative and extensive query logs of similar queries may not be available for interpretations. In our approach, we use the context information around a concept across ontologies to interpret the query. We also discuss query interpretation work in the context of *implicit* and *explicit* query interpretation. Sawant and Chakrabarti [29] propose *implicit* generative and discriminative formulations for joint query interpretation and response ranking in keyword-based searches across web documents. Agarwal *et al.* [1] use probabilistic modeling techniques to mine query templates from query logs for query interpretation. While these approaches may work well for large-scale unstructured data, they may not work in our problem of searching over structured ontologies with low number of redundancies. Our technique of interpreting relations among keywords in the query by using a rich ontology structure is different from the rest of the approaches.

Search Result Diversification. There are two main approaches to diversification: (1) *implicit* ones that assume that similar documents that cover similar intent/aspects of the query should be demoted to achieve diversified ranking (maximum marginal relevance, or MMR [8]); and (2) those that explicitly model query aspects by sub-queries and maximize the coverage of selected documents with respect to these aspects [10,21,28]. These approaches are applied to the unstructured text document search. We believe the diversification techniques have not been designed for the structured data setting of ontologies. Herzig *et al.* [18] propose language model (LM) approach for consolidating entity search results to reduce redundancy by grouping similar entities. However, their approach does not consider diversity of intent capture in the top-k results. While in our explicit diversification approach, we eliminate redundancy and also capture the fine-grained intents in the top-k search results.

3   Overall Approach

Given a multi-word query Q that consists of m keywords, $Q = \{k_1, k_2, \ldots, k_m\}$ on a search space of diverse web ontologies $\mathcal{O} = \{O_1, O_2, \ldots, O_n\}$, the goal is to retrieve relevant (named) concepts $R = \{R_1, R_2, \ldots, R_p\}$ across these ontologies. We retrieve concept results R by interpreting relations among keywords in the query via the context of a concept (class). Given an ontology, O_g ($g = 1, 2, \ldots, n$), the *entities* of O_g include named concepts and named (object-, datatype-, or annotation) properties that are declared in O_g. Given an axiom $a \in O_g$ (logical or annotation), let $\mathtt{sig}(a)$ represent the signature (the set of entities) in a. $\mathtt{sig}(\cdot)$ is extended naturally to apply to sets of axioms also. For an entity e, let $\mathtt{annotations}(e)$ represent the values of annotation axioms on e. These annotation axioms include $\mathtt{rdfs:label}$, $\mathtt{rdfs:comments}$, $\mathtt{rdfs:isDefinedBy}$, $\mathtt{rdfs:seeAlso}$ as well as those defined in other widely-used vocabularies such as SKOS.

A keyword query is interpreted using the *context* of each concept. The context of a given concept C_j across ontologies $\mathcal{O}$ is defined as the set of annotation values of the concept and of the entities that co-occur with C_j in some axioms in an ontology.

$$Ax_{C_j} = \{a|\, a \in O \wedge C_j \in \mathtt{sig}(a)\} \cup$$
$$\{a|\, a \text{ is } A \sqsubseteq B \wedge \{A,B\} \subseteq \mathtt{sig}(O) \wedge o \vDash a \wedge (C_j = A \vee C_j = B)\} \cup$$
$$\{a|\, a \text{ is } A \equiv B \wedge \{A,B\} \subseteq \mathtt{sig}(O) \wedge o \vDash a \wedge (C_j = A \vee C_j = B\}$$
$$Px_{C_j} = \{a \,|\, a \in O \wedge C_j \in \mathtt{sig}(a) \text{ where } a \text{ is an object-, datatype-}$$
$$\text{or annotation property axiom}\}$$
$$Context(C_j) = \{\mathtt{annotations}(C_j)\} \cup$$
$$\{\mathtt{annotations}(e) \,|\, e \in \mathtt{sig}(Ax_{C_j} \cup Px_{C_j})\}$$

where Ax_{C_j} and Px_{C_j} are sets of class-axioms and property-axioms that are relevant to C_j, respectively. Note that Ax_{C_j} includes $\mathtt{subClassOf}$ and $\mathtt{EquivalentClasses}$ axioms a that are entailed by an ontology (i.e., $O \vDash a$), where both concepts are named concepts (i.e., A and B), and one of them is C_j. These additional, inferred axioms are obtained through reasoning. $Context(C_j)$ consist of its annotation values ($\mathtt{annotations}(C_j)$) and the annotation values of the set of entities that are relevant to C_j ($\mathtt{annotations}(e)|e \in \mathtt{sig}(Ax_{C_j} \cup Px_{C_j})$).

We further employ search result diversification to cover maximum user intents in the top-k search results. We pose our search result diversification problem as a an optimization problem, in which the objective is to maximize the relevance of a result, while minimizing the redundancy among the results. Given a ranked set R of relevant concepts for Q, the goal is to select the subset of concepts $C_s \subseteq R$ that are most relevant to the query and diverse among C_s. Along the lines of the MMR [8] framework, our diversification optimization model is:

$$C^* = \arg\max_{C_i \in R \backslash C_s}((1 - \lambda) \times \mathcal{S}(C_i) + \lambda \times \mathcal{D}(C_i, C_s)) \tag{1}$$

where $\mathcal{S}(C_i)$ is the relevance score of concept C_i, and $\mathcal{D}(C_i, C_s)$ is the diversification score of C_i. $\mathcal{S}(C_i)$ is obtained by using LTR algorithms. $\mathcal{D}(C_i, C_s)$ is estimated using the diversity function in which.C_i is compared with each of the concepts in C_s. The diversity parameter, $\lambda \in [0,1]$, is the tuning parameter that draws a balance between the relevance and the diversity of a concept.

4   The Concept Search Framework

Figure 1 depicts the high-level architecture of our search framework. The components at the bottom are constructed offline, whereas the computations at the

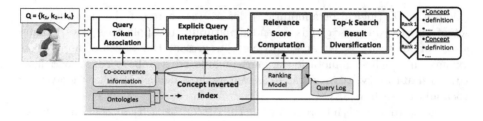

Fig. 1. The high-level architecture of our concept search framework.

top are performed online for each new query. The concept inverted index I is
built offline using $Context(C_j)$, that is relevant to each C_j as defined in Eq. 1
(Sect. 3). Each class and property axiom in Ax_{C_j} and Px_{C_j}, which is relevant
to each C_j in the ontology corpus is indexed as a field such as `rdfs:label`,
`rdfs:comments`, `rdfs:isDefinedBy`. Since the probability of having more than
two words together in the ontology corpus is small, we set the shingle size (number of co-occurring words used in co-occurrence computation) to two. In addition, we store the term-vectors for performing co-occurrence computation. We
perform natural language processing (NLP) techniques such as tokenization and
stemming using the Lucene standard analyzer in order to store the context information in the inverted index.

4.1 The Concept Search Procedure

Given a query Q, the search proceeds by finding concepts with human-readable
label $L(C_j)$ or $class\text{-}name(C_j)$ (fragment of the URI) that match exactly with
the query Q terms as a phrase (line 3 in Procedure CS). We define $L(C_j)$:

$$L(C_j) =\{l \mid l \in \texttt{rdfs:label}(C_j) \vee l \in \texttt{skos:prefLabel}(C_j)\} \quad (2)$$

The lexical co-occurrence LC among keywords in a query is evaluated using
Pearson's Chi-squared test (line 6). A Chi-squared value that is greater than
3.841 implies that the keywords co-occur with 95 % confidence. The Pearson's
Chi-squared test returns a set of all co-occurring terms, C_{terms} in the query.
We use C_{terms} for explicit query interpretation in procedure QI (line 7, further
described in Sect. 4.2). QI generates direct and inferred *parses* for the query
using context of concepts. The parses return a set of concepts as search results.
Feature vectors (fv's) are then built for these results to obtain relevance by using
LTR model (line 8). An LTR ranking model that is trained offline is applied in
order to obtain the relevance score of each search result (line 9, further described
in Sect. 4.3). Finally, the search results are diversified to capture fine-grained user
intents (line 10, further described in Sect. 4.4).

4.2 Explicit Query Interpretation

Our *explicit* query interpretation approach generates interpretations by analyzing the interrelationships among the keywords along with the inherently rich

---

Procedure $CS(Q, \mathcal{C})$

Data: Query $Q = \{k_1, k_2, \ldots, k_m\}$
Data: Number of results to be returned, k
Data: Concepts across ontologies, $C = \{C_1, C_2, \ldots, C_l\}$
Data: An LTR ranking model trained offline, $rankingModel$
Result: SearchResults

1 $SearchResults \leftarrow \emptyset$;
2 **foreach** $C_j \in C$ **do**
3   **if** $IsExactMatch(L(C_j), Q)$ *or* $IsExactMatch(class\text{-}name(C_j), Q)$ **then**
4          $SearchResults \leftarrow SearchResults \cup \{C_j\}$;

5 **if** $|Q| \geq 2$ **then**
6      $C_{terms} = \{C^t \mid C^t \subseteq Q \wedge C^t = \{k_i, k_{i+1}\} \wedge LC(k_i, k_{i+1}) > 3.841\}$;
7      $SearchResults \leftarrow SearchResults \cup QI(C_{terms}, Q)$;

8 $fv \leftarrow BuildFV(Q, SearchResults)$;
9 $SearchResults \leftarrow relevance(fv, SearchResults, rankingModel)$;
10 $SearchResults \leftarrow diversify(SearchResults, k)$;
11 **return** $SearchResults$;

---

structure and semantics of ontologies. Our *explicit* approach embeds a precise understanding of how each search result is obtained. In the procedure *QI*, we use the set of co-occurring terms C_{terms} and all the individual keywords in the query. For each co-occurring terms pair $C^t \in C_{terms}$, we search for the set of classes $CtermClasses$ for which $IsExactMatch(L(C_j), C^t)$ is true (line 5). We implement the direct and inferred parse on each of the concepts C_j in $CtermClasses$. The direct and inferred parse returns the set of relevant concept results (*SearchResults*) for the query (line 6–7). If the *SearchResults* found using direct and inferred parse for co-occurring tokens are less than the threshold (set to 50), we search for a set of classes $StermClasses$ in order to match each keyword $S^t \in Q$, for which $IsExactMatch(L(C_j), S^t)$ is true (line 10). We implement the direct and inferred parse on each of the classes C_j in $StermClasses$ to obtain relevant concept results (*SearchResults*) for the query (line 11–12). In addition we also return the set of classes $CtermClasses$ and $StermClasses$ as *SearchResults* if the *SearchResults* found using direct and inferred parse are less than the threshold (line 14–15).

Direct Parse: The *direct parse* (DP) returns sets of concepts (*SearchResults*). DP analyzes the relation among the query keywords by using the context of a concept and is defined as:

$$DP(tC, S) = \{C_j \mid C_j \in tC \wedge sim(Context(C_j), S) > 0\} \qquad (3)$$

where tC is either $CtermClasses$ or $StermClasses$, and $sim(Context(C_j), S)$ is calculated using the Jaccard similarity measure.

We explain DP with an example for a query "Myocardial infarction causes" in Fig. 2. The query contains the co-occurring terms pair C^t,

---

Procedure $QI(C_{terms}, Q)$

    Data: Co-occurring terms C_{terms}, Q
    Result: SearchResults

1   $SearchResults \leftarrow \emptyset$;
2   $CtermClasses \leftarrow \emptyset$;
3   $StermClasses \leftarrow \emptyset$;
4   **foreach** $C^t \in C_{terms}$ **do**
5      $CtermClasses \leftarrow CtermClasses \cup \{C_j | IsExactMatch(L(C_j), C^t))\}$;
6      $SearchResults \leftarrow SearchResults \cup DP(CtermClasses, Q \setminus C^t)$;
7      $SearchResults \leftarrow SearchResults \cup IP(CtermClasses, Q \setminus C^t)$;

8   **if** $|SearchResults| \leq th$ **then**
9   **foreach** $S^t \in Q$ **do**
10        $StermClasses = StermClasses \cup \{C_j | IsExactMatch(L(C_j)), S^t)\}$;
11        $SearchResults \leftarrow SearchResults \cup DP(StermClasses, Q \setminus S^t)$;
12        $SearchResults \leftarrow SearchResults \cup IP(StermClasses, Q \setminus S^t)$;

13 **if** $|SearchResults| \leq th$ **then**
14      $SearchResults \leftarrow SearchResults \cup CtermClasses$;
15      $SearchResults \leftarrow SearchResults \cup StermClasses$;

16 **return** $SearchResults$;

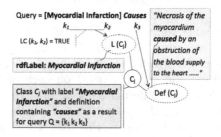

Fig. 2. Direct parse

"Myocardial infarction". The keywords "Myocardial infarction" appear as the label of some concept C_j in our concept index i.e., $IsExactMatch$ $(L(C_j), C^t) = true$. If the context of the same concept C_j contains "causes" (hence $sim(Context(C_j), S) > 0$ is satisfied), then C_j with label 'Myocardial infarction" will be returned as a search result.

Inferred Parse: The *inferred parse* (IP) returns a set of other concepts that do not directly appear in the query, but rather indirectly through SubClassOf or EquivalentClasses axioms. The IP is defined as:

$$OC(C_j) = \{C_k \mid \text{SubClassOf}(C_j, C_k)\} \cup \qquad (4)$$
$$\{C_k \mid \text{EquivalentClasses}(C_j, C_k)\}$$

$$IP(tC, S) = \bigcup_{C_j \in tC} \{C_k \mid C_k \in OC(C_j) \wedge sim(Context(C_k), S) > 0\} \qquad (5)$$

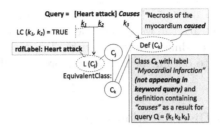

Fig. 3. Inferred parse

where $OC(C_j)$ is a collection of all other classes that are indirectly related to C_j through either `SubClassOf` or `EquivalentClasses` axioms, and IP is similarly constructed from classes in $OC(C_j)$. Here tC, sim are defined in the same way as explained in Eq. 4 of the definition of direct parse.

An example of IP is shown in Fig. 3. Consider the query "`Heart attack causes`". The keywords "`Heart attack`" co-occur and appear as a label of some concept, C_j, in our concept index that is, $IsExactMatch(L(C_j), C^t) = true$. However, the keyword "`causes`" is not present in the context of C_j. An equivalent class of C_k, `Myocardial_infarction`, has the context that contains "`causes`". This is interpreted as an inferred relation between "`Heart attack`" (co-occurring terms) and "`causes`" (single term) and the class `Myocardial_infarction` with `causes` as its relevant property will be returned as a search result.

4.3 Relevance Score Computation

The relevance score for the search results are computed by the ranking model that is built using learning to rank (LTR) algorithms [25]. LTR algorithms are supervised machine learning algorithms. Training data for the ranking model is generated from a query log, in which feature vectors (FV's) are generated for each combination of query and a result. Components of such FV's are ranking features, which are obtained using ISUB [30] (12 features) and Jaccard (12 features) similarity, between query terms and concept context fields in the index. The twelve features are computed as the ISUB similarity of query with `rdfs:label`, `rdfs:isDefinedBy`, `skos:prefLabel`, `rdfs:comments`, `rdfs:seeAlso`, synonym, dataproperty, objectpropertydomain, objectpropertyrange, `SuperClassOf`, `SubClassOf`, `EquivalentClasses`, respectively. Similar features are obtained using the Jaccard similarity measure.

The training data of FV's are used to build LTR models, by employing the RankLib implementation[2]. We use the pairwise RankNet algorithm because the highest normalized distributive cumulative gain (NDCG) value was obtained from this algorithm among the pairwise algorithms by using the query log test data. The RankNet parameters that are used are: the number of epochs to train = 100; the number of hidden layers = 1; number of hidden nodes per layer = 10;

---

[2] http://people.cs.umass.edu/~vdang/ranklib.html.

and the learning rate = 0.00005. The LTR model is trained using 70 % of the log queries in order to learn the weights of the features and 30 % of the queries are used for testing (excluding training queries). The model is then applied to search results in order to obtain the relevance scores for all the concept search results.

4.4 Search Result Diversification

A keyword query may have diverse possible search intents. Search result diversification aims to retrieve k items that are the subset of all relevant results that contain the most relevant and the most diverse intent results. We use the relevance score that is obtained by the LTR algorithm for search result diversification. Search results are diversified by capturing fine-grained query intents explicitly, using the context of concepts.

Baseline Approach. The baseline *Implicit* approach assumes that similar search results map to the same query intent. Such results should be demoted in order to achieve diversified ranking. Maximum marginal relevance MMR [8] is a canonical technique from the implicit approach. The implicit diversity function is defined as follows:

$$\mathcal{D}(C_i, C_s) = \sum_{C_j \in C_s} (1 - SC(C_i, C_j)) \tag{6}$$

We calculate the similarity $SC(C_i, C_j)$ among two concepts by comparing their respective context similarity. For example, the context information that is captured in $\mathtt{sig}(a)$, in which a is a $\mathtt{SubClassOf}$ axiom of the concept C_i is compared with similar information of the other C_j. We use the greedy algorithm [13] by substituting Eq. 6 in Eq. 1 of Sect. 3 in order to implement the implicit diversification.

$$C^* = \arg\max_{C_i \in R \setminus C_s} ((1 - \lambda) \times \mathcal{S}(C_i) + \lambda \times (\sum_{C_j \in C_s} (1 - SC(C_i, C_j)))) \tag{7}$$

We iteratively select the best concept result with the highest LTR score ($\mathcal{S}(C_i)$) from R which can maximize the diversity of the selected concepts C_s. The iterative process is repeated until top-k results ($|C_s| = k$) are obtained.

Fine-Grained Explicit Diversification. *Explicit* approaches [10,28] directly map search results to query intents. Diversified ranking is achieved by selecting results that maximize coverage with respect to query intents. Existing explicit diversification approaches obtain query intents from commercial search engines such as Google, which may not be useful in our setting because they are independent of the ontology corpus. Our explicit diversification is based on fine-grained intents that are captured by the *contextual* information around a concept. We make use of super-class and subclass relations of the returned concepts to generate search intents. Super-classes of concepts cover more generic intents, while subclasses of concepts generate more specific intents.

More specifically, for a query, Q, and the set of most relevant concepts R returned by the LTR model, two levels of intents are generated. Firstly, the *top-level intents* consists of all of the super-classes of concepts in R. Secondly, the *sub-level intents* consists of all the subclasses of the super-classes of R.

Top-level intent diversity: The *top-level intents* are represented as I. Along the lines of the work by Hu et al. [21], we define diversity of the top-level intents as follows:

$$\mathcal{D}(C_i, C_s, I) = \sum_{x \in I} \left[p(C_i|x) \times p(x|Q) \times \prod_{C_j \in C_s} (1 - p(C_j|x)) \right] \tag{8}$$

$p(C_i|x)$, is the probability that C_i satisfies the top-level intent x, and I represents the set of the top-level intents of Q. $p(C_i|x) = sim(L(C_i), L(x))$ is estimated as the Jaccard similarity between the labels $L(C_i)$ of a concept and the intent $L(x)$, in which $L(C_i)$ and $L(x)$ are defined by Eq. 3 in Sect. 4.1.

$p(x|Q)$, which is the probability of x for the given query Q, is estimated by assuming uniform probability distribution $p(x|Q) = \frac{1}{|I|}$. Uniform intent distribution has been demonstrated to be the most useful [28].

$(1 - p(C_j|x))$ is the probability that C_j does not satisfy intent x, which indicates that x is less substantially covered and should have higher "priority" in getting more results. The product $\prod_{C_j \in C_s} (1 - p(C_j|x))$ estimates the probability that all concepts C_s, that are selected by the LTR model fail to satisfy intent x.

After summing over all query intents, and after being weighted by $p(x|Q)$, the diversity measure in Eq. 8 is the probability that C_i covers the search intents I while the existing list C_s, fails to satisfy them.

Sub-level intent diversity: Each of the top-level intents $x \in I$ is subdivided into sub-level intents S. The sub-level intents are represented as S_x in which x is a top-level intent.

$$\mathcal{D}(C_i, C_s, S) = \sum_{x \in I} \sum_{y \in S_x} \left[p(C_i|y) \times p(y|Q) \times \prod_{C_j \in C_s} (1 - p(C_j|y)) \right] \tag{9}$$

where $p(C_i|y)$ estimates the probabilities that concept C_i satisfies the sub-level intent y, and $p(y|Q)$ is the probability of each of the subclass level intents y for the given query Q. The probability of each of the sub-level intents, y, for query, Q, $p(y|Q)$, is estimated by assuming uniform probability for sub-level intents, $p(y|Q) = \frac{1}{|I| \times |S_x|}$.

By combining the diversity of the top-level and sub-level intents, our fine-grained explicit diversification is estimated as follows

$$\mathcal{D}(C_i, C_s) = \gamma \times \mathcal{D}(C_i, C_s, I) + (1 - \gamma) \times \mathcal{D}(C_i, C_s, S) \tag{10}$$

where γ is the tuning parameter for the top-level and the sub-level depending on the granularity of the diversification.

By plugging Eq. 10 into Eq. 1, our diversification optimization model is

$$C^* = \arg \max_{C_i \in R \backslash C_s} ((1 - \lambda) \times \mathcal{S}(C_i) + \lambda(\gamma \times \mathcal{D}(C_i, C_s, I) + (1 - \gamma) \times \mathcal{D}(C_i, C_s, S)))$$
(11)

where λ is the diversity parameter and γ is the intent parameter for the top-level and sub-level intents.

The model considers the relevance between the concept results C_s and query Q and the diversity among concepts in C_s. Using a greedy algorithm [13], it iteratively selects the next best concept that is relevant to query Q which maximizes the diversity of selected concepts C_s.

5   Evaluation

We compare our system with the search function of two large, widely-used, and openly available ontology repositories, the Bio-medical domain BioPortal,[3] and the generic Falcons[4]. A summary of the two repositories can be found in Table 1. A separate inverted index was built for each of the BioPortal and Falcons repositories respectively.

We evaluate our system's performance in terms of relevance only (query interpretation, Sect. 5.1), as well as relevance and diversity (search diversification, Sect. 5.2). Standard information retrieval (IR) ranking measures [25], mean reciprocal rank (MRR) and normalized distributive cumulative gain (NDCG) are used to evaluate our query interpretation approach. The standard search diversification metric of normalized cumulative gain-intent aware (NDCG-IA) [2] is used for evaluation of our diversification technique. Our evaluation dataset is publicly available.[5]

Table 1. A summary of the BioPortal and Falcons repositories.

Repository	Type	# ontologies	# concepts	# axioms
BioPortal	Domain-specific	296	2,062,080	9,221,087
Falcons	Generic	294,504	804,380	2,566,921

5.1   Query Interpretation Evaluation

Comparison with BioPortal. The BioPortal query log (July 2012 to July 2014) contains more than 2,000 real-world queries as well as click-through data. Among these queries, more than 50 % are multiple-token queries.

---

[3] http://bioportal.bioontology.org/.
[4] http://ws.nju.edu.cn/falcons/conceptsearch/.
[5] https://dx.doi.org/10.4225/03/57218DB2399B9.

Table 2. Comparison with BioPortal.

Measure	Multi-token		Single-token	
	BioPortal	Ours	BioPortal	Ours
NDCG	0.61	**0.72**	0.62	**0.63**
MRR	0.42	**0.60**	0.49	**0.51**

Table 3. Comparison of *implicit* and *explicit* with BioPortal for multi-token queries.

Measure	BioPortal	Ours	
		Implicit	Explicit
NDCG	0.61	0.69	**0.72**
MRR	0.42	0.52	**0.60**

Comparison with BioPortal Average Values. We present average NDCG and MRR for our approach vis-a-vis BioPortal for multi-token and single-token queries in Table 2. Our system significantly outperforms BioPortal for multi-token queries, and both systems demonstrate comparable performances in single-token queries.

Query-Wise Comparison with BioPortal. For each query, we calculated the difference between the NDCG values obtained by our system and BioPortal. Of the 2,000 queries, the NDCG values for 1,000 queries (>50 %) are better in our system, and more than 700 queries (>35 %) have the same level of performance. The number of queries in which BioPortal performs better is 300 (<15 %). Our system performs better (>50 %) due to effective use of context information in query interpretation. The level of performance is the same for the queries (>35 %) in which the keywords match the class label exactly. The lower performance (<15 %) may be due to unavailability of context information in the ontologies. The better performance of BioPortal in these (<15 %) queries can be attributed to their statistical consideration of ontology popularity in ranking search results.

Comparison with Implicit Query Interpretation. We evaluated *explicit* and *implicit* implementations of query interpretation on the BioPortal dataset. The feature-based implicit model was trained using query logs. Explicit techniques can be more useful in searches over structured data with a low number of redundancies due to the structuredness of the corpora; we also confirm this experimentally. A comparison of the implicit, explicit query interpretation approaches and BioPortal can be found in the Table 3.

Comparison with Falcons. We compared our system vis-a-vis the Falcons search engine [27] in order to explore the generic applicability of our approach.

Table 4. Comparison with Falcons for multi-token queries.

Measure	Falcons	Ours
NDCG	0.54	**0.79**
MRR	0.49	**0.78**

We performed a human-based evaluation in this experiment for better evaluation accuracy and to eliminate noise in automatic clicks. We performed an evaluation on 102 queries that were obtained from two years of TREC web track competitions.[6] This TREC dataset does not contain single token queries. Our system was evaluated by 30 human users who were undergraduate, graduate, and postgraduate students and had a high level of Web search experiences. Each of the 102 queries was evaluated by at least three of the users. We recorded the binary relevance judgment for the same set of queries for each result on both systems. The performance was evaluated using standard information retrieval measures of NDCG and MRR. Again, our system outperforms Falcons.

Comparison with Average Values of Falcons. We present average NDCG and MRR for our approach vis-a-vis Falcons for multi-token queries in Table 4.

Query-Wise Comparison with Falcons. We calculated the difference of the NDCG and P@k (precision at k) values of our system (with QI) in comparison with Falcons, for the same set of queries. Our NDCG performance was better for >66 % of the queries. It was at par in >25 % and lower in <8 % of the queries. We recorded the top-k ($k = 1, 3, 5$) P@k (precision at k^{th} position) results of our system and Falcons. Of all the queries, the performance of our system in P@1, P@3, and P@5 respectively was better than Falcons in >50 %, >60 %, and >70 % respectively, the same as Falcons in >40 %, >30 %, and >20 % respectively, and lower than Falcons for <10 % for all P@k. The average precision (AP) was calculated by taking an average of P@1, P@3, and P@5 for each query. The positive difference for AP for >70 % queries indicates the better overall performance of our approach.

The subsequent indicative queries give a fair idea of our performance. A query, *standard axioms of set theory*, has co-occurring keywords *set theory* and the keywords *standard axioms* appears in the context of the *set theory* class. The same query failed to return any results in the Falcons system. For another query, *machine learning algorithms*, Falcons failed to return relevant results while our system returned relevant result such as *machine learning program*, and *machine learning topic*.

5.2   Evaluation of Diversification

Search result diversification was evaluated using a variation of NDCG which is known as the intent-aware normalized cumulative gain measure (NDCG-IA) [2]

---

[6] http://trec.nist.gov/data/webmain.html.

Table 5. Comparison of our *implicit* and *explicit* diversification techniques with Bio-Portal and Falcons on two separate indices. The best NDCG-IA value in each comparison is highlighted in **bold**

Measure	BioPortal	Ours		Falcons	Ours	
		Implicit	Explicit		Implicit	Explicit
NDCG-IA	0.66	0.75	**0.83**	0.47	0.73	**0.77**

on the BioPortal and Falcons dataset. We have implemented the implicit diversification approach as defined in Eq. 7 as a baseline, and the explicit diversification approach as defined in the Eq. 11. We set the diversity parameter λ to 0.5, and assigned equal probability to diversity and relevance. Similarly, we set the intent level parameter γ to 0.5, assigning equal priority to top-level and sub-level intents. Table 5 compares the NDCG-IA values produced by our explicit fine-grained diversification method with the implicit diversification baseline, as well as BioPortal and Falcons. Note that separate indices are constructed for the comparison with BioPortal and Falcons.

Comparison with BioPortal. We conducted experiments with 52 queries randomly selected from the BioPortal query log in order to evaluate the effectiveness of intent-capture in our concept search results. Each query was evaluated by three users with a basic level of bio domain expertise and a high level of web search experience. We designed an interface for evaluating our diversification approach. The evaluation interface provided the users with list of intents for search results. The users selected intent for each search result that was used for computing NDCG-IA values.

Comparison of NDCG-IA Values. We report the average NDCG-IA values for the top-10 results of baseline implicit diversification and explicit diversification vis-a-vis BioPortal for multi-token queries in the left part of Table 5. Of the total queries, 70 % fared well with diversification, 25 % were the same as the baseline and 5 % performed lower. The better results for diversification are due to the use to explicit intent capture in our approach. For example, a query *Myocardial infarction* captures the following intents-*myocardial infarction definition*, *myocardial infarction symptoms*, *myocardial infarction types*, and *myocardial infarction causes* in the top-k results. On the other hand, the implicit approach removes redundancy but may not address the specific user intents, whereas, the BioPortal captures the intents in their results but not in the top-k.

Comparison with Falcons. We conducted experiments with 50 queries that were randomly selected from the TREC competitions on the Falcons dataset in order to evaluate the effectiveness of intent-capture in our results. Each query was evaluated by three users. The users were graduate students and had a high level of search experience. NDCG-IA was computed for the intents assigned by the user during evaluation.

Comparison of NDCG-IA Values. We present the average NDCG-IA values of baseline implicit diversification and explicit diversification vis-a-vis Falcons, in the right part of Table 5. Of the total queries, 75 % fared well with diversification, 10 % were the same as the baseline and 15 % performed lower. Our explicit diversification techniques effectively captures the fine grained intents. For example, *natural language processing applications* captures diverse intents such as the *linguistic translation process, linguistic topic*, and *artificial intelligence* in our approach, while Falcons returns search results that repeat the single intent in the top-k results.

5.3   Discussion

Our comprehensive log-based and human-based evaluation includes domain-specific and generic ontologies. Our system demonstrated better performance in both settings using standard information retrieval measures, indicating the effectiveness of our framework, especially in multi-token queries. Relation among the keywords in the multi-token queries is effectively captured in our approach. All of our experiments presented in this section (for both multi-token and single-token queries) were found to be statistically significant using the Wilcoxon signed-rank test with p-value <0.0001. Our system's effectiveness can be attributed to the following factors: (1) Co-occurrence is prevalent among multi-token queries (>50 % queries). (2) Contexts of concepts facilitate effective query interpretation. (3) Our search result diversification methods effectively captures fine-grained intents in top-k results for multi-token queries. As a result, our system seldom returns null or irrelevant results.

6   Conclusion

In this paper we present a novel and effective concept search framework that balances relevance and diversity. We propose to construct *contexts* for concepts, and use these contexts to (1) interpret user queries and (2) capture fine-grained search intents. The effectiveness of our context-based query interpretation and search result diversification techniques is demonstrated through a comprehensive evaluation against two concept search systems, BioPortal and Falcons. Our evaluation shows that our concept search framework significantly outperforms both systems on widely-used IR metrics.

Our work opens up several directions for further research. Our approach of *explicit* query interpretation can be improved by incorporating user involvement in the customization of the search. Our explicit diversification formulation can be improved by using proportionality-based optimization techniques. Finally, implementing the applicability of concept search in applications such as ontology population may be useful.

Acknowledgments. We thank Prof. Mark Musen and Prof. Paul Alexander of Stanford University for sharing the BioPortal query log. We would like to thank Prof. Gong Cheng for his advice and guidance on working with Falcons.

References

1. Agarwal, G., Kabra, G., Chang, K.C.C.: Towards rich query interpretation: walking back and forth for mining query templates. In: WWW 2010 (2010)
2. Agrawal, R., Gollapudi, S., Halverson, A., Ieong, S.: Diversifying search results. In: WSDM 2009, pp. 5–14. ACM, New York (2009)
3. Alani, H., Brewster, C., Shadbolt, N.R.: Ranking ontologies with AKTiveRank. In: Cruz, I., Decker, S., Allemang, D., Preist, C., Schwabe, D., Mika, P., Uschold, M., Aroyo, L.M. (eds.) ISWC 2006. LNCS, vol. 4273, pp. 1–15. Springer, Heidelberg (2006)
4. Atemezing, G.A., Troncy, R.: Information content based ranking metric for linked open vocabularies. In: SEMANTICS 2014, pp. 53–56 (2014)
5. Blanco, R., Mika, P., Vigna, S.: Effective and efficient entity search in RDF data. In: Aroyo, L., Welty, C., Alani, H., Taylor, J., Bernstein, A., Kagal, L., Noy, N., Blomqvist, E. (eds.) ISWC 2011, Part I. LNCS, vol. 7031, pp. 83–97. Springer, Heidelberg (2011)
6. Butt, A.S., Haller, A., Xie, L.: Ontology search: an empirical evaluation. In: Mika, P., et al. (eds.) ISWC 2014, Part II. LNCS, vol. 8797, pp. 130–147. Springer, Heidelberg (2014)
7. Butt, A.S., Haller, A., Xie, L.: Relationship-based top-K concept retrieval for ontology search. In: Janowicz, K., Schlobach, S., Lambrix, P., Hyvönen, E. (eds.) EKAW 2014. LNCS, vol. 8876, pp. 485–502. Springer, Heidelberg (2014)
8. Carbonell, J., Goldstein, J.: The use of MMR, diversity-based reranking for reordering documents and producing summaries. In: SIGIR 1998 (1998)
9. Dali, L., Fortuna, B., Duc, T.T., Mladenić, D.: Query-independent learning to rank for RDF entity search. In: Simperl, E., Cimiano, P., Polleres, A., Corcho, O., Presutti, V. (eds.) ESWC 2012. LNCS, vol. 7295, pp. 484–498. Springer, Heidelberg (2012)
10. Dang, V., Croft, W.B.: Diversity by proportionality: an election-based approach to search result diversification. In: SIGIR 2012, pp. 65–74. ACM, New York (2012)
11. d'Aquin, M., Motta, E.: Watson, more than a semantic web search engine. Semant. Web 2(1), 55–63 (2011)
12. Ding, L., Pan, R., Finin, T.W., Joshi, A., Peng, Y., Kolari, P.: Finding and ranking knowledge on the semantic web. In: Gil, Y., Motta, E., Benjamins, V.R., Musen, M.A. (eds.) ISWC 2005. LNCS, vol. 3729, pp. 156–170. Springer, Heidelberg (2005)
13. Drosou, M., Pitoura, E.: Search result diversification. SIGMOD Rec. 39, 41–47 (2010)
14. Fagin, R., Kimelfeld, B., Li, Y., Raghavan, S., Vaithyanathan, S.: Understanding queries in a search database system. In: PODS 2010, pp. 273–284 (2010)
15. Finin, T., Peng, Y., Scott, R., Joel, C., Joshi, S.A., Reddivari, P., Pan, R., Doshi, V., Ding, L.: Swoogle: a search and metadata engine for the semantic web. In: CIKM 2014 (2004)
16. Fu, H., Anyanwu, K.: Effectively interpreting keyword queries on RDF databases with a rear view. In: Aroyo, L., Welty, C., Alani, H., Taylor, J., Bernstein, A., Kagal, L., Noy, N., Blomqvist, E. (eds.) ISWC 2011, Part I. LNCS, vol. 7031, pp. 193–208. Springer, Heidelberg (2011)
17. Gavankar, C., Li, Y.F., Ramakrishnan, G.: Context-driven concept search across web ontologies using keyword queries. In: K-CAP 2015, pp. 20:1–20:4. ACM (2015)
18. Herzig, D.M., Mika, P., Blanco, R., Tran, T.: Federated entity search using on-the-fly consolidation. In: Alani, H., et al. (eds.) ISWC 2013, Part I. LNCS, vol. 8218, pp. 167–183. Springer, Heidelberg (2013)

19. Hogan, A., Harth, A., Decker, S.: Reconrank: a scalable ranking method for semantic web data with context. In: 2nd Workshop on Scalable Semantic Web Knowledge Base Systems (2006)

20. Hogan, A., Harth, A., Umbrich, J., Kinsella, S., Polleres, A., Decker, S.: Searching and browsing linked data with SWSE: the semantic web search engine. J. Web Semant. **9**(4), 365–401 (2011)

21. Hu, S., Dou, Z., Wang, X., Sakai, T., Wen, J.R.: Search result diversification based on hierarchical intents. In: CIKM 2015, pp. 63–72. ACM (2015)

22. Kleinberg, J.M.: Authoritative sources in a hyperlinked environment. J. ACM **46**, 604–632 (1999)

23. Kollia, I., Glimm, B., Horrocks, I.: SPARQL query answering over OWL ontologies. In: Antoniou, G., Grobelnik, M., Simperl, E., Parsia, B., Plexousakis, D., De Leenheer, P., Pan, J. (eds.) ESWC 2011, Part I. LNCS, vol. 6643, pp. 382–396. Springer, Heidelberg (2011)

24. Konrath, M., Gottron, T., Staab, S., Scherp, A.: Schemex - efficient construction of a data catalogue by stream-based indexing of linked data. J. Web Sem. **16**, 52–58 (2012)

25. Liu, T.Y.: Learning to Rank for Information Retrieval. Springer, Berlin (2011)

26. Noy, N.F., Alexander, P.R., Harpaz, R., Whetzel, P.L., Fergerson, R.W., Musen, M.A.: Getting lucky in ontology search: a data-driven evaluation framework for ontology ranking. In: Alani, H., et al. (eds.) ISWC 2013, Part I. LNCS, vol. 8218, pp. 444–459. Springer, Heidelberg (2013)

27. Qu, Y., Cheng, G.: Falcons concept search: a practical search engine for web ontologies. IEEE Trans. Syst. Man Cybern. Part A **41**(4), 810–816 (2011)

28. Santos, R.L., Macdonald, C., Ounis, I.: Exploiting query reformulations for web search result diversification. In: WWW 2010, pp. 881–890. ACM, New York (2010)

29. Sawant, U., Chakrabarti, S.: Learning joint query interpretation and response ranking. In: Proceedings of the 22nd International Conference on World Wide Web (2013)

30. Stoilos, G., Stamou, G., Kollias, S.D.: A string metric for ontology alignment. In: Gil, Y., Motta, E., Benjamins, V.R., Musen, M.A. (eds.) ISWC 2005. LNCS, vol. 3729, pp. 624–637. Springer, Heidelberg (2005)

31. Tran, T., Wang, H., Rudolph, S., Cimiano, P.: Top-k exploration of query candidates for efficient keyword search on graph-shaped (RDF) data. In: ICDE 2009, pp. 405–416 (2009)

32. Tummarello, G., Delbru, R., Oren, E.: Sindice.com: weaving the open linked data. In: Aberer, K., et al. (eds.) ASWC 2007 and ISWC 2007. LNCS, vol. 4825, pp. 552–565. Springer, Heidelberg (2007)

Predicting Energy Consumption of Ontology Reasoning over Mobile Devices

Isa Guclu[1(✉)], Yuan-Fang Li[2], Jeff Z. Pan[1], and Martin J. Kollingbaum[1]

[1] University of Aberdeen, Aberdeen, UK
r01ig15@abdn.ac.uk
[2] Monash University, Melbourne, Australia

Abstract. The unprecedented growth in mobile devices, combined with advances in Semantic Web (SW) Technologies, has given birth to opportunities for more intelligent systems on-the-go. Limited resources of mobile devices demand approaches that make mobile reasoning more applicable. While Mobile-Cloud integration is a promising method for harnessing the power of semantic technologies in the mobile infrastructure, it is an open question how to decide when to reason over ontologies on mobile devices. In this paper, we introduce an energy consumption prediction mechanism for ontology reasoning on mobile devices that allows an analysis of the feasibility of performing an ontology reasoning on a mobile device with respect to energy consumption. The developed prediction model contributes to mobile–cloud integration and helps to improve further developments in semantic reasoning in general.

Keywords: Energy · Semantic web · Ontology reasoning · Mobile computing · Prediction · Random forests

1 Introduction

Server and desktop machines have been the main environment for ontology reasoning in assisting knowledge management so far. With the rapid improvement of hardware capabilities, as well as software developments in mobile devices (e.g., smartphones, tablets, PDAs, smartwatches), semantic reasoners start to become adopted [24] in mobile environments. Mobile applications (apps) that use semantic technologies, such as for the integration with diverse data sources and knowledge due to inferences made during semantic reasoning, are also being developed. However, this potential seems not fully utilized yet.

According to Yus and Pappachan's research [25] on semantic mobile apps, 23 out of 36 apps implement a client-server architecture, where the mobile app is used as an interface for the results processed by a server and just 6 apps utilize a semantic reasoner directly on the device to infer facts. According to their observation, the use of Semantic Web technologies on mobile devices is on the rise and there is a need for the development of more tools to facilitate this growth [25]. Groth [6] asserts that, while there has been a large amount of effort in Semantic

© Springer International Publishing AG 2016
P. Groth et al. (Eds.): ISWC 2016, Part I, LNCS 9981, pp. 289–304, 2016.
DOI: 10.1007/978-3-319-46523-4_18

Web Services, even going as far as developing a standard for describing those services, we have not seen a corresponding take-up in using these languages to enable the execution of actions either on the Web or in the real world.

Our goal is to make semantic technologies more feasible for a new era of mobile and cloud computing by building an energy prediction mechanism that will guide us "to what extent ontology reasoning can be made on mobile devices". *Mobile-Cloud Integration* can significantly enhance the capabilities and benefits of semantic technologies. For a successful mobile-cloud integration, a mechanism is needed that will (1) predict the cost of data processing (including loading, parsing, reasoning, query answering) on a mobile device itself in terms of time and energy consumption, (2) predict the cost of data processing on the cloud, and (3) ultimately determine where data processing should be conducted in an optimal way.

In this paper, we focus on the *Prediction of Energy Consumption* aspect of ontology reasoning *on the mobile front*, using statistical methods and execution data collected during experiments. We present an energy consumption prediction mechanism that predicts how much energy a new ontology will consume and whether this ontology may be processed within a predefined time, using previous reasoning results and specific metrics for ontologies [12].

We focus on the energy consumption aspect of semantic data processing, because, as [15,19] pointed out, *energy consumption* is a principal design concern for mobile platforms, rather than just a desirable attribute. Our investigation (see Sect. 5.2) shows that it *cannot* be assumed that the energy consumption of a reasoning process on a mobile device correlates with its time consumption. The main contributions of this paper can be summarised as follows:

1. We show that metrics of ontologies are very effective for accurate prediction (having R^2 between 0.8985 and 0.9859, and a maximum $RMSE$ of 10.86) of energy consumption of ontology reasoning on the Android platform, as validated by our comprehensive evaluation.
2. A comprehensive dataset ontologies in OWL 2 EL profile (a tractable profile in OWL 2) is made available for assessing and improving the performance of reasoning algorithms in terms of energy consumption.

2   Related Work and Background

Kleemann [14] discusses resource limitations in terms of computing power, memory and energy and presents a study for the development of a reasoner suitable for resource constrained environments such as mobile devices. Cerri et al. [3] propose the "knowledge in the cloud" approach, extending "data in the cloud" with support for handling semantic information, such as organising and finding it efficiently and providing reasoning and quality support. Despite presenting an efficient approach for harnessing the power of the cloud, this study is limited to cloud computing and doesn't take into account the capabilities of mobile devices. Rietveld and Schlobach [20] present a study about how the constraints in computing environments influence SW applications. In their study, they take *battery power* as one constraint, however, no deeper study is provided about the relationship between the energy need of the application and the structure of semantic data. Corradi et al. [4] propose an architecture and describe a prototype system for a mobile–cloud support of semantically enriched speech recognition in social care. In their approach, they move resource-demanding tasks that consume a high amount of energy on a mobile device to the cloud computing infrastructure. Hogan et al. [11] discuss scalability issues of reasoning and propose an approach for making the processing of a billion triples of open-domain Linked Data feasible. While they contribute to the feasibility of semantic data processing with regard to complexity, energy consumption of these approaches haven't been investigated.

Metrics of ontologies have been used for assessing the quality [2], complexity [26], cohesion [23], population task [16] and time consumption [13] of ontology reasoning. Hasan and Gandon [8] implemented a machine learning approach for predicting the performance of SPARQL queries using previous execution data. These investigations targeted server machines and efficient results were obtained. In our investigation, we are going to make use of metrics to deal with the energy bottleneck of mobile devices.

2.1 Electric Power and Energy Consumed

(Electric) Power (P) is the rate of doing work, measured in *watts*. The electric power in watts produced by an electric current I passing through an electric potential (voltage) difference of V is,

$$P = V * I, \quad (watts = volts * amperes). \tag{1}$$

Energy (E) is equal to the power *(P)* times the time period *(t)* is,

$$E_J = P_W * t_s, \quad (joules = watts * seconds). \tag{2}$$

We measure *Energy Consumed*, in *watt-seconds (Ws.)*, which is equal to joules.

2.2 Measuring Energy Consumption Programmatically

Various techniques [7,22] have been used to measure and predict energy consumption on mobile devices. For measuring energy consumed in reasoning, Patton and McGuinness propose a power benchmark [18] using a physical device setup that consists of a power monitor and a notebook computer to collect data. Because of the difficulty of implementing hardware-dependent (requiring any extra equipment not natively available on the mobile device) techniques to a solution that is desired to be applicable to all mobile devices, we searched for a software-based technique that can be programmatically implemented.

We adopted the Power Consumption Benchmark Framework [21] proposed by Valincius et al., which is hardware-independent and easily programmable[1].

---

[1] https://github.com/evalincius/PowerBenchMark, https://github.com/evalincius/HermitOWLAPI.

Energy is calculated using the properties in Android's `BatteryManager` class, `BATTERY PROPERTY CURRENT NOW` and `EXTRA VOLTAGE`, *Current* and *Voltage* are retrieved (see Eq. (1)). Valincius et al. measured total energy consumption with interval of 1 s (see Eq. (2)).

Observing and measuring overall energy consumption and battery drainage of the mobile device in 1 sec intervals poses a problem – a measurement with a resolution of 1 s shows the energy consumed by the mobile device during this second, independent of whether during such a measuring interval the processing of an ontology lasts 1 ms or 1000 ms. In order to get a more accurate measurement of how much energy the processing of a single ontology consumes, we, therefore, increased the precision of measurements by shortening the interval to 100 ms. In order to do this, we recorded the value of $\frac{V*I}{10}$ with intervals of 100 ms from the start of the data processing to the end. The cumulation of these values constitutes the ***total energy consumed***. With that, we reach a precision of 100 ms. With this method, an ontology processed in 1 ms is measured to consume the energy calculated for 1 interval of 100 ms. And, an ontology processed in 1000 ms is measured to consume the energy calculated for 10 intervals of 100 ms cumulatively.

2.3   Ontology Metrics

To be able to capture the complexity of ontologies thoroughly, we have adopted the set of 91 metrics proposed by Kang et al. [12,13]. These metrics include the number of general class inclusions, number of individuals, and the count of additional types of logical axioms (including reflexive properties, irreflexive properties and domain/range axioms). There are 24 ontology-level metrics to measure the overall size and complexity of an ontology, 15 class-level metrics to measure characteristics of OWL classes in an ontology, 22 anonymous class expression metrics to capture different types of class axioms, 30 property definition and axiom metrics to capture different types of property declarations and axioms. The complexity of all the metrics calculation algorithms is polynomial [13] in the size of the graph representation of the ontology.

2.4   Statistical Methods for Energy Prediction

We use a series of statistical methods for our energy prediction. *Regression Analysis* [9] is a statistical tool for the investigation of relationships between variables using some predictor variables and an output variable. We have built a regression model in which metrics are the predictor variables and the overall energy consumption of processing an ontology is the output variable. The output variable is denoted by Y, and the set of predictors by a vector X ($X_1, X_2, \ldots, X_n$, where n is the number of predictor variables). A regression model is formalized as $Y \approx f(X, \beta)$, where β is the *unknown parameters*, X is the *independent variables* and Y is the *dependent variable*. *Classification* identifies to which of a set of categories a new observation belongs, on the basis of a training set of data containing observations (or instances) whose category membership is known. In our classification model, metrics are the predictor variables and the output variable states either "able to process the ontology in 100 s " or "not able to process the ontology in 100 s". *Random Forest* [1] is an ensemble learning method for classification, regression and other tasks that operate by constructing a multitude of decision trees at training time and outputting the class that is the mode of the classes (classification) or mean prediction (regression) of the individual trees. In this paper, we train Random Forest-based classification models to predict whether an ontology can be processed within a predefined time and Random Forests-based regression models to predict energy consumption for an ontology using the power benchmark introduced above and syntactic metrics as features. A *Moving Average* is a calculation to analyse data points by creating series of averages of different subsets of the full data set in statistics. A moving average is commonly used with time series data to smooth out short-term fluctuations and highlight longer-term trends or cycles. We will use moving average to see whether there is a trend in the energy consumption while the battery level is decreasing from 100 % to 1 %. *The Coefficient of Variation* (CV), also known as "relative variability", is a standardized measure of dispersion of a probability distribution or frequency distribution. It is often expressed as a percentage, and is defined as the ratio of the standard deviation to the mean (or its absolute value). In our work, we will use CV for examining the variability of the energy measurement results from 100 % to 1 % battery level.

3   Our Approach

Making an energy prediction mechanism is a challenging task. Firstly, as detailed in Sect. 5.2, there *may not be always* a linear relation between time and energy consumption of ontology reasoning on every device. Hence, prediction models for reasoning time, and those done in a desktop/server environment (such as [12,13]), cannot be re-used as-is. Secondly, trying to model all the variables of real-world environments for energy consumption prediction, especially for a mobile device, is *very difficult*. Adapting to improvements in mobile environments is another complication for developing predication models, as changes to operating systems or in the utilisation of the CPU may render existing models obsolete.

In addressing these challenges, we developed prediction models by using a programmable (and hardware-independent) energy measurement tool ("Power Consumption Benchmark Framework" [21] proposed by Valincius et al.) and we use metrics that provide us with a numerical representation of particular properties of an ontology and use this information as our data source, in order to deal with the complexity of the semantic web and the uncertainty of internal and external influences on measuring energy consumption during ontology reasoning processes on mobile devices. We have chosen ontologies[2] in EL profile which were used at the ORE 2014 (The OWL Reasoner Evaluation Workshop 2014). We used two substantially different devices for our experiments. The measurement and prediction results will also provide an opportunity to

---
[2] https://zenodo.org/record/10791.

identify unforeseeable effects due to changes in the environment. For example, we observed that energy consumption may not always be correlated with reasoning time during our experiments with Machine2. Our case scenario (work-flow of the mechanism) has the following steps:

1. The mobile device asks the server whether it shall try to perform a reasoning task locally by sending the IRI of the ontology.
2. The server takes measures of the ontology according to the ontology metrics discussed in Sect. 2, and applies those measurement results to the classification model, which is trained to predict whether this ontology can be processed on this mobile device with a chosen reasoner using the metrics of the ontology, and will return either:
   (a) *Positive*: "this ontology can be processed within the predefined time limit (100 s in our experiment), and (using the regression model for predicting the energy consumption) it will consume this amount of energy"; or
   (b) *Negative*: "it cannot be processed in the predefined time".
3. If the mobile device gets a positive result from the server, it will then analyse the remaining energy available on the device and decide whether to proceed locally or in the cloud. If the mobile device receives a negative result, it will wait for the cloud to perform the reasoning task and return the result. For experimental purposes, we will process all the ontologies on the device. If the process exceeds the predefined time limit (100 s in our experiments), the process will be terminated.
4. The server will be informed whether the reasoning finishes with success within the set time limit.
5. The data collected about energy and time consumption will be used to improve our model to produce better prediction results.

Our approach regards mobile device as a "black-box" and accepts all its internal/external influences over data processing as the nature of it. We gather the execution data produced by the device and make inferences using this data with prediction data. We measure overall energy consumed (including loading/parsing of ontology, classification of the ontology TBox and executing the SPARQL query to retrieve the classification result) during the processing of ontologies in EL profile on each mobile device-reasoner pair. We describe experiments with a particular query answering task (explained in Sect. 5.1) by sending the same SPARQL query to the two reasoners we investigate, in order to get results for subsumption reasoning. Experiment results of these ontologies are used to make a prediction model and predict energy consumption of a new ontology on the same device-reasoner pair. This prediction mechanism is validated by the statistical results obtained from experiments.

In our experiments, we have implemented a separate model for each device-reasoner pair to see its validity in that scope. We are planning to work on one model for classification and one model for regression of all device-reasoner pairs as future work.

4   Experimental Setup

For calculating the error rate of our classification model, we divide wrong predictions by total predictions. For deciding whether our regression model is acceptable to describe the relation between the variables and the result obtained from the model, we have referred to R^2 and $RMSE$. The coefficient of determination (R^2) is a key output of regression analysis, which indicates the extent to which the dependent variable is predictable. An R^2 of 0.95 means that 95 percent of the variance in dependent variable can be predictable from independent variables. The Root Mean Squared Error $(RMSE)$ is simply the square root of the mean/average of the square of all of the error. $RMSE$ represents the sample standard deviation of the differences between predicted and observed values.

4.1   Data Collection

*Reasoners:*We have used HermiT [5], a DL reasoner, and TrOWL, an EL reasoner, (version 1.5, ported on Android) as testing reasoners. We implemented an android-ported version[3] of HermiT provided by Yus et al. [24], as the desktop version could not be directly supported by Android Runtime (ART).

Ontologies: The ORE2014 Reasoner Competition Dataset is chosen as the dataset for our experiments. The OWL 2 EL Profile [10] is chosen, because the computational complexities of ontology consistency, class expression subsumption, and instance checking are all *PTIME-Complete* [17] and both reasoners support it natively. From 16,555 ontologies, 8,805 ontologies, which are in EL profile, were filtered. The RDF/XML format was used in the experiments; however, the validity of our prediction mechanism does not depend on a particular input format. As we have built an extendible prediction mechanism, in the future, other formats can be introduced easily. Being aware of the RAM limitation (and reasoning limitation as a consequence), we ordered ontologies according to their file sizes and started with the ones with a smaller file size. Each device-reasoner pair is analysed with the ontologies it could process within mobile-specific and time limitations. There are 17 cases of exceptions throughout the experiment, which contain 14 "InconsistentOntology" exceptions, 2 "ConcurrentModification" exceptions and 1 error for receiving a voltage value of zero (0) from the operating system. Details of the exceptions are available online[4]. The "0 Voltage" problem occurred just once during the processing of ca. 8000 ontologies. Our point of view, therefore, is that such a low frequency of occurrence of these errors do not invalidate our results. These exceptional 17 cases are excluded in the model generation.

Mobile Devices: We used two mobile devices (Machine1 and Machine2) that have substantially different hardware specifications[5]. Machine1 had the Android 5.1.1

---

[3] https://github.com/evalincius/Hermit_1.3.8_android.

[4] https://github.com/IsaGuclu/PredictionOfEnergy/blob/master/Exceptions.xlsx.

[5] Unofficial   comparison: http://www.phonearena.com/phones/compare/ Samsung-Galaxy-S6,Sony-Xperia-Z3-Compact/phones/8997,8744.

as the OS and Machine2 had Android 6.0.1 as the OS. To avoid the side-effects of other services and processes, we uninstalled apps that could be uninstalled, closed all services and GSM connection and opened the Wi-Fi connection in all experiments to enable TBox retrieval from the internet if needed. For avoiding side-effects of the sensors, we closed location services and kept the device in a fixed place to avoid triggering sensors, e.g., accelerometer, gyro, proximity, compass, barometer, etc. We closed all sort of energy saving utilities in the settings of the machine to have near standard conditions in experiments. Ontologies are run on the same machine sequentially.

Data Preprocessing: Before training every model, to avoid misleading consequences, predictor metrics with zero standard deviation are discarded.

In the experiments with the TrOWL EL Reasoner, 61 of the metrics have been chosen for training the classification model and 60 of the metrics for the regression model. In the experiments with the HermiT Reasoner, 58 of the metrics have been chosen for training the classification model and 57 of the metrics for the regression model.

Prediction Model Construction: For the 1st prediction ("Will this ontology be reasoned in 100 s on this device-reasoner?"), a random forests based classification model is implemented. For the 2nd prediction ("How much energy will this ontology consume on this device-reasoner?"), a random forests based regression model is implemented. Standard 10-fold cross-validation is performed to ensure the generalizability of models.

5 Results and Evaluation

Before starting experiments, we had questions about how accurately we could collect energy consumption data from a mobile device by taking the measures explained above to eliminate the side-effects of mobile devices. The voltage provided by the battery continuously decreases during each reasoning activity. Heating may have adverse effect on computations as the CPU may slow itself when it reaches some threshold. As the Wi-Fi connection is open throughout the reasoning process, there would be some effects of OS-based or manufacturer specific apps on measurements. If measurement results change through the battery level from 100 % to 1 %, this will make a generalizable approach impossible. This made us investigate the *standard error of the mean caused by these (and those that we may not foresee) side-effects in our experiments.*

Experiments were started with fully charged Machine1 and Machine2. We repeatedly reasoned over the same ontology[6] using the TrOWL EL reasoner until the battery was completely drained. In this experiment, we made the following observations. In *Machine1* (Fig. 1), the average energy consumption for an ontology reasoning task is 151.16 Ws., with a standard deviation of 5.91 Ws. The average duration of the reasoning is 74.06 s and standard deviation is 1.73 s.

---

[6] https://github.com/IsaGuclu/PredictionOfEnergy/blob/master/
From100To1PercentBatteryLevel/approximated_00518.owl_RDFXML.owl.

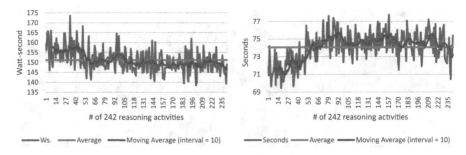

Fig. 1. Machine1 (Android 5.1.1) - Energy/Time consumption with battery level from 100% to 1%. *This figure illustrates (1) the energy/time consumed in experiments, (2) the average energy/time consumption of all experiments, (3) moving average of energy/time consumption with interval of 10.*

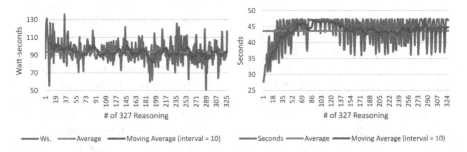

Fig. 2. Machine2 (Android 6.0.1) - Energy/Time consumption with battery level from 100% to 1%. *This figure illustrates (1) the energy/time consumed in experiments, (2) the average energy/time consumption of all experiments, (3) moving average of energy/time consumption with interval of 10.*

We found 3.91% as the CV (standard deviation of energy consumed divided by the average of energy consumed) of energy measurement for this *machine-reasoner pair*. To see whether this result is generalizable, we made the same experiment with a substantially different machine (i.e., Machine2). In *Machine2* (Fig. 2), the average energy consumption of the ontology is 93.76 Ws., with a standard deviation of 11.92 Ws. The average duration of the reasoning is 43.64 s and the standard deviation is 4.34 s. We found 12.71% as the CV of energy measurement for this *machine-reasoner pair*. This result made us search for the reason of such a difference. One of the biggest differences between the two machines is that Machine2 has a CPU which has one 2.1 GHz. quad-core processor and one 1.5 GHz. quad-core processor, but Machine1 has one 2.5 GHz. quad-core processor. To see what kind of a behaviour does the CPU have during our experiments, we used Usemon(CPU Usage Monitor)[7] and three sample execution of Machine2 is illustrated with Fig. 3. Figure 3 shows observations made on Machine2, where different cores, which have different clock-pulses, are used

---

[7] https://play.google.com/store/apps/details?id=com.iattilagy.usemon.

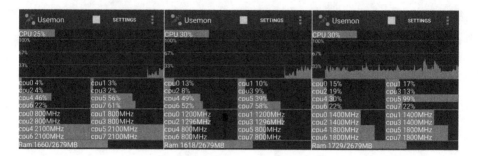

Fig. 3. 3 sample CPU utilization graph of Machine2 during reasoning activities. *This figure illustrates 3 sample attitude of Machine2 during processing ontologies.*

during execution. This feature of Machine2 adds one new dimension for predicting. In the first execution, the faster core (2.1 Ghz) makes the processing and results in a shorter time than the average. In the second execution, the slower core (1.5 Ghz) processes the ontology and results in a longer time than the average. In the third execution, faster core executes the reasoning at a slower speed while the other core is used by other processes. This execution finishes in a longer time than the first execution, but in a shorter time than the second execution. This changeability of cores makes the processing longer or shorter. The OS may decide to use the faster or slower core according to its own decision parameters. And this decision will affect time/energy consumption. We accept this internal effect as a nature of this "device" and continue.

To see whether there is a trend in energy consumption of the battery in relation with the remaining battery level, we implemented moving average over the energy consumption with interval of 10 executions, which is illustrated in Figs. 1 and 2. In Machine1, we see that there is a trend of consuming less energy especially when the battery level is less than 50 %. We accept that there is slight trend (probability) when the battery level is low, and it may result in lower energy consumption with this device-reasoner pair. We searched for whether there is the same trend in Machine2 parallel to Machine1. Making the same experiment using the Machine2, we could not find a concrete trend parallel to Machine1. In our work, we assume that our power benchmark will measure the energy, regardless of the battery level, *within the error rate defined.* Seeing this difference in the results of two different machines, we conclude that *it is very difficult to make a generalizable model that can be applied to all devices.* Thus, regarding "each device" as a black-box in analysing would be more practical.

Machine2 has given us an opportunity to see whether there is a "linear relation" between energy and time, as its energy and time consumption results are varied in time and energy dimension. We ordered execution results of Machine2 according to time consumption and divided into 5 groups as illustrated in Table 1.

According to Table 1, in the 1st group (66 executions with *lowest* time), average time consumption of group is 16.60 % less than average time consumption of all executions. But, energy consumption is not less than general, 1.24 % more than

Table 1. Energy-Time Consumption Relation of Machine2. *Distribution of energy and time consumption of reasoning same ontology from 100 % to 1 % battery level is illustrated.* 1st *column shows average (Avg.) time consumption of group.* 3rd *column shows Avg. energy consumption.*

Group	Avg. sec. of the group	Avg. sec. of all	Avg. Ws. of group	Avg. Ws. of all
1	36.397	43.644	94.951	93.785
2	42.124	43.644	90.737	93.785
3	45.446	43.644	94.043	93.785
4	47.035	43.644	93.754	93.785
5	47.273	43.644	95.258	93.785

general. In the 5th group (66 executions with *highest* time), average time consumption of group is 8.32 % more than average time. Whereas, average energy consumption of 5th group is 1.57 % more than the general. Observing the 1st and 5th group has a difference of 24.92 % average time consumption, but 0.33 % of difference in average energy consumption, we conclude that *we could not find a linear relation between time and energy consumption in this device-reasoner pair.*

Hardware doesn't influence the validity of the mechanism but shows varied results which makes us observe the effects of this hardware. For example, we reached the observation that energy consumption may not always be correlated with time consumption (as in Machine2) with help of this model while questioning why there was a higher variance in predictions of Machine2.

5.1 Experiments

Experiment results and source codes are accessible[8]. A re-run requires the preparation of an application development environment, the recompilation of the code and, finally, the generation of predictions in R. The reasoners TrOWL and HermiT are not part of our contribution, we therefore provide the scripts for running the experiments only. While working with TrOWL on Machine1, we observed that ontologies with the file size (in OWL Functional syntax) between 3000 KB and 3999 KB, 29 of 223 (13 %) could be processed within 100 s. Between 4000 KB and 4999 KB, it was about 2.99 % (5 of 167). Seeing this result, we limited our work for TrOWL within the dataset with the file size between 10 KB and 4999 KB (8281 ontologies). While working with HermiT on Machine1, we observed that ontologies with the file size between 500 KB and 599 KB, about 12.01 % (15 of 124) could be processed within 100 s. Seeing this result, we limited our work for HermiT within the dataset with the file size between 10 KB and 599 KB (6487 ontologies). We regard mobile devices as a black-box and do not search for the reasons of the peaks in energy consumption as in Figs. 1 and 2, whether it is because of OS services or manufacturer specific apps or anything we may not

---

[8] https://github.com/IsaGuclu/PredictionOfEnergy.

foresee, as this is the nature of mobile devices to run with this kind of internal (or external) influences. We preferred this overall approach as we are focusing on the energy consumption of the reasoning activity from a holistic perspective. We will not compare lower levels of reasoners but energy consumption in total. Reasoning experiment of one ontology is in this order:

1. The Counter starts calculating time and the energy. The Counter gets the average voltage and current from the OS, measuring the energy consumed in intervals of 100 ms.
2. The reasoner is called to load the ontology and the following query is sent:

```
prefix rdfs:<http://www.w3.org/2000/01/rdf-schema#>
select * where {?X rdfs:subClassOf ?Y}
```

   We describe experiments with a particular query answering task by sending the same SPARQL query to the two reasoners we investigate, in order to get results for subsumption reasoning.
3. When the request from the reasoner is provided with success and the query result is parsed, the Counter is stopped.

5.2   Results

After training the classification model with the data provided by the previous executions, we predicted whether a new ontology can be processed within the predefined time (100 s) or not, applying 10-fold cross validation. The results are illustrated in Table 2. This table illustrates successful and wrong predictions of the mechanism. "Positive" denotes reasoning CAN be made on mobile device. "Negative" denotes reasoning CANNOT be made on mobile device. For example, in a "Successful-Positive Prediction", it is predicted that reasoning can be accomplished on mobile device and it is observed so. In a "Wrong-Negative Prediction", it is predicted that reasoning cannot be accomplished on mobile device, but just opposite is observed, it could be processed on mobile device. As shown in Table 2, with TrOWL, the error rate of the 1st prediction in Machine1 is 0.52 % and in Machine2 is 1.51 %. With HermiT, the error rate of the 1st prediction in Machine1 is 0.76 % and in Machine2 is 1.86 %. Working on the ontologies which resulted in wrong predictions, deeper analysis can be made about the energy prediction, but we leave this analysis as a future work. After 1st prediction, we focussed on the prediction of the energy consumption. We trained our regression

Table 2. Classification model assessment.

	Machine1				Machine2			
	Successful prediction		Wrong prediction		Successful prediction		Wrong prediction	
	Positive	*Negative*	*Positive*	*Negative*	*Positive*	*Negative*	*Positive*	*Negative*
HermiT	90.38 %	8.87 %	0.48 %	0.28 %	92.55 %	6.59 %	0.52 %	0.34 %
TrOWL	91.76 %	7.72 %	0.36 %	0.16 %	95.40 %	3.08 %	0.72 %	0.79 %

Table 3. Regression model assessment

	Machine1		Machine2	
	$R^2(\%)$	$RMSE$	$R^2(\%)$	$RMSE$
HermiT	94.05	5.43	89.85	10.35
TrOWL	98.59	4.62	95.64	10.86

Table 4. Percentage of error rates according to actual energy consumption

	Machine1		Machine2	
Group	Hermit	TrOWL	Hermit	TrOWL
1 (Up to 1 Ws.)	20.21 %	18.88 %	43.50 %	50.20 %
2 (1–5 Ws.)	24.90 %	13.88 %	56.33 %	47.38 %
3 (5–10 Ws.)	25.51 %	16.35 %	58.88 %	33.93 %
4 (10–50 Ws.)	21.48 %	15.66 %	40.59 %	28.40 %
5 (50– Ws.)	14.80 %	5.43 %	21.11 %	14.06 %
General	21.46 %	14.11 %	45.12 %	40.00 %

model with the data provided by the previous executions and predicted how much energy will a new ontology consume, applying 10-fold cross validation. The results are illustrated in Table 3.

In Table 3, R^2 and $RMSE$ values as obtained from the prediction models are shown. Making more observations with different device-reasoner pairs will enhance the precision of the model, which we plan to do in future work.

To see the percentage of this error in prediction according to the amount of actual energy consumptions, we have grouped ontologies according to actual energy consumptions and obtained average percentage of error in prediction according to amount of actual energy consumption, as illustrated in Table 4.

From Table 4, we make the following observations. Machine1, which produces less varied energy consumption results, has less error rate in all groups of actual energy consumption. Whereas, Machine2, which produces more varied energy consumption results, has more error rate in all groups of actual energy consumption.

When the variation of energy consumption of the device-reasoner is lower, percentage of error rate is lower too. This encourages us to obtain more accurate execution data for training our model. Because, the more we can standardize our results for training, the more precise our prediction will be.

The random forests based regression model makes predictions in a very balanced way. As our energy measurement interval is 100 ms, we were expecting that there would be high percentages of error in predictions of 1st group of which the reasoning finishes less than a second. We find the difference of error rate between the 1st group and general acceptable.

These energy prediction results are obtained after predicting whether an ontology can be processed within 100 s with an accuracy of over 98 %. When energy consumption is very small, the energy prediction model can predict with an accuracy of nearly 80 % in Machine1 and nearly 50 % in Machine2. With the increase in reasoning time, the accuracy in Machine1 reaches about 90 % and the accuracy in Machine2 reaches about 80 %.

From all the experiments we have done, we are concluding that:

1. Treating the device (including OS, manufacturer specific apps and hardware specifications, etc.) as a black-box with the reasoner, we obtained affirmative results, indicating that the classification and regression models generated with this approach show a good measure for validity to describe the relation between energy consumption and structure of the ontology.
2. The classification models (which predict whether the ontology will be processed in the predefined time (100 s)) achieve very low error rates. It validates the feasibility and practicality of our approach, as it can be applied to minimize the risk of Out of memory (OOM) exceptions and general uncertainty about whether an ontology can be processed on a mobile device.
3. Using structure of the ontology (metrics) and previous ontology reasoning energy consumption data, actual energy consumption of a new ontology can be predicted with high accuracy. When the execution time of the ontology increases and standardized training data can be supplied, this accuracy reach 94.57 % as in Machine1 with TrOWL reasoner.
4. Patton and McGuinness had hypothesized that the amount of energy used for reasoning would be linearly related to the amount of time required to perform the reasoning, in their power consumption benchmark [18] for reasoners over mobile devices. Seeing experiment results with Machine2 about energy–time relation, we observe that energy consumption is not always parallel to the time consumption and this hypothesis is limited to old CPUs with standard speed. As the device contains many internal (OS policies-services-apps, manufacturer specific apps, hardware specifications, etc.) and external (movement of the user, bandwidth change, temperature, etc.) influences, instead of trying to sort out every variable and their weight in the energy consumption, using a holistic approach and collecting more and more data will be a more effective way for obtaining a more precise energy prediction mechanism. We conclude that the relation between time and energy is *changeable* according to hardware and software specifications of the device and this necessitates making separate prediction mechanisms for *time* and *energy* consumption.

6   Conclusion

Mobile devices, such as smartphones and tablets, have markedly different performance characteristics and requirements, most prominently limited energy, which poses a significant challenge for deploying computation-intensive tasks, such as ontology reasoning on mobile devices. In this paper, we developed statistical methods that predict energy consumption of ontology reasoning on various mobile devices, using different reasoners and ontologies in the OWL 2 EL profile. Our main contributions include the following. Firstly, high prediction accuracy is achieved for our random forest-based regression models with R^2 of 90 % or higher. It is also observed that the prediction error rate is the lowest for ontologies with the highest actual energy consumption, showing that our prediction models are accurate *when it matters*. Our approach is hardware independent, i.e. hardware specification is not used as a parameter of our prediction model, thus our approach can be applied to devices other than the two that we tested. Secondly, we observe that a linear relation between time and energy consumption on a mobile device is not a valid assumption, especially with new hardware (CPU's containing cores with different speed) and software (multi-threading) improvements. Thirdly, the comprehensive dataset used in our evaluation has been made available to allow for reproducibility and encourage further investigation.

Our plan for future work is to improve our approach and make it applicable to real-world scenarios. First, we will extend our experiments with more devices and combine all models of different device-reasoner pairs into one comprehensive, general model. Second, we are planning to implement this approach in the Android version of TrOWL reasoner and empowering this prediction mechanism by collecting data from devices using this implementation. Third, we will build an optimisation mechanism that will manage the integration of mobile-cloud using this approach with user preferences taken into account.

Acknowledgments. This work is partially funded by the EU IAPP K-Drive project (286348). We would like to thanks Prof. Breiman for making the Fortran code of his Random Forests algorithm available and thank Edgaras Valincius for providing us with his codes in energy measurement over mobile devices.

References

1. Breiman, L.: Random forests. Mach. Learn. **45**, 5–32 (2001)
2. Burton-Jones, A., Storey, V.C., Sugumaran, V., Ahluwalia, P.: A semiotic metrics suite for assessing the quality of ontologies. Data Knowl. Eng. **55**, 84–102 (2005)
3. Cerri, D., et al.: Towards knowledge in the cloud. In: Meersman, R., Tari, Z., Herrero, P. (eds.) OTM-WS 2008. LNCS, vol. 5333, pp. 986–995. Springer, Heidelberg (2008)
4. Corradi, A., Destro, M., Foschini, L., Kotoulas, S., Lopez, V., Montanari, R.: Mobile cloud support for semantic-enriched speech recognition in social care. IEEE Trans. Cloud Comput. **PP**(99), 1 (2016). doi:10.1109/TCC.2016.2570757
5. Glimm, B., Horrocks, I., Motik, B., Stoilos, G., Wang, Z.: HermiT: an OWL 2 reasoner. J. Autom. Reasoning **53**, 245–269 (2014)
6. Groth, P.: The rise of the verb. In: SW2022 (2012)
7. Gurun, S., Krintz, C., Wolski, R.: NWSLite : a light-weight prediction utility for mobile devices. In: MOBISYS (2004)
8. Hasan, R., Gandon, F.L.: A machine learning approach to SPARQL query performance prediction. In: WEBI (2014)
9. Hastie, T., Tibshirani, R., Friedman, J.: The Elements of Statistical Learning. Springer, New York (2009)

10. Hitzler, P., Krötzsch, M., Parsia, B., Patel-Schneider, P.F., Rudolph, S.: OWL 2 web ontology language primer. W3C Recommendation **27**(1), 123 (2009)
11. Hogan, A., Pan, J.Z., Polleres, A., Ren, Y.: Scalable OWL 2 reasoning for linked data. In: Polleres, A., d'Amato, C., Arenas, M., Handschuh, S., Kroner, P., Ossowski, S., Patel-Schneider, P. (eds.) Reasoning Web 2011. LNCS, vol. 6848, pp. 250–325. Springer, Heidelberg (2011)
12. Kang, Y.-B., Li, Y.-F., Krishnaswamy, S.: Predicting reasoning performance using ontology metrics. In: Cudré-Mauroux, P., et al. (eds.) ISWC 2012, Part I. LNCS, vol. 7649, pp. 198–214. Springer, Heidelberg (2012)
13. Kang, Y.-B., Pan, J.Z., Krishnaswamy, S., Sawangphol, W., Li, Y.-F.: How long will it take? Accurate prediction of ontology reasoning performance. In: AAAI (2014)
14. Kleemann, T.: Towards mobile reasoning. In: DLOG (2006)
15. Li, Y.-F., Pan, J.Z., Krishnaswamy, S., Hauswirth, M., Nguyen, H.H.: The ubiquitous semantic web: promises, progress and challenges. Int. J. Semant. Web Inf. Syst. **10**, 1–16 (2014)
16. Maynard, D., Peters, W., Li, Y.: Metrics for evaluation of ontology-based information extraction. In: WWW Workshop on Evaluation of Ontologies for the Web (2006)
17. Motik, B., Cuenca Grau, B., Horrocks, I., Wu, Z., Fokoue, A., Lutz, C., Hitzler, P., Krötzsch, M., Parsia, B., Patel-Schneider, P., et al.: OWL 2 web ontology language: profiles. In: W3C Recommendation, 27 October 2009 (2012)
18. Patton, E.W., McGuinness, D.L.: A power consumption benchmark for reasoners on mobile devices. In: Mika, P., et al. (eds.) ISWC 2014, Part I. LNCS, vol. 8796, pp. 409–424. Springer, Heidelberg (2014)
19. Piccolo, L.S.G., Baranauskas, M.C.C., Fernández, M., Alani, H., Liddo, A.D.: Energy consumption awareness in the workplace,: technical artefacts and practices. In: IHC (2014)
20. Rietveld, L., Schlobach, S.: Semantic web in a constrained environment. In: Proceedings of the Downscaling the Semantic Web Workshop (ESWC) (2012)
21. Valincius, E., Nguyen, H.H., Pan, J.Z.: A power consumption benchmark framework for ontology reasoning on android devices. In: ORE (2015)
22. Wen, Y., Wolski, R., Krintz, C.: Online prediction of battery lifetime for embedded and mobile devices. In: Falsafi, B., VijayKumar, T.N. (eds.) PACS 2003. LNCS, vol. 3164, pp. 57–72. Springer, Heidelberg (2005)
23. Yao, H., Orme, A.M., Etzkorn, L.: Cohesion metrics for ontology design and application. J. Comput. Sci. **1**, 107–113 (2005)
24. Yus, R., Bobed, C., Esteban, G., Bobillo, F., Mena, E.: Android goes semantic: DL reasoners on smartphones. In: ORE (2013)
25. Yus, R., Pappachan, P.: Are apps going semantic? A systematic review of semantic mobile applications. In: SEMWEB (2015)
26. Zhang, H., Li, Y.-F., Tan, H.B.K.: Measuring design complexity of semantic web ontologies. J. Syst. Softw. **83**, 803–814 (2010)

Walking Without a Map: Ranking-Based Traversal for Querying Linked Data

Olaf Hartig[1(✉)] and M. Tamer Özsu[2]

[1] Department of Computer and Information Science (IDA), Linköping University,
Linköping, Sweden
olaf.hartig@liu.se
[2] Cheriton School of Computer Science, University of Waterloo, Waterloo, Canada
tamer.ozsu@uwaterloo.ca

Abstract. The traversal-based approach to execute queries over Linked Data on the WWW fetches data by traversing data links and, thus, is able to make use of up-to-date data from initially unknown data sources. While the downside of this approach is the delay before the query engine completes a query execution, user perceived response time may be improved significantly by returning as many elements of the result set as soon as possible. To this end, the query engine requires a traversal strategy that enables the engine to fetch result-relevant data as early as possible. The challenge for such a strategy is that the query engine does not know a priori which of the data sources discovered during the query execution will contain result-relevant data. In this paper, we investigate 14 different approaches to rank traversal steps and achieve a variety of traversal strategies. We experimentally study their impact on response times and compare them to a baseline that resembles a breadth-first traversal. While our experiments show that some of the approaches can achieve noteworthy improvements over the baseline in a significant number of cases, we also observe that for every approach, there is a non-negligible chance to achieve response times that are worse than the baseline.

1 Introduction

The availability of large amounts of Linked Data on the World Wide Web (WWW) presents an exciting opportunity for building applications that use the data and its cross-dataset connections in innovative ways. This possibility has spawned research interest in approaches to enable such applications to query Linked Data [4,7]. A well-understood approach to this end is to populate a centralized repository of Linked Data copied from the WWW. By using such a repository it is possible to provide almost instant query results. This capability comes at the cost of setting up and maintaining the centralized repository. Further limitations of this approach are that query results may not reflect the most recent status of the copied data, new data and data sources cannot be exploited, and legal issues may prevent storing a copy of some of the data in the repository.

© Springer International Publishing AG 2016
P. Groth et al. (Eds.): ISWC 2016, Part I, LNCS 9981, pp. 305–324, 2016.
DOI: 10.1007/978-3-319-46523-4_19

To address these limitations a number of works adopt an alternative view on querying Linked Data: The idea is to conceive the Web of Linked Data itself as a distributed database in which URI lookups are used to access data sources at query execution time [8,11–13,16,17]. A particularly interesting approach is *traversal-based query execution* which intertwines the query execution process with a traversal of data links [8,11–13]. This approach can discover initially unknown data and data sources on the fly, and it can be used to start querying right away (without first having to populate a repository of data). An inherent downside, however, is the delay before the data retrieval process terminates and a complete query result can be returned to the user. Nonetheless, users may want to start receiving elements of the query result set as soon as possible. The following example shows that the user experience may be improved significantly by query optimization approaches that aim to reduce the *response times* of query executions, that is, the times required to find a particular number of result elements (as opposed to the overall time required to complete the query execution).

Example 1. Consider the following SPARQL query from the FedBench benchmark [15].

```
SELECT * WHERE { ?person nyt:latest_use ?mentionInNYT . ?person owl:sameAs ?chancellor .
            ?chancellor dct:subject <http://dbpedia.org/resource/Category:Chancellors_of_Germany> }
```

We used the URI at the end of this query as a starting point for a traversal-based execution of the query over the WWW (under c_{Match}-bag-semantics; cf. Sect. 2). For this execution we used a randomized traversal strategy; that is, we prioritized the retrieval of Linked Data by using randomly chosen lookup priorities for all URIs that are discovered and need to be looked up during the execution process. By repeating this query execution five times, for each of these executions, we measured an overall execution time of 8.9 min (because all five executions eventually retrieve the same set of documents, which always requires almost the same amount of time). However, due to the random prioritization, the documents always arrive in a completely different order, which affects the time until all the data has been retrieved that is needed to compute any particular result element: In the best of the five cases, a first element of the result set can be returned after 9 s, that is, 1.7 % of the overall query execution time; on average however the five executions require 3.1 min (34.8 %) to return a first result element, and the standard deviation of this average is as high as 1.3 min (14.6 %).

The example illustrates that there exists a huge potential for optimizing the response times of traversal-based query executions (i.e., returning result elements as early as possible) and that these response times may vary significantly depending on the strategy chosen to traverse the queried Web of Linked Data. A desirable traversal strategy is one that prioritizes the lookup of URIs such that it discovers as early as possible the result-relevant documents (whose data can be used to compute at least one of the elements of the query result). Then, as soon as these documents arrive, a pipelined result construction process can compute and output result elements. The primary challenge in this context is that the URIs to be looked up are discovered only recursively, and we cannot assume up-to-date a priori information about what URIs will be discovered and which of the discovered URIs allow us to retrieve documents that are result-relevant. Given these issues, an investigation of possible approaches to prioritize URI lookups and their impact on the response times is an open research problem that is important for improving the user experience of applications that can be built on the traversal-based paradigm.

In this paper, we focus on this problem. To this end, we identify a diverse set of *14 different approaches to prioritize URI lookups* during a traversal-based query execution. None of these approaches assumes any a priori information about the queried Web. Then, as our main contribution, we conduct an experimental analysis to study the effects that each of these prioritization approaches can have on the response times of traversal-based query executions. This analysis is based on a comprehensive set of structurally diverse test Webs. We show that some of the approaches can achieve significant improvements over a breadth-first search baseline approach that looks up URIs on a first-come, first-served basis. However, we also observe that, even for the most promising ones of our approaches, there is a non-negligible number of cases in which they perform worse than the baseline. Before describing the approaches (Sect. 4) and discussing our experiments in more detail (Sects. 5 and 6), we briefly review the state of the art in querying Linked Data on the Web and elaborate more on the focus of our work.

2 Linked Data Query Processing

The prevalent query language used in existing work on querying Linked Data on the WWW is the basic fragment of SPARQL. Approaches to evaluate such basic graph patterns (BGPs) over Linked Data can be classified into traversal-based, index-based, and hybrid [7,11]. All these approaches compute a query result based on Linked Data that they retrieve *by looking up URIs during the query execution process*. Their strategy to select these URIs is where the approaches differ.

Traversal-based approaches perform a recursive URI lookup process during which they incrementally discover further URIs that can be selected for lookup. Existing work in this context focuses on techniques to implement such a traversal-based query execution [8,12,13]; additionally, as a well-defined foundation for these approaches, we have proposed a family of *reachability-based query semantics* for SPARQL that restrict the scope of any query to a query-specific *reachable subweb* [6]. To this end, the specification of any query in this context includes a set of seed URIs (in addition to the query pattern). Then, a document in the queried Web of Linked Data is defined to be *reachable* (and, thus, part of the reachable subweb) if it can be retrieved by looking up either a seed URI—in which case we call it a *seed document*—or a URI u such that (i) u occurs in an RDF triple of some other reachable document and (ii) u meets a particular *reachability condition* specified by the given reachability-based query semantics.

For instance, such a condition may require that the triple in which the URI is found is a matching triple for any of the triple patterns in the given query. Our earlier work formalizes this condition in a reachability-based query semantics that we call c_{Match}-*semantics* [6].

Index-based approaches use a pre-populated index whose entries are URIs that can be looked up to retrieve Linked Data. Then, for any given query, such an approach uses its index to select a set of URIs whose lookup will result in retrieving query-relevant data. By relying on their index, index-based approaches fail to exploit query-relevant data added to indexed documents after building the index, and they are unaware of new documents. Existing work on such approaches focuses on different ways to construct the corresponding index [17], on techniques to leverage such an index [17], and on ranking functions that prioritize the lookup of the selected URIs in order to reduce response times [11,17]. The latter aims to achieve the same objective as our work in this paper. However, the ranking functions proposed for index-based approaches rely on statistical metadata that has been added to the index. For our work on traversal-based query executions we do not assume an a priori availability of any metadata whatsoever.

The only ***hybrid approach*** that has been proposed in the literature so far exploits an index to populate a prioritized list of seed URIs; additional URIs discovered during a subsequent traversal-based execution are then integrated into the list [11]. To this end, discovered URIs that are in the index (but have not been selected initially) are prioritized based on a ranking function that uses information from the index. For any URI for which no index entry exists, the approach simply uses as priority the number of retrieved Linked Data documents that mention the URI in some of their RDF triples (i.e., the number of known incoming links). One of the prioritization approaches that we analyze in this paper resembles the latter strategy (cf. Sect. 4.1).

3   Focus of Our Work

As discussed in the previous section, the prioritization of URI lookups is an idea that has been shown to be suitable to improve the response time of queries over the Web of Linked Data. However, the only systematic analyses of approaches that implement this idea focus on index-based query executions [11,17]. The approaches proposed in this context cannot be used for a traversal-based execution because they rely on statistical metadata that may be recorded when building an index but that is not a priori available to a (pure) traversal-based query execution system (which also rules out these approaches for non-indexed URIs in a hybrid system). Therefore, the overall goal of the work presented in this paper is to investigate URI prioritization approaches that can be used to reduce the response times of traversal-based query executions.

For this work we make minimal assumptions about how traversal-based query execution is implemented, which ensures independence of the peculiarities of any particular implementation techniques (such as those proposed in earlier work [8,12,13]). That is, we only make the general assumption that traversal-based query engines consist of a data retrieval component (DR-component) and a result construction component (RC-component), and these two components operate in parallel to execute a query as follows.

The DR-component receives Linked Data by looking up URIs. To this end, the component is equipped with a *lookup queue* that is initialized with the seed URIs of the given query. The component may use multiple *URI lookup threads*. Whenever such a thread is free, it obtains the next URI from the queue, looks up this URI on the Web, and scans the RDF triples that are contained in the document retrieved by the lookup. This scan has two goals: First, the triples may contain new URIs that can be scheduled for lookup. However, the lookup threads do not necessarily have to add all new URIs to the lookup queue. Instead, the DR-component can support an arbitrary reachability-based query semantics. That is, the component may schedule only those URIs for lookup that satisfy the reachability condition specified by the semantics. By doing so, the lookup threads incrementally discover (and retrieve) the specific reachable subweb that the given query semantics defines as the scope of the query. Hence, all triples scanned by the lookup threads—and only these—have to be considered to compute the sound and complete query result. Consequently, the second goal of scanning these triples is to identify triples that match a triple pattern in the given query. Any such matching triple is then sent to the RC-component, which starts processing them as soon as they arrive.

Regarding the RC-component we make only three assumptions: (i) it uses the incoming matching triples to compute the final query result, (ii) it processes intermediate results in a push-based manner, and (iii) as soon as an element of the final query result is ready, it is sent to the output. For techniques to implement such a push-based RC-component we refer to the literature [12,13] and to the extended version of this paper [10].

The whole query execution process continues until the DR-component has accessed all data from the query-specific reachable subweb and the RC-component has finished processing the resulting intermediate solutions. If the queried Web is distributed over a comparably slow network such as the Internet (as we assume in this paper), it is not surprising to observe that data retrieval is the dominating factor for query execution times. In fact, as we verify experimentally in the extended version of this paper, due to this dominance of data retrieval, the execution times of traversal-based query executions *over the WWW* are not affected at all by the order in which URIs are looked up [10]. For the same reason, however, the URI lookup order has a crucial impact on the times required to find a specific number of result elements (as demonstrated by Example 1). Therefore, when aiming to minimize such response times, a suitable approach to prioritize URI lookups is of critical importance. We study 14 candidates in this paper.

In this study we focus on conjunctive queries (represented by BGPs) under the bag version of the aforementioned c_{Match}-semantics . This semantics is the most prominent reachability-based query semantics supported by the traversal-based approaches studied in the literature [8,12,13]. While, in theory, there exist an infinite number of other reachability-based query semantics and our experiments can be repeated for any of them, we conjecture that the results will be similar to ours because none of the approaches studied in this paper makes use of anything specific to c_{Match}-semantics. Using the bag version of c_{Match}-semantics allows us to focus on a notion of the response time optimization problem that is isolated from the additional challenge of avoiding the discovery of duplicates (which is an additional aspect of response time optimization under set semantics and worth studying as an extension of our work).

As a final caveat before going into the details, we emphasize the following limitations of our study: We ignore factors that may impact the response times of traversal-based queries but that cannot be controlled by a system that executes such queries (e.g., varying latencies when accessing different Web servers). Moreover, we focus only on approaches that do *not* assume any a priori information about the queried Web of Linked Data. That is, the topology of the Web or statistics about the data therein is unknown at the beginning of any query execution. This focus also excludes approaches that aim to leverage such information collected during earlier query executions (of course, studying such approaches is an interesting direction for future work). Similarly, we ignore the possibility to cache documents for subsequent query executions. While caching can reduce the time to execute subsequent queries [5], this reduction comes at the cost of potentially outdated results. However, studying approaches to balance the performance vs. freshness trade-off in this context is another interesting direction for future work.

4   Approaches to Prioritize URI Lookups

A variety of approaches to prioritize URI lookups are possible. In this section, we identify different classes of such approaches. Figure 1 illustrates our taxonomy. All these approaches assume that the lookup queue of the DR-component is maintained as a priority queue. Priorities are denoted by numbers; the greater the number, the higher the priority. URIs that are queued with the same priority are handled in

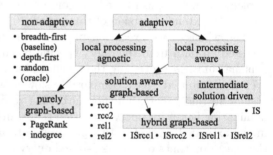

Fig. 1. Approaches to prioritize URI lookups

a first-come, first-served manner (after all higher priority URIs have been looked up).

A first class includes *non-adaptive approaches* that determine a *fixed* priority for each URI when the URI is added to the lookup queue. A trivial example is to treat all URIs equal, which resembles a breadth-first traversal. We consider this approach as our baseline. In the extended version of this paper we also discuss depth-first and random as alternative non-adaptive approaches (Example 1 uses the latter). These turn out to be unsuitable for reducing the response times of traversal-based query executions [10].

4.1 Purely Graph-Based Approaches

In contrast to non-adaptive approaches, *adaptive approaches* may reprioritize queued URIs. A first class of such approaches is based on the idea of applying a vertex scoring method to a directed graph that represents the topology of the queried Web as discovered during the data retrieval process. Each vertex in this graph corresponds either to a retrieved document or to a queued URI. Each directed edge between two document vertices represents a data link that is established by URIs that occur in some RDF triple in the source document and that turned out to resolve to the target document when looked up. Directed edges from a document vertex to a URI vertex represent data links to documents that are yet to be retrieved. Obviously, such a graph is an incomplete model of the topology of the queried Web. However, as a side-effect of the data retrieval process, the DR-component obtains increasingly more information about the topology and, thus, can augment its model continuously. That is, any URI vertex becomes a document vertex after the corresponding URI has been looked up. If such a lookup results in discovering new URIs for the lookup queue, new URI vertices and connecting edges can be added to the graph. Similarly, new edges can be added if a retrieved document mentions URIs that either are already queued for lookup or have already been looked up.

Given such a graph, it is possible to apply a vertex scoring method and use the score of each URI vertex as the priority of the corresponding URI in the lookup queue. Whenever the DR-component extends the graph after completing some URI lookup, the vertex scores can be recomputed, and the priorities can be adapted accordingly.

A multitude of different vertex scoring methods exist. We select PageRank and indegree-based scoring as two examples for our study. PageRank is a well-known method that uses an iterative algorithm to determine a notion of importance of vertices [14]. Indegree-based scoring is a less complex method that simply uses the number of incoming edges as the score of a vertex. Hereafter, we refer to the two resulting URI prioritization approaches as PageRankand indegree, respectively. We note that the latter approach is equivalent to the only existing proposal to prioritize URI lookups during traversal-based query executions [11]. However, its effectiveness has not been studied so far.

4.2 Solution-Aware Graph-Based Approaches

We now turn to *local processing aware approaches* that aim to leverage runtime information about the result construction process in the RC-component. To enable an implementation of these approaches, the traversal-based query execution engine must be extended with a *feedback channel* from the RC-component to the DR-component. Then, specific information required to prioritize URI lookups can be sent over this channel.

Given the possibility to obtain runtime information from the RC-component, we now can define graph-based URI prioritization approaches for which we use vertex scoring methods that leverage such runtime information . In this paper we focus on methods that are based on the number of solutions that retrieved documents have contributed to.

To enable the application of such methods, intermediate solutions must be augmented with provenance annotations. In particular, each intermediate solution must be annotated with a set of all documents that contributed a matching triple to the construction of that intermediate solution. To this end, before sending a matching triple to the RC-component, the DR-component augments this triple with metadata that identifies the source document of the triple. This document becomes the provenance of the initial intermediate solution that the RC-component generates from the matching triple. When two intermediate solutions are joined in the RC-component, the union of their provenance annotations becomes the provenance of the resulting intermediate solution. Then, whenever an intermediate solution has been completed into a solution that is ready to be sent to the output, the RC-component uses the feedback channel to send the provenance annotation of this solution to the DR-component. The DR-component uses these annotations to maintain a *result contribution counter (RCC)* for every document vertex in the Web graph model that the component builds incrementally as described in Sect. 4.1. This counter represents the number of solutions that the document represented by the vertex has contributed to so far, which may increase as the query execution progresses.

Given these counters, we define four vertex scoring functions that can be applied to the Web graph model. Informally, for each vertex $v \in V$ in such a graph $G = (V, E)$, the *rcc-1 score* of v, denoted by $\text{rccScore}_1(v)$, is the sum of the (current) RCCs of all document vertices in the in-neighborhood of v; and the *rel–1 score* of v, denoted by $\text{relScore}_1(v)$, is the number of document vertices in the in-neighborhood of v whose RCC is greater than 1. Similarly, the *rcc-2 score* and *rel-2 score* of v, denoted by $\text{rccScore}_2(v)$ and $\text{relScore}_2(v)$, respectively, focus on the 2-step in-neighborhood. To define these scores formally, let $\text{in}_k(v)$ denote the set of vertices in the k-step in-neighborhood of v, and, if v is a document vertex, let $\text{rcc}(v)$ be its (current) RCC. Then, for each vertex $v \in V$ and $k \in \{1, 2\}$, the scoring functions are defined as follows:

$$\text{rccScore}_k(v) = \sum_{v' \in \text{in}_k(v)} \text{rcc}(v'), \quad \text{relScore}_k(v) = \left| \{v' \in \text{in}_k(v) \mid \text{rcc}(v') > 0\} \right|.$$

These vertex scoring functions can be used by a graph-based approach to prioritize URI lookups (in the same manner as the PageRank and indegree approaches use the PageRank algorithm and indegree-based scoring, respectively). Hereafter, we refer to the four resulting URI prioritization approaches as rcc1, rcc2, rel1, and rel2, respectively.

4.3   Intermediate Solution Driven Approaches

An alternative class of local processing aware approaches use the aforementioned feedback channel to obtain information about all the intermediate solution mappings sent between operators in the RC-component. We focus on one such approach, denoted by IS, that assigns an initial priority of 0 to any new URI added to the lookup queue, and it reprioritizes queued URIs based on the following two assumptions:

A1: The greater the number of operators that have already processed a given solution mapping μ, the more likely it is that this intermediate solution μ can be completed into a solution μ' that covers the whole query and, thus, can be sent to the output.

A2: The documents that can be retrieved by looking up the URIs mentioned in a given (intermediate) solution mapping μ are the documents that are most likely to contain matching triples needed for completing μ into a solution.

Recall that the objective is to return solutions as early as possible. Hence, by assumption A2, it seems reasonable to increase the priority of a URI in the lookup queue if the URI is mentioned in an intermediate solution. Furthermore, by assumption A1, such an increase should be proportional to the number of operators that have already processed the intermediate solution. Then, to implement the IS approach, intermediate solutions do not only have to be sent between operators in the RC-component, but they also have to be sent over the feedback channel to the DR-component—after annotating them with the number of operators that contributed to their construction. Given an intermediate solution mapping μ with such a number, say $opcnt$, the IS approach iterates over all variables that are bound by μ. For each such variable $?v \in \text{vars}(\mu)$, if μ binds the variable to a URI (i.e., $\mu(?v)$ is a URI) and this URI is queued for lookup with a priority value that is smaller than $opcnt$, then IS increases the priority of this URI to $opcnt$.

4.4 Hybrid Local Processing Aware Approaches

The idea of the IS approach (cf. Sect. 4.3) can be combined with the solution-aware graph-based approaches (cf. Sect. 4.2). To this end, the DR-component has to obtain via the feedback channel both the provenance annotation of each solution and all intermediate solution mappings. Based on the former, the DR-component increases the RCCs of document vertices in the Web graph model (as described in Sect. 4.2). The intermediate solutions are used to maintain an additional number for every URI that is queued for lookup; this number represents the maximum of the $opcnt$ values of all the intermediate solutions that bind some variable to the URI. Hence, initially (i.e., when the URI is added to the lookup queue) this number is 0, and it may increase as the DR-component gets to see more and more intermediate solutions via the feedback channel. Observe that this number is always equal to the lookup priority that the IS approach would ascribe to the URI. Therefore, we call this number the *IS-score* of the URI.

Given such IS-scores, we consider four different approaches to prioritize URI lookups, each of which uses one of the RCC-based vertex scoring functions introduced in Sect. 4.2. We call these approaches isrcc1, isrcc2, isrel1, and isrel2 (the name indicates the vertex scoring function used). Each of them determines the priority of a queued URI by multiplying the current IS-score of the URI by the current vertex score that their vertex scoring function returns for the corresponding URI vertex. Whenever the DR-component increases the IS-score of a URI, or the vertex score of the corresponding URI vertex changes , then the lookup priority of that URI is adapted accordingly.

4.5 Oracle Approach

We also want to gain an understanding of what response times a traversal-based query execution system could achieve if it had complete information of the queried Web (which is impossible in practice). To this end, we developed another approach assuming an *oracle* that, for each reachable document, knows (i) the URIs leading to the document and (ii) the final RCC of the document (i.e., the number of solutions of the complete query result that are based on matching triples from the document). Then, this oracle approach uses as priority of a URI lookup the final RCC of the document that will be retrieved by this lookup. As a consequence, retrieving documents with a greater final RCC has a higher priority. Clearly, without a priori information about the queried Web, a traversal-based system can determine such final RCCs only after retrieving all reachable documents —which is when it is too late to start prioritizing URI lookups. Hence, the oracle approach cannot be used in practice. However, for our experiments we performed a baseline-based "dry run" of our test queries and collected the information necessary to determine the RCCs that are required to execute the queries using the oracle approach.

5   Experimental Setup

In this section we specify the setup of our experiments. Although the execution of queries over Linked Data on the WWW is the main use case for the concepts in this paper, the WWW is not a controlled environment to run experiments on. For this reason, we set up a simulation environment consisting of two identical machines, each with an   Athlon 64 X2 dual core CPU, and 3.6 GB of main memory. Both machines use an Ubuntu 12.04 LTS operating system with Sun Java 1.6.0 and are connected via a fast university network. One machine runs a Tomcat server (7.0.26) with a Java servlet that can simulate different Webs of Linked Data (one at a time); the documents of these Webs are materialized on the machine's hard disk. The other machine executes queries over such a simulated Web by using an in-memory, Java implementation of a traversal-based query engine. To rule out any effects of parallelized URI lookups as a factor that may influence our measurements we set up the system to use a single lookup thread.

In the following, we specify the Webs of Linked Data simulated for our experiments, the corresponding test queries, and the metrics that we use. Software and data required for our experiments are available at http://squin.org/experiments/ISWC2016/.

5.1   Test Webs

The goal of our experiments is to investigate how the different URI prioritization approaches impact the response times of traversal-based query executions. This impact (as well as the chance to observe it) may be highly dependent on how the queried Web of Linked Data is structured and how data is distributed. Therefore, we generated multiple test Webs for our experiments. To be able to meaningfully compare measurements across our test Webs, we used the same base dataset for generating these Webs.

We selected as base dataset the set of RDF triples that the data generator of the Berlin SPARQL Benchmark (BSBM) suite [1] produces for a scaling factor of 200. This dataset, hereafter denoted by G_{base}, consists of 75,150 RDF triples and describes 7,329 entities, each of which is identified by a unique URI. Let U_{base} denote the set consisting of these 7,329 URIs. Hence, the subject of any triple $\langle s, p, o \rangle \in G_{base}$ is such a URI (i.e., $s \in U_{base}$), and the object o either is a literal or also a URI in U_{base}.

Every test Web that we generated from this base dataset consists of 7,329 documents, each of which is associated with a different URI in U_{base}. To distribute the triples of G_{base} over these documents, we partitioned G_{base} into 7,329 potentially overlapping subsets (one for each document). First, we always placed any base dataset triple whose object is a literal into the subset of the document for the subject of that triple. Next, for any of the other base dataset triples $\langle s, p, o \rangle \in G_{base}$ (whose object o is a URI in U_{base}), we considered three options: placing the triple (i) into both the documents for s and for o—which establishes a bidirectional data link between both documents, (ii) into the document for s only—which establishes a data link from that document to the document for o, or (iii) into the document for o only—which establishes a data link to the document for s. It is easy to see that choosing among these three options impacts the link structure of the resulting test Web (note that the choice may differ for each triple).

We exploited this property to systematically generate test Webs with different link structures. That is, we applied a random-based approach that, for every generated test Web, uses a particular pair of probabilities (ϕ_1, ϕ_2) as follows: For every base dataset triple $\langle s, p, o \rangle \in G_{base}$ with $o \in U_{base}$, we chose the first option with a probability of ϕ_1; otherwise, ϕ_2 is the (conditional) probability of choosing the second option over the third. To cover the whole space of possible link structures in a systematic manner, we have used each of the twelve pairs $(\phi_1, \phi_2) \in \{0, 0.33, 0.66\} \times \{0, 0.33, 0.66, 1\}$ to generate twelve test Webs $W_{test}^{0,0}, \dots, W_{test}^{66,100}$, and we complemented them with the test Web W_{test}^{100} that we generated using probability $\phi_1 = 1$ (in which case ϕ_2 is irrelevant) .

While these 13 test Webs cover a wide range of possible link structures, we are also interested in an additional test Web whose link structure is most representative of real Linked Data on the WWW. To identify a corresponding pair of probabilities (ϕ_1, ϕ_2) we analyzed the 2011 Billion Triple Challenge dataset [3]. For this corpus of real Linked Data we identified a ϕ_1 of 0.62 and a ϕ_2 of 0.47. Given this pair of probabilities, we used our base dataset to generate another test Web, $W_{\text{test}}^{62,47}$. In this paper we discuss primarily the measurements obtained by querying this test Web. However, for our analysis we also queried the other test Webs; the measurements of all these query executions contribute to our empirical comparison of the URI prioritization approaches (cf. Sect. 6.6).

We emphasize that a systematic creation of test Webs with different link structures as achieved by the given, random-based approach requires a base dataset that has a high degree of structuredness, which is the case for our BSBM dataset [2]. On the other hand, even if the base dataset is highly structured, our random-based approach ensures that the documents in each generated test Web (except W_{test}^{100}) contain data with varying degrees of structuredness, which reflects most of the Linked Data on the WWW [2].

5.2   Test Queries

For our experiments we use six SPARQL basic graph patterns (BGPs) under c_{Match}-bag-semantics (cf. Sect. 2); as seed URIs, we use all URIs in the given BGP, respectively. These queries, denoted by Q1 to Q6, are listed in the extended version of this paper [10].

We created these six queries so that they satisfy the following three requirements: First, each of these queries can be executed over all our test Webs. Second, the queries differ w.r.t. their syntactical structure (shape, size, etc.). Third, to avoid favoring any particular traversal strategy, the reachable subwebs induced by the queries differ along various dimensions. For instance, Table 1 lists several properties of the six query-specific reachable subwebs of test Web $W_{\text{test}}^{62,47}$. These properties are the number of reachable documents, the number of edges between these documents in the link graph of the reachable subweb, the number of strongly connected components and the diameter of the link graph, the number of reachable documents that are *result-relevant* (i.e., their data is required for at least one solution of the corresponding query result), the percentage of reachable documents that are result-relevant, the mean lengths of the shortest paths (in the link graph) from seed documents to these relevant documents, the lengths of the shortest and the longest of these shortest paths, and similar statistics for the reachable documents that are not result-relevant. Additionally, Table 1 lists the cardinality of the corresponding query results. We emphasize

Table 1. Statistics about the reachable subwebs of test queries Q1–Q6 over test Web $W_{\text{test}}^{62,47}$.

Query	link graph of reachable subweb				result-relevant reachable documents					res.-irrel. reach. documents			result
	# of docs	# of edges	str. conn. components	dia-meter	# of docs	% of all reach.docs	shortest paths from seeds mean (st.dev)	min	max	shortest paths from seeds mean (st.dev)	min	max	cardi-nality
Q1	3818	10007	413	8	572	15.0%	1.12 (±0.43)	1	3	1.69 (±0.93)	1	3	2481
Q2	214	627	8	15	22	10.3%	2.34 (±1.70)	1	8	5.04 (±1.40)	2	8	34
Q3	234	410	57	6	3	1.3%	1.41 (±0.50)	1	2	2.74 (±0.53)	1	3	4
Q4	1098	7805	36	12	43	3.9%	1.38 (±0.73)	1	3	3.49 (±0.98)	1	5	804
Q5	333	2340	14	10	36	10.8%	2.21 (±0.78)	1	4	3.83 (±0.37)	3	5	116
Q6	2232	6417	88	45	12	0.5%	2.40 (±0.78)	1	4	4.08 (±1.34)	1	8	28

that a computation of any of the properties in Table 1 requires information that a traversal-based system discovers only during query execution. Hence, such statistics can be computed only after *completing* a traversal-based query execution and, thus, they cannot be used for query optimization.

By comparing the properties in Table 1, it can be observed that our six test queries induce a very diverse set of reachable subwebs of test Web $W_{\text{test}}^{62,47}$. In an earlier, more detailed analysis of these queries we make the same observation for the other 13 test Webs [9]. Moreover, if we consider each query in separation and compare its reachable subwebs across the different test Webs, we observe a similarly high diversity [9]. Hence, these six queries in combination with all 14 test Webs represent a broad spectrum of test cases. That is, we have some test cases that reflect interlinkage characteristics of a real snapshot of Linked Data on the WWW (i.e., $W_{\text{test}}^{62,47}$) and others that systematically cover other possible interlinkage characteristics ($W_{\text{test}}^{0,0}$ ... W_{test}^{100}).

5.3   Metrics

For each solution that our test system computes during a query execution, it measures and records the fraction of the overall execution time after which the solution becomes available. An example of such numbers for the first reported solution are the percentages given in Example 1. For our analysis we focus primarily on the two extreme cases: the relative response times for a first solution and the relative response times for the last solution. The former is interesting because it identifies the time after which users can start looking over some output for their query; the latter marks the availability of the complete result (even if the system cannot verify the completeness at this point). Hence, we define two metrics. Let *exec* be a query execution; let t_{start}, t_{end}, t_{1st}, and t_{last} be the points in time when *exec* starts, ends, returns a first solution, and returns the last solution, respectively. The *relative first-solution response time* (*relRT1st*) and the *relative complete-result response time* (*relRTCmpl*) of *exec* are defined as follows:

$$\text{relRT1st} = \frac{t_{\text{1st}} - t_{\text{start}}}{t_{\text{end}} - t_{\text{start}}} \quad \text{and} \quad \text{relRTCmpl} = \frac{t_{\text{last}} - t_{\text{start}}}{t_{\text{end}} - t_{\text{start}}}.$$

We can use such relative metrics for our study because, for each query, the overall query execution time is always the same, independent of the URI prioritization approach (cf. Sect. 3). The advantage of relative metrics is that they directly show the differences in response times that can be achieved by different URI prioritization approaches *relative to each other*. Measuring absolute times—such as the times that Example 1 provides in addition to the percentages—would not provide any additional insight for such a comparison. Moreover, absolute times that we may measure in our simulation environment are mostly a function of how fast our simulation server responds to URI requests. Hence, such absolute times in our simulation would be quite different from what could be measured for queries over the "real" Web of Linked Data (such as in Example 1).

To increase the confidence in our measurements we repeat every query execution five times and report the geometric mean of the measurements obtained by the five executions. The confidence intervals (i.e., error bars) in the charts in this paper represent one standard deviation. To avoid measuring artifacts of concurrent query executions we execute queries sequentially. To also exclude possible interference between subsequent query executions we stop and restart the system between any two executions.

6   Experimental Results

To experimentally analyze the URI prioritization approaches introduced in Sect. 4 we used each of these approaches for traversal-based query executions over our test Webs. The charts in Fig. 2 illustrate the mean relRT1st and the mean relRTCmpl measured for the query executions over test Web $W_{\text{test}}^{62,47}$ (in some cases the bars for relRT1st are too small to be seen). For instance, the leftmost bars in Fig. 2(a) indicate that the baseline executions of query Q1 returned a first solution of the query result after 26.5 % of the overall query execution time, and it took them about 99 % of the time to complete the query result. In this section, we discuss these measurements, as well as further noteworthy behavior as observed for query executions over the other test Webs. The discussion is organized based on the classification of URI prioritization approaches as introduced in Sect. 4 (Fig. 1). However, we begin with some general observations.

6.1   General Observations

A first, unsurprising observation is that, in almost all cases, none of the approaches achieves response times that are smaller than the response times achieved by the oracle approach. However, we also notice a few (minor) exceptions. These exceptions can be explained by the fact that—independent of what URI prioritization approach is applied—the DR-component discovers the URIs to be looked up only gradually. Then, by *greedily* ordering the currently available URIs (based on our pre-computed RCCs), the oracle approach may only achieve a local optimum but not the global one.

Ignoring the oracle approach for a moment, we note that approaches that achieve a good relRT1st for a query do not necessarily also achieve a good relRTCmpl for that query.

Another general observation is that, by using different URI prioritization approaches to execute the same query over the same test Web, the number of intermediate solutions processed by our system can vary significantly, and so does the number of priority changes initiated by the adaptive approaches. These variances indicate that the amount of computation within our system can differ considerably depending on which URI prioritization approach is used. Nonetheless, in all our experiments the overall time to execute the same query over the same test Web is always almost identical for the different approaches! This fact again illustrates the dominance of the data retrieval fraction of query

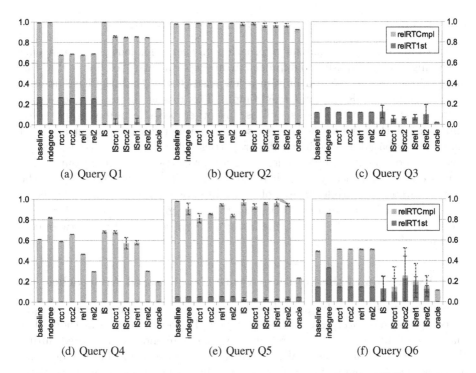

Fig. 2. Relative response times for queries Q1 to Q6 over test Web $W_{test}^{62,47}$ as achieved by employing the different approaches to prioritize URI lookups.

execution time (cf. Sect. 3) and, thus, is a strong verification of the comparability of our relative measurements. The only exception is the PageRank approach for which query execution times range from 120 % to 320 % of the execution times measured for the other approaches. Hence, in contrast to the additional computation required for each of the other approaches, the frequent execution of the iterative PageRank algorithm becomes a non-negligible overhead. As a result, the PageRank approach cannot compete with the other approaches and, thus, we ignore it in the remainder.

Finally, in Fig. 2(c) (for query Q3 over test Web $W_{test}^{62,47}$), we note that, for all approaches, the differences between the time needed to return a first solution and the time to return the last solution are insignificant. We explain this phenomenon as follows: Only three of the 234 reachable documents for Q3 over $W_{test}^{62,47}$ contribute to the query result and this result consists of four solutions (cf. Table 1). It turns out that the computation of each of these four solutions requires data from each of the three result-relevant documents. Hence, only after (and as soon as) the last of these three documents has been retrieved, the system can compute and return all four solutions.

6.2   Evaluation of the Purely Graph-Based Approaches

After ruling out the PageRank approach (cf. Sect. 6.1), indegree is the only remaining purely graph-based approach in our experiments. We observe that this approach is often worse and only in a few cases better than the baseline approach (for both relRT1st and relRTCmpl). The reason for the negative performance of this approach—as well as any other possible purely graph-based approach—is that the applied vertex scoring method rates document and URI vertices only based on graph-specific properties, whereas the result-relevance of reachable documents is independent of such properties. In fact, in our earlier work we show empirically that there does not exist a correlation between the result-relevance—or irrelevance— of reachable documents and the indegree of the corresponding document vertices in the Web graph model (similarly, for the PageRank, the HITS scores, the k-step Markov score, and the betweenness centrality) [9].

6.3   Evaluation of the Solution-Aware Graph-Based Approaches

In contrast to the purely graph-based approaches, the solution-aware graph-based approaches (rcc1, rcc2, rel1, and rel2) employ vertex scoring methods that make use of information about result-relevant documents as discovered during the query execution process. We notice that, until such information becomes available (that is, not before a first query solution has been computed), these methods rate all vertices equal. As a consequence, all URIs added to the lookup queue have the same priority and are processed in the order in which they are discovered. Hence, until a first solution has been computed, the solution-aware graph-based approaches behave like the baseline approach. Therefore, these approaches always achieve the same relRT1st as the baseline.

Once a first set of result-relevant documents can be identified, the solution-aware graph-based approaches begin leveraging this information. As a result, for several query executions in our experiments, these approaches achieve a relRTCmpl that is significantly lower than the baseline. Moreover, for the majority of query executions for which this is not the case, the relRTCmpl achieved by the solution-aware graph-based approaches is comparable to the baseline. In the following, we identify characteristics of reachable subwebs that are beneficial for our four solution-aware graph-based approaches (for a more detailed discussion refer to the extended version of this paper [10]).

A necessary (but not necessarily sufficient) characteristic is that every reachable document that is result-relevant must have at least one in-neighbor that is also result-relevant (for rel2 and rcc2 it may also be a 2-step in-neighbor). However, even if the in-neighborhood of a relevant document d contains some other relevant documents, the solution-aware graph-based approaches can increase the retrieval priority of document d only if the relevance of at least one of these other documents, say d', is discovered before the retrieval of d. This is possible only if the relevance of d' can be attributed to its contribution to some query solution whose computation does not require document d.

Hence, an early discovery of a few first solutions may increase chances that the solution-aware graph-based approaches retrieve all relevant documents early, which then leads to smaller complete-result response times (relRTCmpl). However, there are also cases in which the identification of relevant documents may mislead these approaches; in particular, if some relevant documents link to many irrelevant documents. A special case that is particularly worse for rcc1 and rcc2 is the existence of a result-relevant document d with an RCC that is significantly higher than the RCCs of the other relevant documents in the corresponding subweb; such a high RCC may dominate the RCC-based scores in the in-neighborhood of document d. The Q4-specific reachable subweb of test Web $W_{\text{test}}^{62,47}$ is an example of such a case (cf. Fig. 2(d)).

6.4   Evaluation of the Intermediate Solution Driven Approaches

Intermediate solution driven approaches (including the hybrid approaches analyzed in the next section) use information about all intermediate solutions sent between operators in the RC-component. Regarding these approaches, we notice a high variation in our measurements (observe the error bars in Fig. 2). We attribute this variation to the multithreaded execution of all operators in the RC-component of our traversal-based query engine (which we describe in detail in the extended version of this paper [10]). Due to multithreading, the exact order in which intermediate solutions appear in the RC-component and are sent to the DR-component is nondeterministic. As a result, the intermediate solution driven adaptation of the priorities of URIs that are queued for lookup becomes nondeterministic. Then, due to this nondeterminism, the order in which reachable documents are retrieved may differ for repeated executions with the same prioritization approach. Such differences may cause different response times because the retrieval order of documents determines which intermediate solutions can be generated at which point during the query execution process.

Irrespective of the variations, our measurements indicate that, in a number of cases, the IS approach can achieve an advantage over the baseline approach. For instance, compare the relRT1st values in Fig. 2(a) or the relRTCmpl values in Fig. 2(f). However, there also exist a significant number of cases in which IS performs worse than the baseline approach (e.g., query Q4 over test Web $W_{\text{test}}^{62,47}$; cf. Fig. 2(d)).

6.5   Evaluation of the Hybrid Approaches

For the hybrid approaches (isrcc1, isrcc2, isrel1, isrel2) we first notice that they all achieve similar response times in many cases. More importantly, however, these response times are comparable, or at least close, to the best of either the response times achieved by the solution-aware graph-based approaches or the response times of the IS executions.

A typical example are the executions of Q1 over test Web $W_{\text{test}}^{62,47}$ (cf. Fig. 2(a)). On one hand, the hybrids achieve complete-result response times (relRTCmpl) for this query that are smaller than the baseline—which is also the case for the solution-aware graph-based approaches but not for the IS-based executions. On the other hand, instead of also achieving first-solution response times (relRT1st) as achieved by the solution-aware graph-based approaches (which are as high as the baseline), the hybrids achieve a relRT1st that is as small as what the IS-based executions achieve. Regarding relRT1st we recall that the solution-aware graph-based approaches cannot be better than the baseline. The latter observation shows that this is not the case for the hybrid approaches. On the contrary, even if each hybrid approach is based on a solution-aware graph-based approach, their combination with intermediate solution driven functionality enables the hybrid approaches to outperform the baseline in terms of relRT1st.

6.6   Comparison

Our measurements show that there is no clear winner among the URI prioritization approaches studied in this paper. Instead, for each approach, there exist cases in which the approach is better than the baseline and cases in which the approach is worse.

Table 2 quantifies these cases; that is, the table lists the percentage of cases in which the response times achieved by each approach are at least 10 % better (resp. 10 % worse) than the baseline. For this comparison, we consider the executions of all six test queries *over all 14 test Webs* (i.e., 84 cases for each approach), and we use the threshold of 10 % to focus only on noteworthy differences to the baseline. In addition to relRT1st and relRTCmpl, the table also covers *relative 50 % response time* (*relRT50*); that is, the fraction of the overall execution time after which 50 % of all solutions of the corresponding query result have been computed.

approach	relRT1st worse	relRT1st better	relRT50 worse	relRT50 better	relRTCmpl worse	relRTCmpl better
DFS	23.2%	26.1%	58.9%	17.8%	53.6%	10.1%
random	13.0%	27.5%	58.9%	8.2%	59.4%	8.7%
indegree	21.7%	21.7%	65.8%	4.1%	50.7%	5.8%
rcc1	0.0%	0.0%	4.1%	1.4%	7.2%	24.6%
rcc2	0.0%	0.0%	2.7%	2.7%	4.1%	20.3%
rel1	0.0%	0.0%	5.5%	1.4%	11.6%	29.0%
rel2	0.0%	0.0%	11.0%	0.0%	2.9%	26.1%
IS	7.2%	31.9%	15.1%	27.4%	26.1%	10.1%
isrcc1	2.9%	30.4%	5.5%	26.0%	14.5%	18.8%
isrcc2	5.8%	33.3%	5.5%	24.7%	13.0%	26.1%
isrel1	0.0%	33.3%	2.7%	24.7%	15.9%	26.1%
isrel2	2.9%	31.9%	4.1%	23.3%	11.6%	26.1%
oracle	0.0%	35.3%	0.0%	41.2%	0.0%	64.7%

Table 2. Percentage of cases in which the approaches achieve response times that are at least 10 % worse (resp. 10 % better) than the baseline

For both relRT1st and relRT50, we observe that isrel1 is the best of the approaches tested (ignoring the oracle approach which cannot be used in practice; cf. Sect. 4.5). Although the other intermediate solution driven approaches (IS, isrel2, isrcc1, isrcc2) have a similarly high number of cases in which they are at least 10 % better than the baseline, these approaches have a higher number of cases in which they are at least 10 % worse. We also notice that, as discussed in Sect. 6.3, for relRT1st, the solution-aware graph-based approaches (rcc1, rcc2, rel1, rel2) behave like the baseline.

For relRTCmpl, we observe some differences. The hybrid approaches (isrcc1, ..., isrel2) still have a comparably high number of cases in which they are at least 10 % better than the baseline, but they also have a significant number of noteworthy cases in which they are worse. IS has an even higher number of such worse cases. In contrast, the solution-aware graph-based approaches are more suitable, with rel2 being the best choice.

In summary, to return some solutions of query results as early as possible, isrel1 appears to be the most suitable choice among the approaches studied in this paper. However, if the objective is to reduce complete-result response times (relRTCmpl), the solution-aware graph-based approaches are usually better suited; in particular, rel2. In the extended version of the paper we additionally show that, by and large, these general findings are independent of whether the queried Web is densely populated with bidirectional data links (i.e., $\phi_1 \geq 0.66$) or sparse (i.e., $\phi_1 \leq 0.33$) [10].

7 Conclusions

This is the first paper that studies the problem of optimizing the response times of traversal-based query executions over Linked Data. In particular, we focus on the fundamental problem of fetching result-relevant data as early as possible. To this end, we introduce heuristics-based approaches to prioritize URI lookups during data retrieval and analyze their impact on response times. For this experimental analysis we use a broad range of simulated, structurally diverse Webs of Linked Data. One of these test Webs reflects interlinkage characteristics of a real snapshot of Linked Data on the WWW, and the others systematically cover other possible interlinkage characteristics as may reflect other Webs of Linked Data (e.g., within the intranet of an enterprise).

Our experiments show that some of the approaches can achieve noteworthy improvements over the baseline in a significant number of cases. However, even for the best URI prioritization approaches in this paper, there exist cases in which the baseline approach achieves better response times. Moreover, a comparison to the oracle approach shows that there is further room for improvement. A promising direction of future work are approaches that collect statistics during (traversal-based) query executions and leverage these statistics to optimize the response times for subsequent queries.

References

1. Bizer, C., Schultz, A.: The Berlin SPARQL benchmark. Sem. Web Inf. Sys. **5**(2), 1–24 (2009)
2. Duan, S., Kementsietsidis, A., Srinivas, K., Udrea, O.: Apples and oranges: a comparison of RDF benchmarks and real RDF datasets. In: Proceedings of the ACM SIGMOD (2011)
3. Harth, A.: Billion triples challenge data set (2011). http://km.aifb.kit.edu/projects/btc-2011
4. Harth, A., Hose, K., Schenkel, R. (eds.): Linked Data Management. Chapman & Hall, London (2014)

5. Hartig, O.: How caching improves efficiency and result completeness for querying linked data. In: Proceedings of the 4th Linked Data on the Web Workshop (LDOW) (2011)
6. Hartig, O.: SPARQL for a web of linked data: semantics and computability. In: Simperl, E., Cimiano, P., Polleres, A., Corcho, O., Presutti, V. (eds.) ESWC 2012. LNCS, vol. 7295, pp. 8–23. Springer, Heidelberg (2012)
7. Hartig, O.: An overview on execution strategies for linked data queries. Datenbank-Spektrum 13(2), 89–99 (2013)
8. Hartig, O., Bizer, C., Freytag, J.-C.: Executing SPARQL queries over the web of linked data. In: Bernstein, A., Karger, D.R., Heath, T., Feigenbaum, L., Maynard, D., Motta, E., Thirunarayan, K. (eds.) ISWC 2009. LNCS, vol. 5823, pp. 293–309. Springer, Heidelberg (2009)
9. Hartig, O., Özsu, M.T.: Reachable subwebs for traversal-based query execution. In: Proceedings of the 23rd International World Wide Web Conference (WWW) (2014)
10. Hartig, O., Özsu, M.T.: Walking without a map: optimizing response times of traversal-based linked data queries (extended version). CoRR abs/1607.01046 (2016)
11. Ladwig, G., Tran, T.: Linked data query processing strategies. In: Patel-Schneider, P.F., Pan, Y., Hitzler, P., Mika, P., Zhang, L., Pan, J.Z., Horrocks, I., Glimm, B. (eds.) ISWC 2010, Part I. LNCS, vol. 6496, pp. 453–469. Springer, Heidelberg (2010)
12. Ladwig, G., Tran, T.: SIHJoin: querying remote and local linked data. In: Antoniou, G., Grobelnik, M., Simperl, E., Parsia, B., Plexousakis, D., De Leenheer, P., Pan, J. (eds.) ESWC 2011, Part I. LNCS, vol. 6643, pp. 139–153. Springer, Heidelberg (2011)
13. Miranker, D.P., Depena, R.K., Jung, H., Sequeda, J.F., Reyna, C.: Diamond: a SPARQL query engine, for linked data based on the rete match. In: Proceedings of the AImWB (2012)
14. Page, L., Brin, S., Motwani, R., Winograd, T.: The pagerank citation ranking: bringing order to the web. Technical report 1999-66, Stanford InfoLab, November 1999
15. Schmidt, M., Görlitz, O., Haase, P., Ladwig, G., Schwarte, A., Tran, T.: FedBench: a benchmark suite for federated semantic data query processing. In: Aroyo, L., Welty, C., Alani, H., Taylor, J., Bernstein, A., Kagal, L., Noy, N., Blomqvist, E. (eds.) ISWC 2011, Part I. LNCS, vol. 7031, pp. 585–600. Springer, Heidelberg (2011)
16. Umbrich, J., Hogan, A., Polleres, A., Decker, S.: Link traversal querying for a diverse web of data. Semant. Web J. 6(6), 585–624 (2015)
17. Umbrich, J., Hose, K., Karnstedt, M., Harth, A., Polleres, A.: Comparing data summaries for processing live queries over linked data. World Wide Web 14(5–6), 495–544 (2011)

CubeQA—Question Answering on RDF Data Cubes

Konrad Höffner[1]([✉]), Jens Lehmann[2,3], and Ricardo Usbeck[1]

[1] University of Leipzig, Institute of Computer Science, AKSW Group,
Augustusplatz 10, 04109 Leipzig, Germany
{hoeffner,usbeck}@informatik.uni-leipzig.de
[2] Computer Science Institute, University of Bonn, Bonn, Germany
jens.lehmann@cs.uni-bonn.de
[3] Knowledge Discovery Department, Fraunhofer IAIS, Sankt Augustin, Germany
jens.lehmann@iais.fraunhofer.de

Abstract. Statistical data in the form of RDF Data Cubes is becoming increasingly valuable as it influences decisions in areas such as health care, policy and finance. While a growing amount is becoming freely available through the open data movement, this data is opaque to laypersons. Semantic Question Answering (SQA) technologies provide intuitive access via free-form natural language queries but general SQA systems cannot process RDF Data Cubes. On the intersection between RDF Data Cubes and SQA, we create a new subfield of SQA, called RDCQA. We create an RDQCA benchmark as task 3 of the QALD-6 evaluation challenge, to stimulate further research and enable quantitative comparison between RDCQA systems. We design and evaluate the domain independent CubeQA algorithm, which is the first RDCQA system and achieves a global F_1 score of 0.43 on the QALD6T3-test benchmark, showing that RDCQA is feasible.

1 Introduction

Statistical data influences decisions in domains such as health care, policy, governmental decision making and finance. The general public is increasingly interested in accessing such open information [19]. This coincides with the open data movement and has led to an increased availability of statistical government data in the form of data cubes. Initiatives that publish those statistics include OpenSpending[1] and World Bank Open Data[2]. However, this type of data is multidimensional, numerical and often voluminous, and thus not easily approachable for laypersons.

While singular data points can be queried using a tabular and faceted browsing interfaces offered by those initiatives, common questions often require [12] the combination and processing of many different datapoints. This processing can be performed by specialized tools but they require knowledge of a specific

---

[1] http://openspending.org/.
[2] http://data.worldbank.org/.

© Springer International Publishing AG 2016
P. Groth et al. (Eds.): ISWC 2016, Part I, LNCS 9981, pp. 325–340, 2016.
DOI: 10.1007/978-3-319-46523-4_20

vocabulary or a formal query language and are thus also hindering the access for laypersons. To provide a more intuitive interface, we present the CubeQA approach.

Our contributions are as follows: (1) To the best of our knowledge, we are the first to tackle the intersection between Data Cubes, Question Answering (QA) and RDF, creating a new subfield of QA, which we call RDCQA. (2) We stimulate further research with the creation of the QALD-6 Task 3 benchmark (QALD6T3): *Statistical Question Answering over RDF datacubes*. This enables quantitative comparison between RDCQA systems, see Sect. 4.2. Moreover, the introduction of this task has led to the development of another system, QA^3 ("QA cube"), and the extension of the SPARQL query builder Sparklis (see Sect. 5). (3) With the CubeQA algorithm, which achieves a global F_1 score of 0.43 on QALD6T3-test (see Sect. 4), we show that RDCQA is feasible.

The rest of the paper is structured as follows: Section 2 introduces the preliminaries. Section 3 defines the CubeQA algorithm. Section 4 presents our benchmark and evaluates the CubeQA algorithm. Section 5 summarizes general SQA approaches and work in progress on RDCQA. We summarize our contributions in Sect. 6 and present challenges to be addressed by future work.

2   Preliminaries

Unlike common data representations such as tables or relational databases, the *data cube*[3] formalism adequately represents multidimensional, numerical data. A data cube is a multidimensional array of *cells*. Each cell is uniquely identified by its associated dimension values and contains one or more numeric *measurement* values. Data cubes are often *sparse*, i.e., for most combinations of dimension values there is no cell in the cube. Data cubes, such as in Fig. 2a, allow the following operations supported by CubeQA:

1. *Dicing* a data cube creates a subcube by constraining a dimension to a subset of its values, see Fig. 2b.
2. *Slicing* a data cube reduces its dimensionality by one by constraining a dimension to one specific value, see Fig. 2c.
3. *Rolling Up* a data cube means summarizing measure values along a dimension, such as a sum, count, or arithmetic mean. A roll-up of Fig. 2c answers Fig. 1.

Definition 1. *We define* Question Answering (QA) [11] *as users (1) asking questions in natural language (NL) (2) using their own terminology to which they (3) receive a concise answer. In* Semantic Question Answering (SQA), *the*

How much did the Philippines receive in the years of 2007 to 2008?

Fig. 1. The running example used throughout this paper

---

[3] also *OLAP* cube or *hypercube*.

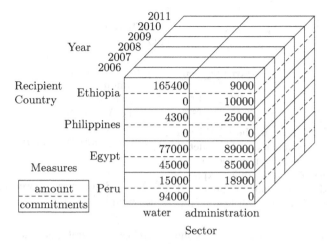

(a) Simplified excerpt of the LinkedSpending RDC *Finland Aid Data* for Fig. 1. Measure units are provided by the *currency* attribute in each cell (omitted for brevity).

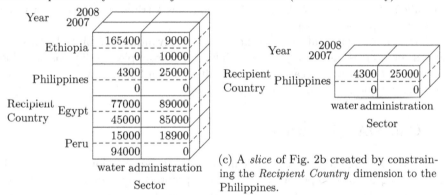

(b) A *dice* of Fig. 2a created by constraining the *year* dimension to 2007 and 2008.

(c) A *slice* of Fig. 2b created by constraining the *Recipient Country* dimension to the Philippines.

Fig. 2. Example of a data cube and its operations.

natural language question is transformed into a formal query on an RDF knowledge base, commonly using SPARQL.

The *RDF Data Cube (RDC) Vocabulary* models both the schema and the observations of a data cube to RDF (see Fig. 3a). Each data cube, an instance of **qb:DataSet**, has an attached schema, the *data structure definition*, which specifies the *component properties*[4]. A component property is either a dimension, an attribute or a measure. *Measures*, of which there has to be at least one, represent the measured quantities, while the dimensions and the optional attributes provide context. Because the RDC vocabulary is focused on statistical data, its cells

---

[4] qb:ComponentProperty in Fig. 3a, not to be confused with rdf:Property.

328 K. Höffner et al.

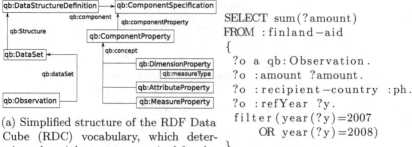

(a) Simplified structure of the RDF Data Cube (RDC) vocabulary, which determines the triple patterns required for the SPARQL query. A more detailed explanation of the RDC vocabulary is presented in [12]. Figure originally published in [13].

```
SELECT sum(?amount)
FROM :finland-aid
{
    ?o a qb:Observation.
    ?o :amount ?amount.
    ?o :recipient-country :ph.
    ?o :refYear ?y.
    filter(year(?y)=2007
        OR year(?y)=2008)
}
```

(b) A SPARQL query answering the running example (Fig. 1).

Fig. 3. The RDC vocabulary and an exemplary SPARQL query for Fig. 1.

are called *observations*. Each observation contains a triple that specifies a value for each component property[5].

RDCs allow data cubes to profit from the advantages of Linked Data [3], such as ontologies, reasoning and interlinking. For example, the dimension value for *recipient country* :ph can be linked to dbpedia:Philippines, whose properties can then be queried from DBpedia [16].

3 CubeQA Algorithm

The CubeQA algorithm converts a natural language question to a SPARQL query using a linear pipeline. Its first step, preprocessing, indexes the target datasets, extracts simple constraints and creates the parse tree used by the following steps. Next, the matching step recursively traverses the parse tree downwards until it identifies reference candidates at each branch. Starting at those candidates, the combination step merges those candidates upwards until it creates a final template in the root of the parse tree. Finally, in the execution step, the template is converted to a SPARQL query that is executed to generate the result set containing the answer.

3.1 Preprocessing

Number Normalization. First, numbers are normalized, for example "5 thousand" to "5000", as the other components do not recognize numbers in words.

Keyphrase Detection. In this step, phrases referring to data cube operations are detected. These operations are typically referenced by certain keyphrases and are thus detected using regular expressions during preprocessing, see Table 1.

---

[5] except attributes, which may also apply to the whole data cube.

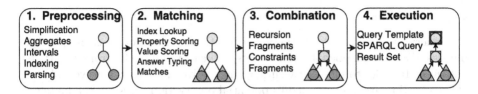

Fig. 4. The CubeQA pipeline

Those keyphrases, as well as the following entity recognition and disambiguation steps, are domain independent, so that CubeQA can be used with any set of RDCs. Nevertheless, the keyphrase to operation mapping can be extended for a specialized vocabulary to increase the recall on a particular domain.

Table 1. Data cube operations extracted in the preprocessing step

Element	SPARQL	Example phrase
Dice	Filter	2007 to 2008
Roll-up	Aggregate	In total
Slice	Filter	In 2008
Modifier	ORDER LIMIT	The 5 highest amounts

If a roll-up is not explicitly expressed in the question, a default aggregation is assumed for some answer types. A SPARQL aggregate rolls up all dimensions that are not bound by query variables. The roll-up aggregates *sum, arithmetic mean* and *count* are handled differently then *minimum* and *maximum*. The former aggregates return new values, so that they can be mapped to the SPARQL aggregation keywords SUM, AVG and COUNT. Minimum and maximum, however, choose a value among the existing ones, which allows identification of a cell and thus of a different component value. For example, in "Which company has the highest research spending?", the user probably asks for the total, which can be achieved by a roll-up with addition followed by selecting the company with the highest sum.

Dataset Detection. CubeQA uses a dataset index that is initialized once with the set of available RDCs. It is implemented as a Lucene index with fields for the labels, comments and property labels of each RDC. The dataset name alone is not sufficient because questions (Definition 1) often do not refer to the dataset. For example, in Fig. 1, the dataset *finland-aid* is not mentioned but "the Philippines", "2007" and "2008" will all be found by the index.

Parsing. At the end of the preprocessing step, a syntactic parse tree is generated for the modified question. This tree structure is traversed for matching nodes as described in Sect. 3.2.

3.2   Matching

Our query model starts with the whole RDC. The question is then split into phrases that are mapped to constraints, which exclude cells from the target datacube. In order to increase accuracy and resolve ambiguity, however, phrases of the question are first mapped to potential observation values in the matching step, based on the following definitions:

Definition 2 (RDF Knowledge Base (KB)). *Let U be a set of URIs, $U = I \cup P \cup C$, where I is are the instances, P the properties and C the classes. Let $L \subset \Sigma^*$ be a set of literals, where Σ is the unicode alphabet. We define a KB k as a set of triples $k \subseteq U \times P \times (U \cup L)$ (we disregard blanknodes). In the context of this work, the KB contains the triples of a set of RDCs.*

Definition 3 (Values). *The* values *of a component property p for an RDC c and a KB k are defined as: $V_{p,c,k} = \{v | \exists o : \{(o,p,v), (o, qb{:}dataset, c), (o, rdf{:}type, qb{:}Observation)\} \subset k\}$. The numerical values $D_{p,c,k}$ are the values representing numbers, converted to double values. The temporal values $T_{p,c,k}$ are the union of all values representing dates or years, converted to time intervals $\tau(v) : T_{p,c,k} = \bigcup_{v \in V_{p,c,k}} \tau(v)$.*

Example: `:ph` $\in V_{\{}$`:recipient-country, finland-aid, linkedspending`$\}$.

Definition 4 (Scorer). *A* scorer *for a component property p is represented formally by a partial function $d_p : \Sigma^* \to (L \cup U) \times (0, 1]$, Table 3 shows the three types of scorers, which are assigned to a property based on its range (see Table 2). Informally, the scorer of p returns the value with the closest match to a given phrase and its estimated probability.*

Table 2. Component property scorer and answer type assignment. Integers include datatypes derived by constraint

Range	Scorer	Answer type
xsd:integer	Numeric	Countable
xsd:float, xsd:double	Numeric	Uncountable
xsd:gYear, xsd:date	Temporal	Temporal
No match	String	Entity

For the query template, the scorer results are converted to constraints. A naive approach is to create a value reference for the highest scored property of a phrase but this penalizes short phrases and suffers from ambiguity as it does not take the context of the phrase into account. Accordingly, CubeQA inserts an intermediate step: the *match*, which represents the possible references to component properties and their values.

Table 3. Definitions of the different types of scorers. The String Scorer uses both the Levenshtein distance, to quickly find candidates, and bigrams, for a more accurate scoring. All three scorers are partial functions whose result is undefined if no value is found. Only String Scorers return scores < 1, as they can correct for typographical errors in the input, while Numeric and Temporal Scorers are either undefined or return the input number, respectively time interval, with a score of 1. Type casting and conversion is omitted for brevity, e.g. in Fig. 1 the phrase "Philippines" is equated to the language tagged label `"Philippines"@en` and the phrase "2007" to the year $2007\,\hat{}\,\hat{}\,xsd{:}gYear$.

Type	Scoring function $d_p(a)$
String	$(\arg\max_{b\in\beta_p(a)}(\mathrm{ngram}(a,b)), \max_{b\in\beta_p(a)}(\mathrm{ngram}(a,b)),$ where $\beta_p(a) = \{b \in V_{p,c,k} \mid \mathrm{lev}(a,b) \le 2\}$, using n-gram similarity [15] (ngram, $n{=}2$) and a Levenshtein-Automaton [21] (lev)
Numeric	$(a,1)$, if $a \in [\min(D_{p,c,k}), \max(D_{p,c,k})]$, otherwise undefined
Temporal	$(a,1)$, if $T_{p,c,k} \cap \tau(a) \ne \emptyset$, otherwise undefined,

Definition 5 (Match). *A match m is represented formally by a pair (ρ, γ), where ρ is the partial property scoring function, $\rho : P \to (0,1]$ and γ is the partial value scoring function, $\gamma : P \to (L \cup U) \times (0,1]$.*

3.3 Combining Matches to Constraints

The recursive combination process is used because (1) it favours longer phrases over shorter ones, giving increased coverage of the question and (2) it favours combination of phrases that are nearby in the question parse tree.

Definition 6 (Constraint). *A constraint c is represented by a triple (G, ω, λ), where:*

- *G is a set of SPARQL triple patterns and filters as defined in [20]*
- *ω is an optional order by modifier, $\omega \in (\{ASC, DESC\} \times P) \cup \{null\}$*
- *λ is an optional limit modifier, $\lambda \in \mathbb{N}^+ \cup \{null\}$*

Constraints are based on three different criteria:

1. A **Value Constraint** can be applied to any component property to confine it to an exact value, which can be a string, a number or a URI. It consists of a single SPARQL triple pattern: $c_v = (\{(?o, p, v), (?o, qb{:}DataSet, d), (?o, a, qb{:}Observation)\}, null, null)\}$, with $p \in P$ and $v \in L \cup U$.
2. An **Interval Constraint** confines a value to a numeric or temporal interval. Accordingly, it can only apply to a component property whose range is an XSD numeric or temporal data type. It consists of a SPARQL triple pattern and a filter: $c_i = (\{?o\ p\ ?x,\ \texttt{filter}(?x > x_1)\ \texttt{AND}\ (?x < x_2)\}, null, null)$, with $p \in P$, the lower limit x_1 and an upper limit x_2. Example: $(\{?o\ \texttt{:refYear}\ ?y,\ \texttt{filter(year(?y)>=2007}\ \texttt{AND}\ \texttt{year(?y)<=2008)}\}, null, null)$. Closed or half-bounded intervals are defined analogously.

3. **Top/Bottom n Constraints** place an upper limit on the number of selected cells. They consist of three parts: The order (ascending or descending), the limit and the numeric component property whose values imply the order. Formally, $c_t = (\emptyset, (\text{DESC}, p), n), c_b = (\emptyset, (\text{ASC}, p), n)$

To identify Value Constraints, each component property has a scorer (Definition 4), which tries to find a value similar to an input phrase. For example, "How much total aid was given to the regional FLEG programme in Mekong?", could refer to a dimension "programme" with a value of "FLEG" and a dimension "region" with a value of "Mekong". Equally possible would be a dataset description of "aid to Mekong" and a dimension "target" with a value of "FLEG programme". The other types of constraints are matched in the preprocessing step because they are identified by certain keyphrases, such as "the 5 highest X".

In the example question, "How much did the Philippines receive in the year of 2007?", there are multiple candidates for the number "2007". The candidates can be disambiguated using the property scoring function of the "year" node by upward combination. As a *match* only holds the information collected from a single node in the question parse tree, there is additional information needed to represent a whole subtree. This extended representation is called a *fragment* and holds: (1) multiple matches collected in the recursive merge and (2) constraints extracted from fitting matches.

Definition 7 (Fragment). *Formally, a fragment f is a pair (M, R), where M is a set of matches (see Definition 5) and R is a set of constraints.*

Algorithm 1 describes the process that combines the fragments of a list of child nodes into the fragment for their parent node.

3.4 Execution

Algorithm 1 combines the fragments of child nodes to create a fragment for the parent node. When this recursive process reaches the root node, Algorithm 2 transforms the fragment that results from the successive combination up to that point into a template (see Definition 8). All leftover value references whose property has not been referenced yet over a certain score threshold are transformed into additional Value Constraints. Other name and value references are discarded. All constraints, as well as the aggregate, if available, are then used to construct a SPARQL *select* query.

Definition 8 (Template). *A template t is a tuple (R, a, α), where R is as defined in Definition 7, $a \in P$ is the answer property and α is an optional aggregate function, $\alpha(X) \in \{\min(X), \max(X), \sum_{x \in X} x, |X|, \sum_{x \in X} \frac{x}{|X|}, null\}$.*

Next, the values of the answer properties are requested. If the set of answer properties is empty, the default measure of the dataset is used as an answer property to determine the properties. Executing the SPARQL query on the target knowledge base results in the set of answers requested by the user. The algorithm

---

Algorithm 1. Fragment Combination

Input: A list of fragments $F = \{(M_1, R_1), \ldots, (M_n, R_n)\}$, with $M_i = (\rho_i, \gamma_i)$
Output: The combined fragment $f = (M, R)$
$R \leftarrow \cup_{i=1}^{n} R_i$;
$P' \leftarrow (P \setminus \delta(R)) \cap \bigcup_{i=1}^{n} (\mathrm{dom}(\rho_i) \cup \bigcup_{x \in \mathrm{dom}(\gamma_i)} \pi_1(x))$;
$M \leftarrow \{M_1, \ldots, M_n\}\}$;
foreach $p' \in P'$ **do**
$\quad m_{\mathrm{property}} \leftarrow \arg\max_{(\rho, \gamma) \in M'} \rho(p')$;
$\quad m_{\mathrm{value}} \leftarrow \arg\max_{(\rho, \gamma) \in M' \setminus \{m_{\mathrm{property}}\}} \gamma(p')]$;
$\quad g \leftarrow ?\mathrm{o}\ p\ \pi_2(m_{\mathrm{value}})(p).$;
$\quad R \leftarrow R \cup \{(\{g\}, \mathrm{null}, \mathrm{null})\}$;
$\quad M \leftarrow M \setminus \{m_{\mathrm{property}}, m_{\mathrm{value}}\}$;
return (M, R)

$\pi_i(t)$ is the projection on the i-th element of the tuple t. The domain dom(f) is the set of all elements for which the (partial) function f is defined. $\delta(R)$ is the set of all component properties that occur in at least one triple pattern in R.

implementation is publicly available under an open license at (link temporarily removed for anonymity). The algorithm implementation is publicly available under an open license at https://github.com/AKSW/cubeqa.

4 Evaluation

4.1 Research Questions

The goal of the evaluation was to obtain answers to the following research questions: Q1: Is CubeQA powerful enough to be practically useful on challenging statistical questions? Q2: Is there a tendency towards either high precision or recall? Q3: How do other RDCQA systems perform? Q4: What types of errors occur? How frequently are they? What are the reasons?

4.2 Experimental Setup and Benchmark

As there was no existing benchmark for RDCQA, we created a benchmark based on a statistical question corpus [12] and included it in the QALD-6[6] evaluation challenge. We used the existing corpus and significantly extended it to 100 questions, forming the training set QALD6T3-train. While keeping a similar structure, we adapted it to 50 of the, at this time, 983 financial datasets of Linked-Spending [13]. Chosen are the first 50 datasets that are manually confirmed as English from a list of all datasets. The list was sorted in descending order by their proportion of English labels (having at least 100 labels) as determined by automatic language detection. The datasets contain in total 158 dimensions, 81 measures, 176 attributes, 950149 observations and 16359532 triples (Table 4).

---

[6] http://www.sc.cit-ec.uni-bielefeld.de/qald/.

Algorithm 2. Fragment to Template Conversion.

Input: A fragment $f = \{(M, R)\}$, an optional aggregate function α identified in preprocessing. The set of expected answer types E is defined in Table 4. answerType(p) is defined by Table 2.

Output: A template $t = (R', a, \alpha')$

$R' = R$;

$P' \leftarrow (P \setminus A \setminus \delta(R))$;

foreach $(\rho, \gamma) \in M'$ **do**

$\quad p_{\max} = \arg\max_{p \in (\mathrm{dom}(\gamma) \cap P')}(\pi_2(\gamma(p)))$;

$\quad$ **if** $(p_{max} \neq null) \wedge (\pi_2(\gamma(p_{max})) \leq \theta)$ **then**

$\quad\quad R' \leftarrow R' \cup (\{?o\; p_{\max}\; \pi_1(\gamma(p_{\max})).\}, null, null)$;

$A \leftarrow \bigcup_{(\rho,\gamma)\in M} dom(\rho)$;

$A' \leftarrow \{p \in A | answerType(p) \in E\}$;

if $A' = \emptyset$ **then**

$\quad a \leftarrow$ DEFAULT_MEASURE;

else

$\quad a \leftarrow \arg\max_{(\rho,\gamma)\in M), p \in A'} \rho(p)$;

return (R', a, α)

Using the same 50 datasets, the test set QALD6T3-test was created in the same way, but with slightly less complex questions. The questions, correct SPARQL queries, correct answers and our evaluation results are available online.[7]

QALD6T3 provides several challenges that are supported by the CubeQA algorithm. These are implied aggregations, intervals, implied or differently referenced measures and numerical values that are contained in several component properties. It also includes questions that require features not provided by CubeQA, such as SPARQL subqueries. The performance of CubeQA on the benchmark is measured as follows: Given C the correct set of resources and O the output of the algorithm, we define precision $p = \frac{|C \cap O|}{|O|}$, recall $r = \frac{|C \cap O|}{|C|}$ and the F_1-score $F_1 = 2\frac{pr}{p+r}$. The average global F_1 score calculates $p = 0$ for empty answers.

Results. Of the 100 questions, 82 resulted in a nonempty answer, with an average precision of 0.401, a recall of 0.324 and an F_1 score of 0.392. Expected Answer Typing positively impacts the performance, as its removal results in a significant decrease in all three scores. Due to the cube index, many questions can be answered even if they do not specify their target dataset. With all the 50 datasets as candidates, the performance drops even more than without using answer typing, but the index chooses the dataset correctly for the majority of the questions (74 of 100). Answering the 100 questions on a PC with an Intel Core i5-3230M CPU, hosting both the SPARQL endpoint and the system implementation, took

---

Table 4. Mapping m of a question word to a set of expected answer types E, along with the frequency of each question word in the benchmark. When the question word is unknown or not found, or the unspecific "what" is used, all 6 answer types are possible.

Question word	Expected answer type	f
What	Uncountable, countable, count, temporal, location, entity	35
How much	Uncountable	33
Which	Temporal, location, entity	19
How many	Countable, count	6
When	Temporal	4
None, other	Uncountable, countable, count, temporal, location, entity	3
Total		100

87.45 s, 63.29 s and 63.33 s on three consecutive runs with preexisting index structures.[8] Table 5 shows the runtime distributions for the core tasks. Without preexisting index structures, the runs took 228.11 s, 228.77 s and 224.73 s, respectively.

4.3 Research Question Summary

A brief summary of the initial research questions is as follows: Q1: CubeQA is sufficiently powerful to be applied on challenging questions over statistical data and we believe it will be a strong baseline for future research. Q2: Precision is higher than recall, similar to general SQA systems, on QALD6T3-train but similar on QALD6T3-test. Q3: CubeQA achieves a global F_1 score of 44%, surpassed by QA^3 with 53%, using a template-base algorithm. Q4: The most common cause for problems is ambiguity, followed by the lexical gap and query structure.

4.4 Discussion

Comparison. We believe that CubeQA will be a strong baseline in this new research subfield. As QALD6T3 was launched prior to submitting this publication to attract further research, two additional systems emerged: the yet unpublished QA^3 RDCQA system and the Sparklis [8] query builder (see Table 6). A query builder lets the user construct queries visually by selecting and combining SPARQL features and knowledge base resources. It enables users to create SPARQL queries and, if they build those queries correctly, achieves high accuracies. As such it occupies a middle ground, both in accuracy and usability, between RDCQA and manually creating SPARQL queries. QA^3 achieves a 9%

---

[8] The higher initial time is assumed to be caused by cache warmup both in the system and the SPARQL endpoint.

Table 5. Runtimes and error causes.

task	$t(ms)$
SPARQL	25489
scoring	23326
index lookup	16070
parsing	5066
detectors	466
answer typing	13
total	61673

error cause	n
ambiguity	30
lexical gap	18
query structure	17
unknown	1
no error	34
total errors	66

(a) Runtimes of the core tasks on QALD6T3-train with preexisting cache structures. SPARQL querying, scoring and index lookup are intersecting and not all tasks are measured, so that the times do not add up to the total.

(b) Categorization of errors from the different benchmark questions (at most one error per question), including the categories automatically excluded before the evaluation.

higher f-score than CubeQA but due to its purely template-base approach, it is unclear how it performs on open domain questions.

Limitations. CubeQA does not support query structures that require SPARQL subqueries, express negations of facts or unions of concepts. Ambiguities and lexical gaps are hard challenges that are not solved yet [14]. Nevertheless, they occur in almost every question and must be adressed by every SQA system to avoid massive penalties to precision and recall. Table 5 categorizes the different errors that prevented CubeQA from returning a correct result to a question.

The most common cause is ambiguity, which mainly results from a high number of similar resources or equal numbers in the observation values. In benchmark question 86, "How much was budgeted for general services for the Office of the President of Sierra Leone in 2013?", two different properties contain the literal "Office of the President". Because only the property value and not the property name is referenced, the algorithm cannot determine which property is correct. SQA systems like TBSL [26] resolve ambiguity by template scoring, so that the user chooses among the top n, where candidate combinations are ranked highest that maximize textual and semantic relatedness between the candidates [22]. But this approach is not applicable to RDCQA because of the RDC meta model, where component properties are not directly connected.

Instead, CubeQA relies on references consisting of a name reference as well as a value reference, as in "the year 2008", where the name-value pair with the maximal score product of the name reference and the value reference is chosen. In case such a two-part reference does not occur, it is alleviated by giving temporal dimensions priority to others. For example, "2008" gets mapped to the year, if it exists, rather than the more improbable measurement value.

Table 6. QALD-6T3 performance of, indicated by average precision (over defined values), recall, and global F_1-score, rounded to 2 decimal places. The training set is used for evaluation, as it contains 100 harder questions compared to 50 in the test set. The correct target RDC was predefined for the training set, as the cube index is evaluated separately, with 74 of 100 correct choices.

Algorithm	Benchmark	$\varnothing p$	$\varnothing r$	$\varnothing F_1$
CubeQA	Train	0.40	0.32	0.32
QA^3	Test	0.59	0.62	0.53
CubeQA	Test	0.49	0.41	0.44
Sparklis	Test	0.96	0.94	0.95

The second most common cause the lexical gap, where a reference could not be mapped to an entity due to the differences in surface forms. It is caused, among others, by different capitalization, typing errors, word transpositions ("extended amount", "amount extended") and different word forms ("committed", "commitments"). Another issue with the lexical gap is that a measurement can be referenced using a quantity reference ("amount"), a unit ("How many dollars are given"), or the type ("aid"), of which only the first one guarantees a match. Thus, CubeQA matches the range of a property as well as a its label. The RDC vocabulary provides sdmx-attribute:unitMeasure to specify units of measurement, but it does not support multiple measures so that the fallback has the same effect. In case of future vocabulary specification updates, we plan to integrate measurement units into our approach.

All of those, except typing errors, occur in the benchmark. As these mentioned causes occur in document retrieval and Web search as well, full text indexes have been developed that robustly handle those problems. The employed Lucene index cannot overcome the lexical gap in some cases, which are not recognized by the stemmer and where the edit distance is too large for the fuzzy index as well. Sometimes a concept is implicitly required but there is no explicit reference at all. Implicit references are part of future work and include aggregates.

5   Related Work

SQA in general is an active and established area of research with too many systems to cite individually but *surveys* [2,6,9,14,17] give a qualitative overview of the field. Also, *evaluation campaigns* present quantitative comparisons with benchmarks on either general tasks like QALD [5] or specialized tasks like BioASQ [25]. RDCQA has not existed until recently, but non-semantic QA is implemented by Wolfram—Alpha, which queries several structured sources using the computational platform *Mathematica* [27], but the source code and algorithm are not published. We inspired the RDCQA sub-field by discussing RDCs in relation to SQA and by categorizing of a statistical question corpus [12]. Next, we developed CubeQA and QALD6-T3 to stimulate further research, which led to the development of QA^3 and Sparklis (see Sect. 4.4).

CubeQA uses time intervals for handling dates, similar to the system in [24] that uses the *Clinical Narrative Temporal Relation Ontology (CNTRO)* to incorporate the time dimension in answering clinical questions. The ontology is based on Allen's Interval Based Temporal Logic [1] but it represents time points as well. The framework includes a reasoner for time inference, for example based on the transitivity of the *before* and *after* relations. The time dimension is used there to identify the direction of possible causality between different events.

Furthermore, CubeQA generates query templates recursively, which is similarly employed by Intui2 [7], which uses DBpedia and is based on *synfragments*, minimal parse subtrees of a question, that are combined based on syntactic and semantic characteristics to create the final query.

The motivation to develop RDCQA algorithms and their benefit rises with the quantity, quality and significance of available RDCs. On the flipside, we expect that the emergence and improvement of RDCQA algorithms increases the value of RDCs. Because of this interdependence, we summarize efforts to improve the quality of, create and publish RDF in general and RDCs in particular: RDCs are usually created by transforming databases or other structured data sources using either custom software or mapping languages like R2RML[9] and SML [23]. Eurostat—Linked Data[10] transforms tabular data of Eurostat[11], providing statistics for comparing the European countries and regions. Linked-Spending [13] uses the OpenSpending JSON API to provide finance data from countries around the world. The most widely used statistical data format is SDMX (Statistical Data and Metadata eXchange), which can be transformed to RDCs using SDMX-ML [4]. A systematic review of Linked Data quality [28] provides a qualitative analysis over established approaches, tools and metrics.

6   Conclusions and Future Work

We introduce RDCQA and design the CubeQA algorithm, provide a benchmark based on real data, and evaluate the results. In future work, we plan to continue contributing to the yearly QALD evaluation campaign by providing progressively more challenging benchmarks. The next iteration of CubeQA will answer questions that require the consolidation of several RDCs. We will also investigate how to integrate RDCQA techniques with SQA frameworks, such as OpenQA [18], so that all-purpose systems can also answer questions on RDCs. On the flipside, we also plan to integrate general SQA into RDCQA, to answer questions on RDCs that require world knowledge. We also identified the following improvements:

- Implement selection filters as logical formula of constraints instead of flat sets, including negations and unions.

---

[9] https://www.w3.org/TR/r2rml.
[10] http://eurostat.linked-statistics.org/.
[11] http://ec.europa.eu/eurostat.

- Support SPARQL subqueries to handle nested information dependencies.
- Support languages other than English using language detection components as well as fitting parsers, indexes and preprocessing templates.
- Incorporate measurement units if the RDC vocabulary adds support for them for multiple measures. For elaborate phrase patterns, like "How many people live in" for "population", there are pattern libraries like BOA [10] which need to be adapted to statistical data by retraining on a comprehensive statistical question corpus.

Overall, we believe to have opened a novel research subfield within SQA, which will increase in importance due to the rise of both the volume of statistical data and the usage of QA approaches in everyday life.

Acknowledgment. This work was supported by a grant from the EU H2020 Framework Programme provided for the project HOBBIT (GA no. 688227).

References

1. Allen, J.F.: Maintaining knowledge about temporal intervals. Commun. ACM **26**(11), 832–843 (1983)
2. Athenikos, S., Han, H.: Biomedical question answering: a survey. Comput. Meth. Programs Biomed. **99**(1), 1–24 (2010)
3. Berners-Lee, T.: Linked Data-Design issues, W3C design issue (2009)
4. Capadisli, S., Auer, S., Ngonga Ngomo, A.C.: Linked SDMX data. Semant.Web J. **6**(2), 105–112 (2015)
5. Cimiano, P., Lopez, V., Unger, C., Cabrio, E., Ngonga Ngomo, A.-C., Walter, S.: Multilingual question answering over linked data (QALD-3): lab overview. In: Forner, P., Müller, H., Paredes, R., Rosso, P., Stein, B. (eds.) CLEF 2013. LNCS, vol. 8138, pp. 321–332. Springer, Heidelberg (2013)
6. Cimiano, P., Minock, M.: Natural language interfaces: what Is the problem? – A data-driven quantitative analysis. In: Horacek, H., Métais, E., Muñoz, R., Wolska, M. (eds.) NLDB 2009. LNCS, vol. 5723, pp. 192–206. Springer, Heidelberg (2010)
7. Dima, C.: Intui2: A prototype system for question answering over linked data. In: Forner, P., Navigli, R., Tufis, D. (eds.) Question Answering over Linked Data (QALD-3), CLEF 2013 Evaluation Labs and Workshop, Online Working Notes (2013)
8. Ferré, S.: Sparklis: an expressive query builder for SPARQL endpoints with guidance in natural language. Semant. Web J. (2015)
9. Freitas, A., Curry, E., Oliveira, J., O'Riain, S.: Querying heterogeneous datasets on the Linked Data Web: challenges, approaches, and trends. IEEE Internet Comput. **16**(1), 24–33 (2012)
10. Gerber, D., Ngomo, A.-C.N.: Extracting multilingual natural-language patterns for RDF predicates. In: ten Teije, A., Völker, J., Handschuh, S., Stuckenschmidt, H., d'Acquin, M., Nikolov, A., Aussenac-Gilles, N., Hernandez, N. (eds.) EKAW 2012. LNCS, vol. 7603, pp. 87–96. Springer, Heidelberg (2012)
11. Hirschman, L., Gaizauskas, R.: Natural language question answering: the view from here. Nat. Lang. Eng. **7**(4), 275–300 (2001)

12. Höffner, K., Lehmann, J.: Towards question answering on statistical linked data. In: Proceedings of the 10th International Conference on Semantic Systems, SEM 2014, pp. 61–64. ACM (2014)
13. Höffner, K., Martin, M., Lehmann, J.: LinkedSpending: OpenSpending becomes Linked Open Data. Semant. Web J. **7**, 95–104 (2015)
14. Höffner, K., Walter, S., Marx, E., Usbeck, R., Lehmann, J., Ngonga Ngomo, A.C.: Survey on challenges of Question Answering in the Semantic Web. Semant. Web J. (2016, submitted)
15. Kondrak, G.: N-Gram similarity and distance. In: Consens, M.P., Navarro, G. (eds.) SPIRE 2005. LNCS, vol. 3772, pp. 115–126. Springer, Heidelberg (2005)
16. Lehmann, J., Bizer, C., Kobilarov, G., Auer, S., Becker, C., Cyganiak, R., Hellmann, S.: DBpedia - a crystallization point for the web of data. J. Web Semant. **7**(3), 154–165 (2009)
17. Lopez, V., Uren, V., Sabou, M., Motta, E.: Is question answering fit for the semantic web? A survey. Semant. Web J. **2**(2), 125–155 (2011)
18. Marx, E., Usbeck, R., Ngomo Ngonga, A.C., Höffner, K., Lehmann, J., Auer, S.: Towards an open Question Answering architecture. In: SEMANTiCS 2014 (2014)
19. Piotrowski, S.J., Van Ryzin, G.G.: Citizen attitudes toward transparency in local government. Am. Rev. public Adm. **37**(3), 306–323 (2007)
20. Prud'hommeaux, E., Seaborne, A.: SPARQL query language for RDF. W3C Recommendation (2008)
21. Schulz, K.U., Mihov, S.: Fast string correction with Levenshtein automata. Int. J. Doc. Anal. Recogn. **5**(1), 67–85 (2002)
22. Shekarpour, S., Ngonga Ngomo, A.C., Auer, S.: Query segmentation and resource disambiguation leveraging background knowledge. In: Proceedings of WoLE Workshop (2012)
23. Stadler, C., Unbehauen, J., Westphal, P., Sherif, M.A., Lehmann, J.: Simplified RDB2RDF mapping. In: Proceedings of the 8th Workshop on Linked Data on the Web (LDOW2015), Florence, Italy (2015)
24. Tao, C., Solbrig, H.R., Sharma, D.K., Wei, W.-Q., Savova, G.K., Chute, C.G.: Time-oriented question answering from clinical narratives using semantic-web techniques. In: Patel-Schneider, P.F., Pan, Y., Hitzler, P., Mika, P., Zhang, L., Pan, J.Z., Horrocks, I., Glimm, B. (eds.) ISWC 2010, Part II. LNCS, vol. 6497, pp. 241–256. Springer, Heidelberg (2010)
25. Tsatsaronis, G., Schroeder, M., Paliouras, G., Almirantis, Y., Androutsopoulos, I., Gaussier, E., Gallinari, P., Artieres, T., Alvers, M.R., Zschunke, M., et al.: BioASQ: a challenge on large-scale biomedical semantic indexing and Question Answering. In: 2012 AAAI Fall Symposium Series (2012)
26. Unger, C., Bühmann, L., Lehmann, J., Ngonga Ngomo, A.C., Gerber, D., Cimiano, P.: Template-based Question Answering over RDF data. In: Proceedings of the 21st International Conference on World Wide Web, pp. 639–648 (2012)
27. Wolfram, S.: The Mathematica Book, vol. 100, pp. 61820–67237. Cambridge University Press and Wolfram Research Inc., New York (2000)
28. Zaveri, A., Rula, A., Maurino, A., Pietrobon, R., Lehmann, J., Auer, S.: Quality assessment for Linked Data: a survey. Semant. Web J. **7**, 63–93 (2015)

Optimizing Aggregate SPARQL Queries Using Materialized RDF Views

Dilshod Ibragimov[1,2]($\boxtimes$), Katja Hose[2], Torben Bach Pedersen[2],
and Esteban Zimányi[1]

[1] Université Libre de Bruxelles, Brussels, Belgium
{dibragim,ezimanyi}@ulb.ac.be
[2] Aalborg University, Aalborg, Denmark
{diib,khose,tbp}@cs.aau.dk

Abstract. During recent years, more and more data has been published as native RDF datasets. In this setup, both the size of the datasets and the need to process aggregate queries represent challenges for standard SPARQL query processing techniques. To overcome these limitations, materialized views can be created and used as a source of precomputed partial results during query processing. However, materialized view techniques as proposed for relational databases do not support RDF specifics, such as incompleteness and the need to support implicit (derived) information. To overcome these challenges, this paper proposes MARVEL (MAterialized Rdf Views with Entailment and incompLetness). The approach consists of a view selection algorithm based on an associated RDF-specific cost model, a view definition syntax, and an algorithm for rewriting SPARQL queries using materialized RDF views. The experimental evaluation shows that MARVEL can improve query response time by more than an order of magnitude while effectively handling RDF specifics.

1 Introduction

The growing popularity of the Semantic Web encourages data providers to publish RDF data as Linked Open Data, freely accessible, and queryable via SPARQL endpoints [25]. Some of these datasets consist of billions of triples. In a business use case, the data provided by these sources can be applied in the context of On-Line Analytical Processing (OLAP) on RDF data [5] or provide valuable insight when combined with internal (production) data and help facilitate well-informed decisions by non-expert users [1].

In this context, new requirements and challenges for RDF analytics emerge. Traditionally, OLAP on RDF data was done by extracting multidimensional data from the Semantic Web and inserting it into relational data warehouses [19]. This approach, however, is not applicable to autonomous and highly volatile data on the Web, since changes in the sources may lead to changes in the structure of the data warehouse (new tables or columns might have to be created) and will impact the entire Extract-Transform-Load process that needs to reflect the changes.

© Springer International Publishing AG 2016
P. Groth et al. (Eds.): ISWC 2016, Part I, LNCS 9981, pp. 341–359, 2016.
DOI: 10.1007/978-3-319-46523-4_21

In comparison to relational systems, native RDF systems are better at handling the graph-structured RDF model and other RDF specifics. For example, RDF systems support triples with *blank* nodes (triples with unknown components) whereas relational systems require all attributes to either have some value or *null*. Additionally, RDF systems support entailment, i.e., new information can be derived from the data using RDF semantics while standard relational databases are limited to explicit data.

Processing analytical queries in the context of Linked Data and federations of SPARQL endpoints has been studied in [15,16]. However, performing aggregate queries on large graphs in SPARQL endpoints is costly, especially if RDF specifics need to be taken into account. Thus, triple stores need to employ special techniques to speed up aggregate query execution. One of these techniques is to use materialized views – named queries whose results are physically stored in the system. These aggregated query results can then be used for answering subsequent analytical queries. Materialized views are typically much smaller in size than the original data and can be processed faster.

In this paper, we consider the problem of using materialized views in the form of RDF graphs to speed up analytical SPARQL queries. Our approach (MARVEL) focuses on the issues of selecting RDF views for materialization and rewriting SPARQL aggregate queries using these views. In particular, the contributions of this paper are:

- A cost model and an algorithm for selecting an appropriate set of views to materialize in consideration of RDF specifics
- A SPARQL syntax for defining aggregate views
- An algorithm for rewriting SPARQL queries using materialized RDF views

Our experimental evaluation shows that our techniques lead to gains in performance of up to an order of magnitude.

The remainder of this paper is structured as follows. Section 2 discusses related work. Section 3 introduces the used RDF and SPARQL notation and describes the representation of multidimensional data in RDF. Section 4 specifies the cost model for view selection, and Sect. 5 describes query rewriting. We then evaluate MARVEL in Sect. 6, and Sect. 7 concludes the paper with an outlook to future work.

2   Related Work

Answering queries using views is a complex problem that has been extensively studied in the context of relational databases [13]. However, as discussed in [13,22], aggregate queries add additional complexity to the problem.

In relational systems, the literature proposes semantic approaches for rewriting queries [22] as well as syntactic transformations [11]. However, SPARQL query rewriting is more complex. The results for views defined as *SELECT* queries represent solutions in tabular form, so that the solutions need to be converted afterwards into triples for further storage, thus making a view definition in SPARQL more complex and precluding view expansion (replacing the view by its definition).

Another problem in this context is to decide which views to materialize in order to minimize the average response time for a query. [14] addresses this problem in relational systems by proposing a cost model leading to a trade-off between space consumption and query response time for an arbitrary set of queries. [23] provides a method to generate views for a given set of select-project-join queries in a data warehouse by detecting and exploiting common sub-expressions in a set of queries. [2] further optimizes the view selection by automatically selecting an appropriate set of views based on the query workload and view materialization costs. However, these approaches have been developed in the context of relational systems and, therefore, do not take into account RDF specifics such as entailment, the different data organization (triples vs. tuples), the *graph*-like structure of the stored data, etc.

The literature proposes some approaches for answering SPARQL queries using views. [4] proposes a system that analyzes whether query execution can be sped up by using precomputed partial results for conjunctive queries. The system also reduces the number of joins between tables of a back-end relational database system. While [4] examines core system improvements, [18] considers SPARQL query rewriting algorithms over a number of virtual SPARQL views. The algorithm proposed in [18] also removes redundant triple patterns coming from the same view and eliminates rewritings with empty results. Unlike [18], [9] examines materialized views. Based on a cost model and a set of user defined queries, [9] proposes an algorithm to identify a set of candidate views for materialization that also account for implicit triples. However, these approaches [4,9,18] focus on conjunctive queries only. The complexity of loosing the multiplicity on grouping attributes (by grouping on attribute X, we loose the multiplicity of X in data) and aggregating other attributes is not addressed by these solutions.

The performance gain of RDF aggregate views has been empirically evaluated in [17], where views are constructed manually and fully match the predefined set of queries. Hence, the paper empirically evaluates the performance gain of RDF views but does not propose any algorithm for query rewriting and view selection.

Algorithms that use the materialized result of an RDF analytical query to compute the answer to a subsequent query are proposed in [3]. The answer is computed based on the intermediate results of the original analytical query. However, the approach does not propose any algorithm for view selection. It is applicable for the subsequent queries and not to an arbitrary set of queries.

Although several approaches consider answering queries over RDF views [4,9,18], none of them considers analytical queries and aggregation. In this paper, we address this problem in consideration of RDF specifics such as entailment and data organization in the form of triples, and taking into account the graph structure of the stored data. In particular, this paper proposes techniques for cost-based selection of materialized views for aggregate queries, query rewriting techniques, and a syntax for defining such views.

3   RDF Graphs and Aggregate Queries

The notation we use in this paper follows [21, 25] and is based on three disjoint
sets: blank nodes B, literals L, and IRIs I (together BLI). An RDF triple
$(s, p, o) \in BI \times I \times BLI$ connects a subject s through property p to object o.
An RDF dataset (G) consists of a finite set of triples and is often represented
as a graph. Queries are based on graph patterns that are matched against G.
A Basic Graph Pattern (BGP) consists of a set of triple patterns of the form
$(IV) \times (IV) \times (LIV)$, where V ($V \cap BLI = \emptyset$) is a set of query variables. Variable
names begin with a question mark symbol, e.g., $?x$. In graph notation, a BGP
can be represented as a directed labeled multi-graph whose nodes N correspond
to subjects and objects in the triple patterns. The set of edges E contains one
edge for each triple pattern and the property as its label. In data analytics, graph
patterns have a special, *rooted* pattern [5]. A BGP is rooted in node $n \in N$ iff
any node $x \in N$ is reachable from n following directed edges in the graph.

The most common SPARQL [25] aggregate queries conform to the form
SELECT RD *WHERE* GP *GROUP BY* GRP, where RD is the result description based on a subset of variables in the graph pattern GP. GP defines a BGP
and optionally uses functions, such as *assignment* (e.g., *BIND*) and constraints
(e.g., *FILTER*). GRP defines a set of grouping variables, whereas RD contains
selection description variables as well as aggregation variables with corresponding aggregate functions. In this paper, we consider the standard aggregate functions *COUNT, SUM, AVG, MIN,* and *MAX*.

Data that SPARQL analytical queries are typically evaluated on can be represented in an n-dimensional space, called a *data cube*. The data cube is defined
by *dimensions* (perspectives used to analyze the data) and *observations* (facts).
Dimensions are structured in *hierarchies* to allow analyses at different aggregation levels. Hierarchy level instances are called *members*. Observations (cells of
a data cube) have associated values called *measures*, which can be aggregated.

An example of data with hierarchical dimensions is the utilities
consumption data from electricity
and gas meters for Scottish Government buildings[1] enhanced in the
Date dimension. The data is available as energy usage over a daily
period. Figure 1 sketches the data

Fig. 1. Representing observations in RDF

with two hierarchical dimensions: Building, Locality, and Region are hierarchy
levels in the Geography dimension, Report Date, Month, and Year are hierarchy levels in the Date dimension, and Data represents an observation with
the utility consumption measure. A dataset with such observations is stored in a
SPARQL endpoint to enable analytical querying with grouping on different hierarchy levels. For example, the following query computes the daily consumption
of electricity in each city in September 2015:

---

[1] http://cofog01.data.scotland.gov.uk/id/dataset/golspie/utilities.

```
SELECT ?dt ?plc (SUM(?v) as ?value) WHERE {
  ?slc gol:refBuilding ?bld ; gol:reportDateTime ?tm ; qb:observation ?ob .
  ?ob  gol:utilityConsumption ?v . ?bld org:siteAddress/vc:adr/vc:locality ?plc .
  ?mn skos:narrower ?tm . ?mn gol:value ?mVal . FILTER (?mVal = 'September 2015')
} GROUP BY ?tm ?plc
```

Listing 1.1. Example query with grouping and aggregation

Based on the multidimensional data model, we can define traditional OLAP operations over the data such as *roll-up*, *drill-down*, *slicing*, and *dicing*. Intuitively, the slice operator fixes a single value for a level of a dimension to define a subcube with one dimension less. Dice uses a Boolean condition and returns a new cube containing only the cells satisfying the condition. Roll-up aggregates measure values at a coarser granularity for a given dimension while drill-down disaggregates previously summarized data to a child level in order to obtain measure values at a finer granularity. In SPARQL, slice and dice can be achieved by adding a constraint function (like *FILTER*) to the graph pattern, while roll-up and drill-down can be achieved by removing/adding connected triple patterns to the existing graph pattern of the query.

4   View Materialization in MARVEL

A high number of triples needs to be processed for evaluating OLAP queries on a dataset. This imposes high execution costs, especially when the amount of data increases. To enable scalable processing, we propose RDF-specific techniques to select a set of materialized views that can be used to evaluate queries more efficiently. We define a materialized view as a named graph described by a query whose results are physically stored in a triple store. Given a query, the system checks whether the query can be answered based on the available materialized views. As materialized views are typically smaller than the original/raw data, this can yield a significant performance boost. Precalculating all possible aggregations over all dimension levels is usually infeasible as it requires much more space than the raw data [24]. Thus, it is important to find an appropriate set of materialized views to minimize the total query response time.

4.1   Creating Materialized RDF Views

Views used for rewriting conjunctive queries can be defined by *CONSTRUCT* queries with *WHERE* clauses [18]. Views defining queries for aggregate views are more complex since these views group and aggregate the original data. Grouping and aggregation are achieved by using *SELECT* queries. However, *SELECT* queries return data in a tabular format, not triples. Thus, the *CONSTRUCT* clause needs to define a new graph structure and triples for the obtained results. As only the combination of values for variables in the *GROUP BY* clause of a *SELECT* query is unique, we can use these values to construct the new triples.

Listing 1.2 gives an example of such a query, where the *SELECT* query aggregates utility consumptions by City and Date. We use the *IRI* and *STRAFTER* functions to create a resource identifier *id* based on the unique combination of City and Date. The *CONSTRUCT* clause then creates triples by connecting the *id* to the resulting aggregate and grouping values. The algorithm for constructing such view queries is similar to the algorithm described in [17].

```
CONSTRUCT { ?id gol:reportDate ?date ; gol:reportLocality ?vCity ;
     gol:utilityConsumption ?value . } WHERE {
  SELECT ?id ?date ?vCity (SUM(?cons) as ?value)
  WHERE { ?fact gol:refBuilding ?bld ; gol:reportDateTime ?date ;
  qb:observation ?data . ?data  gol:utilityConsumption ?cons .
  ?bld org:siteAddress/vc:adr/vc:locality ?vCity .
  BIND(IRI('http://ex.org/id#', CONCAT(STRAFTER(STR(?dt), 'http://'),
  STRAFTER(STR(?vCity), 'http://'))) AS ?id). } GROUP BY ?id ?date ?vCity }
```

Listing 1.2. Query to construct materialized view

4.2   Data Cube Lattice

To represent dependencies between views, we use the notion of a data cube lattice. The data cube lattice is, essentially, a schema of a data cube with connected nodes, where a node represents an aggregation by a given combination of dimensions. Nodes are connected if a node j can be computed from another node i and the number of grouping attributes of i corresponds to the number of attributes of j plus one. A view is defined by a query with the same grouping as in the corresponding node. For example, in case of 3 dimensions, Part (P), Customer (C) and Date (D), possible nodes (grouping combinations) are PCD, PC, PD, CD, P, C, D and *All* (all values are grouped into one group). In our example, the view corresponding to node PC can be computed from the view corresponding to node PCD. We denote this dependence relation as $PC \preceq PCD$ and refer to view PCD as the *ancestor* of view PC. In the presence of dimension hierarchies, the total number of different lattice nodes is $\prod_{i=1}^{k}(h_i + 1)$, where h_i represents the number of hierarchy levels in dimension i and $(h_i + 1)$ accounts for the top level *All*.

We use the data cube lattice since it formalizes which views (nodes) can be used to evaluate a particular query. Given a query grouping (*GROUP BY*), the lattice node with the exact same grouping (and its ancestors) can be used. Since these views are smaller in size than the raw data, calculating the answer from the views will be cheaper than calculating it from the raw data. Thus, to answer user queries we need to find an appropriate set of views so that the multidimensional queries posed against the data can be mapped to one of these views.

The data cube lattice has originally been proposed for selecting aggregate views in a relational framework [14]. This framework considers data that is complete and complies with a predefined schema, and therefore cannot be directly applied to RDF graphs that lack these characteristics. Additionally, RDF data may be *incomplete*. For instance, the canonicalized Ontology Infobox dataset from the DBpedia Download 3.8 contains birth place information for 266,205 persons (either as a country, a city or village, or both). However, out of 266,205 records, 16,351 records contain information only about the country of birth. Thus, the information *available* in the source may not contain the information that holds in the world (Open World Assumption) and, therefore, should *ideally* be present in the source. Accordingly, an incomplete data source is defined as a pair $\Omega = (G_a, G_i)$ of two graphs, where $G_a \subseteq G_i$. G_a corresponds to the available graph and G_i is the ideal graph [6]. Thus, a view is *incomplete* if its defining query over the available graph does not produce the same results as the defining query over the ideal graph: $[q_v]G_a \neq [q_v]G_i$.

Such incomplete views may not be used to answer queries involving the grouping over a higher hierarchy level than in the view. In the above example, the aggregation over the city of birth is incomplete and the city level view, due to incompleteness, cannot be used to roll-up to the country level even though the relationship City → Country between the levels holds. Instead, the aggregation over the country level needs to be computed from the raw data taking into account the derived information that connects cities of birth to the countries.

In summary, we need to account for the graph-like structure of the stored data, presence of implicit knowledge in data, and incompleteness of views for RDF data cubes. Therefore, we propose MARVEL – a novel aggregate view selection approach that, unlike earlier approaches, supports RDF-specific requirements.

4.3   MARVEL Cost Model

MARVEL assumes that RDF data are stored as triples. Thus, the cost of answering an aggregate SPARQL query in a generic RDF store is defined as the number of triples contained in the materialized view used to answer the query. This cost model is simple and works for the general case. More complex models that account for algorithms and auxiliary structures of a particular triple store are certainly possible.

The number of triples to represent an observation in an RDF view is $(n + m)$ where n is the number of dimensions and m is the number of measures. Thus, the size of a view w is equal to $Size(w) = (n + m) * N$, where N is the number of observations. This number is used to calculate the benefit of materializing the view. Note that the size of w serves as the cost of v if v is computed from w: $Cost(v) = Size(w)$. View sizes can be estimated using VoID statistics and cardinality estimation techniques [12], using a small representative subset and, in some cases, with $COUNT$ queries.

Let B_w be the benefit of view w. For every view v such that $v \preceq w$ the benefit of view w relative to v is calculated as $B_{w,v} = (Cost(v) - Size(w))$ if $Cost(v) > Size(w)$ and $B_{w,v} = 0$ otherwise. The difference between the current cost of view v and the possible cost of v (if the materialized view w is used to compute view v) contributes to the benefit of view w. We sum up the benefits for all appropriate views to receive the full benefit of view w: $B_w = \sum B_{w,v_i}$ for all i such that $v_i \preceq w$. Note that this value of benefit is absolute. If the storage space is limited, the benefit of each view per *unit space* can be considered instead. In this case, the value of the benefit is calculated by dividing the absolute benefit of the view by its size: $B'_w = \frac{B_w}{Size(w)}$.

In addition, our cost model needs to account for RDF specifics, such as incomplete views and complex and indirect hierarchies. We use an annotated QB4OLAP schema [7] to describe the dataset and extend it with information about the completeness of levels, the patterns for defining hierarchy steps (which predicates are used), the types of hierarchy levels, etc. This schema reflects the source data structure and does not require adding any triples to the graph. The schema is also used to define aggregate queries for the view selection process (Sect. 4.5). We chose QB4OLAP since unlike alternatives, such as AnS [5] and QB (http://www.w3.org/TR/vocab-data-cube/), QB4OLAP allows to define multidimensional concepts such as dimensions, levels, members, roll-up relations, complex hierarchies (e.g., ragged, recursive), etc.

For example, for a birth place dimension we can specify that the roll-up to the Country level should be calculated from both the City and the Person levels since for some people we might only know the birth country, whereas for others we know the city. When a hierarchy level is computed from several ancestor levels, we say that the view corresponding to this level should be calculated from a *set* of views (to avoid double-counting in such cases, MARVEL uses the *MINUS* statement). We denote this dependence relation as $w \preceq \{v_i, \ldots, v_n\}$, where w is the current view and $\{v_i, \ldots, v_n\}$ are the ancestor views. In general, we can distinguish the following roll-up cases:

- **Single path roll-up**: a view w can be derived from any of the views $v_1, \ldots, v_n$, i.e., $\exists w, v_1, \ldots, v_n$ such that $w \preceq v_i$ and $v_i \nmid v_j$ for $i, j = \{1, \ldots, n\}$
- **Multiple path roll-up**: a view w can be derived from the union of views $v_1 \cup \cdots \cup v_n$ while deriving w from any single v_i will be incomplete: $\exists w, v_1, \ldots, v_n$ such that $w \preceq \{v_i, \ldots, v_n\}$, $w \nmid v_i$, and $v_i \nmid v_j$ for $i, j = \{1, \ldots, n\}$

However, before selecting the views to materialize we should take into account implicit triples since they are considered to be part of the graph.

4.4   RDF Entailment

Accounting for implicit triples in views is necessary for returning a complete answer. The W3C RDF Recommendation (http://www.w3.org/RDF/) defines a number of entailment patterns which lead to deriving implicit triples from RDF datasets. RDF Schema (RDFS) entailment patterns are particularly interesting since RDFS encodes the domain semantics.

Aggregate queries are designed to run only on available correct data; computing the sum over a set of unknown values, for instance, would not yield any useful results. Hence, in this paper we focus on deriving implicit information based on existing data and specified semantics only. Deriving information unknown due to the Open World Assumption or adding missing information using logical rules are orthogonal problems that are difficult to solve in general [8] and therefore beyond the scope of this paper.

There are two main methods for processing queries when considering RDF entailment. In the *dataset saturation* approach, all implicit triples are material-

ized and added to the dataset. While requiring more space and complex maintenance, this method benefits from applying plain query evaluation techniques to
compute the answer. *Query reformulation*, on the other hand, leaves the dataset
unchanged but reformulates a query to a union of queries and increases the
overhead during query evaluation.

MARVEL uses the RDFS entailment regime during view materialization. The
system reformulates queries to materialize the complete answer of a view, which
allows us to leave a dataset unchanged but still account for implicit triples in
query answers. Taking into account that the evaluation of the view query takes
place once and the results are reused for other queries, we believe that this
overhead is justified.

4.5   MARVEL View Selection Algorithm

Given the open nature of SPARQL endpoints, we assume that all groupings
in user queries are equally likely. Algorithm 1 outlines the method for selecting
materialized views in MARVEL; the goal is to materialize N views with the
maximum benefit, regardless of their size.

Input: Set of views W, cube schema S, number of needed views N
Output: Selected views W'
1 $W' = \emptyset$ -- set of selected views ;
2 **while** $|W'| \leq N$ **do**
3   RecalculateViewCosts(W) ;
4       $\{V \times B\} = \emptyset$ -- set of views together with the benefit ;
5   **foreach** *view* $w \in W$ **do**
6          $\lfloor \{V \times B\} = \{V \times B\} \cup (w,$ CalculateBenefit(w)) ;
7   **foreach** $\{w_1...w_n\}$ *for which* $\exists v$ *such that* $v \preceq \{w_1...w_n\}$ *(according to S)* **do**
8          $\lfloor \{V \times B\} = \{V \times B\} \cup (\{w_1...w_n\},$ CalculateBenefit($\{w_1...w_n\}$)) ;
9   Select (set of) views w from $\{V \times B\}$ for which B_w is MAX and
        $|w| \leq (N - |W'|)$;
10      $\lfloor W' = W' \cup w$; $W = W \setminus w$;
11 **return** W' ;

Algorithm 1. Algorithm for selecting views to materialize in MARVEL

Given all views as candidates, we start by assigning each view initial costs
corresponding to the size of the original dataset (line 3). View costs are recalcu-
lated in each iteration to take previously selected view(s) into account. Then, we
compute the benefit of a candidate view for the cases when it is used to derive a
full answer (single path roll-up) to another view in the cube lattice (line 6). The
benefit of the candidate view is computed according to the cost model defined in
Sect. 4.3. The same algorithm is applied when a view should be computed from
a set of views (multiple path roll-up – line 8). In these cases, all the views in the
set are considered together.

Having calculated the benefit of the views, the algorithm selects the view with the maximum benefit and adds it to the set of views proposed for materialization. This process is repeated until we have identified N views.

5   Query Rewriting in MARVEL

There are several aspects that complicate the problem of rewriting queries over SQL aggregate views. First, in SPARQL a user query and a view definition may use different variables to refer to the same entity. Thus, the query rewriting algorithms require variable mapping to rewrite a query. A variable mapping maps elements of a triple pattern in the BGP of the view to the same elements of a triple pattern in the BGP of the query. Second, the algorithms need to match the new graph structure that is formed by the $CONSTRUCT$ query of the view to the graph patterns of the user query and possibly aggregate and group these data anew. Third, complex and indirect hierarchies present in RDF data complicate query rewriting and need to be taken into consideration.

The rewriting algorithms proposed in [9,18] target conjunctive queries and do not consider grouping and aggregation of data. Therefore, we built upon these algorithms and developed an algorithm to rewrite aggregate queries that identifies the views which may be used for query rewriting and selects the one with the least computational cost.

For ease of explanation, we split the algorithm used in MARVEL for aggregate query rewriting using views into two parts: an algorithm for identifying the best view for rewriting (Algorithm 2) and a query rewriting algorithm (Algorithm 3). In the algorithms, we need to look for dimension *roll-up paths* (RUPs), i.e., path-shaped joins of triple patterns of the form $\{(root, p_1, o_1), (s_2, p_2, o_2), \ldots, (s_n, p_n, d)\}$ where $root$ is the root of the BGP, p_x is a predicate from the set of hierarchy steps defined for hierarchies in a cube schema, and triple patterns in the path are joined by subject and object values, e.g., $o_{x-1} = s_x$. We denote such a RUP as $\delta_{p_{dim}}(d_i)$ where p_{dim} is a predicate connecting the root variable to the first variable in the roll-up path and d_i represents the last variable in the path. These algorithms use $\gamma(agg_N)$ and $\gamma(g_N)$ to denote sets of triple patterns in the CONSTRUCT clause $CnPtrn$ $\{(s, p_{C1}^V, g_1), \ldots, (s, p_{Cn}^V, g_n), (s, p_{Cm}^V, agg_m), \ldots, (s, p_{Ck}^V, agg_k)\}$ describing the results of aggregation, e.g., (s, p_{Cx}^V, agg_x), and grouping, e.g., (s, p_{Cx}^V, g_x).

The first step in Algorithm 2 is to replace all literals and IRIs in the user query with variables and corresponding $FILTER$ statements (line 2): $(?s, p, \#o) \rightarrow (?s, p, ?o)$. $FILTER(?o = \#o)$. We do this to make graph patterns of views and queries more compatible with each other, since the graph patterns in the aggregated views should not contain literals. This may also potentially increase the number of candidate views since we may now use the views grouping by the hierarchy level of the replaced literal and then apply restrictions imposed by the $FILTER$ statement.

To make the user query and the view query more compatible, we rename all variable names in the user query to the corresponding variable names in a

---

Input: Set of materialized views MV, query Q, data cube schema S
Output: Selected view w
1 $W = \emptyset$ -- Set of candidate views ;
2 $Q = ReplaceLiteralsAndURI(Q)$;
3 **foreach** $v \in MV$ **do**
4 $\quad Q = RenameVariables(Q, v)$;
5 $\quad \{d_1^Q, \ldots d_n^Q\} = FindMinimalRUP(Q)$;
6 $\quad \{d_1^v, \ldots d_n^v\} = FindMinimalRUP(v)$;
7 $\quad$ let $\{hlvl(d_1)^Q \ldots hlvl(d_n)^Q\}$ be a set of hierarchy levels of Q defined in S ;
8 $\quad$ let $\{hlvl(d_1)^v \ldots hlvl(d_m)^v\}$ be a set of hierarchy levels of v defined in S ;
9 $\quad agg^Q = \{\varphi(o_1), ..., \varphi(o_n)\}$ -- All aggregate expressions in Q ;
10 $\quad agg^v = \{\varphi(o_1), ..., \varphi(o_m)\}$ -- All aggregate expressions in v ;
11 $\quad$ **if** $agg^Q \subseteq agg^v$ and $(\{hlvl(d_1)^Q \ldots hlvl(d_n)^Q\} \preceq \{hlvl(d_1)^v \ldots hlvl(d_m)^v\})$
   such that $hlvl(d_i)^Q \preceq hlvl(d_i)^v$ for all i **then**
12 $\quad\quad W = W \cup v$;

13 **return** $w \in W$ with minimal costs ;

Algorithm 2. Algorithm for selecting a candidate view

---

view (line 4). We start from the root variable and replace all occurrences of this variable name in the user query with the name that is used in the view query. We then continue renaming variables that are directly connected to the previously renamed variables. We continue until we have renamed all corresponding variables in the user query.

Afterwards, for each dimension of the query graph pattern we define the appropriate roll-up path that the candidate view should have (lines 5–6). This path depends on the conditions (*FILTER* statements) and/or grouping related to the corresponding hierarchy and is the minimum of both; we take the roll-up paths to variables in *FILTER* and *GROUP BY* for the same dimension and keep only the triple patterns that are the same in both – common RUP. For example, if the query groups by regions of a country but the *FILTER* statement restricts the returned values to only some cities (*Region* $\preceq$ *City*), the required level of the hierarchy in the view should not be higher than the City level.

Then, we identify the hierarchy levels for all dimensions in the query and all dimensions in a view and compare them. We check that the hierarchy levels of all dimensions defined in the view do not exceed the needed hierarchy levels of the query and that the set of aggregate expressions defined in a view may be used to compute the aggregations defined in the query. The views complying with these conditions are added to the set of candidate views (line 12). Out of these views we select one with the least cost for answering the query (line 13).

Let us consider an example. Given the materialized view described in Listing 1.2 and the query of Listing 1.1, the system renames all variables in the query to the corresponding variable names in the view (i.e. *?place* → *?vCity*; *?fact* → *?obs*; *?val* → *?cons*) and defines the roll-up paths for the dimensions in the query (i.e. (*?fact gol:refBuilding/org:siteAddress/vc:adr/vc:locality ?vCity*) and (*?fact gol:reportDateTime ?date*)). Note that the roll-up path in the Date dimension contains the Date level and not the Month level since the query groups by dates. Then the system identifies the roll-up paths for the dimensions in the view (i.e. (*?fact gol:refBuilding/org:siteAddress/vc:adr/vc:locality ?vCity*) and (*?fact gol:reportDateTime ?date*)) and compares them. The system also identifies aggregation expressions in the query and the view (*?fact qb:observation/gol:utilityConsumption ?cons, (SUM(?cons) as ?value)*). Since the view contains the same aggregate expression and all necessary dimensions and the hierarchy levels of the dimensions in the view do not exceed those in the query, this view is added to the set of candidate views.

Given one of the collected views, MARVEL uses Algorithm 3 to rewrite a query. For every dimension in the query we identify the common roll-up path in the query and the view. In the rewritten query Q', these triple patterns will be replaced by the triple patterns from the *CONSTRUCT* clause of the view $(\gamma^V(c^V))$. The remaining triple patterns belonging to the dimensions $(\Delta(d^Q))$ remain unchanged (lines 4–11).

Input: View v, query Q
Output: Rewritten query Q'
1 $GP' = \emptyset; RD' = \emptyset; GBD = vars^Q_{GRP};$
2 let Φ^Q be *assignment* and *constraint* functions of Q ;
3 $GBGP' = \emptyset;$ -- A graph pattern of *GRAPH* statement ;
4 $qDims = \{\delta_p(d^Q)\ldots\}$ -- Set of RUP in query Q ;
5 $vDims = \{\delta_p(d^v)\ldots\}$ -- Set of RUP in view v ;
6 **foreach** $\delta_p(d^Q) \in qDims$ **do**
7      $\delta_p(c^Q) = \delta_p(d^Q) \cap \delta_p(d^v)$ -- Common RUP in Q and v ;
8      $\Delta(d^Q) = \delta_p(d^Q) \setminus \delta_p(c^Q)$ -- Remaining part of a RUP (remaining triple patterns) in Q after subtracting the part in common with v;
9   let $\gamma^v(c^v)$ be a triple pattern $\in CnPtrn$ such that $\gamma^v(c^v)$ represents $\delta_p(c^v)$;
10     $GP' = GP' \cup \Delta(d^Q); GBGP' = GBGP' \cup \gamma^v(c^v);$
11     $RD' = RD' \cup \{d^Q\};$
12 $agg^Q = \{\varphi(o_1), \ldots, \varphi(o_n)\}$ -- Aggregate expressions in Q over variables $\{o_1 \ldots o_n\}$;
13 $agg^v = \{\varphi(o_1), \ldots, \varphi(o_m)\}$ -- Aggregate expressions in v over variables $\{o_1 \ldots o_m\}$;
14 **foreach** $\varphi^Q(x) \in agg^Q$ **do**
15   let $\gamma^v(x)$ be a triple pattern $\in CnPtrn$ such that $\gamma^v(x)$ represents $\varphi^v(x) \in agg^v$ and $\varphi^v = \varphi^Q$;
16     $GBGP' = GBGP' \cup \gamma^v(x)$;
17     $RD' = RD' \cup \{f'(\gamma^v(x))\}$ where f' is a rewrite of the aggregate function φ ;
18 $GP' = GBGP' \cup GP' \cup \Phi^Q);$
19 $Q' = SELECT\ RD'\ WHERE\ GP'\ GROUP\ BY\ GBD$;
20 **return** Q' ;

Algorithm 3. Algorithm for query rewriting using a view

Afterwards, the algorithm compares the aggregate functions of the query and the *SELECT* clause of the view and identifies those that are needed for rewriting. We add the corresponding triple pattern from the *CONSTRUCT* clause and rewrite the aggregate functions to account for the type of the function (algebraic or distributive) (lines 12–17). *GROUP BY* and *ORDER BY* clauses do not change. Additionally, the triple patterns of the *CONSTRUCT* clause will be placed inside the *GRAPH* statement of the SPARQL query to account for the different storage of the view triple patterns (lines 10, 16, 18).

```
SELECT ?date ?vCity (SUM(?value) as ?aggValue)
FROM <http://data.gov.uk>  FROM NAMED <http://data.gov.uk/matview1>
WHERE { GRAPH <http://data.gov.uk/matview1> { ?id gol:reportDate ?date;
    gol:reportLocality ?vCity; gol:utilityConsumption ?value. }
  ?month skos:narrower ?date . ?month gol:value ?mVal .
  FILTER (?mVal = 'September 2015')  } GROUP BY ?date ?vCity
```

Listing 1.3. Rewritten query

Listing 1.3 shows the result of rewriting the query from Listing 1.1 using the view from Listing 1.2. The algorithm identifies common roll-up paths for the two dimensions in the view and in the query: *?fact gol:refBuilding/org:siteAddress/ vc:adr/vc:locality ?vCity* and *?fact gol:reportDateTime ?date*. The system replaces these triple patterns with the triple pattern from the *CONSTRUCT* clause and puts these replaced triple patterns inside the *GRAPH* statement. The remaining triple patterns in the Date dimension (*?month skos:narrower ?date .* and *?month gol:value ?mVal .*) are added to the query graph pattern outside the *GRAPH* statement. The aggregate function is rewritten; since *SUM* is a distributive function, it is rewritten using the same aggregation (*SUM*). All assignment and constraint functions (e.g., *FILTER*) are copied to the rewritten query.

6   Evaluation

To evaluate the performance gain for queries executed over materialized views against the queries over the raw data, we implemented MARVEL using the .NET Framework 4.0 and the dotNetRDF (http://dotnetrdf.org/) API with Virtuoso v07.10.3207 as triple store. The times reported in this section represent total response time, i.e., they include query rewriting and query execution. All queries were executed 5 times following a single warm-up run. The average runtime is reported for all queries. The triple store was installed on a machine running 64-bit Ubuntu 14.04 LTS with CPU Intel(R) Core(TM) i7-950, 24GB RAM, 600GB HDD.

6.1   Datasets and Queries

Unfortunately, none of the benchmarks for SPARQL queries are applicable to our setup. Data generators for benchmarks produce a complete set of data; they do not have an option to withhold some data and generate instead the implicit data that can be used to derive the missing information. Furthermore, existing benchmarks either do not define analytical SPARQL queries or do not require

RDFS entailment to answer these queries. Therefore, we decided to test our approach on 2 different datasets and adapt the data generators and queries to our needs. All queries, schemas, and datasets are available at http://extbi.cs.aau.dk/aggview.

LUBM [10] uses an ontology in the university domain. We decided to build our data cube and corresponding queries on the information related to courses. Inspired by [3], we defined 6 analytical SPARQL queries (using $COUNT$) involving grouping over several classification dimensions. These queries compute the number of courses offered by departments, number of courses taught by professors in each department, number of graduate courses in each department, etc. The data cube schema is defined in QB4OLAP, specifying 3 dimensions (Student, Staff, and Course), hierarchy levels, and steps between the levels. In total, the schema contains 183 triples.

We changed the data generator to omit some information that relates staff to courses. Instead, we introduced information about the department that offers these courses (*lubm:offeringDepartment*). In this

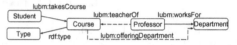

Fig. 2. Excerpt of an altered LUBM schema

case, the roll-up path Course → Staff → Department needs to be complemented by the roll-up path Course → Department and the aggregation of courses by Department cannot be answered by the results of the aggregation by Staff. A simplified schema of the data structure is presented in Fig. 2. We generated 3 datasets containing 30, 100, and 300 universities (4, 13.5 and 40M triples). We applied Algorithm 1 to select a set of views providing a good performance gain for answering user queries. The execution of the algorithm on a data cube lattice with 60 nodes and known view sizes took 213 ms.

To choose which views to materialize, we ran MARVEL's view selection algorithm and measured (i) the total query response time for all queries in the cube using materialized views whenever possible and (ii) the total space these views require. The unit in which we measured both space and time consumption is the number of triples. The results for the first 25 views sorted by their benefit are presented in Fig. 4a. Based on these results we decided to materialize the first 5 views where the benefit in total response time for the views is good compared to the growth in space consumption for storing these views.

Fig. 3. SSB Dataset in QB4OLAP format

Selecting more views substantially increases the used space while the total query time does not decrease significantly. Generating the views took 2:49, 8:38, and 18:47 min with 5, 15, and 33M triples in all views.

In our experiments we also used the Star Schema Benchmark (SSB) [20], originally designed for aggregate queries in relational database systems. This benchmark is well-known in the database community and was chosen for its well-defined testbed and its simple design.

The data in the SSB benchmark represent sales in a retail company; each transaction is defined as an observation described by 4 dimensions (Parts, Customers, Suppliers, and Dates). We translated the data into the RDF multidimensional representation (QB4OLAP) introducing incompleteness to this dataset as well, as illustrated in Fig. 3. An observation is connected to dimensions (objects) via certain predicates. Every connected dimension object is in turn defined as a path-shaped subgraph. Hierarchies in dimensions are connected via the *skos:broader* predicate. Measures (represented as rectangles in Fig. 3) are directly connected to observations. We changed the data generator to omit some information that relates suppliers to their corresponding cities in the Supplier dimension (and parts to their brands in the Part dimension). Instead, we connected suppliers with missing city information directly to their respective nations (*ssb:s_nation*) and parts with missing brand information directly to the categories (*ssb:p_category*). Thus, in the roll-up path Supplier → City → Nation → Region the City level is incomplete. The Part dimension is affected in the level Brand (Part → Brand → Category → Manufacturer). We used scaling factors 1 to 3 to obtain datasets of different sizes (122 to 365M triples).

SSB defines 13 classic data warehouse queries that are typical in business intelligence scenarios. We converted all 13 queries into SPARQL. Then we applied Algorithm 1 to select a set of materialized views. The execution of the algorithm on a cube lattice with 500 nodes and

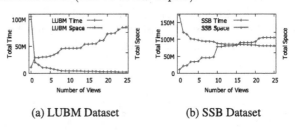

(a) LUBM Dataset   (b) SSB Dataset

Fig. 4. Time and space vs number of views

known view sizes took 11.8 seconds. We then conducted the same time and space analysis as described above (Fig. 4b). We identified and materialized 6 views with the maximum benefit and stored these views in named graphs. Generating the views took 20:42, 43:21, and 59:48 min with 104, 191 and 277M triples in all views.

6.2 Query Evaluation

LUBM. Figure 5 shows the results of executing the LUBM queries for 3 scale factors – queries with similar runtimes are grouped into separate graphs for better visualization. For queries over raw data we materialized the implicit triples and saved them to the dataset to avoid the entailment during query execution. Note that the performance gain for queries over materialized views becomes more evident with the growth in the volume of data, due to the growing difference in

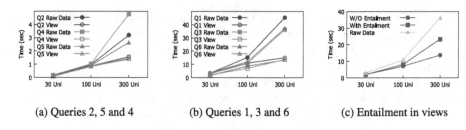

(a) Queries 2, 5 and 4   (b) Queries 1, 3 and 6   (c) Entailment in views

Fig. 5. Execution times of LUBM queries over raw data and views

their sizes. For scale factor 3 the execution of the queries over materialized views is on average 3 times faster.

We also compared the performance of the queries over views that take implicit triples into account and those that do not. Query 3 requests information on the number of courses taken by research assistants whose advisors are professors. We materialized 2 views: one takes into account that all professor ranks are subclasses of the more general class Professor and the other view does not. The execution of Query 3 over the view with implicit information for scale factor 3 was 1.7 times faster than the execution over the other view (Fig. 5c).

SSB. Given the set of materialized views, MARVEL was able to rewrite 10 out of the 13 queries. The other 3 benchmark queries (Q1, Q2, and Q3) apply restrictions on measures. Since the views group by dimensions and only store aggregates over the measures, these queries cannot be evaluated on any aggregate view.

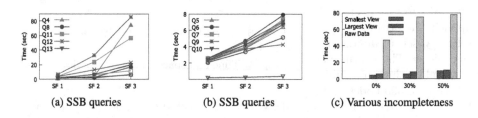

(a) SSB queries   (b) SSB queries   (c) Various incompleteness

Fig. 6. Execution times of SSB queries over raw data and views

Figure 6 shows the runtime of the queries evaluated on the original datasets and on the views (dashed lines of the same colors indicate the execution times over views). Our results for scale factor 3 show that evaluating queries using views is on average 5 times faster (up to 18 times faster for Query 10). This can be explained by the decreased size of the data and the availability of partial results.

We also compared the performance gain for queries over views with different
levels of incompleteness. For scale factor 3, we generated datasets with 0 %, 30 %,
and 50 % levels of incompleteness and identified a set of views for every dataset.
In each case, the set of materialized views is different due to the difference in the
size and the benefit of the views. We then evaluated the execution of Query 4
over the raw data and over the largest and the smallest view. The slight increase
in the query execution time over the raw data for incomplete datasets is caused
by a rewriting of the query into a more complex query. The results show that in
all cases the evaluation of queries over views is far more beneficial (on average
11 times more beneficial – Fig. 6c).

Additionally, we compared the performance gain of MARVEL to the approach in [3] which materializes partial results of user queries to answer subsequent queries. We used the original (non-modified) LUBM dataset containing
approx. 100M triples, analytical queries, and views introduced in the technical
report of [3]. The execution times for the queries over the original data and views
are reported in Fig. 7. As shown in the figure, MARVEL is on average more than
twice as fast as partial result materialization [3]. This can be explained by the
difference in the size of the data – partial results contain identifiers for facts
while our materialized views contain aggregated data only.

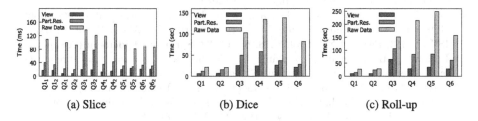

Fig. 7. Comparison with results from [3]

In summary, the experimental results show that MARVEL accounts for RDF-
specific requirements and finds an appropriate set of views that provide a good
balance between the benefit of the views and their storage space. The rewriting
algorithm of MARVEL is able to rewrite analytical SPARQL queries based on a
set of materialized views. The experiments also show that evaluating queries over
materialized views is on average 3–11 times faster than evaluating the queries
over raw data.

7 Conclusion and Future Work

In this paper, we have addressed the problem of selecting a set of aggregate RDF
views to materialize and proposed a cost model and techniques for choosing these
views. The selected materialized views account for implicit triples present in the
dataset. The paper also proposes a SPARQL syntax for defining RDF views and
an algorithm for rewriting user queries given a set of materialized RDF views.

A comprehensive experimental evaluation showed the efficiency and scalability of MARVEL resulting in 3–10 times speedup in query execution. In future work, we plan to investigate algorithms for incrementally maintaining the materialized views in the presence of updates.

Acknowledgments. This research is partially funded by the Erasmus Mundus Joint Doctorate in "Information Technologies for Business Intelligence – Doctoral College".

References

1. Abelló, A., Romero, O., Pedersen, T.B., Berlanga, R., Nebot, V., Aramburu, M., Simitsis, A.: Using semantic web technologies for exploratory OLAP: a survey. TKDE **27**(2), 571–588 (2015)
2. Agrawal, S., Chaudhuri, S., Narasayya, V.R.: Automated selection of materialized views and indexes in SQL databases. In: VLDB, pp. 496–505 (2000)
3. Azirani, E.A., Goasdoué, F., Manolescu, I., Roatis, A.: Efficient OLAP operations for RDF analytics. In: ICDE Workshops, pp. 71–76 (2015)
4. Castillo, R., Rothe, C., Leser, U., RDFMatView: indexing RDF data using materialized SPARQL queries. In: SSWS (2010)
5. Colazzo, D., Goasdoué, F., Manolescu, I., Roatis, A.: RDF analytics: lenses over semantic graphs. In: WWW, pp. 467–478 (2014)
6. Darari, F., Nutt, W., Pirrò, G., Razniewski, S.: Completeness statements about RDF data sources and their use for query answering. In: Alani, H., et al. (eds.) ISWC 2013, Part I. LNCS, vol. 8218, pp. 66–83. Springer, Heidelberg (2013)
7. Etcheverry, L., Vaisman, A., Zimányi, E.: Modeling and querying data warehouses on the semantic web using QB4OLAP. In: Bellatreche, L., Mohania, M.K. (eds.) DaWaK 2014. LNCS, vol. 8646, pp. 45–56. Springer, Heidelberg (2014)
8. Galárraga, L., Teflioudi, C., Hose, K., Suchanek, F.: Fast rule mining in ontological knowledge bases with AMIE+. VLDB J. **24**(6), 707–730 (2015)
9. Goasdoué, F., Karanasos, K., Leblay, J., Manolescu, I.: View selection in semantic web databases. PVLDB **5**(2), 97–108 (2011)
10. Guo, Y., Pan, Z., Heflin, J.: LUBM: a benchmark for OWL knowledge base systems. J. Web Semant. **3**(2–3), 158–182 (2005)
11. Gupta, A., Harinarayan, V., Quass, D.: Aggregate-query processing in data warehousing environments. In: VLDB, pp. 358–369 (1995)
12. Hagedorn, S., Hose, K., Sattler, K-U., Umbrich, J.: Resource planning for SPARQL query execution on data sharing platforms. In: COLD (2014)
13. Halevy, A.: Answering queries using views: a survey. VLDB J. **10**(4), 270–294 (2001)
14. Harinarayan, V., Rajaraman, A., Ullman, J.: Implementing data cubes efficiently. ACM SIGMOD **25**, 205–216 (1996)
15. Ibragimov, D., Hose, K., Pedersen, T.B., Zimányi, E.: Towards exploratory OLAP over linked open data – a case study. In: Castellanos, M., Dayal, U., Pedersen, T.B., Tatbul, N. (eds.) BIRTE 2013 and 2014. LNBIP, vol. 206, pp. 114–132. Springer, Heidelberg (2015)
16. Ibragimov, D., Hose, K., Pedersen, T.B., Zimányi, E.: Processing aggregate queries in a federation of SPARQL endpoints. In: Gandon, F., Sabou, M., Sack, H., d'Amato, C., Cudré-Mauroux, P., Zimmermann, A. (eds.) ESWC 2015. LNCS, vol. 9088, pp. 269–285. Springer, Heidelberg (2015)

17. Kämpgen, B., Harth, A.: No size fits all – running the star schema benchmark with SPARQL and RDF aggregate views. In: Cimiano, P., Corcho, O., Presutti, V., Hollink, L., Rudolph, S. (eds.) ESWC 2013. LNCS, vol. 7882, pp. 290–304. Springer, Heidelberg (2013)
18. Le, W., Duan, S., Kementsietsidis, A., Li, F., Wang, M.: Rewriting queries on SPARQL views. In: WWW, pp. 655–664 (2011)
19. Nebot, V., Berlanga, R.: Building data warehouses with semantic web data. DSS **52**(4), 853–868 (2012)
20. O'Neil, P., O'Neil, E.J., Chen, X.: The star schema benchmark (SSB). Technical report, UMass/Boston, June 2009
21. Pérez, J., Arenas, M., Gutierrez, C.: Semantics and complexity of SPARQL. In: Cruz, I., Decker, S., Allemang, D., Preist, C., Schwabe, D., Mika, P., Uschold, M., Aroyo, L.M. (eds.) ISWC 2006. LNCS, vol. 4273, pp. 30–43. Springer, Heidelberg (2006)
22. Srivastava, D., Dar, S., Jagadish, H., Levy, A.: Answering queries with aggregation using views. In: VLDB, pp. 318–329 (1996)
23. Theodoratos, D., Ligoudistianos, S., Sellis, T.K.: View selection for designing the global data warehouse. DKE **39**(3), 219–240 (2001)
24. Vaisman, A., Zimányi, E.: Data Warehouse Systems - Design and Implementation. Springer, Berlin (2014)
25. WWW Consortium. SPARQL 1.1 Query Language (W3C Recommendation 21 March 2013). http://www.w3.org/TR/sparql11-query/

Algebraic Calculi for Weighted Ontology Alignments

Armen Inants[(✉)], Manuel Atencia, and Jérôme Euzenat

Inria and University Grenoble Alpes, Grenoble, France
{Armen.Inants,Manuel.Atencia,Jerome.Euzenat}@inria.fr

Abstract. Alignments between ontologies usually come with numerical attributes expressing the confidence of each correspondence. Semantics supporting such confidences must generalise the semantics of alignments without confidence. There exists a semantics which satisfies this but introduces a discontinuity between weighted and non-weighted interpretations. Moreover, it does not provide a calculus for reasoning with weighted ontology alignments. This paper introduces a calculus for such alignments. It is given by an infinite relation-type algebra, the elements of which are weighted taxonomic relations. In addition, it approximates the non-weighted case in a continuous manner.

Keywords: Weighted ontology alignment · Algebraic reasoning · Qualitative calculi

1 Introduction

Ontology alignments are used for facilitating the integration of semantically related ontologies [8]. They are sets of correspondences relating entities from two ontologies using semantic relations such as equivalence ($\equiv$), subsumption ($\sqsubseteq, \sqsupseteq$) and disjointness ($\perp$). Very often, these correspondences are coupled with weights in $[0, 1]$. The intended meaning of these weights is a degree of confidence on the correspondence, i.e. a measure of how much we can trust that the correspondence is true. For example, the correspondence (AssociateProfessor, SeniorLecturer, $\sqsubseteq, 0.9$) states that the class AssociateProfessor is subsumed by the class SeniorLecturer with a confidence degree of 0.9, and, therefore, one should trust that this subsumption is true. The automatic treatment of ontology alignments calls for a calculus for reasoning with weighted correspondences. However, such a calculus has not been proposed yet.

In previous work, we advocated the algebraic approach to reasoning with ontology alignments [6,13]. An algebraic calculus of alignments is given by an algebra of ontology alignment relations. In this paper, we show how to compose *weighted* ontology alignment relations, based on their algebraic semantics.

Previous work introduced a formal semantics for weighted ontology alignments [1]. A weighted correspondence between two classes C and D is written

© Springer International Publishing AG 2016
P. Groth et al. (Eds.): ISWC 2016, Part I, LNCS 9981, pp. 360–375, 2016.
DOI: 10.1007/978-3-319-46523-4_22

$C\ r_{[a,b]}\ D$ where $r \in \{\sqsubseteq, \equiv, \sqsupseteq, \perp\}$ and a, b are real numbers in $[0, 1]$. The semantics is based on the classification interpretation of alignments, when a common finite set of instances is classified under classes of different ontologies. The weighted correspondence $C \sqsubseteq_{[a,b]} D$, for example, is interpreted as "the proportion of instances classified under C that are classified under D lies in the interval $[a, b]$." Although [1] provides some entailment rules for reasoning with weighted correspondences, none of these rules allows to compose alignment relations. In addition, the current semantics has some shortcomings, discussed below.

First, although the interval $[a, b]$ may be any subinterval of $[0, 1]$, in practice we are mostly interested in intervals of the form $[a, 1]$. Think, for example, of the previous correspondence (AssociateProfessor, SeniorLecturer, $\sqsubseteq, 0.9$). This should be translated into AssociateProfessor $\sqsubseteq_{[0.9,1]}$ SeniorLecturer and not AssociateProfessor $\sqsubseteq_{[0.9,0.9]}$ SeniorLecturer. Indeed, the latter is interpreted as "(exactly) 90 % of the associate professors are senior lecturers" from which it follows that the crisp subsumption is not true. However, the former is interpreted as "at least 90 % of the associate professors are senior lecturers" which leaves room for the possibility that the crisp subsumption is true. In general, $C\ r_{[a,b]}\ D \models \neg(C\ r\ D)$ if $b < 1$. Furthermore, from a theoretical point of view, if we restrict to $[a, 1]$ intervals, then weighted relations can be seen as relaxed crisp relations, i.e., $C\ r\ D \models C\ r_{[a,1]}\ D$, or equivalently $r \models r_{[a,1]}$. In what follows, r^a will replace $r_{[a,1]}$.

Second, one would expect that $\sqsubseteq^1 \models \neg\perp$. However, with the current semantics of the disjointness relation, this is not the case. Let us illustrate this with an example. Consider the classes BrazilianSnakes and VenomousSnakes, and imagine that 100 snakes are classified under these two classes, and that from these 100 snakes, 10 are Brazilian, and all of them are venomous. Thus, BrazilianSnakes $\sqsubseteq_{[1,1]}$ VenomousSnakes and BrazilianSnakes $\sqsupseteq_{[0.1,0.1]}$ VenomousSnakes. The weight of the equivalence relation is the harmonic mean of 1 and 0.1, i.e. BrazilianSnakes $\equiv_{[0.2,0.2]}$ VenomousSnakes, and the weight of the disjointness relation is 1 minus the harmonic mean, i.e. BrazilianSnakes $\perp_{[0.8,0.8]}$ VenomousSnakes.

Finally, although in the crisp case equivalence entails subsumption, i.e. $\equiv \models \sqsubseteq$, this does not hold in general for weighted correspondences, that is, from equivalence with a confidence interval $[a, 1]$ one cannot entail subsumption with (at least) the same confidence: $\equiv^a \not\models \sqsubseteq^a$ and $\equiv^a \not\models \sqsupseteq^a$ for any $a \in (0, 1)$. This becomes evident in the previous example, since, although BrazilianSnakes $\equiv_{[0.2,0.2]}$ VenomousSnakes, BrazilianSnakes $\sqsupseteq_{[0.1,0.1]}$ VenomousSnakes.

This weighted semantics is a generalization of the crisp or Boolean semantics: if all weights are 1, then the semantics is exactly the crisp semantics. However, the way it approaches the crisp semantics is, in a sense which will be explained in this paper, discontinuous: as close as the weighted semantics approaches the crisp one, these two properties ($\sqsubseteq^1 \models \neg \perp$ and $\equiv \models \sqsubseteq$) do not hold, but as soon as all weights are 1, they do.

In this paper, we propose a calculus for reasoning with weighted alignments based on the semantics that overcomes the shortcomings explained above.

The paper is structured as follows. Section 2 introduces the state of the art and other related work. Section 3 contains some mathematical notions and preliminary results, upon which the developments of this paper are based. The key notion that we employ is that of a (relational) constraint language. Section 4 introduces the constraint language QTAX of *quantified taxonomic relations*. We show that both crisp and weighted taxonomic relations can be expressed in QTAX. In Sect. 5, we specify a sublanguage of QTAX consisting of the relaxed taxonomic relations. We compare the revisited semantics of $\equiv^a$ and $\perp^a$ with the old one and discuss its advantages. In Sect. 6, we develop the calculus of relaxed taxonomic relations. Section 7 discusses how this calculus can be used to reason with weighted ontology alignments. Finally, Sect. 8 summarizes the results and provides some concluding remarks.

2 Related Work

Different semantics to weighted ontology alignments have been proposed [1,16].

[16] relies on tightly integrated description logics programs, i.e., pairs of description logic T-boxes and answer set programs. In that work, weights are interpreted as probabilistic distributions over *models*. We here concentrate on extensional interpretations.

The semantics proposed in [1] is based on a classificational interpretation of alignments: if O_1 and O_2 are two ontologies used to classify a common set X, then correspondences between O_1 and O_2 are interpreted as encoding how elements of X classified in the concepts of O_1 are re-classified in the concepts of O_2, and weights are interpreted to measure how precise and complete re-classifications are. Syntactically, a weighted correspondence between ontologies O_1 and O_2, expressed in a description logic [2], is an expression of the form:

$$1 : C \; r_{[a,b]} \; 2 : D,$$

such that C and D are concepts in O_1 and O_2 respectively, $r \in \{\sqsubseteq, \sqsupseteq, \equiv, \perp\}$ and $a, b \in [0, 1]$ $(a \le b)$.

The semantics of such correspondences is based on pairs of description logic interpretations $\mathcal{I}_1 = (U_1, \cdot^{\mathcal{I}_1})$ and $\mathcal{I}_2 = (U_2, \cdot^{\mathcal{I}_2})$ of O_1 and O_2 respectively. A pair of interpretations is a model of a weighted correspondence if the degree ds_X that can be computed from the interpretations lies within the interval $[a, b]$ assigned to the correspondence. The degrees are defined as follows:

$$\mathrm{ds}_X(\mathcal{I}_1, \mathcal{I}_2, C \sqsubseteq D) = \frac{|C_X^{\mathcal{I}_1} \cap D_X^{\mathcal{I}_2}|}{|C_X^{\mathcal{I}_1}|}$$

$$\mathrm{ds}_X(\mathcal{I}_1, \mathcal{I}_2, C \sqsupseteq D) = \frac{|C_X^{\mathcal{I}_1} \cap D_X^{\mathcal{I}_2}|}{|D_X^{\mathcal{I}_2}|}$$

$$\mathrm{ds}_X(\mathcal{I}_1, \mathcal{I}_2, C \equiv D) = \frac{2 \times \mathrm{ds}_X(\mathcal{I}_1, \mathcal{I}_2, C \sqsubseteq D) \times \mathrm{ds}_X(\mathcal{I}_1, \mathcal{I}_2, C \sqsupseteq D)}{\mathrm{ds}_X(\mathcal{I}_1, \mathcal{I}_2 C, \sqsubseteq D) + \mathrm{ds}_X(\mathcal{I}_1, \mathcal{I}_2, C \sqsupseteq D)}$$

$$\mathrm{ds}_X(\mathcal{I}_1, \mathcal{I}_2, C \perp D) = 1 - \mathrm{ds}_X(\mathcal{I}_1, \mathcal{I}_2, C \equiv D)$$

The interpretation of $\sqsubseteq$ and $\sqsupseteq$ are expressed as the proportion of common individuals in the class interpretations, which can also be interpreted as the probability of reclassification of individuals [1]. This is justifiable by extensional practice of ontology matching. The interpretation of $\equiv$ mitigates the impact of these two through the use of the F-measure between them, while $\perp$ is interpreted as the complement of equivalence. This semantics of weighted alignments approximates classical crisp semantics [4,18] in the sense that if all weights are assigned 1 or 0, i.e., $[1,1]$ or $[0,0]$, then the models are those of the crisp semantics.

However, as mentioned in the introduction, this semantics has some undesirable consequences and we will show how they can be addressed. For that purpose, we will reconsider it in the framework of algebras of relations.

The algebraic approach to reasoning with relational assertions, which we adopt in this paper, comes from the domain of qualitative spatial and temporal reasoning. This approach may also be applied to reasoning with aligned ontologies [6,13,15] and was extended to support relations between different kinds of entities [12]. The central notion is that of a *qualitative calculus* [5,14], which is a finite symbolic algebra used for constraint-based reasoning based on the path-consistency method. There exist reasoning toolboxes which support qualitative calculi [9,17]. The only principal difference of $\mathbb{A}_{\mathsf{QTAX}}$ from qualitative calculi is that it contains infinitely many relations. This may call for adjustments to existing reasoning algorithms.

3 Preliminaries

Here we introduce the notion of constraint languages for relations (Sect. 3.1) and the algebras of relations that they generate (Sect. 3.2).

3.1 Constraint Languages

Constraint languages are a mathematical framework for defining semantics of relational assertions. A (relational) constraint language is given by a collection of relation symbols and their interpretations. We use the formal definition of a constraint language as a relational structure in the model-theoretic sense [11].

Definition 1 (Constraint language). A *relational signature* is a set σ of relation symbols (also called predicate symbols), each with an associated finite arity. A *(relational) constraint language* over σ, or shortly a *σ-language*, is a tuple $\Gamma = (\sigma, U, \cdot^\Gamma)$, where σ is a relational signature, U is a set called the *universe* and $\cdot^\Gamma$ is the *interpretation* function defined on σ, which maps each relation symbol with arity n to an n-ary relation over U.

In this paper we confine ourselves to binary constraint languages, i.e., those that consist of binary relations.

Given a constraint language $\Gamma = (\sigma, U, \cdot^\Gamma)$, we say that R is a Γ-*relation*, if R is equal to r^Γ for some relation symbol $r \in \sigma$. We may write $R \in \Gamma$, meaning that R is a Γ-relation. When the interpretation of relation symbols in σ is clear from the context, we will specify a constraint language over a finite signature as $\Gamma = (U;\ r_1, r_2, \ldots, r_n)$, where U is the universe and $r_1, r_2, \ldots, r_n$ are the relation symbols.

Example 1. The constraint language of base taxonomic relations between sets. baseTAX5 $= (U;\ \equiv, \sqsubset, \sqsupset, \emptyset, \perp)$, where U is some powerset and $\emptyset$ the partial overlap relation symbol.

Constraint languages can be compared in terms of granularity. We start with a general definition of granularity relations [7].

Definition 2 (Granularity). Let $\mathcal{X}$ and $\mathcal{Y}$ be two collections of sets. $\mathcal{X}$ is said to be

- *finer* than $\mathcal{Y}$ if, for every $X \in \mathcal{X}$, there exists $Y \in \mathcal{Y}$ such that $X \subseteq Y$;
- *coarser* than $\mathcal{Y}$ if, for every $X \in \mathcal{X}$, there exists $\mathcal{Y}_0 \subseteq \mathcal{Y}$ such that $X = \cup \mathcal{Y}_0$;
- a *refinement* of $\mathcal{Y}$, if $\mathcal{X}$ is finer than $\mathcal{Y}$ and $\mathcal{Y}$ is coarser than $\mathcal{X}$.

The relations "finer than", "coarser than" and "refinement of" are transitive. A σ-language Γ is said to be *finer* than, *coarser* than, or a *refinement* of a σ'-language Γ', if so is the set of Γ-relations w.r.t. the set of Γ'-relations.

Definition 3 (Disjunctive Expansion). Let $\Gamma = (\sigma, U, \cdot^{\Gamma})$ be a constraint language. The *disjunctive expansion* of Γ is the constraint language $\Gamma_{\vee} = (\hat{\sigma}, U, \cdot^{\Gamma_{\vee}})$, where $\hat{\sigma}$ consists of all subsets of σ ($\hat{\sigma} = \wp(\sigma)$) and, for every $r \in \hat{\sigma}$, $r^{\Gamma_{\vee}} = \cup\{r_0^{\Gamma}\ :\ r_0 \in r\}$.

The signature of $\Gamma_{\vee}$ can be also defined, following the logical notation, as the set of all disjunctions of relation symbols from σ. For the signature of $\Gamma_{\vee}$ we will use the set-theoretic notation with one reservation: we will identify a singleton set $\{r\} \in \wp(\sigma)$ with the element $r \in \sigma$. Thus, for $r \in \sigma$ we may also write that $r \in \wp(\sigma)$. If $r \in \sigma$, then the relation $r^{\Gamma_{\vee}}$ is called a *base $\Gamma_{\vee}$-relation*. If $r \subseteq \sigma$, then $r^{\Gamma_{\vee}}$ is said to be a *disjunctive $\Gamma_{\vee}$-relation*.

Example 2. The disjunctive expansion of baseTAX5 (Example 1) is called the *constraint language of taxonomic relations between sets*, denoted as TAX5. Among the disjunctive TAX5-relations is subsumption and its converse: $\sqsubseteq = \{\sqsubset, \equiv\}$ and $\sqsupseteq = \{\sqsupset, \equiv\}$.

We will usually assume that different relation symbols correspond to different relations. In these cases, for a binary relation $R \in \Gamma$, by R^{σ} we will denote the relation symbol $r \in \sigma$, for which $r^{\Gamma} = R$. If $R \in \Gamma_{\vee}$, then $R^{\sigma} := \{r \in \sigma\ :\ r^{\Gamma} \subseteq R\}$.

3.2   Algebras Generated by Constraint Languages

If a constraint language Γ is closed under all intersections (finite or infinite) and contains the universal relation, then we can define weak composition of Γ-relations as follows: for $R, S \in \Gamma$, their *weak composition* is defined as $R \diamond_{\Gamma} S =$

$\cap\{T \in \Gamma \ : \ R \circ S \subseteq T\}$. (When it causes no ambiguity, we will write $\diamond$ instead of $\diamond_\Gamma$.) Likewise, *weak converse* is defined as $R^\smile = \cap\{T \in \Gamma \ : \ R^{-1} \subseteq T\}$. The operations of weak composition and weak converse are naturally induced on the relation symbols: $r \diamond s = (r^\Gamma \diamond s^\Gamma)^\sigma$ and $r^\smile = ((r^\Gamma)^\smile)^\sigma$.

A more specific and well-studied case is when a constraint language is obtained by the disjunctive expansion of a *partition scheme*. The notion of a partition scheme was introduced in [14] and then extended in [5]. We refer to the former definition as strong partition schemes and to the latter as abstract partition schemes.

Definition 4 (Partition scheme). Let X be some nonempty set and $\mathcal{P}$ a set of its subsets. $\mathcal{P}$ is said to be a *partition* of X if each element of X belongs to one and only one element of $\mathcal{P}$. A constraint language $\Gamma = (\sigma, U, \cdot^\Gamma)$ is said to be an *(abstract) partition scheme*, if Γ-relations make up a partition of $U \times U$. In this case Γ-relations are also said to be *jointly exhaustive and pairwise disjoint* (JEPD) on U. An abstract partition scheme Γ is said to be *strong*, if it is closed under converse and contains the identity relation over U.

The signature of the disjunctive expansion $\Gamma_\vee$ of a constraint language $\Gamma = (\sigma, U, \cdot^\Gamma)$ is a powerset algebra, hence a complete atomic Boolean algebra [10]. If Γ is an abstract partition scheme, then $\Gamma_\vee$ is closed under intersection and contains the universal relation $U \times U$. Thus, there are two additional operations on $\wp(\sigma)$: namely, weak composition and weak converse. The algebra $\mathbb{A}_\Gamma = (\wp(\sigma), \cup, \cap, -, \varnothing, \sigma, \diamond, ^\smile)$ is said to be generated by the abstract partition scheme Γ. The algebra $\mathbb{A}_\Gamma$ provides a symbolic calculus of Γ-relations.

Example 3. The constraint language baseTAX5 is a partition scheme only if its universe U does not contain the empty set. Then it generates an algebra A5, which is specified in [6]. If the universe U contains the empty set, then the relations of baseTAX5 are not pairwise disjoint any more. In that case it takes 8 base relations to refine baseTAX5 into a partition scheme (for more details see [12,13]).

Proposition 1 establishes an important property of algebras generated by partition schemes, which says that it is enough to define weak composition and weak converse on atoms.

Proposition 1 ([12]). *Let Γ be an arbitrary (finite or infinite) abstract partition scheme over a set U. Then weak composition and weak converse operations of $\mathbb{A}_\Gamma$ are completely additive, i.e., they completely distribute over the union.*

In addition to the algebraic method for reasoning with constraint languages, there are other approaches coming from recent research in CSP [3]. The main advantage of the algebraic approach is that it is polynomial (cubic) time. The disadvantage is that its reasoning capabilities vary from one constraint language to another and in many cases are rather limited.

4   The Constraint Language of Quantified Taxonomic Relations

In this section, we consider the universe of all finite sets and define a constraint language, called QTAX, of cardinality-based binary relation over this universe. We show that QTAX contains the crisp taxonomic relations (TAX5) and also the weighted taxonomic relations introduced in [1].

Let D be some countably infinite set, we consider the set of nonempty finite subsets of D as the *universe* and denote it as $\mathcal{U}_D$, or simply $\mathcal{U}$:

$$\mathcal{U}_D = \{X \ : \ X \subseteq D \text{ and } 0 < |X| < \omega\},$$

where ω is the first uncountable ordinal number. The set of all rational numbers from 0 to 1 will be denoted as $[0,1]_\mathbb{Q}$. We define a binary relational signature σ as a set of ordered pairs (α, β), where $\alpha, \beta \in [0,1]_\mathbb{Q}$. Further, we define a σ-language Δ on the universe $\mathcal{U}$ as follows:

$$(\alpha, \beta)^\Delta = \left\{ (X, Y) \in \mathcal{U} \times \mathcal{U} \ : \ \frac{|X \cap Y|}{|X|} = \alpha \text{ and } \frac{|X \cap Y|}{|Y|} = \beta \right\}.$$

Clearly, if $\alpha = 0$ and $\beta \neq 0$, or $\alpha \neq 0$ and $\beta = 0$, then $(\alpha, \beta)^\Delta = \varnothing$. This means that the relation symbols $(0, \beta)$ or $(\alpha, 0)$, in which $\alpha, \beta \neq 0$, are synonyms and all denote the empty relation; we will exclude such relation symbols from consideration. For the rest of σ-symbols we will say (α, β) is equal to (α', β') iff $\alpha = \alpha'$ and $\beta = \beta'$.

We denote the *disjunctive expansion* of Δ as QTAX and call it *the constraint language of quantified taxonomic relations*. A base QTAX-relation can be visually represented as a point on the unit square of α, β parameters (Fig. 1a), which we will call the (α, β)-*space*. A disjunctive QTAX-relation correspond then to a regions of the (α, β)-space, as shown in Fig. 1b.

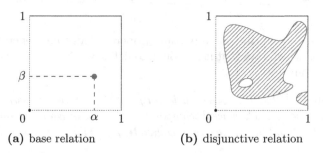

(a) base relation   (b) disjunctive relation

Fig. 1. Visual representation of QTAX-relations on the (α, β)-space.

Recall the constraint language TAX5 of taxonomic relations considered in Example 2. Proposition 2 shows that, if defined on the same universe, QTAX is a refinement of TAX5.

Proposition 2. *QTAX is a refinement of TAX5.*

Proof (Sketch). Figure 2a shows that each taxonomic relation can be presented as a disjunction of quantified taxonomic relations. This means that TAX5 is coarser than QTAX. QTAX is also finer than TAX5, because the latter contains the universal relation of the former. Hence, QTAX is a refinement of TAX5.

TAX-relation	QTAX-relation
$\equiv$	$(1,1)$
$\sqsupset$	$\{(\alpha,1) \ : \ 0 < \alpha < 1\}$
$\sqsubset$	$\{(1,\beta) \ : \ 0 < \beta < 1\}$
$\emptyset$	$\{(\alpha,\beta) \ : \ 0 < \alpha, \beta < 1\}$
$\perp$	$(0,0)$

(a) signature mapping

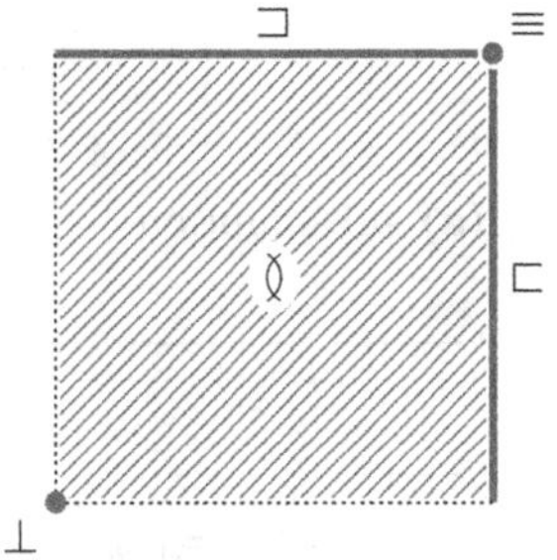

(b) visualization on the (α,β)-space

Fig. 2. The constraint language TAX5 of taxonomic relations is a sublanguage of the constraint language of quantified taxonomic relations QTAX.

The base taxonomic relations are visualized on the (α,β)-space in Fig. 2b. The weighted taxonomic relations $r_{[a,b]}$ (Sect. 2) can also be expressed in QTAX, as shown in Table 1. Figure 3 visualizes these relations on the (α,β)-space.

Table 1. Weighted taxonomic relations $r_{[a,b]}$ expressed in the constraint language QTAX.

Weighted taxonomic relation	QTAX-relation
$\sqsubseteq_{[a,b]}$	$\{(\alpha,\beta) \in \sigma \ : \ a \leq \alpha \leq b\}$
$\sqsupseteq_{[a,b]}$	$\{(\alpha,\beta) \in \sigma \ : \ a \leq \beta \leq b\}$
$\equiv_{[a,b]}$	$\left\{(\alpha,\beta) \in \sigma \ : \ a \leq \frac{2\alpha\beta}{\alpha+\beta} \leq b\right\}$
$\perp_{[a,b]}$	$\left\{(\alpha,\beta) \in \sigma \ : \ a \leq 1 - \frac{2\alpha\beta}{\alpha+\beta} \leq b\right\}$

Proposition 3 says that base QTAX-relations make up a strong partition scheme, thus they generate an algebra $\mathbb{A}_{\text{QTAX}}$.

Proposition 3. Δ *is an infinite strong partition scheme.*

Proof (Sketch). First, any $\alpha, \beta \in [0,1]_{\mathbb{Q}}$, such that α and β are either both zero or both nonzero, the relation $(\alpha,\beta)^{\Delta}$ is not empty. Further, it is easy to check that Δ-relations are jointly exhaustive and pairwise disjoint. Finally, it remains to check that Δ is closed under converse and contains the identity relation. Indeed, $((\alpha,\beta)^{\Delta})^{-1} = (\beta,\alpha)^{\Delta}$ and $(1,1)^{\Delta} = Id_{\mathcal{U}}$.

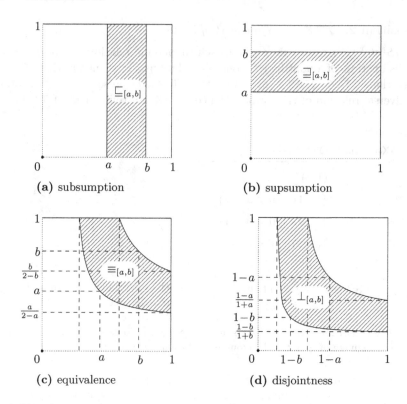

Fig. 3. Visualization of weighted relations $r_{[a,b]}$ (in the sense of [1]) on the (α, β)-space.

5   The Relaxed Taxonomic Relations

In this section, we discuss the shortcomings of weighted equivalence and disjointness and propose different semantics for these relations. The revisited set of weighted relations constitutes a sublanguage of QTAX, called the constraint language of *relaxed taxonomic relations*. We compare the relaxed semantics of equivalence and disjointness with the former one and discuss its advantages.

As mentioned in the introduction, in a weighted relation $r_{[a,b]}$, if the upper bound b of the confidence interval $[a, b]$ is less than 1, then $r_{[a,b]}$ negates the crisp relation r (in symbols, $r_{[a,b]} \models \neg r$), which is counter-intuitive. This issue can be solved by confining to confidence intervals $[a, 1]$, in which the upper bound is always 1, as shown in Fig. 4.

We denote the relations $r_{[a,1]}$ as r^a and call them *relaxed taxonomic relations*, since they are weaker than r, i.e., $r \models r^a$ for any $r \in \{\equiv, \sqsubseteq, \sqsupseteq, \bot\}$ and any $a \in [0, 1]$. The semantics of relaxed equivalence $\equiv^a$ and relaxed disjointness $\bot^a$, proposed in [1], has some shortcomings. First, the "equivalence entails subsumption" property, which holds for crisp equivalence and crisp subsumption (in symbols, $\equiv \models \sqsubseteq$), is not preserved by their relaxed counterparts. That is, from equivalence with a confidence interval $[a, 1]$ one cannot entail subsumption

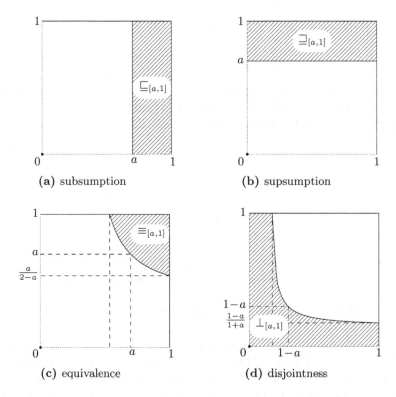

Fig. 4. Visualization of weighted relations $r_{[a,1]}$ (in the sense of [1]) on the (α, β)-space.

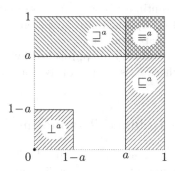

Fig. 5. Visualization of base relaxed taxonomic relations on the (α, β)-space.

with (at least) the same confidence: $\equiv^a \not\models \sqsubseteq^a$, for any $0 < a < 1$. Second, one would intuitively expect the relaxed disjointness and subsumption to be mutually exclusive, as it is the case with the crisp relations. However, this property does not hold either: for any $0 < a < 1$, the assertions $A \perp^a B$ and $A \sqsubseteq^a B$ do not contradict each other.

We overcome these drawbacks by refining the semantics of relaxed equivalence and disjointness as follows:

$$\equiv^a = \{(\alpha,\beta) \in \sigma \ : \ \alpha,\beta \geq a\}, \qquad \perp^a = \{(\alpha,\beta) \in \sigma \ : \ \alpha,\beta \leq 1-a\}.$$

These relations are visualized in Fig. 5. From this definition it follows that $\equiv^a$ is the intersection of $\sqsubseteq^a$ and $\sqsupseteq^a$. Moreover, relaxed disjointness $\perp^a$ does not overlap with relaxed subsumption $\sqsubseteq^a$ for any $a > 0.5$.

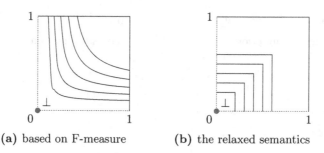

(a) based on F-measure   (b) the relaxed semantics

Fig. 6. Comparison of semantics for weighted disjointness.

It is now time to justify the discontinuity observed in the weighted semantics of [1] with the help of QTAX. The crisp semantics of $\perp$ is the $(0, 0)$ point. The weighted semantics approaches it, but because it is the result of using the F-measure it always preserves the possibility that the segments $(0, 1)$ and $(1, 0)$ denote $\perp$ because F-measure$(1, 0) = $ F-measure$(0, 1) = 0$. This is what is shown in Fig. 6a. Hence, the discontinuity comes from preserving these segments — and the points $(0, 1)$ and $(1, 0)$ which are in the interpretation of $\sqsubseteq$ and $\sqsupseteq$ — whatever closed the weights are from crisp.

This is different when relations are approached by reducing a distance. This is illustrated in Fig. 6b where $\perp$ is continuously approximated through α with the Manhattan distance.

6   The Calculus of Relaxed Taxonomic Relations

In this section, we define the algebraic calculus of QTAX, which allows for composing the relaxed taxonomic relations (Sect. 6.1) and introduce two algebras which can be used for reasoning with alignments (Sect. 6.2).

6.1   Composition of Relaxed Taxonomic Relations

Composition in QTAX distributes over union (Proposition 1). Thus, to compose two relaxed taxonomic relations, one has to compose pairwise all constituent base relations.

$$r^a \diamond s^b = \bigcup_{\substack{(\alpha,\beta)\in r \\ (\alpha',\beta')\in s}} (\alpha,\beta) \diamond (\alpha',\beta').$$

Before providing the formula for composing base QTAX-relations, let us introduce abbreviations for some relation symbols in $\wp(\sigma)$.

$$\mathsf{INT}(\alpha_0, \alpha_1, k) := \{(\alpha, k\alpha) \ : \ \alpha_0 \leq \alpha \leq \alpha_1\}$$
$$\mathsf{REC}(\alpha_0, \beta_0, \alpha_1, \beta_1) := \{(\alpha, \beta) \ : \ \alpha_0 \leq \alpha \leq \alpha_1 \text{ and } \beta_0 \leq \beta \leq \beta_1\}$$

The relation symbols $\mathsf{INT}(\alpha_0, \alpha_1, k)$, where $\alpha_0 \leq \alpha_1 \in [0,1]_{\mathbb{Q}}$ and $0 < k\alpha_1 \leq 1$, correspond to intervals on the (α, β)-space, as shown in Fig. 7a. We call them *interval relations* (not to confuse with Allen's temporal intervals). On the (α, β)-space these relations lie on a line which passes through the point $(0, 0)$. The relation symbols $\mathsf{REC}(\alpha_0, \beta_0, \alpha_1, \beta_1)$, where $\alpha_0, \beta_0, \alpha_1, \beta_1 \in [0,1]_{\mathbb{Q}}$, correspond to rectangles on the (α, β)-space, the edges of which are parallel to those of the unit square (Fig. 7b). We call them *rectangle relations*.

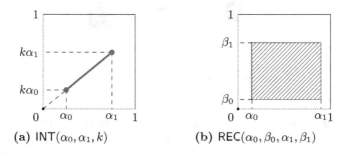

(a) $\mathsf{INT}(\alpha_0, \alpha_1, k)$ (b) $\mathsf{REC}(\alpha_0, \beta_0, \alpha_1, \beta_1)$

Fig. 7. Visual representation of interval and rectangle QTAX-relations.

Now we can formulate the main result. The composition of two base QTAX-relations is either a rectangle relation, if one of the base relations is the disjointness (Theorem 1). Otherwise, the composition is an interval relation.

Theorem 1.

$$(\alpha, \beta) \diamond (\alpha', \beta') = \begin{cases} \mathsf{REC}(0, 0, 1, 1 - \beta'), & \text{if } \alpha, \beta = 0, \\ \mathsf{REC}(0, 0, 1 - \alpha, 1), & \text{if } \alpha', \beta' = 0, \\ \mathsf{INT}(\alpha_0'', \alpha_1'', \frac{\beta\beta'}{\alpha\alpha'}), & \text{if } \alpha, \beta, \alpha', \beta' \neq 0, \end{cases}$$

where

$$\alpha_0'' = \frac{\alpha}{\beta} max\,(\alpha' + \beta - 1, 0)\,,$$

$$\alpha_1'' = min\left[1, \frac{\alpha\alpha'}{\beta\beta'}, \alpha\left(min(1, \frac{\alpha'}{\beta}) + min(\frac{\alpha'}{\beta}\frac{1 - \beta'}{\beta'}, \frac{1 - \alpha}{\alpha})\right)\right]$$

Proof. The proof can be found in [12].

6.2 Approximation and Parametrization of QTAX relations

$\mathbb{A}_{\mathsf{QTAX}}$ is an algebra of relation symbols and not of actual binary relations. A relation symbol is a set of pairs (α, β), where $\alpha, \beta \in [0, 1]_{\mathbb{Q}}$.

Composition of some relaxed taxonomic relations is visually represented in Fig. 8. In general, the composition of such relations is not a relaxed taxonomic relation, but some "irregular" QTAX-relation represented by the black area. However, it can be always approximated by a rectangle relation, and in some cases even by another relaxed taxonomic relation, as shown in Fig. 8 by the grayed area. The composition of relaxed equivalences $\equiv^{0.6}$ and $\equiv^{0.8}$ has a shape close to a rectangle. The REC-approximation of composition is $\equiv^{0.4}$.

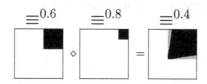

Fig. 8. REC-approximation of composition.

All rectangle relations plus the empty relation are closed under intersection and contain the universal relation. Thus, weak composition $\diamond_{\mathsf{REC}}$ is a valid operation on the REC sublanguage of QTAX. The operation $\diamond_{\mathsf{REC}}$ can be specified based on numeric evaluation of a set of (α, β) relation symbols which constitute the composition in QTAX. A union of rectangle relations may not be a rectangle relation, but can always be approximated by one. This defines the operation of *weak union* on REC, denoted as $\cup_w$. The rectangle relations, together with operations of weak composition, converse intersection and weak union, form an algebra $\mathbb{A}_{\mathsf{REC}}$:

$$\mathbb{A}_{\mathsf{REC}} = (R, \cup_w, \cap, \varnothing, \mathsf{REC}(0,0,1,1), \diamond_{\mathsf{REC}}, {}^{\smallsmile}), \tag{6.1}$$

where $R = \{\mathsf{REC}(\alpha_0, \beta_0, \alpha_1, \beta_1) \ : \ \alpha_0 \leq \alpha_1, \beta_0 \leq \beta_1 \in [0,1]_{\mathbb{Q}}\} \cup \{\varnothing\}$. A general formula for composing relaxed equivalence relations is the following:

$$\equiv^x \diamond_{\mathsf{REC}} \equiv^y = \equiv^{max(0,\ x+y-1)} \tag{6.2}$$

Similar formulas can be obtained for other pairs of relaxed taxonomic relations.

Another approach to make $\mathbb{A}_{\mathsf{QTAX}}$ computationally feasible is to discretize the (α, β)-space as an $n \times n$ matrix and thus obtain a finite algebra $\mathbb{A}_{\mathsf{QTAX}}^n$. This approach was used for computing the composition of relaxed taxonomic relations in Table 2.

Table 2. Composition of relaxed taxonomic relations visualized on the (α, β)-space.

i	j	$(\equiv^i \diamond \equiv^j)$	$(\equiv^i \diamond \sqsubseteq^j)$	$(\equiv^i \diamond \sqsupseteq^j)$	$(\equiv^i \diamond \perp^j)$	$(\sqsubseteq^i \diamond \sqsubseteq^j)$	$(\sqsubseteq^i \diamond \perp^j)$
0.5	0.5						
0.5	0.6						
0.5	0.7						
0.5	0.8						
0.5	0.9						
0.5	1						
0.6	0.5						
0.6	0.6						
0.6	0.7						
0.6	0.8						
0.6	0.9						
0.6	1						
0.7	0.5						
0.7	0.6						
0.7	0.7						
0.7	0.8						
0.7	0.9						
0.7	1						
0.8	0.5						
0.8	0.6						
0.8	0.7						
0.8	0.8						
0.8	0.9						
0.8	1						
0.9	0.5						
0.9	0.6						
0.9	0.7						
0.9	0.8						
0.9	0.9						
0.9	1						
1	0.5						
1	0.6						
1	0.7						
1	0.8						
1	0.9						
1	1						

7   Application to Reasoning with Ontology Alignments

The relaxed semantics of taxonomic relations can be used by ontology matchers. Some matchers induce relations between classes based on the instance-level data. Since the semantic web is an open environment with potentially invalid data, many instance-based matchers induce a relation between two concepts, if it holds for *most* instances of these concepts. The level of fault-tolerance is usually set by a threshold. This threshold may be expressed as the weight of an ontology alignment relation, in compliance with the relaxed semantics.

To reason with weighted ontology alignments, both algebras $\mathbb{A}_{\mathsf{REC}}$ or $\mathbb{A}_{\mathsf{QTAX}}^n$ can be used. The algebra $\mathbb{A}_{\mathsf{REC}}$ contains infinitely many relations, but is computationally feasible, since REC-relations are finitely parametrized. However, using $\mathbb{A}_{\mathsf{REC}}$ for automated reasoning requires adjustments to the existing reasoning algorithms, which are designed for finite algebras. The algebras $\mathbb{A}_{\mathsf{QTAX}}^n$ are finite, thus can be used with existing reasoning tools that support qualitative calculi.

8   Summary and Conclusion

Weights in ontology alignments have been widely adopted. This paper shows how to define algebraic calculi which can be used for expressing both the relation and the weight of correspondences. Its goal is to be able to provide sound compositional reasoning for alignments.

We introduced the $\mathbb{A}_{\mathsf{QTAX}}$ calculus of relaxed taxonomic relations generalising the previous weighted semantics as well as the semantics of crisp relations. We provided a semantics that overcomes the problems identified and, in particular, discontinuity. $\mathbb{A}_{\mathsf{QTAX}}$ composition is not computationally feasible, however we discussed two different ways to make it computationally feasible: $\mathbb{A}_{\mathsf{REC}}$ based on rectangular approximation of these relations and $\mathbb{A}_{\mathsf{QTAX}}^n$ based on a discretization of the (α, β)-space.

On the one hand, this proposal provides a way to reason by composition with weighted alignment that is well grounded and can compose any relation. On the other hand, [1] gave rules for reasoning with concept constructors which are absent here. It would be worth studying if such rules still holds and can be generalised to the new context.

Acknowledgement. This research has been partially supported by the joint NSFC-ANR Lindicle project (12-IS01-0002) with Tsinghua university.

References

1. Atencia, M., Borgida, A., Euzenat, J., Ghidini, C., Serafini, L.: A formal semantics for weighted ontology mappings. In: Cudré-Mauroux, P., et al. (eds.) ISWC 2012. LNCS, vol. 7649, pp. 17–33. Springer, Heidelberg (2012). doi:10.1007/978-3-642-35176-1_2

2. Baader, F., Calvanese, D., McGuinness, D.L., Nardi, D., Patel-Schneider, P.F. (eds.): The description logic handbook: theory, implementation, and applications. Cambridge University Press, New York (2003)
3. Bodirsky, M., Dalmau, V.: Datalog and constraint satisfaction with infinite templates. J. Comput. Syst. Sci. **79**(1), 79–100 (2013)
4. Borgida, A., Serafini, L.: Distributed description logics: assimilating information from peer sources. In: Spaccapietra, S., March, S., Aberer, K. (eds.) Journal on Data Semantics I. LNCS, vol. 2800, pp. 153–184. Springer, Heidelberg (2003)
5. Dylla, F., Mossakowski, T., Schneider, T., Wolter, D.: Algebraic properties of qualitative spatio-temporal calculi. In: Tenbrink, T., Stell, J., Galton, A., Wood, Z. (eds.) COSIT 2013. LNCS, vol. 8116, pp. 516–536. Springer, Heidelberg (2013). doi:10.1007/978-3-319-01790-7_28
6. Euzenat, J.: Algebras of ontology alignment relations. In: Sheth, A., Staab, S., Dean, M., Paolucci, M., Maynard, D., Finin, T., Thirunarayan, K. (eds.) ISWC 2008. LNCS, vol. 5318, pp. 387–402. Springer, Heidelberg (2008). doi:10.1007/978-3-540-88564-1_25
7. Euzenat, J.: An algebraic approach to granularity in qualitative time, space representation. In: Proceedings of the IJCAI-95, pp. 894–900 (1995)
8. Euzenat, J., Shvaiko, P.: Ontology Matching, 2nd edn. Springer, Heidelberg (2013)
9. Gantner, Z., Westphal, M., Wölfl, S.: GQR-a fast reasoner for binary qualitative constraint calculi. In: Proceedings of the AAAI, vol. 8 (2008)
10. Givant, S., Halmos, P.: Introduction to Boolean Algebras. Springer Undergraduate Texts in Mathematics and Technology. Springer, New York (2009)
11. Hodges, W.: Model Theory Encyclopedia of Mathematics and its Applications. Cambridge University Press, Cambridge (1993)
12. Inants, A.: Qualitative calculi with heterogeneous universes. Ph.D. thesis. University of Grenoble Alpes (2016)
13. Inants, A., Euzenat, J.: An algebra of qualitative taxonomical relations for ontology alignments. In: Arenas, M., Corcho, O., Simperl, E., Strohmaier, M., d'Aquin, M., Srinivas, K., Groth, P., Dumontier, M., Heflin, J., Thirunarayan, K., Staab, S. (eds.) ISWC 2015. LNCS, vol. 9366, pp. 253–268. Springer, Heidelberg (2015). doi:10.1007/978-3-319-25007-6_15
14. Ligozat, G., Renz, J.: What is a qualitative calculus? A general framework. In: Zhang, C., W. Guesgen, H., Yeap, W.-K. (eds.) PRICAI 2004. LNAI, vol. 3157, pp. 53–64. Springer, Heidelberg (2004). doi:10.1007/978-3-540-28633-2_8
15. Locoro, A., David, J., Euzenat, J.: Context-based matching: design of a flexible framework and experiment. J. Data Semant. **3**(1), 25–46 (2014)
16. Lukasiewicz, T., Predoiu, L., Stuckenschmidt, H.: Tightly integrated probabilistic description logic programs for representing ontology mappings. Ann. Math. Artif. Intell. **63**(3–4), 385–425 (2011)
17. Wolter, D.: SparQ - a spatial reasoning toolbox. In: AAAI Spring Symposium: Benchmarking of Qualitative Spatial and Temporal Reasoning Systems, p. 53 (2009)
18. Zimmermann, A., Euzenat, J.: Three semantics for distributed systems and their relations with alignment composition. In: Cruz, I., Decker, S., Allemang, D., Preist, C., Schwabe, D., Mika, P., Uschold, M., Aroyo, L.M. (eds.) ISWC 2006. LNCS, vol. 4273, pp. 16–29. Springer, Heidelberg (2006)

Ontologies for Knowledge Graphs:
Breaking the Rules

Markus Krötzsch[✉] and Veronika Thost

Center for Advancing Electronics Dresden (cfaed), TU Dresden, Dresden, Germany
{markus.kroetzsch,veronika.thost}@tu-dresden.de

Abstract. Large-scale knowledge graphs (KGs) are widely used in industry and academia, and provide excellent use-cases for ontologies. We find, however, that popular ontology languages, such as OWL and Datalog, cannot express even the most basic relationships on the normalised data format of KGs. Existential rules are more powerful, but may make reasoning undecidable. Normalising them to suit KGs often also destroys syntactic restrictions that ensure decidability and low complexity. We study this issue for several classes of existential rules and derive new syntactic criteria to recognise well-behaved rule-based ontologies over KGs.

1 Introduction

Graph-based representations are playing a major role in modern knowledge management. Their simple, highly normalised data models can accommodate a huge variety of different information sources, and led to large-scale *knowledge graphs* (KGs) in industry (e.g., at Google and Facebook); on the Web (e.g., Freebase [6] and Wikidata [26]); and in research (e.g., YAGO2 [16] and Bio2RDF [5]).

Not all data is graph-shaped, but it is usually easy to translate into this format using well-known methods. For example, the W3C *RDB to RDF Mapping Language* provides mappings from relational databases to RDF graphs [13]. Relational tuples with three or more values are represented by introducing new graph nodes, to which the individual values of the tuple are then connected directly. For example, the tuple spouse(ann, jo, 2013), stating that Ann married Jo in

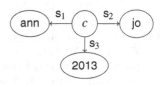

Fig. 1. Tuple as Graph

2013, may be represented by the graph in Fig. 1, where c is a fresh element introduced for this tuple, and s_1 to s_3 are binary edge labels used for all tuples of the spouse relation.

In this way, KGs unify data formats, so that many heterogeneous datasets can be managed in a single system. Unfortunately, however, syntactic alignment is not the same as semantic integration. The KG's flexibility and lack of schematic constraints lead to conceptual heterogeneity, which reduces the KG's utility. This is a traditional data integration problem, and ontologies promise to solve it in an interoperable and declarative fashion [19]. Indeed, ontologies can be used to model semantic relationships between different structures, so that a coherent global view can be obtained.

P. Groth et al. (Eds.): ISWC 2016, Part I, LNCS 9981, pp. 376–392, 2016.
DOI: 10.1007/978-3-319-46523-4_23

It therefore comes as a surprise that ontologies are so rarely used with KGs. A closer look reveals why: modern ontology languages cannot express even the simplest relationships on KG models. In our example, a natural relationship to model would be that marriage is symmetric, so that we can infer spouse(jo, ann, 2013). In a KG, this fact would again be represented by a structure as in Fig. 1, but with Ann and Jo switched, and – importantly – using a fresh auxiliary node in place of c. This entailment could be expressed by the following logical axiom:

$$\forall x, y_1, y_2, y_3. \ s_1(x, y_1) \wedge s_2(x, y_2) \wedge s_3(x, y_3) \ \rightarrow \ \exists v. \ s_1(v, y_2) \wedge s_2(v, y_1) \wedge s_3(v, y_3). \quad (1)$$

Two ontology languages proposed for information integration in databases are *global-as-view* and *local-as-view* mappings [19]. Neither can express (1), since they support only single atoms on the source and on the target side, respectively. *Datalog*, a popular language for defining recursive views, cannot express (1) either, since it lacks existential quantification in conclusions of rules. Another very popular ontology language is OWL [22], which was specifically built for use with RDF graphs. However, even OWL cannot express (1): it supports rules with existential quantifiers, but only with exactly one universally quantified variable occurring in both premise and conclusion.

This problem is not specific to our particular example. KGs occur in many formats, which are rarely as simple as RDF. It is, e.g., common to associate additional information with edges. Examples are validity times in YAGO2, statement qualifiers in Wikidata, and arbitrary edge attributes in Property Graphs (a very popular data model in graph databases). If we want to represent such data in a simple relational form that is compatible with first-order predicate logic, we arrive at encodings as in Fig. 1.

So how can we realise ontology-based information integration on KGs? Formula (1) is in fact what is called a *tuple-generating dependency* in databases [1] and an *existential rule* in AI [2]. While query answering over such rules is undecidable, many decidable fragments have been proposed (see overviews [2], [8], and [11]). These rules use a relational model, and they can be translated to a KG setting just like facts. For example, rule (1) could be the result of translating $\forall y_1, y_2, y_3. \text{spouse}(y_1, y_2, y_3) \rightarrow \text{spouse}(y_2, y_1, y_3)$. However, this changes the rules' syntax and semantics, and it destroys known criteria that guarantee decidability or complexity.

We therefore ask to which extent known decidable fragments of existential rules are applicable to KGs, and we propose alternative definitions where necessary, to recover desirable properties. Our main results are:

– We show that *acyclicity* criteria and related complexities are generally preserved when transforming rules to KGs, and we identify a restricted class of acyclic rules that comprises transformed Datalog and retains its complexity.
– We show that the transformation destroys other basic syntactic criteria such as *linearity* and *guardedness*, though it preserves the underlying semantic notions (FO-rewritability and tree-like model property).
– We propose a new way of *denormalising* KG rules, based on the intuition that several edges can be grouped into "objects", and we exhibit cases for which this approach succeeds in producing rule sets that fall into known decidable classes.
– We introduce a notion of *incidental functional dependency*, which we use to extend our denormalisation to wider classes of rules, and we exhibit a sound procedure for computing such dependencies.

In all cases, we develop criteria that significantly generalise the motivating scenario of translating relational ontologies to KGs. In practice, it is more realistic to assume that ontologies are constructed over KGs directly. In this case, one cannot expect rules to have a regular structure as obtained by a rigid syntactic transformation, but patterns guaranteeing decidability and complexity bounds might still be identifiable.

Full proofs are available in an extended technical report [18].

2 Preliminaries

We briefly introduce essential notation and define the important notion of *graph normalisation*. We consider a standard language of first-order predicate logic, using *predicates* p of *arity* $\mathrm{ar}(p)$, *variables*, and *constants*. A *term* is a constant or variable. Finite lists of variables etc. are denoted in bold, e.g., x. We use the standard predicate logic definitions of *atom* and *formula*. An *existential rule* (or simply *rule*) is a formula of form $\forall x, y.\varphi[x, y] \rightarrow \exists v.\psi[x, v]$ where φ and ψ are conjunctions of atoms, called the *body* and *head* of the rule, respectively. Rules without existentially qualified variables are *Datalog rules*. We usually omit the universal quantifiers when writing rules.

We separate input relations (EDB) from derived relations (IDB). Formally, for a set of rules $\mathbb{P}$, the predicate symbols that occur in the head of some rule are called *intensional* (or *IDB*); other predicates are called *extensional* (or *EDB*). A *fact* is an atom that contains no variables. A *database* $\mathbb{D}$ is a set of facts over EDB predicates. A *conjunctive query* (CQ) is a formula $\exists y.\varphi[x, y]$, where φ is a conjunction of atoms. A *Boolean CQ* (BCQ) is a CQ without free variables.

We only consider rules without constants. They can be simulated as usual, by replacing every constant a in a rule by a new variable x_a, adding the atom $O_a(x_a)$ to the body, and extending the database to include a single fact $O_a(a)$.

Rules and databases can be evaluated under a first-order logic semantics, and we use $\models$ to denote the usual first-order entailment relation between (sets of) formulae. CQ answering over existential rules can be reduced to BCQ entailment, i.e., the problem of deciding if $\mathbb{D}, \mathbb{P} \models \exists y.\varphi$ holds for a given BCQ $\exists y.\varphi$, database $\mathbb{D}$, and set of rules $\mathbb{P}$ [1]. This is undecidable in general, but many special classes of rule sets have been identified where decidability is recovered; we will see several examples later.

We now formalise the standard transformation of n-ary facts into directed graphs that was given in the introduction, and extend it to rules over n-ary predicates.

Definition 1. *For every predicate p, let $p_1, \ldots, p_{\mathrm{ar}(p)}$ be fresh binary predicates. Given an atom $p(t)$ and a term s, the* graph normalisation $\mathrm{GN}(s, p(t))$ *is the set $\{p_1(s, t_1), \ldots, p_{\mathrm{ar}(p)}(s, t_n)\}$ of binary atoms. For a database $\mathbb{D}$, define $\mathrm{GN}(\mathbb{D})$ to be the union of the sets $\mathrm{GN}(c_A, A)$ for all facts $A \in \mathbb{D}$ where c_A is a fresh constant for A. For a rule $\rho = B_1 \wedge \ldots \wedge B_m \rightarrow \exists v.H_1 \wedge \ldots \wedge H_\ell$, let $\mathrm{GN}(\rho)$ be the rule $\bigwedge_{i=1}^{m} \mathrm{GN}(z_i, B_i) \rightarrow \exists v.\exists w. \bigwedge_{j=1}^{\ell} \mathrm{GN}(w_j, H_j)$ using fresh variables z and w. For a set of rules $\mathbb{P}$, let $\mathrm{GN}(\mathbb{P}) := \bigcup_{\rho \in \mathbb{P}} \mathrm{GN}(\rho)$.*

Example 1. Consider a database about PhD graduates and theses with facts of the form sup(*person, supervisor*) and phd(*person, thesis title, date*). We can express that every supervisor of a PhD graduate also has a PhD, using P for inferred (IDB) PhD relations:

$$\mathsf{phd}(x, y_1, y_2) \rightarrow \mathsf{P}(x, y_1, y_2) \tag{2}$$

$$\mathsf{P}(x_1, y_1, y_2) \wedge \mathsf{sup}(x_1, x_2) \rightarrow \exists v_1, v_2.\mathsf{P}(x_2, v_1, v_2) \tag{3}$$

The graph normalisation of this rule set is as follows:

$$\mathsf{phd}_1(z, x) \wedge \mathsf{phd}_2(z, y_1) \wedge \mathsf{phd}_3(z, y_2) \rightarrow \exists v.\mathsf{P}_1(v, x) \wedge \mathsf{P}_2(v, y_1) \wedge \mathsf{P}_3(v, y_2) \tag{4}$$

$$\mathsf{P}_1(z_1, x_1) \wedge \mathsf{P}_2(z_1, y_1) \wedge \mathsf{P}_3(z_1, y_2) \wedge \mathsf{sup}_1(z_2, x_1) \wedge \mathsf{sup}_2(z_2, x_2) \tag{5}$$
$$\rightarrow \exists v, v_1, v_2.\mathsf{P}_1(v, x_2) \wedge \mathsf{P}_2(v, v_1) \wedge \mathsf{P}_3(v, v_2)$$

3 Acyclicity

Sets of existential rules may require models to be infinite. An immediate approach for ensuring decidability is to consider criteria that guarantee the existence of a finite universal model, which can be fully computed and used to answer queries. This led to many so-called *acyclicity* criteria [11]. We review one of the simplest cases, *weak acyclicity*.

Definition 2. *A* position *in a predicate p is a pair $\langle p, i \rangle$, where $i \in \{1, \ldots, ar(p)\}$. The* dependency graph *G of a rule set $\mathbb{P}$ is defined as follows. The vertices of G are all positions of predicates in $\mathbb{P}$. For every rule $\varphi[\boldsymbol{x}, \boldsymbol{y}] \rightarrow \exists v.\psi[\boldsymbol{x}, \boldsymbol{v}] \in \mathbb{P}$: (1) G has an edge from $\langle p, i \rangle$ to $\langle q, j \rangle$ if $x \in \boldsymbol{x}$ occurs at position $\langle p, i \rangle$ in φ and at $\langle q, j \rangle$ in ψ; (2) G has a* special *edge from $\langle p, i \rangle$ to $\langle q, j \rangle$ if $x \in \boldsymbol{x}$ occurs at position $\langle p, i \rangle$ in φ and there is an existentially quantified variable $v \in \boldsymbol{v}$ at $\langle q, j \rangle$ in ψ.*

$\mathbb{P}$ *is* weakly acyclic *if its dependency graph does not contain a directed cycle that involves a special edge.*

Theorem 1. *If $\mathbb{P}$ is weakly acyclic, then so is $\mathsf{GN}(\mathbb{P})$. Analogous preservation properties hold for rule sets that are jointly acyclic, super-weakly acyclic, model-faithful acyclic, or that have an acyclic graph of rule dependencies.*

While most acyclicity notions are thus preserved, this is not a general rule: *model-summarising acyclicity* (MSA) might be destroyed by graph normalisation [18].

BCQ entailment for acyclic rule sets is 2ExpTime-complete [11]. Datalog, however, enjoys a lower ExpTime-complete complexity [12], so Theorem 1 does not yield tight complexity estimates there. ExpTime complexity bounds for acyclic rules were given for rule sets where the maximal length of paths in a (slightly different) type of dependency graph is bounded [17, Theorem 5]. This condition is implied by the following property:

Theorem 2. *If $\mathbb{P}$ is a set of Datalog rules, then the dependency graph of $\mathsf{GN}(\mathbb{P})$ is such that every path contains at most one special edge.*

The number of special edges on paths can therefore be used to recognise (generalisations of) graph-normalised Datalog for which CQ answering is in ExpTime.

4 Beyond Acyclicity

Acyclicity is only one of several approaches for determining that reasoning is decidable for a given set of existential rules. It turns out, however, that other syntactic criteria are not as robust when applying graph normalisation to a set of rules, although one can show that essential semantic properties are preserved.

Baget et al. have identified several general classes of rule sets for which reasoning is decidable [2]. Acyclic rule sets are a typical form of *finite expansion set* (fes), which have a finite universal model. Rule sets without this property may still have an infinite universal model that is sufficiently "regular" to be presented finitely. This is the case if there is a universal model of bounded treewidth, leading to *bounded treewidth sets* (bts). A third general class of practical importance are *finite unification sets* (fus), corresponding to the class of first-order rewritable rule sets for which conjunctive queries (CQs) can be rewritten into finite unions of CQs (UCQs).

All of these abstract properties are preserved during graph normalisation. For fes and bts, this can be shown by noting that any (universal) model of $\mathbb{P}$ can be transformed into a (universal) model of $\mathsf{GN}(\mathbb{P})$ by treating it like an (infinite) database and applying $\mathsf{GN}(\cdot)$. For fus, the result follows since we can apply graph normalisation to the UCQ rewriting to obtain a valid rewriting for $\mathsf{GN}(\mathbb{P})$.

Theorem 3. *If $\mathbb{P}$ is fes/bts/fus, then $\mathsf{GN}(\mathbb{P})$ is fes/bts/fus.*

However, membership in these abstract classes is undecidable, so we need simpler sufficient conditions in practice. We disregard fes here, since it is already covered in Sect. 3. For bts, an easy-to-check criterion is (frontier) *guardedness* [2]:

Definition 3. *A rule $\varphi[x, y] \to \exists v.\psi[x, v]$ is* frontier guarded *if φ contains an atom that contains all variables of x. A rule set $\mathbb{P}$ is* frontier guarded *if all of its rules are.*

Frontier guarded rule sets are bts, and, by Theorem 3, so are their graph normalisations. Unfortunately, this is not easy to recognise, since frontier guardedness is often destroyed when breaking apart body atoms during graph normalisation. For instance, the original rules in Example 1 are frontier guarded, but the normalised rule (4) is not. The only general criterion that could recognise bts in normalised rules is *greedy bts* [4]; but a procedure for recognising this criterion has not been proposed yet, and the problem is generally assumed to be of very high complexity.

The situation is similar for fus. One of the simplest syntactic conditions for this case is *linearity* (a.k.a. *atomic hypothesis* [2]):

Definition 4. *An existential rule is* linear *if its body consists of a single atom. A rule set $\mathbb{P}$ is* linear *if all of its rules are.*

Again, this condition is clearly not preserved by graph normalisation. For example, rule (2) is linear while rule (4) is not.

Towards a way of recognising fus and bts rules even after graph normalisation, we look for ways to undo this transformation, i.e., to *denormalise* the graph. A natural approach of reversing the transformation from $p(x)$ to $p_1(z, x_1) \wedge \ldots \wedge p_n(z, x_n)$ is to group atoms by their first variable z. We may think of such groups of atoms as *objects* (as in object-oriented programming), motivating the following terminology.

Definition 5. *Consider a rule $\varphi \rightarrow \exists v.\psi$. An object in φ (or ψ) is a maximal conjunction of atoms of the form $p_1(z, x_1) \wedge \ldots \wedge p_n(z, x_n)$ that occur in φ (or ψ), where neither variables x_i nor predicates p_i need to be mutually distinct. We call z object variable, $p_1, \ldots, p_n$ attributes, and $x_1, \ldots, x_n$ values of the object. The interface of the object is the set of variables $y \subseteq \{x_1, \ldots, x_n\}$ occurring in atoms in $\varphi \rightarrow \exists v.\psi$ that do not belong to the object.*

Note that each object is confined to either body or head, but cannot span both. In general, several attributes of an object may share a value, and several objects may use the same attributes. The definition therefore generalises the specific conjunctions of binary attributes introduced in graph normalisation. Existential rules may be thought of as "creating" new objects when using existential object variables. It is suggestive to use objects for defining KG versions of the above criteria:

Definition 6. *A rule $\varphi[x, y] \rightarrow \exists v.\psi[x, v]$ over binary predicates is pseudo KG linear if φ consists of a single object. It is pseudo KG frontier guarded if φ contains an object ξ where all variables of x occur in. A rule is KG linear (KG frontier guarded) if it is pseudo KG linear (pseudo KG frontier guarded), and no object variable occurs as a value in any object.*

The "pseudo" versions of the above notions are not enough to obtain the desired properties, as the following example illustrates.

Example 2. The following rules are pseudo KG frontier guarded:

$$p(z, x) \rightarrow P(z, x) \tag{6}$$

$$P(z, x) \rightarrow \exists w_1, w_2.H(z, w_1) \wedge V(z, w_2) \tag{7}$$

$$H(z, y_1) \wedge V(z, y_2) \rightarrow \exists v, w.P(v, w) \wedge H(y_2, v) \wedge V(y_1, v) \tag{8}$$

where p is EDB and the other predicates are IDB. However, the rules are not bts, since applying them to the database with fact $p(a, b)$ leads to models in which V and H form (possibly among other things) an infinite grid – a structure of unbounded treewidth.

5 Graph Denormalisation

To understand how and when our intuition of "objects" can be used to recognise rules with good properties, we introduce a systematic process for *denormalising* rules. Its goal is to replace objects $p_1(z, x_1) \wedge \ldots \wedge p_n(z, x_n)$ by single atoms of the form $D(z, x')$, while preserving semantics. D is a new predicate for this specific object. Note that x' can be limited to the interface of the object with its rule. For example, rule (5) contains the object $P_1(z_1, x_1) \wedge P_2(z_1, y_1) \wedge P_3(z_1, y_2)$, but y_1 and y_2 do not occur in any other object in body or head. One could therefore replace this object by $D_P(z_1, x_1)$, and add a *defining rule*

$$P_1(z_1, x_1) \wedge P_2(z_1, y_1) \wedge P_3(z_1, y_2) \rightarrow D_P(z_1, x_1) \tag{9}$$

to preserve semantics. We do not need the reverse implication, since D is used in the body only. The defining rule is essential to ensure completeness, but it is still in a normalised syntactic form that is usually not acceptable. To address this, we eliminate defining rules by rewriting them using resolution ("backward chaining"). We define this here for the special case of rewriting defining rules for single objects:

Definition 7. *Consider rules $\rho_1 : \varphi_1 \wedge \bar{\varphi}_1 \to D(z, \boldsymbol{x})$ where $\varphi_1 \wedge \bar{\varphi}_1$ is a single object, and $\rho_2 : \varphi_2 \to \exists \boldsymbol{v}.(\psi_2 \wedge \bar{\psi}_2) \wedge \xi$ where $\psi_2 \wedge \bar{\psi}_2$ is a single object, so that ρ_1 and ρ_2 do not share variables. If there is a substitution θ that maps variables of ρ_1 to variables of ρ_2 such that $\bar{\varphi}_1\theta = \bar{\psi}_2$, and $\varphi_1\theta$ does not contain any variables from $\boldsymbol{v}$, then the rule $\varphi_1\theta \wedge \varphi_2 \to \exists \boldsymbol{v}.D(z, \boldsymbol{x})\theta \wedge \xi$ is a rewriting of ρ_1 using ρ_2. We also consider rewritings of rules that share variables, assuming that variables are renamed apart before rewriting.*

Notice that we do not require $\bar{\varphi}_1$ to be the maximal part of the body object for which a rewriting is possible, as is common in (Boolean) conjunctive query rewriting [2]. Doing so would be incomplete, since we need to derive all possible bindings for $D(x, y)$, which may require different parts to be unified with different rule heads. On the other hand, it is sufficient for our purposes to weaken the result by omitting the remaining head object parts ψ_2.

Example 3. Rewriting rule (9) with rules (4) and (5) yields two rules

$$\mathsf{phd}_1(z, x) \wedge \mathsf{phd}_2(z, y_1) \wedge \mathsf{phd}_3(z, y_2) \to \exists v.D_P(v, x) \tag{10}$$

$$P_1(z_1, x_1) \wedge P_2(z_1, y_1) \wedge P_3(z_1, y_2) \wedge \mathsf{sup}_1(z_2, x_1) \wedge \mathsf{sup}_2(z_2, x_2) \to \exists v.D_P(v, x_2). \tag{11}$$

Since the P_i are IDB predicates that only follow from rules (4) and (5), this represents all possible ways to infer new information using rule (9), and we can omit the latter. The bodies of rules (10) and (11) can be denormalised by adding further auxiliary predicates:

$$D_{\mathsf{phd}}(z, x, y_1, y_2) \to \exists v.D_P(v, x) \tag{12}$$

$$D_P(z_1, x_1) \wedge D_{\mathsf{sup}}(z_2, x_1, x_2) \to \exists v.D_P(v, x_2) \tag{13}$$

where D_{phd} and D_{sup} are EDB predicates that need to be defined by denormalising the database, and D can be re-used. We have therefore found a way of expressing (9) in terms of denormalised rules.

Our basic denormalisation algorithm needs to rewrite defining rules exhaustively, and might require to rewrite the same rule several times using its own rewritings, with variables renamed to avoid clashes. For a rule ρ_1 and rule set $\mathbb{P}$, we therefore define $\mathtt{rewrite}(\rho_1, \mathbb{P})$ to be the result (least fixed point) of the following recursive process:

- Initialise $\mathtt{rewrite}(\rho_1, \mathbb{P}) := \mathbb{P}$.
- Add to $\mathtt{rewrite}(\rho_1, \mathbb{P})$ every rewriting of ρ_1 using some rule in $\mathtt{rewrite}(\rho_1, \mathbb{P})$.
- Repeat the previous step until no further changes occur.

This approach terminates and $\mathtt{rewrite}(\rho_1, \mathbb{P})$ is finite since each new rewriting contains fewer head objects than the rule used to obtain it. In particular, only rules with more than a single head object may ever require multiple rewritings.[1]

---

[1] For existential rules, replacing $\varphi \to \psi_1 \wedge \psi_2$ by two rules $\varphi \to \psi_1$ and $\varphi \to \psi_2$ is only correct if ψ_1 and ψ_2 do not share existential variables. Rules with multiple head objects are therefore unavoidable in general. Inseparable parts of rule heads are called *pieces* [2].

Algorithm 1. Generic denormalisation algorithm

Input : rule set $\mathbb{P}$; database $\mathbb{D}$
Output: denormalised rule set $\mathsf{Result}_{\mathbb{P}}$ and denormalised database $\mathsf{Result}_{\mathbb{D}}$

1 $\mathsf{Todo} := \{\varphi \to D_\varphi(z, x) \mid \varphi \text{ an object with object term } z \text{ and interface } x \text{ in a rule body of } \mathbb{P}\}$
2 $\mathsf{Done} := \emptyset$
3 $\mathsf{Rules} := \mathbb{P}$
4 **while** *there is some rule* $\rho \in \mathsf{Todo}$ **do**
5      $\mathsf{Todo} := \mathsf{Todo} \setminus \{\rho\}$
6      $\mathsf{Done} := \mathsf{Done} \cup \{\rho\}$
7   **foreach** $(\varphi \to \exists v.\psi) \in \texttt{rewrite}(\rho, \mathsf{Rules})$ **do**
8   **foreach** *body object* $\xi[z, x]$ *with object term* z *and interface* x *in* $\varphi \to \exists v.\psi$ **do**
9   **if** *there is* $\xi'[z', x'] \to D(z', x') \in \mathsf{Done}$ *such that* $\xi[z, x] \equiv \xi'[z', x']$ **then**
10   replace $\xi[z, x]$ in φ by $\xi'[z, x]$
11   **else**
12                 $\mathsf{Todo} := \mathsf{Todo} \cup \{\xi[z, x] \to D(z, x)\}$ for a fresh predicate D
13   **end**
14   **end**
15         $\mathsf{Rules} := \mathsf{Rules} \cup \{\varphi \to \exists v.\psi\}$
16   **end**
17 **end**
18 $\mathsf{Result}_{\mathbb{P}} := \mathsf{Rules}$ with each body object replaced by its predicate as defined in Done
19 $\mathsf{Result}_{\mathbb{D}} := $ set of all facts $D(c, d_1, \ldots, d_n)$ for which $\mathbb{D}, \mathsf{Done} \models D(c, d_1, \ldots, d_n)$
20 **return** $\langle \mathsf{Result}_{\mathbb{P}}, \mathsf{Result}_{\mathbb{D}} \rangle$

Algorithm 1 shows the main part of our procedure, which makes use of some additional notation explained shortly. The algorithm recursively uses rewriting to eliminate defining rules for all (body) objects that are to be denormalised. Todo and Done are sets of defining rules that still need to be rewritten and that already have been rewritten, respectively. Rules is a set of rules obtained from the rewriting. The defining rules needed for the body objects that occur in Rules are always found in Todo $\cup$ Done.

Initially, Rules are the input rules and Todo are the defining rules for their body objects. For each rule in Todo (Line 4), we consider each rewriting using Rules (Line 7) for being added to Rules (Line 15). First, however, we ensure that every body object of newly rewritten rules is defined (Line 8): either we already defined an equivalent object before (Line 9) that we can reuse, or we add a new object definition to Todo (Line 12).

By $\xi[z, x] \equiv \xi'[z', x']$ in Line 9, we express that the two conjunctions are equivalent conjunctive queries, i.e., there is a bijection $\{z\} \cup x \to \{z'\} \cup x'$ that extends to a homomorphism from ξ to ξ', and whose inverse extends to a homomorphism from ξ' to ξ [1]. Checking this could be NP-hard in general, but is possible in subpolynomial time for our special (star-shaped) object conjunctions. By $\xi'[z, x]$ in Line 10, we mean ξ' with $\{z'\} \cup x'$ replaced by $\{z\} \cup x$ according to the bijection that shows equivalence.

If the algorithm terminates, we return the rewritten rules Rules with all body objects replaced using the newly defined D-atoms, and the set of all denormalised facts that follow from the input database. Note that the heads of rules in Rules may already contain denormalisation atoms $D(z, x)$, while the bodies remain normalised during the rewriting.

In Line 19, we do not need to consider rules in Done that contain IDB predicates in their body, so this database denormalisation is simply conjunctive query answering.

Example 4. Applying Algorithm 1 to Example 1, Todo initially contains three defining rules: rule (9), rule $\mathsf{phd}_1(z, x) \wedge \mathsf{phd}_2(z, y_1) \wedge \mathsf{phd}_3(z, y_2) \rightarrow \mathsf{D}_{\mathsf{phd}}(z, x, y_1, y_2)$, and rule $\mathsf{sup}_1(z_2, x_1) \wedge \mathsf{sup}_2(z_2, x_2) \rightarrow \mathsf{D}_{\mathsf{sup}}(z_2, x_1, x_2)$. The latter two rules contain only EDB predicates in their bodies and therefore have no rewritings: they are moved to Done without adding rules to Rules or Todo. Rule (9) has two rewritings (10) and (11), with the same body objects as the original rule set: all of them are equivalent to objects in Done and can be reused. The algorithm terminates to return four rules: (12) and (13), and analogous denormalisations of the original rules (4) and (5).

Theorem 4. *Consider a database $\mathbb{D}$ and a rule set $\mathbb{P}$, such that Algorithm 1 terminates and returns $\langle \mathsf{Result}_\mathbb{P}, \mathsf{Result}_\mathbb{D} \rangle$. For any Boolean conjunctive query $\exists v.\varphi[v]$, we have that $\mathbb{D}, \mathbb{P} \models \exists v.\varphi[v]$ iff $\mathsf{Result}_\mathbb{D}, \mathsf{Result}_\mathbb{P} \models \exists v.\varphi[v]$.*

As usual, this result extends to non-Boolean CQ answering [1]. To prove Theorem 4, one can show the following invariant to hold before and after every execution of the while loop: $\mathbb{D}, \mathbb{P} \models \exists v.\varphi[v]$ iff $\mathbb{D}, \mathsf{Result}_\mathbb{D}, \mathsf{Result}_\mathbb{P} \models \exists v.\varphi[v]$, where $\mathsf{Result}_\mathbb{P}$ and $\mathsf{Result}_\mathbb{D}$ are obtained as in Lines 18 and 19 using the current Done. Showing this to hold when the program terminates successfully shows the claim, since $\mathbb{D}$ can be omitted as the rules in $\mathsf{Result}_\mathbb{P}$ do not use any EDB predicates from $\mathbb{D}$.

6 Termination of Denormalisation

Although the results of Algorithm 1 are correct, it may happen that the computation does not terminate at all, even in cases where an acceptable rewriting would exist.

Example 5. Consider the rule

$$\mathsf{s}(z_1, x_1) \wedge \mathsf{C}(z_1, x_2) \wedge \mathsf{q}(z_2, x_1) \wedge \mathsf{r}(z_2, x_2) \rightarrow \mathsf{C}(z_1, x_1) \tag{14}$$

where s, q, and r are EDB predicates. There are two body objects in (14), where only the first needs rewriting. Rewriting the rule $\mathsf{s}(z_1, x_1) \wedge \mathsf{C}(z_1, x_2) \rightarrow \mathsf{D}(z_1, x_1, x_2)$ with (14) leads to a new rule $\mathsf{s}(z_1, x_1) \wedge \mathsf{s}(z_1, x_2) \wedge \mathsf{C}(z_1, x_3) \wedge \mathsf{q}(z_2, x_2) \wedge \mathsf{r}(z_2, x_3) \rightarrow \mathsf{D}(z_1, x_1, x_2)$. This rule introduces a new object for object variable z_1. Since the interface now contains three variables $\{x_1, x_2, x_3\}$, it cannot be equivalent to the previous object. A new defining rule is added to Todo, which will subsequently be rewritten to $\mathsf{s}(z_1, x_1) \wedge \mathsf{s}(z_1, x_2) \wedge \mathsf{s}(z_1, x_3) \wedge \mathsf{C}(z_1, x_4) \wedge \mathsf{q}(z_2, x_3) \wedge \mathsf{r}(z_2, x_4) \rightarrow \mathsf{D}'(z_1, x_1, x_2, x_3)$. The algorithm therefore does not terminate, and indeed the generated rules are necessary to retain completeness.

As in this example, non-termination of Algorithm 1 is always associated with objects of growing interface. Indeed, for a fixed interface, there are only finitely many non-equivalent objects, so termination is guaranteed. While general (query) rewriting techniques in existential rules tend to have undecidable termination problems, our specific approach allows us to get a more favourable result:

Theorem 5. *It is P-complete to decide if Algorithm 1 terminates on a given set of rules. For rule sets that do not contain head atoms of the form $p(x, v)$, where x is a universally quantified variable and v is existentially quantified, the problem becomes NL-complete.*

To see why this is the case, let us first observe that non-termination is only caused by rules that use object variables in frontier positions:

Proposition 1. *If object variables do not occur in the frontier of any rule in* $\mathbb{P}$, *then Algorithm 1 terminates on input* $\mathbb{P}$. *In particular, this occurs if* $\mathbb{P}$ *is of the form* GN($\mathbb{P}'$).

Indeed, consider a rewriting step as in Definition 7 where we rewrite ρ_1 using ρ_2. If the object variable z in ρ_1 is mapped to an existential variable in ρ_2, i.e., $z\theta \in \nu$, then no atom of the object in ρ_1 can occur in the body of the rewriting, i.e., φ_1 is empty. Otherwise, there would be an existential (object) variable in the body, which is not allowed by Definition 7. Hence, the body of the rewriting is φ_2, and no new objects are introduced. If all rules are of this form, the overall number of objects that need to be processed is finite and the algorithm must terminate.

Coming back to Theorem 5, we can therefore see that only rewritings using rules with object variables in the frontier need to be considered (we call the associated objects *body frontier object* and *head frontier object*). For investigating termination, we can restrict to "minimal" rewritings that affect only one value y in the rewritten object, i.e., where $\bar{\varphi}_1$ from Definition 7 has the form $p_1(z, y) \wedge \ldots \wedge p_k(z, y)$.

In the (simpler) case that head frontier objects do not have any existentially quantified values, it is even enough to rewrite single attribute-value pairs. A rule with body frontier object $p_1(z, y_1) \wedge \ldots \wedge p_n(z, y_n)$ and head frontier object $q_1(z, x_1) \wedge \ldots \wedge q_m(z, x_m)$ thus gives rise to "replacement rules" of the form $q_i(z, x_i) \mapsto p_j(z, y_j)$ that specify how objects might be rewritten using this rule. This defines a graph on attribute-value pairs of $\mathbb{P}$. Non-termination can be shown to occur exactly if this graph has a cycle along which the interface of the object has increased.

For the latter, we trace the size of the rewritten object's interface during rewriting. Every rewriting with a frontier object may increase or decrease the interface. An increase may occur if the body frontier object contains at least two values in its interface (one interface value preserves size: it is either the frontier value that was unified in the rewriting, or there is no frontier value and the rewritten value was mapped to an existential variable and thereby eliminated). Rule (14), for example, has two interface values, x_1 and x_2, causing non-termination. We can keep track of the interface size in logarithmic space. Cycle detection in the above graph is possible in NL. This shows membership. Hardness is also shown by exploiting the relationship to cycle detection.

Using our understanding of interface-increasing rules as a cause for nontermination, we can also generalise Proposition 1:

Theorem 6. *If every body frontier object that occurs in some rule of* $\mathbb{P}$ *has an interface of size* ≤ 2, *then Algorithm 1 terminates on* $\mathbb{P}$.

We have only shown the NL-part of Theorem 5 yet. The general case with existential values is more complicated and we just give the key ideas of the proof in [18]. The problem is that existential values can only be used for rewriting if all attributes of the rewritten object value are found in the head. Hence, it is not enough to trace single attribute-value pairs. P-hardness is shown by reduction from propositional Horn logic entailment, where we encode propositional rules $a \wedge b \rightarrow c$ as $p_a(x, y) \wedge p_b(x, y) \rightarrow p_c(x, y)$ and true propositions a as $t(x, y) \rightarrow p_a(x, y)$. Finally, we add a rule $p_c(x, y) \wedge p_c(x, z) \rightarrow \exists v.t(x, v)$, where c is a proposition. One can show that Algorithm 1 terminates on the resulting rule set if and only if c is *not* entailed from the Horn rules. Membership can use a similar cycle-detection approach, but the construction of the underlying graph now runs in P.

Even Theorem 6 does not guarantee termination for KG linear rules, and indeed our approach may not terminate in this case. To fix this, we need to observe that we can simplify rewriting if all rules contain only one object in their body: using the notation of Definition 7, a *linear rewriting* of rule ρ_1 using ρ_2 is the rule $\varphi_1\theta \wedge \varphi_2 \rightarrow \exists v.\mathsf{D}(x,y)\theta$. In words: we are reducing the head to contain only the denormalisation atom, and no other atoms. It is easy to check that the procedure remains complete for KG linear rules.

Theorem 7. *If $\mathbb{P}$ is KG linear, then Algorithm 1, modified to use linear rewriting of rules, terminates and returns a rule set $\mathsf{Result}_{\mathbb{P}}$ that is linear.*

It is not hard to see that rewritings of KG linear rules must also be KG linear, showing the second part of the claim. Termination follows since the interface of KG linear rules as obtained during rewriting is bounded by the size of the frontier, which cannot increase when using linear rewriting.

Finally, we remark that our denormalisation shares some similarities with CQ rewriting for existential rules, which is known to be semi-decidable: there is an algorithm that terminates and returns a finite rewriting of a BCQ over a set of rules whenever such a rewriting exists [2]. One may wonder if we could achieve a similar behaviour for Algorithm 1, extending it so that termination is semi-decidable and the algorithm is guaranteed to produce a denormalisation for, e.g., all rule sets that are fus. However, under our assumption that EDB and IDB predicates are separated, the rewritability of BCQs is in fact no longer semi-decidable, not even for plain Datalog. Similar observations have been made for the closely related problem of Datalog *predicate boundedness* [10]. Hence, there is no hope of finding an algorithm that will always compute a denormalisation whenever one exists, even if we cannot decide if this will eventually happen or not. In exchange for this inconvenience, our algorithm also benefits from the separation of IDB and EDB predicates, as it enables us to eliminate defining rules after rewriting them in all possible ways – since IDB predicates cannot occur in the database, this preserves inferences, although it is not semantically equivalent in first-order logic.

7   Frontier Guardedness and Functional Attributes

Our denormalisation procedure can also be applied to KG frontier guarded rules.

Theorem 8. *If $\mathbb{P}$ is KG frontier guarded and Algorithm 1 terminates on $\mathbb{P}$, then the denormalised rule set $\mathsf{Result}_{\mathbb{P}}$ is frontier guarded.*

This follows since a KG frontier guarded rule can only have one object variable in its frontier, so that the object in this case must be the guard. Rewriting therefore can only increase the size of the guard, preserving frontier guardedness.

Theorem 8 is still weaker than Theorem 7, since it does not guarantee termination as in the case of KG linear rules. To compensate, we add another mechanism for making termination more likely, following our intuition of viewing conjunctions as "objects".

In typical objects, attributes often can have at most one value. This holds for all objects created when normalising rules. Making this restriction formal could also ensure termination, since the size of each object would be bounded, and the number of possible objects finite. Example 5 shows how a non-terminating case might violate this. The constraint that attributes have at most one value is captured by functional dependencies:

Definition 8. *A functional dependency (FD) for attribute p is a rule $p(z, x_1) \wedge p(z, x_2) \rightarrow x_1 \approx x_2$, where $\approx$ is a special predicate that is interpreted as identity relation in all models: $\approx^I = \{\langle \delta, \delta \rangle \mid \delta \in \Delta^I\}$. The functional dependency is an EDB-FD if p is an EDB predicate, and an IDB-FD otherwise.*

We use built-in equality in this definition, making FDs a special case of *equality generating dependencies* (egds) [1]. Alternatively, $\approx$ could also be axiomatised using Datalog, which turns FDs into regular Datalog rules and $\approx$ into a regular predicate.

Intuitively, we want functional dependencies to apply to some attributes. However, we cannot just introduce FDs as additional rules: query answering is undecidable for the combination of (frontier) guarded existential rules and FDs [15]. Conversely, it is not true that the given rule set *entails* any IDB-FDs, even if some EDB-FDs are guaranteed to hold in the database. Indeed, any model of a set of rules can be extended by interpreting each IDB predicate as a maximal relation (i.e., as an arity-fold cross-product of the domain), resulting in a model that refutes all possible IDB-FDs. Therefore, rather than *asserted* or *entailed* FDs, we are interested in FDs that are *incidental*:

Definition 9. *Consider a set $\mathbb{P}$ of rules and a set $\mathbb{F}$ of EDB-FDs. An IDB-FD for attribute p is incidental to $\mathbb{P}$ and $\mathbb{F}$ if, for all databases $\mathbb{D}$ with $\mathbb{D} \models \mathbb{F}$ and for all BCQs φ, we have that $\mathbb{D}, \mathbb{P} \models \varphi$ iff $\mathbb{D}, \mathbb{P} \cup \{p(z, x_1) \wedge p(z, x_2) \rightarrow x_1 \approx x_2\} \models \varphi$. The set of all FDs incidental to $\mathbb{P}$ and $\mathbb{F}$ is denoted IDP$(\mathbb{P}, \mathbb{F})$.*

In other words, an FD is incidental if we might as well assert it without affecting the answer to any conjunctive query.

Given a set $\mathbb{F}$ of FDs and a conjunction φ of binary atoms of the form $p(x, y)$, we write $\mathbb{F}(\varphi)$ for the conjunction obtained by identifying variables in φ until all FDs in $\mathbb{F}$ are satisfied. This is unique up to renaming of variables. Moreover, let $\theta_{\mathbb{F}(\varphi)}$ denote a corresponding substitution such that $\mathbb{F}(\varphi) = \varphi\theta_{\mathbb{F}(\varphi)}$. For our simple attribute dependencies, this can be computed in polynomial time. Using this notation, we can extend Algorithm 1 to take a given set of FDs into account:

Definition 10. *Let Algorithm $1_{\mathbb{F}}$ be the modification of Algorithm 1 that takes an additional set $\mathbb{F}$ of FDs as an input, and that replaces the rewriting $\varphi \rightarrow \exists v.\psi$ after Line 7 by $\mathbb{F}(\varphi) \rightarrow \exists v.\psi\theta_{\mathbb{F}(\varphi)}$, i.e., which factorises each rewriting using the given FDs before continuing.*

This may help to achieve termination, since the application of FDs may decrease the size of objects to be rewritten next. Our approach shares some ideas with the use of database constraints for optimising query rewriting [23], but the details are different.

Example 6. Consider again the rule of Example 5, and assume that we know that attribute s is functional. Algorithm $1_{\mathbb{F}}$ will again obtain the rewriting $s(z_1, x_1) \wedge s(z_1, x_2) \wedge C(z_1, x_3) \wedge q(z_2, x_2) \wedge r(z_2, x_3) \rightarrow D(z_1, x_1, x_2)$. Denoting the body of this rewriting by φ, we find that $\theta_{\mathbb{F}(\varphi)} = \{x_2 \mapsto x_1\}$, so that the rewriting becomes $s(z_1, x_1) \wedge C(z_1, x_3) \wedge q(z_2, x_1) \wedge r(z_2, x_3) \rightarrow D(z_1, x_1, x_1)$. The object for variable z_1 now is equivalent to the object that has been rewritten in the first step, and so can be replaced by $D(z_1, x_1, x_3)$. The algorithm terminates.

8   Obtaining Incidental FDs

The improved denormalisation of Definition 10 hinges upon the availability of a suitable set of functional dependencies. For EDB predicates, these might be obtained from constraints that have been declared explicitly for the underlying database, or they might even be determined to simply hold in the given data. Example 6 shows that this can already help. In general, however, we would also like to use incidental IDB-FDs. This section therefore asks how they can be computed.

Our first result is negative: it is impossible to determine all incidental FDs even for very restricted subsets of Datalog. This can be shown by reducing from the undecidable problem of deciding non-emptiness of the intersection of two context-free grammars.

Theorem 9. *For a set $\mathbb{P}$ of Datalog rules containing only binary predicates and no constants, a set $\mathbb{F}$ of EDB-FDs, and an IDB-FD σ, it is undecidable if $\sigma \in \mathsf{IDP}(\mathbb{P}, \mathbb{F})$.*

We therefore have to be content with a sound but incomplete algorithm for computing incidental FDs. We use a top-down approach that initially assumes all possible FDs to hold, and then checks which of them might be violated when applying rules, until a fixed point has been reached. This approach is closely related to a work of Sagiv [24, Sect. 9] where the author checks if a given set of existential rules is *preserved non-recursively* by a given Datalog program. We extend this idea from Datalog to existential rules and from non-recursive to (a form of) recursive preservation. For simplicity, we give the algorithm only for checking FD preservation, but it is not hard to extend it to arbitrary rules. We also remark that Theorem 9 settles an open question of Sagiv [24].

Our algorithm tries to discover a violation of an FD by considering a situation where the premise holds (expressed as a CQ $p(z, x_1) \wedge p(z, x_2)$), and then checking all possible ways to derive this situation in one step, using rewriting. If any of the rewritten queries is such that the FD does not follow from the FDs assumed to far, the FD is eliminated.

To check functionality in the presence of existential quantifiers, we first replace existential variables by Skolem terms. The actual check then has to be based on a rewriting of $p(z, x_1) \wedge p(z, x_2)$ where both atoms have been rewritten, which we ensure by renaming the predicates. For the next definition, recall that rewriting conjunctive queries can be achieved like rewriting rules in Definition 7 but dropping the head in all rewritings.

Definition 11. *The Skolemisation of rule $\varphi[\boldsymbol{x},\boldsymbol{y}] \rightarrow \exists \boldsymbol{v}.\psi[\boldsymbol{x},\boldsymbol{v}]$ is the rule $\varphi[\boldsymbol{x},\boldsymbol{y}] \rightarrow \psi'[\boldsymbol{x}]$ where ψ' is obtained from ψ by replacing each $v \in \boldsymbol{v}$ by a term $f_v(\boldsymbol{x})$, where f_v is a freshly introduced function symbol. The Skolemisation of all rules in $\mathbb{P}$ is denoted $\mathsf{skolem}(\mathbb{P})$.*

For a conjunction of atoms φ, let $\hat{\varphi}$ be φ with all predicates p replaced by fresh predicates $\hat{p}$. For a rule set $\mathbb{P}$, let $\hat{\mathbb{P}}$ be the set $\{\varphi \rightarrow \exists \boldsymbol{v}.\hat{\psi} \mid \varphi \rightarrow \exists \boldsymbol{v}.\psi \in \mathbb{P}\}$. The one-step rewriting $\mathsf{os\text{-}rewrite}(\varphi, \mathbb{P})$ is the set of all conjunctions obtained by exhaustively rewriting $\hat{\varphi}$ using rules in $\mathsf{skolem}(\hat{\mathbb{P}})$, and where no predicate from $\hat{\varphi}$ occurs.

Algorithm 2. Algorithm for computing some incidental FDs

Input : rule set $\mathbb{P}$; set $\mathbb{F}$ of EDB-FDs
Output: set $\mathbb{F}_{IDB}$ of incidental IDB-FDs

1 $\mathbb{F}_{IDB} := \{p(z,x_1) \wedge p(z,x_2) \rightarrow x_1 \approx x_2 \mid p$ an IDB predicate$\}$

2 **repeat**

3   **foreach** $p(z,x_1) \wedge p(z,x_2) \rightarrow x_1 \approx x_2 \in \mathbb{F}_{IDB}$ **do**

4   **foreach** $\varphi \in$ os-rewrite$(p(z,x_1) \wedge p(z,x_2), \mathbb{P})$ **do**

5              $y_i :=$ the variable that x_i has been mapped to for the rewriting φ $(i \in \{1,2\})$

6   **if** $y_1 \theta_{(\mathbb{F} \cup \mathbb{F}_{IDB})(\varphi)} \neq y_2 \theta_{(\mathbb{F} \cup \mathbb{F}_{IDB})(\varphi)}$ **then**

7                  $\mathbb{F}_{IDB} := \mathbb{F}_{IDB} \setminus \{p(z,x_1) \wedge p(z,x_2) \rightarrow x_1 \approx x_2\}$

8   **break** // continue with next FD in Line 3

9   **end**

10   **end**

11   **end**

12 **until** $\mathbb{F}_{IDB}$ *has not changed in previous iteration*

13 **return** $\mathbb{F}_{IDB}$

The result of os-rewrite is finite, since heads and bodies of $\hat{\mathbb{P}}$ do not share predicates. Our procedure is given in Algorithm 2. It proceeds as explained above checking, given a pair of IDB atoms, every possible derivation for a potential violation of an FD. A violation is detected if two values of an attribute are not necessarily equal based on the current FDs (Line 6). Note that φ may not contain x_1 and/or x_2 since they may be unified during rewriting. We therefore consider the values y_i they have been mapped to (Line 5). As a special case, y_i can be Skolem terms, which typically causes the FD to be violated, unless both x_1 and x_2 are rewritten together and replaced by the same term.

Note that the check in Line 5 uses the set $\mathbb{F}_{IDB}$, including the FD that is just checked. Intuitively speaking, this is correct since the rewriting approach searches for the first step (in a bottom-up derivation) where an FD would be violated. Initially, when all IDB predicates are empty, all FDs hold.

Theorem 10. *For inputs $\mathbb{P}$ and $\mathbb{F}$, Algorithm 2 returns a set $\mathbb{F}_{IDB} \subseteq$ IDP$(\mathbb{P}, \mathbb{F})$ after polynomial time.*

While the algorithm must be incomplete, and in particular cannot detect all FDs for the rules used for our proof of Theorem 9, it can detect many cases of FDs.

Example 7. Consider the following rules, with EDB predicates p and s:

$$\mathsf{p}(x,y) \wedge \mathsf{s}(x,y) \rightarrow \mathsf{Q}(x,y) \tag{15}$$

$$\mathsf{s}(x,y) \rightarrow \exists v,w.\mathsf{Q}(v,w) \wedge \mathsf{R}(x,v) \wedge \mathsf{R}(x,w) \tag{16}$$

Assume that p is functional. Algorithm 2 first checks the IDB-FD for Q by rewriting $\hat{\mathsf{Q}}(z,x_1) \wedge \hat{\mathsf{Q}}(z,x_2)$. We can rewrite the first atom using rule (15) (mapping z to x and x_1 to y) to obtain $\mathsf{p}(x,y) \wedge \mathsf{s}(x,y) \wedge \hat{\mathsf{Q}}(x,x_2)$. Rewriting $\hat{\mathsf{Q}}(x,x_2)$ using rule (15) with variables renamed to x' and y', we get $p(x, y) \wedge s(x, y) \wedge p(x, y') \wedge s(x, y')$. Hence $y_1 = y$ and $y_2 = y'$ in Line 5, and these variables are identified since p is functional.

Rewriting $\hat{Q}(z, x_1) \wedge \hat{Q}(z, x_2)$ using rule (16) for both atoms, we obtain $s(x, y) \wedge s(x, y')$, with original variables replaced by $\{z \mapsto x, x_1 \mapsto f_w(x), x_2 \mapsto f_w(x)\}$ where $f_v(x)$ and $f_w(x)$ are Skolem terms. Again, the FD is preserved. As it is not possible to rewrite one atom with rule (15) and the other with rule (16), we find that Q is functional.

In contrast, functionality for R is violated, since we cannot identify $f_v(x)$ and $f_w(x)$.

9   Discussion and Outlook

Our central observation is that support for ontological modelling and reasoning over knowledge graphs (KGs) is severely lacking. Ontology language features needed for KGs are not supported by mainstream approaches such as OWL and Datalog, and take us outside of known decidable classes of existential rules. Practical tools and methods for modelling and reasoning are even further away. A lot of research is still to be done.

Our work is a first step into this field, focussing on basic language definitions and decidability properties. A core concept of our work is to view some conjunctive patterns as *objects* with *attributes* and *values*, such that existential quantification plays the role of object creation. This leads to a very natural view on existential rules, but it also extends to the data, where objects correspond to groups of triples. We believe that such grouping might also help to improve performance of reasoning with KG-based rules.

Each decidability criterion (acyclicity/fes, bts, rewritability/fus) calls for a different reasoning procedure. For the types of acyclicity we mention, any bottom-up forward chaining inference engine will terminate, even if rules are Skolemised. Rule engines in RDF stores (e.g., Jena) or logic programming tools (e.g., DLV) could be used. Linear rules (and fus in general) are supported by backward-chaining reasoners such as Graal [3]. Interestingly, reasoners for fes and fus do not need to know if and why the rules meet the criteria – it is enough if they do. In particular, rules do not have to be denormalised for reasoning. Denormalisation is only needed to find out which tool to use.

Tools for guarded rules and bts seem to be missing today. They could be implemented by augmenting bottom-up reasoners with additional blocking conditions to ensure termination. Similar ideas are used successfully in OWL reasoning, but generalising them to arbitrary rules will require further research and engineering. Our work may motivate such research by identifying a wider class of rules that would benefit from this.

There are too many connections to other recent works to list, but we highlight some. Ontologies for non-classical data models are currently also studied for key-value stores [21] and for the object database MongoDB [7]. A rule language for declarative programming on KGs was recently proposed in Google's Yedalog [9], and several new rule-based reasoners now support RDF graphs [20,25]. There are numerous works on decidable classes of existential rules. We covered essential approaches, but there remain many others, such as *warded* [14] or *sticky* rules [8], that deserve investigation for KGs.

This diversity of works witnesses a huge current interest in practical data models and rule-based ontologies, but many further works will still be needed for bringing KG-based ontologies to the level of maturity that past semantic technologies have acquired.

Acknowledgements. This work is partly supported by the German Research Foundation (DFG) in CRC 912 (HAEC) and in Emmy Noether grant KR 4381/1-1.

References

1. Abiteboul, S., Hull, R., Vianu, V.: Foundations of Databases. Addison Wesley, Redwood City (1994)
2. Baget, J.F., Leclère, M., Mugnier, M.L., Salvat, E.: On rules with existential variables: walking the decidability line. Artif. Intell. **175**(9–10), 1620–1654 (2011)
3. Baget, J.-F., Leclère, M., Mugnier, M.-L., Rocher, S., Sipieter, C.: Graal: a toolkit for query answering with existential rules. In: Bassiliades, N., Gottlob, G., Sadri, F., Paschke, A., Roman, D. (eds.) RuleML 2015. LNCS, vol. 9202, pp. 328–344. Springer, Heidelberg (2015)
4. Baget, J., Mugnier, M., Rudolph, S., Thomazo, M.: Walking the complexity lines for generalized guarded existential rules. In: Proceedings of the 22nd International Joint Conference on Artificial Intelligence (IJCAI 2011), pp. 712–717 (2011)
5. Belleau, F., Nolin, M., Tourigny, N., Rigault, P., Morissette, J.: Bio2RDF: towards a mashup to build bioinformatics knowledge systems. J. Biomed. Inf. **41**(5), 706–716 (2008)
6. Bollacker, K., Evans, C., Paritosh, P., Sturge, T., Taylor, J.: Freebase: a collaboratively created graph database for structuring human knowledge. In: Proceedings of the 2008 ACM SIGMOD International Conference on Management of Data, pp. 1247–1250. ACM (2008)
7. Botoeva, E., Calvanese, D., Cogrel, B., Rezk, M., Xiao, G.: OBDA beyond relational DBs: a study for MongoDB. In: Proceedings of the 29th International Workshop on Description Logics (DL 2016) (2016)
8. Calì, A., Gottlob, G., Pieris, A.: Towards more expressive ontology languages: the query answering problem. J. Artif. Intell. **193**, 87–128 (2012)
9. Chin, B., von Dincklage, D., Ercegovac, V., Hawkins, P., Miller, M.S., Och, F., Olston, C., Pereira, F.: Yedalog: exploring knowledge at scale. In: 1st Summit on Advances in Programming Languages (SNAPL 2015), pp. 63–78 (2015)
10. Cosmadakis, S.S., Gaifman, H., Kanellakis, P.C., Vardi, M.Y.: Decidable optimization problems for database logic programs (preliminary report). In: Simon, J. (ed.) Proceedings of the 20th Annual ACM Symposium on Theory of Computing (STOC 1988), pp. 477–490. ACM (1988)
11. Cuenca Grau, B., Horrocks, I., Krötzsch, M., Kupke, C., Magka, D., Motik, B., Wang, Z.: Acyclicity notions for existential rules and their application to query answering in ontologies. J. Artif. Intell. Res. **47**, 741–808 (2013)
12. Dantsin, E., Eiter, T., Gottlob, G., Voronkov, A.: Complexity and expressive power of logic programming. ACM Comput. Surv. **33**(3), 374–425 (2001)
13. Das, S., Sundara, S., Cyganiak, R. (eds.): R2RML: RDB to RDF Mapping Language. W3C Recommendation (2012). https://www.w3.org/TR/r2rml/
14. Gottlob, G., Pieris, A.: Beyond SPARQL under OWL 2 QL entailment regime: rules to the rescue. In: Proceedings of the 24th International Joint Conference on Artificial Intelligence (IJCAI 2015), pp. 2999–3007 (2015)
15. Grädel, E.: On the restraining power of guards. J. Symb. Log. **64**(4), 1719–1742 (1999)
16. Hoffart, J., Suchanek, F.M., Berberich, K., Weikum, G.: YAGO2: a spatially and temporally enhanced knowledge base from Wikipedia. J. Artif. Intell. **194**, 28–61 (2013)

17. Krötzsch, M., Rudolph, S.: Extending decidable existential rules by joining acyclicity and guardedness. In: Proceedings of 22nd International Joint Conference on Artificial Intelligence (IJCAI 2011), pp. 963–968 (2011)
18. Krötzsch, M., Thost, V.: Ontologies for knowledge graphs: breaking the rules. Extended technical report, TU Dresden, April 2016. https://iccl.inf.tu-dresden.de/web/TR3029/en
19. Lenzerini, M.: Data integration: a theoretical perspective. In: Popa, L. (ed.) Proceedings of the 21st Symposium on Principles of Database Systems (PODS 2002), pp. 233–246. ACM (2002)
20. Motik, B., Nenov, Y., Piro, R., Horrocks, I., Olteanu, D.: Parallel materialisation of Datalog programs in centralised, main-memory RDF systems. In: Proceedings of the 28th AAAI Conference on Artificial Intelligence, pp. 129–137 (2014)
21. Mugnier, M.L., Rousset, M.C., Ulliana, F.: Ontology-mediated queries for NOSQL databases. In: Proceedings of the 30th AAAI Conference on Artificial Intelligence (2016)
22. W3C OWL Working Group: OWL 2 Web Ontology Language: Document Overview. W3C Recommendation, 27 October 2009. http://www.w3.org/TR/owl2-overview/
23. Rodriguez-Muro, M., Kontchakov, R., Zakharyaschev, M.: Query rewriting and optimisation with database dependencies in ontop. In: Proceedings of the 26th International Workshop on Description Logics (2013)
24. Sagiv, Y.: Optimizing Datalog programs. Technical report CS-TR-86-1132, Stanford University, Department of Computer Science (1986). http://i.stanford.edu/TR/CS-TR-86-1132.html
25. Urbani, J., Jacobs, C., Krötzsch, M.: Column-oriented Datalog materialization for large knowledge graphs. In: Proceedings of the 30th AAAI Conference on Artificial Intelligence, pp. 258–264 (2016)
26. Vrandečić, D., Krötzsch, M.: Wikidata: a free collaborative knowledgebase. Commun. ACM **57**(10), 78–85 (2014)

An Extensible Linear Approach for Holistic Ontology Matching

Imen Megdiche[(✉)], Olivier Teste[(✉)], and Cassia Trojahn[(✉)]

Institut de Recherche en Informatique de Toulouse (UMR 5505), Toulouse, France
{Imen.Megdiche,Olivier.Teste,Cassia.Trojahn}@irit.fr

Abstract. Resolving the semantic heterogeneity in the semantic web requires finding correspondences between ontologies describing resources. In particular, with the explosive growth of data sets in the Linked Open Data, linking multiple vocabularies and ontologies simultaneously, known as holistic matching problem, becomes necessary. Currently, most state-of-the-art matching approaches are limited to pairwise matching. In this paper, we propose a holistic ontology matching approach that is modeled through a linear program extending the maximum-weighted graph matching problem with linear constraints (cardinality, structural, and coherence constraints). Our approach guarantees the optimal solution with mostly coherent alignments. To evaluate our proposal, we discuss the results of experiments performed on the Conference track of the OAEI 2015, under both holistic and pairwise matching settings.

Keywords: Ontology matching · Holistic matching · Linear approach

1 Introduction

Ontology matching is an essential task in the management of the semantic heterogeneity problem in several scientific disciplines and applied fields, notably to support data exchange, schema/ontology evolution, data integration, and data linkage. The typically high degree of semantic heterogeneity reflected in different ontologies makes this task an inherently complex task [21]. Several approaches for automatic or semi-automatic ontology matching have emerged [6] in the literature, which exploit in many different ways the knowledge encoded within each ontology when identifying correspondences between their features or structures.

Despite the different proposals in the field, most ontology matching approaches have been designed to deal with pairs of ontologies, a task so-called *pairwise matching*. However, with the continuously increasing amount of data sources being produced by the Linked Open Data community, designing solutions to deal with the simultaneously matching of different schemas and ontologies is becoming necessary [19,27]. This task is called *holistic ontology matching* [21]. The holistic ontology matching problem is one of the key challenges proposed in [19] in its future research agenda. The proposal of the paper falls within the scope of holistic approaches.

© Springer International Publishing AG 2016
P. Groth et al. (Eds.): ISWC 2016, Part I, LNCS 9981, pp. 393–410, 2016.
DOI: 10.1007/978-3-319-46523-4_24

Broadly speaking, the matching process takes as input a set of ontologies, denoted by Ω, and determines as output a set of correspondences, called alignment. The *pairwise ontology matching* process takes as input two ontologies, $\Omega = \{O_1, O_2\}$, and determines as output a set of correspondences denoted as $A = \{c_1, c_2, ..., c_x\}$. A correspondence c_i can be defined as $<e_1, e_2, r, n>$, such that: e_1 and e_2 are ontology entities (e.g. properties, classes, instances) of O_1 and O_2, respectively; r is a relation holding between e_1 and e_2 (usually, $\equiv$, $\sqsupseteq$, $\bot$, $\sqcap$); and n is a confidence measure in the $[0, 1]$ range assigning a degree of trust on the correspondence. The correspondence $<e_1, e_2, r, n>$ asserts that the relation r holds between the ontology entities e_1 and e_2 with confidence n. The higher the confidence value, the higher the likelihood that the relation holds. Within an individual mapping entity, one or more O_1 entities can match with one or more O_2 entities. Alignments have different cardinalities; we distinguish 1:1 (one-to-one), 1:m (one-to-many), n:1 (many-to-one) or n:m (many-to-many). An alignment may be a simple alignment 1:1, or a multiple alignment 1:n or n:1, and n:m.

The *holistic ontology matching* process extends the ontology pairwise matching using a set $\Omega = \{O_1, ..., O_N\}$ of ontologies with $N \geq 2$. For instance, if $\Omega = \{O_1, O_2, O_3\}$, then the alignment is defined as $A = A_{12} \cup A_{13} \cup A_{23}$ where

- $A_{12} = \{<e_1, e_2, r_{12}, c_{12}> \,|e_1 \in O_1 \wedge e_2 \in O_2\}$,
- $A_{13} = \{<e_1, e_3, r_{13}, c_{13}> \,|e_1 \in O_1 \wedge e_3 \in O_3\}$,
- $A_{23} = \{<e_2, e_3, r_{23}, c_{23}> \,|e_2 \in O_2 \wedge e_3 \in O_3\}$.

Triple correspondences between entities of O_1, O_2, and O_3 can be deduced from A by detecting *cliques*; e.g., each subset of adjacent correspondences $< e_1, e_2, r, c_{12}>$, $<e_1, e_3, r, c_{13}>$ and $<e_2, e_3, r, c_{23}>$.

The main limitation of the pairwise approaches regard to the holistic approaches is that in the former, A is considered as a local solution depending of the order with which the ontology matching is carried out; e.g. $A_{12} \cup A_{(12)3} \neq A_{13} \cup A_{(13)2} \neq A_{23} \cup A_{(23)1}$. Thus the set of correspondences in A differs according to the order users apply the ontology matching pairwise approach. Our holistic approach resolves the problem globally thus the solution is unique and considered as a global solution.

In this paper, we tackle the challenges of providing an extensible holistic ontology matching solution at schema-level. We provide an holistic approach which is able to link multiple ontologies simultaneously from Ω with $N \geq 2$. The approach guarantees to find always the same A global optimal solution. Our solution is extensible to operate with simple and multiple correspondences. To identify the best correspondences, a normalized degree of similarity between 0 and 1 is calculated using various similarity metrics. We develop a linear program based on an extension of the maximum-weighted graph matching problem [23], which is solved in polynomial time [15]. Our linear program encompasses different constraints related to the ontology matching problem. The constraints are used to guarantee the structural coherence between matched ontologies.

The main contributions of this paper are as follows:

- We provide an efficient approach to determine holistic correspondences between multiple ontologies. We model the approach within a linear program by reducing the ontology matching problem to the maximum-weighted graph matching problem, which is solvable in polynomial time.
- The approach is extensible with different structural similarity strategies and several linear constraints, which ensure mostly coherent alignments. We provide four constraints allowing the matching of classes and properties between ontologies.
- This approach extends a contribution [1] in the field of schema matching, especially designed to hierarchical schema structures like XML. The flexibility of the employed technique has allowed us to adapt the previous model with new constraints in order to take into account the specificities of the ontology matching problem.

The rest of the paper is organised as follows. Section 2 discusses related work. Section 3 presents our extensible linear approach for matching multiple ontologies. Section 4 discusses the experiments conducted on the *Conference* track of the Ontology Alignment Evaluation Initiative Campaign (OAEI) 2015, under both pairwise and holistic settings. Finally, Sect. 5 concludes the paper and discusses future directions.

2   Related Work

This paper concerns the problem of holistic ontology matching, which is modelled through the maximum-weighted graph matching problem with constraints and techniques from the combinatorial optimisation field.

Graph-Based Approaches. In [28], an association graph is built from two input ontologies, where nodes represent candidate correspondences and edges as affinities between them. The selection of correspondences is formulated as a node ranking in the association graph using a Markov random walk process [3]. An iterative matcher (GMO) using bipartite graphs to represent ontologies is proposed in [11]. It computes structural similarities between entities by recursively propagating their similarities in the graphs. A similar representation is adopted by OLA [7], where the selection of alignments is reduced to a weighted bipartite graph matching problem. This approach models structural similarity computation as a set of equations of the different properties of ontologies.

Combinatorial Optimisation Strategies. S-Match [8] reduces the semantic matching to the propositional validity problem, which is theoretically a co-NP hard problem. The elements of schemes are translated into logical formulas and the matching consists of resolving propositional formula constructed between entities. Similarity Flooding (SF) [18] reduces the selection of correspondences to the stable marriage problem, which returns a local optimal solution. SF proposes a graph-based structural-matcher which propagates similarities between neighbourhood nodes until a fixed point computation. CODI [12] implements the probabilistic markov logical framework, transforming the matching problem into a maximum-a-posteriori (MAP) optimization problem which is equivalent to Max-Sat problem (NP-hard). Recently, [20] proposes a multi-cultural taxonomies matching that is modelled as a combinatorial optimisation problem using integer linear programming and quadratic programming. Mamba [17] is another system applying a combinatorial optimization approach with constraints and Markov Logic.

Holistic Approaches. While state-of-the-art matching proposals mainly focus on pairwise matching, most works on holistic matching give special attention to pairwise-attribute matching. In [9], a probabilistic framework for hidden model discovery is used for determining an underlying unified model capturing the correspondences between attributes in different schemes. Given the input schemas as observations, it reconstructs the hidden generative distribution by selecting consistent models with highest probability. For dealing with complex attribute correspondences, [10] exploit co-occurrence information across schemes and a correlation mining approach. It is based on the observations that frequent attribute co-presence indicates a synonym relationship and rare ones indicates a grouping relationship. This approach has been extended in [24] improving accuracy and efficiency, by reducing the number of synonymous candidates (assuming that two attributes co-present in the same schema cannot be synonymous candidates). [22] present an approach for incrementally merging 2-way schemes and generating an integrated one by clustering the nodes based on linguistic similarity and a tree mining technique. Under a different perspective, [27] proposes a holistic matching approach for aligning large ontologies from different domains, by grouping concepts in topics that are aligned locally. The topic identification is based on TF-IDF applied on Wikipedia pages related to concepts, resulting in a category trees (forests), while the similarity of topics is based on Jaccard, resulting in a graph containing topically related forest nodes. The correspondences between forests are determined using a tree overlap measure, before applying logical reasoning for removing conflicting correspondences.

Discussion. While the alignment selection strategy in [28] is based on paths in the graph, we reduce the selection to the maximum-weighted bipartite graph matching (MWGM) problem like OLA and we adopt a different structural similarity strategy from [11]. The complexity of MWGM with linear programming is known to be polynomial [23] even with the simplex algorithm [23]. Compared to OLA we do not compute structural similarities but encode structural properties as linear constraints. As CODI, we perform both structural matching (without additional structural similarity computation) and alignment extraction phases. Compared to CODI, we consider disjointness for all types of entities. Unlike CODI whose pairwise approach is reduced to a NP-Hard problem, our solution extends a polynomial problem in both pairwise and holistic versions. While MAMBA can be reduced to an NP-Hard problem, our approach is reduced to a polynomial problem. In a holistic and monolingual setting, we apply a combinatorial optimisation problem using linear programming, as done in [20] in pairwise. The constraints proposed by [20] for multiple correspondences can be simply added to our model to enhance the matching of multiple correspondences in the relaxed version of our model (i.e. with relaxed decision variables). While most holistic approaches focus on pairwise-attribute matching [9,10,24,27], our approach is not restricted to attributes. These holistic approaches handle simple attributes compared to the more structured schemes of ontologies. Differently from [27], we do not perform cross-domain holistic matching. Compared to [10], our approach can also return simple and multiple correspondences. Finally, as some pairwise matchers [13,14], we adopt constraints that reduce the possibility of generating incoherent alignments. In that sense, an interesting direction concerns applying repair techniques in holistic ontology matching [5].

3 Extensible Holistic Approach

3.1 Global Overview

Our approach is based on a well-known combinatorial optimisation problem, the maximum-weighted graph matching (MWGM) problem [23]. The idea consists in generalizing the pairwise matching on a set of N input ontologies through generic decision variables and generic linear constraints modelled in a linear program. The MWGM problem aims at finding a set of disjoint edges having the maximum weights in a weighted graph G. Here, we reduce the ontology matching to the MWGM problem[1]. Indeed, we consider that G expresses the potential candidate correspondences between the input ontologies and has (i) three types of nodes representing classes, object and data properties, and (ii) edges representing virtual connections between the same types of nodes (i.e. classes related to classes, object properties to object properties and data properties to data properties). These edges have weights represent similarities between the nodes and can be establish using different strategies. In our approach, the similarities are calculated in a pre-processing step (Sect. 3.2). Given this reduction, searching simple correspondences (1:1) with a maximum weight on similarities is equivalent to find a set of disjoint edges with a maximum weight in the MWGM problem.

Our approach processes simultaneously $N \geq 2$ input ontologies. It involves a pre-processing step and a processing step. In the pre-processing step, we apply element-level matchers and then aggregate the results in order to produce similarities between the entities of the ontologies. In the processing step, we instantiate the different elements of the linear program (decision variables and linear constraints) and then resolve the model by using the CPLEX solver[2].

We will use the following notations in the remainder of this paper:

- $N = |\Omega|$ is the number of input ontologies;
- i, j are internal identifiers of the ontologies O_i and O_j;

---

[1] Note that we do not transform an OWL ontology into a graph but represent all entities as nodes with connections between them representing candidate correspondences.

[2] http://www-01.ibm.com/software/commerce/optimization/cplex-optimizer/.

- $\{k, l\}$, $\{m, n\}$, $\{q, r\}$ refer respectively to the order of classes, object properties, data properties in the ontology (the order refers to an internal identifier of the entity in the set of ordered entities of the same type);
- C_i, OP_i, DP_i refer respectively to the set of classes, object properties and data properties in the ontology O_i;
- nbC_i, $nbDP_i$, $nbOP_i$ refer respectively to the cardinality of classes ($|C_i|$), the cardinality of object properties ($|OP_i|$) and the cardinality of data properties ($|DP_i|$) in O_i;
- c_{i_k} is the class of order k in the ontology O_i;
- op_{i_m} is the object property of order m in the ontology O_i;
- dp_{i_q} is the data property of order q in the ontology O_i;

Running Example. In order to illustrate our approach, we have chosen three ontologies from the OAEI Conference track [26]. These ontologies are *Cmt*, *Sigkdd* and *Conf-of*. For the sake of brevity, we present only some fragments of these ontologies as depicted in Fig. 1. This example will be used in the remainder of the paper.

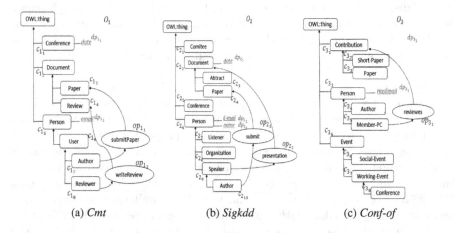

(a) *Cmt*   (b) *Sigkdd*   (c) *Conf-of*

Fig. 1. Example of three ontologies from conference track in OAEI

The objective of our model is to resolve simultaneously the set of alignments given the ontologies O_1 (Cmt), O_2 (Sigdkk) and O_3 (Conf-of). It will resolve in a unique run the alignments for A_{12}, A_{23} and A_{13}. As depicted in Fig. 1, O_1 is composed of $nbC_1 = 8$ classes, $nbOP_1 = 2$ object properties and $nbDP_1 = 2$ data properties. $C_1 = \{c_{1_1}, c_{1_2}, \ldots, c_{1_8}\}$, $OP_1 = \{op_{1_1}, op_{1_2}\}$, $DP_1 = \{dp_{1_1}, dp_{1_2}\}$.

3.2   Pre-processing Step

Our linear program takes as input a set of $N \geq 2$ ontologies $O_i = C_i \cup OP_i \cup DP_i$, $i \in [1, N]$, and a set of $N(N-1)/2$ similarity matrices representing the average results of different element-level matchers. These matrices are computed between each pair of ontologies for classes, object properties and data properties. For instance, sim_{i_k,j_l} denotes a similarity measure computed between the classes c_{i_k} and c_{j_l}, which belong respectively to ontology O_i and O_j. We have selected a restrictive set of element-level matchers according to their time performance and to their quality in the recent comparative study of [25]. The selected metrics are as follows: (1) from the *character-based category* metrics we have chosen ISUB and 3-gram to compute similarity between tokens and we have applied the generalized Mongue-Elkan method on these metrics to get the similarity between entities, (2) from the *token-based category*, we have applied Jaccard and (3) from the *language-based category* we applied Lin measure. These metrics are aggregated according to the average function in order to keep a balanced result.

3.3 Linear Program

In this section, we describe the formalization of our linear program for holistic matching named LPHOM. The formalization is generalizable for $N \geq 2$ graphs.

Decision Variables. Our model is composed of three types of binary decision variables referring respectively to the three simple types of alignments in ontologies:

- The first type refers to the possible correspondences between classes. For each O_i and O_j, $\forall i \in [1, N-1]$, $j \in [i+1, N]$, x_{i_k,j_l} is a binary decision variable equals to 1 if the class c_{i_k} in the ontology O_i aligns with the class c_{j_l} in O_j and 0 otherwise. ontology O_j and 0 otherwise.
- The second type refers to the possible correspondences between object properties. For each O_i and O_j, $\forall i \in [1, N-1]$, $j \in [i+1, N]$, y_{i_m,j_n} is a binary decision variable equals to 1 if the object property op_{i_m} in the ontology O_i aligns with the object property op_{j_n} in the ontology O_j and 0 otherwise.
- The third type refers to the possible correspondences between data properties. For each O_i and O_j, $\forall i \in [1, N-1]$, $j \in [i+1, N]$, z_{i_q,j_r} is a binary decision variable equals to 1 if the data property dp_{i_q} in the ontology O_i aligns with the data property dp_{j_r} in the ontology O_j and 0 otherwise.

Example 1. For the concept c_{1_1} in the ontology O_1, we have the following decision variables: $x_{1_1,2_1}, x_{1_1,2_2} \ldots, x_{1_1,2_9}, x_{1_1,3_1}, \ldots, x_{1_1,3_8}$. For the object property op_{1_1} in the ontology O_1, we have the following decision variables: $y_{1_1,2_1}, y_{1_1,2_2}, y_{1_1,3_1}$. For the data property dp_{1_1} in the ontology O_1, we have the following decision variables: $z_{1_1,2_1}, \ldots, z_{1_1,3_1}$.

Linear Constraints. LPHOM involves four types of constraints:

- Constraints of type $C1$ express the matching cardinality, we apply this type of constraint on classes, object properties and data properties;
- Constraints of type $C2$ allow reducing the incoherences by limiting the correspondences to non-disjoint entities;

- Constraints of type $C3$ express restrictions in aligning object properties considering that classes represent ranges and domains of object properties;
- Constraint of type $C4$ express the relationships between data properties and classes by mean of involving the domain restrictions of data properties. We have not considered ranges because they are less restrictive than domains.

In the following, we detail and illustrate each constraint. For binary decision variables, we propose to use this classical *C1 constraint* in order to resolve 1:1 alignments. This constraint is equivalent to resolve a set of disjoint edges in the MWGM problem.

Definition 1 (C1 Constraint). We define a C1 constraint for each type of decision variables. Each class c_{i_k} (respectively object property op_{i_m}, data property dp_{i_q}) in the ontology O_i could match with at most one class c_{j_l} (respectively object property op_{j_n}, data property dp_{j_r}) in the ontology O_j, $\forall i \times j \in [1, N-1] \times [i+1, N]$. These constraints are defined as follows:

- C1 constraints for classes are : $\sum_{l=1}^{nbC_j} x_{i_k,j_l} \leq 1, \quad \forall k \in [1, nbC_i]$
- C1 constraints for object properties are : $\sum_{n=1}^{nbOP_j} y_{i_m,j_n} \leq 1, \quad \forall m \in [1, nbOP_i]$
- C1 constraints for data properties are: $\sum_{r=1}^{nbDP_j} z_{i_q,j_r} \leq 1, \quad \forall q \in [1, nbDP_i]$

Example 2. Applying C1 for object properties in O_1, O_2 and O_3 generates the following constraints:

$y_{1_1,2_1} + y_{1_1,2_2} \leq 1$; $y_{1_2,2_1} + y_{1_2,2_2} \leq 1$; $y_{1_1,2_1} + y_{1_2,2_1} \leq 1$; $y_{1_1,2_2} + y_{1_2,2_2} \leq 1$
$y_{2_1,3_1} + y_{2_2,3_1} \leq 1$; $y_{1_1,3_1} + y_{1_2,3_1} \leq 1$

The *C2 constraint* aims at reducing the possibility of producing incoherent alignments by considering the disjointness between entities. If we suppose that we have two disjoint classes c_{i_k} and $c_{i_{k'}}$ in the ontology O_i ($c_{i_k} \sqsubseteq \neg c_{i_{k'}}$) so each class c_{j_l} in the ontology O_j should align either with c_{i_k} or $c_{i_{k'}}$. By this mean, we take into consideration the disjointness between classes, object and data properties.

Definition 2 (C2 Constraint). For each pair of ontologies O_i, O_j $\forall i \times j \in [1, N-1] \times [i+1, N]$ such as $i \neq j$, we define C2 constraint for each type of decision variables:

- For disjoint classes, $\forall k, k' \in [1, nbC_i]$ $\forall l \in [1, nbC_j]$, C2 constraint is defined as follows: $x_{i_k,j_l} + x_{i_{k'},j_l} \leq 1$
- For disjoint object properties, $\forall m, m' \in [1, nbOP_i]$ $\forall n \in [1, nbOP_j]$, C2 constraint is defined as follows: $y_{i_m,j_n} + x_{i_{m'},j_n} \leq 1$
- For disjoint data properties, $\forall q, q' \in [1, nbDP_i]$ $\forall r \in [1, nbDP_j]$, C2 constraint is as follows: $z_{i_q,j_r} + x_{i_{q'},j_r} \leq 1$

Example 3. In O_1, *Person* is disjoint with *Document*, $(c_{1_5} \sqsubseteq \neg c_{1_2})$. A part of instantiated C2 constraints is as following:
$x_{1_5,2_l} + x_{1_2,2_1} \leq 1$; $x_{1_5,2_2} + x_{1_2,2_2} \leq 1$; $x_{1_5,3_l} + x_{1_2,3_1} \leq 1$; $x_{1_5,3_2} + x_{1_2,3_2} \leq 1$

The *C3 constraint* takes the advantage of the restrictions of domain and range of each object property in order to make a sense between aligned object properties and aligned classes. We have noticed that when some object properties are aligned, we have either domains aligned or ranges aligned or both of them aligned. The following constraint aims to guide alignments according to this observation. If we suppose that we have some object property op_{i_m} in the ontology O_i and some other object property op_{j_n} in the ontology O_j, such that $T \sqsubseteq \forall op_{i_m}{}^-.c_{i_{k'}}$ and $T \sqsubseteq \forall op_{i_m}.c_{i_{k''}}$ and $T \sqsubseteq \forall op_{j_n}{}^-.c_{j_{l'}}$ and $T \sqsubseteq \forall op_{j_n}.c_{j_{l''}}$. Supposing that op_{i_m} aligns with op_{j_n} so either $c_{i_{k'}}$ aligns with $c_{j_{l'}}$ (i.e. domain of op_{i_m} aligns with domain of op_{j_n}) or $c_{i_{k''}}$ aligns with $c_{j_{l''}}$ (i.e. range of op_{i_m} aligns with range of op_{j_n}) or both of them.

Definition 3 (C3 Constraint). For each pair of ontologies O_i, O_j $\forall i \times j \in [1, N-1] \times [i+1, N]$ such as $i \neq j$ and $\forall m \in [1, nbOP_i]$ $\forall k', k'' \in [1, nbC_i]$ and $\forall n \in [1, nbOP_j]$ $\forall l', l'' \in [1, nbC_j]$, we express C3 constraints as follows:

$$y_{i_m,j_n} \leq x_{i_{k'},j_{l'}} + x_{i_{k''},j_{l''}}$$

Example 4. In O_1 and O_2, the properties *submit* and *submitPaper* are similar. By applying the constraint C3 between these object properties we obtain: $y_{1_1,2_1} \leq x_{1_7,2_{10}} + x_{1_3,2_4}$. This constraint leads to aligning both domains and ranges. We can also observe that due to the similarity between *reviews* in O_3 and *writeReview* in O_1 we obtain *Member_PC* aligned to *Reviewer* by the following constraint: $y_{3_1,1_2} \leq x_{3_5,1_8} + x_{3_1,1_4}$.

Finally, for the *C4 constraint*, we investigate the domains of the data properties. The idea consists of making classes be aligned when data properties gets aligned. If we suppose that some data property dp_{i_q} in the ontology O_i get aligned with another data property dp_{j_r} in the ontology O_j, such that $T \sqsubseteq \forall dp_{i_q}{}^-.c_{i_{k'}}$ and $T \sqsubseteq \forall dp_{j_r}{}^-.c_{j_{l'}}$ so the class $c_{i_{k'}}$ in ontology O_i will align with the class $c_{j_{l'}}$ in ontology O_j.

Definition 4 (C4 Constraint). For each pair of ontologies O_i, O_j $\forall i \times j \in [1, N-1] \times [i+1, N]$ such as $i \neq j$ and $\forall q \in [1, nbOP_i]$ $\forall k' \in [1, nbC_i]$ and $\forall r \in [1, nbOP_j]$ $\forall l' \in [1, nbC_j]$, C4 constraints are defined as follows:

$$z_{i_q,j_r} \leq x_{i_{k'},j_{l'}}$$

Example 5. We can illustrate the constraint C4 through the similar data properties *hasEmail* in O_3 and *email* in ontology O_1: $z_{3_1,1_2} \leq x_{3_3,1_5}$ as *hasEmail* and *email* are similar, their domains, which are also similar will also be aligned.

We summarize our linear program for holistic ontology matching (LPHMO) as depicted in Fig. 2. We emphasize that our model focuses on 1:1 alignments by using binary decision variables. We must however also point out that by relaxing the decision variables in the [0, 1] interval, this model is able to find n:m alignments. Moreover, we have to emphasize too that by using thresholds for entities similarities, we reduce significantly the size of the generated problem.

$$\max \sum_{i=1}^{N-1} \sum_{j=i+1}^{N} \sum_{k=1}^{nbC_i} \sum_{l=1}^{nbC_j} sim_{i_k,j_l} \, x_{i_k,j_l} + \sum_{m=1}^{nbOP_i} \sum_{n=1}^{nbOP_j} sim_{i_m,j_n} \, y_{i_m,j_n} + \sum_{q=1}^{nbDP_i} \sum_{r=1}^{nbDP_j} sim_{i_q,j_r} \, z_{i_q,j_r}$$

$s.t.$ $\displaystyle\sum_{l=1}^{nbC_j} x_{i_k,j_l} \leq 1, \; \forall k \in [1,nbC_i]$ (C1 Classes)

$\qquad\qquad\qquad \forall i \in [1,N-1], \, j \in [i+1,N]$

$\displaystyle\sum_{n=1}^{nbOP_j} y_{i_m,j_n} \leq 1, \; \forall m \in [1,nbOP_i]$ (C1 Object Properties)

$\qquad\qquad\qquad \forall i \in [1,N-1], \, j \in [i+1,N]$

$\displaystyle\sum_{r=1}^{nbDP_j} z_{i_q,j_r} \leq 1, \; \forall q \in [1,nbDP_i]$ (C1 Data Properties)

$\qquad\qquad\qquad \forall i \in [1,N-1], \, j \in [i+1,N]$

$x_{i_k,j_l} + x_{i_{k'},j_l} \leq 1$ (C2 Classes)

$\qquad \forall i \in [1,N-1], \, j \in [i+1,N]$
$\qquad \forall k,k' \in [1,nbC_i], \, \forall l \in [1,nbC_j]$

$y_{i_m,j_n} + x_{i_{m'},j_n} \leq 1$ (C2 Object Properties)

$\qquad \forall i \in [1,N-1], \, j \in [i+1,N]$
$\qquad \forall m,m' \in [1,nbDP_i], \, \forall n \in [1,nbDP_j]$

$z_{i_q,j_r} + x_{i_{q'},j_r} \leq 1$ (C2 Data Properties)

$\qquad \forall i \in [1,N-1], \, j \in [i+1,N]$
$\qquad \forall q,q' \in [1,nbDP_i], \, \forall r \in [1,nbDP_j]$

$y_{i_m,j_n} \leq x_{i_{k'},j_{l'}} + x_{i_{k''},j_{r}}$ (C3)

$\qquad \forall i \in [1,N-1], \, j \in [i+1,N]$
$\qquad \forall m \in [1,nbOP_i], \, \forall n \in [1,nbOP_j]$
$\qquad \forall k',k'' \in [1,nbC_i], \, \forall l',l'' \in [1,nbC_j]$

$z_{i_q,j_r} \leq x_{i_{k'},j_{l'}}$ (C4)

$\qquad \forall i \in [1,N-1], \, j \in [i+1,N]$
$\qquad \forall q \in [1,nbDP_i], \, \forall r \in [1,nbDP_j]$
$\qquad \forall k' \in [1,nbC_i], \, \forall l' \in [1,nbC_j]$

$x_{i_k,j_l} \in \{0,1\} \; \forall i \in [1,N-1], \, j \in [i+1,N]$

$\qquad\qquad \forall k \in [1,nbC_i], \, \forall l \in [1,nbC_j]$

$y_{i_m,j_n} \in \{0,1\} \; \forall i \in [1,N-1], \, j \in [i+1,N]$

$\qquad\qquad \forall m \in [1,nbOP_i], \, \forall n \in [1,nbOP_j]$

$z_{i_q,j_r} \in \{0,1\} \; \forall i \in [1,N-1], \, j \in [i+1,N]$

$\qquad\qquad \forall q \in [1,nbDP_i], \, \forall r \in [1,nbDP_j]$

Fig. 2. LPHOM Linear model

4   Experimental Evaluation

In the following we present the results of our approach in both pairwise and holistic matching settings. For both settings, our approach has been evaluated for similarities higher than a fixed threshold equals to 0.65 for both classes and properties. Furthermore, all generated correspondences have a confidence degree of 1.0.

4.1   OAEI Conference Data Set

The evaluation of LPHOM is carried out using the OAEI *Conference* track[3]. The intent of this track is to provide expressive and real-world matching problems over expressive ontologies covering the same domain [2]. This data set is composed of 16 ontologies covering the domain of conference organization and a subset of 21 reference alignments involving 7 ontologies. The track evaluation is based on crisp reference alignments (RA1) and two other entailed alignments (RA2 and RAR2) deduced from RA1. Our evaluation is restricted to the RA1 alignments as they are the only publicly available set. RA1 is divided into three sub-evaluations, as follows:

- In RA1-M1 only alignments between classes are evaluated;
- In RA1-M2 only alignments between properties (object and data) are evaluated;
- In RA1-M3 both alignments between classes and properties are evaluated.

4.2   Pairwise Matching Evaluation

Here, we compare the results of our approach with the results of the 14 matchers participating in the 2015 OAEI campaign. These results have been obtained from the Web page describing the results of the campaign[4]. With exception of MAMBA, that applies an optimization method with constraints and Markov Logic, these matchers apply different matching strategies than us. For example, AML is based o lexical similarities, external resources and alignment coherence; XMAP applies both lexical and structural contexts and exploits external resources; LogMap applies consistency and locality principles while their variants LogMap-C further implements the conservativity principle and LogMapLite essentially applies string matching techniques; GMAP uses a sum-product network encoding the similarities on individuals and disjointness axioms and a noisy-or model encoding probabilistic matching rules; RSDLWB exploits lexical and structural heuristics and machine learning on statistical patterns and DKP-AOM is based on linguistic, synonym and axiomatic based alignment. Although MAMBA is very performing in this track, it can not deal with more than 1000 classes [17], contrarily to LPHOM. (e.g., for the anatomy track, 2744 and 3305

---

[3] http://oaei.ontologymatching.org/2015/conference/.
[4] http://oaei.ontologymatching.org/2015/conference/eval.html.

	P	F0.5	F1	F2	R
AML	0.83	0.8	0.76	0.72	0.7
Mamba	0.84	0.8	0.74	0.69	0.66
XMAP	0.86	0.8	0.73	0.67	0.63
GMAP	0.76	0.75	0.73	0.72	0.71
LogMap	0.82	0.78	0.73	0.68	0.65
LogMap-C	0.84	0.78	0.71	0.65	0.62
our approach	0.76	0.73	0.69	0.66	0.64
DKP-AOM	0.84	0.77	0.69	0.63	0.59
Edna	0.88	0.78	0.67	0.59	0.54
COMMAND	0.84	0.76	0.66	0.58	0.54
RSDLWB	0.88	0.78	0.66	0.58	0.53
LogMapLite	0.84	0.76	0.66	0.58	0.54
ServOMBI	0.64	0.64	0.64	0.65	0.65
StringEquiv	0.88	0.76	0.64	0.55	0.5
Lily	0.59	0.6	0.61	0.62	0.63
CroMacher	0.72	0.67	0.6	0.54	0.51
JarvisOM	0.88	0.73	0.59	0.49	0.44

(a) RA1-M1

	P	F0.5	F1	F2	R
Mamba	0.89	0.79	0.67	0.59	0.54
AML	0.89	0.78	0.58	0.46	0.41
LogMap-C	1	0.51	0.39	0.32	0.28
LogMap	0.65	0.5	0.39	0.31	0.28
CroMatcher	0.62	0.31	0.34	0.37	0.39
JarvisOM	0.3	0.31	0.34	0.37	0.39
GMAP	0.3	0.46	0.31	0.23	0.2
our approach	0.23	0.24	0.25	0.26	0.26
LogMapLite	0.29	0.27	0.25	0.23	0.22
ServOMBI	0.29	0.27	0.24	0.21	0.2
XMAP	0.67	0.37	0.22	0.15	0.13
Edna	0.24	0.19	0.15	0.12	0.11
COMMAND	0.18	0.11	0.07	0.05	0.04
RSDLWB	0.03	0.04	0.05	0.1	0.24
StringEquiv	0.08	0.05	0.03	0.02	0.02

(b) RA1-M2

Fig. 3. Average results for RA1-M1 and RA1-M2 in the conference track. Results are ranked according to the F1-Measure

classes, LPHOM spent about 36sec with a F-measure of 0,76). However, it is out of the scope of this paper to provide a deep analysis of the results obtained by each tool.

The evaluation is based on Precision (P), Recall (R), F1-measure (F1), F2-measure (F2) and F0.5-measure (F0.5) computed for the threshold that provides the highest average F1-measure computed for each matcher. F1-measure is the harmonic mean of precision and recall. F2-measure weights recall higher than precision and F0.5-measure weights precision higher than recall.

RA1-M1. For this evaluation, we have evaluated LPHOM with the constraints exclusively dedicated to classes (C1 and C2). The average results for the 21 pairs of alignments are summarize in the table of Fig. 3a. We observe that our results are situated in the middle, we are better than 8 participants but lower than the other 6 participants. The best approaches benefits from more elaborate strategies and external resources to compute similarities. Even if our approach uses simple average similarities between known measures in the literature, we can observe that the strategy to find the best set of alignments checking coherence seems returning very good results on recall. These results are slightly closer, see even better than the recall of XMap and LogMap-C participants.

RA1-M2. Here, only one type of constraints exclusively dedicated to properties is applied (C1 on data properties and on object properties). As shown in the table of Fig. 3b, we observe that except Mamba and AML perform well in this task, all the other approaches have difficulties in aligning properties. The results of our approach are once again in the middle. We have noticed that the chosen threshold (65 %) applied on properties is not a very good compromise for this task. We have observed several properties having similarities equal to 0 (according to our measures), that we have not been able to capture. The results of baseline approaches Edna and StringEquiv confirm our observations, since these approaches uses very high similarity threshold.

RA1-M3. Finally, Table 1 summarises our results compared to the results of the other participants for the evaluation on both classes and proprieties. We have

Table 1. Average results for RA1-M3 in the conference track

	Precision	F0.5-Measure	F1-Measure	F2-Measure	Recall
AML	0.84	0.8	0.74	0.69	0.66
AML (semantic)	0.84	0.79	0.79	0.69	0.67
Mamba	0.83	0.78	0.72	0.67	0.64
Mamba(semantic)	0.84	0.79	0.79	0.68	0.66
XMAP	0.85	0.77	0.68	0.6	0.56
XMAP (semantic)	0.87	0.79	0.79	0.62	0.58
LogMap	0.8	0.75	0.68	0.62	0.59
LogMap (semantic)	0.82	0. 77	0.77	0.65	0.62
LogMap-C	0.82	0.75	0.67	0.61	0.57
LogMap-C (semantic)	0.83	0.77	0.77	0.63	0.6
GMAP	0.66	0.66	0.65	0.65	0.65
GMAP (semantic)	0.68	0.68	0.68	0.69	0.7
DKP-AOM	0.84	0.74	0.63	0.54	0.5
DKP-AOM (semantic)	0.86	0.76	0.76	0.56	0.52
Our approach	0.65	0.63	0.61	0.59	0.58
Our approach (semantic)	0.65	0.65	0.66	0.66	0.67
ServOMBI	0.61	0.6	0.59	0.59	0.58
ServOMBI (semantic)	0.58	0.6	0.6	0.69	0.73
COMMAND	0.78	0.69	0.59	0.51	0.47
COMMAND (semantic)	0.6	0.6	0.6	0.63	0.65
LogMapLite	0.73	0.67	0.59	0.53	0.5
LogMapLite (semantic)	0.75	0.7	0.7	0.58	0.56
Edna	0.79	0.7	0.59	0.51	0.47
Lily	0.59	0.58	0.56	0.54	0.53
Lily (semantic)	0.58	0.58	0.58	0.61	0.62
StringEquiv	0.8	0.68	0.56	0.47	0.43
CroMatcher	0.59	0.57	0.54	0.52	0.5
CroMatcher (semantic)	0.61	0.59	0.59	0.54	0.53
JarvisOM	0.84	0.67	0.51	0.42	0.37
JarvisOM (semantic)	0.84	0.69	0.69	0.45	0.41
RSDLWB	0.25	0.28	0.33	0.41	0.49
RSDLWB (semantic)	0.32	0.36	0.36	0.59	0.76

evaluated our model with all the constraints (C1, C2, C3, C4). Our approach keep a stable rank compared to other approaches. We notice that GMAP or RSDLWB have non stable positions through the evaluations Using all constraints seems advantageous for recall more than for precision because of the noise caused by the false positive aligned properties.

Semantic Evaluation. As we have observed that our generated alignments seem semantically close to the crisp reference, we have evaluated our results and those of the other approaches, using the semantic measures [4] (Table 1). Indeed, our results are semantically very interesting. In particular, we observe an improvement in the recall, which is equivalent to the recall of AML and MAMBA. We note also that the semantic evaluation reveals a slight change in the ranking of systems.

To sum up, our approach reaches promising results for its first comparison with regard to the pairwise ontology matching problem. Our model is more efficient when we use all the proposed constraints (RA1-M3). The interaction between constraints leads to semantically significant results closer to gold references which are illustrated by a good recall on semantic distances. The constraints proposed for reducing incoherence are experimentally efficient. We applied the ALCOMO [16] to evaluate if there is incoherence in our results and we get the following average results (for the 21 combinations we removed between 3 and 0 correspondences per alignment): (1) for RA1-M1 we have 0.95 removed correspondences, (2) for RA1-M2 we have 0 removed correspondences and (3) for RA1-M3 we have 0.85 removed correspondences.

Finally, the average runtime of LPHOM (pre-processing, linear program generation and resolution), over 21 pairs of the conference track was 2.84 s using the different types of measures and 0.24 s using only the token-based measure Jaccard.

4.3   Holistic Matching Evaluation

The ontology matching field lacks in benchmarks dedicated to the evaluation of holistic ontology matching. In order to be able to evaluate our holistic approach, we analyse:

- the differences between cliques manually deduced from reference alignments and the cliques generated by our holistic approach (remember that cliques define correspondences between N ontologies, which have to be matched);
- the differences between the results of pairwise and holistic matching settings.

In the following, we denote a clique as $Cl_i = <e_1, \ldots, e_N>$, such as each e_j belongs to ontology O_j. In the first part of this evaluation, we compare the cliques generated by LPHOM with the cliques that we have manually identified from the reference alignments involving 3 ontologies. For the 7 available ontologies in the Conference Track, which are classified into types (Tool, Insider and Web), we selected 3 ontologies from the 'Tool' type (Cmt, Conf-Of, Edas). In order to maximize the chance to have cliques in the reference alignments, we have tried to find $N \geq 2$ ontologies of the same type. The only combination of ontologies verifying that was Cmt, Conf-Of and Edas, for which the reference alignments are available. Given O_1 (Cmt), O_2 (Conf-Of) and O_3 (Edas), we have manually identified the following cliques from the reference alignment:

- $Cl_1(reference)$ =<$author_1, author_2, author_3$>
- $Cl_2(reference)$ =<$hasBeenAssigned_1, reviwes_2, isReviewing_3$>
- $Cl_3(reference)$ =<$person_1, person_2, person_3$>
- $Cl_4(reference)$ =<$hasAuthor_1, writtenBy_2, isWrittenBy_3$>

Applying our approach, we have found the following cliques:

- Cl_1 =<$author_1, author_2, author_3$>
- Cl_2 =<$paper_1, paper_2, paper_3$>
- Cl_3 =<$person_1, person_2, person_3$>
- Cl_4 =<$hasAuthor_1, writtenBy_2, isWrittenBy_3$>
- Cl_5 =<$writePaper_1, writes_2, hasRelatedPaper_3$>
- Cl_6 =<$email_1, hasEmail_2, hasEmail_3$>

We first notice that cliques Cl_1, Cl_3 and Cl_4 are the same as the cliques identified in the reference alignments whereas the clique $Cl_2(reference)$ has not been found by our approach. However, our model has found three other significant cliques Cl_2, Cl_5 and Cl_6. Cl_2 is composed of the same concept *Paper* occurring in all ontologies. In the reference alignments, the correspondences in which *Paper* occur does not form a clique. We emphasize here the benefit of holistic matching which inspects simultaneously all ontologies. The Cl_5 clique is particularly interesting since that the properties of Cl_5 are the inverse of the properties of Cl_4. Finally, Cl_6 is composed of similar data properties which is also relevant and strangely not provided in the reference alignments.

We also analyse the differences between the results of pairwise and holistic matching settings, applied on the example of the Fig. 1 (O_1 is Cmt, O_2 is Sigkdd and O_3 is Conf-Of). The holistic approach discovers simultaneously alignments for N ontologies, from all combinations of pairs of input ontologies. The resulting alignments are collected from a simultaneous resolution of A_{12}, A_{13} and A_{23}. Here we focus on main differences occurring in the alignments:

- If we match O_1 and O_2 by producing A_{12}, then we match with O_3 by producing $A_{(12)3}$, we get the following alignments:
  $A_{12} = \{<Conference, ConferenceHall, \equiv, 0.63>, <ConferenceMember, Conference, \equiv, 0.66>, <Paper, Paper, \equiv, 1>\}$
  and $A_{(12)3} = \{<Conference, Conference, \equiv, 1>, <Paper, Paper, \equiv, 1>\}$
- If we produce A_{13}, then we produce $A_{(13)2}$, we get the following alignments:
  $A_{13} = \{< Paper, ShortPaper, \equiv, 0.63 >, < PaperFullVersion, Paper, \equiv, 0.66>, <Conference, Conference, \equiv, 1>\}$
  and $A_{(13)2} = \{< Conference, Conference, \equiv, 1 >, < ShortPaper, AuthorOfPaper, \equiv, 0.5>, <Paper, Paper, \equiv, 1>\}$

Applying the holistic matching for O_1, O_2 and O_3, we get the following alignments:

- $A_{12} = \{<Conference, Conference, \equiv, 1>, <Paper, Paper, \equiv, 1>\}$
- $A_{23} = \{<Conference, Conference, \equiv, 1>, <Paper, Paper, \equiv, 1>\}$
- $A_{13} = \{<Conference, Conference, \equiv, 1>, <Paper, Paper, \equiv, 1>\}$

From these alignments, two cliques are deduced:

- $Cl_1 = <Paper_1, Paper_2, Paper_3>$
- $Cl_2 = <Conference_1, Conference_2, Conference_3>$

To sum up, the results presented in this section show the subtleties between a local and global investigations on $N \geq 2$ ontologies, which confirm the usefulness of holistic approaches for ontology matching.

5   Conclusion and Future Work

In this paper, we have presented an extensible linear model named LPHOM performing holistic ontology matching. The main contribution of this approach consists in allowing simultaneous matching of multiple ontologies. We model the approach within a linear program by reducing the ontology matching problem to the maximum-weighted graph matching problem, which is solvable in polynomial time. Our approach is extensible with different linear constraints handling classes and properties of ontologies. These constraints are used to reduce the logical incoherence in generated alignments, what is not done systematically by all matching systems. We experimented LPHOM on the OAEI Conference set on both pairwise and holistic settings. For future work, we intend to deeply study the similarity computation of entities with more accurate external resources. With respect to the constraints, we plan to add the constraint that classes can also match with properties and other hypothesis concerning incoherence. We also intend to extend our evaluation on the whole set of Conference and other data sets. Finally, we plan to extend the approach to deal with holistic instance matching and larger data sets.

References

1. Berro, A., Megdiche, I., Teste, O.: A linear program for holistic matching: assessment on schema matching benchmark. In: Proceedings of the International Conference on Database and Expert Systems Applications, pp. 383–398 (2015)
2. Cheatham, M., Hitzler, P.: Conference v2.0: an uncertain version of the OAEI conference benchmark. In: Mika, P., Tudorache, T., Bernstein, A., Welty, C., Knoblock, C., Vrandečić, D., Groth, P., Noy, N., Janowicz, K., Goble, C. (eds.) ISWC 2014, Part II. LNCS, vol. 8797, pp. 33–48. Springer, Heidelberg (2014)
3. Cho, M., Lee, J., Lee, K.M.: Reweighted random walks for graph matching. In: Proceedings of the 11th European Conference on Computer Vision, pp. 492–505 (2010)

4. Euzenat, J.: Semantic precision and recall for ontology alignment evaluation. In: Proceedings of the 20th International Joint Conference on Artificial Intelligence, pp. 348–353 (2007)
5. Euzenat, J.: Revision in networks of ontologies. Artif. Intell. **228**, 195–216 (2015)
6. Euzenat, J., Shvaiko, P.: Ontology Matching, 2nd edn. Springer, Heidelberg (2013)
7. Euzenat, J., Valtchev, P.: Similarity-based ontology alignment in OWL-Lite. In: Proceedings of the 16th European Conference on Artificial Intelligence, pp. 333–337 (2004)
8. Giunchiglia, F., Yatskevich, M., Shvaiko, P.: Semantic matching: algorithms and implementation. In: Spaccapietra, S., et al. (eds.) Journal on Data Semantics IX. LNCS, vol. 4601, pp. 1–38. Springer, Heidelberg (2007)
9. He, B., Chang, K.C.-C.: Statistical schema matching across web query interfaces. In: Proceedings of the 2003 ACM SIGMOD International Conference on Management of Data, SIGMOD 2003, pp. 217–228. ACM (2003)
10. He, B., Chang, KC.-C., Han, J.: Discovering complex matchings across web query interfaces: a correlation mining approach. In: Proceedings of the 20th International Conference on Knowledge Discovery and Data Mining, pp. 148–157 (2004)
11. Hu, W., Jian, N., Qu, Y., Wang, Y.: GMO: a graph matching for ontologies. In: K-Cap. 2005 Workshop on Integrating Ontologies 2005, pp. 43–50 (2005)
12. Huber, J., Sztyler, T., Nößner, J., Meilicke, C.: CODI: combinatorial optimization for data integration: results for OAEI 2011. In: Proceedings of the 6th International Workshop on Ontology Matching (2011)
13. Jean-Mary, Y., Shironoshita, E., Kabuka, M.: Ontology matching with semantic verification. Web Semant. Sci. Serv. Agents World Wide Web **7**(3), 235–251 (2009)
14. Jiménez-Ruiz, E., Cuenca Grau, B.: LogMap: logic-based and scalable ontology matching. In: Aroyo, L., Welty, C., Alani, H., Taylor, J., Bernstein, A., Kagal, L., Noy, N., Blomqvist, E. (eds.) ISWC 2011, Part I. LNCS, vol. 7031, pp. 273–288. Springer, Heidelberg (2011)
15. Karmarkar, N.: A new polynomial-time algorithm for linear programming. Combinatorica **4**(4), 373–395 (1984)
16. Meilicke, C.: Alignment incoherence in ontology matching. Ph.D. thesis, University of Mannheim (2011)
17. Meilicke, C.: MAMBA - results for the OAEI 2015. In: Proceedings of the 10th International Workshop on Ontology Matching, pp. 181–184 (2015)
18. Melnik, S., Garcia-Molina, H., Rahm, E.: Similarity flooding: a versatile graph matching algorithm and its application to schema matching. In: Proceedings of the 18th International Conference on Data Engineering. IEEE Computer Society (2002)
19. Otero-Cerdeira, L., Rodrguez-Martnez, F.J., Gmez-Rodrguez, A.: Ontology matching: a literature review. Expert Syst. Appl. **42**(2), 949–971 (2015)
20. Prytkova, N., Weikum, G., Spaniol, M.: Aligning multi-cultural knowledge taxonomies by combinatorial optimization. In: Proceedings of the 24th International Conference on World Wide Web, pp. 93–94. ACM (2015)
21. Rahm, E.: Towards large-scale schema and ontology matching. In: Bellahsene, Z., Bonifati, A., Rahm, E. (eds.) Schema Matching and Mapping, pp. 3–27. Springer, Heidelberg (2011)
22. Saleem, K., Bellahsene, Z., Hunt, E.: PORSCHE: Performance ORiented SCHEma mediation. Inf. Syst. **33**(7–8), 637–657 (2008)
23. Schrijver, A.: Combinatorial Optimization - Polyhedra and Efficiency. Springer, Heidelberg (2003)

24. Su, W., Wang, J., Lochovsky, F.H.: Holistic schema matching for web query inter-faces. In: Ioannidis, Y., Scholl, M.H., Schmidt, J.W., Matthes, F., Hatzopoulos, M., Böhm, K., Kemper, A., Grust, T., Böhm, C. (eds.) EDBT 2006. LNCS, vol. 3896, pp. 77–94. Springer, Heidelberg (2006)
25. Sun, Y., Ma, L., Shuang, W.: A comparative evaluation of string similarity metrics for ontology alignment. J. Inf. Comput. Sci. **12**(3), 957–964 (2015)
26. Svatek, V., Berka, P.: Ontofarm: towards an experimental collection of parallel ontologies. In: Poster Session at International Semantic Web Conference (2005)
27. Naumann, F., Gruetze, T., Böhm, C.: Holistic and scalable ontology alignment for linked open data. In: Proceedings of the 5th Linked Data on the Web Workshop at the 21th International World Wide Web Conference, April 2012
28. Xiang, C., Chang, B., Sui, Z.: An ontology matching approach based on affinity-preserving random walks. In: Proceedings of the 24th International Conference on Artificial Intelligence, pp. 1471–1477 (2015)

Semantic Sensitive Simultaneous Tensor Factorization

Makoto Nakatsuji[(✉)]

NTT Resonant Inc., Granparktower, 3-4-1 Shibaura, Minato-ku,
Tokyo 108-0023, Japan
nakatuji@nttr.co.jp

Abstract. The semantics distributed over large-scale knowledge bases can be used to intermediate heterogeneous users' activity logs created in services; such information can be used to improve applications that can help users to decide the next activities/services. Since user activities can be represented in terms of relationships involving three or more things (e.g. a user tags movie items on a webpage), tensors are an attractive approach to represent them. The recently introduced Semantic Sensitive Tensor Factorization (SSTF) is promising as it achieves high accuracy in predicting users' activities by basing tensor factorization on the semantics behind objects (e.g. item categories). However, SSTF currently focuses on the factorization of a tensor for a single service and thus has two problems: (1) *the balance problem* occurs when handling heterogeneous datasets simultaneously, and (2) *the sparsity problem* triggered by insufficient observations within a single service. Our solution, Semantic Sensitive Simultaneous Tensor Factorization (S^3TF), tackles the problems by: (1) Creating tensors for individual services and factorizing them simultaneously; it does not force the creation of a tensor from multiple services and factorize the single tensor. This avoids the low prediction accuracy caused by the balance problem. (2) Utilizing shared semantics behind distributed activity logs and assigning semantic bias to each tensor factorization. This avoids the sparsity problem by sharing semantics among services. Experiments using real-world datasets show that S^3TF achieves higher accuracy in rating prediction than the current best tensor method. It also extracts implicit relationships across services in the feature spaces by simultaneous factorization with shared semantics.

1 Introduction

Recently, many large-scale knowledge bases (KBs) have been constructed, including academic projects such as YAGO [8], DBpedia [2], and Elementary/ Deep-Dive [15], and commercial projects, such as those by Google [6] and Walmart [4]. These knowledge repositories hold millions of facts about the world, such as information about people, places, and things. Such information is deemed essential for improving AI applications that require machines to recognize and understand queries and their semantics in search or question answering systems. The applications include Google search and IBM's Watson, as well as smart

© Springer International Publishing AG 2016
P. Groth et al. (Eds.): ISWC 2016, Part I, LNCS 9981, pp. 411–427, 2016.
DOI: 10.1007/978-3-319-46523-4_25

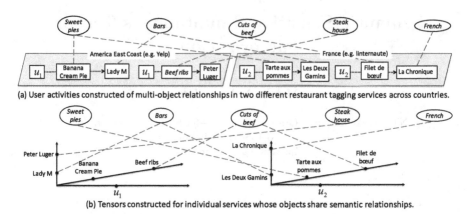

(a) User activities constructed of multi-object relationships in two different restaurant tagging services across countries.

(b) Tensors constructed for individual services whose objects share semantic relationships.

Fig. 1. Creating tensors for individual services whose objects are linked by semantics

mobile assistants such as Apple's Siri and NTT docomo's Shabette-Concier [5]. They now assist users to acquire meaningful knowledge in their daily activities; e.g. looking up an actor's birthday by question-answering systems or searching restaurants near the user's current location by smart mobile assistants.

The KBs can also be used to provide background knowledge that is shared by the different services [2]. Thus, beyond the above described usages of facts stored in the KBs, the semantics in those bases can be effectively used for mediating distributed users' activity logs in different services. Thus they have the potential to let AI applications assist users to decide next activities across services by analyzing heterogeneous users' logs distributed across services. In this paper, we assume services are different with each other if they do not share any objects, e.g. users, venues, or reviews. For example, in Fig. 1, US restaurant review service, Yelp[1], and French one, linternaute[2], are quite different services.

Tensor factorization methods have become popular for analyzing users' activities, since users' activities can be represented in terms of relationships involving three or more things (e.g. when a user tags venues on a webpage) [9,11,13,17,20]. Among the proposals made to date, Bayesian Probabilistic Tensor Factorization (BPTF) [20] is promising because of its efficient sampling of large-scale datasets and simple parameter settings. Semantic Sensitive Tensor Factorization (SSTF) [11,13] extends BPTF and applies semantic knowledge in the form of vocabularies/taxonomies extracted from Linked Open Data (LOD) to tensor factorization to solve the sparsity problem caused by sparse observation of objects. By incorporating the semantics behind objects, SSTF achieves the best rating prediction accuracy among the existing tensor factorization methods [13].

However, SSTF cannot enhance prediction accuracy across services for two reasons: (1) SSTF suffers from the *balance problem* that arises when handling heterogeneous datasets, e.g. the predictions for the smaller services are greatly

---

[1] http://www.yelp.com.
[2] http://www.linternaute.com/restaurant/.

biased by the predictions for the larger services [10]. Even if we merge user activity logs across services based on objects that appear across services, SSTF prediction results are poor when faced with merged logs. (2) SSTF focuses on only the factorization of a tensor representing users' activities within a single service and cannot solve *the sparsity problem*. Even if the logs in different services share some *semantic* relationships, SSTF can not make use of them.

We think that a tensor factorization method that uses the semantics in the KBs to intermediate different services is needed since LOD project aims to mediate distributed data in different services [1]. Thus, this is an important goal for the Semantic Web community. For example, we can simultaneously analyze logs in an American restaurant review service and those in an equivalent Japan service by using semantics even if they share no users, restaurant venues, and review descriptions. As a result, we can improve the prediction accuracy of the individual services, extract the implicit relationships across services, and recommend good Japanese restaurants to users in the United States (and vice verse). So, this paper enhances SSTF and proposes Semantic Sensitive Simultaneous Tensor Factorization (S^3TF) that simultaneously factorizes tensors created for different services by relying on the semantics shared among services. It overcomes the above mentioned problems by taking the following two ideas:

(1) It creates tensors for individual services whose objects are linked by semantics. This means that S^3TF does not force a tensor to be created from multiple services and then factorize that single tensor to make predictions. Below, for ease of understanding, this paper uses the scenario in which there are two different restaurant review services in different countries (they share no users, restaurants, or food reviews); e.g. Yelp and linternaute in Fig. 1. Figure 1(a) presents an example of users' activities involving three objects: a user who assigned tags about impressive foods served by restaurants with ratings on those relationships. The restaurants and foods are linked by the semantics from the KB. In the figure, say American user u_1 assigned tag "Banana cream pie" to restaurant "Lady M". French user u_2 assigned tag "Tarte aux pommes" to restaurant "Les Deux Gamins". In Fig. 1(b), S^3TF creates tensors for two different services while sharing semantic classes; e.g. Food "Banana cream pie" is linked with food class "Sweet pies" and restaurant "Lady M" is linked with restaurant class "Bars" in a tensor for "America East Coast". Food "Tarte aux pommes" is linked with the food class "Sweet pies" and restaurant "Les Deux Gamins" is linked with the restaurant class "Bars" in a tensor for "French". As a result, S^3TF can factorize those individual tensors "individually" while sharing semantics across tensors. This solves the *balance problem*.

(2) It uses the shared semantics present in distributed services and uses the semantics to bias the latent features learned in each service's tensor factorization. Thus, it can avoid *the sparsity problem* of tensor factorization, by using not only the semantics shared within a service but also those shared among services. This has another effect: the semantic biases are shared in latent features for the tensors of individual services and thus S^3TF can extract

the implicit relationships among services present in the latent features. For example, in Fig. 1(a), user u_1 and u_2 share no foods and no restaurants with each other, though they may share almost the same tendencies in food choice (e.g. they both tend to eat "Sweet pies" at "Bars" and "Cuts of beef" at nice restaurants). If such multi-object relationships are sparsely observed in each country, they cannot be well predicted by current tensor factorization methods because of the sparsity problem. S^3TF solves this by using the shared semantics among services. It propagates observations for "Banana cream pie" and "Tarte aux pommes" to the class "Sweet pie" as well as the observations for "Lady M" and "Les Deux Gamins" to the class "Bar". It then applies the semantic biases from food class "Sweet pie" to "Banana cream pie" as well as those from restaurant class "Bars" to restaurant "Lady M" when the tensor for United States is factorized. It also applies semantic biases from food class "Sweet pie" to "Tarte aux pommes" as well as those from restaurant class "Bars" to restaurant "Les Deux Gamins" when the tensor for France is factorized. In this way, S^3TF solves the sparsity problem by using the semantics shared across services. It also can find the implicit relationships from the latent features (e.g. the relationships shared by users u_1 and u_2 described above) by the mediation provided by the shared semantics.

We evaluated S^3TF using restaurant review datasets across countries. The reviews do not share any users, restaurant venues, or review descriptions as the languages are different. Thus, they are considered to be different services. The results show that S^3TF outperforms the previous methods including SSTF by sharing the semantics behind venues and review descriptions across services.

The paper is organized as follows: Sect. 2 describes related works while Sect. 3 introduces the background of this paper. Section 4 explains our method and Sect. 5 evaluates it. Finally, Sect. 6 concludes the paper.

2   Related Work

Tensor factorization methods have recently been used in various applications such as recommendation systems [11,17] and LOD analyses [7,14]. For example, [14] proposed methods that use tensor factorization to analyze huge volumes of LOD datasets in a reasonable amount of time. They, however, did not use the simultaneous tensor factorization approach and thus could not explicitly incorporate the semantic relationships behind multi-object relationships into the tensor factorization; in particular, they failed to use taxonomical relationships behind multi-object relationships such as "subClassOf" and "subGenreOf", which are often seen in LOD datasets. A recent proposal, SSTF [11,13], solves the sparsity problem by providing semantic bias from KBs to the feature vectors for sparse objects in multi-object relationships. SSTF was, however, not designed to perform cross-domain analysis even though LOD can be effectively used for mediating distributed objects in different services [2]. Generalized Coupled Tensor Factorization (GCTF) methods [22] and recent Non-negative Multiple

Tensor Factorization (NMTF) [19] try to incorporate extra information into tensor factorization by simultaneously factorizing observed tensors and matrices representing extra information. They, however, do not focus on handling semantics behind objects while factorizing tensors created for different services. Furthermore, according to the evaluations in [13], they have much worse performance than SSTF.

Other than tensor methods, [23] applies embedding models including heterogeneous network embedding and deep learning embedding to automatically extract semantic representations from the KB. Then it jointly learns the latent representations in collaborative filtering as well as items' semantic representations from the KB. There are, however, no embedding methods that analyze different services by using shared KBs.

Recent semantic web studies try to find missing links between entities [21] or find an explanation on a pair of entities in KBs [16]. [12, 18] incorporate semantic categories of items into the model and improve the recommendation accuracies. They, however, do not focus on the analysis of users' activities across services and find implicit relationships between entities by the above mentioned analysis.

3   Preliminary

Here, we explain Bayesian Probabilistic Tensor Factorization (BPTF) since S^3TF was implemented within the BPTF framework due to its efficiency with simple parameter settings.

This paper deals with the relationships formed by user u_m, venue v_n, and tag t_k. A third-order tensor $\mathcal{R}$ is used to model the relationships among objects from sets of users, venues, and tags. Here, the (m, n, k)-th element $r_{m,n,k}$ indicates the m-th user's rating of the n-th venue with the k-th tag. Tensor factorization assigns a D-dimensional latent feature vector to each user, venue, and tag, denoted as $\mathbf{u}_m$, $\mathbf{v}_n$, and $\mathbf{t}_k$, respectively. Here, $\mathbf{u}_m$ is an M-length, $\mathbf{v}_n$ is an N-length, and $\mathbf{t}_k$ is a K-length "column" vector. Accordingly, each element $r_{m,n,k}$ in $\mathcal{R}$ can be approximated as the inner-product of the three vectors as follows:

$$r_{m,n,k} \approx \langle \mathbf{u}_m, \mathbf{v}_n, \mathbf{t}_k \rangle \equiv \sum_{d=1}^{D} u_{m,d} \cdot v_{n,d} \cdot t_{k,d} \qquad (1)$$

where index d represents the d-th "row" element of each vector.

BPTF [20] models tensor factorization over a generative probabilistic model for ratings with Gaussian/Wishart priors over parameters. The Wishart distribution is most commonly used as the conjugate prior for the precision matrix of a Gaussian distribution.

We denote the matrix representations of $\mathbf{u}_m$, $\mathbf{v}_n$, and $\mathbf{t}_k$ as $\mathbf{U} \equiv [\mathbf{u}_1, \mathbf{u}_2, \ldots, \mathbf{u}_M]$, $\mathbf{V} \equiv [\mathbf{v}_1, \mathbf{v}_2, \ldots, \mathbf{v}_N]$, and $\mathbf{T} \equiv [\mathbf{t}_1, \mathbf{t}_2, \ldots, \mathbf{t}_K]$. To account for randomness in ratings, BPTF uses the following probabilistic model for generating ratings:

$$\mathcal{R}|\mathbf{U}, \mathbf{V}, \mathbf{T} \sim \prod_{m=1}^{M} \prod_{n=1}^{N} \prod_{k=1}^{K} \mathcal{N}(\langle \mathbf{u}_m, \mathbf{v}_n, \mathbf{t}_k \rangle, \alpha^{-1}).$$

This represents the conditional distribution of $\mathcal{R}$ given $\mathbf{U}$, $\mathbf{V}$, and $\mathbf{T}$ in terms of Gaussian distributions, each with means of $\langle \mathbf{u}_m, \mathbf{v}_n, \mathbf{t}_k \rangle$ and precision α.

The generative process of BPTF requires parameters μ_0, β_0, $\mathbf{W}_0$, ν_0, $\tilde{W}_0$, $\tilde{A}$, and $\tilde{\nu}_0$ in the hyper-priors, which should reflect prior knowledge about a specific problem and are treated as constants during training. The process is as follows:

1. Generate $\Lambda_{\mathbf{U}}$, $\Lambda_{\mathbf{V}}$, and $\Lambda_{\mathbf{T}} \sim \mathcal{W}(\Lambda|\mathbf{W}_0, \nu_0)$, where $\Lambda_{\mathbf{U}}$, $\Lambda_{\mathbf{V}}$, and $\Lambda_{\mathbf{T}}$ are the precision matrices (a precision matrix is the inverse of a covariance matrix) for Gaussians. $\mathcal{W}(\Lambda|\mathbf{W}_0, \nu_0)$ is the Wishart distribution of a $D \times D$ random matrix Λ with ν_0 degrees of freedom and a $D \times D$ scale matrix $\mathbf{W}_0$: $\mathcal{W}(\Lambda|\mathbf{W}_0, \nu_0) = \frac{|\Lambda|^{(\nu_0 - D - 1)/2}}{c} \exp(-\frac{Tr(\mathbf{W}_0^{-1}\Lambda)}{2})$, where C is a constant.
2. Generate $\mu_{\mathbf{U}} \sim \mathcal{N}(\mu_0, (\beta_0 \Lambda_{\mathbf{U}})^{-1})$, where $\mu_{\mathbf{U}}$ is used as the mean vector for a Gaussian. Similarly, generate $\mu_{\mathbf{V}} \sim \mathcal{N}(\mu_0, (\beta_0 \Lambda_{\mathbf{V}})^{-1})$ and $\mu_{\mathbf{T}} \sim \mathcal{N}(\mu_0, (\beta_0 \Lambda_{\mathbf{T}})^{-1})$, where $\mu_{\mathbf{V}}$ and $\mu_{\mathbf{T}}$ are mean vectors for Gaussians.
3. Generate $\alpha \sim \mathcal{W}(\tilde{A}|\tilde{W}_0, \tilde{\nu}_0)$.
4. For each $m \in (1 \ldots M)$, generate $\mathbf{u}_m \sim \mathcal{N}(\mu_{\mathbf{U}}, \Lambda_{\mathbf{U}}^{-1})$.
5. For each $n \in (1 \ldots N)$, generate $\mathbf{v}_n \sim \mathcal{N}(\mu_{\mathbf{V}}, \Lambda_{\mathbf{V}}^{-1})$.
6. For each $k \in (1 \ldots K)$, generate $\mathbf{t}_k \sim \mathcal{N}(\mu_{\mathbf{T}}, \Lambda_{\mathbf{T}}^{-1})$.
7. For each non-missing entry (m, n, k), generate $r_{m,n,k} \sim \mathcal{N}(\langle \mathbf{u}_m, \mathbf{v}_n, \mathbf{t}_k \rangle, \alpha^{-1})$.

Parameters μ_0, β_0, $\mathbf{W}_0$, ν_0, $\tilde{W}_0$, $\tilde{A}$, and $\tilde{\nu}_0$ should be set properly according to the objective dataset; fortunately, varying their values, has little impact on the final prediction [20].

BPTF views the hyper-parameters α, $\Theta_{\mathbf{U}} \equiv \{\mu_{\mathbf{U}}, \Lambda_{\mathbf{U}}\}$, $\Theta_{\mathbf{V}} \equiv \{\mu_{\mathbf{V}}, \Lambda_{\mathbf{V}}\}$, and $\Theta_{\mathbf{T}} \equiv \{\mu_{\mathbf{T}}, \Lambda_{\mathbf{T}}\}$ as random variables, yielding a predictive distribution for unobserved ratings $\hat{\mathcal{R}}$, which, for observable tensor $\mathcal{R}$, is given by:

$$p(\hat{\mathcal{R}}|\mathcal{R}) = \int p(\hat{\mathcal{R}}|\mathbf{U}, \mathbf{V}, \mathbf{T}, \alpha)$$
$$p(\mathbf{U}, \mathbf{V}, \mathbf{T}, \alpha, \Theta_{\mathbf{U}}, \Theta_{\mathbf{V}}, \Theta_{\mathbf{T}}|\mathcal{R})d\{\mathbf{U}, \mathbf{V}, \mathbf{T}, \alpha, \Theta_{\mathbf{U}}, \Theta_{\mathbf{V}}, \Theta_{\mathbf{T}}\}. \quad (2)$$

BPTF computes the expectation of $p(\hat{\mathcal{R}}|\mathbf{U}, \mathbf{V}, \mathbf{T}, \alpha)$ over the posterior distribution $p(\mathbf{U}, \mathbf{V}, \mathbf{T}, \alpha, \Theta_{\mathbf{U}}, \Theta_{\mathbf{V}}, \Theta_{\mathbf{T}}|\mathcal{R})$; it approximates the expectation by averaging samples drawn from the posterior distribution. Since the posterior is too complex to be directly sampled, it applies the Markov Chain Monte Carlo (MCMC) indirect sampling technique to infer the predictive distribution for unobserved ratings $\hat{\mathcal{R}}$ (see [20] for details on the inference algorithm of BPTF).

The time and space complexities of BPTF are $O(\#nz \times D^2 + (M + N + K) \times D^3)$. $\#nz$ is the number of observation entries, and M, N, and K are all much greater than D. BPTF can also compute feature vectors in parallel while avoiding fine parameter tuning during factorization.

4   Method

We now explain S^{3}TF. We first explain how to create augmented tensors, which share semantics among services, from individual services' tensors. Table 1 summarizes the notations used by our method.

Table 1. Definition of main symbols

Symbols	Definitions
$\mathcal{R}^i$	Tensor that includes ratings by users of venues with tags for the i-th service
α^i	Observation precision for $\mathcal{R}^i$
$\mathbf{u}_m^i$	m-th user feature vector for i-th service
$\mathbf{v}_n^i$	n-th venue feature vector for i-th service
$\mathbf{t}_k^i$	k-th tag feature vector for i-th service
$\mathbf{U}^i$	Matrix representation of $\mathbf{u}_m^i$ for i-th service
$\mathbf{V}^i$	Matrix representation of $\mathbf{v}_n^i$ for i-th service
$\mathbf{T}^i$	Matrix representation of $\mathbf{t}_k^i$ for i-th service
X	Number of services
$\mathbb{V}_s^i$	Set of the most sparse venues for the i-th service
$\mathbb{T}_s^i$	Set of the most sparse tags for the i-th service
$\mathcal{A}^v$	The augmented tensor that includes the classes of sparse venues in all services
$\mathcal{A}^t$	The augmented tensor that includes classes of sparse tags in all services
$\mathbf{c}_j^v$	j-th semantically biased venue feature vector from $\mathcal{A}^v$
$\mathbf{c}_j^t$	j-th semantically biased tag feature vector from $\mathcal{A}^t$
$\mathbf{C}^v$	Matrix representation of $\mathbf{c}_j^v$
$\mathbf{C}^t$	Matrix representation of $\mathbf{c}_j^t$
S^v	Number of classes that include sparse venues in all services
S^t	Number of classes that include sparse tags in all services
$f(o)$	Function that returns the classes of object o
δ	Parameter that adjusts the number of the most sparsely observed objects in each service

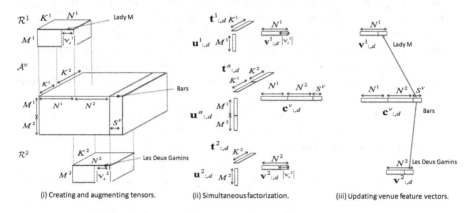

(i) Creating and augmenting tensors.   (ii) Simultaneous factorization.   (iii) Updating venue feature vectors.

Fig. 2. Examples of our factorization process

4.1   Creating Augmented Tensors

Following SSTF, S³TF creates the augmented tensor $\mathcal{A}^v$ that has all the observations across X services (those services do not share any object) as well as the observations for sparsely observed venues lifted in the augmented venue classes. The classes are chosen from shared KBs such as DBPedia and Freebase, and thus they are shared among services; e.g. for restaurant review services, the types of restaurants and the food categories are listed in DBPedia or Freebase in detail.

First, S³TF extracts the observations for sparsely observed venues. Here, the set of sparse venues for the i-th service $(1 \leq i \leq X)$, denoted as $\mathbb{V}_s^i$, is defined as the group of the most sparsely observed venues, v_s^is, among all venues in the i-th service. We set a 0/1 flag to indicate the existence of relationships composed of user u_m^i, venue v_n^i, and tag t_k^i as $o_{m,n,k}^i$. Then, $\mathbb{V}_s^i$ is computed as follows:

(1) S³TF first sorts the venues from the rarest to the most common in the i-th service $(1 \leq i \leq X)$ and creates a list of venues: $\{v_{s(1)}^i, v_{s(2)}^i, \ldots, v_{s(N^i-1)}^i, v_{s(N^i)}^i\}$ where N^i is the number of venues in the i-th service. For example, $v_{s(2)}^i$ is not less sparsely observed than $v_{s(1)}^i$.
(2) It iterates the following step (3) from $j = 1$ to $j = N^i$.
(3) If it satisfies the following equation, S³TF adds the j-th sparse venue $v_{s(j)}^i$ to set $\mathbb{V}_s^i$: $(|\mathbb{V}_s^i| / \sum_{m,n,k} o_{m,n,k}^i) < \delta$ where $\mathbb{V}_s^i$ initially does not have any venues and $|\mathbb{V}_s^i|$ is the number of venues in set $\mathbb{V}_s^i$. If not, it stops the iterations and returns the set $\mathbb{V}_s^i$ as the most sparsely observed venues in the i-th service. Here, δ is a parameter used to determine the number of sparse venues in $\mathbb{V}_s^i$. Typically, we set δ to range from 0.05 to 0.20 in accordance with the long-tail characteristic such that sparse venues account for 5–20% of all observations [13].

Second, S³TF constructs the augmented tensor $\mathcal{A}^v$ as follows:.

(1) S³TF inserts the multi-object relationship composed of user u_m^i, venue v_n^i, and tag t_k^i, observed in the i-th service, into $\mathcal{A}^v$. Here, the rating $r^i{}_{m,n,k}$ corresponding to the above relationship is inserted into the $((M_1^{i-1} + m), (N_1^{i-1} + n), (K_1^{i-1} + k))$-th element in $\mathcal{A}^v$ where we denote M_1^{i-1}, N_1^{i-1}, and K_1^{i-1} as the sum of number of users, that of venues, and that of tags in services whose identifiers are from 1 to $(i-1)$, respectively. As a result, $\mathcal{A}^v$ has all users, all venues, and all tags in all services. In Fig. 2(i), all observations in $\mathcal{R}^1$ and $\mathcal{R}^2$ are inserted into $\mathcal{A}^v$.
(2) S³TF additionally inserts the multi-object relationships composed of user u_m^i, a class of sparse venue c_j^v, and tag t_k^i into $\mathcal{A}^v$ if v_n^i is included in $\mathbb{V}_s^i$ and c_j^v is one of the classes of v_n^i. Thus, the rating $r^i{}_{m,n,k}$ is inserted into the $((M_1^{i-1} + m), (N_1^X + j), (K_1^{i-1} + k))$-th element in $\mathcal{A}^v$. If sparse venue v_n^i has several classes, S³TF inserts the rating $r^i{}_{m,n,k}$ into all corresponding elements in $\mathcal{A}^v$. In Fig. 2(i), observations for classes for sparse venues ("Lady M" in service 1 and "Les Deux Gamins" in service 2) are added to $\mathcal{A}^v$ (in the elements corresponding to their class "Bars"). Here, the number of classes

that have the sparse venues in all services is denoted as S^v; it is computed as: $S^v = | \bigcup_{V^i_s} f(v^i_s)|_{(1 \leq i \leq X)}$ where $f(v^i_s)$ is a function that returns the classes of sparse venue v^i_s in the i-th service.

The set of sparse tags $\mathbb{T}^i_s$ is defined as the group of the most sparsely observed tags in i-th service and is computed using the same procedure as it creates $\mathbb{V}^i_s$. The augmented tensor for tags $\mathcal{A}^t$ is also computed in the same way as it creates $\mathcal{A}^v$. So we omit the explanations of the procedures for creating those here.

Tensor creation by S^{3}TF has the following two benefits: (1) It solves the balance problem by creating individual tensors for services and so avoids strongly biasing any particular service. (2) It overcomes the sparsity problem by propagating observations in sparse objects to their classes shared among services in the augmented tensor.

4.2   Simultaneously Factorizing Tensors Across Services

S^{3}TF factorizes individual services' tensors and augmented tensors simultaneously. We first explain our approach and then the algorithm.

Approach. S^{3}TF takes the following three techniques in factorizing tensors.

(A) It factorizes individual service tensors $\mathcal{R}^i$s ($1 \leq i \leq X$), and augmented tensors $\mathcal{A}^v$ and $\mathcal{A}^t$ simultaneously. In particular, it creates feature vectors for users, $\mathbf{u}^i_m$s, those for venues, $\mathbf{v}^i_n$s, and those for tags, $\mathbf{c}^t_j$s, by factorizing tensor $\mathcal{R}^i$ for each i-th service as well as feature vectors for their venue classes $\mathbf{c}^v_j$s by $\mathcal{A}^v$ and those for their tag classes $\mathbf{c}^t_j$s by $\mathcal{A}^t$. As a result, S^{3}TF factorizes individual tensors while enabling the semantic biases from $\mathbf{c}^v_j$s and $\mathbf{c}^t_j$s to be shared during the factorization process. This approach to "simultaneously" factorizing individual service tensors solves the balance problem. In the example shown in Fig. 2(ii), $\mathcal{R}^1$, $\mathcal{R}^2$, and $\mathcal{A}^v$ are factorized simultaneously into D-dimensional "row" feature vectors.

(B) It shares feature vectors $\mathbf{u}^i_m$, $\mathbf{v}^i_n$, $\mathbf{t}^i_k$ which are computed by factorizing $\mathcal{R}^i$, in the factorization of augmented tensors $\mathcal{A}^v$ and $\mathcal{A}^t$. This means that it computes the feature matrix for users for the augmented tensor $\mathbf{U}^a$ by joining X numbers of service feature matrices for users, $[\mathbf{U}^1, \ldots, \mathbf{U}^i, \ldots, \mathbf{U}^X]$. Similarly, it computes the feature matrix for venues, $\mathbf{V}^a$, and that for tags, $\mathbf{T}^a$, for the augmented tensor. Then, it computes the feature matrix for venue (or tag) classes by reusing the joined feature matrices $\mathbf{U}^a$ and $\mathbf{T}^a$ (or $\mathbf{U}^a$ and $\mathbf{V}^a$). As a result, it can, during the factorization process, share the tendencies of users' activities across services via those shared parameters. In Fig. 2(ii), $\mathbf{u}^a_{m,d}$ is computed as: $[\mathbf{u}^1_{m,d}, \mathbf{u}^2_{m,d}]$ and $\mathbf{t}^a_{k,d}$ is as: $[\mathbf{t}^1_{k,d}, \mathbf{t}^2_{k,d}]$.

(C) It updates latent feature vectors for sparse venues (or tags) in the i-th service, $\mathbf{v}^i_s$s (or $\mathbf{t}^i_s$s), by incorporating semantic biases from $\mathbf{c}^v_j$s (or $\mathbf{c}^t_j$s) to $\mathbf{v}^i_s$s (or $\mathbf{t}^i_s$s). Here, $\mathbf{c}^v_j$s (or $\mathbf{c}^t_j$s) are feature vectors for classes of the sparse venues v^i_ss (or sparse tags t^i_ss). This process incorporates the semantic tendencies of

users' activities across services captured by idea (B) into each service's factorization; this is useful in solving the sparsity problem. In Fig. 2(iii), each
row vector $\mathbf{c}^v_{:,d}$ has latent features for $(N^1 + N^2)$ venues and for S^v classes.
The features in $\mathbf{c}^v_{:,d}$ share semantic knowledge of sparse venues across services. For example, the feature for "Bars" in $\mathbf{c}^v_{:,d}$ share semantic knowledge
of sparse venues "Lady M" and "Les Deux Gamins" across US restaurant
review service and French one (see also Fig. 1).

Algorithm. Here we explain how to compute the predictive distribution for unobserved ratings. Differently from the BPTF model (see Eq. (2)), S^{3}TF considers
the tensors for individual services and augmented tensors in computing the distribution. Thus, the predictive distribution is computed as follows:

$$p(\hat{\mathcal{R}}|\mathcal{R}, \mathcal{A}^v, \mathcal{A}^t) = \int p(\hat{\mathcal{R}}|\mathbf{U}, \mathbf{V}, \mathbf{T}, \mathbf{C}^v, \mathbf{C}^t, \alpha, \alpha^a)$$
$$p(\mathbf{U}, \mathbf{V}, \mathbf{T}, \mathbf{C}^v, \mathbf{C}^t, \Theta_\mathbf{U}, \Theta_\mathbf{V}, \Theta_{\mathbf{C}^v}, \Theta_{\mathbf{C}^t}, \alpha, \alpha^a|\mathcal{R}, \mathcal{A}^v, \mathcal{A}^t)$$
$$d\{\mathbf{U}, \mathbf{V}, \mathbf{T}, \mathbf{C}^v, \mathbf{C}^t, \Theta_\mathbf{U}, \Theta_\mathbf{V}, \Theta_\mathbf{T}, \Theta_{\mathbf{C}^v}, \Theta_{\mathbf{C}^t}, \alpha, \alpha^a\} \qquad (3)$$

where $\mathcal{R} \equiv \{\mathcal{R}^i\}_{i=1}^X$, $\alpha \equiv \{\alpha^i\}_{i=1}^X$, $\mathbf{U} \equiv \{\mathbf{U}^i\}_{i=1}^X$, $\mathbf{V} \equiv \{\mathbf{V}^i\}_{i=1}^X$, $\mathbf{T} \equiv \{\mathbf{T}^i\}_{i=1}^X$,
$\Theta_\mathbf{U} \equiv \{\Theta_{\mathbf{U}^i}\}_{i=1}^X$, $\Theta_\mathbf{V} \equiv \{\Theta_{\mathbf{V}^i}\}_{i=1}^X$, and $\Theta_\mathbf{T} \equiv \{\Theta_{\mathbf{T}^i}\}_{i=1}^X$.

Equation (3) involves a multi-dimensional integral that cannot be computed analytically. Thus, S^{3}TF views Eq. (3) as the expectation of
$p(\hat{\mathcal{R}}|\mathcal{R}, \mathcal{A}^v, \mathcal{A}^t)$ over the posterior distribution $p(\mathbf{U}, \mathbf{V}, \mathbf{T}, \mathbf{C}^v, \mathbf{C}^t, \Theta_\mathbf{U}, \Theta_\mathbf{V}, \Theta_{\mathbf{C}^v},$
$\Theta_{\mathbf{C}^t}, \alpha, \alpha^a|\mathcal{R}, \mathcal{A}^v, \mathcal{A}^t)$, and approximates the expectation by MCMC with the
Gibbs sampling paradigm. It collects a number of samples, L, to approximate
the integral in Eq. (3) as:

$$p(\hat{\mathcal{R}}|\mathcal{R}, \mathcal{A}^v, \mathcal{A}^t) \approx \sum_{l=1}^L p(\hat{\mathcal{R}}|\mathbf{U}[l], \mathbf{V}[l], \mathbf{T}[l], \mathbf{C}^v[l], \mathbf{C}^t[l], \alpha[l], \alpha^a[l]) \qquad (4)$$

where l represents the l-th sample.

The MCMC procedure is as follows (detail is given in the supplemental material[3]):

(1) Initialize $\mathbf{U}^i[1]$, $\mathbf{V}^i[1]$, and $\mathbf{T}^i[1]$ $(1 \le i \le X)$ for each i-th service as well as
$\mathbf{C}^v[1]$ and $\mathbf{C}^t[1]$ for the augmented tensors by Gaussian distribution as per
BPTF. $\mathbf{C}^v[1]$ and $\mathbf{C}^t[1]$ are used for sharing the semantics across services
(see our approach (A)). Next, it repeats steps (2) to (8) L times.
(2) Samples the hyperparameters for each i-th service as per BPTF i.e.:
  - $\alpha^i[l+1] \sim p(\alpha^i[l]|\mathbf{U}^i[l], \mathbf{V}^i[l], \mathbf{T}^i[l], \mathcal{R}^i)$
  - $\Theta_{\mathbf{U}^i}[l+1] \sim p(\Theta_{\mathbf{U}^i}[l]|\mathbf{U}^i[l])$
  - $\Theta_{\mathbf{V}^i}[l+1] \sim p(\Theta_{\mathbf{V}^i}[l]|\mathbf{V}^i[l])$
  - $\Theta_{\mathbf{T}^i}[l+1] \sim p(\Theta_{\mathbf{T}^i}[l]|\mathbf{T}^i[l])$

  here, $\Theta_\mathbf{X} \equiv \{\mu_\mathbf{X}, \Lambda_\mathbf{X}\}$ and is computed in the same way as BPTF.

---

[3] Please see https://sites.google.com/site/sapplementalfile/appendix-html.

(3) Samples the feature vectors the same way as is done in BPTF:
- $\mathbf{u}_m^i[l+1] \sim p(\mathbf{u}_m^i|\mathbf{V}^i[l], \mathbf{T}^i[l], \alpha^i[l+1], \Theta_{U^i}[l+1], \mathcal{R}^i)$
- $\mathbf{v}_n^i[l+1] \sim p(\mathbf{v}_n^i|\mathbf{U}^i[l+1], \mathbf{T}^i[l], \alpha^i[l+1], \Theta_{V^i}[l+1], \mathcal{R}^i)$
- $\mathbf{t}_k^i[l+1] \sim p(\mathbf{t}_k^i|\mathbf{U}^i[l+1], \mathbf{V}^i[l+1], \alpha^i[l+1], \Theta_{T^i}[l+1], \mathcal{R}^i)$

(4) Joins the feature matrices in services in order to reuse them as the feature matrices for the augmented tensors as (see our approach (B)):
- $\mathbf{U}^a[l+1] = [\mathbf{U}^1[l+1], \cdots, \mathbf{U}^i[l+1], \cdots, \mathbf{U}^X[l+1]]$
- $\mathbf{V}^a[l+1] = [\mathbf{V}^1[l+1], \cdots, \mathbf{V}^i[l+1], \cdots, \mathbf{V}^X[l+1]]$
- $\mathbf{T}^a[l+1] = [\mathbf{T}^1[l+1], \cdots, \mathbf{T}^i[l+1], \cdots, \mathbf{T}^X[l+1]]$

(5) Samples the hyperparameters for the augmented tensors similarly:
- $\alpha^a[l+1] \sim p(\alpha^a[l]|\mathbf{U}^a[l+1], \mathbf{V}^a[l+1], \mathbf{T}^a[l+1], \mathcal{R}^a)$
- $\Theta_{C^v}[l+1] \sim p(\Theta_{C^v}[l]|\mathbf{C}^v[l])$
- $\Theta_{C^t}[l+1] \sim p(\Theta_{C^t}[l]|\mathbf{C}^t[l])$

(6) Samples the semantically-biased feature vectors by using $\alpha^a[l+1]$, $\mathbf{U}^a[l+1]$, $\mathbf{V}^a[l+1]$, and $\mathbf{T}^a[l+1]$ as follows (see our approach (B)):
- $\mathbf{c}_j^v[l+1] \sim p(\mathbf{c}_j^v|\mathbf{U}^a[l+1], \mathbf{T}^a[l+1], \alpha^a[l+1], \Theta_{C^v}[l+1], \mathcal{A}^v)$
- $\mathbf{c}_j^t[l+1] \sim p(\mathbf{c}_j^t|\mathbf{U}^a[l+1], \mathbf{V}^a[l+1], \alpha^a[l+1], \Theta_{C^t}[l+1], \mathcal{A}^t)$

(7) Samples the unobserved ratings $\hat{r}_{m,n,k}^i[l+1]$ by applying $\mathbf{U}^i[l+1]$, $\mathbf{V}^i[l+1]$, $\mathbf{T}^i[l+1]$, $\mathbf{C}^v[l+1]$, $\mathbf{C}^t[l+1]$, $\alpha^i[l+1]$ to equation (4).

(8) Updates $\mathbf{v}_n^i[l+1]$ as follows and uses it in the next iteration (see our approach (C)):

$$\mathbf{v}_n^i = \begin{cases} \frac{1}{2}\left(\mathbf{v}_n^i + \frac{\sum_{\mathbf{c}_j^v \in f(v_n^i)} \mathbf{c}_j^v}{|f(v_n^i)|}\right) & (v_n^i \in \mathbb{V}_s^i) \\ \mathbf{v}_n^i & (\text{otherwise}) \end{cases} \tag{5}$$

Updates $\mathbf{t}_k^i[l+1]$ similarly (we halt the explanation here).

The complexity of S^3TF in each MCMC iteration is $O(\#nz \times D^2 + (M_1^X + N_1^X + K_1^X + S^V + S^T) \times D^3)$. Because the first term is much larger than the rest, the computation time is almost the same as that of BPTF. Parameter δ and parameters for factorization can be easily set based on the long-tail characteristic and the full Bayesian treatment inherited by the BPTF framework, respectively. S^3TF is faster than SSTF when analyzing X numbers of services since S^3TF creates and factorizes only one set of augmented tensors ($\mathcal{A}^v$ and $\mathcal{A}^t$) for all services while SSTF needs X sets of augmented tensors.

5   Evaluation

The method's accuracy was confirmed by evaluations.

5.1   Dataset

We used the Yelp ratings/reviews[4] together with DBPedia [2] food vocabularies. Yelp datasets contain user-made ratings of restaurant venues and user reviews

---

[4] Available at http://www.yelp.com/dataset_challenge/.

of venues across four countries (United Kingdom (UK), United States (US)[5], Canada[6], and Germany). The logs of users who are included in several countries are excluded from the datasets. Thus we can consider the datasets of individual countries are made from different services. Food vocabularies are extracted from food ontology[7] and categories are extracted from DBPedia article categories. We first extracted English food entries and then translated them into French or German by using BabelNet[8], which is a multilingual encyclopedic dictionary based on Wikipedia entries. Thus, the resulting food entries share the same categories. We then extracted tags from the reviews that match the instances in a DBPedia food vocabulary entry as was done in [13]. Consequently, we extracted 988, 1,100, 1,388, and 435 tags for UK, US, Canada, Germany, respectively. We used the genre vocabulary provided by Yelp as the venue vocabulary, it has 179 venue classes. The tag vocabulary provided by DBPedia has 1,358 food classes. The size of the user-venue-tag tensors in UK, US, Canada, and Germany were $2,052 \times 1,398 \times 988$, $10,736 \times 1,554 \times 1,100$, $10,700 \times 3,085 \times 1,388$, and $286 \times 332 \times 435$, respectively. The numbers of ratings in those countries were 54,774, 118,012, 172,182, and 3,062, respectively. The ratings range from 1 to 5.

5.2 Comparison Methods

We compared the accuracy of the following six methods:

1. *NMTF* [19], which utilizes the auxiliary information like GCTF. It factorizes the target tensors (user-item-tag tensors created for each countries) and auxiliary matrices (item-class matrix and tag-class matrix) simultaneously.
2. *BPTF* proposed by [20].
3. *SSTF*, which applies Semantic Sensitive Tensor Factorization proposed by [13] to the observed relationships in each service.
4. *SSTF_all*, which combines observed relationships in different services to create a merged tensor and factorizes the merged tensor by SSTF.
5. $S^3\,TFT$, which utilizes only the tag vocabulary.
6. $S^3\,TFV$, which utilizes only the venue vocabulary.
7. $S^3\,TF$, which is our proposal.

5.3 Methodology and Parameter Setup

We split each dataset into two halves; a training set that holds reviews entered in the first half period of all logs and a test set consisting of the reviews entered in the last half. We then performed evaluations for the two-joint combinations (total 6) of those sets to check the repeatability of results. Following the evaluation

---

[5] We focused on restauran reviews for Midwestern United States to efficiently perform evaluations.

[6] The Canada dataset includes venues located in the Quebec area, so the languages used in the reviews are written in French or English.

[7] http://dbpedia.org/ontology/Food/.

[8] http://babelnet.org.

methodology used in previous studies [3,11,13,20], we computed Root Mean Square Error (RMSE), which is computed by $\sqrt{(\sum_{i=1}^{n}(P_i - R_i)^2)/n}$, where n is the number of entries in the test set, and P_i and R_i are the predicted and actual ratings of the i-th entry, respectively. The RMSE is more appropriate to represent model performance than the Mean Absolute Error (MAE) when the error distribution is expected to be Gaussian. We varied D from 5 to 20 for each method, and set the optimum value to 20 since it gave the highest accuracies for all methods. We set the iteration count, L, to 100 since all methods could converge with this setting. δ was set to 0.8 following [13].

5.4 Results

We first investigated the sparseness of objects observed. Figure 3 plots the distribution of venue frequencies observed in the UK dataset. From this figure, we can confirm that venue observation frequencies exhibit the long-tail characteristic. Thus, observations of multi-object relationships become very sparse with respect to the possible combinations of observed objects. The distributions of other datasets showed the same tendencies. Thus, a solution to the sparsity problem across services is required. Figure 4 presents the accuracy of the UK dataset on the simultaneous factorization on UK and US datasets when the number of iterations, L, was changed. This confirms that the accuracy of S^3TF saturated before $L = 100$. Results on other datasets showed similar tendencies.

We then compared the accuracy of the methods for the simultaneous factorizations on the six datasets. The results shown in Table 2 are the average RMSE values computed for each country. They show that $SSFT$ has better accuracy than $BPFT$. This is because $SSFT$ uses the semantics shared within a single

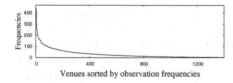

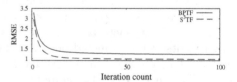

Fig. 3. Distribution of venue frequencies

Fig. 4. Accuracy vs. iteration count

Table 2. Comparing RMSE values of the methods

	$NMTF$	$BPTF$	$SSTF$	$SSTF\_all$	$S^3\ TFT$	$S^3\ TFV$	$S^3\ TF$
UK	1.7192	1.0063	0.9928	0.9960	0.9967	0.9594	*0.9501*
US	1.9011	1.2303	1.2176	1.2267	1.1939	1.1733	*1.1727*
Canada	1.8723	1.1853	1.1431	1.1655	1.1219	1.1331	*1.1215*
German	1.8923	1.3266	1.2789	1.2868	1.2744	1.2847	*1.2527*

service (e.g. within a service in US) and thus solves the sparsity problem. *SSTF* has better accuracy by *SSTF_all* though *SSTF_all* uses the entire logs. This is because *SSTF_all* creates a tensor by mixing the heterogeneous datasets in different countries and thus suffers from the balance problem. S^3 *TFT* and S^3 *TFV* had better performance than *BPTF* or *SSTF* since S^3 *TFT* and S^3 *TFV* can use the shared semantics on venues and those on tags across services, respectively. Finally, S^3 *TF*, which utilizes the semantic knowledge across services while performing coupled analysis of two tensors, yielded higher accuracy than the current best method, *SSTF*, with the statistical significance of $\alpha < 0.05$.

The RMSEs of *NMTF* are much worse than those of $S^3 TF$. This is mainly because: (1) *NMTF* straightforwardly combines different relationships, i.e., rating relationships among users, items, and tags, link relationships among items and their classes, and link relationships among tags and their classes. Thus, it suffers from the balance problem. (2) *NMTF* uses the KL divergence for optimizing the predictions since its authors are interested in "discrete value observations such as stars in product reviews", as described in [19]. Our datasets are those they are interested in; however, exponential family distributions like Poisson distribution do not fit our rating datasets so well.

Table 3. Computation time (seconds) when $L = 100$

German × UK			German × US		
BPTF	*SSTF*	$S^3 TF$	*BPTF*	*SSTF*	$S^3 TF$
63	113	109	85	142	134

Table 3 presents the computation times of *BPTF*, *SSTF*, and S^3 *TF* when simultaneously factorizing tensor for German and that for UK datasets as well as simultaneously factorizing tensor for German and that for US datasets. All experiments were conducted on a Linux 3.33 GHz Intel Xeon (24 cores) server with 192 GB of main memory. All methods were implemented with Matlab and GCC. We can see that the computation time of S^3 *TF* is shorter than that of *SSTF*. Furthermore, we can set L smaller than 100 (see Fig. 4). Thus, we can conclude that S^3 *TF* can compute more accurate predictions quickly; it works better than *SSTF* and *BPTF* on real applications.

Table 4. Prediction examples for US (the upper row) and German (the lower row)

Training dataset			Rating predictions by *SSTF* (S) and $S^3 TF$ (S^3)				
Tag in review sentence	Item/genre	Rating	Tag in review sentence	Item/genre	S	S^3	Actual
Berry *streusel* is tasty	A/Bakeries	5.0	*Bratwurst* was incredible	B/American	3.8	4.7	5.0
A *enchilada* is perfect	C/Tex-Mex	4.0	An amazing *pretzel* roll	D/Breakfast	3.8	4.8	5.0
Ich genoss *Marzipan*	E/Bakeries	5.0	*Burrito* war wirklich gut	F/Bars	3.1	3.7	4.0
nachos ist wertvoll	G/Bars	4.0	*Schnitzel* ist lecker	H/German	3.9	3.4	3.0

We then show the examples of the differences between the predictions output by *SSTF* and S^3TF in Table 4. The columns "S", "S^3", and "Actual" list prediction values by *SSTF*, those by S^3TF, and actual ratings given by users as found in the test dataset, respectively. In the US dataset, the combination of tag "streusel" at Bakeries "A" and "enchilada" at Tex-Mex restaurant "C" were highly rated in the training set. In the test set, the combination of tag "bratwurst" at American restaurant "B" and "pretzel" at Breakfast restaurant "D" were highly rated. The tags "streusel", "bratwurst", and "pretzel" (they are included in "german cuisine" class) are sparsely observed in the US's training set. In the Germany dataset, the combination of tag "marzipan" at Bakeries "E" and "nachos" at Bars "G" were highly rated in the training set. In the test set, the combination of tag "burrito" at Bars "F" and "Schnitzel" at German restaurant "H" were highly rated. The tags "nachos" and "burrito" (they are included in "mexican cuisine" class) are sparsely observed in the German's training set. S^3TF accurately predicted those observations formed by sparse tags since it uses knowledge that the tags "streusel" and "marzipan" both lie in tag class "german cuisine", as well as the knowledge that tags "enchilada" and "nachos" both lie in tag class "mexican cuisine". Thus, S^3TF can use such knowledge that the combinations of "german cuisine" and "mexican cuisine" are often seen in datasets across countries. *SSTF* predictions were inaccurate since they were not based on the semantics behind the objects being rated across services.

We also show the implicit relationships extracted when we factorized three datasets, UK, US, and Canada, simultaneously. The implicit relationships were computed as: (1) The probability that the relationship composed by u_m, v_n, and t_k is included in the i-th dimension is computed by $u_{i,m} \cdot v_{i,n} \cdot t_{i,k}$ where $1 \leq i \leq D$. (2) Each observed relationship is classified into the dimension that gives the highest probability value among all D dimensions. (3) The relationships included in the same dimension are considered to form implicit relationships across services. Figure 5 presents examples as the extraction results. The first line, the second line, and the third line in balloons in the figure indicate the representative venues, venue classes, and foods, respectively. The relationships

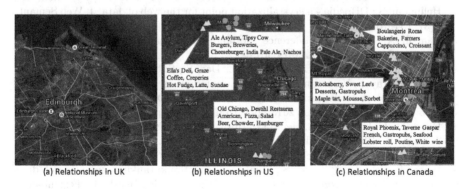

| (a) Relationships in UK | (b) Relationships in US | (c) Relationships in Canada |

Fig. 5. Examples of implicit relationships extracted by S^3TF

in the same dimension are represented by the same marks; circles (1) or triangles (2): (1) This dimension includes several local dishes with alcoholic content across countries. E.g. People in UK who love "Haggis" and drink "Scotch whisky" are implicitly related to those in US who love "Cheeseburger" and drink "Indian pale ale" as well as those in Canada who love "Lobster roll" and drink "White wine". (2) This dimension includes several sweet dishes across countries. E.g. People in UK who love "Shortbread" are implicitly related to those in US who love "Sundae" as well as those in Canada who love "Maple tart". Such implicit relationships can be used to create recommendation lists for users across services. BPTF and SSTF cannot extract such implicit relationships since they cannot use shared semantics, and thus latent features, across services.

6   Conclusion

This is the first study to show how to include the semantics behind objects into tensor factorization and thus analyze users' activities across different services. Semantic-Sensitive Simultaneous Tensor Factorization, S^3TF, proposed here, presents a new research direction to the use of shared semantics for the cross service analysis of users' activities. S^3TF creates individual tensors for different services and links the objects observed in each tensor to the shared semantics. Then, it factorizes the tensors simultaneously while integrating semantic biases into tensor factorization. Experiments using real-world datasets showed that S^3TF achieves much higher accuracy than the current best tensor method and extracts implicit relationships across services during factorization. One interesting future direction is to apply our idea to the recent embedding models (e.g. [23]) and analyze different services simultaneously by using KBs.

References

1. Bizer, C., Heath, T., Berners-Lee, T.: Linked data - the story so far. Int. J. Semant. Web Inf. Syst. **5**(3), 122 (2009)
2. Bizer, C., Lehmann, J., Kobilarov, G., Auer, S., Becker, C., Cyganiak, R., Hellmann, S.: DBpedia - a crystallization point for the web of data. J. Web Semant. **7**(3), 154–165 (2009)
3. Cemgil, A.T.: Bayesian inference for nonnegative matrix factorisation models. Comput. Intell. Neurosci. **2009**, 4:1–4:17 (2009)
4. Deshpande, O., Lamba, D.S., Tourn, M., Das, S., Subramaniam, S., Rajaraman, A., Harinarayan, V., Doan, A.: Building, maintaining, and using knowledge bases: a report from the trenches. In: Proceedings of SIGMOD 2013, pp. 1209–1220 (2013)
5. http://www.ntt.co.jp/csr_e/2013report/pdf/en_19-22.pdf
6. Dong, X., Gabrilovich, E., Heitz, G., Horn, W., Lao, N., Murphy, K., Strohmann, T., Sun, S., Zhang, W.: Knowledge vault: a web-scale approach to probabilistic knowledge fusion In: Proceedings of KDD 2014, pp. 601–610 (2014)
7. Franz, T., Schultz, A., Sizov, S., Staab, S.: TripleRank: ranking semantic web data by tensor decomposition. In: Bernstein, A., Karger, D.R., Heath, T., Feigenbaum, L., Maynard, D., Motta, E., Thirunarayan, K. (eds.) ISWC 2009. LNCS, vol. 5823, pp. 213–228. Springer, Heidelberg (2009)

8. Hoffart, J., Suchanek, F.M., Berberich, K., Weikum, G.: YAGO2: a spatially and temporally enhanced knowledge base from Wikipedia. Artif. Intell. **194**, 28–61 (2013)
9. Karatzoglou, A., Amatriain, X., Baltrunas, L., Oliver, N.: Multiverse recommendation: n-dimensional tensor factorization for context-aware collaborative filtering. In: Proceedings of RecSys 2010, pp. 79–86 (2010)
10. Menon, A.K., Elkan, C.: Link prediction via matrix factorization. In: Gunopulos, D., Hofmann, T., Malerba, D., Vazirgiannis, M. (eds.) ECML PKDD 2011, Part II. LNCS, vol. 6912, pp. 437–452. Springer, Heidelberg (2011)
11. Nakatsuji, M., Fujiwara, Y., Toda, H., Sawada, H., Zheng, J., Hendler, J.A.: Semantic data representation for improving tensor factorization. In: Proceedings of AAAI 2014, pp. 2004–2012 (2014)
12. Nakatsuji, M., Fujiwara, Y., Uchiyama, T., Toda, H.: Collaborative filtering by analyzing dynamic user interests modeled by taxonomy. In: Cudré-Mauroux, P., et al. (eds.) ISWC 2012, Part I. LNCS, vol. 7649, pp. 361–377. Springer, Heidelberg (2012)
13. Nakatsuji, M., Toda, H., Sawada, H., Zheng, J.G., Hendler, J.A.: Semantic sensitive tensor factorization. Artif. Intell. **230**, 224–245 (2016)
14. Nickel, M., Tresp, V., Kriegel, H.-P.: Factorizing YAGO: scalable machine learning for linked data. In: Proceedings of WWW 2012, pp. 271–280 (2012)
15. Niu, F., Zhang, C.R.C., Shavlik, J.W.: Elementary: large-scale knowledge-base construction via machine learning and statistical inference. Int. J. Semant. Web Inf. Syst. **8**(3), 42–73 (2012)
16. Pirr, G.: Explaining and suggesting relatedness in knowledge graphs. In: Proceedings of ISWC 2015, pp. 622–639 (2015)
17. Rendle, S., Schmidt-Thieme, L.: Pairwise interaction tensor factorization for personalized tag recommendation. In: Proceedings of WSDM 2010, pp. 81–90 (2010)
18. Rowe, M.: Transferring semantic categories with vertex kernels: recommendations with semanticSVD++. In: Mika, P., Tudorache, T., Bernstein, A., Welty, C., Knoblock, C., Vrandečić, D., Groth, P., Noy, N., Janowicz, K., Goble, C. (eds.) ISWC 2014, Part I. LNCS, vol. 8796, pp. 341–356. Springer, Heidelberg (2014)
19. Takeuchi, K., Tomioka, R., Ishiguro, K., Kimura, A., Sawada, H.: Non-negative multiple tensor factorization. In: Proceedings of ICDM 2013, pp. 1199–1204 (2013)
20. Xiong, L., Chen, X., Huang, T.-K., Schneider, J.G., Carbonell, J.G.: Temporal collaborative filtering with Bayesian probabilistic tensor factorization. In: Proceedings of SDM 2010, pp. 211–222 (2010)
21. Xu, M., Wang, Z., Bie, R., Li, J., Zheng, C., Ke, W., Zhou, M.: Discovering missing semantic relations between entities in Wikipedia. In: Alani, H., et al. (eds.) ISWC 2013, Part I. LNCS, vol. 8218, pp. 673–686. Springer, Heidelberg (2013)
22. Yilmaz, Y.K., Cemgil, A.-T., Simsekli, U.: Generalised coupled tensor factorisation. In: Proceedings of NIPS 2011 (2011)
23. Zhang, F., Yuan, N.J., Lian, D., Xie, X., Ma, W.-Y.: Collaborative knowledge base embedding for recommender systems. In: Proceedings of KDD 2016 (2016)

Multi-level Semantic Labelling
of Numerical Values

Sebastian Neumaier[1], Jürgen Umbrich[1(✉)], Josiane Xavier Parreira[2],
and Axel Polleres[1]

[1] Vienna University of Economics and Business, Vienna, Austria
juergen.umbrich@wu.ac.at
[2] Siemens AG Österreich, Vienna, Austria

Abstract. With the success of Open Data a huge amount of tabular data sources became available that could potentially be mapped
and linked into the Web of (Linked) Data. Most existing approaches
to "semantically label" such tabular data rely on mappings of textual
information to classes, properties, or instances in RDF knowledge bases
in order to link – and eventually transform – tabular data into RDF.
However, as we will illustrate, Open Data tables typically contain a large
portion of numerical columns and/or non-textual headers; therefore solutions that solely focus on textual "cues" are only partially applicable for
mapping such data sources. We propose an approach to find and rank
candidates of semantic labels and context descriptions for a given bag
of numerical values. To this end, we apply a hierarchical clustering over
information taken from DBpedia to build a background knowledge graph
of possible "semantic contexts" for bags of numerical values, over which
we perform a nearest neighbour search to rank the most likely candidates. Our evaluation shows that our approach can assign fine-grained
semantic labels, when there is enough supporting evidence in the background knowledge graph. In other cases, our approach can nevertheless
assign high level contexts to the data, which could potentially be used
in combination with other approaches to narrow down the search space
of possible labels.

1 Introduction

With the uptake of the Open Data movement a large number of tabular data
sources become freely available comprising a wide range of domains, such as
finance, mobility, tourism, sports, or cultural heritage, just to name a few. The
published data is a rich corpus that could be mapped and linked into the Web
of Data, but RDF and Linked Data still remain too high an entry barrier in
many cases, such that "3-star Open Data" (cf. http://5stardata.info/) in the form
of tabular CSV data remains the predominant data format of choice in the
majority of Open Data portals [19]. Connecting CSV data to the Web of Linked
Data involves typically two steps, that is, (i) transforming tabular data to RDF
and (ii) mapping, i.e. linking the columns (which adhere to different arbitrary

© Springer International Publishing AG 2016
P. Groth et al. (Eds.): ISWC 2016, Part I, LNCS 9981, pp. 428–445, 2016.
DOI: 10.1007/978-3-319-46523-4_26

schemata) and contents (cell values) of such tabular data sources to existing RDF knowledge bases. While a recent W3C standard [18],[1] provides a straightforward canonical solution for (i), the mapping step (ii) though remains difficult.

Mapping involves to semantically label columns by linking column headers or cell values to either properties or classes in ontologies or instances in knowledge bases, and to determine the relationship between columns [17]. For the semantic labelling, most approaches so far rely on mapping textual values [16,20,21]; these work well e.g. for HTML/Web tables which have rich textual descriptions, as they are published mainly for human consumption. However, in typical Open Data portals many data sources exist where such textual descriptions (such as column headers or cell labels) are missing or cannot be mapped straightforwardly to known concepts or properties using linguistic approaches, particularly when tables contain many numerical columns for which we cannot establish a semantic mapping in such manner. Indeed, a major part of the datasets published in Open Data portals comprise tabular data containing many numerical columns with missing or non human-readable headers (organisational identifiers, sensor codes, internal abbreviations for attributes like "population count", or geo-coding systems for areas instead of their names, e.g. for districts, etc.) [9]. We verified this observation by inspecting 1200 tables collected from the European Open Data portal and the Austrian Government Open Data Portal and attempted to map the header values using the BabelNet service (http://babelnet.org): on average, half of the columns in CSV files served on these portals contain numerical values, only around 20 % of which the header labels could be mapped with the Babel-Net services to known terms and concepts (cf. more details in our evaluation in Sect. 6.3). Therefore, the problem of semantically labelling numerical values, i.e., identifying the most likely property or classes for instances described by a bag of numerical values remains open.

Some early attempts focus on specific "known" numerical datatypes, such as longitude and latitude values [3], or – more generally – on classifying numerical columns using (manually) pre-labelled numeric value sets [11]. To the best of our knowledge, so far no unsupervised approaches have been devised for semantic labelling of numerical value sets. Additionally, the latter approach by Ramnandan et al. only assigns a single predefined semantic label, corresponding to a "property" per column. In the context of RDF, we deem such semantic labelling insufficient in (at least) two aspects: (a) We do not only need to map columns to properties, but to what we will call "contexts", that is property-domain pairs. (b) Since, given the variety and heterogeneity of Open

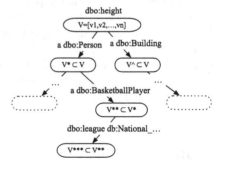

Fig. 1. Hierarchical background knowledge

---

[1] Or, likewise with RDB2RDF direct mapping [2], the basis of [18].

Data, it is likely we cannot rely on a manually curated, pre-defined set of semantic labels. Therefore, there is a need to build a hierarchical "background knowledge graph" of semantic labels in an unsupervised manner, cf. Fig. 1. As an example for (a), we do not only want to label a bag of numerical values as *height*, but instead we want to identify that the values represent the *heights of basketball players who played in the NBA*, or that the values represent the *heights of buildings*.

Even if we cannot identify such precise labels, we still want to assign the most likely contexts the values belong to, e.g. *height of a person*. To this end, and in order to achieve (b), we automatically generate a hierarchical background knowledge base of contexts from DBpedia. Different than previous approaches that assign a single label to a bag of values, we assign different labels/contexts, with different confidence values. This way, our approach could potentially be combined with textual labelling techniques for further label refinement, which is left for future work. In this particular paper, we focus on the following concrete contributions:

1. We propose a hierarchical clustering over an RDF knowledge base to build a background knowledge graph containing information about typical numerical representatives of contexts, i.e., grouped by properties and their shared domain (subject) pairs, e.g. city temperatures, peoples ages, longitude and latitudes of cities.
2. We perform a k-nearest neighbours search and aggregate the results of semantically label numerical values at different levels in our knowledge graph.
3. We evaluate our approach by cross-validating over a sample of DBpedia data generated from the most widely used numeric properties and their associated domain concepts.
4. We test our approach "in the wild" on tabular data extracted from Open Data portals and report valuable insights and upcoming challenges which we have to tackle in order to successfully label data from the Open Data domain.

In the remainder of this paper, after an overview of related works (Sect. 2), we describe our overall approach (Sect. 3). Next, we present the construction of the background knowledge graph from DBpedia in Sect. 4, as well as the actual semantic labelling of a column (i.e., a bag of numeric values) in Sect. 5. Finally, we present the evaluation results of the efficiency of different background graph construction strategies and our experiments with attempting to find matching columns in Open Data, Sect. 6. We conclude with a summary and ideas for future work (Sect. 7).

2   Related Work

There exists an extensive body of research in the Semantic Web community to derive semantic labels for attributes in structured data sources (such as columns in tables) which are used to (i) map the schema of the data source to ontologies or existing semantic models or (ii) categorise the content of a data source (e.g. a table talking about politicians, i.e., in our case mapping the rows of a table into classes). The majority of these approaches [1,5,11,12,17,20,21] assume well-formed relational tables, rely on textual information, such as table headers and string cell values in the data sources, and apply common, text-based entity linkage techniques for the mapping (see [24] for a good survey). Moreover, typical approaches for semantic labelling such as [1,20,21] recover the semantics of Web tables by considering as additional information the, again textual, "surrounding" (section headers, paragraphs) of the table and leverage a database of class labels and relationships automatically extracted from the Web. Note, that in contrast to our concept of "context" of a column, the labels here are one-dimensional. In summary, the main focus of all these works is on textual relations inside the tables and in their surroundings. Techniques for recovering numerical relationships are often left for future work. As for used techniques, while these are out of the scope of our paper, many advanced textual entity recognition and linkage techniques are implemented in the Babelnet system [10], as we highlighted in the previous section these techniques are not necessarily applicable to a large portion of (numerical) Open Data. In contrast, our approach assumes that we only have a bag of numerical values available, in the worst case lacking any other rich textual information.

Most closely related to our efforts is the work by Ramnandan et al. [11], where the authors proposed to semantically label tuples of attribute-value pairs (textual and numerical). The semantic labelling of numerical values is achieved by analysing the distribution of the values and compare it to known and labelled distributions given as input by using statistical hypothesis testing. In contrast to their approach, we build a knowledge hierarchy and annotate sources not only with a single label but with a possible type and shared property-object pairs. Also complementary to our efforts is the work of Cruz et al. [3] which focus on detecting geolocation information in tables and apply heuristics specifically for numerical longitude and latitude values.

Outside the area of semantic labeling as such, but as an inspiration for our approach, the authors of [6,22] developed approaches to detect natural errors/outliers in RDF knowledge bases and automatically clustered candidate sets from the RDF knowledge base they want to analyse by grouping numerical values of a selected property by their types. We use a similar approach to build our background knowledge: we also group the subjects (and their corresponding values) by their types. However, we use a more fine-grained notion of "type", not only considering named classes but also "subtypes" defined in terms of shared property-object pairs.

While our present work explicitly focuses on instance sets labeling in the absence of a schema, previous work that addressed the automatic labeling problem using different combinations of instance and schema matching are relevant and will be considered in future extensions of our work [8,13,23].

3   Approach

Next, we outline the steps of our approach of finding the most likely semantic label and to determine the context in which a bag of numerical values are derived. In the following we formally define our notation and state the problem.

We denote a bag of numerical values annotated by a given label and context description $<l, c>$ as $\mathcal{V}^{<l,c>} = \{v_1, v_2, \cdots, v_n\}$, with $v_i \in \mathbb{R}$. Similar to [11], we define a semantic label l as an attribute of a set of values, which can potentially appear in different contexts. In this work, the semantic label l is a property from an ontology. However, this could be generalised. The concept of context description corresponds to a set of attribute-value pairs which explain/describe the commonalities of the values in $\mathcal{V}^{<l,c>}$. As such, one can assume that the set of input values $\mathcal{V}^{<l,c>}$ are the result of applying a query over a knowledge base ($\mathcal{V}^{<l,c>} = \mathcal{Q}(KB) = \{v_1, v_2, \cdots, v_k\}$) with the semantic label and the set of attribute value pairs as filter attributes of the query. For instance, the following SPARQL query returns the set of values labelled with *height* and sharing the attribute-value pair *a basketball player*:

```
SELECT ?v WHERE {[a dbo:BasketballPlayer] dbp:height ?v.}
```

Numerical values for a semantic label can appear in different contexts. For instance, values can represent the *height* of a building, mountain or a person. Even further, we might find values representing the height of basketball players that played in the NBA. We model this observation in form of a tree for each label l. The root node in such a tree corresponds to the set of all values which fulfill the property l. The remaining nodes of the tree represent further semantic information for this values, i.e., a shared context in the form of attribute-value pair. Edges in the tree are subset-relations between these values, directed from the superset to the subset. For instance, considering the semantic label *height*, the root node could have child nodes corresponding to the context *a mountain* and *a person*.

The background knowledge can be constructed in an either top-down or bottom-up approach The former starts with the root node of the graph and then detects subsets while the latter starts with leaf nodes which are then combined into parent/super nodes. The top-down approach is suitable for building the context graph from RDF knowledge bases and requires to start with a set of entities which are described by several attribute-value pairs. Next, we can group such entities by attributes which have numerical values, and then detect subgroups of entities with shared attribute-value pairs. We will show in the next section how we can build the background knowledge graph from an RDF knowledge base.

The bottom-up approach is more suitable for building the background knowledge from a set of CSV files. We first find a set of annotated numerical value triples $\{(v_1, l_1, c_1), (v_2, l_1, c_2), \cdots, (v_n, l_m, c_n)\}$, each consisting of a set of numerical values v_i, a label l_j and a context c_i. An input triple (v, l, c) can be extracted from a numerical column which was either manually or automatically annotated with semantic labels (e.g. based on the column header). The possible context

Table 1. Example table

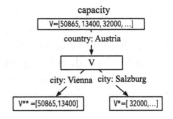

Fig. 2. Resulting tree

name	capacity	city	country
Ernst Happel...	50865	Vienna	Austria
Franz Horr Stadium	13400	Vienna	Austria
Red Bull Arena	32000	Salzburg	Austria
...	...	...	...

information can be modelled from column headers, the author or title of the table, or shared attributes within the table. For instance, take the example table in Table 1 and the numerical column `capacity`, as context we could extract that the numerical values describe an attribute of entities which are of type *football stadium*. Further, all values share the attribute-value pair *country: Austria*. Additionally, we could build a subset of values with the common context *city: Vienna* and another subgroup with the context *city: Salzburg* (cf. Fig. 2). The resulting background knowledge can be exploited by machine learning algorithms or statistical methods to predict the most likely label and context for a given bag of input values. We will outline how we apply a nearest neighbour search approach to derive the most likely label and context pair for a set of values in Sect. 5.

4   Background Knowledge Graph Construction from DBpedia

In this section, we outline our automatic top-down approach to build a background knowledge base from RDF data. To do so, we execute the following steps:

1. We extract all RDF properties which have numerical values as their objects and group the subjects by their numerical properties. These properties are used as labels. We derive the list of RDF properties which have numerical values as their range; the following SPARQL query could be used, cf. [6], however, we note that this query does not return results on the live DBpedia SPARQL endpoint due to timeouts:

```
SELECT ?p, COUNT(DISTINCT ?o) AS ?cnt
WHERE {?s ?p ?o. FILTER (isNumeric(?o))} GROUP BY ?p
```

Another approach would be to directly query the vocabularies if we know that the RDF KB contains OWL vocabulary listing all datatype properties. We resorted to just filtering triples of the DBpedia dump with numeric objects, sorting them by property and counting via a script.

2. Next (in another pass/sorting), we collect/group by subjects in the different property groups the values of the numerical properties *l*. For "typing" of these subjects we collect property-object pairs – what we call context – for which

the object is an RDF resource (an IRI); this includes `rdf:type-`*Class* triples, but also others, e.g. `dbo:locatedIn-dbr:Japan`.

3. Next, we also extract and materialise the OWL class hierarchy for the *Classes*. This can be done directly by extracting the `rdfs:subClassOf` hierarchy from the DBpedia ontology for these *Classes*; we will use this *type hierarchy* to further enrich our background graph collecting contexts.

After grouping the entities by the selected context labels we construct our background knowledge graph as follows: An abstraction of our graph is depicted on the left hand side of Fig. 3: the graph consists of multiple trees, each tree corresponding to a property. The root node of such a tree is labelled by the property and contains the bag (i.e., multiset) of all numerical values of this property.

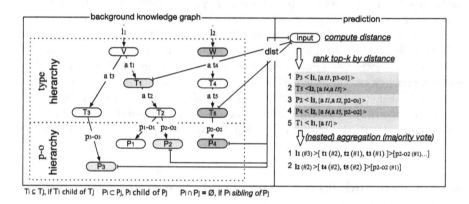

Fig. 3. background knowledge and prediction

4. The first "layer" of our knowledge graph is the so-called type hierarchy which represents the `rdfs:subClassOf` relation for all available types of the triples for property l from the *type hierarchy*. Since subjects can be of more than one type, the sibling nodes in this layer can share values from the same triples. In order to not keep too fine grained, rare classes, we filter by discarding types with less than δ instances (e.g., property-class combinations with less than 50 instances).

5. Next, we construct the second layer, termed *p-o* hierarchy for the identified non-`rdf:type` property-object pairs to further refine out context structure, beyond classes, using a divisive hierarchical clustering approach. We start with one node/group and split/compute sub-contexts recursively as we move down the hierarchy, to further refine the type hierarchy. In order to decide how to split a node, we impose the following requirements for possible candidates:
   (a) **constrain property-object:** we use the same constraint as [6] that subjects in a candidate node share the same property-object pair.

(b) **constrain size:** again, in order to avoid too fine-grained subdivision, the size of a candidate node has to be larger than 1 % of the parent node size (or, resp. larger than δ) and smaller than 99 % of the parent node size.

Once the set of possible sub-contexts is computed, we sort the candidates by their distance to the parent node in descending order. Details on the distance measures used to compare bags of numerical values are given in Sect. 5.1. To guarantee a high diversity as well as disjointedness of the sub-contexts within the hierarchy, we select the candidate with the biggest distance first, and then subsequently the non-overlapping sub-contexts from the list with decreasing parent distance. Additionally, the disjointedness requirement also helps to limit the number of sub groups. We recursively perform the above steps for the new selected groups. Consequently, shared property-object pairs of a node on the *p-o* hierarchy are encoded in the path to the resp. *p-o* node.

Node type terminology: Regarding terminology, we refer to the *exact type* of a context graph node as the lowest type node in the path to a *p-o* node. For instance, considering node P_3 in our example in Fig. 3, the exact type would be T_2. As a *super type*, we consider all type nodes on the path between the exact type node and the *p-o* node (e.g. T_1 would be a super type of node P_3). Eventually, the *root type* of a node, is the highest type node on the path to the *p-o* node (e.g. T_1 is the root type of P_3).

5 Prediction Approach

We use nearest neighbours classification over our background knowledge graph to predict the most likely "semantic context" for a given bag of numerical values. Given an input bag, we compute the distance between the values to all context nodes in our background knowledge graph and return the resp. contexts in ascending order of distance. Ideally, the node with the closest distance is the most likely semantic context/description for the input values. However, obviously numerical values for different types and properties might share the same value range and distribution and so we cannot even expect that the correct semantic description is always the top ranked result. As such, we also provide aggregation functions for predicates, type and *p-o* nodes over the top-k results. The idea is similar to the K-nearest neighbour classification for which the classification of an object is based on a majority vote over the top-k neighbour contexts.

5.1 Distance Measures

An important part for any prediction algorithm, be it based on machine learning or statistical methods, is the distance measure to determine how closely related two items (e.g. feature vectors) are. We consider two distance measures, namely (i) the euclidean distance between two feature vectors and (ii) the distribution similarity between two bags of numerical values.

Euclidean Distance Between Descriptive Features. The first distance function is the euclidean distance between two numerical n-dimensional feature vectors. For our use case we consider the following features for the vectors:

- *min and max value:* The range of minimum and maximum values is an important feature which allows us to easily discard "out of scope" labels or contexts. For instance, the heights of humans might have a maximum range of 213 cm which distinguishes it from buildings which have much higher max height.
- *5 % and 95 % quantile*: Due to the fact that minimum and maximum values as features are prone to outliers and errors in the set of values we also consider quantiles and inter-quantile ranges, e.g., using 5 %- and 95 %-quantile instead of min and max as features in a feature vector [6].
- *Additional descriptive statistics (mean, stddev):* Additionally, descriptive features such as the mean and the standard deviation of a set of values give better results for values which are within the same range but follow different distributions.

Distribution Similarity. Another distance measure is the similarity of two distributions of numeric values. This approach was already successfully used in a similar setup by Ramnadan et al. [11]. The authors also showed in their evaluation that the Kolmogorov-Smirnov (KS) test performs best for this particular setup compared to tests such as Welch's *t*-test or Mann-Whitney's U-test.

Kolmogorov-Smirnov (KS) distance: The KS test is a non-parametric test which quantifies the distance between two empirical distribution functions with the advantage of making no assumptions about the distribution of the data. As a distance measure between two samples, the KS test computes the KS-statistic D for two given cumulative distribution functions F_1 and F_2 in the following way:

$$D = \sup_{x} |F_1(x) - F_2(x)| \tag{1}$$

where *sup* is the supremum of the distances. If two samples are equally distributed, i.e., the two bags hold the same numeric values, then the statistic D converges to 0.

5.2 Aggregation Function

As in the K-nearest neighbour classification, we also aggregate the top-k nearest neighbours by their properties, types and property object pairs. This allows us to classify the input values at several levels:

Before we apply the specific voting function, we aggregate the neighbours for the following different levels:

- `property level`: aggregation of the top-k neighbours by their properties
- `exact type level`: aggregation of the top-k neighbours by their exact type
- `root type level`: aggregation of the top-k neighbours by their root type

- **all types level**: aggregation of the top-k neighbours by each of their types (including the exact and all super types)
- *p-o* **level**: aggregation of the top-k neighbours by each of their *p-o* nodes

We consider the following two aggregation functions:

- *Majority vote:* This is the standard method for the K-nearest neighbour classification for which the input values are classified based on a majority vote over the k nearest neighbours. Therefore, given an aggregation level, we rank the aggregated results (e.g. properties) based on the appearance in the top-k neighbours. Consider the right part of Fig. 3 in which we illustrate such a ranking process. For instance, the property aggregation would rank p_1 higher than p_2 since p_1 appears three times in comparison to p_2 which only appears 2 times.
- *Aggregated distance:* Our second aggregation function, we rank the aggregated results not by the number of their appearances, but compute the average distance. For instance, we would compute the distance for p_1 in Fig. 3 by averaging the distance of node P_3, P_2 and T_1.

In addition to the aggregation of properties, types and property-object pairs, we can also perform a nested level aggregation. For instance, we could aggregate first on the property level and then inside each property on the type level. An example for the nested aggregation based on the majority vote is depict in Fig. 3; the most likely type for p_1 would be t_1 with 2 votes, followed by t_2 and t_3.

6   Evaluation and Experiments

We have implemented a prototype system in Python to evaluate our approach with different functions. As a dataset to construct our background knowledge we use the DBpedia 3.9 dump.[2] The aim of our evaluation is twofold: We first automatically evaluate the accuracy of our prediction functions with different setups of the background knowledge in a controlled environment by splitting the DBpedia data into a test and training dataset. Secondly, we manually test our approach over Open Data CSV files to gain first insights for future directions, whether there is a chance to label tabular columns outside of DBpedia.

6.1   Background Knowledge Construction

We selected 50 of the the most frequently used numerical DBpedia properties to build our background knowledge for both evaluation scenarios:[3] we excluded properties which clearly indicate internal DBpedia ids only, such as `dbo:wikiPageRevisionID` as well as properties which are not directly in the root

---

[2] http://downloads.dbpedia.org/3.9/en/mappingbased_properties_en.nt.bz2,   last accessed 2016-04-28.

[3] The full list of properties is online at http://data.wu.ac.at/iswc2016_numlabels/properties.html.

path of the http://dbpedia.org/ontology/ prefix. Figure 4 plots in the left figure the 5 % to 95 % inter-quantile ranges of our selected properties (in logarithmic scale) and in the right figure the total number of numeric values for each property. The range plot visualises the overlap of numerical values for our different properties and the quantiles are used to smoothen the ranges and eliminate possible outliers. About 60 % of the properties have values within the range 0–1000 and about 90 % within 0–2000.[4] The shortest range has the property `dbp:displacement` (inter-quantile range of 0.0058) and the maximum range of 2.56 billion has the property `dbp:areaTotal`. Regarding the total number of values, the longest bar, with 421k values, corresponds to the `dbp:years` property and the shortest to `dbp:width` (9.6k values).

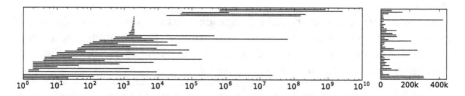

Fig. 4. 5 %–95 % inter-quantile ranges and number of values of training properties

We built three versions of our background knowledge graph to better understand the impact of the three different distance functions. One function is based on the Kolmogorov - Smirnov distribution test, and two based on the euclidean distance over feature vectors. The first type of vector uses the minimum and maximum of the values as features while the other uses the 5 % and 95 % quantile as features. We add to both vectors the mean and standard deviation as additional dimensions. Table 2 gives an overview of our three knowledge bases together with the number of nodes, the construction time of the background knowledge graph and the average prediction time for a given set of values (based on our evaluation runs).

In addition we added our average prediction times for the different setups. However, please note that we did not optimize our system wrt. runtimes.

Table 2. Setup of our three background knowledge graphs

ID	Distance measure	Nodes	Build time	Avg. pred. time
KS	Kolmogorov-Smirnov test	11431	30 m	2.5 s
FV1	(min, max, mean, std)	11432	24 m	2.3 s
FV2	(5-q, 95-q, mean, std)	11432	38 m	4.6 s

---

[4] Note, that around 30 % of the properties have values in the range of 1000–2000 and mainly describe years.

In future work we plan the improvement of these prediction times in order to provide our algorithm as a live service. All evaluation are conducted on a machine with 30 GB of RAM.

6.2   Model Evaluation

Our first experiment is designed to obtain the performance characteristics of our prediction for different distance functions.

Test and training data selection: To get an unbiased assessment we randomly assigned 20 % of the subjects for each property as test data and the remaining subjects are used to build the knowledge bases. The test data is further processed to find suitable test groups. To build those test-groups, we proceed in a similar manner as for the construction of the background knowledge base. That is, we analogously built type hierarchy and *p-o* hierarchy for per property, however, this time without imposing any constraints and creating all possible test contexts and sub-contexts. Eventually, we randomly select the leaf nodes of this "test context graph" and the respective numerical value bags as test data. This process ensures that we select context nodes which are not necessarily contained 1-to-1 in the background knowledge graph.

Evaluating Distance Functions. Our first evaluation aims to (i) test the impact of the distance function for the prediction and (ii) to select the best setup for further tests. We set up an initial experiment by randomly selecting a maximum of 50 leave nodes from each property tree in our test dataset; resulting in 1787 test nodes.

Table 3. Accuracy in % for different distance functions

	FV1			FV2			KS		
	Top-1	Top-5	Top-10	Top-1	Top-5	Top-10	Top-1	Top-5	Top-10
exact	2.5	8.2	8.2	2.5	8.2	8.2	**12.3**	**41.8**	**47.9**
prop	45.4	60.3	60.3	45.4	60.3	60.3	**57.1**	**74.1**	**79.8**
type	11.3	24.9	24.9	11.3	24.9	24.9	**16.1**	**43.9**	**56.0**
stype	24.9	41.1	41.1	24.9	41.1	41.1	**35.8**	**58.6**	**67.5**

To initially measure the accuracy of the top-k neighbours, we introduce the following evaluation measures:

- exact: the top-k neighbours contain the *correct node* in the graph, that is, the test node and predicted node share the same property, type and *p-o* pairs.
- prop: the top-k neighbours contain the *correct property/label*
- type: the top-k neighbours contain the *correct type*
- stype: the top-k neighbours contain the *correct super type* of the test node

The results in Table 3 show the accuracy for different metrics for the top-k neighbours, with the best results marked bold. We can clearly see that the Kolmogorov-Smirnov based distance function (KS) outperforms for all metrics the feature vectors based functions in terms of prediction accuracy. The initial results show that our approach already predicts the correct property of the input values in the top-10 neighbours for 79 % of all test and the right type in 56 % of the cases. Based on the clear results, we decided to use the prediction approach based on Kolmogorov-Smirnov distance function in the remaining evaluation.

Large-Scale Model Validation: The next experiment focuses on the evaluation of the different aggregation functions and levels. We randomly sampled 33657 test nodes by selecting a maximum of 20 % of the leave nodes for each property in our test data set. The test data is ~18 times larger than in the previous experiment and 3 times the size of our training nodes. In addition, only 9 % of the test context nodes are contained 1-to-1. This allows us to study our approach for input data for which we have only partial evidences available. We evaluate the accuracy for the different levels by measuring if the top-k aggregated results contain the correct property, type, parent types or any p-o context of the test instance.

Table 4 summarises the accuracy (in %) over 33k test instances for two aggregation functions over the top-k nearest neighbours. Overall, the results show a high prediction accuracy of over 92 % across all different levels for the top-10 aggregated results using the top-50 closest neighbours. For the root-type prediction, we observe the highest accuracy within the top-5 aggregated results. Regarding test nodes for which we have only partial information available, our approach can still predict the correct property, (parent) type and even some of the shared p-o pairs. Our results also show that doubling the number of neighbours significantly improves the prediction accuracy by up to 15 %. Considering the two aggregation functions, we see that ranking the results based on majority votes performs slightly better than using the average distance, with the

Table 4. Accuracy in % for different aggregation levels and functions (maj. = majority vote, avg. = average distance)

Top-k		prop		type		all-types		root-type		p-o level	
Neigh.	Agg. results	Maj.	Avg.	Maj.	Avg.	Maj.	Avg.	Maj.	Avg.	Maj.	Avg.
25	1	59.3	34.5	64.7	57.8	64.7	57.8	66.4	69.2	20.4	24.9
	5	87.9	82.9	91.4	85.3	91.4	85.3	94.7	94.7	75.8	66.2
	10	98.5	98.5	94.7	94.7	94.7	94.7	94.7	94.7	83.8	74.0
50	1	57.4	23.7	66.4	37.6	66.4	37.6	66.7	70.7	20.4	24.9
	5	98.4	83.7	93.2	65.4	93.2	65.4	96.3	96.3	75.8	66.2
	10	99.3	99.3	96.3	96.1	96.3	96.1	96.3	96.3	83.8	74.0

biggest impact for the *p-o* level aggregation. Interestingly, inspecting the top-1 aggregated results using the 50 nearest neighbours, we see that the `root-type` accuracy is lower than the `all types` accuracy. This is a false negative classification which can happen if there exists more than k results with equal votes or average distances. In such case, we rank the results in alphabetical order and return only the top-k, leading to a cutoff of possible correct results.

Looking at the top-10 of the aggregated results, we correctly predicated 99.5 % of the properties, 96.3 % of the exact and parent types and 92 % of the *p-o* pairs. These results are encouraging to use our approach for labelling numerical columns in tabular data, especially since we can also partially label values for which we do not have full evidences in our background knowledge graph.

6.3   Semantic Labelling of Numerical Columns in Open Data Tables

Eventually, we study how our approach performs for numerical columns in Open Data tables. We have to emphasise upfront, that this experiment is of rather exploratory than quantitative nature, since - due to the heterogeneity of data typically published in Open Data portals vs. DBpedia, we could not expect a lot of exact matches.

To conduct our experiment, we downloaded and parsed in total 1343 CSV files from two Open Data portals, namely the Austrian Open Government Data portal (AT[5]) and the European Open Data portal (EU[6]). We used the standard Python CSV parser to analyse the tables for missing header rows and performed a simple datatype detection to identify numerical columns. In order to get insights into the descriptiveness of these headers we tried to map header labels to BabelNet [10] in a non-restrictive manner: we performed a simple preprocessing on the headers (splitting on underscores and camel case) and retrieved all possible mappings from the BabelNet API.

Table 5 shows some basic statistics of the CSV tables in the two portals. An interesting observation is that the AT portal has an average number of 20 columns per table with an average of 8 numerical columns, while the EU portal has larger tables with an average of 4 out of 20 columns being numerical. Regarding the descriptiveness of possible column headers, we observed that 28 % of the tables have missing header rows. Eventually, we extracted headers from 7714 out of around 10K numerical columns and used the BabelNet service to retrieve possible mappings. We received only 1472 columns mappings to BabelNet concepts or instances, confirming our assumption that many headers in Open Data CSV files cannot easily be semantically mapped.

Exploratory Experiments: We used the numerical columns from our CSV corpus as input for our system and manually study select columns to gain first insights. Initially, we ranked the columns by their average distance over the 50

---

[5] http://data.gv.at/.
[6] http://open-data.europa.eu/.

Table 5. Header mapping of CSVs in Open Data portals

Portal	Tables	*cols*	*num.cols*	w/o Header	Num. H	Mapped
AT	968	13	8	154	6,482	1,323
EU	357	20	4	223	1,233	349

nearest neighbours and inspected the top-100 columns for each portal. We share the interesting results for tables and columns online.[7]

Our first observation observation is related to the time coverage of numerical values and the difference between Open Data and DBpedia. For instance, the Austrian Open Data portal hosts tables as specific as numbers of cars per brands per district in Vienna, or current (every 15 min) weather data from different weather stations in Austria. We do not expect matches for such specific numbers or even for temperature values if they are given in the form of timelines. In contrast, DBpedia typically has numeric values only for "current" or "latest" for many properties, taking population numbers of settlements as an example. Still, we are curious to see what the method would return and partially could explore interesting findings.

Another observation is that our knowledge base does not cover some of the domains and attributes of the numerical Open Data columns. For instance, many columns describe "counts" or "statistics". Examples for such count columns are the number of registered car model per district, the count of tourists grouped by their nationality, month of year and country/region they visit or the count of valid or invalid votes for an election. Examples for statistics are election results or the percentage of registered people for different age groups and districts in a city. For instance, take the 14th ranked Column#14[8] which describes election results divided by different regions, with a non-descriptive header UNG. (we assume this means "invalid votes"). The second-ranked property is *population-Total* which is arguable a related labelling, since election results are basically sub-populations of different regions. Looking at the results of the type aggregation for this column, we find five times the type *Settlement* within the first ten neighbours, which further indicates that the values rise from (sub-)populations. Similarly, Column#1[9] holds counts of car models grouped by regions which our algorithm again labelled as population. This shows clearly that to label Open Data columns we need a very broad coverage of numerical domains in our background knowledge.

We also aggregate the results across columns to identify the "domain" of a table using the top-10 results of our all-types level aggregation and manually inspected some results. Again, we ranked the tables based on their average distances across all their numerical columns. For instance, consider the second

---

[7] http://data.wu.ac.at/iswc2016_numlabels/.

[8] http://data.wu.ac.at/iswc2016_numlabels/submission/col14.html.

[9] http://data.wu.ac.at/iswc2016_numlabels/submission/col1.html.

ranked Table#2[10] which consists of multiple columns which describe population counts for different districts. Aggregating and ranking the types across these columns results in the types *Town* and *PopulatedPlace* which proved to be right.

Discussion of Findings: While the findings did not yet provide, clear and convincing matches, we could collect valuable insights from this results on challenges to be tackled in future work:

- *Dealing with timeline data:* To correctly handle timeline data, we first need to be able to detect the time dependency and than regroup or transform the table.
- *Domain specific background knowledge:* Open Data contains many tabular data which is similar in itself, but not necessarily matching DBpedia categories and values reported there, e.g. reports for spendings/budget election results, tourism or population demographics. Our results highlight the limited coverage of DBpedia, which was also observed in the work from Ritze et al. [14]. Therefore, we have to gradually enrich the background knowledge graph from categories learned from Open Data tables themselves.
- *Aggregating column scores:* While single columns provided partially bad recognition, in some cases combined recognition of columns revealed interesting combinations.
- *Combine with existing complementary approaches:* Lastly, while we deliberately left it out of scope in this paper, linguistic cues could and probably should be used in combination with our numerical methods as an additional cue to gradually improve labelling/matching capabilities, as we explore and collect more Open Data sources.

7 Conclusions and Future Work

To the best of our knowledge, this is the first work addressing semantic labelling of numerical values by applying k-nearest neighbours search over a background knowledge graph, which is constructed in an unsupervised manner using hierarchical clustering. Our evaluation shows that we can assign fine-grained semantic labels when there is enough supporting evidence in our background knowledge graph. In other cases, our approach can nevertheless assign high level contexts to the data. Given a bag of numerical values, we correctly identified in 99.5 % of the test cases the properties, in 96.3 % the exact or parent type, and in 92 % the shared property-object pairs. Despite the simplicity of our solution, we can confirm that a knowledge base can be harnessed to perform automatic semantic labelling of datasets with promising results.

The obtained results are encouraging for labelling numerical columns in tabular data. A first feasibility evaluation using numerical columns in Open Data CSV files showed that further research is needed to extend our knowledge graph to cater for the specifics of the Open Data domain, such as addressing timeliness.

---

[10] http://data.wu.ac.at/iswc2016_numlabels/submission/tab2.html.

In future work, we plan to extend our background knowledge, using more properties from DBpedia and combining it with knowledge from other RDF datasets, such as WikiData or eurostats. To achieve better results in such combined datasets the handling of units of measurement and the time dimension is a vital extension of our system [23]. Complementary, we will focus on building the background knowledge in a bottom-up approach from information extracted out of CSV files as outlined in this work. We will also investigate performance optimization techniques, since our prediction time increases linearly with the number of context nodes. For example, we will explore range indices or pre-filtering to reduce the search space in the context graph. Another direction is to exploit our system in combination with other approaches for labelling tables based on textual header; e.g., [8] nicely complements our approach: Halevy et al. group together semantically related attributes and relate them to corresponding classes.

We believe that our approach can provide important clues about the context of numerical values which can be exploited in other domains, e.g., as input for ontology alignment between two different RDF datasets [7,15] or as input for computing the relatedness between tables such as used in [4].

Acknowledgments. This work has been supported by the Austrian Research Promotion Agency (FFG) under the project ADEQUATe (grant no. 849982).

References

1. Adelfio, M.D., Samet, H.: Schema extraction for tabular data on the web. Proc. VLDB Endow. **6**(6), 421–432 (2013)
2. Arenas, M., Bertails, A., Prud'hommeaux, E., Sequeda, J.: A direct mapping of relational data to RDF, W3C Recommendation, September 2012. http://www.w3.org/TR/rdb-direct-mapping/
3. Cruz, I.F., Ganesh, V.R., Mirrezaei, S.I.: Semantic extraction of geographic data from web tables for big data integration. In: Proceedings of the 7th Workshop on Geographic Information Retrieval, GIR 2013, pp. 19–26. ACM, New York (2013)
4. Das Sarma, A., Fang, L., Gupta, N., Halevy, A., Lee, H., Wu, F., Xin, R., Yu, C.: Finding related tables. In: Proceedings of the 2012 ACM SIGMOD International Conference on Management of Data, pp. 817–828. ACM (2012)
5. Ermilov, I., Auer, S., Stadler, C.: User-driven semantic mapping of tabular data. In: Proceedings of the 9th International Conference on Semantic Systems, I-SEMANTICS 2013, pp. 105–112. ACM, New York (2013)
6. Fleischhacker, D., Paulheim, H., Bryl, V., Völker, J., Bizer, C.: Detecting errors in numerical linked data using cross-checked outlier detection. In: Mika, P., et al. (eds.) ISWC 2014, Part I. LNCS, vol. 8796, pp. 357–372. Springer, Heidelberg (2014)
7. Gal, A., Roitman, H., Sagi, T.: From diversity-based prediction to better ontology & schema matching. In: Proceedings of the 25th International Conference on World Wide Web, WWW 2016, Montreal, Canada, pp. 1145–1155 (2016)
8. Halevy, A.Y., Noy, N.F., Sarawagi, S., Whang, S.E., Yu, X.: Discovering structure in the universe of attribute names. In: Proceedings of the 25th International Conference on World Wide Web, WWW 2016, Montreal, Canada, pp. 939–949 (2016)

9. Lopez, V., Kotoulas, S., Sbodio, M.L., Stephenson, M., Gkoulalas-Divanis, A., Aonghusa, P.M.: QuerioCity: a linked data platform for urban information management. In: Cudré-Mauroux, P., et al. (eds.) ISWC 2012, Part II. LNCS, vol. 7650, pp. 148–163. Springer, Heidelberg (2012)
10. Navigli, R., Ponzetto, S.P.: BabelNet: the automatic construction, evaluation and application of a wide-coverage multilingual semantic network. Artif. Intell. **193**, 217–250 (2012)
11. Ramnandan, S.K., Mittal, A., Knoblock, C.A., Szekely, P.: Assigning semantic labels to data sources. In: Gandon, F., Sabou, M., Sack, H., d'Amato, C., Cudré-Mauroux, P., Zimmermann, A. (eds.) ESWC 2015. LNCS, vol. 9088, pp. 403–417. Springer, Heidelberg (2015)
12. Rastan, R.: Towards generic framework for tabular data extraction and management in documents. In: Proceedings of the Sixth Workshop on Ph.D. Students in Information and Knowledge Management, PIKM 2013, pp. 3–10. ACM, New York (2013)
13. Ritze, D., Lehmberg, O., Bizer, C.: Matching HTML tables to DBpedia. In: Proceedings of the 5th International Conference on Web Intelligence, Mining and Semantics, WIMS 2015, Larnaca, Cyprus, pp. 10:1–10:6 (2015)
14. Ritze, D., Lehmberg, O., Oulabi, Y., Bizer, C.: Profiling the potential of web tables for augmenting cross-domain knowledge bases. In: Proceedings of the 25th International Conference on World Wide Web, WWW 2016, Montreal, Canada, pp. 251–261 (2016)
15. Rong, S., Niu, X., Xiang, E.W., Wang, H., Yang, Q., Yu, Y.: A machine learning approach for instance matching based on similarity metrics. In: Cudré-Mauroux, P., et al. (eds.) ISWC 2012, Part I. LNCS, vol. 7649, pp. 460–475. Springer, Heidelberg (2012)
16. Syed, Z., Finin, T., Mulwad, V., Joshi, A.: Exploiting a web of semantic data for interpreting tables. In: Proceedings of the Second Web Science Conference, April 2010
17. Taheriyan, M., Knoblock, C.A., Szekely, P., Ambite, J.L.: A scalable approach to learn semantic models of structured sources. In: Proceedings of the 8th IEEE International Conference on Semantic Computing (ICSC 2014) (2014)
18. Tandy, J., Herman, I., Kellogg, G.: Generating RDF from tabular data on the web, W3C Recommendation, December 2015. https://www.w3.org/TR/csv2rdf/
19. Umbrich, J., Neumaier, S., Polleres, A.: Quality assessment & evolution of open data portals. In: IEEE International Conference on Open and Big Data, Rome, Italy, August 2015
20. Venetis, P., Halevy, A.Y., Madhavan, J., Pasca, M., Shen, W., Wu, F., Miao, G., Wu, C.: Recovering semantics of tables on the web. PVLDB **4**(9), 528–538 (2011)
21. Wang, J., Wang, H., Wang, Z., Zhu, K.Q.: Understanding tables on the web. In: Atzeni, P., Cheung, D., Ram, S. (eds.) ER 2012 Main Conference 2012. LNCS, vol. 7532, pp. 141–155. Springer, Heidelberg (2012)
22. Wienand, D., Paulheim, H.: Detecting incorrect numerical data in DBpedia. In: Presutti, V., d'Amato, C., Gandon, F., d'Aquin, M., Staab, S., Tordai, A. (eds.) ESWC 2014. LNCS, vol. 8465, pp. 504–518. Springer, Heidelberg (2014)
23. Zhang, M., Chakrabarti, K.: Infogather+: semantic matching and annotation of numeric and time-varying attributes in web tables. In: Proceedings of the 2013 ACM SIGMOD International Conference on Management of Data, pp. 145–156. ACM, New York (2013)
24. Zhang, Z.: Towards efficient and effective semantic table interpretation. In: Mika, P., et al. (eds.) ISWC 2014, Part I. LNCS, vol. 8796, pp. 487–502. Springer, Heidelberg (2014)

Semantic Labeling:
A Domain-Independent Approach

Minh Pham$^{(\boxtimes)}$, Suresh Alse, Craig A. Knoblock, and Pedro Szekely

University of Southern California, Los Angeles, USA
{minhpham,alse}@usc.edu, {knoblock,pszekely}@isi.edu

Abstract. Semantic labeling is the process of mapping attributes in data sources to classes in an ontology and is a necessary step in heterogeneous data integration. Variations in data formats, attribute names and even ranges of values of data make this a very challenging task. In this paper, we present a novel domain-independent approach to automatic semantic labeling that uses machine learning techniques. Previous approaches use machine learning to learn a model that extracts features related to the data of a domain, which requires the model to be re-trained for every new domain. Our solution uses similarity metrics as features to compare against labeled domain data and learns a matching function to infer the correct semantic labels for data. Since our approach depends on the learned similarity metrics but not the data itself, it is domain-independent and only needs to be trained once to work effectively across multiple domains. In our evaluation, our approach achieves higher accuracy than other approaches, even when the learned models are trained on domains other than the test domain.

1 Introduction

Mapping attributes in data sources to a domain ontology is a necessary step in integrating different sources and mapping them to a domain ontology. The problem, which we call semantic labeling, requires annotating source attributes with classes and properties of ontologies. There has been a number of studies conducted to automate the process since labeling attributes manually is laborious and requires a sufficient amount of domain knowledge. However, automatic semantic labeling is difficult to perform accurately for several reasons. First, people have different ways to represent data of same labels. Table 1 shows different formats that *PlayerPosition* can be found in soccer data. On the other hand, data from different labels can be very similar. For example, data of *NumberOfGoalsScores* and *NumberOfGamesPlayed* in soccer data are very similar because both of them are in numeric format with values ranged mainly from 0 to 50. Therefore, a good semantic labeling approach needs to deal with two different issues: to distinguish similar labels and to recognize the same labels from different data, both of which generally make the problem very hard.

To address these issues, we present a domain-independent machine learning approach for semantic labeling. Our contribution is a novel way of using machine

© Springer International Publishing AG 2016
P. Groth et al. (Eds.): ISWC 2016, Part I, LNCS 9981, pp. 446–462, 2016.
DOI: 10.1007/978-3-319-46523-4_27

Table 1. Different representations of *PlayerPosition*

Code	Abbreviation	Full form
1	GK	Goalkeeper
2	DF	Defender
3	MF	Midfieder
4	FW	Forward

learning to solve semantic labeling as a combination of many binary classification sub-problems. Our machine learning model uses similarity metrics as features and learns a matching function to determine whether attributes have the same labels to infer the correct semantic labels. Because the matching function is not related to specific labels, our model is independent from labels and thus independent from the domain ontologies.

We evaluate our approach on many datasets from different domains. When the machine learning models are trained on another domain, the system achieves an average mean reciprocal rank (MRR) [3] over 80 % on 4 datasets. The results are even better if models are trained on the same domain. We also run experiments on the T2D Gold Standard data and achieve a higher F1-measure compared to the property-matching approach in the T2K system [12].

2 Motivating Example

In this section, we provide an example to explain the problem of mapping source attributes to semantic types in a domain ontology. Suppose that we want to map attributes in a data source named *WC2014* (Table 2), which contains information about players of national teams in World Cup 2014, to the DBpedia ontology. First, we define our target label, which we call *semantic type*, as a pair of values consisting of a domain class and one of its properties *<class, property>*. For example, in Table 2, the correct semantic types of column *player*, *height* and *position* are *<dbo:SoccerPlayer, dbo:birthName>*, *<dbo:SoccerPlayer, dbo:height>* and *<dbo:SoccerPlayer, dbo:draftPosition>*. Semantic labeling systems attempt to automatically identify these mappings. However, this cannot be done without knowing about these semantic types in a domain.

Table 2. Sample data from World Cup 2014 players (WC2014)

Player	Height	Position
Alan PULIDO	176	Forward
Robin VAN PERSIE	186	Forward
Miiko ALBORNOZ	180	Defender
Marouane FELLAINI	194	Midfielder

Table 3. Sample data from England Premier League (EPL)

First name <*SoccerPlayer, birthName*>	Position <*SoccerPlayer, draftPosition*>	Height <*SoccerPlayer, height*>
Hazard, Eden	Midfielder	172
Cahill, Gary	Defender	191
Felliani, Marouane	Midfielder	194
Oezil, Mesut	Midfielder	180

Therefore, the semantic labeling problem refers to a situation where we have already mapped one or more sources to a common ontology and we want to label new sources using the same ontology. For example, we have the data source *EPL* containing information about all England Premier League players and it is already labeled with DBpedia semantic types (Table 3). Since we have information about the DBpedia ontology from the *EPL* source, we can label source attributes of *WC2014* based on this information. There are different ways to leverage domain data from labeled sources for semantic labeling. Previous work uses labeled sources such as *EPL* as training data to learn the characteristic of data in different attributes. Table 4 shows some feature values extracted from <*dbo:SoccerPlayer, dbo:birthName*> data. In our approach, we use *EPL* as our base data and compare attributes in *WC2014* with attributes in *EPL*. If these two attributes are similar such as column *first name* in EPL and column *player* in WC2014, we conclude that they have the same semantic types. Because we know that the semantic type of *first name* is <*dbo:SoccerPlayer, dbo:birthName*>, we infer that the semantic type of *player* is also <*dbo:SoccerPlayer, dbo:birthName*>.

Table 4. Some feature values extracted from <*SoccerPlayer, birthName*>

Feature	Value
All capitalized token	1
Starts with char C	0.25
Num. len.	0

The main difference between our approach and previous work is when faced with unseen semantic types. For example, consider the case where we have another labeled source named *BGL* containing information about players in Germany Bundesliga League (Table 5). *BGL* contains a column *salary* which is labeled as <*dbo:Person, dbo:salary*> - an unseen semantic type. In previous approaches, learned models need to be retrained to capture the data characteristic of <*dbo:Person, dbo:salary*> and this needs to be repeated for every unseen semantic type. There are a huge number of data sources and semantic types, which makes the possibility of facing new semantic types very high and it is time-consuming to retrain the learning models each time. For our approach, we just need to store data with the new semantic types for later comparison with unlabeled attributes.

Table 5. Sample data from Germany Bundesliga League (GBL)

Name <*SoccerPlayer, birthName*>	Salary <*SoccerPlayer, salary*>
Neuer; Manuel	150,000
Boateng; Jerome	90,000
Dante	100,000

3   Approach

In this section, we explain our approach to determine similarities between unlabeled and labeled attributes and use machine learning techniques to find the correct semantic type. Section 3.1 describes various similarity metrics that we use as our features and how we compute them. Section 3.2 describes details of how we use machine learning for semantic labeling problem.

3.1   Similarity Metrics

In our approach, we exploit different similarity metrics that measure how attributes are similar to others. In this section, we describe these similarity metrics and explain how they can help in semantic labeling.

Attribute Name Similarity. In relational databases, web tables or spreadsheets, tabular structures usually have titles for each column. We consider these titles as attribute names and use them to compare similarities between two attributes.

Definition 1. *Given two attributes named a and b, we have A and B as sets of character tri-grams extracted from a and b. The **attribute name similarity** is calculated using Jaccard similarity* [8] *as follows:*

$$S(a, b) = \frac{|A \cap B|}{|A \cup B|} \qquad (1)$$

In data sources, people usually name attributes based on the meaning of the data so that similarity in attribute names provides a good indication of the similarity in semantic types. However, as attribute names usually correspond only to ontology properties, using attribute names as the only metric can lead to false positives in labeling. For example, a column named *name* can refer to <*dbo:Person, dbo:birthName*> or <*dbo:SportsTeam, dbp:clubName*> depending on the sources. Collecting data sources from the web can also result in missing or noisy attribute names, which provide no information about the attributes.

Value Similarity. Value similarity is the most common similarity metric, which is widely used in different matching systems. In semantic labeling, attribute values play an important role in identifying attributes that have the same semantic types because they usually contain similar values. In our approach, we compute three different value similarity metrics: Jaccard similarity and TF-IDF cosine similarity for textual data, as well as a modified version of Jaccard similarity for numeric values.

Definition 2. *Given two attributes named a and b with v_a and v_b as the corresponding sets of values, the* **textual Jaccard similarity** *[8] is computed as follows:*

$$S(a, b) = \frac{|v_a \cap v_b|}{|v_a \cup v_b|} \tag{2}$$

Definition 3. *Given set of attributes $\{a_1, a_2, \ldots, a_n\}$ with a corresponding sets of values $\{v_1, v_2, \ldots, v_n\}$, the* **TF-IDF cosine similarity** *[8] is computed using the following steps:*

1. *We concatenate the values in $\{v_1, v_2, \ldots, v_n\}$ by attribute to generate a set of documents: $\{D_1, D_2, \ldots, D_n\}$*
2. *For a document D_i, we calculate the corresponding TF-IDF vector W_i*
3. *We compute TF-IDF cosine similarity between two attributes a and b:*

$$S(a, b) = \frac{W_a \cdot W_b}{|W_a| \times |W_b|} \tag{3}$$

For numeric attributes, set-based similarity metrics such as Jaccard and cosine similarity do not work effectively because numeric data have continuous ranges of values. Therefore, we customize Jaccard similarity to work with range of values instead of sets of values.

Definition 4. *Given two attributes named a and b with v_a and v_b as the corresponding sets of values, the* **numeric Jaccard similarity** *is computed as follows:*

$$S(a, b) = \frac{min(max(v_a), max(v_b)) - max(min(v_a), min(v_b))}{max(max(v_a), max(v_b)) - min(min(v_a), min(v_b))} \tag{4}$$

For example, the numeric Jaccard similarity s of two attributes with values in range [1912, 1980] and [1940, 2000] is computed as follows:

$$s = \frac{1980 - 1940}{2000 - 1912} = 0.45. \tag{5}$$

To reduce sensitivity to outliers, we only use the subsets containing values from first quartile to third quartile instead of the whole set of values in attributes.

Distribution Similarity. For numeric data, there are semantic types that we are unable to distinguish by using value similarity because they have the same range of values. However, since they have different underlying meanings, their distribution of values may be different. For example, consider the example about *NumberOfGoalsScored* and *NumberOfGamesPlayed* in Sect. 1. Although they have the same range of values, *NumberOfGoalsScored* has skewed distribution because the high values are mostly distributed to *Forwards* and *Midfielders* while *NumberOfGamesPlayed* is more likely to follow a near-uniform distribution.

Therefore, we analyze the distribution of numeric values contained in the attributes using statistical hypothesis testing as one of the similarity metrics. For statistical hypothesis testing used in our approach, the null hypothesis is that the two sets of values are drawn from a same population (distribution), which may indicate that they come from a same semantic type. We use Kolmogorov-Smirnov test (KS test)[6] as our statistical hypothesis test based on evaluation of different statistical tests in Ramnandan et al.'s research [11].

Histogram Similarity. For textual data, normal statistical hypothesis testing cannot be applied because there is no order in textual values. Moreover, we cannot use traditional correlation methods such as mutual information or KL-divergence since we are comparing attributes that do not appear in the same source. Therefore, we calculate value histograms in textual attributes and compare their histograms instead. The statistical hypothesis tests used for the histogram case is the Mann-Whitney test (MW test) [6]. The reason we use MW test instead of KS test is that histograms are not ordinal and using methods that compare two empirical value distributions such as KS test is not suitable. Mann-Whitney test computes distribution distances based on medians and, thus, is more appropriate to use for histograms.

When comparing a textual attribute with a numeric attribute, we also transform numeric data into histogram form and use MW test to compute histogram similarity. For the example of *PlayerPosition* in Table 1, even though they have different representations, they have similar histogram forms because every position usually have similar frequencies over the different sources of data. For instance, because every soccer team usually has 1 goalkeeper, 4 defenders, 4 midfielders and 2 forwards, the histogram frequencies are likely to be $[\frac{1}{11}, \frac{4}{11}, \frac{4}{11}, \frac{2}{11}]$.

Mixtures of Numeric and Textual Data. As we have described above, there are similarity measures that can only applied for the textual part of attribute values while some others only work on numeric parts (Table 6). Because textual similarity metrics are more important when comparing attributes with mostly text and numeric similarities are more important for attributes with numeric data, we need to adjust the values of these similarity measures based on the fraction of textual and numeric values contained in attributes.

Table 6. Similarity feature vector

Feature name	Explanantion	Applied data types
ATT NAME	Jaccard similarity for attribute names	All
TEXT JACCARD	Jaccard similarity for textual data	Textual
TF-IDF COSINE	TF-IDF cosine similarity for textual data	Textual
NUM JACCARD	Modified Jaccard similarity for numeric data	Numeric
NUM KS	KS statistical test for numeric data	Numeric
MW HISTOGRAM	MW test for histogram	All

Given r_1 and r_2 are fractions of textual data in the pair of attributes, the adjusted value of textual similarity value is computed as follows:

$$v_{adjusted} = \frac{(r_1 * r_2)}{(r_1 + r_2)} * v_{original} \qquad (6)$$

On the other hand, the adjusted value of numeric similarity value is computed as follows:

$$v_{adjusted} = \frac{[(1 - r_1) * (1 - r_2)]}{[(1 - r_1) + (1 - r_2)]} * v_{original} \qquad (7)$$

The adjusted value is the product of the harmonic mean over r_1, r_2 and the original value. The reason for using harmonic mean follows the intuition that the corresponding similarity values are more reliable when two attributes have similar fractions of textual data or numeric data and vice versa.

3.2   Semantic Labeling

The overall framework is illustrated in Fig. 1. The input of our system is an unlabeled attribute and a set of labeled attributes as domain data and the output is a set of top-k semantic types corresponding to the unlabeled attribute.

Overall Approach. Given a set of attributes $\{a_1, a_2, \ldots a_n\}$, we compute M-dimensional feature vectors f_{ij} $(i \neq j)$. Each dimension k corresponds to a similarity metric, so $f[k]$ represents how similar attributes a_i and a_j are under metric k.

During training we label each f_{ij} as True/False, where True means that attributes a_i and a_j have the same semantic type and vice versa. To set up a new domain, we store a set of labeled attributes $\{a_1, a_2, \ldots a_n\}$ as domain data and use them to compare against new attributes to infer the semantic types.

Given a new attribute a_0, the algorithm computes f_{0j} for all j $(j \neq 0)$, and uses the learned classifier to label each f_{0j} as True/False. If the label of f_{0j} is yes, the algorithm says that the semantic type of a_0 is the semantic type that was recorded for a_j. From that, we can conclude the semantic type of a_0.

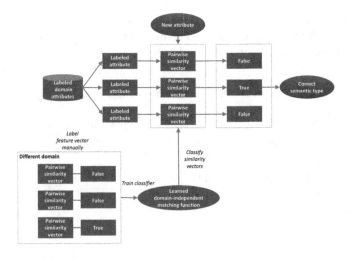

Fig. 1. Overall framework of our semantic labeling system

Previous approaches, tried to predict the semantic label of a_0 based on characteristic of recorded a_i. In contrast, our approach learns a classifier over similarity vectors. It is domain-independent because classification does not depend on the values in attributes, but rather on the similarity scores of multiple metrics between the attributes.

Since there are no constraint on the number of True labels for each attribute, we develop a ranking method and only take the top-k results of semantic types. The ranking algorithm uses the predicted probabilities of the True class in classification as the confidence scores and ranks the candidate semantic types based on that.

Classifiers for Semantic Labeling. To choose the best classifier for semantic labeling, we ran experiments on various of classifiers and compare the results. Because we use class probabilities of classifiers as confidence scores, classifiers need to have class probabilities calculated from the feature vector in order to be applicable. Therefore, we only consider Logistic Regression and Random Forests [2]. Details of the experiments are described in Sect. 4.2. According to the results from Tables 8, 9 and 10, Logistic Regression achieves the best performance and thus is the selected classifier in our system.

4   Evaluation

In our experiments, we use four different datasets: city [11], weather [1], museum [14], and soccer. The soccer data set was created to provide a wide variety of semantic types and consists of numerous real-world data sets about soccer. The purpose of using many datasets from different domains is to evaluate our classifiers when applying a single learned classifier to multiple domains. Table 7 shows the overall information about these data sets. The datasets and code used in our experiments have been published online[1].

Table 7. Data sets from different domains in experiments

Data set	No. sources	No. semantic types	No. attributes
Museum	29	20	217
City	10	52	520
Soccer	12	14	97
Weather	4	11	44

4.1 Experimental Setup

In this section, we evaluate the performance of our system, which is called DSL (Domain-independent Semantic Labeler). The evaluation metric that we measure is the mean reciprocal rank (MRR) [3]. The details of the experimental setup is as follows:

1. Choose a labeling dataset A.
2. Suppose A consists of n sources $\{s_1, s_2, \ldots, s_n\}$, choose the number of labeled sources m in the dataset $(m < n)$.
3. For every source s_i in A, perform semantic labeling using m labeled sources from s_{i+1} to s_{m+i+1}.

For example, the soccer dataset has 12 sources. If we have one labeled source, we label s_1 with labeled data from s_2, label s_2 with labeled data from s_3 and so on. Likewise, if we have five labeled sources, we label s_1 with labeled data in set of sources $s_2, s_3, \ldots, s_6$ and continue through the entire data set.

For classifier training data, we follow the same process as above but we manually label the computed feature vectors generated instead of running semantic labeling. To assure that classifier training data is disjoint from labeling data, if labeling dataset and training dataset are the same, we choose distinct labeled sources for each process.

4.2 Classifier Analysis

In this experiment, we evaluate 2 classifiers: Logistic Regression and Random Forests to choose the best classifier for semantic labeling.

Tables 8, 9 and 10 lists results of two classifiers when being trained and tested on different datasets. We use city, museum and soccer datasets to train Logistic

---

[1] https://github.com/minhptx/iswc-2016-semantic-labeling.git.

Table 8. MRR scores of different classifiers when training on soccer

	Soccer	Museum	City	Weather
Logistic Regression	0.814	0.863	0.944	0.951
Random Forests	0.794	0.799	0.947	0.86

Table 9. MRR scores of different classifiers when training on museum

	Soccer	Museum	City	Weather
Logistic Regression	0.815	0.845	0.940	0.951
Random Forests	0.820	0.778	0.830	0.898

Table 10. MRR scores of different classifiers when training on city

	Soccer	Museum	City	Weather
Logistic Regression	0.782	0.807	0.965	0.955
Random Forests	0.802	0.728	0.912	0.807

Regression and Random Forest since we can generate a sufficient amount of samples for training data. For semantic labeling, we use all 4 datasets: soccer, museum, city and weather with the numbers of labeled sources is 50 % of the total numbers of sources in these datasets.

Overall, Logistic Regression achieves a comparable performance to Random Forests, which is a surprising result, because Random Forests have been shown to be the better classifiers in other research. However, because of the issue where we need to use class probabilities as confidence scores, the results can be explained.

Logistic Regression class probabilities are computed using the following function:

$$P(y = 1|x) = sigmoid(w^T x) \tag{8}$$

where x is the feature vector and w are its coefficients. Because $P(y = 1|x)$ is a monotonically increasing function of $w^T x$, $P(y = 1|x)$ increases when $w^T x$ increases. Thus, feature vectors with higher similarity values have higher class probabilities in Logistic Regression models.

Random Forests, on the other hand, calculate class probabilities based on fraction of samples of the same class in decision tree leaves. As long as the values are higher than splitting values in decision trees, feature vectors are split to the same branches and are likely to receive similar class probabilities. Therefore, using class probabilities of Random Forests as confidence scores performs worse.

Since the labeling accuracy of Logistic Regression and Random Forests are comparable, we consider the training time and labeling time of each classifier as additional measurements. Table 11 lists average system training time and labeling time of these classifiers.

Table 11. Training and labeling time of different classifiers

	Training time	Labeling time
Logistic Regression	144 s	0.31 s
Random Forests	157 s	0.36 s

The results in Table 11 show that Logistic Regression has a smaller training and labeling time. Although the differences are minor, it provides an advantage, especially in real-world scenarios with large amounts of data. Using Logistic Regression also provides more meaningful insights of features because of its linear combination compared with a randomized algorithm as Random Forests. Therefore, we use Logistic Regression as the classifier for the remaining experiments.

4.3   Feature Analysis

In machine learning classifiers, different features have different degrees of influence on the classification results. To analyze the importance of features in our similarity vectors, we train Logistic Regression on different datasets and extract coefficients of features. Table 12 shows coefficients of features when Logistic Regression models are trained on city, museum and soccer data.

Table 12. Coefficients of features in Logistic Regression classifier

Feature	Train on soccer	Train on museum	Train on city
ATT NAME	4.41	6.08	0
TEXT JACCARD	1.88	0.88	9.16
TEXT TF-IDF	3.91	1.03	3.20
NUM JACCARD	4.21	3.28	12.68
NUM KS	1.78	0.78	7.25
MW HISTOGRAM	0.32	1.14	3.83

In general, all of our similarity features have positive correlation with the classification results, which means that higher values in these similarity metrics results in higher probabilities that the attributes have the same semantic type. As we can see from the results, value similarity features play the most important role in Logistic Regression classifiers regardless of training domain. Attribute names similarity has a good impact on soccer and museum data but not in city because city dataset does not have headers or titles for attributes. Distribution and histogram similarity metrics have higher coefficients in city data because city dataset contains mostly numeric attributes.

In conclusion, we have demonstrated that our similarity features contribute to the similarity in the semantic types of attributes. However, the importance of features in the learned classifiers can vary according to the training data as shown in Table 12.

4.4   Semantic Labeling

In this experiment, we evaluate performance of DSL (Domain-independent Semantic Labeler) in comparison with SemanticTyper [11]. To follow real-world scenarios where labeled sources are hard to find and manually labeling sources is tedious, our experiments run on configuration with only 1 to 5 labeled data sources for every dataset (Weather dataset has only 4 sources so the maximum number of labeled sources is 3). For DSL, we follow the setup in Sect. 4.1 while having soccer, city and museum as our classifier training dataset iteratively. For SemanticTyper, the MRR scores reported are the MRR scores when being trained on the testing domains. The weather domain is only used in semantic labeling because it cannot provide a sufficient number of feature vectors for training classifiers.

The results in Tables 13, 14, 15 and 16 show that our approach outperforms SemanticTyper in all four evaluation datasets. Although there are slight changes in performance when the classifiers are trained on different domains, the changes are not significantly different and it shows that our approach is robust across multiple data datasets. According to the table, training the classifier from the same domain, which provides more information about the characteristic of data in domains, slightly improves the accuracy of the classifier.

Table 13. MRR scores of DSL and SemanticTyper on soccer dataset

Number of labeled sources	1	2	3	4	5
DSL (train on soccer)	0.625	0.782	0.777	0.800	0.815
DSL (train on city)	0.601	0.785	0.788	0.808	0.820
DSL (train on museum)	0.600	0.781	0.788	0.808	0.810
SemanticTyper	0.608	0.711	0.720	0.720	0.732

Table 14. MRR scores of DSL and SemanticTyper on museum dataset

Number of labeled sources	1	2	3	4	5
DSL (trained on soccer)	0.471	0.665	0.719	0.755	0.790
DSL (trained on museum)	0.463	0.652	0.709	0.752	0.792
DSL (trained on city)	0.472	0.659	0.706	0.713	0.730
SemanticTyper	0.491	0.615	0.656	0.699	0.697

Table 15. MRR scores of DSL and SemanticTyper on city dataset

Number of labeled sources	1	2	3	4	5
DSL (trained on soccer)	0.913	0.932	0.932	0.941	0.945
DSL (trained on museum)	0.912	0.927	0.928	0.941	0.944
DSL (trained on city)	0.914	0.928	0.930	0.939	0.944
SemanticTyper	0.856	0.893	0.893	0.913	0.919

Table 16. MRR scores of DSL and SemanticTyper on weather dataset

Number of labeled sources	1	2	3
DSL (trained on soccer)	0.899	0.951	0.977
DSL (trained on museum)	0.899	0.951	0.977
DSL (trained on city)	0.902	0.955	0.977
SemanticTyper	0.852	0.920	0.955

We also evaluate our system on the T2D Gold Standard dataset[2] and compare our result with T2K system's approach for properties matching [12]. As described in Ritze's work, labeled sources are extracted from the DBpedia ontology. After that, they divided the T2D Gold Standard dataset into two equal-sized parts: an optimization set and an evaluation set. The optimization set is used to optimize the essential parameters for the system and the result are evaluated on the evaluation set. Although we are unable to reconstruct the exact experiment, we approximated the result by using the following configuration as an alternative:

1. Collect DBpedia ontology data in table format as labeled sources.
2. For every attribute in the ontology, extract only 1000 first values as the set of values for the attribute.
3. Train the classifiers on combination of soccer, museum and city datasets to enrich the training data.
4. Test semantic labeling (properties matching) on the entire T2D Gold Standard dataset.

Table 17. MRR scores of DSL and T2K on T2D Gold Standard dataset

DSL	T2K (evaluation)	T2K (optimization)
0.773	0.730	0.700

Table 17 shows the results of DSL in comparisons with T2K. Although our approach is not optimized on the optimization set as T2K, we achieve a better

---

[2] http://webdatacommons.org/webtables/goldstandard.html.

accuracy on the dataset. Moreover, our classifiers have been trained on different domains and we only use 1000 values for every attribute as domain data instead of the entire set of values. Because we exploit more similarity features, our approach achieves better discriminative ability for the various semantic types. The evaluation also shows that we have a robust, domain-independent system that only needs to be trained once before using it for semantic labeling in a wide range of domains.

5 Related Work

Ramnandan et al. [11] describe an approach that captures and compares distributions and properties of data corresponding to semantic types as a whole. They apply heuristic rules to separate numeric and textual data and then use TF-IDF and KS as measures to compare the data. In our approach, we use more similarity features besides of TF-IDF and KS test, which enables our system to better discriminate between semantic types. Our similarity metrics can be applied to both textual and numeric attributes by the method described in Sect. 3.1.

Ritze et al. [12] propose a new approach for annotating HTML tables with DBpedia classes, properties, and entities. Their system, which is named T2K, use metrics like Jaccard, Levenshtein and deviation similarity to match attributes to properties and values. T2K also uses a iterative process to adjust property weights and filter the candidate sets until the similarity values converge. The system provides good results in entity and class matching but not in property matching. Since they exploit only value similarity for textual data and numeric similarity for numeric data for property matching, they face the same limitation as Ramnandan's work and achieve a lower performance compared to DSL.

A number of approaches have used probabilistic graphical models to solve the problem of semantic labeling. Goel et al. [4] exploit the underlying relationships between attributes and values with attribute characteristics as features and use Conditional Random Fields (CRF) to label attributes. They assign semantic types to every value in an attribute and then combine these semantic types to infer the semantic type for the whole attribute. Limaye et al. [7] use probabilistic graphical models in a broader problem as they annotate tables on the web by entities for cells, types for columns, and relationships for binary relations between columns. They exploit two feature functions that describe the dependency of column type with its values and header. The labels of all columns are then assigned simultaneously using a message passing algorithm to maximize the potential function formulated by features and their weights. Mulwad et al. [9] extend the work of Limaye et al. by proposing a novel *Semantic Message Passing* algorithm that uses Linked Open Data (LOD) knowledge to improve the existing semantic message algorithm. These approaches require the probabilistic graphical models to be retrained when handling new semantic types. The reason for this is that their feature weights are calculated associated with labels and need to be re-estimated for new semantic types. Also, graphical models do not scale well as the number of semantic types increases because of the explosion of different enumerations in the search space.

Mulwad [10] also extend their work into a full system with multiple functions. They incorporated probabilistic semantic labeling with domain knowledge processing and data cleaning to produce a domain-independent semantic labeling system. However, their domain independence is limited in that it requires users to provide domain knowledge or apply preprocessing modules. In our semantic labeling system, the process is automatic and the domain-independent learning models only require a small amount of domain data to perform well on semantic labeling.

Venetis et al. [15] present an approach to annotate tables on the web by leveraging existing data on the web. An isA database in the form of {instance, class} is extracted from the web using linguistic patterns and is used to produce column labels. The column labels are assigned by a maximum likelihood estimator that assigns a column with a class label that maximize the fraction of column values in that label. Syed et al. [13] use Semantic Web data to infer the semantic models of tables. They annotate the table columns by using the column names if available and values inside the columns to build a query to Wikitology. After that, columns are mapped to classes returned in the query result. Both the work of Venetis et al. and Syed et al. extract a huge amount of data from various sources to estimate the probability that a value belongs to a semantic type. Thus, their approach is restricted to domains where online data is widely available. In our approach, our learning model is not domain-specific and thus, we can use any domain as our training data and the system can still label data from other domains effectively.

Gunaratna et al. [5] address a related problem, which is called entity class resolution. Entity class resolution is similar to semantic labeling except that their targets are entity classes instead of semantic types. Their system, FACES, applies natural language processing (NLP) techniques to identify focus terms and uses text similarities to compare focus terms with entity class names in the ontology. Although FACES' approach works well in text documents because it is easy to detect focus terms in grammatical documents, it cannot be applied to most of web data such web tables, spreadsheets, or RDF stores because data values are mostly unstructured and do not follow grammar rules such as numbers and named entity mentions. In contrast, our approach does not rely on NLP algorithms so that it can perform effectively in noisy data sources from the web.

6   Conclusion and Future Work

In this paper, we presented a novel domain-independent approach for semantic labeling that leverages similarity measures and machine learning techniques. In our system, we capture the patterns of matching decisions given the similarity scores between unlabeled attributes and labeled data to find the correct semantic types. The approach allows us to train the machine learning model only once and use it in multiple different domains. Moreover, our similarity features are independent within a semantic type and across other semantic types. We can compute feature vectors using a parallel and distributed implementation which reduces the running time while maintaining labeling accuracy.

In the future, we plan to exploit transfer learning to incorporate some specific information about the domain data to adjust the weights of our features. For example, if a domain contains mostly numeric data, we may give more weight to numeric features. In view of the machine learning model, we can leverage data from Linked Open Data to enrich our learning models. In this way, the model can have information about many difficult cases and, therefore, it will be more likely to generalize well. Finally, although our approach allows new semantic types to be easily integrated, it lacks the ability to detect whether the true semantic type exists in the labeled data. This inability can lead to incorrect mappings in unseen cases and decrease the overall system accuracy. One of the directions of future work is to have the machine learning model detect these cases.

Acknowledgments. This material is based upon work supported in part by the United States Air Force and the Defense Advanced Research Projects Agency (DARPA) under Contract No. FA8750-16-C-0045. Any opinions, findings and conclusions or recommendations expressed in this material are those of the author(s) and do not necessarily reflect the views of the United States Air Force and DARPA.

References

1. Ambite, J.L., Darbha, S., Goel, A., Knoblock, C.A., Lerman, K., Parundekar, R., Russ, T.: Automatically constructing semantic web services from online sources. In: Bernstein, A., Karger, D.R., Heath, T., Feigenbaum, L., Maynard, D., Motta, E., Thirunarayan, K. (eds.) ISWC 2009. LNCS, vol. 5823, pp. 17–32. Springer, Heidelberg (2009)
2. Breiman, L.: Random forests. Mach. Learn. **45**, 5–32 (2001)
3. Craswell, N.: Mean reciprocal rank. In: Liu, L., Özsu, M.T. (eds.) Encyclopedia of Database Systems, p. 1703. Springer, Heidelberg (2009)
4. Goel, A., Knoblock, C.A., Lerman, K.: Exploiting structure within data for accurate labeling using conditional random fields. In: Proceedings of the 14th International Conference on Artificial Intelligence (ICAI), vol. 69 (2012)
5. Gunaratna, K., Thirunarayan, K., Sheth, A., Cheng, G.: Gleaning types for literals in RDF triples with application to entity summarization. In: Sack, H., Blomqvist, E., d'Aquin, M., Ghidini, C., Ponzetto, S.P., Lange, C. (eds.) ESWC 2016. LNCS, vol. 9678, pp. 85–100. Springer, Heidelberg (2016). doi:10.1007/978-3-319-34129-3_6
6. Lehmann, E.L., Romano, J.P.: Testing Statistical Hypotheses. (Springer Texts in Statistics). Springer, New York (2005)
7. Limaye, G., Sarawagi, S., Chakrabarti, S.: Annotating and searching web tables using entities, types and relationships. Proc. VLDB Endow. **3**, 1338–1347 (2010)
8. Manning, C.D., Raghavan, P., Schtze, H.: Introduction to Information Retrieval. Cambridge University Press, New York (2008)
9. Mulwad, V., Finin, T., Joshi, A.: Semantic message passing for generating linked data from tables. In: Alani, H., et al. (eds.) ISWC 2013, Part I. LNCS, vol. 8218, pp. 363–378. Springer, Heidelberg (2013)
10. Mulwad, V.V.: TABEL - a domain independent and extensible framework for inferring the semantics of tables. Ph.D. thesis, University of Maryland, Baltimore County (2015)

11. Ramnandan, S.K., Mittal, A., Knoblock, C.A., Szekely, P.: Assigning semantic labels to data sources. In: Gandon, F., Sabou, M., Sack, H., d'Amato, C., Cudré-Mauroux, P., Zimmermann, A. (eds.) ESWC 2015. LNCS, vol. 9088, pp. 403–417. Springer, Heidelberg (2015)
12. Ritze, D., Lehmberg, O., Bizer, C.: Matching HTML tables to DBpedia. In: Proceedings of the 5th International Conference on Web Intelligence, Mining and Semantics, WIMS 2015, pp. 10:1–10:6. ACM, New York (2015)
13. Syed, Z., Finin, T., Mulwad, V., Joshi, A.: Exploiting a web of semantic data for interpreting tables. In: Proceedings of the Second Web Science Conference (2010)
14. Taheriyan, M., Knoblock, C.A., Szekely, P., Ambite, J.L.: Learning the semantics of structured data sources. Web Semant.: Sci. Serv. Agents World Wide Web **37**, 152–169 (2016)
15. Venetis, P., Halevy, A., Madhavan, J., Paca, M., Shen, W., Wu, F., Miao, G., Wu, C.: Recovering semantics of tables on the web. Proc. VLDB Endow. **4**, 528–538 (2011)

Exploiting Emergent Schemas to Make RDF Systems More Efficient

Minh-Duc Pham[(✉)] and Peter Boncz

CWI, Amsterdam, The Netherlands
{duc,boncz}@cwi.nl

Abstract. We build on our earlier finding that more than 95 % of the triples in actual RDF triple graphs have a remarkably tabular structure, whose schema does not necessarily follow from explicit metadata such as ontologies, but for which an RDF store can automatically derive by looking at the data using so-called "emergent schema" detection techniques. In this paper we investigate how computers and in particular RDF stores can take advantage from this emergent schema to more compactly store RDF data and more efficiently optimize and execute SPARQL queries. To this end, we contribute techniques for efficient emergent schema aware RDF storage and new query operator algorithms for emergent schema aware scans and joins. In all, these techniques allow RDF schema processors fully catch up with relational database techniques in terms of rich physical database design options and efficiency, without requiring a rigid upfront schema structure definition.

1 Emergent Schema Introduction

In previous work [15], we introduced *emergent schemas*: finding that >95 % of triples in all LOD datasets we tested, including noisy data such as Web-Data Commons and DBpedia, conform to a small relational tabular schema. We provided techniques to automatically and at little computational cost find this "emergent" schema, and also to give the found columns, tables, and "foreign key" relationships between them short *human-readable labels*. This label-finding, and in fact the whole process of emergent schema detection, exploits not only value distributions and connection patterns between the triples, but also additional clues provided by RDF ontologies and vocabularies.

A significant insight from that paper is that relational and semantic practitioners give different meanings to the word "schema". It is thus a misfortune that these two communities are often distinguished from each other by their different attitude to this ambiguous concept of "schema" – the semantic approach supposedly requiring no upfront schema ("schema-last") as opposed to relational databases only working with a rigid upfront schema ("schema-first").

Semantic schemas, primarily ontologies and vocabularies, aim at modeling a knowledge universe in order to allow diverse current and future users to denote these concepts in a universally understood way in many different contexts. Relational database schemas, on the other hand, model the structure of one particular

© Springer International Publishing AG 2016
P. Groth et al. (Eds.): ISWC 2016, Part I, LNCS 9981, pp. 463–479, 2016.
DOI: 10.1007/978-3-319-46523-4_28

dataset (i.e., a database), and are not designed with a purpose of re-use in different contexts. Both purposes are useful: relational database systems would be easier to integrate with each other if the semantics of a table, a column and even individual primary key values (URIs) would be well-defined and exchangeable. Semantic data applications would benefit from knowledge of the actual patterns of co-occurring triples in the LOD dataset one tries to query, e.g. allowing users to more easily formulate SPARQL queries with a non-empty result (this often results from using a non-occurring property in a triple pattern).

In [15], we observed *partial* and *mixed* usage of ontology classes across LOD datasets: even if there is an ontology closely related to the data, only a small part of its class attributes actually occur as triple properties (partial use), and typically many of the occurring attributes come from different ontologies (mixed use). DBpedia on average populates <30 % of the class attributes it defines [15], and each actually occurring class contains attributes imported from no less than 7 other ontologies on average. This is not necessarily bad design, rather good re-use (e.g. foaf), but it underlines the point that any single ontology class is a poor descriptor of the actual structure of the data (i.e., a "relational" schema). Emergent schemas are helpful for human RDF users, but in this paper, we investigate how RDF stores can exploit emergent schemas for efficiency.

We address three important problems faced by RDF stores. The first and foremost problem is the high execution cost resulting from the large amount of self-joins that the typical SPARQL processor (based on some form of triple table storage) must perform: one join per additional triple pattern in the query. It has been noted [7] that SPARQL queries very often contain star-patterns (triple patterns that share a common subject variable), and if the properties of the patterns in these stars reference attributes from the same "table", the equivalent relational query can be solved with a table scan, not requiring *any* join. Our work achieves the same reduction of the amount of joins for SPARQL.

The second problem we solve is the low quality of SPARQL query optimization. Query optimization complexity is *exponential* in the amount of joins [17]. In queries with more than 12 joins or so, optimizers cannot analyze the full search space anymore, potentially missing the best plan. Note that SPARQL query plans typically have F times more joins than equivalent SQL plans. Here F is the average size of a star pattern[1]. This leads to a 3^F times larger search space. Additionally, query optimizers depend on cost models for comparing the quality of query plan candidates, and these cost models assume independence of (join) predicates. In case of star patterns on "tables", however, the selectivity of the predicates is heavily correlated (e.g. subjects that have an ISBN property, typically instances of the class Book, have a much higher join hit ratio with AuthoredBy triples than the *independence assumption* would lead to predict) which means that the cost model is often wrong. Taken together, this causes the quality of SPARQL query optimization to be significantly lower than in SQL.

---

[1] A query of X stars has $X \times F$ triple patterns, so needs $P_1 = X \times F - 1$ joins. When each star is collapsed into one tablescan, just $P_2 = (X - 1)$ joins remain: $\frac{P_1}{P_2} \geq F$ times.

Our work eliminates many joins, making query optimization exponentially easier, and eliminates the biggest source of correlations that disturb cost modeling (joins between attributes from the same table).

The third problem we address is that mission-critical applications that depend on database performance can be optimized by database administrators using a plethora of physical design options in relational systems, yet RDF system administrators lack all of this. A simple example are *clustered indexes* that store a table with many attributes in the value order of one or more sort key attributes. For instance, in a data warehouse one may store sales records ordered by Region first and ProductType second – since this accelerates queries that select on a particular product or region. Please note that not only the Region and Product-Type properties are stored in this order, but *all* attributes of the sales table, which are typically retrieved together in queries (i.e. via a star pattern). A similar relational physical design optimization is table-partitioning or even database cracking [9]. Up until this paper, one cannot even think of the RDF equivalent of these, as table clustering and partitioning implies an understanding of the structure of an RDF graph. Emergent schemas allow to leave the "pile of triples" quagmire, so one can enter structured data management territory where advanced physical design techniques become applicable.

In all, we believe our work brings RDF datastores on par with SQL stores in terms of performance, without losing any of the flexibility offered by the RDF model, thus without introducing a need to create upfront or enforce subsequently any explicit relational schema.

2   Emergent Schema Aware RDF Storage

The original emergent schema work allows to store and query RDF data with SQL systems, but in that case the SQL query answers account for only those "regular" triples that fit in the relational tables. In this work, our target is to answer SPARQL queries over 100 % of the triples correctly, but still improve the efficiency of SPARQL systems by exploiting the emergent schema.

RDF systems store triple tables T in multiple orders of Subject (S), Property (P) and Object (O), among which typically T_{PSO} ("column-wise"), T_{SPO} ("row-wise") and either T_{OSP} or T_{OPS} ("value-indexed") – or even all permutations.[2]

In our proposal, RDF systems storage should become emergent schema aware by only changing the T_{PSO} representation. Instead of having a single T_{PSO} triple table, it gets stored as a set of wide relational tables in a *column*-store – we use MonetDB here. These tables represent only the regular triples, the remaining <5 % of "exception" triples that do not fit the schema (or were updated recently) remain in a smaller PSO table T_{pso}. Thus, T_{PSO} is replaced by the union of a smaller T_{pso} table and a set of relational tables.

---

[2] To support named RDF graphs, the triples are usually extended to quads. Our approach trivially extends to that but we discuss triple storage here for brevity.

Relational storage of triple data has been proposed before (e.g. property tables [20]), though these prior approaches advocated an explicit and human-controlled mapping to a relational schema, rather than a transparent, adaptive and automatic approach, as we do. While such relational RDF approaches have performance advantages, they remained vulnerable in case SPARQL queries *do not* consist mainly of star patterns and in particular when they have triple patterns where the P is a variable. This would mean that many, if not all, relational tables could contribute to a query result, leading to huge generated SQL queries which bring the underlying SQL technology to its knees.

Our proposal hides relational storage behind T_{PSO}, and has as advantage that SPARQL query execution can always fall back on existing mechanisms – typically MergeJoins between scans of T_{SPO}, T_{PSO} and T_{OPS}. Our approach at no loss of flexibility, just makes T_{PSO} storage more compact as we will discuss here, and creates opportunities for better handling of star patterns, both in query optimization and query execution, as discussed in the following sections.

Formal Definition. Given the RDF triple dataset $\Delta = \{t|t = (t_S, t_P, t_O)\}$, an emergent schema $(\Delta, \mathcal{E}, \mu)$ specifies the set $\mathcal{E}$ of emergent tables T_k, and mapping μ from triples in Δ to emergent tables in $\mathcal{E}$. A common idea we apply is rather than storing URIs as some kind of string, to represent them as an OID (object identifier) – in practice as a large 64-bit integer. The RDF system maintains a dictionary $\mathcal{D} : OID \rightarrow URI$ elsewhere. We use this $\mathcal{D}$ dictionary creatively, adapting it to the emergent schema.

Definition 1. *Emergent tables ($\mathcal{E} = \{T_1, ..\}$): Let $s, p_1, p_2, \ldots, p_n$ be subject and properties with associated data types OID and $D_1, D_2, \ldots, D_n$, then $T_k = (T_k.s:OID, T_k.p_1:D_1, T_k.p_2:D_2, \ldots, T_k.p_n:D_n)$ is an emergent table where $T_k.p_j$ is a column corresponding to the property p_j and $T_k.s$ is the subject column.*

Definition 2. *Dense subject columns: $T_k.s$ consists of densely ascending numeric values $\beta_k, .. \beta_k + |T_k| - 1$, so s is something like an array index, and we denote $T_k[s].p$ as the cell of row s and column p. For each T_k its base OID $\beta_k = k * 2^{40}$. By choosing β_k to be sufficiently apart, in practice the values of column $T_i.s$ and $T_j.s$ never overlap when $i \neq j$.[3]*

Definition 3. *Triple-Table mapping ($\mu : \Delta \rightarrow \mathcal{E}$): For each table cell $T_k[s].p_j$ with non-NULL value o, $\exists (s, p_j, o) \in \Delta$ and $\mu(s, p_j, o) = T_k$. These triples we call "regular" triples. All other triples $t \in \Delta$ are called "exception" triples and $\mu(t) = T_{pso}$. In fact T_{pso} is exactly the collection of these exception triples.*

The emergent schema detection algorithm [15] assigns each subject to at most 1 emergent table – our storage exploits this by manipulating the URI dictionary $\mathcal{D}$ so that it gives dense numbers to all subjects s assigned to the same T_k.

---

[3] In our current implementation with 64-bit OIDs we thus can support up to 2^{16} emergent tables with each up to $2^{40} = 1$ trillion subjects, still leaving the highest 8 bits free, which are used for type information – see footnote 4.

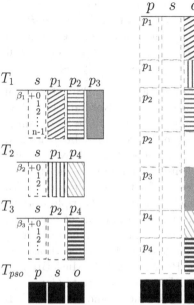

T_1 s p_1 p_2 p_3

T_2 s p_1 p_4

T_3 s p_2 p_4

T_{pso} p s o

		Except%	Null%	Compr
Synthetic RDF datasets				
LUBM		0.0%	6.0%	1.8x
BSBM		0.0%	4.2%	2.5x
SP2Bench		0.4%	5.2%	2.0x
LDBC SNB		0.0%	12.2%	2.0x
RDF datasets with Relational Roots				
MusicBrz		0.4%	3.9%	2.2x
Eurostat		0.5%	3.8%	1.4x
DBLP		0.4%	12.6%	1.7x
PubMed		0.3%	15.3%	1.9x
Native RDF datasets				
WebData		7.5%	42.7%	1.4x
DBpedia		3.8%	32.2%	1.4x

Fig. 1. Columnar storage of emergent tables T_k and exception table T_{pso}

Fig. 2. PSO as view $P_{PSO} \cup T_{pso}$

Fig. 3. Exception percentage, NULL percentage and Compression Factor achieved by Emergent Table-aware PSO storage, over normal PSO storage

Columnar Relational Storage. On the physical level of bytes stored on disk, columnar databases can be thought of as storing all data of one column consecutively. Column-wise data generally compresses better than row-wise data because data from the same distribution appears consecutively, and column-stores exploit this by having advanced data compression methods built-in in their storage and query execution infrastructure. In particular, the dense property of the columns $T_k.s$ will cause column-stores to compress it down to virtually nothing, using a combination of delta encoding (the difference between subsequent values is always 1) and run-length encoding (RLE), encoding these subsequent 1's in just a single run. Our evaluation platform MonetDB supports densely ascending OIDs natively with its VOID (virtual OID) type, that requires no storage.

Figure 1 shows an example of representing RDF triples using the emergent tables $\{T_1, T_2, T_3\}$ and the triple table of exception data T_{pso} (in black, below). We have drawn the subject columns $T_k.s$ transparent and with dotted lines to indicate that there is no physical storage needed for them.

For each individual property column $T_k.p_j$, we can define a triple table view $P_{j,k} = (p_j, T_k.s, T_k.o)$, the first column being a constant value (p_j) which thanks to RLE compression requires negligible storage and the other two reusing storage from emergent table T_k. If we concatenate these views $P_{j,k}$ ordered by j and k, we obtain table $P_{PSO} = \cup_{j,k} P_{j,k}$. This P_{PSO} is shown in Fig. 2. Note that P_{PSO} is simply a re-arrangement of the columns $T_k.p_j$. Thus, with emergent schema aware storage, one can always access the data P_{PSO} as if it were a PSO table at no additional cost.[4] In the following, we show this cost is actually less.

Space Usage Analysis. P_{PSO} storage is more efficient than PSO storage in an efficient columnar RDF store such as Virtuoso would be. Normally in a PSO table, the P is highly repetitive and will be compressed away. The S column is ascending, so delta-compression will apply. However, it would not be dense and it will take some storage ($\log_2(W)$ bits per triple, where W is the average gap width between successive s values[5]) – while a dense S column takes no storage.

Compressing-away the S column is only possible for the regular part P_{PSO}, whereas the exception triples in T_{pso} must fall back to normal PSO triple storage. However, the left table column of Fig. 3 shows that the amount of exception triples is negligible anyway – it is almost 0 in synthetic RDF data (stemming from the LUBM, BSBM, SP2Bench and LDBC Social Network Benchmark), as well as in RDF data with relational roots (EuroStat, PubMed, DBLP, MusicBrainz), and is limited to <10 % in more noisy "native" RDF data (WebData Commons and DBpedia). A more serious threat to storage efficiency could be the NULL values that emergent tables introduce, which are table cells for which no triple exists. In the middle column we see that the first-generation RDF benchmarks (LUBM, BSBM, SP2Bench) ignore the issue of missing values. The more recent LDBC Social Network benchmark better models data with relational roots where this percentage is roughly 15 %. Webdata Commons, which consists of crawled RDFa, has most NULL values (42 %) and DBpedia roughly one third. We note that the percentage of NULLs is a consequence of the emergent table algorithm trying to create a compact schema that consists of relatively few tables. This process makes it merge initial tables of property-combinations into tables that store the union of those properties: less, wider, tables means more NULLs. If human understandability were not a goal of emergent table detection, parameters could be changed to let it generate more tables with less NULLs. Still, space saving is not really an argument for doing so, as the rightmost table column of Fig. 3 shows that emergent table storage is overall at least a factor 1.4 more compact than default PSO storage.

Query Processing Microbenchmark. While the emergent schema can be physically viewed as a compressed PSO representation, we now will argue that every use a RDF store will give to a PSO table can be supported at least as efficiently on emergent table aware storage.

Typically, the PSO table is used for three access patterns during SPARQL processing: (i) Scanning all the triples of a particular property p (i.e., p is known), (ii) Scanning with a particular property p and a range of object value (i.e., p is known + condition on o), and (iii) Having a subset of S as the input for the scan

---

[4] SQL-based SPARQL systems (MonetDB, Virtuoso) still allow SQL on T_k tables.
[5] $W = \frac{1}{n-1} \sum_1^{n-1} (s_{i+1} - s_i)$ where s_i is the subject OID at row i (table with n rows).

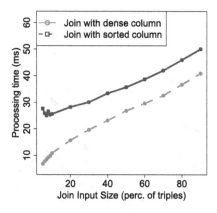

Fig. 4. PSO join performance vs input size (no exceptions)

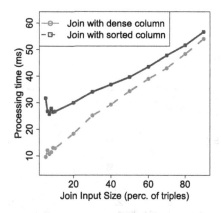

Fig. 5. PSO join performance vs input size (with exceptions)

on a certain p value (i.e., typically s is sorted, and the system performs a filtering MergeJoin). The first and the second access patterns can be processed on the emergent schema in the similar way as with the original PSO representation by using a UNION operator: $\sigma(pso, p, o) = \sigma(P_{PSO}, p, o) \cup \sigma(T_{pso}, p, o)$.

The third access pattern, which is a JOIN with s candidate OIDs is very common in SPARQL queries with star patterns. We test two different cases: with and without of exceptions (i.e. T_{pso}).

Without T_{pso}. In this case, the JOIN can be pushed through the P_{PSO} view and is simply the UNION of JOINs between the s candidates and dense $T_k.s$ columns in each emergent table T_k. MonetDB supports joins into VOID columns very efficiently, essentially this is sequential array lookup.

We conducted a micro-benchmark to compare the emergent schema aware performance with normal PSO access. It executes the JOIN between a set of $I.s$ input OIDs with two different $T_k.s$ columns: a dense column and a sorted (but non-dense) column; in both cases retrieving the $T_k.o$ object values. The benchmark data is extracted from the subjects corresponding to the Offer entities in BSBM benchmark, containing $\approx$5.7 million triples. Each JOIN is executed 10 times and the minimum running time is recorded. Figure 4 shows that dense OID joins are 3 times faster on small inputs: array lookup is faster than MergeJoin.

With T_{pso}. Handling exception data requires merging the result produced by the JOIN between input $(I.s)$ and the dense S column of emergent table $T_k.s$ with the result produced by the JOIN between $I.s$ and the exception table $T_{pso}.s$ – the latter requires an actual MergeJoin. We implemented an algorithm that performs both tasks simultaneously. In order to form the JOIN result between $I.s$ with both $T_k.s$ and $T_{pso}.s$ simultaneously, we modify the original MergeJoin algorithm by checking for each new index of $I.s$, whether the current element from $I.s$ belongs to the dense range of $T_k.s$.

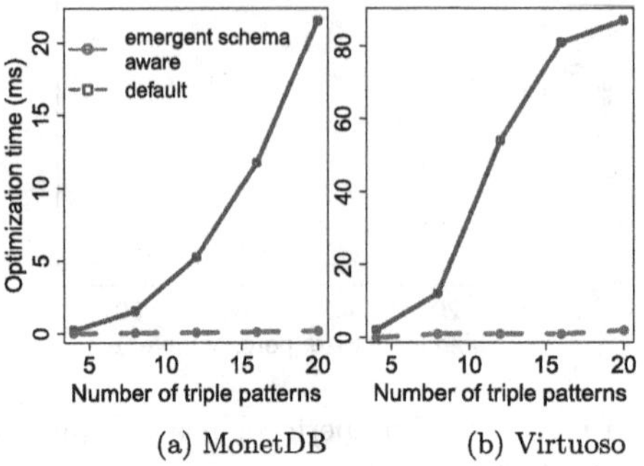

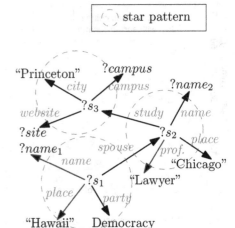

Fig. 6. Optimization time as a function of query size (#triple patterns)

Fig. 7. Example SPARQL graph with three star patterns

We conducted another micro-benchmark using the same 5.7 million triples. The exception data is created by uniformly sampling 3 % of the regular data (BSBM itself is perfectly tabular and has no exceptions). We note that 3 % is already more than the average percentage of exception data in all our tested datasets. The list of input $I.s$ candidates is also generated by sampling from 5 % to 90 % of the regular data. Figure 5 shows that the performance of the JOIN operator on the emergent schema still outperforms that on the original PSO representation even though it needs to handle exception data.

The conclusion of this section is that emergent schema aware storage reduces space by 1.4 times, provides faster PSO access, and importantly hides the relational table storage from the SPARQL processor – such that query patterns that would be troublesome for property tables (e.g. unbound property variables) can still be executed without complication. We take further advantage of the emergent schema in many common query plans, as described next.

3   Emergent Schema Aware SPARQL Optimization

The core of each SPARQL query is a set of (s,p,o) triple patterns, in which s, p, o are either literal values or *variables*. Viewing each pattern as a property-labeled edge between a subject and object, these triples form a *SPARQL graph*. We *group* these triple patterns, where originally each triple pattern is a group of one.

Definition 4. *Star Pattern* ($\rho = (\$s, p_1, o_1), (\$s, p_2, o_2), \ldots$): *A star pattern is a collection of* more than one *triple patterns from the query, that each have a* constant *property* p_i *and an* identical *subject variable* $\$s$.

To exploit the emergent schema, we identify star patterns in the query and at the query optimization, group query's triple patterns by each star. Joins are needed only between these triple pattern groups. Each group will be handled by one table scan subplan that uses a new "$RDFscan$" operator described further on. SPARQL query optimization then largely becomes a *join reordering* problem. The complexity of join reordering is exponential in the number of joins.

To show the effects on query optimization performance, we created a micro-benchmark that forms queries consisting of (small) stars of size $F = 4$. The smallest query is a single star, followed by one with two stars that are connected by sharing the same variable for an object in the first star and the subject of the star, etc. (hence queries have 4, 8, 12, 16 and 20 triple patterns). Our optimization identifies these stars, hence after grouping star patterns their join graph reduces to 0, 1, 2 and 3 joins respectively. We ran the resulting queries through MonetDB and Virtuoso and measured *only* query optimization time. Figure 6 shows that emergent schema aware SPARQL query optimization becomes orders of magnitude faster thanks to its simplification of the join ordering problem. The flattening Virtuoso default line beyond 15 patterns suggests that with large amount of joins, it stops to fully traverse the search space using cutoffs, introducing the risk of choosing a sub-optimal plan.

4   Emergent Schema Aware SPARQL Execution

The basic idea of emergent schema aware query execution is to handle a complete star pattern ρ with one relational table scan($T_i, [p_1, p_2, ..]$) on the emergent table T_i with whose properties p_i from ρ. Assuming a SQL-based SPARQL engine, as is the case in Virtuoso and MonetDB, it is crucial to rely on the existing relational table scan infrastructure, so that advanced relational access paths (clustered indexes, partitioned tables, cracking [9]) get seamlessly re-used.

In case of multiple emergent tables matching star pattern ρ, the *scan plan* (denoted ϑ_ρ) we generate consists of the UNION of such table scans. In ϑ_ρ we also push-down certain relational operators (at least simple filters) below these UNIONs – a standard relational optimization. This push-down means that selections are executed before the UNIONs and optimized relational access methods can be used to e.g. perform IndexScans. For space reasons we cannot go into all details, although we should mention that OPTIONAL triple patterns in ρ are marked and can be ignored in the generated scans (because missing property values are already represented as NULL in the relational tables). Another detail is that on top of ϑ_ρ, we must introduce a Project operator to cast SQL literal types to a special SPARQL value type, that allows multiple literal types as well as URIs to be present in one binding column.[6] Executing (pushed-down) filter operations while values are still SQL literals allows to avoid most casting effort, since after selections much fewer tuples remain.

---

[6] In our MonetDB implementation, the 64-bits OID that encodes (subject) URIs, also encodes literals by using other patterns in its highest 8 bits.

T_1

s	p_1	p_2	p_3
100	11	2	5
101	13	4	6
102	14		5
103	9	6	
104	15	8	5

T_2

s	p_1	p_3	p_4
200	11	7	1
201		5	2
202	13	9	3

T_{pso}

p	s	o
p_1	0	20
p_1	1	9
p_1	201	15
p_2	0	8
p_2	102	6
p_2	201	4
p_3	0	5
p_6	6	7

Result

s	o_1	o_2
100	11	2
104	15	8
0	20	8
102	14	6
201	15	4

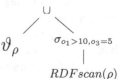

Fig. 8. Example RDF data and expected query result. (Color figure online)

Fig. 9. Query plan for handing exception

This whole approach will still only create bindings for the "regular" triples. To generate the 100 % correct SPARQL result, we introduce an operator called **RDFscan**, that produces *only* the missing bindings. The basic idea is to put another UNION on top of the scan plan ϑ_ρ that adds the RDFscan(ρ) bindings to the output stream, as shown in Fig. 9. Unlike normal scans, we cannot push down filters below the RDFscan - hence these selections remain placed above it, at least until optimization 1 (see later).

Generating Exception Bindings. Correctly generating all result bindings that SPARQL semantics expect is non-trivial, since the exception triples in T_{pso} when combined with *any* emergent table T_k (not only those covering ρ) could produce valid bindings. Consider the example SPARQL query, consisting of a single star pattern and two selections ($o_1 > 10$, $o_3 = 5$):

```
SELECT ?s ?o1 ?o2 WHERE { ?s  p1 ?o1 .
                          ?s  p2 ?o2 .
                          ?s  p3    5 . FILTER (?o1 > 10) }
```

Figure 8 shows the expected result of this query on an example data. (For a better view of the example, we assume s base OID of T_1, T_2 are 100, 200, respectively). In this result, the first two tuples come from the regular triples while the last three tuples is the combination of triples stored in T_{pso} table (i.e., in red color) with those stored in tables T_1 and T_2.

Basic Approach. RDFscan returns all the bindings for a star pattern, in which each binding is generated by at least one irregular triple (the *missing* bindings). Formally, given a star pattern $\rho = \{(s, p_i, o_i), i = 1, .., k\}$, the RDF dataset Δ, the output of the RDFscan operator for this star pattern is defined as:

$$RDFscan(\rho) = \{(s, o_1, \ldots, o_k)\} | (s, p_i, o_i) \in \Delta \wedge (\exists i : (s, p_i, o_i) \in T_{pso}) \quad (1)$$

RDFscan generates the "exception" bindings in 2 steps:

E_1

s	o_1
0	20
1	9
201	15

E_3

s	o_3
0	5

E_2

s	o_2
0	8
102	6
201	4

Output(S_1)

s	o_1	o_2	o_3
0	20	8	5
1	9		
102		6	
201	15	4	

Fig. 10. Step 1

E_1

s	o_1
0	20
201	15

Output(S_1)

s	o_1	o_2	o_3
0	20	8	5
102		6	
201	15	4	

Fig. 12. Step 1 output with pushing down Selection predicates

Require: S_1: Step 1 output
    $lstP$: List of required properties
    $\mathcal{E}$: Emergent tables
Ensure: T_{fin}: Merging results
1: **for** each tuple $t=(s, o_1,...,o_k)$ in S_1 **do**
2:    $id, r = getT\_row(t.s)$ # table & row id
3:    $accept = true$
4:    **for** each p_i in $lstP$ **do**
5:       **if** $t.o_i =$ null & $\mathcal{E}[id][r].p_i =$ null **then**
6:          $accept = false$
7:          Continue next tuple
8:       **else**
9:          store_cand($bind$, $t.o_i$, $\mathcal{E}[id][r].p_i$)
10:      **end if**
11:   **end for**
12:   **if** $accept = true$ **then**
13:      append(T_{fin}, $bind$)
14:   **end if**
15: **end for**

Fig. 11. Merge-exception-regular algorithm

Step 1: Get all possible bindings $(s, o_1, \ldots, o_k)$ where each o_i stems from triple $(s, p_i, o_i) \in T_{pso}$ (for those p_i from ρ), or $o_i =$ NULL if such a triple does not exist, with the constraint that at least one of the object values o_i is non-NULL. *Step 2:* Merge each binding $(s, o_1, \ldots, o_k)$ with the emergent table T_k corresponding to s ($\beta_k \le s < \beta_k + |T_k|$) to produce output bindings for *RDFscan*.

Step 1 is implemented by first extracting the set E_i of all $\{(s, o_i)\}$ corresponding to each property p_i from the T_{pso}: $E_i = \sigma_{p=p_i}(T_{pso})$. Then, it returns the output, S_1, by performing a relational OuterJoin on s between all E_i. We note that, as T_{pso} table is sorted by p, extracting E_i from T_{pso} can be done with no cost by reading a slice of T_{pso} from the starting row of p_i and the ending row of p_i (the information on starting, ending rows of each p in T_{pso} table is preloaded before any query processing). Furthermore, as for each p in T_{pso}, $\{(s, o)\}$ are sorted according to s, E_i are also sorted by s. Thus, the full OuterJoin of all E_i can be efficiently done by using a multi-way sort merging algorithm. Figure 10 demonstrates Step 1 for the example query.

Step 2 merges each tuple in S_1 with a tuple of the same s in the regular table in order to form the final output of *RDFscan*. For example, the 4th tuple of S_1 (201, 15, 4, *null*) merged with the 2nd tuple of T_2 (201, *null*, 5, 2) returns a valid binding (201, 15, 4) for the (s, o_1, o_2) of the example query. Figure 11 shows the detailed algorithm of Step 2. For each tuple t in S_1, it first extracts the corresponding regular table and row Id of the current $t.s$ from encoded information inside each s OID (Line 2). Then, for each property p_i, the algorithm will check whether there is any non-NULL object value appearing in either t

(i.e., $t.o_i$) or the regular column p_i (i.e., $\mathcal{E}\,[id]\,[r]\,.p_i$) (Line 5). If yes, the non-NULL value will be placed in the binding for p_i (Line 9). Otherwise, if both of the values are NULL, there will be no valid binding for the current checking tuple t. Finally, the binding that has non-NULL object values for all non-optional properties will be appended to the output table T_{fin}.

Optimization 1: Selection Push-Down. Pushing selection predicates down in the query plan is an important query optimization technique to apply filters as early as possible. This technique can be applied to RDFscan when there is any selection predicate on the object values of the input star pattern (e.g., $o_1 > 10$, $o_3 = 5$ in the example query). Specifically, we push the selection predicates down in Step 1 of the RDFscan operator to reduce the size of each set E_i (i.e., $\sigma_{p=p_i}(T_{pso})$), accordingly returning a smaller output S_1 of this step. Formally, given λ_i being a selection predicate on the object o_i, the set E_i of $\{(s, o_i)\}$ from T_{pso}) is computed as: $E_i = \sigma_{p=p_i, \lambda_i}(T_{pso})$. In the example query, $E_1 = \sigma_{p=p_1, o_1>10}(T_{pso})$. Figure 12 shows that the size of E_1 and the output S_1 are reduced after applying the selection pushdown optimization, which thus improves the processing time of RDFscan operator.

Optimization 2: Early Check for Missing Property. If a regular table T_k does not have p_i in its list of columns, to produce a valid binding by merging a tuple t of S_1 (i.e., output of Step 1) and T, the exception object value $t.o_i$ must be non-NULL. Thus, we can quickly check whether t is an invalid candidate without looking into the tuple from T_k by verifying whether t contains non-NULL object values for all missing columns of T_k. We implement this by modifying the algorithm for Step 2. Before considering the object values of all properties from both exception and regular data (Line 4), we first check exception object value $t.o_i$ of each missing property to prune the tuple if any $t.o_i$ is NULL. Then, we continue the original algorithm with the remaining properties.

Optimization 3: Prune Non-matching Tables. The exception table T_{pso} mostly contains triples whose subject was mapped to some emergent table. For example, the triple $(201, p_2, 4)$ refers to the emergent table T_2 because $s \geq 200 = \beta_2$. During the emergent schema exploration process [15] this triple was temporarily stored in the *initial* emergent table T_2', but was then moved to T_{pso} during the so-called "schema and instance filtering" step. This filtering moves not only triples but also whole columns from initial emergent tables to T_{pso}, in order to derive a compact and precise emergent schema. Assume column p_2 was removed from T_2 during schema filtering. We observe that before filtering, *all* triples (*regular + exception triples*) of subject s were part of the initial emergent table which means that had a particular set of properties. Accordingly, if C is the set of columns of an initial emergent table T' and if C does not contain the set of properties in ρ, there cannot be a matching subject with all properties of ρ stemming from T' even with the help of T_{pso}. This observation can be exploited to prune all subject ranges corresponding to (initial) emergent tables that cannot have any matching for ρ from the pass over T_{pso}.

Specifically, we pre-store, for each emergent table, its set of columns C before schema and instance filtering was applied during emergent schema detection. Then, given the input star pattern ρ, the possible matching tables for ρ are those tables whose set of columns C contain all properties in ρ. Finally, Step 1 is optimized by removing from E_i all the triples that the subject does not refer to any of the matching tables.

5 Performance Evaluation

We tested with both synthetic and real RDF datasets BSBM [1], LUBM [8], LDBC-SNB [6] and DBpedia (DBPSB) [12]; and their respective query workloads. For BSBM, we also include its relational version, namely BSBM-SQL, in order to compare the performance of the RDF store against a SQL system (i.e., *MonetDB-SQL*). We used datasets of 100 million triples for LUBM and BSBM, and scale factor 3 ($\approx$200 million triples) for LDBC-SNB. The experiments were conducted on a Linux 4.3 machine with Intel Core i7 3.4 GHz CPU and 16 GBytes RAM. All approaches are implemented in the RDF experimental branch of MonetDB.

Query Workload. For BSBM, we use the SELECT queries from Explore workload (ignoring the queries with DESCRIBE and CONSTRUCT). For LUBM, we use its published queries and rewrite some queries (i.e., Q4, Q7, Q8, Q9, Q10, Q13) that requires certain ontology reasoning capabilities in order to account

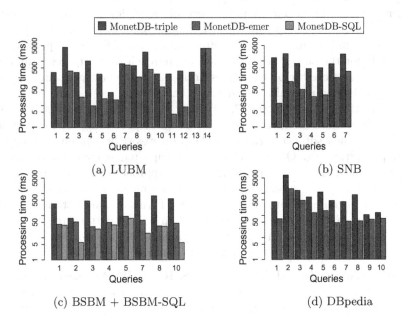

Fig. 13. Query processing time: Emergent schema-based vs triple-based

Table 1. Properties of DBpedia queries

Queries	Q1	Q2	Q3	Q4	Q5	Q6	Q7	Q8	Q9	Q10
Operators: OPTIONAL, FILTER, UNION	-	-	O	-	U	F	-	F,U	O,F,U	O
Modifiers: Distinct, Limit, ORDER	D	D	D,L,O	D	D,L	D	D,L	-	-	D
# of triple pattern	4	5	5	3	10	3	6	4	6	7
# constraints on O?	1	0	1	1	2	2	1	4	2	0
Has multi-valued prop.?	√	√	√	√	√	√	√	-	√	-

for the ontology rules and implicit relationships. For LDBC-SNB, we use its short read queries workload. DBPSB exploits the actual query logs of the DBpedia SPARQL endpoints to build a set of templates for the query workload. Using these templates, we create 10 non-empty result queries w.r.t DBpedia 3.9 dataset[7]. Table 1 shows the features of tested DBpedia queries. In Figs. 13, 14 and 15, X-axis holds query-numbers: 1 means Q1. For each benchmark query we run three times and record the last query execution (i.e., Hot run).

Emergent Schema Aware vs Triple-Based RDF Stores. We perform the benchmarks against two different approaches of MonetDB RDF store: the original triple-based store (*MonetDB-triple*) and the emergent schema-based store (*MonetDB-emer*).

Figure 13 shows the query processing time using two approaches over four benchmarks. For BSBM and LDBC-SNB, the emergent schema aware approach significantly outperforms the triple-based approach in all the queries, by up to two orders of magnitude faster (i.e., Q1 SNB). In a real workload such as DBpedia where there is significant amount of exception triples, our approach is still much faster (note: logscale) by up to more than an order of magnitude (Q8). We also note that multi-valued properties appear in most of DBpedia queries, and this is costly for the emergent schema aware approach as it requires additional MergeJoins to retrieve the object values. In Fig. 13d, the best-performing query Q8 is the one having no multi-valued property.

For LUBM, a few queries (i.e. 7, 14) show comparable processing times for triple-table based and emergent schema aware query processing. The underlying reason is that each subject variable in these queries only contains one or two common properties (e.g., Q14 only contains one triple pattern with the properties rdf:type). Thus, the emergent schema aware approach will not improve the query execution time – however as the optimization does not trigger then it also does not degrade performance in absence of fruitful star patterns. For the queries having *discriminative* properties [15] in a star pattern (e.g., Q4, 11, 12), the emergent schema aware approach significantly outperforms the original triple-based version, by up to two order of magnitude (i.e., Q4).

Emergent Schema-Based RDF Store vs RDBMS. As shown in Fig. 13c, the emergent schema aware SPARQL processing (*MonetDB-emer*) provides comparable performance on most queries (i.e., Q1, Q3, Q4, Q5, Q8) compared to

---

[7] The detailed DBpedia queries can be found at goo.gl/RxzOmy.

MonetDB-SQL. In other queries (Queries 7,10), the emergent schema aware approach also significantly reduces the performance gap between SPARQL and SQL, from almost two orders of magnitude slower (*MonetDB-triple* vs *MonetDB-SQL*) to a factor of 3.8 (*MonetDB-emer* vs *MonetDB-SQL*).

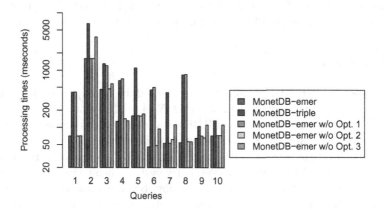

Fig. 14. Query processing with/with-out optimizations

RDFscan Optimizations. Figure 14 shows the effects of each of the three described RDFscan optimization by running the DBpedia benchmark without using with each of them. All optimizations have positive effects, though in different queries, and the longer running queries show stronger effects. Selection push-down (Opt. 1) has most influence, while the early check in T_{pso} to see if it delivers missing properties has the least influence. Obviously, selection push-down does not give any performance boost when there is no constraint on the object variables in the queries (e.g., Query 2). For queries having constraints on the object variables, which are quite common in any query workload, it does speed up query processing by up to a factor of 24 (i.e., Q8).

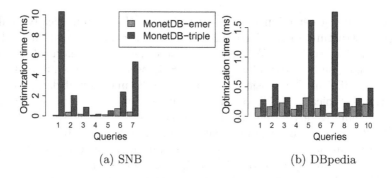

(a) SNB   (b) DBpedia

Fig. 15. Optimization time: Emergent schema-based vs triple-based

Query Optimization Time. Figure 15 shows query optimization time on
LDBC-SNB and DBPSB (due to lack of space, we omit similar results for BSBM
and LUBM). For all queries, the emergent schema aware approach significantly
lowers optimization time, by even up to two orders of magnitude (Q1 SNB)
or a factor of 37 (Q7 DBPSB). Note also that due to the smaller plan space
and strong reduction of join correlations, query optimization also qualitatively
improves, a claim supported by its performance improvements across the board.

6 Related Work

Most state-of-the-art RDF systems store their data in triple- or quad-tables
creating indexes on multiple orders of S,P,O [5,14,16,19]. However, according to
[7,15], these approaches have several RDF data management problems including
unpredictably bad query plans and low storage locality.

Structure-aware storage was first exploited in RDF stores with the "property
tables" approach [4,10,18,20]. However, early systems using this approach [4,20]
do not support automatic structure recognition, but rely on a database adminis-
trator doing the table modeling manually. Automatic recognition is introduced
in some newer systems [10,11,18], however unlike emergent schemas these structures are not apt for human usage, nor did these papers research in depth integration with relational systems in terms of storage, access methods or query
optimization. Recently, Bornea et al. [2] built an RDF store, DB2RDF, on top
of a relational system using hash functions to shred RDF data into multiple
multi-column tables. This approach (nor any of the others) allows both SQL
and SPARQL access to the same data, as emergent schemas do. Gubichev et al.
[7] and Neumman et al. [13] use structure recognition to improve join ordering
in SPARQL queries alone. Brodt et al. [3] proposed a new operator, called *Pivot
Index Scan*, to efficient deliver attribute values for a resource (i.e., subject) with
less joins using something similar to a SPO index – as such it does not recognize
structure in RDF to leverage it on the physical level.

7 Conclusion

Emergent Schema detection is a recent technique that automatically analyzes
the actual structure an RDF graph, and creates a compact relational schema
that fits most of the data. We investigate here how these Emergent Schemas,
beyond helping humans to understand a RDF dataset, can be used to make RDF
stores more efficient. The basic idea is to store the majority of data, the "regular" triples (typically >95 % of all data) in relational tables under the hood,
and the remaining "exception" triples in a reduced PSO triple table. This storage still allows to see the relational data as if it were a PSO table, but is in fact
>1.4x more compact and faster to access than a normal PSO table. Furthermore,
we provide a simple optimization heuristic that groups triple patterns by star-
shape. This reduces the complexity of query optimization by often more than
a magnitude, since the size of the join graph is reduced thanks to only joining

these groups. Finally, we contribute the RDFscan algorithm with three important optimizations. It is designed to work in conjunction with relational scans, which perform most of the heavy-lifting, and can benefit from existing physical storage optimizations such as table clustering and partitioning. RDFscan keeps the overhead of generating additional binding results for "exception" triples low, yielding overall speed improvements of 3–10x on a wide variety of datasets and benchmarks, closing the performance gap between SQL and SPARQL.

References

1. Bizer, C., Schultz, A.: IJSWIS. The Berlin SPARQL Benchmark **5**, 1–24 (2009)
2. Bornea, M., et al.: Building an efficient RDF store over a relational database. In: SIGMOD (2013)
3. Brodt, A., et al.: Efficient resource attribute retrieval in RDF triple stores. In: CIKM (2011)
4. Chong, E., et al.: An efficient SQL-based RDF querying scheme. In: VLDB (2005)
5. Erling, O.: Virtuoso, a hybrid RDBMS/graph column store. DEBULL **35**, 3–8 (2012)
6. Erling, O., et al.: The LDBC social network benchmark. In: SIGMOD (2015)
7. Gubichev, A., et al.: Exploiting the query structure for efficient join ordering in SPARQL queries. In: EDBT, pp. 439–450 (2014)
8. Guo, Y., et al.: LUBM: a benchmark for owl knowledge base systems. Web Semant. **3**, 158–182 (2005)
9. Idreos, S., et al.: Database cracking. In: CIDR, Asilomar, California (2007)
10. Levandoski, J., et al.: RDF data-centric storage. In: ICWS (2009)
11. Matono, A., Kojima, I.: Paragraph tables: a storage scheme based on RDF document structure. In: Liddle, S.W., Schewe, K.-D., Tjoa, A.M., Zhou, X. (eds.) DEXA 2012, Part II. LNCS, vol. 7447, pp. 231–247. Springer, Heidelberg (2012)
12. Morsey, M., Lehmann, J., Auer, S., Ngonga Ngomo, A.-C.: DBpedia SPARQL benchmark – performance assessment with real queries on real data. In: Aroyo, L., Welty, C., Alani, H., Taylor, J., Bernstein, A., Kagal, L., Noy, N., Blomqvist, E. (eds.) ISWC 2011, Part I. LNCS, vol. 7031, pp. 454–469. Springer, Heidelberg (2011)
13. Neumann, T., et al.: Characteristic sets: accurate cardinality estimation for RDF queries with multiple joins. In: ICDE (2011)
14. Neumann, T., et al.: RDF-3X: a RISC-style engine for RDF. Proc. VLDB Endow. **1**, 647–659 (2008)
15. Pham, M.D., et al.: Deriving an emergent relational schema from RDF data. In: WWW (2015)
16. Tsialiamanis, P., et al.: Heuristics-based query optimisation for SPARQL. In: EDBT (2012)
17. Ullman, J., Widom, J.: Database Systems: The Complete Book (2008)
18. Wang, Y., Du, X., Lu, J., Wang, X.: FlexTable: using a dynamic relation model to store RDF data. In: Kitagawa, H., Ishikawa, Y., Li, Q., Watanabe, C. (eds.) DASFAA 2010. LNCS, vol. 5981, pp. 580–594. Springer, Heidelberg (2010)
19. Weiss, C., et al.: Hexastore: sextuple indexing for semantic web data management. Proc. VLDB Endow. **1**, 1008–1019 (2008)
20. Wilkinson, K.: Jena property table implementation (2006)

Distributed RDF Query Answering with Dynamic Data Exchange

Anthony Potter$^{(\boxtimes)}$, Boris Motik, Yavor Nenov, and Ian Horrocks

University of Oxford, Oxford, UK
{anthony.potter,boris.motik,yavor.nenov,ian.horrocks}@cs.ox.ac.uk

Abstract. Evaluating joins over RDF data stored in a shared-nothing server cluster is key to processing truly large RDF datasets. To the best of our knowledge, the existing approaches use a variant of the *data exchange operator* that is inserted into the query plan *statically* (i.e., at query compile time) to shuffle data between servers. We argue that such approaches often miss opportunities for local computation, and we present a novel solution to distributed query answering that consists of two main components. First, we present a query answering algorithm based on *dynamic data exchange*, which exploits data locality to maximise the amount of computation on a single server. Second, we present a partitioning algorithm for RDF data based on graph partitioning whose aim is to increase data locality. We have implemented our approach in the RDFox system, and our performance evaluation suggests that our techniques outperform the state of the art by up to an order of magnitude in terms of query evaluation times, network communication, and memory use.

1   Introduction

RDF datasets used in practice are often too large to fit on a single server. For example, in performance-critical applications, it is common to use an in-memory RDF store, but the comparatively high cost of RAM limits the capacity of such systems. Moreover, linked data applications often require integrating several large datasets that cannot be processed jointly even using disk-based systems. To attain scalability sufficient for such applications, numerous approaches for storing and querying RDF data in a shared-nothing server cluster have been developed [7–10, 12, 15, 17–19, 21, 22].

Such approaches typically consist of a query answering algorithm and a data partitioning strategy, both of which must address a specific set of challenges. First, triples participating in a join may be stored on different servers, so network communication during join evaluation should be minimised. Second, to ensure that servers can progress independently of each other, one must minimise synchronisation between the servers. Third, the intermediate results produced during join evaluation often grow with the overall data size and so they may easily exceed the capacity of individual servers.

The Volcano [5] database system was one of the first to address these challenges by introducing the *data exchange operator* that encapsulates the communication

© Springer International Publishing AG 2016
P. Groth et al. (Eds.): ISWC 2016, Part I, LNCS 9981, pp. 480–497, 2016.
DOI: 10.1007/978-3-319-46523-4_29

between query execution processes.[1] Data exchange operators are added into the query plan to move the data within the system in order to ensure that each operator in the query plan receives all the relevant data. Data exchange can be avoided if the data partitioning strategy guarantees that the triples participating in a join are colocated on the same sever. For example, exchange is not needed for subject–subject joins if all triples with the same subject are assigned to the same server. Data partitioning strategies often replicate data across servers to increase the level of guarantees they offer. As we discuss in detail in Sect. 3, all existing distributed RDF systems we are aware of can be seen as using a variant of the data exchange operator, and they aim to balance the trade-off between data replication and data exchange. Moreover, in all of the existing approaches, the decision about when and how to exchange data is made *statically*—that is, at compile time and independently from the data encountered during query evaluation. In Sect. 3 we argue that this can incur a communication cost even when the data is stored in such a way that no communication is needed in principle.

In this paper we present a new approach to query answering in distributed RDF systems. We focus here on *conjunctive* SPARQL queries (i.e., *basic graph patterns* extended with projection), but we believe that our approach can be extended to handle all SPARQL constructs. As is common in the literature, our solution also consists of a query answering algorithm and a data partitioning strategy.

In Sect. 4 we present a novel distributed query answering algorithm that employs *dynamic data exchange*: the decision when and how to exchange data is made during query processing, rather than statically at query compile time. In this way, each join between triples stored on the same server is computed on that server. Unlike in the existing solutions, local computation in our algorithm is independent of any guarantees about data partitioning, and is determined solely by the actual placement of the data. Our algorithm thus gives the data partitioning strategy more freedom regarding data placement. Moreover, our algorithm uses asynchronous communication between servers, ensuring that a server's progress in query evaluation is largely independent of that of other servers. Finally, our algorithm uses a novel technique that limits the amount of memory each server needs to store intermediate results.

In Sect. 5 we present a novel RDF data partitioning method that aims to maximise data locality by using *graph partitioning* [11]—the task of dividing the vertices of a graph into sets while satisfying certain balancing constraints and simultaneously minimising the number of edges between the sets. Graph partitioning has already been used for partitioning RDF data [7,10], but these approaches duplicate data across servers to increase the chance of local processing. In contrast, our approach does not duplicate any data at all, and it uses a special pruning step to reduce the size of the graph being partitioned. Finally, a balanced partition of vertices does not necessarily lead to a balanced partition of triples so, to achieve the latter, we use *weighted* graph partitioning.

---

[1] These processes may be threads within a single server or processes running on different servers, and so intra- and inter-server communication is handled using the same abstraction.

We have implemented our approach in the in-memory RDF management system RDFox [13], and have compared its performance with that of TriAD [7]—a system that was shown to outperform other state of the art distributed RDF systems on a mix of data and query loads. In Sect. 6 we present the results of our evaluation using the LUBM [6] and WatDiv [1] benchmarks. We show that our approach is competitive with TriAD in terms of query evaluation times, network communication, and memory usage; in fact, RDFox often outperforms TriAD by an order of magnitude.

2   Preliminaries

To make this paper self-contained, in this section we recapitulate certain definitions and notation. For f a function, $\mathsf{dom}(f)$ is its domain; for D a set, $f|_D$ is the restriction of f to $D \cap \mathsf{dom}(f)$; and if f' is a function such that $f(x) = f'(x)$ for each $x \in \mathsf{dom}(f) \cap \mathsf{dom}(f')$, then $f \cup f'$ is a function as well.

The vertices of RDF graphs are taken from a countable set of *resources* $\mathcal{R}$ that consists of *IRI references*, *blank nodes*, and *literals*. A *triple* has the form $\langle t_s, t_p, t_o \rangle$, where t_s, t_p, and t_o are resources. An *RDF graph* G is a finite set of triples. The *vocabulary* $\mathsf{voc}(G)$ of G is the set of all resources that occur in G; moreover, for a position $\beta \in \{s, p, o\}$, set $\mathsf{voc}_\beta(G)$ contains each resource r for which a triple $\langle t_s, t_p, t_o \rangle \in G$ exists such that $t_\beta = r$. SPARQL is an expressive language for querying RDF graphs; for example, the following SPARQL query retrieves all people that have a sister:

SELECT ?x WHERE { ?x rdf:type :Person . ?x :hasSister ?y }

SPARQL syntax is verbose, so we use a more compact notation. An *(RDF) term* is a resource or a *variable*. An *atom* (aka *triple pattern*) A is an expression of the form $\langle t_s, t_p, t_o \rangle$, where t_s, t_p, and t_o are terms; thus, each triple is an atom. For A an atom, let $\mathsf{vars}(A)$ be the set of variables occurring in A; and for $\beta \in \{s, p, o\}$, let $\mathsf{term}_\beta(A) = t_\beta$. A *conjunctive query* (CQ) has the form $Q(\vec{x}) = A_1 \wedge \cdots \wedge A_n$, where each A_i is an atom. Our definition of CQs captures *basic graph patterns* with projection in SPARQL; e.g., $Q(x) = \langle x, rdf{:}type, {:}Person \rangle \wedge \langle x, {:}hasSister, y \rangle$ captures the above query. A *subject-join* query is a query where the same term occurs in the subject position of all query atoms; such queries are used very frequently in practice.

Evaluation of CQs on an RDF graph produces partial mappings of variables to resources called *(variable) assignments*. For α a term or an atom and σ an assignment, $\alpha\sigma$ is the result of replacing each variable x in α with $\sigma(x)$. An assignment σ is an *answer* to a CQ $Q(\vec{x}) = A_1 \wedge \cdots \wedge A_n$ on an RDF graph G if an assignment ν exists such that $\sigma = \nu|_{\vec{x}}$, $\mathsf{dom}(\nu) = \mathsf{vars}(A_1) \cup \cdots \cup \mathsf{vars}(A_n)$, and $\{A_1\nu, \ldots, A_n\nu\} \subseteq G$ holds. SPARQL uses *bag* semantics, so $\mathsf{ans}(Q, G)$ is the multiset that contains each answer σ to Q on G with multiplicity equal to the number of such assignments ν.

Finally, we formalise the computational problems we consider in this paper. Let C be a finite set called a *cluster*; each element $k \in C$ is called a *server*.

A *partition* of an RDF graph G is a function $\mathbf{G}$ that assigns to each server $k \in C$ an RDF graph $\mathbf{G}_k$, called a *partition element*, such that $G = \bigcup_{k \in C} \mathbf{G}_k$. Partition $\mathbf{G}$ is *strict* if $\mathbf{G}_k \cap \mathbf{G}_{k'} = \emptyset$ for all $k, k' \in C$ with $k \neq k'$. A *(data) partitioning strategy* takes an RDF graph G and produces a partition $\mathbf{G}$. Given a CQ Q, a *distributed query answering algorithm* computes $\mathsf{ans}(Q, G)$ on a cluster C where each server $k \in C$ stores $\mathbf{G}_k$. An answer σ to Q on G is *local* if $k \in C$ exists such that σ is an answer to Q on $\mathbf{G}_k$.

3   Motivation and Related Work

We now illustrate the difficulties of distributed query answering and present an overview of the existing approaches.

Data Exchange Operator by Example. To make our discussion concrete, let G be the RDF graph from Fig. 1a partitioned to elements $\mathbf{G}_1$ and $\mathbf{G}_2$ by *subject hashing*; resource c is shown in grey because it occurs in both partition elements. Subject hashing is one of the simplest data partitioning strategies that assigns triple $\langle t_s, t_p, t_o \rangle$ to partition element $(h(t_s) \bmod 2) + 1$ for a suitable hash function h. It was initially studied in the YARS2 [9] system, but modern distributed RDF systems use more elaborate strategies.

To understand the main issue that distributed query processing must address, let $Q_1(x, y, z) = \langle x, S, y \rangle \wedge \langle y, R, z \rangle$. Answer $\sigma_1 = \{x \mapsto b, y \mapsto c, z \mapsto e\}$ spans partition elements so servers must exchange intermediate answers to compute σ_1. The Volcano system [5] proposed the solution in form of a data exchange operator that encapsulates all communication between servers in the query pipeline. In particular, variable y occurs in the second atom of Q_1 in the subject position so, for each triple $\langle t_x, S, t_y \rangle$ matching the first atom of Q_1, subject hashing ensures that any join counterparts are found on server $(h(t_y) \bmod 2) + 1$. Thus, we can answer Q_1 using the query plan shown in Fig. 1b, where $\otimes$ is a *data exchange operator* that (i) sends each variable assignment σ from its input to server $(h(\sigma(y)) \bmod 2) + 1$, and (ii) receives variable assignments sent from other servers and forwards them to the parent join operator. Thus, the rest of the query plan is completely isolated from any data exchange issues.

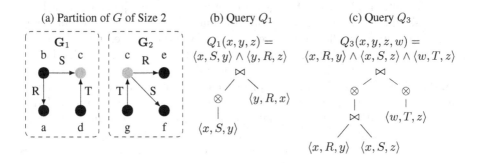

(a) Partition of G of Size 2   (b) Query Q_1 (c) Query Q_3

$Q_1(x, y, z) = \langle x, S, y \rangle \wedge \langle y, R, z \rangle$

$Q_3(x, y, z, w) = \langle x, R, y \rangle \wedge \langle x, S, z \rangle \wedge \langle w, T, z \rangle$

Fig. 1. Example RDF data and query plans

Guarantees about data partitioning can be used to avoid data exchange in some cases. For example, subject hashing ensures that all triples with the same subject are colocated, so subject-join queries can be evaluated without any data exchange. Thus, we can evaluate $Q_2(x, y, z) = \langle x, R, y \rangle \wedge \langle x, S, z \rangle$ independently over $\mathbf{G}_1$ and $\mathbf{G}_2$.

The decision when to introduce the data exchange operators is made *statically* (i.e., at query compile time), which can introduce unnecessary data exchange. For example, let $Q_3(x, y, z, w) = \langle x, R, y \rangle \wedge \langle x, S, z \rangle \wedge \langle w, T, z \rangle$. As in Q_2, we can evaluate the first two atoms locally. In contrast, join variable z occurs in Q_3 only in the object position so, given a value for z, subject hashing does not tell us where to find the relevant triples. Consequently, we need a query plan from Fig. 1c with two data exchange operators that hash their inputs based on the value of z, which allows us to compute answers $\sigma_2 = \{x \mapsto b, y \mapsto a, z \mapsto c, w \mapsto d\}$ and $\sigma_3 = \{x \mapsto b, y \mapsto a, z \mapsto c, w \mapsto g\}$. Note that data exchange is necessary for σ_3; however, σ_2 can be obtained by evaluating Q in $\mathbf{G}_1$, but resource c is hashed to server 2 so σ_2 is unnecessarily computed on server 2.

Data Exchange in Related Approaches. Static data exchange has been extensively used in practice. For example, the map phase in MapReduce [3] assigns to each data record a key that is used to redistribute data records in the shuffle phase; hence, distributed MapReduce-based RDF systems [10,15,17,18] can be seen as using a variant of static data exchange. Moreover, systems such as Sempala [19] implemented on top of big data databases such as Impala and Spark, as well as custom-built systems such as TriAD [7], SemStore [21], and SHAPE [12], use similar ideas. Trinity.RDF [22] uses one master and a number of worker servers: the workers first to compute candidate bindings for each variable using graph exploration, and they then send these bindings to the master to compute the final join, which is a variant of static data exchange.

Some approaches provide stronger locality guarantees by data duplication. For example, Huang et al. [10] distribute the ownership of the resources of G to partition elements using graph partitioning, and they assign each triple to the element that owns the triple's subject. Moreover, they duplicate data using *n-hop duplication*: each server containing a resource r is extended with all triples so that it contains all paths of length n from r. Thus, each query with paths of length less than n can be answered locally, and all other queries are answered using MapReduce. Duplication, however, is costly: for example, query Q_3 needs 2-hop duplication, and Huang et al. [10] show that this can increase the data in the system by a factor of 4.8; this factor is unlikely to scale linearly with the total data size since RDF graphs typically have small diameters. Furthermore, SemStore [21] partitions every rooted subgraph in the original graph. SHAPE [12] partitions subject, object or subject-object groups, extending each group with n-hop duplication and applying optimisations to reduce duplication. Trinity.RDF [22] hashes all triples on subject and object. TriAD [7] first divides resources into groups using graph partitioning, then it computes a *summary* of the input graph by merging all resources in each group, and it assigns groups from the summary to servers by hashing on subject and object. Hashing by subject and object at most doubles the data, and so it is more likely to scale, and it also reduces data exchange: for Q_3, we can use the query plan from Fig. 1c, but without the right-hand data exchange operator.

Our Contribution. In contrast to all of these approaches that use static data exchange, in Sect. 4 we present a novel algorithm for distributed query answering that decides when and how to exchange data independently from any locality guarantees provided by data partitioning. On our example query Q_3, our algorithm computes answer σ_2 on server 1 by discovering that all the data needed for σ_2 is colocated on server 1, and it exchanges only the data necessary for σ_3. Similarly to TriAD, servers in our system exchange messages asynchronously, without coordinating progress through the query plan. This promotes concurrency, but it complicates detecting termination since an idle server can always receive a message from other servers and become busy. We solve this problem using a novel, fully decentralised termination condition. Finally, by processing messages in a specific order we limit the amount of memory needed to store messages.

Although our query answering algorithm does not rely on locality guarantees, ensuring that most answers to a query are local is critical to its efficiency. Thus, in Sect. 5 we present a novel data partitioning strategy based on graph partitioning. Our approach uses no replication, and it produces partition elements that are more balanced in sizes than those produced by related strategies based on graph partitioning [7,10,16].

4   Query Answering Algorithm

We now present our distributed query answering algorithm that uses *dynamic data exchange*. Throughout this section, we fix a cluster C of shared-nothing servers, a strict partition **G** of an RDF graph G distributed over C, and a CQ Q. Our algorithm outputs $\mathsf{ans}(G, Q)$ as pairs $\langle \sigma, m \rangle$ of assignments and multiplicities; each σ can be output several times, but the sum of all m for σ is equal to the multiplicity of σ in $\mathsf{ans}(G, Q)$.

4.1   Intuition

We evaluate Q over G using nested index loop joins: starting with an empty assignment, we recursively extend the assignment by matching the atoms of Q; we call each assignment that matches a prefix of the atoms of Q a *partial answer*. By letting all servers evaluate Q in parallel over their respective partition element, we obtain all local answers to Q without any network communication or synchronisation between the servers. To also obtain answers that are not local, whenever some server k attempts to extend a partial answer σ so that it matches some atom A of Q, the server must take into account that other servers in the cluster may contain facts matching A as well. To identify such situations, server k uses the key notion of *occurrences* that, for any resource r in $\mathbf{G}_k$, allow server

k to identify all servers that contain r. Server k can thus use the occurrences of the resources in A and σ to identify the servers that can potentially extend σ to a match for A, so server k forwards σ only to those servers; each server receiving σ then continues matching the remaining atoms of Q. The occurrences are thus used to avoid sending σ to servers that definitely cannot extend σ to an answer of Q on G, which considerably reduces communication in the cluster.

Consider again our example query Q_3 from Fig. 1c. Evaluating the first two atoms of Q_3 in $\mathbf{G}_1$ left-to-right produces a partial answer $\sigma = \{x \mapsto b, y \mapsto a, z \mapsto c\}$, so we must next evaluate $\langle w, T, z \rangle \sigma = \langle w, T, c \rangle$. By keeping the occurrences of the resources from $\mathbf{G}_1$, server 1 determines that resource c occurs in both $\mathbf{G}_1$ and $\mathbf{G}_2$ so it branches its execution: it continues evaluating the query locally and thus computes σ_2, but it also sends the partial answer σ and atom index 3 to server 2. Upon receiving this message, server 2 continues evaluating the query starting from atom 3 and produces answer σ_3.

Data exchange in our setting is thus dynamic (i.e., it is determined by the occurrences), which allows servers to always compute all local answers locally. Moreover, messages are exchanged asynchronously, without predetermined synchronisation points in the query plan: partial answers can be sent and processed as soon as they are produced, which promotes parallelisation. However, as we shall see, the asynchronous nature of our algorithm makes detecting termination nontrivial.

Our notion of occurrences exhibits two important properties. First, we require each server k to store only the occurrences for the resources that are present in $\mathbf{G}_k$. As we discuss in Sect. 4.3, this complicates determining where to forward partial answers; however, this assumption is critical for scalability because it makes the size of the occurrences at server k proportional to the size of $\mathbf{G}_k$, rather than to the size of G. Second, we track the occurrences of resources for subject, predicate, and object position independently, which we use to further limit communication. For example, if an atom A to be matched next contains a resource r in the subject position, then a partial answer is sent to server k' only if r occurs in $\mathbf{G}_{k'}$ in the subject position. As a consequence of this optimisation, if the data is partitioned such that all triples containing the same resource in the subject are colocated, subject-join queries are answered without any communication.

4.2   Setting

Before presenting our approach in detail, we discuss the assumptions we make on each server in the cluster.

We assume that each server $k \in C$ stores the partition element $\mathbf{G}_k$. For A an atom and X a set of variables, EVALUATE($A, \mathbf{G}_k, X$) evaluates A in $\mathbf{G}_k$ and returns the multiset containing one occurrence of $\rho|_X$ for each assignment ρ with $\mathrm{dom}(\rho) = \mathrm{vars}(A)$ and $A\rho \in \mathbf{G}_k$. For reasons we discuss in Sect. 4.3, this multiset must be represented as a set of pairs $\langle \rho|_X, c \rangle$ where c is the multiplicity of $\rho|_X$.

In addition to $\mathbf{G}_k$, for each $\beta \in \{s, p, o\}$, server k must also store the *occurrences mapping* $\mu_{k,\beta} : \text{voc}_\beta(\mathbf{G}_k) \to 2^C$ that, for each resource $r \in \text{voc}_\beta(\mathbf{G}_k)$, returns the *occurrences* of r as $\mu_{k,\beta}(r) = \{k' \in C \mid r \in \text{voc}_\beta(\mathbf{G}_{k'})\}$.

Finally, we assume that each server can use $\text{SEND}(L, msg)$ to send a message msg to all servers listed in set L. Message delivery must be guaranteed: each sent message must be eventually received and processed; however, we make no assumptions about the order of message delivery, not even for messages sent from the same server. For the moment, we assume that the call always succeeds—that is, each sent message is delivered to all the servers in L in a finite amount of time. In Sect. 4.5, we show how $\text{SEND}(L, msg)$ can be realised so that it handles the case where each server can accept only a bounded number of messages.

4.3   Computing Query Answers

The client can submit Q for processing to any sever k_c in the cluster, and so server k_c becomes the *coordinator* for Q; the client will receive all answers from server k_c. Coordinator processes Q using Algorithm 1. In line 2, the coordinator determines an efficient ordering of the query atoms; this can be done using any of the well-known query planning techniques. In line 4 the coordinator sends the reordered query to all servers; this is done synchronously so that no server starts sending partial answers to servers that have not yet accepted Q. Finally, to start the processing of Q in the cluster, in line 5 the coordinator sends to each server in the cluster the empty partial answer.

Each server $k \in C$ (including the coordinator) accepts Q for processing using procedure $\text{START}(k_c, \vec{x}, A_1, \ldots, A_n)$ from Algorithm 2. The procedure initialises certain local variables, starts a number of message processing threads, and then terminates; all further processing at server k is driven by the messages that the server receives. The server processes messages in lines 15–21. The ANS messages represent partial answers produced at other servers and we discuss them shortly; moreover, the FIN messages are used to detect termination and we discuss them in Sect. 4.4. Each message is associated with a *stage* integer i that satisfies $1 \leq i \leq n + 1$.

Message $\text{ANS}[i, \sigma, m, \lambda_s, \lambda_p, \lambda_o]$ informs a server that σ is a partial answer with multiplicity m. As we discuss later, the algorithm eagerly removes certain variables from partial answers to save bandwidth; thus, although σ does not necessarily cover all the variables of $A_1, \ldots, A_{i-1}$, for each σ there exists an assignment ν that coincides with σ on $\text{dom}(\sigma)$ and that satisfies $\{A_1\nu, \ldots, A_{i-1}\nu\} \subseteq G$. Finally, for $\beta \in \{s, p, o\}$ a position, $\lambda_\beta : \mathcal{R} \to 2^C$ is a partial function that determines the location of certain resources in σ; we discuss the role of λ_β shortly. Such a message is forwarded in line 16 to the MATCHATOM procedure that implements index nested loop join. Line 23 determines the recursion base: if $i = n + 1$, then σ is an answer to Q on G and it is output to the client in line 24. Otherwise, in line 27 atom $A_i\sigma$ is evaluated in $\mathbf{G}_k$ and, for each match ρ, assignment σ is extended with ρ to σ' in line 28 so that the remaining atoms can be evaluated recursively. Due to data distribution, however, recursion may also need to continue on other servers. The set L of relevant servers is identified in lines 29–35 using the following observations.

- If all atoms have been matched, then line 30 ensures that the answer σ' is forwarded to the coordinator so that it can be delivered to the client.
- Otherwise, atom $A_{i+1}\sigma'$ containing a resource r in position β cannot be matched at a server $\ell \in C$ that does not contain r in position β; hence, lines 32–35 determine the servers that contain all resources occurring in $A_{i+1}\sigma'$ at the respective positions.

After the set L of relevant servers has been computed, the computation branches to the servers in $L \setminus \{k\}$ by sending them an ANS message in line 36; and if $k \in L$, processing also continues on server k via a recursive call in line 38.

The Role of λ_s, λ_p, and λ_o. As we have already explained, each server tracks the occurrences only for the resources that it contains, which introduces a complication. For example, consider evaluating query Q_4 over the following partition:

$$Q_4(x, y, z) = \langle x, R, y \rangle \wedge \langle y, S, z \rangle \wedge \langle x, T, z \rangle$$

$$\mathbf{G}_1 = \{\langle a, R, b \rangle, \langle a, T, c \rangle\} \qquad \mathbf{G}_2 = \{\langle b, S, c \rangle\} \qquad \mathbf{G}_3 = \{\langle e, T, f \rangle\}$$

Now let $\sigma' = \{x \mapsto a, y \mapsto b\}$ be the partial answer obtained by matching the first two atoms in $\mathbf{G}_1$ and $\mathbf{G}_2$, respectively, and consider processing in line 28. Then, we have $A_{i+1}\sigma' = \langle a, T, z \rangle$, but resource a does not occur in $\mathbf{G}_2$, and so server 2 has no way of knowing where to forward σ. To remedy this, our algorithm tracks the location of resources matched thus far using partial mappings λ_s, λ_p, and λ_p. When $A_i\sigma$ is matched at server k, the server's mappings $\mu_{k,\beta}$ contain information about each resource r occurring in $A_i\sigma'$; now if r also occurs in $A_j\sigma'$ with $j > i + 1$, then the information about the location of r might be relevant when evaluating such A_j. Therefore, the algorithm records the location of r in λ'_β, which is sent along with partial answers.

Handling Projected Variables. To optimise variable projection, line 26 determines the set X of variables that are needed after A_i. Variables not occurring in X are removed from σ' in line 28 in order to reduce message size. Furthermore, $A_i\sigma$ is evaluated in line 27 using EVALUATE by grouping the resulting assignments X, which can considerably improve performance. For example, let $Q_5(x) = \langle x, R, y \rangle \wedge \langle x, S, z \rangle$, and let $\mathbf{G}_k$ contain triples $\langle a, R, b_i \rangle$ and $\langle a, S, c_j \rangle$ for $1 \leq i \leq u$ and $1 \leq j \leq v$. A naïve evaluation of the index nested loop join requires $u \cdot v$ steps, producing the same number of answer messages. In contrast, our algorithm uses $u + v$ steps: evaluating the first atom returns the pair $\langle \rho = \{x \mapsto a\}, u \rangle$ using u steps, and evaluating the second atom returns the pair $\langle \rho' = \emptyset, v \rangle$ using v steps. In addition, our algorithm sends just one answer message in this case, which is particularly important in a distributed setting.

4.4 Detecting Termination

Termination is detected by tracking the per-server completion of each atom (stage) in the query. In particular, server k can finish processing stage i if (i)

Algorithm 1. Initiating the Query at Coordinator k_c

1: **procedure** ANSWERQUERY(Q)
2:     Reorder the query atoms as $Q(\vec{x}) = A_1 \wedge \cdots \wedge A_n$ to obtain an efficient plan
3:     **for** $k \in C$ **do**
4:         Call START($k_c, \vec{x}, A_1, \ldots, A_n$) on server k synchronously
5:     SEND(C, ANS[$1, \emptyset, 1, \emptyset, \emptyset, \emptyset$])

---

it knows that all servers in C have finished processing stages up to $i - 1$ by receiving the respective FIN messages, and (ii) it has processed all received messages for this stage. At this point, server k sends a FIN message to all other servers informing them that they will not receive further messages from k for stage i. To this end, each server k keeps several counters: P_i and R_i count the ANS messages for stage i that the server processed and received, respectively; and N_i counts the FIN messages that servers have sent to inform k that they have finished processing stage i. Thus, if $N_i = |C|$ holds at server k, then all other servers have finished sending all messages for all stages prior to i and so server k will not get further partial answers to process for stages up to i. If in addition $P_i = R_i$, then server k has finished stage i and it then sends his FIN message for i. Only one thread must detect this condition line 40, which is ensured by SWAP(F_i, *true*): this operation atomically reads F_i, stores *true* into F_i, and returns the previous value of F_i. Hence, this operation returns *false* just once, in which case server k then informs in line 47 all servers (or just the coordinator if $i = n$) of this by sending a message FIN[$i, S_{i,\ell}$], where $S_{i,\ell}$ is the number of ANS messages that server k sent to ℓ for stage i. Server ℓ processes this message in line 19 by incrementing R_i and N_i, which can cause further termination messages to be sent. Since each server sends $|C|$ messages to all other servers per stage, detecting termination requires $\Theta(n|C|^2)$ messages in total.

4.5 Dealing with Finite Resources

Nested index loop joins require just one iterator per query atom, so a query with n atoms can be answered using $O(n)$ memory; this is particularly important when servers store their data in RAM. The algorithm as presented thus far does not have this property: partial answers produced in line 36 must be stored on the sending and/or the receiving server before they are processed. In the worst case, queries can produce exponentially many answers and so the number of messages sent in line 36 can be large; consequently, the cumulative size of all messages sent to a server can exceed the server's capacity. We now describe how our query answering algorithm overcomes this drawback.

To formulate our idea abstractly, we assume that each server in the cluster contains $n + 1$ finite *queues*. Moreover, function PUTINTOQUEUE(ℓ, i, msg) instructs the message passing infrastructure to insert message msg into queue i on server ℓ. The function returns *true* if the infrastructure can guarantee that msg will be placed into the appropriate queue eventually, otherwise it returns

---

Algorithm 2. Processing at Server k

6: **procedure** START($k_c, \vec{x}, A_1, \ldots, A_n$)
7:     **for** $1 \leq i \leq n + 1$ **do**
8:         **for** $\ell \in C$ **do** $S_{i,\ell} := 0$ ▷ # (partial) answers sent to server ℓ for stage i
9:         $P_i := 0$ ▷ # processed (partial) answers for stage i
10:        $R_i := (i = 1 \ ? \ 1 : 0)$ ▷ # received (partial) answers for stage i
11:        $N_i := (i = 1 \ ? \ |C| : 0)$ ▷ # servers finished sending messages for stage i
12:        $F_i := \textit{false}$ ▷ has this server finished stage i?
13:     Start message processing threads that, until **exit** is called, repeatedly
           extract an unprocessed message msg and call PROCESSMESSAGE(msg)

14: **procedure** PROCESSMESSAGE(msg)
15:     **if** $msg = \mathsf{ANS}[i, \sigma, m, \lambda_s, \lambda_p, \lambda_o]$ **then**   ▷ Partial/query answer
16:         MATCHATOM($i, \sigma, m, \lambda_s, \lambda_p, \lambda_o$)
17:         $P_i := P_i + 1$
18:         CHECKTERMINATION(i
19:     **else if** $msg = \mathsf{FIN}[i, m]$) **then**   ▷ Atom/query termination
20:         $R_i := R_i + m, \qquad N_i := N_i + 1$
21:         CHECKTERMINATION(i)

22: **procedure** MATCHATOM($i, \sigma, m, \lambda_s, \lambda_p, \lambda_o$)
23:     **if** $i = n + 1$ **then**
24:         Output answer $\langle \sigma, m \rangle$ to the client
25:     **else**
26:         $X := \vec{x} \cup \mathsf{vars}(A_{i+1}) \cup \cdots \cup \mathsf{vars}(A_n)$
27:         **for each** $\langle \rho, h \rangle \in$ EVALUATE($A_i\sigma, \mathbf{G}_k, X$) **do**
28:             $\sigma' := (\sigma \cup \rho)|_X, \qquad m' := m \cdot h$
29:             **if** $i = n$ **then**
30:                 $L := \{k_c\}, \qquad \lambda_s' := \lambda_p' := \lambda_o' := \emptyset$
31:             **else**
32:                 $L := C$
33:                 **for** $\beta \in \{s, p, o\}$ **do**
34:                     $\lambda_\beta' := (\lambda_\beta \cup \mu_{k,\beta})|_Y$ for $Y = \mathcal{R} \cap \{\mathsf{term}_\beta(A_j\sigma') \mid i + 1 < j \leq n\}$
35:                     **if** $\mathsf{term}_\beta(A_{i+1}\sigma') \in \mathsf{dom}(\lambda_\beta')$ **then** $L := L \cap \lambda_\beta'(\mathsf{term}_\beta(A_{i+1}\sigma'))$
36:             SEND($L \setminus \{k\}, \mathsf{ANS}[i + 1, \sigma', m', \lambda_s', \lambda_p', \lambda_o']$)
37:             **for** $\ell \in L \setminus \{k\}$ **do** $S_{i+1,\ell} := S_{i+1,\ell} + 1$
38:             **if** $k \in L$ **then** MATCHATOM($i + 1, \sigma', m', \lambda_s', \lambda_p', \lambda_o'$)

39: **procedure** CHECKTERMINATION(i)
40:     **if** $P_i = R_i$ and $N_i = |C|$ and SWAP($F_i, true$) = $\textit{false}$ **then**
41:         **if** $i = n + 1$ **then**
42:             Tell client that Q has been answered and **exit**
43:         **else if** $i = n$ **then**
44:             SEND($\{k_c\}, \mathsf{FIN}[i + 1, S_{i+1,k_c}]$)
45:             **if** $k \neq k_c$ **then exit**
46:         **else**
47:             **for** $\ell \in C$ **do** SEND($\{\ell\}, \mathsf{FIN}[i + 1, S_{i+1,\ell}]$)

Algorithm 3. Message Sending for Resource-Constrained Servers
---
48: **procedure** SEND(L, msg)
49:     $i :=$ the stage index that message msg is associated with
50:     **loop**
51:         **for all** $\ell \in L$ **do**
52:             **if** PUTINTOQUEUE(ℓ, i, msg) **then** $L := L \setminus \{\ell\}$
53:         **if** $L = \emptyset$ **then return**
54:         If an unprocessed message for stage j with $j > i$ exists,
                extract one such message msg and call PROCESSMESSAGE(msg)
---

false. Note that the return value of *true* does not imply that the message has actually been delivered; thus, message passing can still be asynchronous. Queues can be implemented in many different ways using common networking infrastructure. For example, TCP uses *sliding window protocol* for congestion control, so one TCP connection could provide a pair of queues. Another solution is to multiplex $n + 1$ queues onto a single TCP connection. Yet another solution is to use explicit signalling: when a server sees that it is running out of queue space, it tells the sender not to send any more data until further notice.

To handle finite resources, our algorithm implements SEND(L, msg) in terms of PUTINTOQUEUE as shown in Algorithm 3: as long as some queue for stage i is blocked, server k keeps processing messages for stages larger than i. This ensures recursion depth of at most $n + 1$, so each server's thread uses $O(n^2)$ memory. To see why Algorithm 2 necessarily terminates, even with queues of bounded size, we make two observations. First, processing a message for stage i only calls PUTINTOQUEUE(ℓ, j, msg) for $j > i$, which fails only if queue j on server ℓ is full. Second, at any given point in time the cluster contains at least one highest-indexed nonempty queue across the cluster. As a result of these observations, messages from the highest-indexed nonempty queue can always be processed. Thus, although individual servers in the cluster can become blocked at different points in time, at least one server in the cluster makes progress at any given point in time, which eventually ensures termination.

The following theorem captures the properties of our algorithm, and its proof is given online at http://www.cs.ox.ac.uk/people/anthony.potter/rdfox-tr.pdf.

Theorem 1. *When Algorithm 1 is applied to a strict partition* **G** *of an RDF graph G distributed over a cluster C of servers where each server has $n + 1$ finite message queues, the following claims hold:*

1. *the coordinator for Q correctly outputs* ans(Q, G),
2. *each server sends $\Theta(n|C|^2)$ FIN messages and then terminates, and*
3. *each server thread uses $O(n^2)$ memory.*

5   Data Partitioning Algorithm

Ensuring that computation is not passed from server to server often is key to ensuring efficiency of our approach. Therefore, in this section we present a new data partitioning strategy based on *weighted graph partitioning* that (i) aims to maximise the number of local answers on common queries, (ii) does not duplicate triples, and (iii) produces partitions balanced in the number of triples. Throughout this section, let G be an RDF graph that we wish to partition into $|C|$ elements. Our algorithm proceeds in three steps.

First, we transform G into an undirected weighted graph (V, E, w) as follows: we define the set of vertices as $V = \text{voc}_s(G)$—that is, V is the set of resources occurring in G in the subject position; we add to the set of edges E an undirected edge $\{s, o\}$ for each triple $\langle s, p, o \rangle \in G$ if $p \neq rdf{:}type$ and o is not a literal (e.g., a string or an integer); and we define the weight $w(r)$ of each resource $r \in V$ as the number of triples in G that contain r in the subject position. Classes and literals often occur in RDF graphs in many triples, and the presence of such hubs can easily confuse partitioners such as METIS, so we *prune* such resources from (V, E, w). As we discuss shortly, this does not affect the performance of distributed query answering for the queries commonly used in practice.

Second, we partition (V, E, w) by *weighted graph partitioning* [11]—that is, we compute a function $\pi : V \to C$ such that (i) the number of edges spanning partitions is minimised, while (ii) the sum of the weights of the vertices assigned to each partition is approximately the same for all partitions.

Third, we compute each partition element by assigning triples based on subject—that is, we assign each triple $\langle s, p, o \rangle \in G$ to partition element $\mathbf{G}_{\pi(s)}$. Note that this ensures no duplication between partition elements.

This data partitioning strategy is tailored to efficiently support common query loads. By analysing more than 3 million real-world SPARQL queries, it was shown [4] that approximately 60 % of joins are subject–subject joins, 35 % are subject–object joins, and less than 5 % are object–object joins. Now pruning classes and literals before graph partitioning makes it more likely that such resources will end up in different partitions; however, this can affect the performance only of object–object joins, which are the least common in practice. In other words, pruning does not affect the performance of 95 % of the queries occurring in practice, but it increases the chance of obtaining a good partition, as well as reduces the size of (V, E, w). Furthermore, by placing all triples with the same subject on a single server in the third step, we can answer the most common type of join without any communication; this includes subject-join queries, which are particularly important in practice. Finally, the weight $w(r)$ of each vertex r in (V, E, w) determines exactly the number of triples are added to $\mathbf{G}_{\pi(r)}$ as a consequence of assigning r to partition $\pi(r)$; since weighted graph partitioning balances the sum of the weights of vertices in each partition, this ensures that the resulting partition elements are balanced in terms of their size. As we experimentally show in Sect. 6, our partitions are indeed much more balanced than the ones produced by the existing approaches based on graph partitioning [7, 10, 16]. This is important because it ensures that the servers in the cluster use roughly the same amount of memory for storing their respective partition elements.

6   Evaluation

We implemented our query answering and data partitioning algorithms in our RDFox system.[2] The authors of TriAD [7] have already shown that their system outperforms Trinity.RDF [22], SHARD [17], H-RDF-3X [10], 4store [8], RDF-3X [14], BitMat [2], and MonetDB [20]; therefore, we have evaluated our approach by comparing it with TriAD only. We have conducted our experiments using the m4.2xlarge servers of the Amazon Elastic Compute Cloud.[3] Each server had 32 GB of RAM and eight virtual cores of 2.4GHz Intel Xeon E5-2676v3 CPUs.

We generated the WatDiv-10K dataset of the WatDiv [1] v0.6 benchmark, and used the 20 basic testing queries, which are divided into four groups: complex (C), snowflake (F), linear (L), and star (S) queries. We also generated the LUBM-10K dataset of the widely used LUBM [6] benchmark. Many of the LUBM queries return no answers if the dataset is not extended via reasoning, so we used the seven queries from [22] that compensate for the lack of reasoning (Q1–Q7), and we manually generated three additional complex queries (Q8–Q10). All queries that we used in the evaluation are given online at http://www.cs.ox.ac.uk/people/anthony.potter/rdfox-tr.pdf.

6.1   Evaluating Query Answering

To evaluate the effectiveness of our distributed query answering algorithm, we compared RDFox and TriAD using a cluster of ten servers. For RDFox, we partitioned the data into ten partition elements as described in Sect. 5. For TriAD, one master server partitioned the data across nine workers using TriAD's summary mode. Both systems produced the answers on one server, but without printing them. For each query, we recorded the wall-clock query time, the total amount of data sent over the network, and the maximum amount of memory used by a server for query processing.

WatDiv-10K results are summarised in Table 1. TriAD threw an exception on queries F4 and S5, which is why the respective entries are empty. Both RDFox and TriAD offer comparable performance for linear and star queries, which in both cases require little network communication. On complex queries, RDFox was faster in two out of three cases despite the fact that TriAD uses a summary graph optimisation [7] to aggressively prune the search space on complex queries. RDFox could process queries F2, F3, and F5 by up to two orders of magnitude quicker and with up to two orders of magnitude less data sent over the network. Moreover, all queries apart from C3 do not return large datasets, so the memory used for query processing was comparable.

LUBM-10K results are summarised in Table 2. RDFox was quicker than TriAD on all queries apart from Q5 and Q8, on which the two systems were roughly comparable. The difference was most pronounced on Q1, Q7, Q9, and Q10, on

---

[2] http://www.cs.ox.ac.uk/isg/tools/RDFox/.

[3] http://aws.amazon.com/ec2/.

Table 1. Query answering results on WatDiv-10K

Query	Answer count	RDFox Time (ms)	Network use (KB)	Max Mem. (MB)	TriAD Time (ms)	Network use (KB)	Max Mem. (MB)
C1	1,504	**148**	9,043	31	248	**3,170**	**27**
C2	288	493	**32,866**	**2**	**343**	45,520	97
C3	42,441,808	**373**	1,190	13	419	**423**	**8**
F1	324	62	4,013	1	**15**	**265**	1
F2	188	**10**	**92**	1	263	11,461	25
F3	865	**15**	**199**	1	208	337	29
F4	2,879	**25**	**471**	1	–	–	–
F5	65	**5**	**61**	1	348	29,900	76
L1	2	**3**	**29**	1	11	227	1
L2	16,132	41	**259**	1	**15**	1,106	1
L3	24	**2**	**20**	1	6	76	1
L4	5,782	14	**92**	1	**5**	299	1
L5	12,957	21	**306**	1	**17**	940	1
S1	12	**5**	**79**	1	41	142	1
S2	6,685	**12**	**183**	1	33	517	1
S3	0	25	**35**	1	**8**	91	1
S4	153	**19**	3,096	1	22	**108**	1
S5	0	**10**	**37**	**1**	–	–	–
S6	453	8	**37**	1	8	151	1
S7	0	**2**	**29**	1	3	58	1

which TriAD used significant amounts of memory. This is because TriAD evaluated queries using bushy query plans consisting of hash joins; for example, on Q10 TriAD used over 6 GB—more than half the amount needed to store the data. In contrast, RDFox uses index nested loop joins that require very little memory: at most 147 MB were used in all cases, mainly to store the messages passed between the servers. Furthermore, on most queries RDFox sent less data over the network, leading us to believe that dynamic data exchange can considerably reduce communication during query processing.

6.2 Effectiveness of Data Partitioning

To evaluate our data partitioning algorithm, we have partitioned our test data into ten elements in four different ways: with both weighted partitioning and pruning as described in Sect. 5, without pruning, without weighted partitioning, and by subject hashing. For each partitioning obtained in this way, Table 3 shows the minimum and maximum number of triples, the average number of resources

Table 2. Query answering results on LUBM-10K

		RDFox			TriAD		
Query	Answer count	Time (ms)	Network use (KB)	Max Mem. (MB)	Time (ms)	Network use (KB)	Max Mem. (MB)
Q1	2,528	**1,927**	**2,261**	**33**	13,410	197,762	1,144
Q2	10,799,863	**701**	150,565	**147**	927	**104,657**	154
Q3	0	**443**	1,809	**1**	771	**466**	708
Q4	10	**4**	**45**	**1**	7	115	1
Q5	10	2	**18**	1	2	63	1
Q6	125	**4**	**39**	1	85	153	1
Q7	439,994	**975**	**10,860**	**8**	7,294	29,592	844
Q8	2,528	1,771	**5,497**	**20**	**1,755**	8,154	232
Q9	4,111,592	**6,281**	**141,603**	**103**	23,711	184,661	3,501
Q10	2,225,206	**1,096**	**38,030**	**29**	33,661	111,571	6,645

Table 3. Comparing the partitioning strategies of RDFox and TriAD

	WatDiv				LUBM			
Partitioning	Min	Max	Avg. Res.	P	Min	Max	Avg. Res.	P
Weighted, pruning	103.1 M	113.0 M	20.9 M	72.1%	126.4 M	138.2 M	32.9 M	0.3%
Weighted, no pruning	102.1 M	113.0 M	21.6 M	72.3%	123.6 M	139.8 M	35.7 M	13.3%
Unweighted, no pruning	22.5 M	410.7 M	18.1 M	63.0%	123.7 M	142.3 M	36.0 M	14.5%
Subject hashing	109.0 M	109.3 M	24.2 M	79.2%	133.3 M	133.7 M	52.5 M	46.8%

Table 4. Comparing the idle memory use of RDFox and TriAD

	WatDiv			LUBM		
System	Mean (GB)	Max (GB)	Sdev (GB)	Mean (GB)	Max (GB)	Sdev (GB)
RDFox	4.39	5.42	0.54	5.49	5.61	0.15
TriAD	9.57	10.99	0.73	12.04	19.26	3.98

per partition, and the average percentage of the resources that occur in more than one partition. In all cases, subject hashing produces very balanced partitions, but the percentage of resources that occur on more than one server is large. Weighted partitioning reduces this percentage on LUBM dramatically to 0.3%. Our partitioning is not as effective on WatDiv, but it still offers some improvement. Partitions are well balanced in all cases, apart from WatDiv with unweighted partitioning: WatDiv contains several hubs, so a balanced number of resources in partitions does not ensure a balanced number of triples.

We also compared the idle memory use (excluding dictionaries) of RDFox and TriAD's workers, in order to indirectly compare the partitioning approaches used by the two systems. Table 4 shows the minimal and maximal memory use per server after the data is loaded, as well as the standard deviation across all

servers. As one can see, RDFox uses about half of the memory of TriAD. We
believe is due to the fact that our partitioning strategy does not duplicate data,
whereas TriAD hashes its groups by subject and object. Furthermore, memory
use per server is more balanced for RDFox, which we believe is due to weighted
graph partitioning.

7 Conclusion

We have presented a new technique for query answering in distributed RDF systems based on dynamic data exchange, which ensures that all local answers to
a query are computed locally and thus reduces the amount of data transferred
between servers. Using index nested loops and message prioritisation, the algorithm is very memory-efficient while still guaranteeing termination. Furthermore,
we have presented an algorithm for partitioning RDF data based on weighted
graph partitioning. The results of our performance evaluation show that our algorithms outperform the state of the art, sometimes by orders of magnitude. In our
future work, we shall focus on adapting the known query planning techniques to
the distributed setting. Moreover, we shall evaluate our approach against modern
big data systems such as Spark and Impala.

Acknowledgments. This work was funded by the EPSRC projects MaSI³, DBOnto,
and ED3, an EPSRC doctoral training grant, and a grant by Roke Manor Research
Ltd.

References

1. Aluç, G., Hartig, O., Özsu, M.T., Daudjee, K.: Diversified stress testing of RDF
   data management systems. In: Mika, P., et al. (eds.) ISWC 2014, Part I. LNCS,
   vol. 8796, pp. 197–212. Springer, Heidelberg (2014)
2. Atre, M., Chaoji, V., Zaki, M.J., Hendler, J.A.: Matrix "Bit" loaded: a scalable
   lightweight join query processor for RDF data. In: Proceedings of WWW, pp.
   41–50 (2010)
3. Dean, J., Ghemawat, S.: MapReduce: a flexible data processing tool. Commun.
   ACM **53**(1), 72–77 (2010)
4. Gallego, M.A., Fernández, J.D., Martínez-Prieto, M.A., de la Fuente, P.: An empirical study of real-world SPARQL queries. CoRR abs/1103.5043 (2011)
5. Graefe, G., Davison, D.L.: Encapsulation of parallelism and architecture-
   independence in extensible database query execution. IEEE Trans. Softw. Eng.
   19(8), 749–764 (1990)
6. Guo, Y., Pan, Z., Heflin, J.: LUBM: a benchmark for OWL knowledge base systems.
   J. Web Semant. **3**(2), 158–182 (2005)
7. Gurajada, S., Seufert, S., Miliaraki, I., Theobald, M.: TriAD: a distributed shared-
   nothing RDF engine based on asynchronous message passing. In: Proceedings SIG-
   MOD, pp. 289–300 (2014)
8. Harris, S., Lamb, N., Shadbolt, N.: 4store: the design and implementation of a
   clustered RDF store. In: Proceedings of SSWS (2009)

9. Harth, A., Umbrich, J., Hogan, A., Decker, S.: YARS2: a federated repository for querying graph structured data from the web. In: Aberer, K., et al. (eds.) ASWC 2007 and ISWC 2007. LNCS, vol. 4825, pp. 211–224. Springer, Heidelberg (2007)

10. Huang, J., Abadi, D.J., Ren, K.: Scalable SPARQL querying of large RDF graphs. PVLDB 4(11), 1123–1134 (2011)

11. Karypis, G., Kumar, V.: A fast and high quality multilevel scheme for partitioning irregular graphs. SIAM J. Sci. Comput. 20(1), 359–392 (1998)

12. Lee, K., Liu, L.: Scaling queries over big RDF graphs with semantic hash partitioning. PVLDB 6(14), 1894–1905 (2013)

13. Nenov, Y., Piro, R., Motik, B., Horrocks, I., Wu, Z., Banerjee, J.: RDFox: a highly-scalable RDF Stored. In: Arenas, M., et al. (eds.) ISWC 2015. LNCS, vol. 9367, pp. 3–20. Springer, Heidelberg (2015). doi:10.1007/978-3-319-25010-6_1

14. Neumann, T., Weikum, G.: The RDF-3X engine for scalable management of RDF data. VLDB J. 19(1), 91–113 (2010)

15. Papailiou, N., Konstantinou, I., Tsoumakos, D., Karras, P., Koziris, N.: H₂RDF+: high-performance distributed joins over large-scale RDF graphs. In: Proceedings of Big Data, pp. 255–263 (2013)

16. Potter, A., Motik, B., Horrocks, I.: Querying distributed RDF graphs: the effects of partitioning. In: Proceedings of SSWS (2014)

17. Rohloff, K., Schantz, R.E.: Clause-iteration with MapReduce to scalably query data graphs in the SHARD graph-store. In: Proceedings of DIDC, pp. 35–44 (2011)

18. Schätzle, A., Przyjaciel-Zablocki, M., Hornung, T., Lausen, G.: PigSPARQL: a SPARQL query processing baseline for big data. In: Proceedings of ISWC (Poster), pp. 241–244 (2013)

19. Schätzle, A., Przyjaciel-Zablockl, M., Neu, A., Lausen, G.: Sempala: interactive SPARQL query processing on hadoop. In: Mika, P., et al. (eds.) ISWC 2014, Part I. LNCS, vol. 8796, pp. 164–179. Springer, Heidelberg (2014)

20. Sidirourgos, L., Gonçalves, R., Kersten, M., Nes, N., Manegold, S.: Column-store support for RDF data management: not all swans are white. PVLDB 1(2), 1553–1563 (2008)

21. Wu, B., Zhou, Y., Yuan, P., Jin, H., Liu, L.: SemStore: a semantic-preserving distributed RDF triple store. In: Proceedings of CIKM, pp. 509–518 (2014)

22. Zeng, K., Yang, J., Wang, H., Shao, B., Wang, Z.: A distributed graph engine for web scale RDF data. PVLDB 6(4), 265–276 (2013)

RDF2Vec: RDF Graph Embeddings
for Data Mining

Petar Ristoski[(⊠)] and Heiko Paulheim

Data and Web Science Group, University of Mannheim, Mannheim, Germany
{petar.ristoski,heiko}@informatik.uni-mannheim.de

Abstract. Linked Open Data has been recognized as a valuable source
for background information in data mining. However, most data mining
tools require features in propositional form, i.e., a vector of nominal or
numerical features associated with an instance, while Linked Open Data
sources are graphs by nature. In this paper, we present RDF2Vec, an approach that uses language modeling approaches for unsupervised feature
extraction from sequences of words, and adapts them to RDF graphs.
We generate sequences by leveraging local information from graph substructures, harvested by Weisfeiler-Lehman Subtree RDF Graph Kernels
and graph walks, and learn latent numerical representations of entities in
RDF graphs. Our evaluation shows that such vector representations outperform existing techniques for the propositionalization of RDF graphs
on a variety of different predictive machine learning tasks, and that feature vector representations of general knowledge graphs such as DBpedia
and Wikidata can be easily reused for different tasks.

Keywords: Graph embeddings · Linked open data · Data mining

1 Introduction

Linked Open Data (LOD) [29] has been recognized as a valuable source of background knowledge in many data mining tasks and knowledge discovery in general
[25]. Augmenting a dataset with features taken from Linked Open Data can, in
many cases, improve the results of a data mining problem at hand, while exter-
nalizing the cost of maintaining that background knowledge [18].

Most data mining algorithms work with a propositional *feature vector* representation of the data, i.e., each instance is represented as a vector of features
$\langle f_1, f_2, \ldots, f_n \rangle$, where the features are either binary (i.e., $f_i \in \{true, false\}$),
numerical (i.e., $f_i \in \mathbb{R}$), or nominal (i.e., $f_i \in S$, where S is a finite set of symbols). LOD, however, comes in the form of *graphs*, connecting resources with
types and relations, backed by a schema or ontology.

Thus, for accessing LOD with existing data mining tools, transformations
have to be performed, which create propositional features from the graphs in
LOD, i.e., a process called *propositionalization* [10]. Usually, binary features
(e.g., **true** if a type or relation exists, **false** otherwise) or numerical features

© Springer International Publishing AG 2016
P. Groth et al. (Eds.): ISWC 2016, Part I, LNCS 9981, pp. 498–514, 2016.
DOI: 10.1007/978-3-319-46523-4_30

(e.g., counting the number of relations of a certain type) are used [20,24]. Other variants, e.g., counting different graph sub-structures are possible [34].

In this work, we adapt language modeling approaches for latent representation of entities in RDF graphs. To do so, we first convert the graph into a set of sequences of entities using two different approaches, i.e., graph walks and Weisfeiler-Lehman Subtree RDF graph kernels. In the second step, we use those sequences to train a neural language model, which estimates the likelihood of a sequence of entities appearing in a graph. Once the training is finished, each entity in the graph is represented as a vector of latent numerical features.

Projecting such latent representations of entities into a lower dimensional feature space shows that semantically similar entities appear closer to each other. We use several RDF graphs and data mining datasets to show that such latent representation of entities have high relevance for different data mining tasks.

The generation of the entities' vectors is task and dataset independent, i.e., once the vectors are generated, they can be used for any given task and any arbitrary algorithm, e.g., SVM, Naive Bayes, Random Forests, Neural Networks, KNN, etc. Also, since all entities are represented in a low dimensional feature space, building machine learning models becomes more efficient. To foster the reuse of the created feature sets, we provide the vector representations of DBpedia and Wikidata entities as ready-to-use files for download.

The rest of this paper is structured as follows. In Sect. 2, we give an overview of related work. In Sect. 3, we introduce our approach, followed by an evaluation in section Sect. 4. We conclude with a summary and an outlook on future work.

2   Related Work

In the recent past, a few approaches for generating data mining features from Linked Open Data have been proposed. Many of those approaches are supervised, i.e., they let the user formulate SPARQL queries, and a fully automatic feature generation is not possible. LiDDM [8] allows the users to declare SPARQL queries for retrieving features from LOD that can be used in different machine learning techniques. Similarly, Cheng et al. [3] propose an approach feature generation after which requires the user to specify SPARQL queries. A similar approach has been used in the RapidMiner[1] semweb plugin [9], which preprocesses RDF data in a way that it can be further processed directly in RapidMiner. Mynarz and Svátek [16] have considered using user specified SPARQL queries in combination with SPARQL aggregates.

FeGeLOD [20] and its successor, the *RapidMiner Linked Open Data Extension* [23], have been the first fully automatic unsupervised approach for enriching data with features that are derived from LOD. The approach uses six different unsupervised feature generation strategies, exploring specific or generic relations. It has been shown that such feature generation strategies can be used in many data mining tasks [21,23].

---

[1] http://www.rapidminer.com/.

A similar problem is handled by *Kernel functions*, which compute the distance between two data instances by counting common substructures in the graphs of the instances, i.e. walks, paths and trees. In the past, many graph kernels have been proposed that are tailored towards specific applications [7], or towards specific semantic representations [5]. Only a few approaches are general enough to be applied on any given RDF data, regardless the data mining task. Lösch et al. [12] introduce two general RDF graph kernels, based on intersection graphs and intersection trees. Later, the intersection tree path kernel was simplified by Vries et al. [33]. In another work, Vries et al. [32,34] introduce an approximation of the state-of-the-art Weisfeiler-Lehman graph kernel algorithm aimed at improving the computation time of the kernel when applied to RDF. Furthermore, the kernel implementation allows for explicit calculation of the instances' feature vectors, instead of pairwise similarities.

Our work is closely related to the approaches DeepWalk [22] and Deep Graph Kernels [35]. DeepWalk uses language modeling approaches to learn social representations of vertices of graphs by modeling short random-walks on large social graphs, like BlogCatalog, Flickr, and YouTube. The Deep Graph Kernel approach extends the DeepWalk approach, by modeling graph substructures, like graphlets, instead of random walks. The approach we propose in this paper differs from these two approaches in several aspects. First, we adapt the language modeling approaches on directed labeled RDF graphs, compared to the undirected graphs used in the approaches. Second, we show that task-independent entity vectors can be generated on large-scale knowledge graphs, which later can be reused on variety of machine learning tasks on different datasets.

3   Approach

In our approach, we adapt neural language models for RDF graph embeddings. Such approaches take advantage of the word order in text documents, explicitly modeling the assumption that closer words in the word sequence are statistically more dependent. In the case of RDF graphs, we consider entities and relations between entities instead of word sequences. Thus, in order to apply such approaches on RDF graph data, we first have to transform the graph data into sequences of entities, which can be considered as sentences. Using those sentences, we can train the same neural language models to represent each entity in the RDF graph as a vector of numerical values in a latent feature space.

3.1   RDF Graph Sub-structures Extraction

We propose two general approaches for converting graphs into a set of sequences of entities, i.e., graph walks and Weisfeiler-Lehman Subtree RDF Graph Kernels.

Definition 1. *An RDF graph is a graph $G = (V, E)$, where V is a set of vertices, and E is a set of directed edges.*

The objective of the conversion functions is for each vertex $v \in V$ to generate a set of sequences S_v, where the first token of each sequence $s \in S_v$ is the vertex v followed by a sequence of tokens, which might be edges, vertices, or any substructure extracted from the RDF graph, in an order that reflects the relations between the vertex v and the rest of the tokens, as well as among those tokens.

Graph Walks. In this approach, for a given graph $G = (V, E)$, for each vertex $v \in V$ we generate all graph walks P_v of depth d rooted in the vertex v. To generate the walks, we use the breadth-first algorithm. In the first iteration, the algorithm generates paths by exploring the direct outgoing edges of the root node v_r. The paths generated after the first iteration will have the following pattern $v_r \rightarrow e_{1i}$, where $i \in E(v_r)$. In the second iteration, for each of the previously explored edges the algorithm visits the connected vertices. The paths generated after the second iteration will follow the following patter $v_r \rightarrow e_{1i} \rightarrow v_{1i}$. The algorithm continues until d iterations are reached. The final set of sequences for the given graph G is the union of the sequences of all the vertices $\bigcup_{v \in V} P_v$.

Weisfeiler-Lehman Subtree RDF Graph Kernels. In this approach, we use the subtree RDF adaptation of the Weisfeiler-Lehman algorithm presented in [32,34]. The Weisfeiler-Lehman Subtree graph kernel is a state-of-the-art, efficient kernel for graph comparison [30]. The kernel computes the number of sub-trees shared between two (or more) graphs by using the Weisfeiler-Lehman test of graph isomorphism. This algorithm creates labels representing subtrees in h iterations.

There are two main modifications of the original Weisfeiler-Lehman graph kernel algorithm in order to be applicable on RDF graphs [34]. First, the RDF graphs have directed edges, which is reflected in the fact that the neighborhood of a vertex v contains only the vertices reachable via outgoing edges. Second, as mentioned in the original algorithm, labels from two iterations can potentially be different while still representing the same subtree. To make sure that this does not happen, the authors in [34] have added tracking of the neighboring labels in the previous iteration, via the multiset of the previous iteration. If the multiset of the current iteration is identical to that of the previous iteration, the label of the previous iteration is reused.

The procedure of converting the RDF graph to a set of sequences of tokens goes as follows: (i) for a given graph $G = (V, E)$, we define the Weisfeiler-Lehman algorithm parameters, i.e., the number of iterations h and the vertex subgraph depth d, which defines the subgraph in which the subtrees will be counted for the given vertex; (ii) after each iteration, for each vertex $v \in V$ of the original graph G, we extract all the paths of depth d within the subgraph of the vertex v on the relabeled graph. We set the original label of the vertex v as the starting token of each path, which is then considered as a sequence of tokens. The sequences after the first iteration will have the following pattern $v_r \rightarrow T_1 \rightarrow T_1 \ldots T_d$, where T_d is a subtree that appears on depth d in the vertex's subgraph; (iii) we repeat step 2 until the maximum iterations h are reached. (iv) The final set of sequences is the union of the sequences of all the vertices in each iteration $\bigcup_{i=1}^{h} \bigcup_{v \in V} P_v$.

3.2   Neural Language Models – Word2vec

Neural language models have been developed in the NLP field as an alternative to represent texts as a bag of words, and hence, a binary feature vector,
where each vector index represents one word. While such approaches are simple
and robust, they suffer from several drawbacks, e.g., high dimensionality and
severe data sparsity, which limits the performances of such techniques. To overcome such limitations, neural language models have been proposed, inducing
low-dimensional, distributed embeddings of words by means of neural networks.
The goal of such approaches is to estimate the likelihood of a specific sequence
of words appearing in a corpus, explicitly modeling the assumption that closer
words in the word sequence are statistically more dependent.

   While some of the initially proposed approaches suffered from inefficient
training of the neural network models, with the recent advancements in the field
several efficient approaches has been proposed. One of the most popular and
widely used is the word2vec neural language model [13,14]. Word2vec is a particularly computationally-efficient two-layer neural net model for learning word
embeddings from raw text. There are two different algorithms, the Continuous
Bag-of-Words model (CBOW) and the Skip-Gram model.

Continuous Bag-of-Words Model. The CBOW model predicts target words
from context words within a given window. The model architecture is shown in
Fig. 1a. The input layer is comprised from all the surrounding words for which the
input vectors are retrieved from the input weight matrix, averaged, and projected
in the projection layer. Then, using the weights from the output weight matrix,
a score for each word in the vocabulary is computed, which is the probability
of the word being a target word. Formally, given a sequence of training words
$w_1, w_2, w_3, \ldots, w_T$, and a context window c, the objective of the CBOW model
is to maximize the average log probability:

$$\frac{1}{T} \sum_{t=1}^{T} log p(w_t | w_{t-c} \ldots w_{t+c}), \tag{1}$$

where the probability $p(w_t | w_{t-c} \ldots w_{t+c})$ is calculated using the softmax
function:

$$p(w_t | w_{t-c} \ldots w_{t+c}) = \frac{exp(\bar{v}^T v'_{w_t})}{\sum_{w=1}^{V} exp(\bar{v}^T v'_w)}, \tag{2}$$

where v'_w is the output vector of the word w, V is the complete vocabulary of
words, and $\bar{v}$ is the averaged input vector of all the context words:

$$\bar{v} = \frac{1}{2c} \sum_{-c \leq j \leq c, j \neq 0} v_{w_{t+j}} \tag{3}$$

Skip-Gram Model. The skip-gram model does the inverse of the CBOW model
and tries to predict the context words from the target words (Fig. 1b). More
formally, given a sequence of training words $w_1, w_2, w_3, \ldots, w_T$, and a context

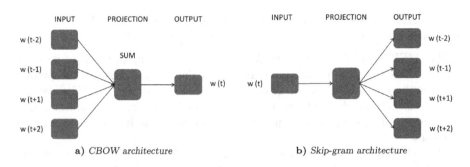

INPUT   PROJECTION   OUTPUT   INPUT   PROJECTION   OUTPUT

Fig. 1. Architecture of the CBOW and Skip-gram model.

window c, the objective of the skip-gram model is to maximize the following average log probability:

$$\frac{1}{T}\sum_{t=1}^{T}\sum_{-c\leq j\leq c, j\neq 0} logp(w_{t+j}|w_t),\tag{4}$$

where the probability $p(w_{t+j}|w_t)$ is calculated using the softmax function:

$$p(w_o|w_i) = \frac{exp(v_{wo}'^T v_{wi})}{\sum_{w=1}^{V} exp(v_w'^T v_{wi})},\tag{5}$$

where v_w and v_w' are the input and the output vector of the word w, and V is the complete vocabulary of words.

In both cases, calculating the softmax function is computationally inefficient, as the cost for computing is proportional to the size of the vocabulary. Therefore, two optimization techniques have been proposed, i.e., hierarchical softmax and negative sampling [14]. Empirical studies haven shown that in most cases negative sampling leads to a better performance than hierarchical softmax, which depends on the selected negative samples, but it has higher runtime.

Once the training is finished, all words (or, in our case, entities) are projected into a lower-dimensional feature space, and semantically similar words (or entities) are positioned close to each other.

4   Evaluation

We evaluate our approach on a number of classification and regression tasks, comparing the results of different feature extraction strategies combined with different learning algorithms.

4.1   Datasets

We evaluate the approach on two types of RDF graphs: (i) small domain-specific RDF datasets and (ii) large cross-domain RDF datasets. More details about the evaluation datasets and how the datasets were generated are presented in [28].

Small RDF Datasets. These datasets are derived from existing RDF datasets, where the value of a certain property is used as a classification target:

- The *AIFB* dataset describes the AIFB research institute in terms of its staff, research groups, and publications. In [1], the dataset was first used to predict the affiliation (i.e., research group) for people in the dataset. The dataset contains 178 members of five research groups, however, the smallest group contains only four people, which is removed from the dataset, leaving four classes.
- The *MUTAG* dataset is distributed as an example dataset for the DL-Learner toolkit[2]. It contains information about 340 complex molecules that are potentially carcinogenic, which is given by the `isMutagenic` property. The molecules can be classified as "mutagenic" or "not mutagenic".
- The *BGS* dataset was created by the British Geological Survey and describes geological measurements in Great Britain[3]. It was used in [33] to predict the lithogenesis property of named rock units. The dataset contains 146 named rock units with a lithogenesis, from which we use the two largest classes.

Large RDF Datasets. As large cross-domain datasets we use DBpedia [11] and Wikidata [31].

We use the English version of the 2015-10 DBpedia dataset, which contains $4,641,890$ instances and $1,369$ mapping-based properties. In our evaluation we only consider object properties, and ignore datatype properties and literals.

For the Wikidata dataset we use the simplified and derived RDF dumps from 2016-03-28[4]. The dataset contains $17,340,659$ entities in total. As for the DBpedia dataset, we only consider object properties, and ignore the data properties and literals.

We use the entity embeddings on five different datasets from different domains, for the tasks of classification and regression. Those five datasets are used to provide classification/regression targets for the large RDF datasets (see Table 1).

- The *Cities* dataset contains a list of cities and their quality of living, as captured by Mercer[5]. We use the dataset both for regression and classification.
- The *Metacritic Movies* dataset is retrieved from Metacritic.com[6], which contains an average rating of all time reviews for a list of movies [26]. The initial dataset contained around $10,000$ movies, from which we selected $1,000$ movies from the top of the list, and $1,000$ movies from the bottom of the list. We use the dataset both for regression and classification.
- Similarly, the *Metacritic Albums* dataset is retrieved from Metacritic.com[7], which contains an average rating of all time reviews for a list of albums [27].

---

[2] http://dl-learner.org.

[3] http://data.bgs.ac.uk/.

[4] http://tools.wmflabs.org/wikidata-exports/rdf/index.php?content=dump_download.php\&dump=20160328.

[5] https://www.imercer.com/content/mobility/quality-of-living-city-rankings.html.

[6] http://www.metacritic.com/browse/movies/score/metascore/all.

[7] http://www.metacritic.com/browse/albums/score/metascore/all.

Table 1. Datasets overview. For each dataset, we depict the number of instances, the machine learning tasks in which the dataset is used (C stands for classification, and R stands for regression) and the source of the dataset

Dataset	# instances	ML task	Original source
Cities	212	R/C (c = 3)	Mercer
Metacritic Albums	1600	R/C (c = 2)	Metacritic
Metacritic Movies	2000	R/C (c = 2)	Metacritic
AAUP	960	R/C (c = 3)	JSE
Forbes	1585	R/C (c = 3)	Forbes
AIFB	176	C (c = 4)	AIFB
MUTAG	340	C (c = 2)	MUTAG
BGS	146	C (c = 2)	BGS

- The *AAUP* (American Association of University Professors) dataset contains a list of universities, including eight target variables describing the salary of different staff at the universities[8]. We use the average salary as a target variable both for regression and classification, discretizing the target variable into "high", "medium" and "low", using equal frequency binning.
- The *Forbes* dataset contains a list of companies including several features of the companies, which was generated from the Forbes list of leading companies 2015[9]. The target is to predict the company's market value as a regression task. To use it for the task of classification we discretize the target variable into "high", "medium", and "low", using equal frequency binning.

4.2 Experimental Setup

The first step of our approach is to convert the RDF graphs into a set of sequences. For each of the small RDF datasets, we first build two corpora of sequences, i.e., the set of sequences generated from graph walks with depth 8 (marked as W2V), and set of sequences generated from Weisfeiler-Lehman sub-tree kernels (marked as K2V). For the Weisfeiler-Lehman algorithm, we use 4 iterations and depth of 2, and after each iteration we extract all walks for each entity with the same depth. We use the corpora of sequences to build both CBOW and Skip-Gram models with the following parameters: window size = 5; number of iterations = 10; negative sampling for optimization; negative samples = 25; with average input vector for CBOW. We experiment with 200 and 500 dimensions for the entities' vectors. The remaining parameters have the default value as proposed in [14].

As the number of generated walks increases exponentially [34] with the graph traversal depth, calculating Weisfeiler-Lehman subtrees RDF kernels, or all graph walks with a given depth d for all of the entities in the large RDF graph

---

[8] http://www.amstat.org/publications/jse/jse_data_archive.htm.

[9] http://www.forbes.com/global2000/list/.

quickly becomes unmanageable. Therefore, to extract the entities embeddings for the large RDF datasets, we use only random graph walks entity sequences. More precisely, we follow the approach presented in [22] to generate limited number of random walks for each entity. For DBpedia, we experiment with 500 walks per entity with depth of 4 and 8, while for Wikidata, we use only 200 walks per entity with depth of 4. Additionally, for each entity in DBpedia and Wikidata, we include all the walks of depth 2, i.e., direct outgoing relations. We use the corpora of sequences to build both CBOW and Skip-Gram models with the following parameters: window size = 5; number of iterations = 5; negative sampling for optimization; negative samples = 25; with average input vector for CBOW. We experiment with 200 and 500 dimensions for the entities' vectors. All the models, as well as the code, are publicly available[10].

We compare our approach to several baselines. For generating the data mining features, we use three strategies that take into account the direct relations to other resources in the graph [20], and two strategies for features derived from graph sub-structures [34]:

- Features derived from specific relations. In the experiments we use the relations *rdf:type* (types), and *dcterms:subject* (categories) for datasets linked to DBpedia.
- Features derived from generic relations, i.e., we generate a feature for each incoming (rel in) or outgoing relation (rel out) of an entity, ignoring the value or target entity of the relation.
- Features derived from generic relations-values, i.e., we generate feature for each incoming (rel-vals in) or outgoing relation (rel-vals out) of an entity including the value of the relation.
- Kernels that count substructures in the RDF graph around the instance node. These substructures are explicitly generated and represented as sparse feature vectors.
  - The Weisfeiler-Lehman (WL) graph kernel for RDF [34] counts full subtrees in the subgraph around the instance node. This kernel has two parameters, the subgraph depth d and the number of iterations h (which determines the depth of the subtrees). We use two pairs of settings, $d = 1, h = 2$ and $d = 2, h = 3$.
  - The Intersection Tree Path kernel for RDF [34] counts the walks in the subtree that spans from the instance node. Only the walks that go through the instance node are considered. We will therefore refer to it as the root Walk Count (WC) kernel. The root WC kernel has one parameter: the length of the paths l, for which we test 2 and 3.

We perform two learning tasks, i.e., classification and regression. For classification tasks, we use Naive Bayes, k-Nearest Neighbors (k = 3), C4.5 decision tree, and Support Vector Machines. For the SVM classifier we optimize the parameter C in the range $\{10^{-3}, 10^{-2}, 0.1, 1, 10, 10^2, 10^3\}$. For regression, we use Linear Regression, M5Rules, and k-Nearest Neighbors (k = 3). We measure accuracy for

---

[10] http://data.dws.informatik.uni-mannheim.de/rdf2vec/.

Table 2. Classification results on the small RDF datasets. The best results are marked in bold. Experiments marked with "\" did not finish within ten days, or have run out of memory

Strategy/ dataset	AIFB				MUTAG				BGS			
	NB	KNN	SVM	C4.5	NB	KNN	SVM	C4.5	NB	KNN	SVM	C4.5
rel in	16.99	47.19	50.70	50.62	\	\	\	\	61.76	54.67	63.76	63.76
rel out	45.07	45.56	50.70	51.76	41.18	54.41	62.94	62.06	54.76	69.05	72.70	69.33
rel in & out	25.59	51.24	50.80	51.80	\	\	\	\	54.76	67.00	72.00	70.00
rel-vals in	73.24	54.54	81.86	80.75	\	\	\	\	79.48	83.52	86.50	68.57
rel-vals out	86.86	55.69	82.39	71.73	62.35	62.06	73.53	62.94	84.95	65.29	83.10	73.38
rel-vals in & out	87.42	57.91	88.57	85.82	\	\	\	\	84.95	70.81	85.80	72.67
WL_1_2	85.69	53.30	92.68	71.08	91.12	62.06	92.59	93.29	85.48	63.62	82.14	75.29
WL_2_2	85.65	65.95	83.43	89.25	70.59	62.06	94.29	93.47	90.33	85.57	91.05	87.67
WC_2	86.24	60.27	75.03	71.05	90.94	62.06	91.76	93.82	84.81	69.00	83.57	76.90
WC_3	86.83	64.18	82.97	71.05	92.00	72.56	86.47	93.82	85.00	67.00	78.71	76.90
W2V CBOW 200	70.00	69.97	79.48	65.33	74.71	72.35	80.29	74.41	56.14	74.00	74.71	67.38
W2V CBOW 500	69.97	69.44	82.88	73.40	75.59	70.59	82.06	72.06	55.43	73.95	74.05	65.86
W2V SG 200	76.76	71.67	87.39	65.36	70.00	71.76	77.94	68.53	66.95	69.10	75.29	71.24
W2V SG 500	76.67	76.18	89.55	71.05	72.35	72.65	78.24	68.24	68.38	71.19	78.10	63.00
K2V CBOW 200	85.16	84.48	87.48	76.08	78.82	69.41	86.47	68.53	93.14	95.57	94.71	88.19
K2V CBOW 500	90.98	88.17	86.83	76.18	80.59	70.88	90.88	66.76	93.48	95.67	94.82	87.26
K2V SG 200	85.65	87.96	90.82	75.26	78.53	69.29	95.88	66.00	91.19	93.24	95.95	87.05
K2V SG 500	88.73	88.66	**93.41**	69.90	82.06	70.29	**96.18**	66.18	91.81	93.19	**96.33**	80.76

classification tasks, and root mean squared error (RMSE) for regression tasks. The results are calculated using stratfied 10-fold cross validation.

The strategies for creating propositional features from Linked Open Data are implemented in the RapidMiner LOD extension[11] [21,23]. The experiments, including the feature generation and the evaluation, were performed using the RapidMiner data analytics platform.[12] The RapidMiner processes and the complete results can be found online.[13]

4.3 Results

The results for the task of classification on the small RDF datasets are given in Table 2. From the results we can observe that the K2V approach outperforms all the other approaches. More precisely, using the skip-gram feature vectors of size 500 in an SVM model provides the best results on all three datasets. The W2V approach on all three datasets performs closely to the standard graph substructure feature generation strategies, but it does not outperform them. K2V outperforms W2V because it is able to capture more complex substructures in the graph, like sub-trees, while W2V focuses only on graph paths.

---

[11] http://dws.informatik.uni-mannheim.de/en/research/rapidminer-lod-extension.
[12] https://rapidminer.com/.
[13] http://data.dws.informatik.uni-mannheim.de/rmlod/LOD_ML_Datasets/.

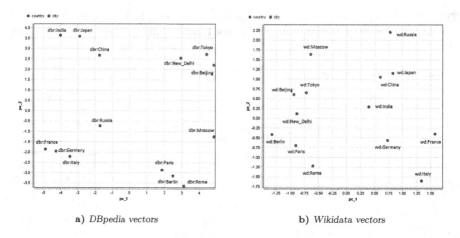

a) *DBpedia vectors*   b) *Wikidata vectors*

Fig. 2. Two-dimensional PCA projection of the 500-dimensional Skip-gram vectors of countries and their capital cities.

The results for the task of classification on the five different datasets using the DBpedia and Wikidata entities' vectors are given in Table 3, and the results for the task of regression on the 5 different dataset using the DBpedia and Wikidata entities' vectors are given in Table 4. We can observe that the latent vectors extracted from DBpedia and Wikidata outperform all of the standard feature generation approaches. In general, the DBpedia vectors work better than the Wikidata vectors, where the skip-gram vectors with size 200 or 500 built on graph walks of depth 8 on most of the datasets lead to the best performances. An exception is the AAUP dataset, where the Wikidata skip-gram 500 vectors outperform the other approaches.

On both tasks, we can observe that the skip-gram vectors perform better than the CBOW vectors. Also, the vectors with higher dimensionality and paths with bigger depth on most of the datasets lead to a better representation of the entities and better performances. However, for the variety of tasks at hand, there is no universal approach, i.e., embedding model and a machine learning method, that consistently outperforms the others.

4.4 Semantics of Vector Representations

To analyze the semantics of the vector representations, we employ Principal Component Analysis (PCA) to project the entities' feature vectors into a two dimensional feature space. We selected seven countries and their capital cities, and visualized their vectors as shown in Fig. 2. Figure 2a shows the corresponding DBpedia vectors, and Fig. 2b shows the corresponding Wikidata vectors. The figure illustrates the ability of the model to automatically organize entities of different types, and preserve the relationship between different entities. For example, we can see that there is a clear separation between the countries and the cities, and the relation "capital" between each pair of country and the

Table 3. Classification results. The first number represents the dimensionality of the vectors, while the second number represent the value for the depth parameter. The best results are marked in bold. Experiments marked with "\" did not finish within ten days, or have run out of memory

Strategy/dataset	Cities				Metacritic Movies				Metacritic Albums				AAUP				Forbes			
	NB	KNN	SVM	C4.5	NB	KNN	SVM	C4.5	NB	KNN	SVM	C4.5	NB	KNN	SVM	C4.5	NB	KNN	SVM	C4.5
Types	55.71	56.17	63.21	59.05	68.00	57.60	71.40	70.00	66.50	50.75	62.31	54.44	41.00	85.62	91.67	92.78	55.08	75.84	75.67	75.85
Categories	55.74	49.98	62.39	56.17	75.25	62.70	76.35	69.50	67.40	54.13	64.50	56.62	48.00	85.83	90.78	91.87	60.38	76.11	75.70	75.70
rel in	60.41	58.46	71.70	60.35	52.75	49.90	60.35	60.10	51.13	62.19	65.25	60.75	45.63	85.94	90.62	92.81	50.24	76.49	75.16	76.10
rel out	47.62	60.00	66.04	56.71	52.90	58.45	66.40	62.70	58.75	63.75	62.25	64.50	41.15	85.83	89.58	91.35	64.73	75.84	75.73	75.92
rel in & out	59.44	58.57	66.04	56.47	52.95	59.30	67.75	62.55	58.69	64.50	67.38	61.56	42.71	85.94	89.67	92.50	22.27	75.96	76.34	75.98
rel-vals in	\	\	\	\	50.60	50.00	50.60	50.00	50.88	52.56	50.81	50.00	54.06	84.69	89.51	\	14.95	76.15	76.97	75.73
rel-vals out	53.79	35.91	55.66	64.13	78.50	54.78	78.71	\	74.06	52.56	76.99	\	57.81	85.73	91.46	91.78	67.09	75.61	75.74	76.74
rel-vals in & out	\	\	\	\	77.90	55.75	77.82	\	74.25	51.25	75.85	\	63.44	84.69	91.56	\	67.20	75.88	75.96	76.75
WL_1.2	70.98	49.31	65.34	75.29	75.45	66.90	79.30	70.80	73.63	64.69	76.25	62.00	58.33	91.04	91.46	92.40	64.17	75.71	75.10	76.59
WL_2.3	65.48	53.29	69.90	69.31	\	\	\	\	\	\	\	\	\	\	\	\	\	\	\	\
WC.2	72.71	47.39	66.48	75.13	75.39	65.89	74.93	69.08	72.00	60.63	76.88	63.69	57.29	90.63	93.44	92.60	64.23	75.77	76.22	76.47
WC.3	65.52	52.36	67.95	65.15	74.25	55.30	78.40	\	72.81	52.87	77.94	\	57.19	90.73	90.94	92.60	64.04	75.65	76.22	76.59
DB2vec CBOW 200 4	59.32	68.84	77.39	64.32	65.60	79.74	82.90	74.33	70.72	71.86	76.36	67.24	73.36	89.65	29.00	92.45	89.38	80.94	76.83	84.81
DB2vec CBOW 500 4	59.32	71.34	76.37	66.34	65.65	79.49	82.75	73.87	69.71	71.93	75.41	65.65	72.71	89.65	29.11	92.01	89.02	80.82	76.95	85.17
DB2vec SG 200 4	60.34	71.82	76.37	65.37	65.25	80.44	83.25	73.87	68.95	73.89	76.11	67.87	71.20	89.65	28.90	92.12	88.78	80.82	77.92	85.77
DB2vec SG 500 4	58.34	72.84	76.87	67.84	65.45	80.14	**83.65**	72.82	70.41	74.34	78.44	67.49	71.19	89.65	28.90	92.23	88.30	80.94	77.25	84.81
DB2vec CBOW 200 8	69.26	69.87	67.32	63.13	57.83	70.08	65.25	67.47	67.91	64.44	72.42	65.39	68.18	85.33	28.90	90.50	77.35	80.34	28.90	85.17
DB2vec CBOW 500 8	62.26	69.87	76.84	63.21	58.78	69.82	69.46	67.67	67.53	65.83	74.26	63.42	62.90	85.22	29.11	90.61	89.86	80.34	78.65	84.81
DB2vec SG 200 8	73.32	75.89	78.92	60.74	79.94	79.49	83.30	75.13	77.25	76.87	**79.72**	69.14	78.53	85.12	29.22	91.04	**90.10**	80.58	78.96	84.68
DB2vec SG 500 8	**89.73**	69.16	84.19	72.25	80.24	78.68	82.80	72.42	73.57	76.30	78.20	68.70	75.07	**94.48**	29.11	94.15	88.53	80.58	77.79	86.38
WD2vec CBOW 200 4	68.76	57.71	75.56	61.37	51.49	48.56	51.04	49.01	50.86	50.29	51.44	50.09	50.54	90.18	89.63	88.83	49.84	81.08	76.77	79.14
WD2vec CBOW 500 4	68.24	57.75	85.56	64.54	49.22	48.56	51.04	50.98	53.08	50.03	52.33	53.28	48.45	90.39	89.74	88.31	51.95	80.74	78.18	80.32
WD2vec SG 200 4	72.58	57.53	75.48	52.32	69.53	70.14	75.39	67.00	60.32	62.03	64.76	58.54	60.87	90.50	89.63	89.98	65.45	81.17	77.74	77.03
WD2vec SG 500 4	83.20	60.72	79.87	61.67	71.10	70.19	76.30	67.31	55.31	58.92	63.42	56.63	55.85	90.60	89.63	87.69	58.95	81.17	79.00	79.56

Table 4. Regression results. The first number represents the dimensionality of the vectors, while the second number represent the value for the depth parameter. The best results are marked in bold. Experiments that did not finish within ten days, or that have run out of memory are marked with "/"

Strategy/dataset	Cities			Metacritic Movies			Metacritic Albums			AAUP			Forbes		
	LR	KNN	M5	LR	KNN	M5	LR	KNN	M5	LR	KNN	M5	LR	KNN	M5
Types	24.30	22.16	18.79	77.80	30.68	22.16	16.45	18.36	13.95	9.83	34.95	6.28	29.22	21.07	18.32
Categories	18.88	22.68	22.32	84.57	23.87	22.50	16.73	16.64	13.95	8.08	34.94	6.16	19.16	21.48	18.39
rel in	49.87	18.53	19.21	22.60	41.40	22.56	13.50	22.06	13.43	9.69	34.98	6.56	27.56	20.93	18.60
rel out	49.87	18.53	19.21	21.45	24.42	20.74	13.32	14.59	13.06	8.82	34.95	6.32	21.73	21.11	18.97
rel in & out	40.80	18.21	18.80	21.45	24.42	20.74	13.33	14.52	12.91	12.97	34.95	6.36	26.44	20.98	19.54
rel-vals in	/	/	/	21.46	24.19	20.43	13.94	23.05	13.95	/	34.96	6.27	/	20.86	19.31
rel-vals out	20.93	23.87	20.97	25.99	32.18	22.93	/	15.28	13.34	/	34.95	6.18	/	20.48	18.37
rel-vals in & out	/	/	/	/	25.37	20.96	/	15.47	13.33	/	34.94	6.18	/	20.20	18.20
WL_1_2	20.21	24.60	20.85	/	21.62	19.84	/	13.99	12.81	/	34.96	6.27	/	19.81	19.49
WL_2_3	17.79	20.42	17.04	/	/	/	/	/	/	/	/	/	/	/	/
WC_2	20.33	25.95	19.55	/	22.80	22.99	/	14.54	12.87	9.12	34.95	6.24	/	20.45	19.26
WC_3	19.51	33.16	19.05	/	23.86	19.19	/	19.51	13.02	/	35.39	6.31	/	20.58	19.04
DB2vec CBOW 200 4	14.37	12.55	14.33	15.90	17.46	15.89	11.79	12.45	11.59	12.13	45.76	12.00	18.32	26.19	17.43
DB2vec CBOW 500 4	14.99	12.46	14.66	15.90	17.45	15.73	11.49	12.60	11.48	12.44	45.67	12.30	18.23	26.27	17.62
DB2vec SG 200 4	13.38	12.54	15.13	15.81	17.07	15.84	11.30	12.36	11.42	12.13	45.72	12.10	17.63	26.13	17.85
DB2vec SG 500 4	14.73	13.25	16.80	15.66	17.14	15.67	11.20	12.11	11.28	12.09	45.76	11.93	18.23	26.09	17.74
DB2vec CBOW 200 8	16.17	17.14	17.56	21.55	23.75	21.46	13.35	15.41	13.43	6.47	55.76	6.47	24.17	26.48	22.61
DB2vec CBOW 500 8	18.13	17.19	18.50	20.77	23.67	20.69	13.20	15.14	13.25	6.54	55.33	6.55	21.16	25.90	20.33
DB2vec SG 200 8	12.85	14.95	12.92	15.15	17.13	**15.12**	10.90	11.43	10.90	6.22	56.95	6.25	18.66	21.20	18.57
DB2vec SG 500 8	11.92	12.67	10.19	15.45	17.80	15.50	**10.89**	11.72	10.97	6.26	56.95	6.29	18.35	21.04	**16.61**
WD2vec CBOW 200 4	20.15	17.52	20.02	23.54	25.90	23.39	14.73	16.12	14.55	16.80	42.61	6.60	27.48	22.60	21.77
WD2vec CBOW 500 4	23.76	18.33	20.39	24.14	22.18	24.56	14.09	16.09	14.00	13.08	42.89	6.08	50.23	21.92	26.66
WD2vec SG 200 4	20.47	18.69	20.72	19.72	21.44	19.10	13.51	13.91	13.67	6.86	42.82	6.52	23.69	21.59	20.49
WD2vec SG 500 4	22.25	19.41	19.23	25.99	21.26	19.19	13.23	14.96	13.25	8.27	42.84	**6.05**	21.98	21.73	21.58

corresponding capital city is preserved. Furthermore, we can observe that more similar entities are positioned closer to each other, e.g., we can see that the countries that are part of the EU are closer to each other, and the same applies for the Asian countries.

4.5 Features Increase Rate

Finally, we conduct a scalability experiment, where we examine how the number of instances affects the number of generated features by each feature generation strategy. For this purpose we use the *Metacritic Movies* dataset. We start with a random sample of 100 instances, and in each next step we add 200 (or 300) unused instances, until the complete dataset is used, i.e., 2,000 instances. The number of generated features for each sub-sample of the dataset using each of the feature generation strategies is shown in Fig. 3.

From the chart, we can observe that the number of generated features sharply increases when adding more samples in the datasets, especially for the strategies based on graph substructures. However, the number of features remains the same when using the RDF2Vec approach, independently of the number of samples in the data. Thus, by design, it scales to larger datasets without increasing the dimensionality of the dataset.

5   Conclusion

In this paper, we have presented RDF2Vec, an approach for learning latent numerical representations of entities in RDF graphs. In this approach, we first convert the RDF graphs in a set of sequences using two strategies, Weisfeiler-Lehman Subtree RDF Graph Kernels and graph walks, which are then used to

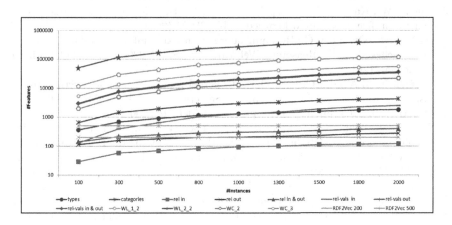

Fig. 3. Features increase rate per strategy (log scale)

build neural language models. The evaluation shows that such entity representations could be used in two different machine learning tasks, outperforming
standard feature generation approaches.

So far we have considered only simple machine learning tasks, i.e., classification and regression, but in the future work we would extend the number of
applications. For example, the latent representation of the entities could be used
for building content-based recommender systems [4]. The approach could also be
used for link predictions, type prediction, graph completion and error detection
in knowledge graphs [19], as shown in [15,17]. Furthermore, we could use this
approach for the task of measuring semantic relatedness between two entities,
which is the basis for numerous tasks in information retrieval, natural language
processing, and Web-based knowledge extractions [6]. To do so, we could easily
calculate the relatedness between two entities as the probability of one entity
being the context of the other entity, using the softmax function given in Eqs. 2
and 5, using the input and output weight matrix of the neural model. Similarly, the approach can be extended for entity summarization, which is also an
important task when consuming and visualizing large quantities of data [2].

Acknowledgments. The work presented in this paper has been partly funded by the
German Research Foundation (DFG) under grant number PA 2373/1-1 (Mine@LOD).

References

1. Bloehdorn, S., Sure, Y.: Kernel methods for mining instance data in ontologies. In:
   Aberer, K., et al. (eds.) ASWC 2007 and ISWC 2007. LNCS, vol. 4825, pp. 58–71.
   Springer, Heidelberg (2007)
2. Cheng, G., Tran, T., Qu, Y.: RELIN: relatedness and informativeness-based centrality for entity summarization. In: Aroyo, L., et al. (eds.) ISWC 2011, Part I.
   LNCS, vol. 7031, pp. 114–129. Springer, Heidelberg (2011)
3. Cheng, W., Kasneci, G., Graepel, T., Stern, D., Herbrich, R.: Automated feature
   generation from structured knowledge. In: CIKM (2011)
4. Di Noia, T., Ostuni, V.C.: Recommender systems and linked open data. In: Faber,
   W., Paschke, A. (eds.) Reasoning Web 2015. LNCS, vol. 9203, pp. 88–113. Springer,
   Heidelberg (2015)
5. Fanizzi, N., d'Amato, C.: A declarative kernel for ALC concept descriptions. In:
   Esposito, F., Raś, Z.W., Malerba, D., Semeraro, G. (eds.) ISMIS 2006. LNCS
   (LNAI), vol. 4203, pp. 322–331. Springer, Heidelberg (2006)
6. Hoffart, J., Seufert, S., Nguyen, D.B., Theobald, M., Weikum, G.: KORE:
   keyphrase overlap relatedness for entity disambiguation. In: Proceedings of the
   21st ACM International Conference on Information and Knowledge Management,
   pp. 545–554. ACM (2012)
7. Huang, Y., Tresp, V., Nickel, M., Kriegel, H.P.: A scalable approach for statistical
   learning in semantic graphs. Semant. Web **5**, 5–22 (2014)
8. Kappara, V.N.P., Ichise, R., Vyas, O.: LiDDM: a data mining system for linked
   data. In: LDOW (2011)
9. Khan, M.A., Grimnes, G.A., Dengel, A.: Two pre-processing operators for
   improved learning from semanticweb data. In: RCOMM (2010)

10. Kramer, S., Lavrač, N., Flach, P.: Propositionalization approaches to relational data mining. In: Džeroski, S., Lavrač, N. (eds.) Relational Data Mining, pp. 262–291. Springer, Berlin (2001)
11. Lehmann, J., Isele, R., Jakob, M., Jentzsch, A., Kontokostas, D., Mendes, P.N., Hellmann, S., Morsey, M., van Kleef, P., Auer, S., Bizer, C.: DBpedia - a large-scale, multilingual knowledge base extracted from Wikipedia. Semant. Web J. (2013)
12. Lösch, U., Bloehdorn, S., Rettinger, A.: Graph kernels for RDF data. In: Simperl, E., Cimiano, P., Polleres, A., Corcho, O., Presutti, V. (eds.) ESWC 2012. LNCS, vol. 7295, pp. 134–148. Springer, Heidelberg (2012)
13. Mikolov, T., Chen, K., Corrado, G., Dean, J.: Efficient estimation of word representations in vector space. arXiv preprint arXiv:1301.3781 (2013)
14. Mikolov, T., Sutskever, I., Chen, K., Corrado, G.S., Dean, J.: Distributed representations of words and phrases and their compositionality. In: Advances in Neural Information Processing Systems, pp. 3111–3119 (2013)
15. Minervini, P., Fanizzi, N., d'Amato, C., Esposito, F.: Scalable learning of entity and predicate embeddings for knowledge graph completion. In: 2015 IEEE 14th International Conference on Machine Learning and Applications (ICMLA), pp. 162–167. IEEE (2015)
16. Mynarz, J., Svátek, V.: Towards a benchmark for LOD-enhanced knowledge discovery from structured data. In: The Second International Workshop on Knowledge Discovery and Data Mining Meets Linked Open Data (2013)
17. Nickel, M., Murphy, K., Tresp, V., Gabrilovich, E.: A review of relational machine learning for knowledge graphs: from multi-relational link prediction to automated knowledge graph construction. arXiv preprint arXiv:1503.00759 (2015)
18. Paulheim, H.: Exploiting linked open data as background knowledge in data mining. In: Workshop on Data Mining on Linked Open Data (2013)
19. Paulheim, H.: Knowlegde graph refinement: a survey of approaches and evaluation methods. Semant. Web J. 1–20 (2016, Preprint)
20. Paulheim, H., Fümkranz, J.: Unsupervised generation of data mining features from linked open data. In: Proceedings of the 2nd International Conference on Web Intelligence, Mining and Semantics, p. 31. ACM (2012)
21. Paulheim, H., Ristoski, P., Mitichkin, E., Bizer, C.: Data mining with background knowledge from the web. In: RapidMiner World 2014 Proceedings, pp.1-14. Shaker, Aachen (2014)
22. Perozzi, B., Al-Rfou, R., Skiena, S.: Deepwalk: online learning of social representations. In: Proceedings of the 20th ACM SIGKDD International Conference on Knowledge Discovery and Data Mining, pp. 701–710. ACM (2014)
23. Ristoski, P., Bizer, C., Paulheim, H.: Mining the web of linked data with rapidminer. Web Semant.: Sci. Serv. Agents World Wide Web 35, 142–151 (2015)
24. Ristoski, P., Paulheim, H.: A comparison of propositionalization strategies for creating features from linked open data. In: Linked Data for Knowledge Discovery (2014)
25. Ristoski, P., Paulheim, H.: Semantic web in data mining and knowledge discovery: a comprehensive survey. Web Semant.: Sci. Serv. Agents World Wide Web 36, 1–22 (2016)
26. Ristoski, P., Paulheim, H., Svátek, V., Zeman, V.: The linked data mining challenge 2015. In: KNOW@LOD (2015)
27. Ristoski, P., Paulheim, H., Svátek, V., Zeman, V.: The linked data mining challenge 2016. In: KNOWLOD (2016)

28. Ristoski, P., de Vries, G.K.D., Paulheim, H.: A collection of benchmark datasets for systematic evaluations of machine learning on the semantic web. In: International Semantic Web Conference. Springer, Berlin (2016, to appear)

29. Schmachtenberg, M., Bizer, C., Paulheim, H.: Adoption of the linked data best practices in different topical domains. In: Mika, P., et al. (eds.) ISWC 2014, Part I. LNCS, vol. 8796, pp. 245–260. Springer, Heidelberg (2014)

30. Shervashidze, N., Schweitzer, P., Van Leeuwen, E.J., Mehlhorn, K., Borgwardt, K.M.: Weisfeiler-Lehman graph kernels. J. Mach. Learn. Res. **12**, 2539–2561 (2011)

31. Vrandečić, D., Krötzsch, M.: Wikidata: a free collaborative knowledgebase. Commun. ACM **57**(10), 78–85 (2014)

32. de Vries, G.K.D.: A fast approximation of the Weisfeiler-Lehman graph kernel for RDF data. In: Blockeel, H., Kersting, K., Nijssen, S., Železný, F. (eds.) ECML PKDD 2013, Part I. LNCS, vol. 8188, pp. 606–621. Springer, Heidelberg (2013)

33. de Vries, G.K.D., de Rooij, S.: A fast and simple graph kernel for RDF. In: DMLOD (2013)

34. de Vries, G.K.D., de Rooij, S.: Substructure counting graph kernels for machine learning from RDF data. Web Semant.: Sci. Serv. Agents World Wide Web **35**, 71–84 (2015)

35. Yanardag, P., Vishwanathan, S.: Deep graph kernels. In: Proceedings of the 21th ACM SIGKDD International Conference on Knowledge Discovery and Data Mining, pp. 1365–1374. ACM (2015)

SPARQL-to-SQL on Internet of Things Databases and Streams

Eugene Siow[✉], Thanassis Tiropanis, and Wendy Hall

Electronics and Computer Science, University of Southampton, Southampton, UK
{eugene.siow,t.tiropanis,wh}@soton.ac.uk

Abstract. To realise a semantic Web of Things, the challenge of achieving efficient Resource Description Format (RDF) storage and SPARQL query performance on Internet of Things (IoT) devices with limited resources has to be addressed. State-of-the-art SPARQL-to-SQL engines have been shown to outperform RDF stores on some benchmarks. In this paper, we describe an optimisation to the SPARQL-to-SQL approach, based on a study of time-series IoT data structures, that employs metadata abstraction and efficient translation by reusing existing SPARQL engines to produce Linked Data 'just-in-time'. We evaluate our approach against RDF stores, state-of-the-art SPARQL-to-SQL engines and streaming SPARQL engines, in the context of IoT data and scenarios. We show that storage efficiency, with succinct row storage, and query performance can be improved from 2 times to 3 orders of magnitude.

Keywords: SPARQL · SQL · Query translation · Analytics · Internet of Things · Web of Things

1 Introduction

The Internet of Things (IoT) envisions a world-wide, interconnected network of smart physical entities with the aim of providing technological and societal benefits [9]. However, as the W3C Web of Things (WoT) Interest Group charter[1] states, the IoT is currently beset by product silos and to unlock its potential, an open ecosystem based upon open standards for identification, discovery and interoperation of services is required.

We see a semantic Web of Things as such an information space, with rich descriptions, shared data models and constructs for interoperability that utilises but is not limited to semantic and web technologies to provide an application layer for IoT applications. As Barnaghi *et al.* [3] have proposed, semantic technologies can serve to facilitate interoperability, data abstraction, access and integration with other cyber, social or physical world data.

The semantic WoT does present a set of unique challenges: handling and storing time-series data as RDF, querying with SPARQL on limited IoT devices and distributed usage scenarios. Buil-Aranda *et al.* [5] have examined traditional

---

[1] https://www.w3.org/2014/12/wot-ig-charter.html.

© Springer International Publishing AG 2016
P. Groth et al. (Eds.): ISWC 2016, Part I, LNCS 9981, pp. 515–531, 2016.
DOI: 10.1007/978-3-319-46523-4_31

SPARQL endpoints on the web and shown that performance for generic queries can vary by up to 3–4 orders of magnitude. Endpoints generally limit or have worsened reliability when issued with a series of non-trivial queries. IoT devices have added resource constraints, however, we argue that time-series IoT data and distribution also present the opportunity for specific optimisation.

The contribution of this paper is to present an optimisation of SPARQL-to-SQL query translation for the particular case of time-series data, both historical and streaming, with a novel approach that uses existing SPARQL engines to resolve Basic Graph Patterns and mappings that allow intermediate nodes of observations to be 'collapsed'. This is advised by our study of IoT schemata which exhibits a flat and wide structure. Our approach compares favourably to native RDF storage, SPARQL-to-SQL engines and RDF stream processing engines deployed on compact, resource-constrained devices, showing 2 times to 3 orders of magnitude performance and storage improvements on published sensor benchmarks and IoT use cases like smart homes.

In Sect. 2, we first study the structure of time-series IoT data which leads us, in Sect. 3, to study related work. We then describe the design and implementation of our approach, that employs metadata abstraction through mappings and SPARQL-to-SQL translation for performance, reusing, at the core, any existing SPARQL engine in Sect. 4. Finally, we evaluate our approach against traditional RDF stores, SPARQL-to-SQL engines and streaming engines using an established benchmark and a common IoT scenario in Sect. 5. Results are presented and discussed in Sect. 6 with the conclusion in Sect. 7.

2   Structure of Internet of Things Data

To investigate the structure of data produced by sensors in the Internet of Things, we collected the schemata of 19,914 unique IoT devices from public data streams on Dweet.io[2] over a one month period in January 2016.

Dweet.io is a cloud platform that supports the publishing of time-series data from IoT devices in JavaScript Object Notation (JSON). The schema represented in JSON can be flat (row-like with a single level of data) or complex (tree-like/hierachical with multiple nested levels of data). It was observed from the schemata, removing the 1542 (7.7 %) that were empty, that 18,280 (99.5 %) of the non-empty schemata were flat while only 92 (0.5 %) were complex.

We also analysed the schemata to investigate how wide the IoT data was. Wideness is defined as the number of properties beside the timestamp and a schema is considered wide if there are 2 or more such properties. We found that 92.2 % of the devices sampled had a schema that was wide. The majority (53.2 %) had 4 properties related to each timestamp. We also obtained a smaller alternative sample of 614 unique devices (over the same period) from Sparkfun[3], that only supports flat schemata, which confirmed that most (76.3 %) IoT devices sampled have wide time-series schemata.

---

[2] http://dweet.io/see.

[3] https://data.sparkfun.com/streams.

We concluded that our sample of over 20,000 unique IoT devices from Dweet.io and Sparkfun contained (1) flat and (2) wide IoT time-series data. It follows that a possible succinct representation of such data is as rows in a relational database with column headings corresponding to properties. SPARQL-to-SQL translation is then a possibility for querying. Investigating column stores was out of the scope of this study, however provides for interesting future work and comparison, as tension between inserts/updates and optimising data structures for reads [15] are reduced for time-series data which are already sorted by time in entry sequence order. The IoT schemata we collected is available on Github[4].

3   Related Work

The fact that we are dealing with time-series sensor data, represented as Linked Data with ontologies like the Semantic Sensor Network (SSN) ontology and Linked Sensor Data [12] for interoperability, prescribes the study of: (i) RDF stores, (ii) R2RML and SPARQL-to-SQL translation with relational databases to improve performance and storage efficiency for time series-data as rows and (iii) streaming engines for efficient processing on real-time streams.

RDF Stores. Virtuoso [8] is based on an Object Relational DBMS optimised for RDF storage while Jena Tuple Database (TDB) is a native Java RDF store using a single table to store triples/quads. Indexes, like the 6 SPO (Subject-Predicate-Object) permutations that Neumann *et al.* [11] propose often improve query performance on tables by reducing scans. TDB creates 3 triple indexes (OSP, POS, SPO) and 6 quad indexes while Virtuoso creates 5 quad indexes (PSOG, POGS, SP, OP, GS; G is graph). Commercial stores like GraphDB, formerly OWLIM, have also shown to perform well on benchmarks [4] with 6 indexes (PSO, POS, entities, classes, predicates, literals). Indexing, however, increases the storage size and memory required to load them.

Relational Databases (SPARL-to-SQL). Efficient SPARQL-to-SQL translation that improves performance and builds on previous literature has been investigated by Rodriguez-Muro and Rezk [14] and Priyatna *et al.* [13] with state-of-the-art engines ontop and morph respectively. Both engines support R2RML[5], a W3C recommendation based on the concept of mapping logical tables in relational databases to RDF via Triples Maps (the subject, predicate and object in a triple can be mapped to columns in a table). They also optimise query translation to remove redundant self-joins. Ontop, which translates mappings and queries to a set of Datalog rules, applies query containment and semantic query optimisation to create efficient SQL queries. However, (1) R2RML is designed for

---

[4] https://github.com/eugenesiow/iotdevices/releases/download/data/dweet_release.
    zip.
[5] http://www.w3.org/TR/r2rml/.

generality rather than abstracting and 'collapsing' (*reducing self joins on identifier columns in tables mapping to IRI templates*) intermediate nodes (Sect. 4) (2) Time-series data can be different from relational data (e.g. does not have primary keys) (3) The round-trip to retrieve database metadata (ontop) could be significant on devices with slower disk/memory access.

Streaming Engines. The C-SPARQL [1] engine supports continuous pull-based SPARQL queries over RDF data streams by using Esper[6], a complex event processing engine, to form windows in which SPARQL queries can be executed on an in-memory RDF model. CQELS [10] is a native RDF stream engine, supporting push and pull queries, that takes a 'white-box' approach for full control over query optimisation and execution. morph-streams, from $SPARQL_{stream}$ [6], supports query rewriting with R2RML mappings and execution with Esper.

4 Designing a SPARQL-to-SQL Engine for the IoT

Based on the ontologies for integrating time-series sensor data, the SSN ontology[7], Semantic Sensor Web and Linked Sensor Data (LSD) [12] mentioned in the previous section, we observe that semantic sensor data is modelled as (1) IoT *device metadata* like the location and specifications of sensors, (2) IoT *observation metadata* like the units of measure and types of observation (3) IoT *observation data* like timestamps and actual readings. Listing 1.1 shows an example division into the 3 categories from the Linked Sensor Data dataset in RDF Turtle.

Listing 1.1. LSD example, rainfall from Station 4UT01 (abbreviated)

```
@prefix ssw:<http://knoesis.wright.edu/ssw/ont/sensor-observation.owl#>
@prefix weather:<http://knoesis.wright.edu/ssw/ont/weather.owl#>
@prefix wgs:<http://www.w3.org/2003/01/geo/wgs84_pos#>
@prefix time:<http://www.w3.org/2006/time#>
@prefix sen:<http://knoesis.wright.edu/ssw/>
sen:System_4UT01 ssw:processLocation                // Device Metadata
  [wgs:lat "40.82944"; wgs:long "-111.88222"].
_:obs a weather:RainfallObservation;                // Observation Metadata
  ssw:observedProperty weather:_Rainfall;
  ssw:procedure sen:System_4UT01;
  ssw:result _:data; ssw:samplingTime _:time.
_:data a ssw:MeasureData;
  ssw:uom weather:degrees.
_:time a time:Instant;
  time:inXSDDateTime "2003-03-31T12:35:00".         // Observation Data
_:data ssw:floatValue "0.1".
```

Although Linked Data as implemented in RDF is flexible and expressive enough to represent both data and metadata as triples as seen in Listing 1.1, however, given the resource constraints of IoT devices, we make these hypotheses:

1. Storing flat and wide IoT observation data as rows is more efficient than storage as RDF as each field value in a row, under a column header, does not require additional subject and predicate terms (Table 1).

---

[6] http://www.espertech.com/products/esper.php.

[7] https://www.w3.org/2005/Incubator/ssn/ssnx/ssn.

Table 1. LSD example, abbreviated row from the Table 4TU01

Time	Rainfall	RelativeHumidity	...
2003-03-31T12:35:00	0.1	37.0	...

2. Queries that retrieve more fields from a row (e.g. Rainfall & RelativeHumidity) will require less joins as compared to RDF stores' and perform better.
3. Most device and observation metadata can be abstracted and stored in-memory, with a mapping language that can express this. Metadata triples can be produced 'just-in-time' and intermediate nodes (e.g. ssw:MeasureData in Listing 1.1), if not projected in queries, can be 'collapsed' (reduces joins in RDF stores and self joins on identifier columns in tables that map to intermediate nodes, e.g. _:obs, _:data and _:time for SPARQL-to-SQL).
4. Efficient queries can be produced without relying on primary keys within time-series data and retrieving database schema from IoT devices.

4.1 sparql2sql and sparql2stream

We present, based on our hypotheses, sparql2sql (translates SPARQL-to-SQL) and sparql2stream (translates SPARQL to Event Processing Language (EPL) for streams) engines. They utilise the same core to provide a holistic approach to SPARQL translation for both historical and streaming IoT datasets.

Firstly, to support SPARQL-to-SQL translation, a mapping for IoT data stored as rows is required. R2RML (as in Sect. 3) is designed for generality rather than for specific IoT time-series data. As such, we propose S2SML in Sect. 4.2, an R2RML-compatible mapping language designed for metadata abstraction, collapsing intermediate nodes and in-memory storage.

Next, in Sect. 4.3, we explain how S2SML mappings can be used to translate SPARQL to SQL, reusing any existing SPARQL engine. Finally, in Sect. 4.4, we show how this applies for SPARQL on streams.

4.2 S2SML Mapping

Sparql2Sql Mapping Language (S2SML) mappings serve the dual purpose of providing bindings from rows and abstracting sensor and observation metadata from observation data stored as rows. Mappings are pure RDF and compatible with R2RML (can be translated to and from). Furthermore, S2SML is also designed to support 'collapsing' intermediate nodes of observation metadata through the use of blank nodes or faux nodes, nodes containing identifiers only created on projection. Listings 1.2 and 1.3 show a comparison of S2SML and R2RML from Listing 1.1. R2RML is more verbose and uses the {time} column for IRI templates, which might not be unique and cannot be 'collapsed' (Sect. 6.2).

Listing 1.2. S2SML

```
_:b a weather:RainfallObservation;
ssw:result _:c.
_:c a ssw:MeasureData;
ssw:floatValue
"4UT01.Rainfall"^^<:LiteralMap>.
```

Listing 1.3. R2RML

```
:t1 a rr:TriplesMap; rr:logicalTable :4UT01;
rr:subjectMap[rr:template "http://...o/{time}";
   rr:class weather:RainfallObservation];
rr:predicateObjectMap[rr:predicate ssw:result;
rr:objectMap[rr:parentTriplesMap :t2]].
:t2 a rr:TriplesMap; rr:logicalTable :4UT01;
rr:subjectMap[rr:template "http://...m/{time}";
rr:class ssw:MeasureData];
rr:predicateObjectMap[rr:predicate ssw:floatValue;
   rr:objectMap[rr:column "Rainfall"]].
```

To define S2SML, we adopt the notation introduced by Chebotko *et al.* [7] where I, B, L denote pairwise disjoint infinite sets of IRIs, blank nodes and literals while I_{map}, L_{map}, F are IRI Map, Literal Map and Faux Node respectively. Examples can be found in Table 2. Combinations of these terms (e.g. $I_{map}IBF$) denote the union of their component sets (e.g. $I_{map} \cup I \cup B \cup F$).

Definition 1 (S2SML Mapping, m). *Given a set of all possible S2SML mappings, M, an S2SML mapping, $m \in M$, is a set of triple tuples, $(s, p, o) \in (I_{map}IBF) \times I \times (I_{map}IBL_{map}LF)$ where s, p and o are subject, predicate and object respectively.*

Table 2. Examples of elements in (s, p, o) sets

Symbol	Name	Example
I	IRI	\<http://knoesis.wright.edu/ssw/ont/weather.owl#degrees\>
I_{map}	IRI Map	\<http://knoesis.wright.edu/ssw/{sensors.sensorName}\>
B	Blank Node	_:bNodeId
L	Literal	"-111.88222"^^ \<xsd:float\>
L_{map}	Literal Map	"readings.temperature"^^ \<s2s:literalMap\>
F	Faux Node	\<http://knoesis.wright.edu/ssw/obs/{readings.uuid}\>

As shown in Table 2, I_{map} are IRI templates that consist of the union of IRI string parts (e.g. http://knoesis.wright.edu/ssw/) and reference bindings to table columns (e.g. {tableName.colName}). L_{map} are RDF literals whose value contains reference bindings to table columns (e.g. "tableName.colName") with a datatype of \<s2s:literalMap\>.

Definition 2 (Faux Node, F). *F is defined as an IRI template that consists of the union of a set of IRI string parts, I_p and a set of placeholders, U_{id}, referencing a table, so that $F = I_p \cup U_{id}$ and $|U_{id}| >= 1, |I_p| >= 1$.*

The example F in Table 2 shows how a placeholder is defined in the format of {tableName.uuid} with keyword '.uuid' identifying this as a Faux node.

Listing 1.2 shows an S2SML mapping of an LSD weather station 4UT01 in Salt Lake City. Observation data is referenced from table columns with Literal Maps, L_{map} (e.g. "4UT01.Rainfall"). Observation metadata which serves to connect nodes (e.g. _:c) is 'collapsed' through the use of blank nodes, B, which in R2RML (Listing 1.3) is mapped to {time} columns. The R2RML specification does support blank nodes but none of the other engines support their use yet. Faux nodes in S2SML are used if there is a possibility that the identifier/intermediate node will be projected in queries (described in Sect. 4.3). Finally, device metadata also contains constant Literals, L (e.g. the latitude of the sensor).

Mapping Closures. IoT devices might also have multiple sensors, each producing a time-series with a corresponding S2SML mapping. In Fig. 1, there might be multiple *observations mappings* each in different *readings* tables and a single sensors mapping and *sensors* table all forming a mapping closure.

Definition 3 (Mapping Closure, M_c). *Given the set of all mappings on a device, $M_d = \{m_d | m_d \in M\}$, where M is a set of all possible S2SML mappings, a mapping closure is the union of all elements in M_d, so $M_c = \bigcup_{m \in M_d} m$.*

Implicit Join Conditions. Observation data that is represented across multiple tables within a mapping closure might need to be joined if matched by a SPARQL query. In R2RML, one or more join conditions (*rr:joinCondition*) may be specified between triple maps of different logical tables.

In S2SML, these join conditions are automatically discovered as they are implicit within mapping closures from IRI template matching involving two or more tables. We define IRI template matching as follows.

Definition 4 (IRI Template Matching). *Let I_p be the set of IRI string parts in an element of I_{map}. I_{map_1} and I_{map_2} are matching if $\bigcup_{i_1 \in I_{p_1}} i_1 = \bigcup_{i_2 \in I_{p_2}} i_2$ and $\forall i_1 \in I_{p_1}, \forall i_2 \in I_{p_2} : pos(i_1) = pos(i_2)$ where $pos(x)$ is a function that returns the position of x within its I_{map}.*

Given matching I_{map}, join conditions can be inferred. Figure 1 shows a mapping closure consisting of a sensor and observation mapping. An IRI map in each of the mappings, *sen:system{sensors.name}* in I_{map_1} and *sen:system{readings.sensor}* in I_{map_2}, fulfil a template matching. A join condition is inferred between the columns *sensors.name* and *readings.sensor* as a result.

Compatibility with R2RML. S2SML is compatible with R2RML as they can be mutually translated without losing expressiveness. Triple Maps are translated to triples based on the elements in Table 2. Table 3 defines additional R2RML predicates and the corresponding S2SML construct. *rr:inverseExpression*, for

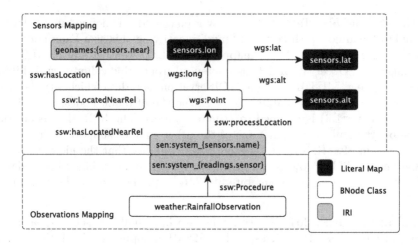

Fig. 1. Graph representation of an implicit join within a mapping closure

example, is encoded within a literal, L_{iv}, with a datatype of <s2s:inverse> and the *rr:column* denoted with double braces {{COL2}}. *rr:sqlQuery* is encoded by generating a context/named graph to group triples produced from that TripleMap and the query is stored in a literal object with context as the subject and <s2s:sqlQuery> as predicate. Faux nodes are translated as IRI templates. A specification of S2SML is available on the sparql2sql wiki on Github.

Table 3. Other R2RML predicates and the corresponding S2SML construct

R2RML predicate	S2SML example
rr:language	"literal"@en
rr:datatype	"literal"^^ <xsd:float>
rr:inverseExpression	"{COL1} = SUBSTRING({{COL2}}, 3)"^^ <s2s:inverse>
rr:class	?s a <ont:class>
rr:sqlQuery	<context1> {<sen:sys_{table.col}> ?p ?o.}
	<context1> s2s:sqlQuery "query"

4.3 Translation

Building a Mapping Closure. Following from Definition 3 of a Mapping Closure, M_c, a translation engine needs to perform, $\bigcup_{m \in M_d} m$, a union of all mappings on a device, M_d. To support template matching with any in-memory RDF store and SPARQL engine, as described in Definition 4, we replace all I_{map} within each mapping m with I_p, the union of IRI string parts, and extract C, the set of table column binding strings. C is then stored within map, m_{join}, with I_p as key and C as value. For example, in Fig. 1, <sen:system_> will replace both <sen:system_{sensors.name}> and <sen:system_{readings.sensor}> while m_{join} will store (<sen:system_>, {sensors.name, readings.sensor}).

SPARQL Algebra and BGP Resolution. A SPARQL query, *sparql*, can be translated by the function $trans(M_c, sparql)$. The first step within *trans* is $algebra(sparql) \rightarrow \propto$, where $\propto$ is a SPARQL algebra expression. For example, SRBench [17] query 6^8, which looks for weather stations that have observed low visibility within an hour of time by projecting stations that have either (by union) low visibility, high rainfall or snowfall observations, has $\propto$ as follows.

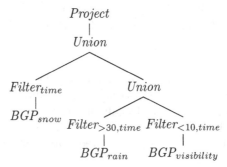

Basic graph patterns (BGPs) are sets of triple patterns within the query. *trans* walks through $\propto$ from the leaf nodes executing function $\sigma(M_c, BGP)$ on each BGP. As the M_c is pure RDF and represents the graph as it is, it can be loaded into an RDF store, ideally, in-memory. A SPARQL *select * query* containing the BGP within its *where* clause can then be executed on the M_c within the store. Literal datatypes are removed from the query and stored in a map. In the above example, BGP_{snow} and $BGP_{visibility}$ return no results for 4UT01 (Listing 1.1) but BGP_{rain} returns a result from σ. Each result from σ is a map of $(vk, vv) \in V \times (I_{map}IBL_{map}LF)$ where V is a variable in a triple pattern. The (vk, vv) maps are passed to the operator, $\propto_{op}$ above in $\propto$. Eventually, an SQL union is performed at the project operator π for all $|\sigma| > 1$. We have implemented a pluggable BGP resolution interface to show various in-memory RDF stores can be supported, with Jena and Sesame as reference examples.

Syntax Translation. *trans* continues its walk from BGP leaf nodes through $\propto$ to the root. At each node, $\propto_{op}$, a syntax translation $syn(SQL_i, (vk, vv)_i, \propto_{op}) \rightarrow (SQL_o, (vk, vv)_o)$ is performed, producing an updated SQL query, SQL_o. In the example, at the $Filter_{>30,time} \propto_{op}$, SQL_i which consists of a blank SQL *where* clause is updated using $(vk, vv)_i$ to translate restrictions on *?time* and *?value* to those with bindings 4UT01.time<...T17:00:00 and 4UT01.Rainfall>30. The SQL *from* clause is also updated with the table 4UT01. An unchanged $(vk, vv)_o$ and the updated SQL_o are output from *syn* and passed upwards.

Table 4 shows a list of common operators $\propto_{op}$ and their corresponding SQL clauses and *syn* descriptions. If an operator uses (vk, vv) for mapping a V and

---

^{8} https://github.com/eugenesiow/sparql2sql/wiki/Q06.

Table 4. Operators $\propto_{op}$ and corresponding SQL Clauses

$\propto_{op}$	SQL clause	Remarks
Project π	Select, From	Restricts relation to subset using (vk, vv)
Extend ρ	Select	Renames an attribute in (vk, vv)
Filter ς	Where, Having, From	Restriction translated using (vk, vv)
Union $\cup$	From	Add unrestricted select of SQL_i in FROM
Group γ	Group By, From	Aggregation translated using (vk, vv)
Slice ς_S	Limit	Add a LIMIT clause
Distinct ς_D	Select	Add a DISTINCT to SELECT clause
Left Join $\bowtie$	Left Join..On, Select	If I add to (vk, vv), else LEFT JOIN

retrieves a I_{map}, L_{map} or F, it adds the table binding to the FROM clause. If there are tables in the FROM without join conditions, a cartesian product (cross join) of two tables is taken. Finally, if faux nodes, F, are encountered in π, an SQL update (UPDATE table SET col=RANDOM_UUID()) is run to generate identifiers and vv in (vk, vv) is updated from {table.uuid} to {table.col}.

4.4 Streaming

The mapping and translation design can be used to translate SPARQL to Event Processing Language (EPL) for streams. Listing 1.4 shows the additional syntax in the SPARQL *from* clause specified in Extended Backus Naur Form.

Listing 1.4. SPARQL FROM Clause Definition for sparql2stream

```
FromClause = FROM NAMED STREAM <StreamIRI> [RANGE Time TimeUnit WindowType]
TimeUnit = ms | s | m | h | d
WindowType = TUMBLING | STEP
```

A *TUMBLING* window is a pull-based buffer that reevaluates at the specified time interval while the *STEP* window is a push-based sliding window extending for the specified time interval into the past. The *syn* function is modified to support EPL as an SQL dialect. Streaming for the IoT is useful for 1) scenarios with high sampling (e.g. accelerometers) or insertion rate (e.g. many sensors to a device/hub) and 2) applications that perform real-time analytics requiring push-based results from queries rather than results at pull intervals.

5   Experiment

To evaluate our approach against RDF stores, SPARQL-to-SQL engines and streaming engines in an WoT context, we selected two unique IoT scenarios using published datasets. Code and experiments can be found on Github[9].

---

[9] https://github.com/eugenesiow/sparql2sql.

Distributed Meteorological System. The first scenario uses Linked Sensor Data with sensor metadata and observation data from about 20,000 weather stations across the United States. In particular, we used the period of the Nevada Blizzard (100k triples) for storage and performance tests and the largest Hurricane Ike period (300k triples) for storage tests. SRBench [17] is an accompanying analytics benchmark for streaming SPARQL queries but can be applied, with similar effect, to SPARQL queries constrained by time. Queries[10] 1 to 10 were used as they involve time-series sensor data while the remaining queries involved integration or federation with DBpedia or Geonames which was not within the scope of the experiment. Queries are available on Github[11]. The experiment simulates a distributed setup as each station's data is stored on an IoT device as RDF or rows with S2SML or R2RML mappings. Queries are broadcast to all stations, total query time was the maximum time as the slowest station was the limiting factor. Due to resource constraints, we assumed broadcast and individual connection times to be similar over a gigabit switch, hence, distributed tests for the 4700+ stations were run in series, recording individual times, averaging over 3 runs and taking the maximum amongst stations for each query.

Smart Home Analytics Benchmark. This scenario uses smart home IoT data collected by Barker *et al.* [2] over 3 months in 2012. 4 queries[12] requiring space-time aggregations with a variety of data for descriptive and diagnostic analytics were devised. (1) hourly aggregation of temperature, (2) daily aggregation of temperature, (3) hourly and room-based aggregation of energy usage and (4) diagnosis of unattended devices through energy usage and motion, aggregating by hour and room. Time taken for queries were averaged over 3 runs.

Environment and Stores. The IoT devices used were Raspberry Pi 2 Model B+s' with 1GB RAM, 900MHz quad-core ARM Cortex-A7 CPU and Class 10 SD Cards, as they are widely available and relatively powerful. 512mb was assigned to the Java Virtual Machine on Raspbian 4.1. Ethernet connections were used between the querying client (i5 3.2GHz, 8GB RAM, hybrid drive) and the Pis'.

RDF stores compared were TDB (Open Source) and GraphDB (Commercial). Virtuoso 7 was not supported on the 32-Bit Raspbian and Virtuoso 6 did not support SPARQL 1.1 time functions like *hours*. H2[13] (disk mode) was used as the relational store for all SPARQL-to-SQL tests. ontop and morph were tested within the limits of query compatibility and a quantitative evaluation of SQL queries and translation time was done. Native SPARQL streaming engine CQELS was compared for push-based performance. As CQELS already benchmarked against C-SPARQL and push results for real-time analytics helped differentiate streams, we did not compare against C-SPARQL.

---

[10] http://www.w3.org/wiki/SRBench.
[11] https://github.com/eugenesiow/sparql2sql/wiki.
[12] https://github.com/eugenesiow/ldanalytics-PiSmartHome/wiki/.
[13] http://www.h2database.com/.

6   Results and Discussion

6.1   Storage Efficiency

Table 5 shows the store sizes of different datasets for the H2, TDB and GraphDB setups. As time-series sensor data benefits from succinct storage as rows, H2 outperformed the RDF stores, which also suffered from greater overheads for multiple stores and indexing [16], from about one to three orders of magnitude.

Table 5. Store size by dataset (in MB)

Dataset	#Store(s)	H2	TDB	GraphDB	Ratio
Nevada Blizzard	4701	90	6162	121694	1:68:1352
Hurricane Ike	12381	761	85274	345004	1:112:453
Smart home	1	135	2103	1221	1:15:9

6.2   Query Performance

Figure 2 shows the performance of SRBench queries on the various stores with the Nevada Blizzard dataset. We see that our sparql2sql approach performs better consistently on all queries with stable average execution times. We argue that this was the result of SQL queries produced not having joins as each station was a single time-series (wide) and intermediate nodes not being projected (could be 'collapsed'). GraphDB generally performed better than the TDB store especially on query 9 due to TDB doing a time consuming join operation in the low-resource environment between two subtrees, WindSpeedObservation and WindDirectionObservation. If queries were executed to retrieve subgraphs individually with TDB, each query cost a 100 times less. Query 4 was similar but with TemperatureObservation and WindSpeedObservation subgraphs instead.

Both ontop (v1.6.1) and morph (v3.5.16), at the time of writing, will only support the aggregation operators required for queries 3 to 9 sans query 6 in future versions. morph was also unable to translate queries 6 and 10 as yet while ontop's SQL query 10 did not return from the H2 store on some stations (e.g. BLSC2). ontop performs better than the RDF stores on queries 2 and 6. Although queries 1 and 2 are similar in purpose, query 2 has an OPTIONAL on the unit of measure term, hence as shown in Table 6, ontop generates different structured queries, explaining the discrepancy in time taken.

We did an additional comparison between SPARQL-to-SQL engines in terms of the structure of queries generated and translation time. Table 6 shows the average translation time, t_{trans} of the 3 engines on the client. The plugin BGP resolution engine for sparql2sql (s2s) used was Jena. Both ontop and morph have additional inference/reasoning features and ontop makes an extra round trip to the Pi to obtain database metadata explaining the longer translation times.

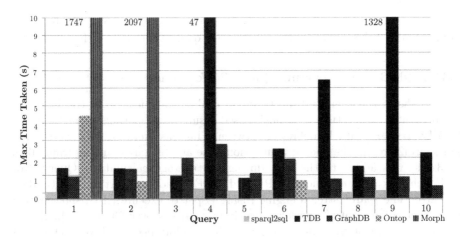

Fig. 2. Max time taken for distributed SRBench queries

In R2RML, as shown in Listing 1.3, in the absence of row identifiers in time-series data, time has to be used in IRI templates for intermediate observation metadata nodes. As timestamps are not unique in LSD (observed from data), they are not suited as a primary key, hence cannot be used to chase equality generating dependencies in the semantic query optimisation ontop does [14]. The resulting queries from ontop and morph both have redundant inner joins on the time column (used to model intermediate IRIs in R2RML).

In the smarthome scenario, sparql2sql query performance on aggregation queries as shown in Fig. 3 is still ahead of the RDF stores. GraphDB also has all-round better performance than TDB. All the queries performed SPARQL 1.1 space-time aggregations, excluding the other SPARQL-to-SQL engines.

Through the experiments, we observe that although other SPARQL-to-SQL engines have reported significant performance improvements over RDF stores on various benchmarks and deployments, there is still room for optimisation for IoT devices and scenarios and perform below RDF stores on Pis' or do not yet support queries relevant to IoT scenarios such as aggregations. sparql2sql with S2SML, utilises the strengths of SPARQL-to-SQL on IoT scenarios and

Table 6. SPARQL-to-SQL translation time and query structure

Q	t_{trans} (ms)			Joins			Join type and structure	
	s2s	Ontop	Morph	s2s	Ontop	Morph	Ontop (qview)	Morph
1	16	702	146	0	6	4	implicit	4 inner
2	17	703	144	0	6	4	5 nested, 1 left outer	4 inner
6	19	703	-	0	5	-	5 implicit	-
10	32	846	-	0	6	-	UNION(2x3 implicit)	-

time-series data and performed better than both RDF stores and SPARQL-to-SQL engines. Table 8 summarises the average query times for all the tests.

6.3 Push-Based Streaming Query Performance

Table 8 shows the average time taken to evaluate a query from the insertion of an event to the return of a push-based result from sparql2stream, t_{s2r} and CQELS, t_{CQELS} with 1 s delays in between. This was averaged over 100 results. For sparql2stream, the one-off translation time at the start (ranging from 16ms to 32ms) was added to the sum during the average calculation. Query 6 of SRBench was omitted due to EPL and CQELS not supporting the UNION operator. The sparql2stream engine (using Esper to execute EPL) showed over two orders of magnitude performance improvements over CQELS. Queries 4, 5 and 9 that involved joining subgraphs (e.g. WindSpeed and WindDirection in 9) and aggregations showed larger differences. It was noted, that although CQELS returned valid results for these queries, they contained an increasing number of duplicates (perhaps from issues in the adaptive implementation) which caused a significant slowdown over time and when averaged over 100 pushes. The experiments are available on Github[14,15].

This ability to answer queries in sub-millisecond average times in a push-based fashion makes sparql2stream a viable option for real-time analytics on IoT devices like medical devices that require reacting instantaneously.

To verify that sparql2stream was able to answer SRBench queries close to the rate they are sent, even at high velocity, we reduced the delay between insertions from 1000 ms to 1 ms and 0.1 ms. Table 8 shows a summary of the average latency (the time from insertion to when query results to be returned) of each query (in ms). We observe that the average latency is slightly higher than the inverse of the rate. The underlying stream engine, Esper, maintains context partition states consisting of aggregation values, partial pattern matches and data

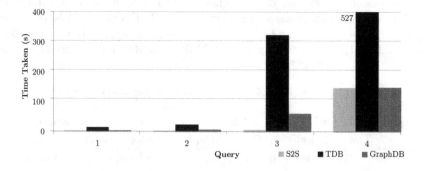

Fig. 3. Average time taken for smarthome analytical queries

---

[14] https://github.com/eugenesiow/cqels.
[15] https://github.com/eugenesiow/sparql2stream.

Table 7. Average query run times (in ms)

SR_{Bench}	t_{s2s}	t_{TDB}	t_{GDB}	t_{ot}	t_{morph}	Ratio	t_{s2r}	t_{CQELS}	Ratio
1	365	1679	1223	4589	1747702	1:5:3:13:4k	0.47	138	1:294
2	415	1651	1627	945	2097159	1:4:4:2:5k	0.46	119	1:261
3	375	1258	2251	-	-	1:3:6	0.66	202	1:306
4	533	47084	3004	-	-	1:88:6	0.67	186k	1:277k
5	415	1119	1404	-	-	1:3:3	0.63	1476k	1:3243k
6	457	2751	2181	987	-	1:6:5:2	-	-	-
7	455	6563	1082	-	-	1:14:2	0.66	2885	1:5245
8	320	1785	1162	-	-	1:6:4	0.67	282	1:426
9	436	1328197	1175	-	-	1:3k:3	0.67	188k	1:280k
10	354	2514	685	-	-	1:7:2	0.73	72	1:98

Smarthome	t_{s2s}	t_{TDB}	t_{GDB}	Ratio	t_{s2r}	t_{CQELS}	Ratio
1	466	13709	3132	1:29:7	0.64	125	1:196
2	2457	21898	6914	1:9:3	0.77	129	1:167
3	4685	322357	59803	1:69:13	0.81	-	-
4	147649	527184	147275	1:4:1	3.78	-	-

Table 8. Average latency (in ms) at different rates

R \ $Q_{\#}$	1	2	3	4	5	7	8	9	10
1	1.300	1.374	1.279	1.303	1.2561	1.268	1.267	1.295	1.255
10	0.155	0.159	0.143	0.161	0.1291	0.137	0.141	0.155	0.129

R = Rate(rows/ms), $Q_{\#}$ = Query number

windows. At high rates, the engine introduces blocking to lock and protect context partition states. However, Fig. 4 shows the effect of this blocking is minimal as the percentage of high latency events is less than 0.3 % (note that x-axis is 99 % to 100 %) across various rates. This comparison which groups messages by latency ranges is also used in the Esper benchmark and by Calbimonte et al. [6].

We also tested the size of data that can fit in-memory for sparql2stream with SRBench Query 8, that uses a long *TUMBLING* window. The engine ran out of memory after 33.5 million insertions. Given a ratio of 1 row to 75 triples within the SSN mapping (each observation type with 10+ triples), by projection, an RDF dataset size of 2.5 billion triples was 'fit' in a IoT devices' memory.

Queries 1 and 2 of the smart home scenario also corroborated the 2 orders of magnitude performance advantage of sparql2stream over CQELS. Queries 3 and 4 were not run on CQELS due to issues with the FILTER operator in the version tested. Query 4 which involved joins on motion and meter streams and an aggregation saw the average latency of sparql2stream increase, though still stay under 4ms. The latency for this query was measured from the insertion time of the last event involved (that trips the push) to that of the push result.

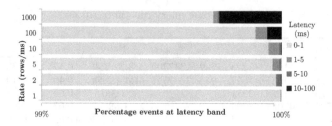

Fig. 4. Percentage latency at various rates for Q1

7   Conclusion

A Web of Things based on open standards and the innovations introduced in the Semantic Web and Linked Data can encourage greater interoperability and bridge product silos. This paper shows how time-series Internet of Things data that is flat and wide, can be stored efficiently as rows on devices with limited resources. By optimising SPARQL-to-SQL translation and 'collapsing' intermediate nodes, performance on smart home monitoring and a distributed meteorological system show storage and query performance improvements that range from 2 times to 3 orders of magnitude. The independence from primary keys and database metadata also resulted in less joins in resultant SQL queries and faster query translation times respectively. Future work will expand experimentation to consider additional datasets, data sizes, queries and include a greater variety of stores and stream processing use cases for time-series data e.g. column stores, stream analytics and compression/approximation.

The limitations of this approach lie in the assumption that the bulk of IoT time-series data is flat and read-only which might change in the future. Exploiting the wideness of time-series data for row access performance is also query dependant. Current state-of-the-art Ontology-Based Data Access (OBDA) systems, which do query translation, support general use cases (web/enterprise relational database mapping) and support reasoning which our approach does not seek to address at the moment.

References

1. Barbieri, D.F., Braga, D., Ceri, S., Valle, E.D., Grossniklaus, M.: Querying RDF streams with C-SPARQL. ACM SIGMOD Rec. **39**(1), 20 (2010)
2. Barker, S., Mishra, A., Irwin, D., Cecchet, E.: Smart*: an open data set and tools for enabling research in sustainable homes. In: Proceedings of the Workshop on Data Mining Applications in Sustainability (2012)
3. Barnaghi, P., Wang, W.: Semantics for the internet of things: early progress and back to the future. Int. J. Semant. Web Inf. Syst. **8**(1), 1–21 (2012)
4. Bishop, B., Kiryakov, A., Ognyanoff, D.: OWLIM: a family of scalable semantic repositories. Semant. Web **2**(1), 33–42 (2011)

5. Buil-Aranda, C., Hogan, A., Umbrich, J., Vandenbussche, P.-Y.: SPARQL web-querying infrastructure: ready for action? In: Alani, H., et al. (eds.) ISWC 2013, Part II. LNCS, vol. 8219, pp. 277–293. Springer, Heidelberg (2013)
6. Calbimonte, J.P., Jeung, H., Corcho, O., Aberer, K.: Enabling query technologies for the semantic sensor web. Int. J. Semant. Web Inf. Syst. **8**(1), 43–63 (2012)
7. Chebotko, A., Lu, S., Fotouhi, F.: Semantics preserving SPARQL-to-SQL translation. Data Knowl. Eng. **68**(10), 973–1000 (2009)
8. Erling, O.: Implementing a sparql compliant RDF triple store using a SQL-ORDBMS. Technical report, OpenLink Software (2001)
9. International Telecommunication Union: Overview of the Internet of things. Technical report (2012)
10. Le-Phuoc, D., Dao-Tran, M., Xavier Parreira, J., Hauswirth, M.: A native and adaptive approach for unified processing of linked streams and linked data. In: Aroyo, L., et al. (eds.) ISWC 2011, Part I. LNCS, vol. 7031, pp. 370–388. Springer, Heidelberg (2011)
11. Neumann, T., Weikum, G.: x-RDF-3X. Proc. VLDB Endow. **3**, 256–263 (2010)
12. Patni, H., Henson, C., Sheth, A.: Linked sensor data. In: Proceedings of the International Symposium on Collaborative Technologies and Systems (2010)
13. Priyatna, F., Corcho, O., Sequeda, J.: Formalisation and experiences of R2RML-based SPARQL to SQL query translation using morph. In: Proceedings of the 23rd International Conference on World Wide Web, pp. 479–489 (2014)
14. Rodriguez-Muro, M., Rezk, M.: Efficient SPARQL-to-SQL with R2RML mappings. Web Semant. Sci. Serv. Agents WWW **33**, 141–169 (2014)
15. Stonebraker, M., Abadi, D., Batkin, A.: C-store: a column-oriented DBMS. In: Proceedings of VLDB, pp. 553–564 (2005)
16. Weiss, C., Karras, P., Bernstein, A.: Hexastore: sextuple indexing for semantic web data management. In: Proceedings of the VLDB Endowment (2008)
17. Zhang, Y., Duc, P.M., Corcho, O., Calbimonte, J.-P.: SRBench: a streaming RDF/SPARQL benchmark. In: Cudré-Mauroux, P., et al. (eds.) ISWC 2012, Part I. LNCS, vol. 7649, pp. 641–657. Springer, Heidelberg (2012)

Can You Imagine... A Language for Combinatorial Creativity?

Fabian M. Suchanek[1]([✉]), Colette Menard[2], Meghyn Bienvenu[3], and Cyril Chapellier[1]

[1] Télécom ParisTech, Paris, France
suchanek@telecom-paristech.fr
[2] STIM, Paris, France
[3] LIRMM Montpellier, Paris, France

Abstract. Combinatorial creativity combines existing concepts in a novel way in order to produce new concepts. For example, we can imagine jewelry that measures blood pressure. For this, we would combine the concept of jewelry with the capabilities of medical devices. In this paper, we concentrate on creating new concepts in the description logic $\mathcal{EL}$. We propose a novel language to this effect, and study its properties and complexity. We show that our language can be used to model existing inventions and (to a limited degree) to generate new concepts.

1 Introduction

What if cars had wings? What if tables could serve as beds? What if spoons could talk? These questions may seem completely absurd. And yet, the following questions are much less absurd: What if phones could go online? What if cars could be used to sleep in them? What if tap water contained medicine? Much of what may seem absurd today may become reality in the future. The field that is concerned with combining components of existing concepts into new concepts is called *combinatorial creativity*[1]. This field serves different purposes. Most prominently, it serves to describe new inventions: a smartphone is a telephone that is connected to the Internet; a Segway is a vehicle with two wheels on the same axis; a Hyperloop is a train without wheels. But combinatorial creativity can also serve to develop new business ideas, to find plots for books or movies, to understand human creativity, to disrupt conventional assumptions, to find design alternatives, or to foster thinking outside the box.

A prominent current of research uses description logics (DLs) to model real-world concepts [2,11], so it is natural to try to use DLs to capture the types of concept modification underlying combinatorial creativity. Suppose for example that cars are defined using the following (simplified) DL axiom:

$$Car \equiv Vehicle \sqcap \exists hasPart.Wheel$$

---

[1] A more constrained subfield of computational creativity in general.

© Springer International Publishing AG 2016
P. Groth et al. (Eds.): ISWC 2016, Part I, LNCS 9981, pp. 532–548, 2016.
DOI: 10.1007/978-3-319-46523-4_32

Now suppose that we wish to consider the concept obtained by *removing the wheels from car*. Any construction of the form $Car \sqcap \neg \exists hasPart.Wheel$ simply leads to a contradiction, i.e., the empty concept $\bot$. Thus, standard DL constructors do not provide any direct means of expressing modifications like taking the concept Car and removing the property $\exists hasPart.Wheel$ from it.

Much work has been done on conceptual blending [1,4,6,7,11,14,20,26], in which two concepts from different thematic areas are blended to create a new concept. However, blending is concerned more with describing analogies and metaphors than with describing modifications of concepts. Blending can, e.g., explain how a human understands an expression such as "sign forest", but it cannot express an atomic operation such as removing the wheels from a car. Non-standard reasoning services for DLs, like semantic matchmaking, consider the problem of modifying concepts to achieve certain objectives, but do not provide a means of expressing explicit updates of concepts. What we would want is a language that allows writing *Remove the wheels from the car*, or: Car *"minus"* $\exists hasPart.Wheel$.

In this paper, we propose a formal language for concept modifications that can serve as a basis for combinatorial creativity. More precisely:

- We define a language that allows modeling the transition from one concept to another one explicitly – by adding, removing, or modifying its constituents.
- We explain the design rationale for our operators, discuss design alternatives, and prove their formal properties.
- We show that our language can be used to describe real-world inventions, and (in a limited manner) to *generate* new concepts.

This paper is structured as follows. We start with a discussion of related work in Sect. 2 and give the preliminaries in Sect. 3. The main part of our paper, Sect. 4, describes our language. Section 5 shows concrete applications of our language. We conclude in Sect. 6.

2   Related Work

Cognitive Sciences. Combinatorial creativity has first been studied in the cognitive sciences [5,8,24]. These analyses center on understanding human cognition and have not led to a formal theory based in logic. The COINVENT project [21] aims to develop a computationally feasible, cognitively-inspired, formal model of concept invention. The project, however, has started only recently, and the model is still in the process of development. Fictional ideation generates new concepts for narratives [16]. In a larger sense, computational creativity is concerned also with creative human-computer interaction, art, figurative language, humor, music, argumentation, generating narratives or poetry, and scientific discovery [25]. We concentrate here on works that come closest to a formal language for describing modifications of concepts.

Conceptual Blending. One of the areas that cognitive science investigates is amalgams and analogies [4]. An *amalgam* of two input concepts is any new concept that combines constraints from abstractions of each of the input concepts. In this way, "a red French sedan" and "a blue German minivan" can be combined to "a red German sedan". *Analogies*, on the other hand, find commonalities between a combined concept (such as "sign forest") and source concepts (such as "forest"). The Structure Mapping Engine [7], likewise, is concerned with analogies. Analogies and amalgams fall into the broader field of conceptual blending [11,14,26]. Recent work in the area of linguistics [1] also discusses conceptual blending, as does the Heuristic-Driven Theory Projection [20]. Closest to our work in conceptual blending is work on upward refinement in the DL $\mathcal{EL}++$ [6]. All of these analyses are centered on describing the space between two concepts. However, they do not give us a language with operators to explicitly modify a single input concept. For example, none of the approaches can express the operation of taking a car, removing a plastic part, and replacing it by an aluminium part.

Modifying DL Concepts. Given a DL description of a product on offer and of a product in demand, semantic matchmaking is concerned with modifying the product in demand so that it matches the product on offer [17]. Work on identifying missing negative constraints also involves generation and manipulation of concepts [9]. However, while concept modification is central to these (and other) works on non-standard reasoning in DLs, they do not provide any means to explicitly express modifications of a concept, like *Remove the wheels from a car*. To the best of our knowledge, the only work that proposes such an operator is Teege [23]. We compare their subtraction operator with our own in Sect. 4.2. Also loosely related is work on belief change in DLs, which aims to modify a knowledge base to consistently incorporate new information, see e.g. [10,18].

3   Preliminaries

Description logics (DLs) are a family of logical formalisms that describe semantic knowledge about a domain in terms of concepts (= classes, unary relations) and roles (=properties, binary relations). For example, the first line in Fig. 1 says that the concept *PlasticRoof* is defined as the intersection of the concept *Roof* and the concept of all those things that are made of plastic. We concentrate here on one particular DL, $\mathcal{EL}$ [2]. We assume fixed sets N_C and N_R of *concept names* and *role names*, respectively. A *concept* is anything of the form

$$\top \mid A \mid C \sqcap D \mid \exists r.C$$

where C and D are concepts, A ranges over N_C, and r ranges over $N_R \cup \{u\}$, where u is the *universal role*. The set of these concepts will be denoted by $\mathcal{L}$.

A concept is *basic* if it is a named concept, $\top$, or an existential concept. By definition, every concept in $\mathcal{L}$ is a conjunction of one or more concepts. We will therefore understand every concept C as a conjunction $C = C_1 \sqcap \cdots \sqcap C_n$. If any C_i is a conjunction, then C_i can be folded into the conjunction. In all of the following, we will therefore assume that every concept is a conjunction of basic

$$PlasticRoof \equiv Roof \sqcap \exists madeOf.Plastic$$
$$Car \equiv Vehicle \sqcap \exists hasPart.PlasticRoof$$
$$\sqcap \exists hasPart.Wheel \sqcap \exists usedFor.Travel$$

Fig. 1. An example terminology $\mathcal{T}$

concepts. To simplify notation, we will talk of "the conjunct C_i of C" to mean that $C = C_1 \sqcap \cdots \sqcap C_n$ and $1 \leq i \leq n$.

DL semantics relies on *interpretations* $\mathcal{I} = (\Delta^{\mathcal{I}}, \cdot^{\mathcal{I}})$, where $\Delta^{\mathcal{I}}$ is a non-empty *domain*, which can be understood as a set of real-world entities, and $\cdot^{\mathcal{I}}$ is an *interpretation function* which maps each $A \in N_C$ to a subset $A^{\mathcal{I}}$ of $\Delta^{\mathcal{I}}$, and each $r \in N_R$ to a subset $r^{\mathcal{I}}$ of $\Delta^{\mathcal{I}} \times \Delta^{\mathcal{I}}$. The universal role u is mapped to $\Delta^{\mathcal{I}} \times \Delta^{\mathcal{I}}$. This interpretation is extended to all concepts as follows: $\top^{\mathcal{I}} = \Delta^{\mathcal{I}}$, $(C_1 \sqcap C_2)^{\mathcal{I}} = C_1^{\mathcal{I}} \cap C_2^{\mathcal{I}}$, and $(\exists r.C)^{\mathcal{I}} = \{x \mid \exists(x,y) \in r^{\mathcal{I}} \text{ such that } y \in C^{\mathcal{I}}\}$. We say that a concept C is *subsumed by* (or *implies*) a concept D, written $C \sqsubseteq D$, if $C^{\mathcal{I}} \subseteq D^{\mathcal{I}}$ in all interpretations $\mathcal{I}$.

Subsumption between $\mathcal{EL}$ concepts can be decided in polynomial time [3] and adding the universal role does not increase the complexity. To test for subsumption between two concepts ($C \sqsubseteq D$), we can employ a polynomial algorithm based upon a syntactic characterization of subsumption [13] that works analogously to the well-known structural subsumption algorithm [2, Sect. 2.3.1] for the $\mathcal{FL}_0$ language. Details can be found in our technical report [22].

Concepts can contain redundant conjuncts, as in $\exists r.A \sqcap \exists r.\top$. We call a concept C *fully reduced* if there does not exist a concept C' such that (1) $C' \sqsubseteq C$, (2) every conjunct of C' appears in C, and (3) C' has less conjuncts than C. By removing redundant conjuncts from C, we can compute a fully reduced concept $red(C)$ equivalent to C. The *normal form* of a concept C, written $norm(C)$, is defined as follows:

- $norm(A) = A$, if A is a named concept or $\top$
- $norm(\exists r.C) = \exists r.norm(C)$
- $norm(C_1 \sqcap \cdots \sqcap C_n) = red(norm(C_1) \sqcap \cdots \sqcap norm(C_n))$

We show in our technical report [22] that the normal form of a concept is unique up to reordering of conjuncts and can be computed in polynomial time.

Ordering. We assume that the set of concept names N_C is ordered by a complete order $\prec_{N_C}$, and the set of relation names N_R is ordered by a complete order $\prec_{N_R}$. Based on these, it is easy to define a complete order $\prec$ on concepts, as follows.

Definition 1 (Order): Given a complete order $\prec_{N_C}$ on concept names $N_C \cup \{\top\}$ and a complete order $\prec_{N_C}$ on relation names $\mathcal{R} \cup \{u\}$, the complete order $\prec$ on minimal concepts $\mathcal{L}$ is defined as follows:

- $A \prec B$ iff $A \prec_{N_C} B$, for $A, B \in N_C$
- $A \prec C$, for $A \in N_C$ and $C \notin N_C$
- $\exists r.C \prec D$ for conjunctions D

- $\exists r.C \prec \exists s.D$ iff $r \prec_{N_R} s$
- $\exists r.C \prec \exists r.D$ iff $C \prec D$
- $C \prec D$ for conjunctions C, D is given by the lexicographical extension of $\prec$.

This order is purely syntactic; it does not relate to concept subsumption. If the conjuncts of a conjunction $C_1 \sqcap \cdots \sqcap C_n$ are written in increasing order $C_1 \prec \cdots \prec C_n$, we will talk of an *ordered conjunction*. In all of the following, we will assume conjunctions to be ordered. For example, from Fig. 1, we will understand that $Vehicle \prec \exists hasPart.PlasticRoof \prec \exists hasPart.Wheel \prec \exists usedFor.Travel$, because the concepts are written in that order. Ordered conjunctions have an important property, which follows from the properties of the normal form:

Property 1 (Ordered Conjunctions in Normal Form): Two ordered conjunctions in normal form are equivalent iff they are identical.

Terminologies. A *concept definition* takes the form $A \equiv C$, where $A \in N_C$ and $C \in \mathcal{L}$. A *terminology* $\mathcal{T}$ is a set of concept definitions, in which no concept name occurs more than once on the left-hand side of a concept definition. Figure 1 shows an example terminology. We will see the terminology as a function $\mathcal{T}$: $N_C \rightarrow \mathcal{L}$, which, given a named concept, replaces it by the right-hand-side of its definition in $\mathcal{T}$. We consider only *acyclic* terminologies, i.e., there are no cyclic dependencies between the concept definitions. This allows us to define the complete recursive *unfolding* $\mathcal{T}^*(\cdot)$, with $\mathcal{T}^*(A) = A$ for concept names that do not have a definition in $\mathcal{T}$, $\mathcal{T}^*(A) = \mathcal{T}^*(\mathcal{T}(A))$ for concept names that have a definition, $\mathcal{T}^*(\exists r.C) = \exists r.\mathcal{T}^*(C)$ and $\mathcal{T}^*(C_1 \sqcap \cdots \sqcap C_n) = \mathcal{T}^*(C_1) \sqcap \cdots \sqcap \mathcal{T}^*(C_n)$. We say that a concept name A is a *declared child* of a concept name B if B appears as a conjunct of the definition of A.

4   Operators for Concept Modification

We will now define the operators of our language for combinatorial creativity. Our operators will work directly on DL concepts. The background terminology $\mathcal{T}$ will play no role in defining the operators, but will serve instead to provide the DL concepts upon which we will apply the operators. This may seem unconventional at first, but perhaps this is only fitting for a paper that treats matters of creativity.

4.1   Addition

Definition 2 (Addition): For two concepts C and D, we define addition as $C + D := norm(C \sqcap D)$.

> *Example 2 (Addition):* In our example from Figure 1, the expression $\mathcal{T}(Car) + (\exists hasPart.Wing \sqcap \exists usedFor.Fly)$ denotes a car with wings that is used for flying. This yields $Vehicle \sqcap \exists hasPart.Wheel \sqcap \exists hasPart.PlasticRoof \sqcap \exists usedFor.Travel \sqcap \exists hasPart.Wing \sqcap \exists usedFor.Fly$.

Addition has the following properties, which follow directly from the properties of $\sqcap$ and the normalization.

Property 2 (Inclusion): If $C \sqsubseteq D$ for two concepts C and D, then $C + D = C$.

Property 3 (Commutative Monoid): For any three concepts C, D, E, the following hold: Addition is closed, $C + D \in \mathcal{L}$. Addition is monotone, $C + D \sqsubseteq C$. Addition is commutative, $C + D = D + C$. Addition is associative, $C + (D + E) = (C + D) + E$. $\top$ is its neutral element. $(+, \top)$ forms a commutative monoid.

4.2 Subtraction

Definition 3 (Subtraction): For a basic concept A, and a concept C that has an ordered normal form $norm(C) = C_1 \sqcap \cdots \sqcap C_n$, we define subtraction as $C - A := norm(C_1 \sqcap \cdots \sqcap C_{j-1} \sqcap C_{j+1} \sqcap \cdots \sqcap C_n)$, where $j = argmin_i\{C_i : C_i \sqsubseteq A\}$. If there is no such j, then $C - A = C$. Subtraction is left-associative. For a conjunction D with $norm(D) = D_1 \sqcap \cdots \sqcap D_m$, subtraction is defined as $C - D = C - D_1 - \cdots - D_m$.

In other words, the subtraction $C - A$ removes the first conjunct of the ordered conjunction C that implies A. If the subtrahent is a conjunction, subtraction removes each conjunct of the subtrahent.

> *Example 3 (Subtraction):* In our example from Figure 1, the expression $\mathcal{T}(Car) - (\exists hasPart.\top \sqcap \exists usedFor.Travel)$ removes the first *hasPart* association and the *usedFor* association. This yields $Vehicle \sqcap \exists hasPart.Wheel$.

The definition of subtraction offers several design choices. We could, e.g., define subtraction simply as the set difference of the conjuncts, without considering subsumption of concepts. However, the proposed definition has the advantage that we can remove a part of a car even if we do not fully specify it, as in $\mathcal{T}(Car) - \exists hasPart.\top$. Another design alternative for a subtraction $C - A$ would be to remove not just the first conjunct from C that implies A, but all conjuncts in C that imply A. However, we can easily express this design alternative in terms of the proposed definition by subtracting A several times, whereas it is not possible to express the subtraction of just one conjunct with an operator that subtracts all implied conjuncts simultaneously. We could define subtraction to remove not the first matching conjunct, but an arbitrary conjunct. This, however, would result in non-determinism. We could also define subtraction so as to return the set of all possible ways of subtracting the subtrahent. However, the result of this operation would be a set, not a concept. Thus, it would not be possible to use this result with the other operators.

Finally, the subtraction of a conjunction $(C - D)$ offers a design alternative. Instead of subtracting each conjunct of D separately, we could remove from C all those conjuncts that are subsumed by the complete concept term D. This, however, would violate the usual set semantics: subtracting a conjunction with more conjuncts would have less effect than subtracting a conjunction with only one conjunct. We could also make subtraction distributive, by defining $C - (A \sqcap B) = (C - A) \sqcap (C - B)$. This, however, would entail that $(A \sqcap B) - (A \sqcap B) = (A \sqcap B)$, which would defeat the purpose of subtraction.

The subtraction operator from Definition 3 has the following properties, which follow from the definition and the normalization.

Property 4 (Monotonicity): For any two concepts C and D, the following holds: Subtraction is closed, $C - D \in \mathcal{L}$. Subtraction is monotone, $C \sqsubseteq C - D$.

Property 5 (Destruction): If $C \sqsubseteq D$ for a basic concept C and a concept D, then $C - D = \top$. In particular, $C - \top = \top$.

Property 6 (Self-Destruction): For any concept C, $C - C = \top$.

Property 7 (Reversibility): If, for two conjunctions in normal form C and D, every conjunct of D appears in C, then $(C - D) + D = C$.

Proof. If C and D are in normal form, and every conjunct of D appears in C, then $C - D$ is obtained by removing every conjunct of D from C (the subsequent normalization will have no effect). Adding back these conjuncts yields C.   □

We can generalize subtraction to remove an arbitrary conjunct among the implied conjuncts, as follows.

Definition 4. (Generalized Subtraction): For a concept C, a basic concept A, and a natural number i, we define

$$C -_1 A := C - A \qquad C -_i A := ((C - A) -_{i-1} A) + (C - (C - A))$$

Property 8. (Generalized Subtraction): The generalized subtraction $C -_i A$ removes the ith conjunct of the normalized ordered conjunction of C that implies A. If no such conjunct exists, then $C -_i A = C$.

Proof. Assume that C has been transformed into an ordered conjunction in normal form. We proceed by induction. For $i = 1$, the claim follows from Definition 3. Now assume that $C -_{i-1} A$ removes the $(i - 1)^{th}$ conjunct of C that implies A. We observe that $C - A$ removes the first such conjunct. Hence, the expression $((C - A) -_{i-1} A)$ removes the first such conjunct and the ith such conjunct. The expression $(C - (C - A))$ yields the first such conjunct.
Hence, $((C - A) -_{i-1} A) + (C - (C - A))$ is C without the ith such conjunct. If the ith such conjunct does not exist, then $((C - A) -_{i-1} A) = C - A$. Then, $C -_i A = (C - A) + (C - (C - A)) = C$.   □

> *Example 4 (Generalized Subtraction):* With generalized subtraction, we can remove, e.g., the second part of our example car from Figure 1, by saying $\mathcal{T}(Car) -_2 \exists hasPart.\top$. This yields $Vehicle \sqcap \exists hasPart.PlasticRoof \sqcap \exists usedFor.Travel$.

Comparison to Related Work. Teege [23, Definition 2.1] defines subtraction as $C - D := max_{\sqsupseteq}\{E \in \mathcal{L} : D \sqcap E \equiv C\}$. We first note that this operator is undefined if $C \not\sqsubseteq D$, while our operator is always defined. Let us now assume that $C \sqsubseteq D$ and that C and D are conjunctions in normal form. If every conjunct of D appears in C, then Property 7 tells us that our operator is equivalent to Teege's. In general, however, the operators are different. Our definition allows removing a conjunct that "matches" the subtrahent, as in $(A \sqcap \exists r.B) - \exists r.T = A$. Teege's operator has a different behavior: $(A \sqcap \exists r.B) - \exists r.T = (A \sqcap \exists r.B)$. This allows our operator to remove the first, second, or n^{th} matching conjunct, while Teege's operator does not offer this functionality. We will show in Sect. 5 how this functionality can be used in practice.

4.3 Succession

Definition 5. (Succession): For an existential concept $\exists r.E$ and a concept C, we define succession as $C \rightarrow \exists r.E := E'$, where $\exists r'.E'$ is the first conjunct of the ordered normal form $norm(C)$ with $\exists r'.E' \sqsubseteq \exists r.E$. If there is no such conjunct, succession is undefined. For a conjunction D with $norm(D) = D_1 \sqcap \cdots \sqcap D_m$, succession is defined as $C \rightarrow D = norm(\sqcap_i(C \rightarrow D_i))$.

Succession finds the first existential conjunct in C that implies $\exists r.E$, and returns the inner concept in that existential conjunct. If succession is used with a conjunction, the operator joins all the target concepts in a conjunction.

> *Example 5 (Succession):* For our example from Figure 1, the expression $T^*(Car) \rightarrow \exists hasPart.(\exists madeOf.Plastic)$ asks for a plastic part of a car. This yields the plastic roof, $Roof \sqcap \exists madeOf.Plastic$.

The definition of succession allows several design choices. The current definition picks the first target concept. We could equally well use the operator to pick one of them at random. This, however, would result in non-determinism. Another design alternative is to return not the first matching conjunct, but the set of all matching conjuncts. Then, however, the result would no longer be a concept, and could no longer be used with the other operators.

We could also combine all target concepts of all matching conjuncts into a conjunction. Then, however, the target concepts could no longer be manipulated individually. If we want to compute such a conjunction nonetheless, we can do so with the current definition of succession. It suffices to extract one concept after the other through generalized succession (see below), and join them by addition.

Succession has the following properties, which follow from Definition 5.

Property 9. (Closedness): For any concepts C and D, $C \rightarrow D$ is either undefined or a concept.

Property 10. (Inclusion): If $C \sqsubseteq D$, then $\exists r.C \rightarrow \exists r.D = C$, for any concepts C, D and role r.

Like for subtraction, we can define a succession for the ith matching conjunct.

Definition 6. (Generalized Succession): For concepts C, an existential concept $\exists r.E$, and a natural number i, we define

$$(C \rightarrow_1 \exists r.E) := C \rightarrow \exists r.E$$
$$(C \rightarrow_i \exists r.E) := (C - \exists r.E) \rightarrow_{i-1} \exists r.E$$

Property 11. (Generalized Succession): The generalized subtraction $C \rightarrow_i A$ returns the inner concept of the ith existential conjunct of the normalized ordered conjunction of C that implies A.

Proof. Assume that C has been transformed into an ordered conjunction in normal form. We proceed by induction. For $i = 1$, the claim follows from Definition 5. Now assume that $C \rightarrow_{i-1} A$ returns the $(i-1)^{th}$ existential conjunct of C that implies A. We observe that $(C - A)$ removes the first such conjunct. Hence, $(C - A) \rightarrow_{i-1} A$ returns the ith such conjunct. □

Example 6 (Generalized Succession): In our example from Figure 1, the expression $\mathcal{T}(Car) \rightarrow_2 \exists hasPart.\top$ retrieves the second part of a car. This yields *Wheel*.

Comparison to Related Work. To the best of our knowledge, the succession operator has no analog in the formalisms of previous work [6,7,17,23].

4.4   Selection and Replacement

We can define a selection operator by using multiple applications of subtraction:

Definition 7. (Selection): For a concept C, a basic concept A, and a natural number i, we define

$$C \uparrow_i A = (C \sqcap \chi) - (C \sqcap \chi -_i A)$$

where χ is a fresh concept name that comes last in the order $\prec_N$.

Property 12. (Selection): The selection $C \uparrow_i A$ returns the ith conjunct of the ordered normal form of C that implies A, or else $\top$.

Proof. Assume that C has been transformed into an ordered conjunction in normal form. If there is no i^{th} conjunct of C that implies A, then $(C \sqcap \chi -_i A) = C \sqcap \chi$, and hence $C \uparrow_i A = \top$. Otherwise, $(C \sqcap \chi -_i A)$ will remove that i^{th} conjunct, and $(C \sqcap \chi) - (C \sqcap \chi -_i A)$ will deliver that conjunct. This holds in particular for the case where $i = 1, A = \top$ and C is basic. In this case, $C \uparrow_1 \top = (C \sqcap \chi) - (C \sqcap \chi - \top) = (C \sqcap \chi) - \chi = C$. Note that without the addition of χ, we would obtain $C \uparrow_1 \top = C - (C - \top) = C - \top = \top$. □

Example 7 (Selection): In our example from Figure 1, we can select the second *hasPart*-conjunct of a car by saying $\mathcal{T}(Car) \uparrow_2 \exists hasPart.\top$. This yields $\exists hasPart.Wheel$.

We can also combine addition and subtraction to define an operator that replaces a certain conjunct, as follows:

Definition 8. (Replacement): For concepts C, P, D with P taking the form $\exists r_1.\exists r_2.\ldots\exists r_n.A$ (with $n \geq 0$, all $r_i \neq u$, and A a concept name), we define

$$C.replace(P, D) := (C - P) + D \qquad \text{for } P \text{ a named concept}$$
$$C.replace(\exists r.P, D) := (C - \exists r.P) + \exists r.((C \to \exists r.P).replace(P, D))$$

Example 8 (Replacement): In our example from Figure 1, we can replace the material of the roof of the car by aluminium by stating

$$\mathcal{T}^*(Car).replace(\exists hasPart.(\exists madeOf.Plastic), Aluminium)$$

This expression first unfolds Car completely w.r.t. $\mathcal{T}$. Then, the replacement operator descends into the *hasPart* conjunct and into the *madeOf* conjunct, where *Plastic* is replaced by *Aluminium*. This yields:

$$Vehicle \sqcap \exists hasPart.(Roof \sqcap \exists madeOf.Aluminium)$$
$$\sqcap \exists hasPart.Wheel \sqcap \exists usedFor.Travel$$

Property 13. (Replacement): The replacement $C.replace(\exists r_1.\exists r_2.\ldots\exists r_n.A, D)$ descends inside the first existential conjunct of $norm(C)$ that implies $\exists r_1.\exists r_2.\ldots\exists r_n.A$ by 'entering' $\exists r_1$, then enters $\exists r_2$ for the first conjunct that implies $\exists r_2.\ldots\exists r_n.A$, continues in this manner until entering $\exists r_n$, and finally replaces the first occurrence of A in the resulting concept by D.

Proof. Let $P = \exists r_1.\exists r_2.\ldots\exists r_n.A$, and assume that C has been transformed into an ordered conjunction in normal form. We proceed by induction. For $n = 0$, we have $C.replace(A, D)$. The first case of Definition 8 will then remove the first conjunct A in C, and then add D. Next assume that the claim holds when the chain of existentials has length at most $n - 1$. The second case of Definition 8 will first remove from C the first conjunct that implies P. Suppose this conjunct is $\exists r_1.F$. Then we will add back the concept $\exists r_1.(F.replace(P, D))$. As F has a chain of $n - 1$ existentials, we can apply the induction hypothesis to infer that $F.replace(P, D)$ is obtained by following the chain $\exists r_2 \ldots \exists r_n$ to reach a concept containing A (always choosing the first $\exists r_i$ conjunct that entails $\exists r_i \ldots \exists r_n.A$) and replacing there the first occurrence of A by D. By adding the prefix $\exists r_1$ to the resulting concept, we obtain a concept with the stated property. $\square$

We define $C.replace^i(P, D)$ as the i-fold application of the operator.

4.5   Tractability and Generality

All of the above operations are defined on concepts without the use of a background theory (TBox). Subtraction, e.g., removes a conjunct that is subsumed

F.M. Suchanek et al.

by a given concept, but this subsumption is purely syntactic and independent of a terminology ($\sqsubseteq$ instead of $\sqsubseteq_{\mathcal{T}}$). If we want a terminology $\mathcal{T}$ to come into play, we have to explicitly unfold a concept by the operator $\mathcal{T}^*(\cdot)$. The reason for this design choice is that for the goal of modifying concepts, it can be counter-intuitive if concepts are automatically expanded. Consider again our example from Fig. 1, and assume that we are looking for the first conjunct in the definition of Car: $\mathcal{T}(Car) \uparrow_1 \top = Vehicle$. If concepts were automatically expanded, then adding a definition of $Vehicle$ to $\mathcal{T}$ would change this result. Moreover, a full recursive concept expansion would be of exponential complexity [2, Sect. 2.2.4.2]. Our framework avoids this complexity.

All of our operations are defined for normal forms. This means that, before applying an operator, its arguments have to be reduced to their normal forms. This guarantees that, for any operator $\otimes$, $(C \otimes D) \equiv (C' \otimes D)$ if $C \equiv C'$. Consider for example the (redundant) concept $C = (\top \sqcap A)$. The term $C - \top$ yields A. Now consider the equivalent concept $C' = A$. Now, $C' - \top$ yields $\top$. To avoid such artifacts, all concepts have to be brought to their normal form before applying the operators. Since two equivalent concepts have the same normal form, we can guarantee that $(C \otimes D) \equiv (C' \otimes D)$, if $C \equiv C'$. Bringing a concept to its normal form can be done by a simple polynomial algorithm [22].

All of our operators are polynomial-time operations. For addition, this is easy to see: it suffices to join the arguments in a conjunction, and to apply the normalization algorithm. For subtraction, we have to run through all conjuncts in the left argument, and perform a (polynomial) subsumption check for each of them. This yields a polynomial algorithm. The same is true for succession. We conclude our discussion of operators by noting that our language can transform any concept into any target concept. This can always be achieved trivially by subtracting all conjuncts of the original concept and adding all conjuncts of the target concept.

5 Evaluation

We conducted three types of evaluations: a case study on a business case of a French start-up; a descriptive study, where we use our language to describe inventions; and a generative study, where we use it to generate new concepts.

5.1 Case Study

As a case study, we consider a business case by Stim[2], a French consulting firm founded in 2013. Stim studies the product or service of the client company, and then uses the C-K method [12] to rethink it and suggest alternatives. We look here at the case of the client Turbomeca[3], a French manufacturer of gas turbine engines for helicopters. The original technology that Turbomeca employed was

---

[2] http://www.wearestim.com/.
[3] http://www.turbomeca.com/.

a motor working with two symmetric turbines, which both run during the whole flight. The suggested alternative was a "hybrid motor", which works with two different turbines that can run and stop during the flight according to power needs. After a feasibility study, Turbomeca is now developing this new motor, which reduces the energy consumption by half.

We focus here on how the transition from the old motor to the new one can be written down explicitly. Assuming suitable sets of named concepts and relations, the original motor looks as follows:

$$Original \equiv Motor \sqcap \exists worksWith.(Turbine \sqcap \exists placed.Left \sqcap \exists runsDuring.Flight)$$
$$\sqcap \exists worksWith.(Turbine \sqcap \exists placed.Right \sqcap \exists runsDuring.Flight)$$

The motor is then modified as follows:

$$Original.replace^2(\exists worksWith.\exists runsDuring.Flight, (Duration \sqcap \exists partOf.Flight))$$

The resulting motor is:

$$Motor \sqcap \exists worksWith.(Turbine \sqcap \exists placed.Right \sqcap \exists runsDuring.(Duration \sqcap \exists partOf.Flight))$$
$$\sqcap \exists worksWith.(Turbine \sqcap \exists placed.Left \sqcap \exists runsDuring.(Duration \sqcap \exists partOf.Flight))$$

This shows that it is possible to model the transition from the old motor to the new one in our language. However, it also shows that we cannot replace both turbines at the same time. We have to apply the *replace* operation twice (by help of the superscript "2", see Sect. 4.4). This can be traced back to a design choice in our language, which allows subtracting the first matching conjunct, but not *all* conjuncts. This choice was made because subtracting all conjuncts can be achieved by successively removing individual conjuncts. What is more, subtraction becomes idempotent after all matching conjuncts have been removed, so that one can simulate the removal of all conjuncts by iterating the removal of individual conjuncts a large number of times (the same is true for *replace()*). We leave such extensions for future work.

5.2   Descriptive Study

We wanted to analyze to what degree real-world inventions can be described in our language. For this purpose, we considered the top 25 inventions of 2015 according to Time Magazine[4]. We asked computer-science students in our department to model, for each invention, a pair of two concepts: the original concept (say, a toy that can talk) and the innovative concept (say, a toy that can engage in a dialog with the child). Each concept had to be drawn as a graph of concept nodes that are connected by role edges. The students were allowed to use any concept and role names. Figure 2 shows the talking toy as an example (in vectorized form).

We translated the graphs to $\mathcal{EL}$, making modifications where necessary. For example, we translated numeric items such as the age range *5–9* into atomic

---

[4] http://time.com/4115398/best-inventions-2015/.

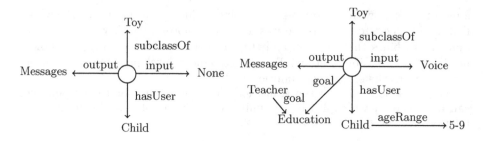

Fig. 2. A talking toy (left) and a toy that can engage in a dialog (right).

concepts (*FiveNine*). We translated incoming edges (as in the *Education* node in Fig. 2 on the right) by introducing new roles (e.g. *isGoalOf*).

Our goal was then to make the transition of the original concept to the new concept explicit, by modeling it in our language. Figure 3 exemplifies this process for the talking toy. In general, we found that the transitions could be modeled with a few operators in our language. The most useful operation proved to be *replace()*. It was used 29 times. The next most frequent operator was addition, with 16 cases. Subtraction was used in 7 cases. We discuss more details of this experiment in our technical report [22].

$((($ *Toy* ⊓ ∃*input.None* ⊓ ∃*output.Messages* ⊓ ∃*hasUser.Child*)
.replace(∃*input.None, Voice)).replace(*∃*hasUser.Child, (Child*⊓∃*ageRange.FiveNine))*
$+$ ∃*goal.(Education* ⊓ ∃*isGoalOf.Teacher))*

$\vdots$

$(((($ *Toy* ⊓ ∃*output.Messages* ⊓ ∃*hasUser.Child* ⊓ ∃*input.Voice)* $-$ ∃*hasUser.Child)*
$+$ ∃*hasUser.((((Toy* ⊓ ∃*output.Messages* ⊓ ∃*hasUser.Child* ⊓ ∃*input.Voice)*
$\rightarrow$ ∃*hasUser.Child)).replace(Child,   (Child ⊓ ∃ageRange.FiveNine))))* $+$
∃*goal.(Education* ⊓ ∃*isGoalOf.Teacher))*

$\vdots$

(Toy ⊓ ∃*output.Messages* ⊓ ∃*input.Voice* ⊓ ∃*hasUser.(Child* ⊓ ∃*ageRange.FiveNine)*
⊓ ∃*goal.(Education* ⊓ ∃*isGoalOf.Teacher))*

Fig. 3. Transition from a talking toy to a toy that engages in dialog. Top line: original concept. Following lines: transition formula. In the middle: an example step from the actual transition. Bottom lines: new concept.

5.3   Generative Study

We want to show that our language can also be used (in a limited manner) to *generate* new concepts. More precisely, our goal is to show that our operators can be used to model something similar to brainstorming.

We use ConceptNet [15], a large knowledge base of commonsense facts. ConceptNet knows, e.g., that cars have wheels, and that they are used for locomotion. Since we are interested in generating new objects, we remove relations that describe words (EtymologicallyDerivedFrom, etc.), as well as relations that describe agents and events. To clean out noise, we also remove all definitions that have 2 or less conjuncts. This leaves us with a terminology $\mathcal{T}$ of 28 relations and 5485 concept definitions, each of which contains 41 conjuncts on average.

To generate new concepts, we use the following formulas. Here, $child_i^{\mathcal{T}}(\cdot)$ retrieves the ith declared child of a concept in $\mathcal{T}$.

1. Addition of a conjunct of a sibling concept:

$$(\mathcal{T}(child_i^{\mathcal{T}}(\mathcal{T}(x) \uparrow_j \top)) \uparrow_k \exists u.\top) - \mathcal{T}(x)$$

   For integers i, j, k and a concept x, this formula selects the jth conjunct of the input concept x. In the example of Fig. 1 with $j = 1$, this yields *Vehicle*. The formula then asks for the ith child of that conjunct in the terminology. This could be, e.g., *Plane* (a sibling of *Car* under *Vehicle*). From this sibling concept, we choose the kth existential conjunct. This could be, e.g., $\exists has.Wing$. We make sure that the chosen conjunct does not already appear in the original concept x. If the result of this operation exists[5], and if the result is not $\top$, we propose to add this conjunct to the original concept.
2. Removal of a conjunct: $\mathcal{T}(x) \uparrow_i \exists u.\top$
   We select the ith existential conjunct of the input concept x. In Fig. 1, we could e.g., choose the 3rd conjunct, $\exists hasPart.Wheel$. Then we propose to remove it.
3. Reversal of a conjunct:

$$\exists HasProperty.(\mathcal{T}(\mathcal{T}(x) \rightarrow_i \exists HasProperty.\top) \rightarrow \exists Antonym.\top)$$

   This formula takes the ith HasProperty conjunct of the input concept x (e.g., $\exists HasProperty.Fast$), and picks its target concept (*Fast*). It expands this target concept from the terminology, and finds its antonym (*Slow*). If the result of this operation exists and is not $\top$, we propose to replace the original conjunct by the conjunct with the antonym.

For each of these cases, we automatically generate a human-readable question such as "Can you imagine a car with wings?" (for Addition), "Can you imagine a car without wheels?" (for Removal), and "Can you imagine a car that is slow?" (for Reversal).

Experiment. We randomly chose 100 concepts from our terminology. Then we applied the above formulas to each of them, increasing each variable i, j, k from 1 to 100. For each formula, we took the first two concepts generated this way (if they exist). Then we asked 9 computer science students to judge the generated concepts. We wanted to know whether the proposed modification is nonsense, already exists in the real world, can be imagined, or is considered creative. Some of concepts in our filtered ConceptNet do not describe objects, but people, actions, or events. We could not filter these out automatically, and hence added an option "This sentence does not describe a physical object". A final option is "I don't understand the sentence".

---

[5] The operation $child_i^{\mathcal{T}}$ may fail if the concept is undefined in the terminology.

Table 1. Human judgement of generated concepts per technique

	Addition		Reversal		Removal	
Already exists	15	(19 %)	7	(41 %)	50	(31 %)
Can be imagined	29	(37 %)	5	(29 %)	66	(40 %)
Funny or creative	10	(12 %)	1	(5 %)	18	(11 %)
Nonsense	23	(29 %)	4	(23 %)	27	(16 %)
Total	77	(100 %)	17	(100 %)	161	(100 %)

In total, we generated 313 new concepts. Of these, 39 did not describe a physical item. 19 used concepts from ConceptNet that the judge did not understand (such as *ny*). For the remaining concepts, Table 1 shows the distribution of human judgements. In up to 29 % of the cases, our techniques return nonsensical concepts, e.g., *A solar wind that is used for learning*. The addition of a sibling conjunct is the most risky technique here. In a large number of cases, our techniques return an existing concept. This is not surprising: it indicates that ConceptNet is incomplete. Interestingly, it also indicates that our techniques actually generated a reasonable concept. 29 %–40 % of the concepts we generate can be imagined, e.g., *A patio that is used for an orchestra to sit*. Finally, around 10 % of our concepts are considered funny or creative. Examples are *Broken glass that does not cut feet*, *A front door without a doorbell*, or *Jelly beans that contain chocolate*.

Many improvements to our generative formulas can be envisaged, but a full investigation is outside the scope of the present paper. Here, we only show that one of the applications of our language is to express formulas that can generate concepts. We leave the study of better concept generation and more extensive experimental evaluation (using e.g., criteria proposed in [19]) for future work.

6   Conclusion

In this paper, we have introduced a formal language for combinatorial creativity. We have justified the choice of our operators and discussed their formal properties. In our experiments, we have shown that our language can be used to describe real-world inventions. In another experiment, we have demonstrated that our language can also be used, to a limited degree, to *generate* new concepts. For future work, we plan to investigate how our language can be used to generate reasonable concepts more systematically, thus working towards the goal of making machines truly creative one day.

Acknowledgments. This research was partially supported by Labex DigiCosme (project ANR-11-LABEX-0045-DIGICOSME) operated by ANR as part of the program "Investissement d'Avenir" Idex Paris-Saclay (ANR-11-IDEX-0003-02).

References

1. Goguen, J.A., Harrell, D.F.: Style: a computational and conceptual blending-based approach. In: Argamon, S., Burns, K., Dubnov, S. (eds.) The Structure of Style. Springer, Heidelberg (2010)
2. Baader, F., Calvanese, D., McGuinness, D.L., Nardi, D., Patel-Schneider, P.F. (eds.): The Description Logic Handbook. Cambridge University Press, Cambridge (2003)
3. Baader, F., Küsters, R., Molitor, R.: Computing least common subsumers in description logics with existential restrictions. In: IJCAI (1999)
4. Besold, T.R., Plaza, E.: Generalize, blend: concept blending based on generalization, analogy, and amalgams. In: ICCC (2015)
5. Boden, M.A., Mind, T.C.: Myths and Mechanisms. Routledge, Abingdon-on-Thames (2004)
6. Confalonieri, R., Schorlemmer, M., Plaza, E., Eppe, M., Kutz, O., Peñaloza, R.: Upward refinement for conceptual blending in description logic: an ASP-based approach and case study in EL++. In: Joint Ontology Workshops (2015)
7. Falkenhainer, B., Forbus, K.D., Gentner, D.: The structure-mapping engine: algorithm and examples. Artif. Intell. **41**, 1–63 (1989)
8. Fauconnier, G., Turner, M.: Conceptual integration networks. Cogn. Sci. **22**(2), 133–187 (1998)
9. Ferré, S., Rudolph, S.: Advocatus diaboli - exploratory enrichment of ontologies with negative constraints. In: EKAW (2012)
10. Giacomo, G.D., Lenzerini, M., Poggi, A., Rosati, R.: On the update of description logic ontologies at the instance level. In: AAAI (2006)
11. Goguen, J.: What is a concept? In: Dau, F., Mugnier, M.-L., Stumme, G. (eds.) ICCS 2005. LNCS (LNAI), vol. 3596, pp. 52–77. Springer, Heidelberg (2005)
12. Hatchuel, A., Weil, B.: C-K design theory: an advanced formulation. Res. Eng. Des. **19**(4), 181 (2009)
13. Konev, B., Ludwig, M., Walther, D., Wolter, F.: The logical difference for the lightweight description logic EL. CoRR, abs/1401.5850 (2014)
14. Kutz, O., Bateman, J., Neuhaus, F., Mossakowski, T., Bhatt, M.: Computational Creativity Research: Towards Creative Machines. Springer, Heidelberg (2015). E Pluribus Unum
15. Liu, H., Singh, P.: Conceptnet. BT Technol. J. **22**(4), 211–226 (2004)
16. Llano, M., Hepworth, R., Colton, S., Charnley, J., Gow, J.: Automating fictional ideation using conceptnet. In: AISB14 Symposium on Computational Creativity (2014)
17. Noia, T.D., Sciascio, E.D., Donini, F.M.: Semantic matchmaking as non-monotonic reasoning: a description logic approach. Artif. Int. Res. **29**, 269–307 (2007)
18. Qi, G., Yang, F.: A survey of revision approaches in description logics. In: Web Reasoning and Rule Systems (2008)
19. Ritchie, G.: Some empirical criteria for attributing creativity to a computer program. Minds Mach. **17**(1), 67–99 (2007)
20. Schmidt, M., Krumnack, U., Gust, H., Kühnberger, K.-U.: Heuristic-driven theory projection: an overview. In: Prade, H., Richard, G. (eds.) Computational Approaches to Analogical Reasoning: Current Trends. Springer, Heidelberg (2014)
21. Schorlemmer, M., Smaill, A., Kühnberger, K.-U., Kutz, O., Colton, S., Cambouropoulos, E., Pease, A.: Coinvent: towards a computational concept invention theory. In ICCC (2014)

22. Suchanek, F.M., Menard, C., Bienvenu, M., Chapellier, C.: A language for combinatorial creativity. Technical report, Telecom ParisTech (2016). https://suchanek.name
23. Teege, G.: Making the difference: a subtraction operation for DL. In: KR (1994)
24. Thagard, P., Stewart, T.C.: The AHA! experience: creativity through emergent binding in neural networks. Cogn. Sci. **35**(1), 1–33 (2011)
25. Toivonen, H., Colton, S., Cook, M., Ventura, D. (eds). International Conference on Computational Creativity (2015)
26. Veale, T., O'Donoghue, D., Keane, M.T.: Computation and blending. Cogn. Ling. **11**(3/4), 253–281 (2000)

Leveraging Linked Data to Discover Semantic Relations Within Data Sources

Mohsen Taheriyan$^{(\boxtimes)}$, Craig A. Knoblock, Pedro Szekely,
and José Luis Ambite

University of Southern California Information Sciences Institute,
Marina del Rey, USA
{mohsen,knoblock,pszekely,ambite}@isi.edu

Abstract. Mapping data to a shared domain ontology is a key step in publishing semantic content on the Web. Most of the work on automatically mapping structured and semi-structured sources to ontologies focuses on semantic labeling, i.e., annotating data fields with ontology classes and/or properties. However, a precise mapping that fully recovers the intended meaning of the data needs to describe the semantic relations between the data fields too. We present a novel approach to automatically discover the semantic relations within a given data source. We mine the small graph patterns occurring in Linked Open Data and combine them to build a graph that will be used to infer semantic relations. We evaluated our approach on datasets from different domains. Mining patterns of maximum length five, our method achieves an average precision of 75 % and recall of 77 % for a dataset with very complex mappings to the domain ontology, increasing up to 86 % and 82 %, respectively, for simpler ontologies and mappings.

Keywords: Semantic model · Semantic relation · Semantic label · Linked data · Semantic web

1 Introduction

A critical task in generating rich semantic content from information sources such as relational databases and spreadsheets is to map the data to a domain ontology. Manually mapping data sources to ontologies is a tedious task. Several approaches have been proposed to automate this process [3,4,6–12,17,18], nonetheless, most approaches focus on *semantic labeling*, annotating data fields, or *source attributes*, with classes and/or properties of a domain ontology. However, a precise mapping needs to describe the semantic relations between the source attributes in addition to their semantic types.

In our earlier Karma work [5], we build a graph from learned semantic types and a domain ontology and use this graph to map a source to the ontology. Since only using the ontology does not necessarily generate accurate models, we had the user in the loop to interactively refine the suggested mappings. Later, we

© Springer International Publishing AG 2016
P. Groth et al. (Eds.): ISWC 2016, Part I, LNCS 9981, pp. 549–565, 2016.
DOI: 10.1007/978-3-319-46523-4_33

title	creation	name
The Island	2009	Walton Ford
Excavation at Night	1908	George Wesley Bellows
Rose Garden	1901	Maria Oakey Dewing

Fig. 1. Sample data from the Crystal Bridges Museum of American Art

introduced an automatic approach that exploits the mappings already created for similar sources in addition to the domain ontology to learn a model for a new unknown source (*target source*) [15]. One limitation of this approach is that the quality of the generated mappings is highly dependent on the availability of the previous mappings. However, in many domains, there are none or very limited number of known mappings available, but we still want to learn the mapping of a new source without requiring the user to map some similar sources first.

We present a novel approach that exploits Linked Open Data (LOD) to automatically infer the semantic relations within a given data source. LOD contains a vast amount of semantic data in many domains that can be used to learn how instances of different classes are linked to each other. The new approach leverages the instance-level data in LOD rather than schema-level mappings as in previous work. Our work in this paper focuses on learning the relationships between the source attributes once they are annotated with semantic labels. The main contribution of our work is leveraging the graph patterns occurring in the linked data to disambiguate the relationships between the source attributes. First, we mine graph patterns with different lengths occurring in the linked data. We combine these patterns into one graph and expand the resulting graph using the paths inferred from the domain ontology. Then, we explore the graph starting from the semantic labels of the source to find the candidate mappings covering all the labels.[1]

We evaluated our approach on different datasets and in different domains. Using patterns of maximum length five, our method achieved an average precision of 75 % and recall of 77 % in inferring the semantic relations within a dataset with complex mappings to the domain ontology, including 13.5 semantic types and 12.5 object properties on average per mapping. The precision and recall are over 80 % for datasets with simpler mappings. Our evaluation shows that longer patterns yield more accurate semantic models.

2 Motivating Example

In this section, we provide an example to explain the problem of inferring semantic relations within structured sources. We want to map a data source containing data about artwork in the Crystal Bridges Museum of American Art (Fig. 1) to the CIDOC-CRM ontology (www.cidoc-crm.org). We formally write the signature of this source as *s(title, creation, name)* where *s* is the name of the source and *title*, *creation*, and *name* are the names of the source attributes (columns).

---

[1] This paper is a significantly extended version of a workshop paper [16].

The first step in mapping the source s to the ontology is to label its attributes with semantic types. We formally define a semantic type to be a pair consisting of a domain class and one of its data properties $\langle class\_uri, property\_uri \rangle$ [6,9]. In this example, the correct semantic types for the columns *title*, *creation*, and *name* are $\langle E35\_Title, label \rangle$, $\langle E52\_Time\text{-}Span, P82\_at\_some\_time\_within \rangle$, and $\langle E82\_Actor\_Appellation, label \rangle$. A mapping that only includes the types of the attributes is not complete because it does not necessarily reveal how the attributes relate to each other. To build a precise mapping, we need a second step that determines how the semantic labels should be connected to capture the intended meaning of the data. Various techniques can be employed to automate the labeling task [4, 6–10, 17, 18], however, in this work, we assume that the labeling step is already done and we focus on inferring the semantic relations.

We use a conceptual graph called *semantic model* to represent the complete set of mappings between the sources and ontologies. In a semantic model, the nodes correspond to ontology classes and the links correspond to ontology properties. Figure 2 shows the correct semantic model of the source s. As we can see in the figure, none of the semantic types corresponding to the source columns are directly connected to each other, which makes the problem of finding the correct semantic model more complex. There are many paths in the CIDOC-CRM ontology connecting the assigned labels. For instance, we can use the classes *E39_Actor* and *E67_Birth* to relate the semantic types *E82_Actor_Appellation* and *E52_Time-Span*:

$(E39\_Actor,\ P1\_is\_identified\_by,\ E82\_Actor\_Appellation)$
$(E39\_Actor,\ P98i\_was\_born,\ E67\_Birth)$
$(E67\_Birth,\ P4\_has\_time\text{-}span,\ E52\_Time\text{-}Span)$

However, this way of modeling does not correctly represent the semantics of this particular data source (Fig. 2 shows the correct model). In general, the ontology defines a large space of possible semantic models and without additional knowledge, we do not know which one is a correct interpretation of the data.

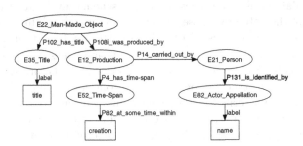

Fig. 2. The semantic model of the source s. Class nodes (ovals) correspond to ontology classes and data nodes (rectangles) correspond to the source attributes

The LOD cloud includes a vast and growing collection of semantic content published by various data providers in many domains. Suppose that some other museums have already mapped their data to the CIDOC-CRM ontology and published it as linked data. Can we exploit the available linked data as background knowledge to infer relationships between attributes of s? The basic idea of our approach is to leverage this linked data to bias the search space to prefer those relationships that are used for related sources. Once we have identified the semantic types of the source attributes, we can search the linked data to find the frequent patterns connecting the corresponding classes. For example, by querying the available linked data corpus, we find that $P131\_is\_identified\_by$ is more popular than $P1\_is\_identified\_by$ to connect instances of $E82\_Actor\_Appellation$ and instances of $E21\_Person$, and this makes sense when we investigate the definitions of these two properties in the ontology. The property $P1\_is\_identified\_by$ describes the naming or identification of any real world item by a name or any other identifier, and $P131\_is\_identified\_by$ is a specialization of $P1\_is\_identified\_by$ that identifies a name used specifically to identify an instance of $E39\_Actor$ (superclass of $E21\_Person$). We can query the linked data to find longer paths between entities. For instance, by inspecting the paths with length two between the instances of $E22\_Man\text{-}Made\_Object$ and $E21\_Person$, we observe that the best way to connect these two classes is through the path: $E22\_Man\text{-}Made\_Object$ $\xrightarrow{P108i\_was\_produced\_by} E12\_Production \xrightarrow{P14\_is\_carried\_out\_by} E21\_Person$.

3   Inferring Semantic Relations

In this section, we explain our approach to automatically deduce the attribute relationships within a data source. The inputs to our approach are the domain ontology, a repository of (RDF) linked data in the same domain, and a data source whose attributes are already labeled with the correct semantic types.[2] The output is a semantic model expressing how the assigned labels are connected.

3.1   Extracting Patterns from Linked Data

Given the available linked data, we mine the schema-level patterns connecting the instances of the ontology classes. Each pattern is a graph in which the nodes correspond to ontology classes and the links correspond to ontology properties. For example, the pattern $c_1 \xrightarrow{p} c_2$ indicates that at least one instance of the class c_1 is connected to an instance of the class c_2 with the property p.

Depending on the structure of the domain ontology, there might be a large number of possible patterns for any given number of ontology classes. Suppose that we want to find the patterns that only include the three classes c_1, c_2, and c_3. Figure 3 exemplifies some of the possible patterns to connect these classes. We define the length of a pattern as the number of links (ontology properties) in a pattern. Thus, the general forms of the patterns with length two will be $c_1 \rightrightarrows c_2$, $c_1 \rightleftarrows c_2$, $c_1 \rightarrow c_2 \rightarrow c_3$ (chain), $c_1 \rightarrow c_2 \leftarrow c_3$ (V-shape), or $c_2 \leftarrow c_1 \rightarrow c_3$ (A-shape).

---

[2] For this paper, we assume that the correct semantic types are given, but our approach can support the more general case where a set of candidate semantic types are assigned to each attribute.

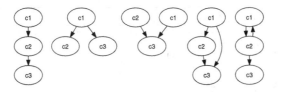

Fig. 3. Sample structures of the patterns connecting the classes c_1, c_2, and c_3

We can write SPARQL queries to extract patterns with different lengths from a triplestore. For example, the following SPARQL query extracts the patterns with length one and their frequencies from the linked data:

```
SELECT DISTINCT ?c1 ?p ?c2 (COUNT(*) as ?count)
WHERE { ?x ?p ?y. ?x rdf:type ?c1. ?y rdf:type ?c2. }
GROUP BY ?c1 ?p ?c2
```

Pattern mining is a preprocessing step in our approach and can be done offline. Yet, using SPARQL queries to extract long patterns from a large number of triples is not efficient. For example, having a Virtuoso repository containing more than three million triples on a Linux machine with a 2.4 GHz Intel Core CPU and 16 GB of RAM, the response time to the query to extract V-shape patterns of length two was roughly 1 h. We were only able to collect a few forms of the patterns with length three, four, and five in a 5-hour timeout. The problem is that the SPARQL queries to extract long patterns include many joins and there is no binding between the variables in the queries and the classes and properties in the ontology. Therefore, we adopt a different approach to extract patterns of length two or more.

Algorithm 1 shows our method to find patterns from a triplestore. This is a recursive algorithm that incrementally generates longer patterns by joining shorter patterns with patterns of length one. The intuition is that we enumerate the candidate patterns and create SPARQL queries whose variables are bound to ontology classes and properties rather than writing a join-heavy query with unbound variables. This technique allows us to exploit the indexes created by triplestore over subjects, predicates, objects, and different combinations of them.

First, we use the SPARQL query shown above to retrieve all the patterns of length one from the linked data. Then, we construct candidate patterns of length two by joining patterns of length one with themselves (join on the shared ontology class). For example, the pattern $c1 \xrightarrow{p_1} c_2$ can be joined with $c_1 \xrightarrow{p_2} c_3$ on c_1 to construct a pattern of length two: $c_2 \xleftarrow{p_1} c_1 \xrightarrow{p_2} c_3$. For each candidate pattern, we query the linked data to see whether an instance of such pattern exists in the data. Once we find the patterns of length two that occur in the linked data, we join them with patterns of length one and check the occurrence of the newly formed patterns (length three) in the triplestore. We can repeat the same process to retrieve longer patterns. To prevent generating duplicate patterns, we perform some optimizations in the code that are not shown in Algorithm 1. Using this algorithm, we could extract all patterns of length 1, 2,

---

Algorithm 1. Extract LD Patterns

Input: LD (the linked data repository), k (maximum length of patterns)
Output: A set of LD patterns
1: $P_1 \leftarrow$ extract patterns of length one from LD
2: $P \leftarrow P_1$, $i \leftarrow 1$
3: **while** $i < k$ **do**
4:     **for** each pattern $p_i \in P_i$ **do**
5:         **for** each ontology class c in p_i **do**
6:             $P_{1,c} \leftarrow$ all the patterns in P_1 that include c
7:             **for** each pattern $p_{1,c} \in P_{1,c}$ **do**
8:                 $p_{join} \leftarrow$ construct a pattern by joining p_i and $p_{1,c}$ on node c
9:                 **if** p_{join} exists in LD **then**
10:                     $P_{i+1} \leftarrow P_{i+1} \cup p_{join}$
11:                 **end if**
12:             **end for**
13:         **end for**
14:     **end for**
15:     $P \leftarrow P \cup P_{i+1}$, $i \leftarrow i + 1$
16: **end while**
    return P

Algorithm 2. Construct Graph G

Input: LD Patterns, Semantic Types, Domain Ontology
Output: Graph G
    ▷ Add LD patterns
1: sort the patterns descending based on their length
2: exclude the patterns contained in longer patterns
3: merge the nodes and links of the remaining patterns into G
    ▷ Add Semantic Types
4: **for** each semantic type $\langle class\_uri, property\_uri \rangle$ **do**
5:     add the class to the graph if it does not exist in G
6: **end for**
    ▷ Add Ontology Paths
7: **for** each pair of classes c_i and c_j in G **do**
8:     find the directed and inherited properties between c_i and c_j in the ontology
9:     add the properties that do not exist in G
10: **end for**
    return G

3, and 4 from the same triplestore in only 10 min, and all patterns of length 5 in less than one hour.

3.2   Merging Linked Data Patterns into a Graph

Once we extract the patterns from the linked data (LD patterns), we combine them into a graph G that will be used to infer the semantic relations. Building the graph has three parts: (1) adding the LD patterns, (2) adding the semantic labels assigned to the source attributes, and (3) expanding the graph with the paths inferred from the ontology.

The graph G is a weighted directed graph in which nodes correspond to ontology classes and links correspond to ontology properties. The algorithm to construct the graph is straightforward (Algorithm 2). First, we add the LD patterns to G. We start from the longer patterns and merge the nodes and links of patterns into G if they are not subgraphs of the patterns already added to the graph (lines 1–3). Next, we add the semantic types of the target source for which we want to learn the semantic model (lines 4–6). As we mentioned before, we assume that the source attributes have already been labeled with semantic types. Each semantic type is a pair consisting of a domain class and one of its data properties $\langle class\_uri, property\_uri \rangle$. If G does not contains any node with the label $class\_uri$, we add a new node to G and label it with $class\_uri$. The final step in building the graph is to find the paths in the ontology that relate the current classes in G. The goal is to connect class nodes of G using the direct paths or the paths inferred through the subclass hierarchy in the ontology.

The links in G are weighted. We adopt a subtle approach to assign weights to the links. There are two types of links; the links that are added from the LD patterns, and the links that do not exist in any pattern and are inferred from the ontology. The weight of the links in the former group has an inverse relation with the frequency of the links. If the number of instances for the pattern $c1 \xrightarrow{p} c_2$ in the linked data is x and the total number of instances of patterns with length one is n, the weight of the link p from c_1 to c_2 in G is computed as $1 - x/n$. Since we are generating minimum-cost models in the next section, this weighting strategy gives more priority to the links occurring more frequently in the linked data. For the links in the second category, the ones added from the ontology, we assign a much higher weight comparing to the links in the LD patterns. The intuition is that the links used by other people to model the data in a domain are more likely to represent the semantics of a given source in the same domain. One reasonable value for the weight of the links added from the ontology is the total number of object properties in the ontology. This value ensures that even a long pattern costs less than a single link that does not exist in any pattern.

The other important feature of the links in the graph is their *tags*. We assign an identifier to each pattern added to the graph and annotate the links with the identifiers of the supporting patterns. Suppose that we are adding two patterns $m_1 : c_1 \xrightarrow{p_1} c_2 \xrightarrow{p_2} c_3$ and $m_2 : c_1 \xrightarrow{p_1} c_2 \xrightarrow{p_3} c_4$ to G. The link p_1 from c_1 to c_2 will be tagged with $\{m_1, m_2\}$, the link p_2 from c_2 to c_3 will have only $\{m_1\}$ as its tag set, and the link p_3 from c_2 to c_4 will be tagged with $\{m_2\}$. We use the link tags later to prioritize the models containing larger segments from the LD patterns.

We provide an example to help the reader to understand our algorithm for creating the graph. Suppose that we are trying to infer a semantic model for the source s in Fig. 1 and we only extract the patterns with length one and two from the available linked data. Table 1 lists the extracted LD patterns. Figure 4 shows the graph constructed using Algorithm 2. The black links are the links added from the LD patterns. The weight of these links is calculated as $1 - (link\ frequency)/(sum\ of\ frequencies)$. For example, the weight of the link $P98i\_was\_born$ from $E21\_Person$ to $E67\_Birth$ is $1 - 5/63 = 0.92$ (total number of instances of the patterns with length one is 63). The black links are also tagged with the identifier of the patterns they originate from. For instance, the link $P98i\_was\_born$ is tagged with m_1 because the pattern m_1 contains this link. Note that only the patterns m_1, m_2, m_3, and m_4 are added to the graph and the rest of the patterns are ignored by the algorithm. This is because the patterns m_5, m_6, m_7, m_8, m_9, and m_{10} are subgraphs of the longer patterns that are already added to the graph (line 2 in Algorithm 2). The blue node $E53\_Title$ does not exist in any pattern, however, it is added to the graph since the semantic type of the column *title* in the source s, $\langle E53\_Title, label \rangle$, contains $E53\_Title$ (line 5 in

Table 1. The sample patterns extracted from the linked data. Each pattern is a set of (c_1, p, c_2) triples where c_1 and c_2 are ontology classes and p is an ontology property. The second and third columns are the length and frequency of patterns (only the frequency of the patterns with length one matters in our algorithm)

Id	Pattern	Len	Freq
m_1	$(E21\_Person, P98i\_was\_born, E67\_Birth)$, $(E67\_Birth, P4\_has\_time\text{-}span, E52\_Time\text{-}Span)$	2	-
m_2	$(E22\_Man\text{-}Made\_Object, P108i\_was\_produced\_by, E12\_Production)$, $(E12\_Production, P4\_has\_time\text{-}span, E52\_Time\text{-}Span)$	2	-
m_3	$(E22\_Man\text{-}Made\_Object, P14\_carried\_out\_by, E39\_Actor)$, $(E39\_Actor, P131\_is\_identified\_by, E82\_Actor\_Appellation)$	2	-
m_4	$(E21\_Person, P131\_is\_identified\_by, E82\_Actor\_Appellation)$	1	12
m_5	$(E22\_Man\text{-}Made\_Object, P108i\_was\_produced\_by, E12\_Production)$	1	8
m_6	$(E12\_Production, P4\_has\_time\text{-}span, E52\_Time\text{-}Span)$	1	3
m_7	$(E22\_Man\text{-}Made\_Object, P14\_carried\_out\_by, E39\_Actor)$	1	10
m_8	$(E39\_Actor, P131\_is\_identified\_by, E82\_Actor\_Appellation)$	1	20
m_9	$(E21\_Person, P98i\_was\_born, E67\_Birth)$	1	5
m_{10}	$(E67\_Birth, P4\_has\_time\text{-}span, E52\_Time\text{-}Span)$	1	5

Algorithm 2). The red links are the links that do not exist in the LD patterns but are inferred from the ontology. For example, the red link $P14\_carried\_out\_by$ is added because $E21\_Person$ is a subclass of the class $E39\_Actor$, which is in turn the range of the object property $P14\_carried\_out\_by$. We assign a high weight to the red links, e.g., the total number of object properties in the ontology (in this example, assume that the ontology consists of 100 object properties).

3.3   Generating and Ranking Semantic Models

The final part of our approach is to compute the semantic model of the source s from the graph. First, we map the semantic types of s to the nodes of the graph. In our example, the semantic types are $\langle E53\_Title, label \rangle$, $\langle E52\_Time\text{-}Span, P82\_at\_some\_time\_within \rangle$, and $\langle E82\_Actor\_Appellation, label \rangle$, and they will be mapped to the nodes $E53\_Title$, $E52\_Time\text{-}Span$, and $E82\_Actor\_Appellation$ of the graph in Fig. 4. Then, we compute the top k trees connecting the mapped nodes based on two metrics: *cost* and *coherence*. Cost of a tree is the sum of the link weights. Because the weights of the links have an inverse relation with their popularity, computing the minimum-cost tree results in selecting more frequent links. However, selecting more popular links does not always yield the correct semantic model. The coherence of a model is another important factor that we need to consider. Coherence in this context means the ratio of the links in a computed tree that belong to the same LD pattern. The coherence metric gives priority to the models that contain longer patterns. For example, a model that includes one pattern with length 3 will be ranked higher than a model including two patterns with length 2, and the latter in turn will be preferred over a model with only one pattern with length 2. Our algorithm prefers the coherence over the cost in choosing the best model. If two models are equivalent as far as coherence, the model with the lowest cost will be ranked higher.

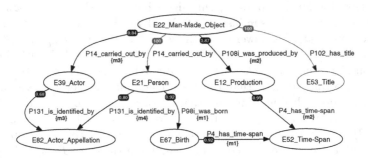

Fig. 4. The graph constructed from the patterns in Table 1, the semantic types of the source s, and an example subset of the CIDOC-CRM ontology

The algorithm that finds the top k trees is a customized version of the BANKS algorithms [1]. The BANKS algorithm computes the top k minimum-cost trees that span a subset of the nodes in a graph (the nodes that the semantic types are mapped to). It creates one iterator for each of the nodes corresponding to the semantic types, and then the iterators follow the incoming links to reach a common ancestor. The BANKS algorithm uses the iterator's distance to its starting point to decide which link should be followed next. To incorporate the coherence into the algorithm, we use a heuristic that prefers the links that are parts of the same pattern even if they have higher weights. In our example, the algorithm creates three iterators, one starting from the node *E53_Title*, one from *E52_Time-Span*, and one from *E82_Actor_Appellation*. Each iterator has a queue consisting of the candidate links to traverse. At each step, the algorithm chooses a link and adds the incoming links of the source node of the selected link to the queue of the corresponding iterator. At the beginning, the candidates are:

e_1 *(itr 1)*: *(E22_Man-Made_Object,P102_has_title,E53_title)*, *distance* $= 100$
e_2 *(itr 2)*: *(E67_Birth,P4_has_time-span,E52_Time-Span)*, *distance* $= 0.92$
e_3 *(itr 2)*: *(E12_Production,P4_has_time-span,E52_Time-Span)*, *distance* $= 0.95$
e_4 *(itr 3)*: *(E21_Person,P131_is_identified_by,E21_Actor_Appellation)*, *distance* $= 0.80$
e_5 *(itr 3)*: *(E39_Actor,P131_is_identified_by,E21_Actor_Appellation)*, *distance* $= 0.68$

The algorithm pulls e_5 from the queue of the third iterator because e_5 is the lowest-cost link in the candidates. Then, it inserts the incoming links of the source node of e_5 (*E39_Actor*) to the queue:

e_6 *(itr 3)*: *(E22_Man-Made_Object,P14_carried_out_by,E39_Actor)*, *distance* $= 1.52$

Now the candidate links to traverse are e_1, e_2, e_3, e_4, and e_6. Although the distance of e_4 to the starting point is less than other links, our algorithm prefers e_6. This is because e_6 is part of the pattern m_3 which includes the previously traversed link e_5. Considering the coherence in traversing the graph forces the algorithm to converge to models that include larger segments from the patterns.

Once we compute top k trees, we rank them first based on their coherence and then their cost. The model in Fig. 5 shows the top semantic model computed by our algorithm for the source s. Our algorithm ranks this model higher than a model that has $E21\_Person$ instead of $E39\_Actor$, because the model in Fig. 5 is more coherent and has lower cost (the weight of the link $P14\_carried\_out\_by$ from $E22\_Man\text{-}Made\_Object$ to $E21\_Person$ is 100, while the link with the same label from $E22\_Man\text{-}Made\_Object$ to $E39\_Actor$ has a weight of 0.84).

It is important to note that the trees containing longer patterns are not necessarily more coherent. Suppose that a source has two semantic labels A and B, and there are only two LD patterns: $p_1 = \{(A, e_1, B)\}$ and $p_2 = \{(A, e_2, B)(B, e_3, C)\}$. After constructing the graph, we compute top k trees based on both coherence and cost. The candidate trees are $T_1 = \{(A, e_1, B)\}$, $T_2 = \{(A, e_2, B)\}$, $T_3 = \{(A, e2, B)(B, e3, C)\}$, and $T_4 = \{(A, e1, B)(B, e3, C)\}$. The trees T_1, T_2, and T_3 have a coherence value equal to 1 (all the links belong to one pattern), while coherence of T_4 is 0.5 (one link is from p_1 and one link is from p_2). Between T_1, T_2, and T_3, our algorithm ranks T_3 third because it is longer (higher cost) and ranks T_1 and T_2 as the top two trees (based on the frequency of e_1 and e_2 in the linked data). Thus, the algorithm does not always prefer the longer patterns.

4   Evaluation

To evaluate our approach, we performed four experiments on different datasets, using different ontologies, and in different domains. The details of the experiments are as follows:

– **Task 1**: The dataset consists of 29 museum data sources in CSV, XML, or JSON format containing data from different art museums in the US. The domain ontology is CIDOC-CRM, and the background linked data is the RDF triples generated from all sources in the dataset except the target source (leave-one-out settings).

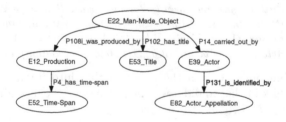

Fig. 5. The top semantic model computed for the source s

- **Task 2**: The dataset and the domain ontology are the same as the ones in Task 1, but we use the linked data published by Smithsonian American Art Museum[3] as the background knowledge. This linked data repository contains more than 3 million triples mapping the artwork collection in Smithsonian museum to the CIDOC-CRM ontology.
- **Task 3**: The dataset is the same as the one in Task 1, but we use a different domain ontology, EDM[4], which is much simpler and less structured than CIDOC-CRM. The goal is to evaluate how our approach performs with respect to different representations of knowledge in a domain. We use a leave-one-out setting is this task.
- **Task 4**: The dataset includes 15 data sources in a domain containing ads for weapons. The ontology is an extension of the schema.org ontology, and the experiment adopts a leave-one-out setting.

Table 2 shows more details of the evaluation tasks. In each task, we applied our approach on the dataset to find the top 50 candidate semantic models for each source. We then ranked the candidate models and compared the first ranked models with the gold standard models created by an expert in the test domain. The datasets including the sources, the domain ontologies, and the gold standard models are available on GitHub.[5] The source code of our approach is integrated into Karma which is available as open source.[6]

We assume that the correct semantic labels for the source attributes are known. The goal is to see how well our approach learns the relationships having the correct semantic types. For example, from the total of 825 links in the ground-truth models in Task 1, 458 of them correspond to data properties connecting the semantic types to the source attributes. Therefore, the 367 internal links are the semantic relations that we want to infer using our approach.

Table 2. The evaluation tasks

	Dataset		Ontology		Ground Truth	
	#sources	#attributes	#classes	#properties	#nodes	#links
Task 1	29	458	147	409	852	825
Task 2	29	458	147	409	852	825
Task 3	29	329	119	351	470	441
Task 4	15	175	736	1081	261	246

We measured the accuracy of the computed semantic models by comparing them with the gold standard models in terms of *precision* and *recall*. Assuming that the correct semantic model of the source s is sm and the semantic model learned by our approach is sm', we define the precision and recall as:

---

[3] http://americanart.si.edu/collections/search/lod/about.
[4] http://pro.europeana.eu/page/edm-documentation.
[5] https://github.com/taheriyan/iswc-2016.
[6] https://github.com/usc-isi-i2/Web-Karma (directory: `karma-research`).

$$precision = \frac{rel(sm) \cap rel(sm')}{rel(sm')}, \quad recall = \frac{rel(sm) \cap rel(sm')}{rel(sm)}$$

where $rel(sm)$ is the set of triples (u, v, e) in which e is a link from the class node u to the class node v in the semantic model sm. Note that we do not consider the links from the classes to the source attributes since they are part of the semantic types that are given as input to our algorithm. Consider the semantic model of the source s in Fig. 2, $rel(sm)=\{$ (E22_Man-Made_Object, P108i_was_produced_by, E12_Production), $\cdots\}$. Assuming that the semantic model in Fig. 5 is what our algorithm infers for s, we will have $precision = 3/5 = 0.6$ and $recall = 3/5 = 0.6$. Note that we are using a strict evaluation metric. The learned semantic model is not semantically wrong because E21_Person is a subclass of E39_Actor according to the CIDOC-CRM ontology. However, our formula treats both (E22_Man-Made_Object, P14_carried_out_by, E39_Actor) and (E39_Actor, P131_is_identified_by, E82_Actor_Appellation) as incorrect links.

Table 3. The evaluation results

	Task 1		Task 2		Task 3		Task 4	
	Precision	Recall	Precision	Recall	Precision	Recall	Precision	Recall
Max length of patterns = 0	0.07	0.05	0.07	0.05	0.01	0.01	0.03	0.02
Max length of patterns = 1	0.60	0.60	0.28	0.29	0.85	0.78	0.84	0.79
Max length of patterns = 2	0.64	0.67	0.53	0.58	0.81	0.81	0.83	0.79
Max length of patterns = 3	0.67	0.68	0.55	0.60	0.84	0.83	0.86	0.81
Max length of patterns = 4	0.74	0.76	0.55	0.60	0.83	0.82	0.86	0.82
Max length of patterns = 5	0.75	0.77	0.61	0.67	0.83	0.82	0.86	0.82

Table 3 shows the average precision and recall for all sources in the evaluation tasks for different maximum lengths of LD patterns (the input parameter k in Algorithm 1). The maximum length 0 means that we did not incorporate the LD patterns and only used the domain ontology to build a graph on top of the semantic labels. An interesting observation is that when we do not use the LD patterns, the precision and recall are close to zero. This low accuracy comes from the fact that in most of the gold standard models, the attributes are not directly connected and there are multiple paths between each pair of classes in the ontology (and thus in our graph), and without additional information we cannot resolve the ambiguity. Leveraging LD patterns as background knowledge yields a remarkable improvement in both precision and recall compared to the case in which we only consider the domain ontology. Since we are using the pattern frequencies in assigning the weights to the links of the graph, using patterns of length one means that we are only taking into account the popularity of the links in computing the semantic models. Leveraging longer patterns improves both precision and recall. This means that considering coherence in addition to the link popularity empowers our approach to derive more accurate models.

As we can see in the table, the precision and recall in Task 2 are lower than Task 1 even though they have the same dataset and ontology. The reason is that Task 2 employs the triples from the Smithsonian museum rather than the triples generated from other sources in the same dataset, and the overlap between these triples and the ground truth models is less than Task 1. The results in Task 3 are better than both Task 1 and Task 2 because the ground-truth models created using EDM are simpler and smaller than the models created using CIDOC-CRM (441 links vs. 825 links). Although CIDOC-CRM is well structured, the level of ambiguity in inferring the relations is more than simpler ontologies with a flat structure because there are many links (and paths) between each pair of classes (many properties are inherited through the class hierarchy). We achieve high precision and recall in Task 4. The ontology in this task is much larger than the ontologies in the other tasks (cf. Table 2), however, the hierarchy level is less than the CIDOC-CRM ontology, and the models are also smaller. We believe that extracting longer patterns (length > 5) will improve the results for Task 1 and Task 2 as many of the semantic models in these tasks include more than 5 internal links. However, as Table 3 suggests, longer patterns do not contribute to the precision and recall for Task 3 and Task 4.

The results show that our method suggests semantic models with high precision and recall for large, real-world datasets even in complex domains. There are several reasons explaining why we did not achieve higher accuracy in Task 1 and Task 2. First and foremost, there were sources in the dataset whose semantic model contained structures that did not exist in any pattern. In other words, some sources did not have much overlap with other sources in the dataset. A simple example of this case is where the semantic model of a source includes an ontology class that does not have any instance in the linked data, and consequently, no LD pattern contains relations of that class. A second important reason is that in some cases, more than one pattern can be used to connect semantic labels. For example, many sources contain two columns labeled with the class *E52_Time-Span*. We have a pattern with length four in which one *E52_Time-Span* is connected to *E12_Production* and the other one is connected to *E67_Birth* (or *E69_Death*). We also have another pattern with length four in which one *E52_Time-Span* is connected to *E67_Birth* and the second one is linked to *E69_Death*. In such situations, our algorithm may select the wrong pattern. Inspecting these cases convinced us that taking into account longer patterns will resolve the issue. For a few sources, the ground-truth semantic model was not a rooted tree, while our algorithm computes only tree candidate models. Therefore, we missed some of the links in our learned semantic models. Finally, our strict evaluation metric penalizes the precision and recall even though some of the learned models are not semantically wrong. For instance, the correct model of one source includes the link *P138_has_representation* from *E22_Man-Made_Object* to *E38_Image*. Our algorithm infers the inverse link *P138i_represents* to connect these two classes because it is more frequently used in other models, and thus, it is more popular in the linked data. Our evaluation considers this link as an incorrect inference. Another example is the one we discussed earlier where our method suggests *E39_Actor* instead of *E21_Person*. We are exploring the possibility of defining a looser evaluation metric that provides more flexible interpretation of "correct" models. Our initial idea is to consider the subclass and subproperty definitions in the ontology. For example, we can give some credit if the algorithm infers a parent property of the one in the ground-truth model (instead of considering it completely wrong).

To compare the work in this paper with our previous approach that exploits the known semantic models [15], we applied the previous approach on the same dataset in Task 1, which resulted in semantic models with 81 % precision and 82 % recall. The reason why the accuracy is lower in the current work is that we only used patterns of maximum length five in our experiment. On the other hand, our previous work exploits complete semantic models of previously modeled sources, which are more coherent than the small graph patterns we used in this work. Our work in this paper complements our previous work in more common and critical case where few, if any, known semantic models are available.

We ran our experiments on a single machine with a Mac OS X operating system and a 2.3 GHz Intel Core i7 CPU. The total time to extract all the patterns of length one, two, three, and four from our Virtuoso repository was less than 10 min, and it was approximately one hour for the patterns of length five. Then, we fed the extracted patterns to Algorithm 2. Let T be the time from combining LD patterns into a graph until generating and ranking candidate semantic models. The average value of T is different when we run the code with different maximum length for patterns. However, the average value of T never exceeds 5 s (for most of the sources, the actual value of T was less than a second).

5   Related Work

There have been many studies to automatically describe the semantics of data sources as a mapping from the source to an ontology. Since the focus of our work is on inferring the semantic relations, we compare our work with the ones that pay attention to inferring semantic relationship and not only semantic labeling.

Limaye et al. [7] use the YAGO ontology to annotate web tables and generate binary relationships using machine learning approaches. Venetis et al. [17] present a scalable approach to describe the semantics of tables on the Web. To recover the semantics of tables, they leverage a database of class labels and relationships automatically extracted from the Web. They attach a class label to a column if a sufficient number of the values in the column are identified with that label in the database of class labels, and analogously for binary relationships. Although these approaches are very useful in publishing semantic data from tables, they are limited in learning the semantics of sources as a united model. Both of these approaches only infer individual binary relationships between pairs of columns. They are not able to find the link between two columns if no relation is directly instantiated between the values of those columns. Our approach can connect one column to another one through a path in the ontology.

Carman and Knoblock [2] use known source descriptions to learn a semantic description that describes the relationship between the inputs and outputs of a source. However, their approach is limited in that it can only learn sources whose models are subsumed by the models of known sources. That is, the description of a new source is a conjunctive *combination* of known source descriptions.

Our work is closely related to other work leveraging the Linked Open Data (LOD) cloud to capture the semantics of sources. Mulwad et al. [8] use *Wikitology* [14], an ontology which combines some existing manually built knowledge systems such as DBpedia and Freebase, to link cells in a table to Wikipedia entities. They query the background LOD to generate initial lists of candidate classes for column headers and cell values and candidate properties for relations between columns. Then, they use a probabilistic graphical model to find the correlation between the columns headers, cell values, and relation assignments. The quality of the semantic data generated by this category of work is highly dependent on how well the data can be linked to the entities in LOD. While for most popular named entities there are good matches in LOD, many tables contain domain-specific information or numeric values (e.g., temperature and age) that cannot be linked to LOD. Moreover, these approaches are only able to identify individual binary relationships between the columns of a table. However, an integrated and united semantic model is more than fragments of binary relationships between the columns. In a complete semantic model, the columns may be connected through a path including the nodes that do not correspond to any column in the table.

The most closely related work to ours is the recent work by Schaible et al. [13]. They extract *schema-level patterns* (SLPs) from LOD and generate a ranked list of vocabulary terms for reuse in modeling tasks. The main difference between their work and our method is the complexity of the patterns. SLPs are (sts, ps, ots) triples where sts and ots are sets of RDF types and ps is a set of RDF properties. For example, the SLP ({*Person,Player*}, {knows},{Person,Coach}) indicates that some instances of *Person* ∩ *Player* are connected to some instances of *Person* ∩ *Coach* via the property *knows*. In our approach, we mine graph patterns that are more complex than SLPs allowing us to automatically compose a complete semantic model for a target source rather than presenting recommendations in an interactive mapping task.

6 Discussion

We presented a novel approach to infer semantic relations within structured sources. Understanding how the source attributes are related is an essential part of building a precise semantic model for a source. Such models are the key ingredients to automatically integrate heterogeneous data sources. They also automate the process of publishing semantic data on the Web. The core idea of our work is to exploit the small graph patterns occurring in the Linked Open Data to hypothesize attribute relationships within a data source.

Manually constructing semantic models, in addition to being time-consuming and error-prone, requires a thorough understanding of the domain ontologies. Tools such as Karma can help users to model data sources through a graphical user interface. Yet, building the models in Karma without any automation requires significant user effort. Incorporating our method in source modeling tools enables them to infer an initial semantic model for the input source that can be transformed to the correct model with only a few user actions.

The evaluation shows that our approach infers the semantic relations with a high precision and recall for a dataset with very complex semantic models (on average 13.5 classes and 12.6 object properties per semantic model). We have shown that we gain higher precision and recall when we apply our method on data sources with simpler models. The results support the theory that more accurate models can be constructed when longer LD patterns are used. We observed that the structure of the patterns also affects the quality of the learned models. For example, using only the chain-shape patterns resulted in more precise models for some of the sources. One direction of our future work is to investigate the correlation between the shape of the LD patterns, the structure of the domain ontology, and the ground-truth semantic models. This can help us to incorporate certain types of patterns when mapping a data source to a domain ontology.

Our work plays a role in helping communities to produce consistent Linked Data so that sources containing the same type of data use the same classes and properties when published in RDF. Often, there are multiple correct ways to model the same type of data. A community is better served when all the data with the same semantics is modeled using the same classes and properties. Our work encourages consistency because our algorithms bias selection of classes and properties towards those used more frequently in existing data.

Acknowledgments. This research was supported in part by the National Science Foundation under Grant No. 1117913 and in part by Defense Advanced Research Projects Agency (DARPA) via AFRL contract numbers FA8750-14-C-0240 and FA8750-16-C-0045. The U.S. Government is authorized to reproduce and distribute reprints for Governmental purposes notwithstanding any copyright annotation thereon. The views and conclusions contained herein are those of the authors and should not be interpreted as necessarily representing the official policies or endorsements, either expressed or implied, of NSF, DARPA, AFRL, or the U.S. Government.

References

1. Bhalotia, G., Hulgeri, A., Nakhe, C., Chakrabarti, S., Sudarshan, S.: Keyword searching and browsing in databases using BANKS. In: Proceedings of the 18th International Conference on Data Engineering, pp. 431–440 (2002)
2. Carman, M.J., Knoblock, C.A.: Learning semantic definitions of online information sources. J. Artif. Intell. Res. **30**(1), 1–50 (2007)
3. Ding, L., DiFranzo, D., Graves, A., Michaelis, J., Li, X., McGuinness, D.L., Hendler, J.A.: TWC Data-gov Corpus: incrementally generating linked government data from data.gov. In: Proceedings of the 19th WWW, pp. 1383–1386 (2010)

4. Han, L., Finin, T.W., Parr, C.S., Sachs, J., Joshi, A.: RDF123: from spreadsheets to RDF. In: Sheth, A.P., Staab, S., Dean, M., Paolucci, M., Maynard, D., Finin, T., Thirunarayan, K. (eds.) ISWC 2008. LNCS, vol. 5318, pp. 451–466. Springer, Heidelberg (2008)
5. Knoblock, C.A., Szekely, P., Ambite, J.L., Goel, A., Gupta, S., Lerman, K., Muslea, M., Taheriyan, M., Mallick, P.: Semi-automatically mapping structured sources into the semantic web. In: Simperl, E., Cimiano, P., Polleres, A., Corcho, O., Presutti, V. (eds.) ESWC 2012. LNCS, vol. 7295, pp. 375–390. Springer, Heidelberg (2012)
6. Ramnandan, S.K., Mittal, A., Knoblock, C.A., Szekely, P.: Assigning semantic labels to data sources. In: Gandon, F., Sabou, M., Sack, H., d'Amato, C., Cudré-Mauroux, P., Zimmermann, A. (eds.) ESWC 2015. LNCS, vol. 9088, pp. 403–417. Springer, Heidelberg (2015)
7. Limaye, G., Sarawagi, S., Chakrabarti, S.: Annotating and searching web tables using entities, types and relationships. PVLDB 3(1), 1338–1347 (2010)
8. Mulwad, V., Finin, T., Joshi, A.: Semantic message passing for generating linked data from tables. In: Alani, H., Kagal, L., Fokoue, A., Groth, P., Biemann, C., Parreira, J.X., Aroyo, L., Noy, N., Welty, C., Janowicz, K. (eds.) ISWC 2013, Part I. LNCS, vol. 8218, pp. 363–378. Springer, Heidelberg (2013)
9. Pham, M., Alse, S., Knoblock, C.A., Szekely, P.: Semantic labeling: a domain-independent approach. In: Proceedings of the 15th International Semantic Web Conference (ISWC) (2016)
10. Polfliet, S., Ichise, R.: Automated mapping generation for converting databases into linked data. In: ISWC Posters & Demos (2010)
11. Sahoo, S.S., Halb, W., Hellmann, S., Idehen Jr., K., T.T., Auer, S., Sequeda, J., Ezzat, A.: A survey of current approaches for mapping of relational databases to RDF. W3C RDB2RDF XG report (2009)
12. Saquicela, V., Vilches-Blazquez, L.M., Corcho, O.: Lightweight semantic annotation of geospatial RESTful services. In: Antoniou, G., Grobelnik, M., Simperl, E., Parsia, B., Plexousakis, D., De Leenheer, P., Pan, J. (eds.) ESWC 2011, Part II. LNCS, vol. 6644, pp. 330–344. Springer, Heidelberg (2011)
13. Schaible, J., Gottron, T., Scherp, A.: TermPicker: enabling the reuse of vocabulary terms by exploiting data from the linked open data cloud. In: Sack, H., Blomqvist, E., d'Aquin, M., Ghidini, C., Ponzetto, S.P., Lange, C. (eds.) ESWC 2016. LNCS, vol. 9678, pp. 101–117. Springer, Heidelberg (2016). doi:10.1007/978-3-319-34129-3_7
14. Syed, Z., Finin, T.: Creating and exploiting a hybrid knowledge base for linked data. In: Filipe, J., Fred, A., Sharp, B. (eds.) ICAART 2010. CCIS, vol. 129, pp. 3–21. Springer, Heidelberg (2011)
15. Taheriyan, M., Knoblock, C., Szekely, P., Ambite, J.L.: Learning the semantics of structured data sources. Web Seman.: Sci., Serv. Agents World Wide Web 37, 152–169 (2016)
16. Taheriyan, M., Knoblock, C., Szekely, P., Ambite, J.L., Chen, Y.: Leveraging linked data to infer semantic relations within structured sources. In: Proceedings of the 6th International Workshop on Consuming Linked Data (COLD) (2015)
17. Venetis, P., Halevy, A., Madhavan, J., Paşca, M., Shen, W., Wu, F., Miao, G., Wu, C.: Recovering semantics of tables on the web. Proc. VLDB Endow. 4(9), 528–538 (2011)
18. Wang, J., Wang, H., Wang, Z., Zhu, K.Q.: Understanding tables on the web. In: Atzeni, P., Cheung, D., Ram, S. (eds.) ER 2012 Main Conference 2012. LNCS, vol. 7532, pp. 141–155. Springer, Heidelberg (2012)

Integrating Medical Scientific Knowledge with the Semantically Quantified Self

Allan Third[1(✉)], George Gkotsis[2], Eleni Kaldoudi[3],
George Drosatos[3], Nick Portokallidis[3], Stefanos Roumeliotis[3],
Kalliopi Pafili[3], and John Domingue[1]

[1] Knowledge Media Institute, Open University, Milton Keynes, UK
{allan.third,john.domingue}@open.ac.uk
[2] King's College London, Biomedical Research Centre Nucleus, London, UK
george.gkotsis@kcl.ac.uk
[3] School of Medicine, Democritus University of Thrace, Alexandroupoli, Greece
kaldoudi@med.duth.gr, gdrosato@ee.duth.gr,
portokallidis@gmail.com, st_roumeliotis@hotmail.com,
kpafili@hotmail.com

Abstract. The assessment of risk in medicine is a crucial task, and depends on scientific knowledge derived by systematic clinical studies on factors affecting health, as well as on particular knowledge about the current status of a particular patient. Existing non-semantic risk prediction tools are typically based on hard-coded scientific knowledge, and only cover a very limited range of patient states. This makes them rapidly out of date, and limited in application, particularly for patients with multiple co-occurring conditions. In this work we propose an integration of Semantic Web and Quantified Self technologies to create a framework for calculating clinical risk predictions for patients based on self-gathered biometric data. This framework relies on generic, reusable ontologies for representing clinical risk, and sensor readings, and reasoning to support the integration of data represented according to these ontologies. The implemented framework shows a wide range of advantages over existing risk calculation.

Keywords: Health · Comorbidities · Risk factor · Scientific modelling · Knowledge capture · Semantics · Ontology · Linked data

1 Introduction

An important task in medicine is the assessment of risk. This depends on scientific knowledge derived by rigorous clinical studies regarding the (quantified) factors affecting clinical changes. Existing risk prediction tools typically only cover a very limited range of patient states, and the scientific knowledge informing the predictions is hardcoded into the tool. This makes them limited in application, particularly for patients with comorbidities (multiple co-occurring conditions), and rapidly out of date. An explicit representation of this knowledge, covering a wide (and, more importantly, expandable) range of risks and outcomes, would enable more sophisticated and maintainable risk prediction, prevention and management.

© Springer International Publishing AG 2016
P. Groth et al. (Eds.): ISWC 2016, Part I, LNCS 9981, pp. 566–580, 2016.
DOI: 10.1007/978-3-319-46523-4_34

In order actually to assess risk for an individual patient, it is necessary to link this generic clinical knowledge of risk to actual data relating to that patient's physical state. Traditionally, a doctor will make specific observations of a patient, and mentally determine the relevant known clinical evidence to make a risk prediction. In recent years, risk calculators based on individual clinical studies have been implemented as, e.g., web tools, where a patient can enter certain observations and be presented with numerical risks. With the advent of "Quantified Self" (QS) devices for low-cost and easy collection of individual physical and emotional data, there is a significant opportunity for personalized predictive medicine to combine this data with up-to-date knowledge of risk.

We present here a framework for calculating clinical risk predictions for patients based on self-gathered biometric data, using Semantic Web technologies at the core. The framework is shown to enable a large body of medical knowledge to be encoded in a common framework, and faithfully applied to QS data to perform automatic risk calculation, providing qualitative and quantitative improvements over the state of the art.

2 Previous Work

Existing algorithms for risk prediction for, e.g., cardiovascular risk, include the Framingham equation [1], the Joint British Societies (JBS) formula [2] and the ASSIGN score [3]. These take account of a limited set of risk factors and possible outcomes, as these have been produced by specific clinical studies – thus can be limited in application. For example, the ASSIGN score is specialized for Scottish populations, and, while Framingham includes diabetes as a risk factor, it is omitted from the JBS formula (diabetic patients are always high-risk). The Framingham equation takes account of 9 different patient observables and predicts the risk of only one outcome. More fundamentally, each of these hardcode the scientific knowledge about risk into the prediction formula itself, thus requiring new versions to be created to accommodate new scientific knowledge. This limited and non-extensible approach motivates our construction of a generic semantic model.

As there is no other model addressing the concept of risk factor, to the best of our knowledge, we compare related work addressing similar concepts and level of abstraction. A number of models have been proposed for capturing various aspects of clinical research at various levels of granularity. In particular, the Ontology-Based eXtensible data model (OBX) [4] has been developed to represent results of clinical research in order to promote data reuse, but does not address the concept of population-level risk factor. Models maintained by the Clinical Data Interchange Standards Consortium (CDISC) [5], and the Ontology of Clinical Research (OCRe) [6] take a more top-down approach to the modelling of clinical research and focus on data interchange formats and on the conceptual modelling of proposed and ongoing clinical trials. Overall, existing models aim to support the process of generating new scientific knowledge in medicine, rather than represent the actual knowledge itself, which is required for the task of risk prediction. In order to capture this scientific knowledge, we have developed an ontology for medical risk factors.

The "Quantified Self" (QS) refers to the use of technology for automated tracking of various measurements related to oneself (e.g., daily step count, distance walked, weight, and so on). Although considered a new trend with the potential to transform healthcare, it has received only a small amount of attention from the Semantic Web community. The MoodMap app [8] represents emotional states using an ontology, in order to support analysis of mood as tracked in the workplace, but does not concern itself with other Quantified Self measurements. An ontology for QS was presented in [9], but this is very high-level and at an early stage, lacking the detail needed to implement a Semantic Web system making use of it. This paper presents a detailed and practical ontology for representing QS measurements semantically, in a way which encourages flexibility and reuse, linking to other concepts related to each measurement, and which is usable in practical systems.

Finally, in order to achieve the integration between medical knowledge and QS data, it is necessary to express rules describing when a particular piece of knowledge is relevant to an individual on the basis of gathered data. There are two main candidate standards for representing rules for the Semantic Web – SWRL [10] and RIF [11] - as well as widely-used systems such as Jena [12]. Unfortunately, none of these rule systems can offer the expressive power needed to describe the conditions necessary to personalize a risk factor description to an individual person's data. In particular, it is common that the conditions under which clinical risks can be identified depend on a range of functions, e.g., body mass index, or the time since the occurrence of a myocardial infarction. This requirement rules out Jena, SWRL, and the Core dialect of RIF. These conditions can often also require disjunction to express correctly – "if estimated glomerular filtration rate is less than 44 OR chronic kidney disease is diagnosed at stage 3 or 4 or 5", and negation ("if the patient is male and does not have a family history of ischemic heart disease"). While the RIF Basic Logic Dialect (RIF-BLD) does support disjunction (where SWRL and Jena do not), and is compatible with OWL [13], it does not support negation. It is therefore necessary to develop a dedicated rules expression format, in a way which is, by design, easily interpreted and evaluated, supports the required logical features, and which allows the contents of rules to be easily authored and understood by clinicians.

The risk ontology implements the model described in [7]. The measurements ontology and data integration via rules are presented for the first time here.

3 Risk Factors

In medicine, risk is the probability of a negative outcome on the health of a population of subjects. The agents responsible for that risk are called risk factors when they aggravate a situation and are used to predict (up to a degree) the occurrence of a condition or deterioration of a patient's health dividing the population into high and low risk groups [14]. The following paragraphs present our model of the concept of **risk factor** in medicine [7] which is shown schematically in Fig. 1.

In general, risk factors can be: environmental (e.g. chemical, physical, mechanical, biological and psychosocial elements that constitute risk factors to public health); demographic (e.g. age, sex, race, location, occupation); genetic; behavioral and lifestyle related (e.g. smoking, overeating, unprotected sexual life, excessive alcohol drinking, drug abuse and sedentary lifestyle); and biomedical (i.e. conditions present in a patient that can influence his/her health by creating or affecting other conditions). Extending work on general risk analysis [15, 16], we can present a risk factor as a triplet, which includes the **source** of the risk, the outcome (**target**) and an expression of their association. The source of the risk is an agent (an event, a condition, a disorder or any other factor) that is shown via empirical studies to be associated with a consequence, that is, the outcome. The outcome itself is a negative health condition or disorder. Most often the outcome itself is found to be a source of another risk factor.

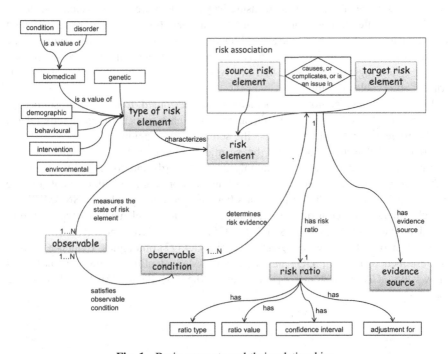

Fig. 1. Basic concepts and their relationships.

Thus in the general case the source and the outcome can both be treated as health related conditions (including disorders). In this work, we collectively refer to both the source and the outcome as **risk elements**. A **risk association** between the source and the outcome is a complex construct which describes the type of relation, the likelihood of an outcome to occur, and the initial conditions under which such likelihood can be estimated. The existence of a risk factor is not a determinant of consequence but the degree of its influence can be statistically calculated. The way to measure the likelihood requires a certain quantitative biomarker and observational studies that statistically calculate a probability. Different study designs and analyses can generate different types of probability measures [17] - a **Risk Ratio** (RR), such as the Relative Risk or Hazard Ratio (HR). A probability determined from a clinical study lies within a confidence interval, and the study design/analysis may have been adjusted, or not, for certain factors (for example, age, sex, and so on). In order to be able properly to represent risk factors, these must be included – especially where the goal is to produce personalized risk calculations.

An event, a condition, a disorder or any other factor becomes a risk source when certain conditions are met. These conditions are associated with one or more **observables**, which is either environmental or a physical or mental property of the patient. Therefore, in order to describe properly a risk association we have to state a specific observable that provides a measure/description of the risk source and the specific condition or value of this observable. For the same risk factor, a number of different risk associations can be measured in the literature, each association corresponding to a different observable or a different **observable condition** or even different combinations of observables corresponding to different concurrent risk sources. The circumstances under which a risk association is relevant to an individual are ascertained via an explicit logical expression that involves observables; this logical expression is termed 'observable condition'.

Finally, risk associations in medicine are determined from clinical studies as reported in evidence based medical literature. Thus, each association is directly related to an **evidence source** which is a specific scientific publication.

To ensure that the model can be seamlessly integrated into existing medical information systems, we adopt commonly used standards and controlled vocabularies in the description of the concepts presented above. For example, risk elements of type biomedical include an ICD-10 [18] classifier, of type demographic, a SNOMED-CT [19] classifier. Other controlled vocabularies used for risk elements of type environmental or intervention include SNOMED-CT, RxNorm [20], and EnvO [21]. Measurements and units follow the QUDT [22] and UO [23] ontologies. Evidence sources are described using their DOI and/or their PubMed identifier, while evidence level follows the OCEBM system [24]. In general, where available UMLS [25] codes are also used.

View	Risk factor	Observable condition ↑₁	Ratio value
⊕	central obesity [is an issue in] acute myocardial infarction	(waist circumference < 102 AND waist circumference ≥ 94) AND sex = 'male'	1.1
⊕	central obesity [is an issue in] acute myocardial infarction	(waist circumference < 88 AND waist circumference ≥ 80) AND sex = 'female'	1.5
⊕	central obesity [is an issue in] acute myocardial infarction	waist circumference ≥ 102 AND sex = 'male'	2.8
⊕	central obesity [is an issue in] acute myocardial infarction	waist circumference ≥ 88 AND sex = 'female'	1.4

Fig. 2. Example of risk associations and corresponding observable conditions

Figure 2 shows the risk associations relevant to the risk factor "central obesity is an issue in acute myocardial infarction", with the risk ratio values associated with patients who satisfy each observable condition, respectively. Although omitted here for reasons of space, each of these ratios is also associated with the original publication providing the evidence for it, as well as a confidence interval and specific ratio type.

4   Measurements and Sensors

The aim of the readings and measurements ontology is to represent the concepts involved in the gathering of data from personal Quantified Self sensors. In particular, it is important to represent details which are common to measurements generically, while allowing details relevant to specific measurement types to be captured also. Crucially for data integration, each measurement should be associated with a canonical type (representing, e.g., "systolic blood pressure") and a unit (e.g., "mmHg"), both preferably denoted by terms in standardized external vocabularies where possible. Figure 3 illustrates the ontology for the CARRE measurements and sensors.

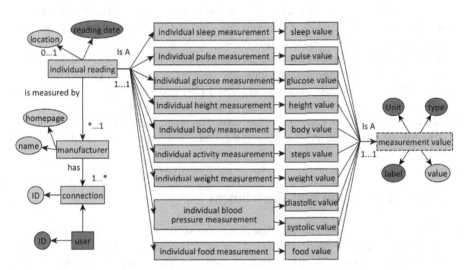

Fig. 3. The CARRE Sensors and Measurements Ontology, including some specific types of measurement to illustrate types of data which can be represented.

A **user** (an individual whose data is being represented using the terms of this ontology) has an identifier and **connections**. A connection represents that user's login details to the cloud data source, usually in practice provided by a device **manufacturer**, which has a name and a website.

Procedurally, data for an individual user is gathered from a manufacturer by means of the connection. Data is in the form of one or more device **readings**. Every device reading must of course have a date at which the reading was taken. Some manufacturers also provide location information in the form of latitude and longitude. A device reading may represent a set of **measurements**, all of which are semantically related. For example, a device reading may originate with the user stepping onto a set of body analysis scales, which can provide measurements of weight, body fat percentage, muscle mass, and so on. A reading may also have a provenance, which at the time of writing is simply whether the measurement came from a device automatically, or was manually entered into a web form by the user, and an actuality: manufacturers may provide actual measurements from devices or users, or goal measurements (e.g., a target weight). Finally, a device reading may be associated with a textual note added by the user.

The device reading class may be sub-classed, for measurements of, for example, activity, weight, blood pressure, and so on. Each of these has properties relating to the type of measurement value represented: for example, an individual blood pressure measurement relates to both systolic and diastolic blood pressure valuess.

Every measurement value has a common structure. A measurement value has a measurement type and a unit, which are its type and unit expressed in an external vocabulary wherever possible, a value which can be an integer, string, floating point value, and so on, and a label, which is a human-readable string.

To ensure that the model can be seamlessly integrated into existing medical information systems, we adopt the commonly used standards and controlled vocabularies in the description of the concepts presented above. The FOAF ontology [28] is extremely widely used and well-known, and allows easy representation of data relating to people. Types of measurement are indicated with respect to the Logical Observation Identifier Names and Codes ontology (LOINC) [29] and the Clinical Measurements Ontology (CMO) [30], with preference given to CMO on the basis of coverage for the set of measurement types currently being used. For units, we use QUDT [22] and the Unit Ontology [23], with preference for the Unit Ontology, again, on the basis of coverage.

5   Data Aggregation and Enrichment

These two ontologies are generic, describing the structure of data relating to measurements and to risk factors. To be useful, we need to populate them with instances of particular measurements and risk factors, respectively. The output of the relevant data aggregation processes is Linked Data, expressed according to the vocabularies defined by the relevant ontology, and stored in an RDF quad-store (Virtuoso, [31]).

Measurement data is subject to some extra constraints compared to the risk factor data. While clinical knowledge relating to risk is generic, and therefore can (and, we would argue, should) be public, measurement data is specific to an individual, and, as personal health-related data, required to be kept private. We thus maintain a separation between them at the quad-store level. Risk data is stored in a (curated, for quality and safety purposes) publicly accessible RDF graph, where measurement data relating to an individual is stored in an authentication-protected RDF graph belonging to that individual, accessible only via HTTPS.

There is a wide range of different wearable and personal sensors available which can, usually via a smartphone connection, automatically upload measurements to a manufacturer service. Such devices exist to measure activity levels (step counts, distance travelled), heart rate, blood pressure, blood oxygen saturation, weight, body fat, and others. In this work we have developed aggregators for data from devices from multiple manufacturers, including Fitbit, Medisana, iHealth, and Withings [32–35]. In each of these cases, the measurements are available for programmatic access via a Web API secured by some variant of the OAuth authentication schemes. Each such

API is supported in the aggregator by a plugin module, which, when supplied with access tokens for a particular user, retrieves that user's measurements, enriches them with RDF, and stores them in the relevant graph in the quad-store. Once set up, measurements are retrieved automatically, according to either the device's or user's chosen sampling interval, unless the user chooses to revoke access.

The risk ontology was populated with scientific information on medical risk factors in the area of cardiorenal disease. Chronic cardiorenal disease is the condition characterized by simultaneous kidney and heart disease while the primarily failing organ may be either the heart or the kidney. The cardio-renal patient (or the person at risk of this condition) presents an interesting case example for exploring risk factors, as (a) it is a complex comorbid condition which involves and is affected by a number of related health disorders as well as lifestyle related factors; (b) chronic cardiorenal disease has an increasing incidence and a number of serious (and of increasing incidence) comorbidities, including diabetes and hypertension, and may lead to serious chronic conditions such as nephrogenic anemia, renal osteodystrophy, peripheral neuropathy, malnutrition, and various systemic diseases (e.g. rheumatoid arthritis, lupus erythematosus); and (c) prevention is of major importance. Good appreciation of risks therefore plays an important role for the various stages of cardiorenal disease evolution, from normal health condition, to chronic disease, to end-stage renal and/or heart failure.

The process of collecting risk factor data begins with a literature review by the medical experts, to identify risk associations and associated entities and properties according to the ontology model. Identified risk factors are recorded in a tabular format, which mirrors the structure of the model, and these are reviewed by multiple clinicians. Observables, evidence sources, risk elements and associations are then translated to RDF.

6 Data Integration

The integration between the medical scientific knowledge and the semantic QS data is achieved using observable conditions. Each specific risk association is associated with a list of relevant observables, and an observable condition written in terms of these observables. Observable conditions are built using two basic types of operators, logical and comparison operators. More precisely, we follow prefix notation syntax for logical operators and infix notation syntax for comparison operators, and support as logical operators the disjunction "OR" and the conjunction "AND", and as comparison operators the equality "=", inequality "!=", greater than ">", greater than or equal to ">=", less than "<" and less than or equal to "<=". We have also identified functions which occur in the current domain of application, and use the idea of "calculated observable" to represent them. For example, "time since myocardial infarction" is calculable given the current date and a (non-calculated) observation of a myocardial infarction event. Generic functions such as averages over time are also important to take account of possible differences in sampling interval in measurement data. These calculated observables can be used in observable conditions.

Figure 4 shows a user-friendly interface that is used to build the observable conditions. This interface is implemented in HTML5, CSS and JavaScript using the AngularJS framework. As output of this expression builder, we support two different formats, an abstract syntax tree format (Fig. 5a) and a simple free text format (Fig. 5b). The first one is more suitable for expression editors and other parsers because it follows formal JSON syntax, and the second is more suitable for humans and evaluation algorithms and tools because it follows formal plain text syntax.

Fig. 4. Web based interface of expression builder.

```
{
    group: {
        operator: "AND",
        rules: [
            {
                group: {
                    operator: "AND",
                    rules: [
                        {
                            condition: "<",
                            field: "OB_80",
                            data: 102
                        },
                        {
                            condition: ">=",
                            field: "OB_80",
                            data: 94
                        }
                    ]
                }
            },
            {
                condition: "=",
                field: "OB_64",
                data: "male"
            }
        ]
    }
}
```

(a)

```
( OB_80 >= 94 AND OB_80 < 102 )
        AND  OB_64 = 'male'
```

(b)

Fig. 5. Observable condition: (a) abstract syntax tree format (b) simple free text format.

The software evaluates these conditions by retrieving the relevant measurement data for the patient in question, and substituting values into the condition expression. The (boolean) result of this evaluation determines whether or not the condition's risk factor applies to that patient, and hence with what particular ratio the patient is at risk of its target.

For example, if we evaluate the expression of Fig. 5 with observable values **waist circumference** (OB_80) equal to **98** and **sex** (OB_64) **"male"**, the expression evaluates to *true*. Referring back to Fig. 2, we can see that this therefore means that with regard to the risk factor "central obesity is an issue in acute myocardial infarction", there is a risk ratio of 1.1 that the central obesity of the patient concerned will be an issue in the probability of acute myocardial infarction.

Data integration of this form remains scalable over large numbers of both risk factors and users, since each observable condition is only ever evaluated with respect to *one* patient at a time, and, for clinical relevance, only ever with regard to (a small set of) that patient's most recent measurements.

7   Evaluation

To test the expressive utility of the risk ontology, as well as to populate it with data for use with QS data, a group of 8 medical doctors (members of the CARRE project team) reviewed current medical literature to identify major risk factors related to cardiorenal syndrome. At this time, 96 different risk factors were identified and described formally. The evidence sources used were 60 scientific publications. The evidence selection methodology and the available descriptions in text (tabular) format are provided in CARRE Deliverable 2.2 available from the project site [26]. A web entry system [27] allows these descriptions to be entered and reviewed, and produces RDF data representing their contents in accordance with the ontology. The manual curation of this data is necessary for regulatory and ethical reasons: as the aim of the system is to be used with patients, it is important to maintain strict quality control.

In addition, 10 project members connected a range of QS devices to the data aggregators and used or wore them to build up a sample corpus of semantically-annotated QS data. The aggregators collected data over a period of at least 12 months for all users (some users wore devices for longer), and stored them as RDF (with an average of 110,483 triples per user, at the time of writing). This length of time allowed the overall physical activity patterns of each user to be determined at different times of year and in different conditions, and thoroughly tested the data aggregators, and, importantly, was able to capture measurements which vary slowly over time, such as body weight. Other measurements, such as blood pressure, do not typically need to be measured over a long period of time to be useful in risk calculations – although it is worth noting that this commonly-held belief may simply result from a lack of data, as the ability to capture such measurements over long periods easily from non-hospitalized subjects is a comparatively recent development.

The rules expression evaluator, which evaluates the observable conditions and calculates a risk ratio for an individual for the target of a particular risk factor, is applied to each user, for each risk association stored in the system. The same calculations were performed manually to check the fidelity of the knowledge capture. (Despite the quantity of potential risk factors, this manual process can be streamlined effectively by discarding all those risk factors which can *never* apply to a particular user – e.g., those which only apply to male populations need never be evaluated for female users.)

The risk ontology population process resulted in 253 respective associations from 96 risk factors. There were 53 involved risk elements, corresponding to a total of 90 different observables. This is an order of magnitude greater than the observables taken account of by existing risk calculators. The automatic calculation of risks agreed with the manual calculation in *every* case. It should be noted that, of course, this assesses solely whether the risk calculations are faithful to the evidence sources, not whether the evidence source itself provides good predictions (already validated via the original clinical systematic review processes) nor what, if any, effect our approach has on user behaviour to minimise risk; this will be the subject of an upcoming randomized controlled trial. For reasons of confidentiality, particularly given the small and potentially deanonymisable set of participants, the QS data cannot be made public. The online tools, however, permit reproducibility testing with new participants.

This process of testing and using the risk ontology resulted in the following qualitative findings, derived via a focus group analysis of the testing participants. The medical experts found the model straightforward to use to describe risk factors. The terminology used was found to be familiar and thus easy to understand and apply to describe risk factors found in the literature and also to read descriptions already produced by colleagues. The only difficulty identified related to expressing accurately and rigorously the observable condition that has to be satisfied in order for a risk association to hold. Initially, medical experts were asked to produce this condition in the conventional way it is written in the literature, using natural language – which was a straightforward task. Subsequently, they were asked to reformat this condition using a logical operator expression (so that this expression can be easily translated to computer readable format). This task proved to be more cumbersome and required 1-2 h training and testing before the medical experts could independently produce correct expressions.

By using standard semantic technologies, it is possible to link both model and data to other clinical models (such as OCRe and OBX trial and data descriptions) and to external sources of data (e.g., environmental risk factors could be linked to open sources of environmental data). In particular, the semantic annotations on observables relating to medical diagnoses have made it possible to integrate the QS aggregation with Personal Health Record systems, by using UMLS to identify relevant medical concepts. Because of the semantic nature of the model, the outputs of risk calculation are also more useful for automated analysis, since it is always clear what a risk ratio value *means* in probability terms.

Nothing in either model is specific to the motivating domain of cardiorenal conditions, and extension to risk factors relating to other domains of medicine is not anticipated to pose any problems; the terminology and working practices with regard to risk calculation are common across medicine. Extending to more 'distant' domains where evidence-based risk calculation is relevant (e.g., climate science) ought also to be practical. The ontology already accommodates different representations of probability, and so could be adapted to those representations suitable to the new domain's conventions. The concept of "observable" is already generic. It would be necessary to extend the notion of evidence, and in particular, evidence quality, which is currently dependent on medical definitions.

The measurement ontology has also proved to be reusable. Having been conceived as a model for capturing numerical time series data from QS devices, it has proved to be conveniently usable without modification to represent qualitative data, such as that relating to diagnoses and the severity of conditions, as well as, in preliminary work, data relating to changes in patient state. For example, if a patient becomes higher risk for a particular outcome, it is proving to be both natural and useful to clinicians to record the state change as an observation of the patient.

While the motivation and initial thinking was focused on factors which increase the probability of negative consequences, the end result is equally as capable of modelling factors which *decrease* those probabilities, or which increase the probability of *positive* consequences. In other words, it is just as straightforward to represent, for example, an intervention with the potential to *lower* a patient's chance of acute myocardial infarction as a risk association with a risk ratio less than 1. It is interesting to note that this flexibility came as something of a surprise to the medical experts on the project – it appears that the linguistic conventions in medical practice around terms such as "risk factor" and "effectiveness of treatment" obscure, to some degree, the common probabilistic structure underneath – and required a shift in philosophical approach from the clinicians to accommodate. In the same way, having to make explicit the observable conditions for grounding risk predictions in data also required a change in thinking, where conditions easily understood by experienced humans need to be spelled out in precise detail in order to be implementable. Both of these changes in thinking were seen as positive by the clinicians involved. While only a qualitative observation of a small number of people, it is perhaps reasonable to expect similar changes in thinking to be necessary for domain experts in other fields where Semantic Web approaches become more practical and applicable to more situations, and it suggests an interesting avenue for future research into the social aspects of the move to data-based approaches.

Another benefit of modelling risks explicitly in this way is that it gives a very easy to follow overview of the field of medicine under consideration, showing at a glance both which risks are increased by multiple factors, which factors lead to multiple risks, as well as which associations have received more (or less) research attention. Figure 6 illustrates a projection of the various risk factors, as captured by the medical experts in the context of our project. Highlighted is the example of age and ischemic heart disease increasing a patient's risk of a stroke. It can also be seen how many risk elements increase the risk of heart failure, and how many new risks appear in obese patients. Again, this is suggestive of an interesting avenue for future research, to see what may be discovered by analysis of the semantic risk data as a whole with regard to the medical research field of which it represents the output. The semantic nature of our representation is likely to be a significant advantage in such research, enabling, as it does, the integration of the wide variety of different data sources which can be relevant to the study of scientific endeavour.

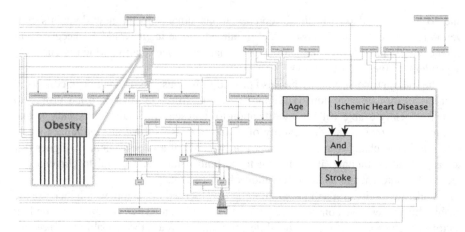

Fig. 6. A visual overview of currently encoded risk factors, with some examples highlighted, available online at http://ontology.carre-project.eu/

8 Conclusion

The risk model presented in this paper enables clinical experts to encode the risk associations between biological, demographic, lifestyle and environmental elements and clinical outcomes in accordance with evidence from the clinical literature. The measurements model enables the automatic capture of Quantified Self data relating to individual patients in a semantically annotated form. The integration of these datasets by means of the "observable condition" rule language makes it possible to compute risks automatically.

Compared to existing risk prediction models, this approach has a significant advantage in being able to be expanded and updated easily as clinical knowledge increases and changes, as well as being transparent and traceable in function and origin. The Semantic Web approach simplifies and encourages the integration of both clinical knowledge and QS data with other sources of relevant data, and, crucially, allows an area of very complex meanings to be expressed in a machine-readable fashion. We have also shown unanticipated extra benefits of having explicit ontological models relating these types of data. In particular, analysis of risk data en masse may provide insight into the current state of overall knowledge regarding a clinical domain, and the process of knowledge capture with clinical experts required some interesting, and positive, changes in thinking and approach, drawing out commonalities and possibilities which had not before been seen. We argue that such insights are likely to be encountered in other complex domains to which Semantic Web techniques are applied.

The work presented here illustrates the value of applying the Semantic Web to Quantified Self and health data, both in and of itself and also as an illustration of using semantics to connect sources of data at very different levels of granularity and acquired through very different methods. The development of the rules language was vital to enabling our results, and we believe it would be beneficial to explore the general question of the use of rules to "bridge" distinct data sources in this way.

Acknowledgments. This work was supported by the FP7-ICT project CARRE (Grant No. 611140), funded in part by the European Commission. We express our gratitude to all project team members for fruitful discussions.

References

1. Sheridan, S., Pignone, M., Mulrow, C.: Framingham-based tools to calculate the global risk of coronary heart disease. J. Gen. Intern. Med. **18**(12), 1039–1052 (2003)
2. Boon, N., Boyle, R., Bradbury, K., Buckley, J., Connolly, S., Craig, S., Wood, D.: Joint British Societies' consensus recommendations for the prevention of cardiovascular disease (JBS3). Heart **100**(Suppl. 2), ii1–ii67 (2014)
3. Woodward, M., Brindle, P., Tunstall-Pedoe, H.: Adding social deprivation and family history to cardiovascular risk assessment: the ASSIGN score from the Scottish Heart Health Extended Cohort (SHHEC). Heart **93**(2), 172–176 (2007)
4. Kong, Y.M., Dahlke, C., Xiang, Q., Qian, Y., Karp, D., Scheuermann, R.H.: Toward an ontology-based framework for clinical research databases. J. Biomed. Inform. **44**(1), 48–58 (2011)
5. CDISC. http://cdisc.org. Accessed 24 July 2015
6. Sim, I., Tu, S.W., Carini, S., Lehmann, H.P., Pollock, B.H., Peleg, M., Wittkowski, K.M.: The Ontology of Clinical Research (OCRe): an informatics foundation for the science of clinical research. J. Biomed. Inform. **52**, 78–91 (2014)
7. Third, A., Kaldoudi, E., Gkotsis, G., Roumeliotis, S., Pafili, K., Domingue, J.: Capturing scientific knowledge on medical risk factors. In: Workshop: 1st International Workshop on Capturing Scientific Knowledge at 8th International Conference on Knowledge Capture, Palisades, NY, USA (2015)
8. Rivera Pelayo, V.: Design and Application of Quantified Self Approaches for Reflective Learning in the Workplace. KIT Scientific Publishing, Karlsruhe (2015)
9. Cena, F., Likavec, S., Rapp, A., Deplano, M., Marcengo, A.: Ontologies for quantified self: a semantic approach. In: HT (Doctoral Consortium/Late-breaking Results/Workshops) (2014)
10. Semantic Web Rules Language. http://www.w3.org/Submission/SWRL/
11. Rules Interchange Format. https://www.w3.org/TR/rif-overview/
12. Reasoners and Rule Engines: Jena Inference Support. https://jena.apache.org/documentation/inference/
13. RIF RDF and OWL compatibility. http://www.w3.org/TR/rif-rdf-owl/
14. Mrazek, P.B., Haggerty, R.J. (eds.): Reducing Risks for Mental Disorders: Frontiers for Preventive Intervention Research: Summary. National Academies Press, Washington, D.C (1994)
15. Kaplan, S.: The words of risk analysis. Risk Anal. **17**(4), 407–417 (1997)
16. Offord, D.R., Kraemer, H.C.: Risk factors and prevention. EBMH **3**, 71 (2000)
17. Crowson, C.S., Therneau, T.M., Matteson, E.L., Gabriel, S.E.: Primer: demystifying risk - understanding and communicating medical risks. Nat. Clin. Pract. Rheumatol. **3**(3), 181–187 (2007)
18. ICD-10: International Classification of Diseases v10, WHO. http://www.who.int/classifications/icd/en/
19. SNOMED-CT: Systemized Nomenclature of Medicine – Clinical Terms, IHTSDO. http://www.ihtsdo.org/snomed-ct/
20. RxNorm: Normalized Names for Clinical Drugs, U.S. National Library of Medicine. http://www.nlm.nih.gov/research/umls/rxnorm/

21. EnvO: Environmental Ontology. http://environmentontology.org/
22. QUDT: Quantity, Unit, Dimension and Type Ontologies. http://qudt.org/
23. UO: The Ontology of Units of Measurement, OBO Foundry Initiative. https://code.google.com/p/unit-ontology/
24. Oxford Centre for Evidence-Based Medicine Levels of Evidence. Produced by Howick, J., Chalmers, I., Glasziou, P., Greenhalgh, T., Heneghan, C., Liberati, A., Moschetti, I., Phillips, B., Thornton, H., Goddard, O., Hodgkinson, M. (2011)
25. UMLS: The Unified Medical Language System, US National Library of Medicine. http://www.nlm.nih.gov/research/umls/
26. CARRE. http://carre-project.eu/
27. CARRE Risk Data Entry system. https://entry.carre-project.eu/
28. FOAF. http://xmlns.com/foaf/spec/
29. LOINC. https://loinc.org
30. CMO. https://bioportal.bioontology.org/ontologies/CMO
31. Virtuoso Universal Server. http://virtuoso.openlinksw.com
32. Fitbit. https://www.fitbit.com
33. Medisana. http://www.medisana.com
34. iHealth. http://www.ihealthlabs.com
35. Withings. http://www.withings.com

Learning to Assess Linked Data Relationships Using Genetic Programming

Ilaria Tiddi[(✉)], Mathieu d'Aquin, and Enrico Motta

Knowledge Media Institute, The Open University, Milton Keynes, UK
{ilaria.tiddi,mathieu.daquin,enrico.motta}@open.ac.uk

Abstract. The goal of this work is to learn a measure supporting the detection of strong relationships between Linked Data entities. Such relationships can be represented as paths of entities and properties, and can be obtained through a blind graph search process traversing Linked Data. The challenge here is therefore the design of a cost-function that is able to detect the strongest relationship between two given entities, by objectively assessing the value of a given path. To achieve this, we use a Genetic Programming approach in a supervised learning method to generate path evaluation functions that compare well with human evaluations. We show how such a cost-function can be generated only using basic topological features of the nodes of the paths as they are being traversed (i.e. without knowledge of the whole graph), and how it can be improved through introducing a very small amount of knowledge about the vocabularies of the properties that connect nodes in the graph.

1 Introduction

The goal of the work here presented is to automatically discover what makes a strong relationship between two entities of the Web of Linked Data. Identifying the strength of the relationship between entities can have many applications, the most common of which is to measure entity relatedness, i.e. identifying how related two entities are. This is a well-known problem for a wide range of tasks, such as text-mining and named-entity disambiguation in Natural Language Processing, or ontology population and query expansion in Semantic Web activities.

From a Web of Data perspective, a relationship can be identified in the graph of Linked Data as a semantic path (expressed as a chain of entities and properties) between two given entities, and graph search techniques can be used to reveal them. When applying such techniques to the Linked Data graph, however, the entities and properties included in the found paths might come from a number of different, unknown data sources. In order to avoid having to index and locally pre-process a necessarily partial subset of the graph, a natural approach is to rely on link traversal, which allows to incrementally and agnostically explore the graph from entity to entity until paths between them are found. In other words, finding relationships between entities in the Linked Data graph requires a uniformed (or blind) search, which does not need pre-computation or

© Springer International Publishing AG 2016
P. Groth et al. (Eds.): ISWC 2016, Part I, LNCS 9981, pp. 581–597, 2016.
DOI: 10.1007/978-3-319-46523-4_35

knowledge over the entire graph. However, to drive such an uniformed search, a function to measure the strength of the explored paths (a "cost-function") is necessary to ensure that only the most promising ones will be followed.

Our goal is therefore to figure out which of (and how) the features of the Linked Data graph along the explored paths could be used by such cost-function. While one could intuitively think that the shortest paths reveal the strongest connections, this assumption does not necessarily hold within the Linked Data space, where entities of different datasets are connected by multiple paths of similar lengths. Our challenge is to find which Linked Data structural information we need in order to design a cost-function that objectively assesses the value of a path. More specifically, we aim at discovering which topological and semantic features of the traversed entities and properties can be used to reveal the strongest relationships.

To answer this question, the approach we propose is to use a supervised method based on Genetic Programming whose scope is to learn the path evaluation function to integrate in a Linked Data blind search. Our idea is that, starting from a randomly generated population of cost-functions created from a set of topological and semantic characteristics of the Linked Data graph, the evolutionary algorithm will reveal which functions best compare with human evaluations, and will show us what is really important to assess strong relationships in Linked Data. The learnt cost-functions are compared and discussed in our experiments, where we show not only that good results are achieved using basic topological features of the nodes of the paths as they are being traversed, but also how those results can be improved through introducing a very small amount of knowledge about the vocabularies used to label the edges connecting the nodes.

2   Related Work

As already mentioned, the goal of our work is to learn a measure to assess strong Linked Data relationships, so that this can be integrated in an uninformed graph search within Linked Data. For this reason, we divided this related work section in three parts. First, we study works that focus on assessing Linked Data entity relatedness, in order to discover which types of interestingness measures have been proposed. Then, we analyse works based on Linked Data traversal, to see how uninformed graph searches can be applied in the context of Linked Data. Finally, we explore works that focused on designing a measure empirically, namely by learning it through Genetic Programming.

Linked Data Entity Relatedness. There is a solid body of literature on entity relatedness, which can be categorised according to the corpus used to assess it [10,18]. Here, we focus mainly of approaches that compute the relatedness based on Linked Data. A first area comprehends Linked Data-based metrics to assess the strength of a relationship between entities quantitatively [10,14, 20,21,24]. They can be divided into entity-based approaches, which compute the similarity between neighbouring concepts based on the entity description (i.e. triples where the entity is involved as a subject or object) [14,21,24], and path-based metrics [10,20], which compute relatedness between concepts that are not directly connected. As in our work, these approaches are motivated by the idea of exploiting the rich resource descriptions existing in (and across) Linked Data. From our perspective, these works are too restrictive for two main reasons: first, they present ad-hoc measures, which have been either manually designed based on the analysed datasets or adapted from existing information theoretical measures; second, the strength of a relationship can only be assessed quantitatively.

A second area includes works that define entity relationships qualitatively, as Linked Data paths or subgraphs. The strongest relationships are identified through information theoretical measures based on node centrality, node frequency or edge informativeness applied on the paths retrieved from one or more Linked Data datasets. We find in this category systems for data visualisation and exploratory searches, such as RelFinder [8], REX [5], Explass [1] and, more recently, Recap [18]. These approaches first identify all possible relationships between two entities, either using SPARQL queries aiming at retrieving paths up to a certain length [8,18], or by extracting them from a pre-computed dataset [1,5], and then rank the results based on some predefined interestingness measures. Pathfinding techniques have also been used to identify entity relationships [3,12,15]: similar to our work is the use of cost-functions based on the Linked Data graph structure to drive the informed searches (e.g. A*, random walks), prioritising nodes and pruning the search space. With that said, their major limitation consist in exploiting Linked Data with an a priori knowledge, either by indexing and pre-processing datasets, or by using queries against SPARQL endpoints, therefore pre-defining the desired portion of data to be analysed.

Linked Data Traversal. The idea behind the Linked Data Traversal (link traversal in short) is to exploit URI dereferencing[1] to discover connections between entities across datasets on-the-fly and in a *follow-your-nose* fashion, so that no or very little domain knowledge has to be introduced. Link traversal relies on the fact that if data are connected (through owl:sameAs, skos:exactMatch, rdfs:seeAlso or simply by vocabulary reuse), then one can naturally span datasources and gather new knowledge serendipitously. Various studies have shown that Linked Data can be traversed agnostically in contexts such as SPARQL query extensions [7] or (cor-)relation explanation [18,22]. So far, however, uninformed graph searches have been only used for Linked Data crawling or indexing [9,20]. To the best of our knowledge, no work has focused on identifying a cost-function suitable to be integrated in an uninformed search over Linked Data.

Genetic Programming. Evolutionary algorithms have proven to perform well in those tasks where it was necessary to identify suitable functions based on a desired output. For instance, Genetic Programming has been successfully applied in Information Retrieval to reveal the most appropriate document ranking functions for search engines [2,4,13,23]. In the Linked Data context, Genetic Pro-

---

[1] Retrieving a representation of a resource identified by a URI.

gramming was used to identify similarity functions for discovering links [16,17], instance clustering [6] or matching [11] across different datasets, but not, prior to the work presented here, to assess relationship strength between Linked Data entities.

3   Motivation, Challenges and Approach

Our motivating scenario is a uniform-cost search process which traverses Linked Data on-the-fly with the aim of identifying the best relationship given two input entities. Uniform-cost search (*ucs*) is an uninformed, non-greedy best-first search strategy where a cost-function $g(n)$ chooses the next node to expand based on the cumulative cost of the edges from the start to that node n. When expanded, new children are created and stored in the queue accordingly. Since the nodes are generated iteratively, *ucs* does not require holding the whole graph in memory, which makes it suitable for large graphs; secondly, $g(n)$ being cumulative (i.e. paths never get shorter as the nodes are added), the search expands nodes in the order of their optimal path, which guarantees search optimality. These characteristics make *ucs* particularly suitable to a context such as Linked Data.

Our process is designed as a bi-directional search, whose aim is to find the path $p = \langle n^l \dots n^r \rangle$ that best represents the relation between two entities n^l and n^r. By "best", it is intended that the relatedness between n^l and n^r, expressed as a score assigned by the cost-function, is maximised. Because *ucs* does guarantee optimality, but its bi-directional version does not, we use a maximum number of node expansions to perform as a termination criterion. An example of such a process, showing how different structural information might be needed to find the best relationship, is presented below. Here, we used entities of the same dataset for clarity purposes but, as demonstrated by the experiments, the process can be applied on entities of two arbitrary datasets.

3.1   Example Scenario

Let us imagine that we want to identify the strongest relationship between two DBpedia entities, e.g. n_1=db:ASongOfIceAndFire(novel) and n_2=db:GOT-TV-series(episodes) of Fig. 1. The process consists in:

(1) *Bi-directional search.* Given the two nodes n^l and n^r, two uniform-cost searches ucs^l and ucs^r are performed simultaneously. Their objective is to iteratively build two search spaces, a left-directed one from n^l and a right-directed one from n^r, to find a common node n^c.
(2) *Entity dereferencing.* Each search space is expanded by dereferencing the entity labelling the next node in the queue, and by finding all the entities that are linked to it. We do consider as "link" any edge of the node, i.e. both incoming and outgoing RDF properties of the dereferenced entity. In the example, n_1 is linked to 5 entities and n_2 to 4. As said, nodes are queued and dereferenced according to their cumulative cost from the start node n^j (with $j \in \{l, r\}$), which guarantees optimality to both ucs^j. This step is repeated until one or more common nodes n^c are found.

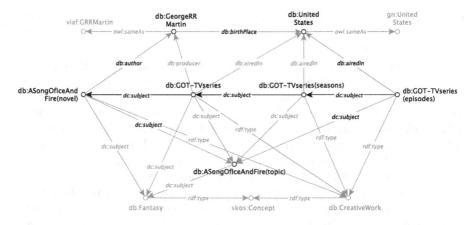

Fig. 1. Paths between db:ASongOfIceAndFire(novel) and db:GOT-TVseries(episodes)

(3) *Path building.* For each common node n^c, we build the two subpaths $p^j = \langle n^j \ldots n^c \rangle$, and then merge them into a path $p = \langle n^l \ldots n^c \ldots n^r \rangle$. Each path then identifies a relationship between the initial two entities. For instance, the graph of Fig. 1 represents all the paths existing between n_1 and n_2 after a few iterations.

(4) *Path scoring.* The cost of each path is evaluated as an approximation (most often, a sum) of the costs of the paths from n^l to n^c and from n^r to n^c. The one with the highest score, highlighted in the Figure, is chosen as the strongest relationship between the initial entities.

3.2   Challenge and Proposed Approach

From the process described above, it becomes clear that the problem to tackle are how to find a good cost-function is necessary to choose among a set of alternative paths between two entities, and how to avoid computational efforts or inconclusive searches. The question arising here is what is the best strategy to find the most representative relationships, and if we can exploit the information in the Web of Data to guide the two searches in Linked Data in the right direction, so that they can quickly get to convergence. When looking at the paths in Fig. 1, an interesting observation can be made: the node corresponding to the entity db:GameOfThrones-TVseries has a lower indegree, which is generally used to measure the authority (its "popularity") of a node, when compared to other nodes, as the ones labelled as ASongOfIceAndFire(topic) or db:UnitedStates. This information could be used to rank nodes so that the path that best specifies the relation between n_1 and n_2 is soon revealed. In other words, the structural features of the graph could be a good insight to drive a blind search in Linked Data. Given this, our challenge is to answer the question: what makes a path strong? Which are the topological or semantic features of a node or an edge, which can be used when deciding if a path is better than another? To reformu-late the problem: can we use the structure of the Linked Data graph to assess relationship strengths?

Our proposition is to use a supervised Genetic Programming (GP) approach to identify the cost-function that best performs in ranking sets of alternative relationship paths. Starting from a random population of cost-functions cre-ated on a set of features related to possible topological or semantic features of the nodes and edges of the path, the evolutionary algorithm will learn the cost-function that best performs when compared to a benchmark of human-evaluated relationship paths. The choice of Genetic Programming over other supervised learning techniques (e.g. SVMs, Neural Networks, Linear Regression or learning-to-rank) is motivated by three main reasons: first, its results are not assessed by comparing directly the path scores, which are hardly comparable to the human rankings provided in the benchmark, but by assessing the ade-quacy of the cost-functions through a fitness function; secondly, these formulas are human-understandable, which means that they can be used to identify the structural features of Linked Data that matter for a successful search; finally, because they are understandable, they can be directly implemented in a graph search mechanism. Additionally, the GP learning process is flexible, so it allows us to easily refine parameters and impose new constraints on the fitness func-tion, and it comfortably deals with wide search spaces, so we can study large populations of possible cost-functions without worrying about scalability issues.

The contributions of this work can be summarised as follows: (i) we present a measure to detect strong entity relationships that can be integrated in unin-formed searches over Linked Data, therefore avoiding data pre-processing; (ii) we demonstrate that such function can be derived empirically, which improves over the state-of-the-art approaches presenting domain-specific or manually-defined measures; (iii) we show that good results are achieved using basic topological features of the nodes of the paths as they are being traversed, and how those results can be improved through introducing a very small amount of knowledge about the vocabularies used to label the edges connecting the nodes.

4   Learning Functions to Evaluate Paths

In this section, we first give and overview of the Genetic Programming framework and then present the supervised approach that we propose to discover the cost-functions to assess entity relationships.

4.1   Genetic Programming Foundations

Inspired by Darwin's theory of evolution, Genetic Programming is an Artifi-cial Intelligence technique that aims at automatically solving problems in which the solution is not known in advance [19]. The general idea is to create a ran-dom population of computer programs, which are the candidate solutions for a problem, that the algorithm stochastic transforms ("evolves") into new, possibly improved, programs. The stochastic process guarantees that the GP proposes diverse solutions to a given problem.

In GP, programs are generally represented as trees of primitive elements, where the internal nodes (mathematical or logical operations) are called functions, while the leaf nodes (constants or variables) are called terminals. A fitness function measures how good each program is with respect to the problem to be solved. Given a population, a new population is created by adding programs using one of the three following genetic operations: (1) reproduction, in which a new child program is generated by copying a randomly selected parent program; (2) crossover, where a child program is generated by combining randomly chosen parts from two randomly selected parent programs; (3) mutation, where a new child program is generated by randomly altering a randomly chosen part of a selected parent. This process is iterated until a termination condition is met: typically either a maximum number of generations is reached, or a satisfying, possibly optimal solution (i.e. a desired fitness) is found. Along with the primitive set, the fitness and the termination condition, a set of parameters such as the population size, the probabilities of performing the genetic operations, the selection methodology or the maximum size for programs need to be decided to control the GP process.

4.2 Preparatory Steps

The described framework can be used to learn the cost-function that best ranks a set of alternative paths between two Linked Data entities. For a better understanding, we invite the reader to use as a reference the graph of Fig. 1 and the three following paths:

$\overline{p_1} =$ db:ASongOfIce AndFire(novel) — db:author→ db:George- RRMartin — db:birthPlace→ db:United States ← db:airedIn — db:GOT-TV series(episodes)

$\overline{p_2} =$ db:ASongOfIce AndFire(novel) ← dc:subject — db:GOT- TVseries ← dc:subject — db:GOT-TV series(seasons) ← dc:subject — db:GOT-TV series(episodes)

$\overline{p_3} =$ db:ASongOfIce AndFire(novel) — dc:subject→ db:ASongOfIce AndFire(topic) ← dc:subject — db:GOT-TV series(episodes)

Process. Let $P_i = \{\overline{p_1}, \ldots, \overline{p}_{|P_i|}\}$ be the set of $|P_i|$ alternative paths between two Linked Data entities, with i being the ith pair in $D = \{P_1, \ldots, P_{|D|}\}$, the set of $|D|$ examples that have been ranked by humans, and $G = \{g_1, \ldots g_{|G|}\}$ a starting population of randomly generated cost-functions g_j. The GP algorithm iteratively evolves the population into a new, possibly improved one, until the stopping condition is met. The evolution consists first in assigning a fitness score to the cost-functions, which in our case reflects how "good" a cost-function is in ranking paths compared to the human evaluators. For instance, assuming 3 users have agreed on ranking the paths as $\overline{p_2}$, $\overline{p_3}$ and $\overline{p_1}$, those functions scoring the them in the same order will obtain the highest fitness. Then, reproduction, mutation and crossover are applied to some randomly (with bias from fitness) chosen individuals, and the generated children are added to the new population. The current population is replaced by the evolved one once they reach the same size, and a new generation starts.

Primitives. Terminals and functions are called primitives. A terminal can be: (i) a constant, i.e. a randomly chosen integer in the set $Z = \{0, \ldots, 1000\}$, or (ii) a combination of an edge weighting function $w(e)$ (with e being the edge) and one aggregator a. We call this combination $a.w$ an aggregated terminal.

Edge weighting functions $w(e)$ assign a weight to each edge of the path, based on the information of its source. We define 10 edge weighting functions, that we divide in topological and semantic terminals. Topological terminals focus on the Linked Data graph structure, and are as follows.

- *Fixed Weight* (1): the edge is assigned a score of 1. This is equivalent to performing a breadth-first search, where nodes are queued and explored in the order they are found.
- *Indegree* (*in*): the edge is weighted according to the number of incoming links of its source. For instance, the edge *db:birthPlace*(db:GeorgeRRMartin, db:UnitedStates) of Fig. 1 has a weight of 2, since the source db:GeorgeRR-Martin has 2 incoming links. This feature is chosen to understand the importance of "authority" nodes, i.e. the ones with many incoming links.
- *Outdegree* (*ou*): the edge is weighted according to the number of outgoing links of its source, e.g. the weight in the previous example is 2. *ou* helps us study the importance of "hub" nodes that point to many other nodes.
- *Degree* (*dg*): an edge is weighted based on the degree of the source, i.e. the sum of *in* and *ou*. To the previous example, *dg* would assign a score of 4.
- *Conditional Degree* (*cd*): the weight attributed to the edge depends on the RDF triple from which the edge has been generated. In fact, each edge $e(u, v)$ is generated from a dereferenced RDF triple, either $\langle u, e, v \rangle$, as in the case of *db:birthPlace*(db:GeorgeRRMartin, db:UnitedStates), or $\langle v, e, u \rangle$, as for *db:pro- ducer*(db:GeorgeRRMartin, db:GOT-TVseries). The *cd* terminal returns either the indegree or the outdegree of the source depending on whether the triple represents a back or a forward link. Therefore, *cd* would return 2 in the former case (the indegree of the node for db:GeorgeRRMartin) and 2 in the latter case (its outdegree). The conditional degree analyses the importance of paths going through large hubs, that are also common to many other paths.

We define semantic terminals those features that are more specific to Linked Data than to common graphs. For that, we first considered the vocabulary usage, then analysed the most frequent RDF properties, as provided by both Linked Open Vocabularies[2] and LODStats[3]. Note that, since we rely upon entity dereferencing to traverse Linked Data, we only considered the most frequent object properties.

- *Namespace Variety* (*ns*): an edge is weighted depending on the number of namespaces of its source node. For instance, the node db:GeorgeRRMartin has the two namespaces *owl:* and *db:* for its three links, while the node db:GOT-TVseries has the 3 namespaces *dc:*, *db:* and *skos:* for its 5 links.

---

[2] http://lov.okfn.org/dataset/lov/terms.
[3] http://lodstats.aksw.org/.

Namespaces variety is intended to analyse the use of vocabularies when semantically describing an entity. While initially we considered incoming and outgoing namespaces separately, we did not find any substantial difference in the process, and eventually reduced the two terminals to one.

- *Type Degree* (*td*): the edge weight depends on the number of *rdf:type* declared for the source entity. For example, db:ASongOfIceAndFire(novel) has a type degree of 1 but, assuming this was declared as a skos:Concept too, its score would be 2. *td* focuses on the taxonomical importance of an entity, with the idea that the more a node is generic (i.e. the entity belongs to many classes), the less informative the path might be. Since *rdf:type* is unidirectional, there is no need to distinguish between in- and outdegree.

- *Topic Outdegree* (*so*): the edge weight is assigned by counting the number of outgoing edges labeled as *dc:subject*, *foaf:primaryTopic* and *skos:broader* of the starting node. The edge *db:author*(db:ASongOfIceAndFire(novel), db:GeorgeRRMartin) has a score of 2. The topic outdegree focuses on authority nodes in topic taxonomies (controlled vocabularies or classification codes).

- *Topic Indegree* (*si*): similarly, the edge weight is assigned by counting the number of incoming *dc:subject*, *foaf:primaryTopic* and *skos:broader* edges. The same edge has a score of 1 in this case. *si* considers hub nodes on controlled vocabularies.

- *Node Equality* (*sa*): the edge is weighted according to how much its source is connected to the external datasets, based on the number of links labeled as *owl:sameAs*, *skos:exactMatch* or *rdf:seeAlso*. For instance, db:UnitedStates is connected to its Geonames[4] corresponding entity gn:6252001 so, according to the *sa* weight, the edge *db:airedIn*(db:UnitedStates, db:GOT-TVseries (episodes)) is scored 1. *sa* considers the importance of the inter-dataset connections. Since those properties are bi-directional, we do not distinguish between in- and outdegree.

Aggregators are functions to combine the weights of edges across the whole path: *sum* returns the sum of the $w(e)$ for each of the l edges of the path; *avg* returns the average edge weight across the path; *min* and *max* the path minimal an maximal $w(e)$, respectively.

To generate an individual, the aggregated terminals are randomly combined through the GP function set, composed of *addition* $x + y$, *multiplication* $x * y$, *division* x/y and *logarithm* $log(x)$. For example, $g_1 = sum.1 + (1/avg.td)$ is interpreted as a function acting almost as a depth-first search, with a small added value from the average type degree of the nodes of the path.

Fitness Evaluation. The fitness of a cost-function is measured with the Normalised Discounted Cumulative Gain (*nDCG*), generally used in Information Retrieval to assess the quality of rankings provided by the web search engines based on the graded relevance of the returned documents[5]. The closer it gets to 1,

---

[4] http://www.geonames.org/.

[5] https://www.kaggle.com/wiki/NormalizedDiscountedCumulativeGain.

the more the engine judges the documents as relevant as the human evaluators did. We apply the same idea by considering a path as a document, therefore evaluating first the DCG for a path p_k at rank k as:

$$DCG(p_k) = rel_1 + \sum_{m=2}^{k} \frac{rel_m}{log_2(m)} \tag{1}$$

where rel_k is the average of the relevance scores given to p_k by human evaluators. The $DCG(p_k)$ is then normalised by comparing it to its ideal score $iDCG(p_k)$, as assessed by the gold standard.

The function $avg(P_i)$ then averages each $nDCG(p_k)$ in the set P_i, so that to obtain the performance of the function for the i-th pair, as in Eq. 2. The overall fitness of a function is obtained by averaging each $avg(P_i)$ of all the $|D|$ pairs of the dataset as in Eq. 3.

$$avg(P_i) = \frac{\sum\limits_{p_k \in P_i} nDCG(p_k)}{|P_i|} \quad (2) \qquad f(g_j) = \frac{\sum\limits_{P_i \in D} avg(P_i)}{|D|} \quad (3)$$

We also add a penalty weight to avoid long and complex cost-functions, by comparing the length l of a function with its ideal length L. The weighted fitness of a function is defined as:

$$fw(g_j) = f(g_j) - (w \times (l - L)^2) \tag{4}$$

where w is the penalty weight.

Genetic Operations. We perform the following genetic operations.

- Reproduction. Given a cost-function parent, a new individual is copied in the new generation without alterations.
- Crossover. Given two parents, two children are generated by swapping two random subtrees of the parents.
- Mutation. Given a selected parent, one node (the mutation point mp) is modified. We designed different kinds of mutations, as in Table 1, depending on the type of mp: if it is a constant x, the node is mutated with a new constant y that is either higher (1) or lower (2) than x in the range of $y = [x - 100, x + 100]$; if mp is an aggregated terminal, the node is mutated by either modifying its aggregator (3), modifying its edge weighting function (4), or by replacing it with a new constant (5); if mp is a function, it can be replaced either with a new constant (7) an aggregated terminal (8), or with a new function (8), in which case we might remove (9) or add (10) a child depending on the arity of the new function.

Training and Testing. We randomly split the dataset into a training set and a test set. Then, we run the GP process on the training set and store a small set

Table 1. Mutation examples for $g_1 = sum.1 + (1/avg.td)$

n	mp	Mutation type	Example
1	Constant	$x < y < x + 100$	$sum.1 + (\mathbf{18}/avg.td)$
2	Constant	$x - 100 < y < x$	$sum.1 + (\mathbf{-18}/avg.td)$
3	Terminal	New a	$sum.1 + (1/\mathbf{max}.td)$
4	Terminal	New w	$sum.1 + (1/avg.\mathbf{in})$
5	Terminal	New x	$sum.1 + (1/\mathbf{20})$
6	Function	New $a.w$	$sum.1 + \mathbf{max.ns}$
7	Function	New x	$sum.1 + \mathbf{40}$
8	Function	New function (same arity)	$sum.1 + (1 \times avg.td)$
9	Function	New function (delete child)	$\mathbf{log}(1/avg.td)$
10	Function	New function (add child)	$\mathbf{min.ou} \times (sum.1 + (1/avg.td))$

of the fittest individuals, i.e. the cost-functions that performed better in ranking paths, while the rest are discarded. Third, the surviving individuals are tested on the test set, and if their fitness is not consistent with the one of the training set, we screen them out. This helps in avoiding overfitting and in obtaining more valid cost-functions. We then keep the best individual of each run.

5 Experiments

The section introduces our experimental scenario, describing the dataset we built and the control parameters for the Genetic Programming learning process. Then, it presents the obtained results, including the discovered cost-function[6].

5.1 Experimental Setting

As previously mentioned, the fitness is assessed on a dataset composed by sets of alternative paths between random pairs of entities. In order to create more variety in the final dataset, so that the learnt functions would not be overfitted to a specific type of data source, we used different types of entities, randomly extracted from different Linked Data sources, namely: (i) 12,630 *events* (from battles to sport to music events) from Yago[7]; (ii) 8,185 *people* from the Virtual International Authority File (VIAF)[8]; (iii) 999 *movies* from the Linked Movie Database[9]; and (iv) 1,174 *countries* and *capitals* from Geonames and the UNESCO[10] datasets.

---

[6] Dataset and results are available online at http://linkedu.eu/dedalo/pathfinding/.
[7] http://yago-knowledge.org.
[8] http://viaf.org/.
[9] http://www.linkedmdb.org/.
[10] http://uis.270a.info/.html.

To make sure to span at least to another dataset when finding paths, therefore guaranteeing more path heterogeneity, we used the DBpedia SPARQL endpoint as a pivot, i.e. we chose a desired `?_class` (event, country, person etc.) and the `?_dataset` we wanted to retrieve it from, and then ran the simple query:

```
select distinct ?same where {
    ?entity a ?_class. # select the entities of a desired class
    ?entity <http://www.w3.org/2002/07/owl#sameAs> ?same.  # get owl:sameAs
    FILTER(strStarts(str(?same), ?_dataset)).  # filter by dataset
}
ORDER BY RAND()  # make sure to get random entities
```

Next, given a random pair, we ran a bi-directional breadth-first search limited to 30 iterations in order to find a set of possible paths between them. Note that other iterations thresholds were also tested (between 20 and 50), and 30 cycles seemed the most reasonable trade-off between missing some relationships and taking more time to obtain almost the same relationships. We discarded the pairs for which no path was found. 8 judges were asked to evaluate each set, assigning the paths *rel* scores between 2 ("highly informative") and 0 ("not informative at all"), and discarded the pairs whose agreement was below 0.1 according to the Fleiss'k rating agreement[11]. The choice of using different scores was motivated to represent the gradation between meaningless relations, valide but weak relations and strong relations. An example of a path to be ranked, showing that the movie "The Skin Game" and the actress Dina Korzun are both based in Europe, is presented in Fig. 2. The final dataset consisted of 100 pairs, whose paths were assigned a score corresponding to the average of the scores given by the users.

Fig. 2. A path example

Finally, Table 2 presents the control parameters we used during the GP process. Because a perfect agreement between the functions with the users could

Table 2. Control parameters for the GP runs

Population size	100 individuals	Reproduction	10 % population size
Max generations	300	Elitism	10 % population size
Termination	Max generation	Penalty weight w	0.001
Selection	5-sized tournament	Ideal length L	3
Crossover rate:	0.65	Validation split	70 % −30 %
Mutation rate:	0.15	Num. individuals (testing)	5

---

[11] https://en.wikipedia.org/wiki/Fleiss%27_kappa.

not be reached, we use a maximum number of generations as a termination condition. In order to not bias the generation process, we also generated trees without limit in depth, and equally distributed the probability of functions, constants and aggregated terminals being chosen. It is worth mentioning that other parameters were also tested but, due to space limitation, we only present the ones giving the best results according to our tests.

5.2 Results

First, we present the results of different runs of the Genetic Programming learning process presented in the previous Section. Table 3 shows the unweighted fitness on training set (f_{tr}) and test set (f_{ts}) of the best cost-functions learnt during 3 different runs (therefore different dataset cuts). We divided the GP runs depending on whether we used topological terminals only (T), or both topological and semantic terminals (S).

Results show that some terminals, i.e. the conditional degree cd, the namespace variety ns and the topic indegree si, are recurrent across different runs of the same block, which demonstrate the stability of our learning process. Given the regularity we noticed of the $min.ns$ aggregated terminal, Table 3 also includes a third block of experiments (N), in which we used only the topological terminals and $min.ns$. We observe that both the T- and the N-functions, based mostly on topological features, have a lower performance when compared to the S-ones, that include semantic terminals too. Nevertheless, the S-functions confirm the importance of $min.ns$.

We then performed a comparative evaluation between the learnt cost-functions and some related work of the literature, namely RelFinder (RF [8]), Recap [18] and the two measures presented by the Everything is connected Engine (EICE [3]) and by Moore et al. (M&V [15]). Figure 3 presents the $avg(P_i)$ score (Y axis) that each of the functions obtained on each of the examples in D (X axis).

Table 3. Best cost-functions for different runs

Run	Fittest individual g_j	f_{tr}	f_{ts}
T$_1$	$log(log(min.cd \times min.cd))/max.cd$	0.79	0.79
T$_2$	$log(min.cd)/(avg.cd + 87)$	0.77	0.78
T$_3$	$min.cd \times (min.cd/max.cd)$	0.78	0.72
N$_1$	$(log((max.ns/max.cd))/avg.ns) + min.ns$	0.82	0.81
N$_2$	$((min.dg/sum.cd)/sum.ou) + min.ns$	0.79	0.77
N$_3$	$min.ns/(log(max.cd)/avg.ns)$	0.83	0.75
S$_1$	$min.ns + (sum.ns/log(log(sum.si)))$	0.88	0.83
S$_2$	$min.ns + (min.cd/log(log(sum.si)))$	0.88	0.86
S$_3$	$min.ns + (log(max.in)/log(log(sum.si)))$	0.87	0.86

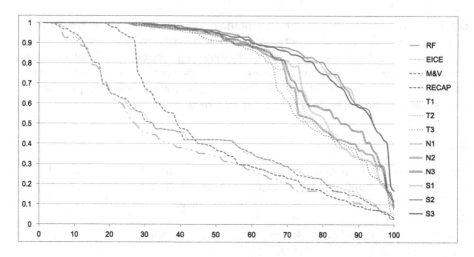

Fig. 3. $avg(P_i)$ of each measure on the full dataset D

What can be noticed from the Figure is a considerable difference between the existing approaches, which are based on ad-hoc information theoretical measures, and the ones that were automatically learnt through Genetic Programming. Indeed, the combination of several topological characteristics sensibly improves a cost-function performance, as demonstrated by the overall fitness of $f(g_i)$ of T_1, T_2 and T_3 (also presented in Table 4, first row). This means that the ranking the T-functions give for a set of path is much more similar to the ones of a human evaluator than the ones attributed by hand-crafted measures. The low performance of the existing measures suggests that they are not suitable to correctly evaluate paths that connect entities across several Linked Data datasets, as the ones we have collected in our experiments. A slight improvement can also be observed with the N-functions: the overall fitness $f(g_i)$ for them improves roughly by 0.02-0.04 when compared to the T-functions. With that said, the Figure clearly shows that adding some semantic information is the key to obtain more precise results, as the S-function overall fitnesses $f(S_1)=0.86$, $f(S_2)=0.88$ and $f(S_3)=0.87$ demonstrate (i.e. fitness improvement is ca. 0.09–0.11).

In Table 4, the cost-functions are compared to the baselines to assess if they perform in a stable way across different datasets sources. We removed, in turn, pairs whose entities belonged to one of the Linked Data sources presented in Sect. 5.1, and then calculated the functions' fitness $f(g_i)$ on the filtered dataset. Results confirm that the S-functions are consistent even with different datasets.

We finally analyse the cost-function that reported the best performance:

$$S_2 = min.ns + \frac{min.cd}{log(log(sum.si))} \qquad (5)$$

and observe that the terminals here included are the same that we had already noted as being the most recurrent among the different runs of Table 3. As can be seen from its shape, this function prioritises paths that:

Table 4. Overall fitness of the functions across datasets. $D\backslash$ indicates from which dataset the entities were removed; s indicates the size of the filtered dataset

$D\backslash$	s	RF	RECAP	EICE	M&V	T_1	T_2	T_3	S_1	S_2	S_3
$\emptyset$ (tot)	100	0.399	0.481	0.449	0.446	0.784	0.763	0.749	0.871	**0.873**	0.865
Geonames	23	0.455	0.457	0.584	0.605	0.756	0.710	0.641	0.909	**0.911**	0.887
Yago	77	0.371	0.513	0.432	0.424	0.819	0.820	0.819	**0.883**	0.881	0.880
VIAF	69	0.378	0.492	0.418	0.414	0.832	0.828	0.801	0.880	**0.888**	0.880
UNESCO	79	0.390	0.457	0.442	0.437	0.784	0.764	0.750	0.858	**0.860**	0.849
LMDB	73	0.439	0.382	0.438	0.435	0.740	0.728	0.705	0.846	**0.847**	0.843

- pass through nodes with rich node descriptions (the higher $min.ns$ is, the more relevant the path is considered);
- do not include high level entities (that have many incoming $dc{:}subject/foaf{:}primaryTopic/skos{:}broader$ links, since many other entities are also of the same category), since the higher $sum.si$ is, the lower the path score is;
- include only specific entities (not hubs) for paths with a small number of topic categories. Indeed, because of the use of the double log function, the ratio between $min.cd$ and $log(log(sum.si))$ is negative if $sum.si$ is lower than 10. However, $min.cd$ becomes a positive factor when $sum.si$ is above 10.

In other words, the function prioritises specific paths (e.g. a movie and a person are based in the same region) to more general paths (e.g. a movie and a person are based in the same country).

6   Conclusions

In this paper, we presented a supervised method based on Genetic Programming in which we learnt a measure to detect strong relationships between entities in the Linked Data graph. Such measure is a cost-function to be used in a blind graph search over Linked Data, in which relationships between entities are identified as Linked Data paths of entities and properties. With the assumption that the topological and semantic structure of Linked Data can be used by a cost-function to identify the strongest connections, we used Genetic Programming to generate a population of cost-functions that was evolved iteratively, based on how well the individuals compared with a human evaluated training data. The results proved our idea that successful path evaluation functions can be built empirically using basic topological features of the nodes traversed by the paths, and that a little knowledge about the vocabularies of the properties connecting nodes in the explored graph is fundamental to obtain the best cost-functions. We analysed the obtained functions to detect which features are important in Linked Data to find the strongest entity relationships, and finally presented the cost-function that we learnt. As future work, we will integrate this cost-function in a Linked Data pathfinding process that can be used in frameworks for on-the-fly knowledge discovery over Linked Data.

References

1. Cheng, G., Zhang, Y., Qu, Y.: Explass: exploring associations between entities via Top-K ontological patterns and facets. In: Mika, P., et al. (eds.) ISWC 2014, Part II. LNCS, vol. 8797, pp. 422–437. Springer, Heidelberg (2014)
2. Cordón, O., Herrera-Viedma, E., López-Pujalte, C., Luque, M., Zarco, C.: A review on the application of evolutionary computation to information retrieval. Int. J. Approx. Reason. **34**(2), 241–264 (2003)
3. De Vocht, L., Coppens, S., Verborgh, R., Vander Sande, M., Mannens, E., Van de Walle, R.: Discovering meaningful connections between resources in the web of data. In: LDOW (2013)
4. Fan, W., Gordon, M.D., Pathak, P.: A generic ranking function discovery framework by genetic programming for information retrieval. Inf. Process. Manag. **40**(4), 587–602 (2004)
5. Fang, L., Sarma, A.D., Yu, C., Bohannon, P.: REX: explaining relationships between entity pairs. Proc. VLDB Endow. **5**(3), 241–252 (2011)
6. Fanizzi, N., d'Amato, C., Esposito, F.: Metric-based stochastic conceptual clustering for ontologies. Inf. Syst. **34**(8), 792–806 (2009). Sixteenth ACM Conference on Information Knowledge and Management (CIKM 2007)
7. Hartig, O.: SQUIN: a traversal based query execution system for the web of linked data. In: Proceedings of the 2013 ACM SIGMOD International Conference on Management of Data, pp. 1081–1084. ACM (2013)
8. Heim, P., Hellmann, S., Lehmann, J., Lohmann, S., Stegemann, T.: RelFinder: revealing relationships in RDF knowledge bases. In: Chua, T.-S., Kompatsiaris, Y., Mérialdo, B., Haas, W., Thallinger, G., Bailer, W. (eds.) SAMT 2009. LNCS, vol. 5887, pp. 182–187. Springer, Heidelberg (2009)
9. Hogan, A., Harth, A., Umbrich, J., Kinsella, S., Polleres, A., Decker, S.: Searching and browsing linked data with SWSE: the semantic web search engine. Web Semant. Sci. Serv. Agents World Wide Web **9**(4), 365–401 (2011)
10. Hulpuş, I., Prangnawarat, C., Hayes, C.: Path-based semantic relatedness on linked data and its use to word and entity disambiguation. In: Arenas, M., et al. (eds.) ISWC 2015. LNCS, vol. 9366. Springer, Heidelberg (2015)
11. Isele, R., Bizer, C.: Learning linkage rules using genetic programming. In: Proceedings of the Sixth International Workshop on Ontology Matching, pp. 13–24 (2011)
12. Kasneci, G., Elbassuoni, S., Weikum, G.: Ming: mining informative entity relationship subgraphs. In: Proceedings of the 18th ACM Conference on Information and knowledge Management, pp. 1653–1656. ACM (2009)
13. Lin, J.Y., Yeh, J.-Y., Liu, C.C.: Learning to rank for information retrieval using layered multi-population genetic programming. In: 2012 IEEE International Conference on Computational Intelligence and Cybernetics (CyberneticsCom), pp. 45–49. IEEE (2012)
14. Meymandpour, R., Davis, J.G.: Linked data informativeness. In: Ishikawa, Y., Li, J., Wang, W., Zhang, R., Zhang, W. (eds.) APWeb 2013. LNCS, vol. 7808, pp. 629–637. Springer, Heidelberg (2013)
15. Moore, J.L., Steinke, F., Tresp, V.: A novel metric for information retrieval in semantic networks. In: García-Castro, R., Fensel, D., Antoniou, G. (eds.) ESWC 2011. LNCS, vol. 7117, pp. 65–79. Springer, Heidelberg (2012)

16. Ngonga Ngomo, A.-C., Lyko, K.: EAGLE: efficient active learning of link specifications using genetic programming. In: Simperl, E., Cimiano, P., Polleres, A., Corcho, O., Presutti, V. (eds.) ESWC 2012. LNCS, vol. 7295, pp. 149–163. Springer, Heidelberg (2012)
17. Nikolov, A., d'Aquin, M., Motta, E.: Unsupervised learning of link discovery configuration. In: Simperl, E., Cimiano, P., Polleres, A., Corcho, O., Presutti, V. (eds.) ESWC 2012. LNCS, vol. 7295, pp. 119–133. Springer, Heidelberg (2012)
18. Pirró, G.: Explaining and suggesting relatedness in knowledge graphs. In: Arenas, M., et al. (eds.) ISWC 2015. LNCS, vol. 9366, pp. 622–639. Springer, Heidelberg (2015)
19. Poli, R., Langdon, W.B., McPhee, N.F., Koza, J.R.: A Field Guide to Genetic Programming. Lulu. com, Raleigh (2008)
20. Schuhmacher, M., Ponzetto, S.P.: Knowledge-based graph document modeling. In: Proceedings of the 7th ACM International Conference on Web Search and Data Mining, pp. 543–552. ACM (2014)
21. Seufert, S., Bedathur, S.J., Hoffart, J., Gubichev, A., Berberich, K.: Efficient computation of relationship-centrality in large entity-relationship graphs. In: International Semantic Web Conference (Posters & Demos), pp. 265–268 (2013)
22. Tiddi, I., d'Aquin, M., Motta, E.: Dedalo: looking for clusters explanations in a labyrinth of linked data. In: Presutti, V., d'Amato, C., Gandon, F., d'Aquin, M., Staab, S., Tordai, A. (eds.) ESWC 2014. LNCS, vol. 8465, pp. 333–348. Springer, Heidelberg (2014)
23. da Torres, R.S., Falcão, A.X., Gonçalves, M.A., Papa, J.P., Zhang, B., Fan, W., Fox, E.A.: A genetic programming framework for content-based image retrieval. Pattern Recogn. **42**(2), 283–292 (2009)
24. Zhou, W., Wang, H., Chao, J., Zhang, W., Yu, Y.: LODDO: using linked open data description overlap to measure semantic relatedness between named entities. In: Pan, J.Z., Chen, H., Kim, H.-G., Li, J., Horrocks, I., Mizoguchi, R., Wu, Z., Wu, Z. (eds.) JIST 2011. LNCS, vol. 7185, pp. 268–283. Springer, Heidelberg (2012)

A Probabilistic Model for Time-Aware Entity Recommendation

Lei Zhang$^{(\boxtimes)}$, Achim Rettinger, and Ji Zhang

Institute AIFB, Karlsruhe Institute of Technology (KIT), Karlsruhe, Germany
l.zhang@kit.edu

Abstract. In recent years, there has been an increasing effort to develop techniques for related entity recommendation, where the task is to retrieve a ranked list of related entities given a keyword query. Another trend in the area of information retrieval (IR) is to take temporal aspects of a given query into account when assessing the relevance of documents. However, while this has become an established functionality in document search engines, the significance of time has not yet been recognized for entity recommendation. In this paper, we address this gap by introducing the task of *time-aware entity recommendation*. We propose the first probabilistic model that takes time-awareness into consideration for entity recommendation by leveraging heterogeneous knowledge of entities extracted from different data sources publicly available on the Web. We extensively evaluate the proposed approach and our experimental results show considerable improvements compared to time-agnostic entity recommendation approaches.

1 Introduction

The ever-increasing quantities of entities in large knowledge bases on the Web, such as Wikipedia, DBpedia and YAGO, pose new challenges but at the same time open up new opportunities of information access on the Web. In recent years, many research activities involving entities have emerged and increasing attention has been devoted to technologies aimed at identifying entities related to a user's information need. *Entity search* has been defined as finding an entity in the knowledge base that is explicitly named in a keyword query [1]. A variant of entity search is *related entity recommendation*, where the goal is to rank relationships between a query entity and other entities in a knowledge base [2,3]. In the context of Web search, *entity recommendation* has been defined as finding the entities related to the entity appearing in a Web search query [4].

On the other hand, temporal dynamics and their impact on information retrieval (IR) have drawn increasing attention in the last decade. In particular, the study of document relevance by taking into account the temporal aspects of a given query is addressed within *temporal IR* [5]. To support a temporal search, a basic solution is to extend keyword search with the creation or publication date of documents, such that search results are restricted to documents from a particular time period given by a time constraint [6,7]. This feature is

© Springer International Publishing AG 2016
P. Groth et al. (Eds.): ISWC 2016, Part I, LNCS 9981, pp. 598–614, 2016.
DOI: 10.1007/978-3-319-46523-4_36

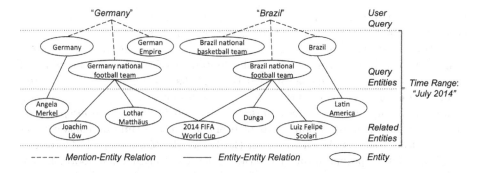

Fig. 1. Examples of the query and related entities for the user query *"Germany Brazil"* and the given time range *"July 2014"*

already available in every major search engine, e.g., Google also allows users to search Web documents using a keyword query and a customized time range. For the effectiveness of temporal IR, the time dimension has been incorporated into retrieval and ranking models, also called *time-aware retrieval and ranking*. More precisely, documents are ranked according to both textual and temporal similarity w.r.t. the given temporal information needs [5].

Inspired by temporal IR, we believe that the time dimension could also have a strong influence on entity recommendation. Existing entity recommendation systems aim to link the initial user query to its related entities in the knowledge base and provide a ranking of them. Typically, this has been done by exploiting the relationships between entities in the knowledge base [2–4]. However, the (temporal) entity importance and relatedness is often significantly impacted by real-world events of interest to users. For example, a sports tournament could drive searches towards the teams and players that participate in the tournament and the acquisition of a company by another company could establish a new relationship between them and thus affect their relatedness. Some efforts have already been devoted to improve the quality of recommendations in particular with respect to data freshness. For example, Sundog [8] uses a stream processing framework for ingesting large quantities of Web search log data at high rates such that it can compute feature values and entity rankings in much less time compared to previous systems, such as Spark [4], and thus can use more recently collected data for the ranking process. However, the time-awareness, which should be a crucial factor in entity recommendation, has still not been addressed.

Let us suppose users issue the keyword query *"Germany Brazil"* (see Fig. 1). Then they are likely looking for related geographic or political entities. However, when additionally specifying the time range *"July 2014"*, their interest is more likely related to the German and Brazilian national football teams during the 2014 FIFA World Cup. Obviously, once time information is available, the goal for a related entity search approach should be to improve entity recommendation such that the ranking of related entities depends not only on entity information in the knowledge base but also on the real-world events taking place in a specific time period. Therefore, it is essential to make *time-awareness* a top priority in entity recommendation when a customized time range is given.

In this paper, we introduce the problem of *time-aware entity recommendation* (TER), which allows users to restrict their interests of entities to a customized time range. In general, the goal of TER is to (1) *disambiguate the query entities* mentioned in the user query and (2) *find the related entities* to the query entities as well as (3) *rank all these query entities and related entities according to time* in order to match information needs of users, where the time dimension plays an important role. As shown in Fig. 1, the keywords *"Germany"* and *"Brazil"* result in different potential query entities. Since Germany_national_football_team and Brazil_national_football_team are of particular interest during the given time range *"July 2014"*, they should more likely be the intended query entities. For each query entity, its related entities will be found through the relations between entities, which can also be influenced by the time dimension. For example, the query entity Brazil_national_football_team results in the related entities Dunga, the current coach of Brazilian national football team, and Luiz_Felipe_Scolari, the coach during 2014 FIFA World Cup. By taking into account the time dimension, Luiz_Felipe_Scolari should be preferred over Dunga since the user requests information from July 2014.

To achieve this, we propose a probabilistic model by decomposing the TER task into several distributions, which reflect heterogeneous entity knowledge including *popularity, temporality, relatedness, mention* and *context*. The parameters of these distributions are then estimated using different real-world data sources, namely Wikipedia[1], Wikilinks[2], Wikipedia page view statistics[3] and a multilingual real-time stream of annotated Web documents. Please note that the data sources used by existing systems are mostly not publicly accessible. Particularly the major Web search engines keep their own usage data, like query terms and search sessions as well as user click logs and entity pane logs, secret, since they are crucial to optimizing their own entity recommendation systems, like the ones of Yahoo! [4,8] and Microsoft [9,10]. In contrast, our approach does not rely on datasets taken from commercial Web search engines, but only resorts to data sources publicly available on the Web.

The main contributions of this paper are: (1) We introduce a formal definition of the TER problem (2) and propose a statistically sound *probabilistic model* that incorporates *heterogeneous entity knowledge* including the *temporal context*. (3) We show how all parameters of our model can be effectively estimated solely based on data sources *publicly available* on the Web. (4) Due to the lack of benchmark datasets for the TER challenge, we have created *new datasets* to enable empirical evaluations and (5) the results show that our approach improves the performance considerably compared to time-agnostic approaches.

The rest of the paper is organized as follows. We present the overall approach, especially the probabilistic model in Sect. 2. Then, we describe the estimation of

---

[1] https://dumps.wikimedia.org/.

[2] http://www.iesl.cs.umass.edu/data/wiki-links/.

[3] https://dumps.wikimedia.org/other/pagecounts-raw/.

model parameters in Sect. 3. The experimental results are discussed in Sect. 4. Finally, we survey the related work in Sect. 5 and conclude in Sect. 6.

2   Approach

We first formally define the *time-aware entity recommendation* (TER) task and then describe the probabilistic model of our approach.

Definition 1 (Time-Aware Entity Recommendation). Given a knowledge base with a set of entities $E = \{e_1, \cdots, e_N\}$, the input is a keyword query q, which refers to one or more entities, and a continuous date range $t = \{d_{start}, \cdots, d_{end}\}$ where $d_{start} \leq d_{end}$, and the output is a ranked list of entities that are related to q, especially within t.

We use DBpedia as the knowledge base in this work, which contains an enormous number of entities in different domains by extracting various kinds of structured information from Wikipedia, where each entity is tied to a Wikipedia article.

2.1   Probabilistic Model

We formalize the TER task as estimating the probability $P(e|q,t)$ of each entity e given a keyword query q and a date range t. The goal is then to find a ranked list of top-k entities e, which maximize the probability $P(e|q,t)$. Based on Bayes' theorem, the probability $P(e|q,t)$ can be rewritten as follows

$$P(e|q,t) = \frac{P(e,q,t)}{P(q,t)} \propto P(e,q,t) \tag{1}$$

where the denominator $P(q,t)$ can be ignored as it does not influence the ranking.

To facilitate the discussion in the following, we first introduce the concepts of *mention* and *context*. For a keyword query q, a *mention* is a term in q that refers to an entity e_q, also called *query entity*, and the *context* of e_q is the set of all other mentions in q except the one for e_q. For each query entity e_q, the keyword query q can be decomposed into the mention and context of e_q, denoted by s_{e_q} and c_{e_q} respectively. For example, given the query entity Germany_national_football_team, the keyword query *"Germany Brazil"* results in the mention *"Germany"* and the context $\{$*"Brazil"*$\}$. Based on that, the joint probability $P(e,q,t)$ is given as

$$P(e,q,t) = \sum_{e_q} P(e_q, e, q, t) = \sum_{e_q} P(e_q, e, s_{e_q}, c_{e_q}, t)$$

$$= \sum_{e_q} P(e)P(t|e)P(e_q|e,t)P(s_{e_q}|e_q,e,t)P(c_{e_q}|e_q,e,t) \tag{2}$$

$$= \sum_{e_q} P(e)P(t|e)P(e_q|e,t)P(s_{e_q}|e_q)P(c_{e_q}|e_q,t) \tag{3}$$

where we assume in (2) s_{e_q} and c_{e_q} are conditionally independent given e_q and t, in (3) s_{e_q} is conditionally independent of e and t given e_q, and c_{e_q} is conditionally independent of e given e_q and t. The intuition behind these assumptions is that a mention s_{e_q} should only rely on the query entity e_q it refers to and a context c_{e_q} that appears together with e_q should depend on both e_q and t.

The main problem is then to estimate the components of $P(e, q, t)$ including the *popularity* model $P(e)$, the *temporality* model $P(t|e)$, the *relatedness* model $P(e_q|e, t)$, the *mention* model $P(s_{e_q}|e_q)$ and the *context* model $P(c_{e_q}|e_q, t)$.

2.2   Data Sources

To derive the estimation of these distributions in our model, we present several publicly available data sources. Based on these data sources, we discuss the details of model parameter estimation in Sect. 3.

Wikipedia and Wikilinks. Wikipedia provides several resources, including article titles, redirect pages and anchor text of hyperlinks, that associate each entity with terms referring to it, also called *surface forms* [11]. Wikilinks [12] also provides surface forms of entities by finding hyperlinks to Wikipedia from a Web crawl and using anchor text as mentions. Based on such sources, we construct a dictionary that maps each surface form to the corresponding entities.

Based on the observation that a more popular entity usually has more pages linking to it, we take link frequency as an indicator of *popularity*. For example, in Wikipedia the famous basketball player Michael Jeffrey Jordan is linked over 10 times more than the Berkeley professor Michael I. Jordan.

Wikipedia link structure has also been used to model *entity relatedness* [13], without considering temporal aspects, where the intuition is that Wikipedia pages containing links to both of the given entities indicate relatedness, while pages with links to only one of the given entities suggest the opposite.

Page View Stream. Wikipedia page view stream provides the number of times a particular Wikipedia page is requested per hour and thus can be treated as a query log of entities. In general, a well-known entity usually gets more page views than the obscure ones, such that the page view frequency also captures the *popularity* of entities.

In addition, an entity is likely to get more page views when an event related to it takes place. For example, during the FIFA World Cup, many participating football teams and players will get more page views. This explains the significant page view spike during an event when the entity receives media coverage, which has been utilized for the event detection task [14]. In this sense, the page view spike captures a user-driven measure of the *temporality* of entities.

Furthermore, an event could result in more page views for all the involved entities. For example, when Facebook acquires WhatsApp, both of them get high page view spikes. Based on this observation, simultaneous page view spikes of entities can help with modeling *the dynamic relatedness* between entities.

Annotated Web Document Stream. Another data source is a real-time aggregated stream of semantically annotated Web documents. We first employ a *news feed aggregator*[4] to acquire a multilingual real-time stream of news articles publicly available on the Web [15], where the enormous number of collected Web documents are in various languages, such as English (50 % of all articles), German (10 %), Spanish (8 %) and Chinese (5 %). Then we employ a cross-lingual semantic annotation system[5] to annotate the multilingual Web documents with DBpedia entities, i.e., to link entity mentions to their referent entities [16]. Based on that, entity co-occurrence statistics extracted from the annotated Web documents can help to identify dynamically related entities and the co-occurrence frequency can be utilized to measure the *dynamic relatedness* between entities w.r.t. a specific time range.

2.3 Candidate Selection

As there are millions of entities in DBpedia, it is extremely time-consuming to calculate $P(e, q, t)$ for all entities. To improve the efficiency of TER, we employ a candidate selection process to filter out the impossible candidates. Given a query q and a date range t, the candidate related entities are generated in three different ways: (1) Based on the dictionary containing entities and their surface forms extracted from Wikipedia and Wikilinks datasets, all query entities, whose mentions can be found in q, are selected as a set of candidates, denoted by E_q. (2) Given the set of subject, predicate and object triples $\{(s, p, o)\}$ in DBpedia, where all subjects and objects are entities, the potential candidate related entities that have a relation to the query entities are identified as $\{e | \exists p : (e, p, e_q), e_q \in E_q\} \cup \{e | \exists p : (e_q, p, e), e_q \in E_q\}$. (3) By analyzing the annotated Web documents, the entities that co-occur with the query entities in the Web documents published during the date range t are also considered as candidate related entities.

3 Model Parameter Estimation

Our probabilistic model is parameterized by $\Phi_e = P(e)$, $\Phi_{t|e} = P(t|e)$, $\Phi_{e'|e,t} = P(e'|e,t)$, $\Phi_{s|e} = P(s|e)$ and $\Phi_{c|e,t} = P(c|e,t)$. In the following, we present the details of parameter estimation based on the introduced data sources.

3.1 Popularity Model Φ_e

The distribution $P(e)$ captures the popularity of entity e. By leveraging both Wikipedia link structure and page view statistics, we first calculate $C(e)$ as

$$C(e) = C_{link}(e) + \beta C_{view}(e) \tag{4}$$

where $C_{link}(e)$ denotes the number of links pointing to e and $C_{view}(e)$ denotes the average number of page views on e per day. While $C_{link}(e)$ represents the prior

---

[4] http://newsfeed.ijs.si/visual_demo/.
[5] http://km.aifb.kit.edu/sites/xlisa/.

popularity of e in Wikipedia, $C_{view}(e)$ captures the popularity of e based on user interests. Due to the different scales of link and page view frequencies, $C_{view}(e)$ is adjusted by a balance parameter $\beta = \frac{\text{total number of links in Wikipedia}}{\text{average number of page views per day}}$, which accounts for the difference in frequencies of Wikipedia links and per-day page views. Then the probability $P(e)$ is estimated as follows

$$P(e) = \frac{\log\left(C(e)\right) + 1}{\sum_{e_i \in W} \log\left(C(e_i)\right) + |W|} \tag{5}$$

where W denotes the set of all entities. The estimation is smoothed using Laplace smoothing for avoiding the zero probability problem.

3.2 Temporality Model $\Phi_{t|e}$

The distribution $P(t|e)$ captures the temporality of entity e w.r.t. date range t. We employ the page view statistics as a proxy for interest of each entity and equate the page view spike with it. For each entity e, we track its per-day page view counts for each date d. Then we compute the mean $\mu(e, d)$ and standard deviation $\sigma(e, d)$ of page views for entity e in a window of n days before d

$$\mu(e, d) = \frac{1}{n} \sum_{d_i=d-n}^{d-1} C(e, d_i) \tag{6}$$

$$\sigma(e, d) = \sqrt{\frac{1}{n} \sum_{d_i=d-n}^{d-1} (C(e, d_i) - \mu(e, d))^2} \tag{7}$$

where $C(e, d_i)$ denotes the number of page views of e on date d_i. Inspired by [17], we calculate the page view spike $S(e, d)$ of entity e on date d as

$$S(e, d) = \begin{cases} \frac{C(e,d)-\mu(e,d)}{\sigma(e,d)} & \text{if } \frac{C(e,d)-\mu(e,d)}{\sigma(e,d)} \geq \kappa, \\ 0 & \text{otherwise} \end{cases} \tag{8}$$

where we assume that only the page view count $C(e, d)$ that is abnormally large compared with the previously seen page views of e, i.e. $\frac{C(e,d)-\mu(e,d)}{\sigma(e,d)} > \kappa$ (κ is a fixed parameter set as 2.5 here), indicates an event and thus will be taken into account to compute the page view spike $S(e, d)$.

Based on the page view spike $S(e, d)$ of entity e for date d, the estimation of $P(d|e)$, which is further smoothed using Laplace smoothing, is given as

$$P(d|e) = \frac{S(e, d) + \kappa}{\sum_{d_i \in T} S(e, d_i) + \kappa|T|} \tag{9}$$

where $|T|$ is the number of days contained in the longest date range T supported by the system, which is set as one year here. Consequently, the probability $P(t|e)$ reflecting events about e happening within t can be calculated as follows (here we assume that the dates within t are independent given the entity e)

$$P(t|e) = \sum_{d_i \in t} P(d_i|e) \tag{10}$$

3.3   Relatedness Model $\Phi_{e'|e,t}$

The distribution $P(e'|e,t)$ models the entity relatedness between e and e' w.r.t. t. To estimate $P(e'|e,t)$, we consider both static and dynamic entity relatedness as

$$P(e'|e,t) = \lambda \frac{R_S(e,e')}{\sum_{e'} R_S(e,e')} + (1-\lambda)\frac{R_D(e,e',t)}{\sum_{e'} R_D(e,e',t)} \tag{11}$$

where $R_S(e,e')$ measures the *static relatedness* between e and e', $R_D(e,e',t)$ measures the *dynamic relatedness* between e and e' w.r.t. t and λ is a parameter, which is set as 0.2 by default and will be discussed in detail in the experiments. For the special case that $e = e'$, we define $P(e'|e,t) = 1$.

For each pair of entities e and e', we calculate their *static relatedness* $R_S(e,e')$ by adopting the Wikipedia link based measure introduced by [13] as

$$R_S(e,e') = 1 - \frac{\log(\max(|E|,|E'|)) - \log(|E \cap E'|)}{\log(|W|) - \log(min(|E|,|E'|))} \tag{12}$$

where E and E' are the sets of entities that link to e and e' respectively, and W is the set of all entities.

In order to measure the *dynamic relatedness* $R_D(e,e',t)$, we propose a novel approach based on *entity co-occurrence* in Web documents and *spike overlap* of page views, which will be discussed in the following.

Entity Co-occurrence. Based on the real-time stream of multilingual Web news articles annotated with entities, we investigate *entity co-occurrence* in the Web documents, which expresses the strength of dynamic entity association. For each pair of e and e' w.r.t. t, we calculate the entity co-occurrence measure $EC(e,e',t)$ by adopting the method of χ^2 hypothesis test introduced by [18] as

$$EC(e,e',t) = \frac{N(t)(C(e,e',t)C(\bar{e},\bar{e'},t) - C(e,\bar{e'},t)C(\bar{e},e',t))^2}{C(e,t)C(e',t)(N(t) - C(e,t))(N(t) - C(e',t))} \tag{13}$$

where $N(t)$ is the total number of Web documents published within the date range t, $C(e,e',t)$ denotes the co-occurrence frequency of e and e' in the Web documents within t, $C(e,t)$ and $C(e',t)$ denote the frequencies of e and e' occurring in the Web documents within t, respectively, and $\bar{e}$, $\bar{e'}$ indicate that e, e' do not occur in Web documents, i.e., $C(\bar{e},\bar{e'},t)$ is the number of documents within t where neither e nor e' occurs, and $C(e,\bar{e'},t)$ $(C(\bar{e},e',t))$ denotes the number of documents within t where e (e') occurs but e' (e) does not.

Spike Overlap. Based on the page view spike of entities, we propose *spike overlap* $SO(e,e',t)$ to affect the dynamic relatedness between entities e and e' w.r.t. t. The intuition is that the page view spike of e and e' on the same date d will contribute to the dynamic relatedness between e and e'. In this regard, we calculate $SO(e,e',t)$ by adopting the weighted Jaccard similarity as

$$SO(e,e',t) = \frac{\sum_{d \in \mathcal{I}} \min\{S(e,d), S(e',d)\}}{\sum_{d \in t} \max\{S(e,d), S(e',d)\}} \tag{14}$$

where $\mathcal{I}$ can be defined as the given date range t, i.e., $\mathcal{I} = t$. However, the above defined measure is only based on page view spikes of entities and thus suffers from the situation that entities with significant page view spike on the same date might not be associated in reality. Therefore, we construct the date set $\mathcal{I}$ as

$$\mathcal{I} = \{d | C(e, e', d) \geq \tau, d \in t\} \tag{15}$$

where the co-occurrence frequency $C(e, e', d)$ of e and e' in the Web documents published on d has to exceed a threshold τ, which helps to determine if the page view spike overlap is more likely to indicate an association between e and e' than just by chance. Based on our observation, it is reasonable to set τ as 10.

By taking both *entity co-occurrence* in Web documents and *spike overlap* of page views into consideration, we calculate the *dynamic relatedness* $R_D(e, e', t)$ between entities e and e' for a specific date range t as follows

$$R_D(e, e', t) = EC(e, e', t) \cdot SO(e, e', t)^2 \tag{16}$$

3.4   Mention Model $\Phi_{s|e}$

The distribution $P(s|e)$ models the likelihood of observing the mention s given the intended entity e. To estimate $P(s|e)$, we employ Wikipedia and Wikilinks datasets and propose a point-wise mutual information (PMI) based method as

$$P(s|e) = \frac{PMI(e, s)}{\sum_{s_i \in S_e} PMI(e, s_i)} \tag{17}$$

where S_e is the set of surface forms of entity e and $PMI(e, s)$ is calculated as

$$PMI(s, e) = \log \frac{P(s, e)}{P(s)P(e)} = \log \frac{C(e, s) \times N}{C(s) \times C(e)} \tag{18}$$

where we have $P(s) = \frac{C(s)}{N}$, $P(e) = \frac{C(e)}{N}$, $P(s, e) = \frac{C(e, s)}{N}$ based on maximum likelihood estimation (MLE), $C(s)$ is the number of links using s as anchor text, $C(e)$ is the number of links pointing to e, $C(e, s)$ is the number of links using s as anchor text pointing to e and N is the total number of links.

3.5   Context Model $\Phi_{c|e,t}$

The probability $P(c|e, t)$ models the likelihood of observing the context c given the query entity e and the date range t. The context c of e contains the surface forms of other entities related to e. Assuming that all surface forms s_c in the context c are independent given e and t, the probability $P(c|e, t)$ is estimated as

$$P(c|e, t) = \prod_{s_c \in c} P(s_c|e, t) \tag{19}$$

The problem remains to estimate $P(s_c|e, t)$, the probability that a surface form s_c appears in the context of e w.r.t. t.

Given the query entity e and date range t, we consider a generation process of the context, where the context model first finds the *related entities* of e w.r.t. t based on the relatedness model, and then generates the surface form s_c of such related entities as the context of e based on the mention model. The form of the context generation for the query entity e and date range t is given as

$$P_R(s_c|e,t) = \sum_{e_{s_c} \in E_{s_c}} P(e_{s_c}, s_c|e,t) = \sum_{e_{s_c} \in E_{s_c}} \underbrace{P(e_{s_c}|e,t)}_{\text{Relatedness}} \underbrace{P(s_c|e_{s_c})}_{\text{Mention}} \qquad (20)$$

where E_{s_c} denotes the set of entities having surface form s_c and we assume that s_c is independent of e and t given e_{s_c}, i.e., $P(s_c|e_{s_c}, e, t) = P(s_c|e_{s_c})$.

The above estimation suffers from the sparse data problem, i.e., some entities are not related to a given query entity e, but might appear as the context of e in the query q, which results in zero probability. Therefore, we perform smoothing by giving some probability mass to such unrelated entities. The general idea is that a surface form s_c of entities that are not related to the query entity e should also be possible to appear in the context of e and can be generated by chance. In this regard, we define the probability $P(s)$ of surface form s, which is built from the *entire collection* of entities and surface forms, as

$$P(s) = \frac{\sum_{e \in E_s} C(e, s)}{\sum_{s_i \in S} \sum_{e_i \in E_{s_i}} C(e_i, s_i)} \qquad (21)$$

where S is the set of all surface forms, E_s is the set of entities having surface form s, and $C(e, s)$ denotes the frequency that s refers to e.

In order to achieve a robust estimation of the context model, we further smooth $P_R(s_c|e,t)$ using $P(s)$ based on Jelinek-Mercer smoothing as follows

$$P(s_c|e,t) = \gamma P_R(s_c|e,t) + (1 - \gamma)P(s_c) \qquad (22)$$

where γ is a tunable parameter that is set to 0.9 by line search in our experiments. This estimation mixes the probability of s_c derived from the related entities of e with the general collection frequency of s_c used to refer to any entities.

4   Evaluation

We now discuss the experiments we have conducted to assess the performance of our approach to TER based on our newly created benchmark datasets.

4.1   Experimental Setup

In our experiments, we employ DBpedia 2014[6] as the knowledge base and the Wikipedia snapshot of June 2014 as the auxiliary data source. Existing datasets for the evaluation of entity recommendation aim to quantify the degree to which

---

[6] http://wiki.dbpedia.org/Downloads2014.

entities are related to the query without involving temporal aspects, which makes such datasets unsuitable for the TER task. There are some studies using a subset of TREC queries for time-aware information retrieval, where the goal is to investigate the user's implicit temporal intent for document retrieval [19,20]. However, such datasets do not contain the time ranges of interest explicitly given by users along with the queries and thus cannot be used for the TER evaluation. Therefore, we have created a new dataset where we asked 6 volunteers, who also serve as judges of the experimental results, to provide information needs of both queries and date ranges. By removing the duplicate ones, it results in a final set of 22 information needs in different domains including Sports, Entertainment, Business, Emergencies, Society, Science and Politics. The datasets used in our experiments are available at http://km.aifb.kit.edu/sites/ter/.

To the best of our knowledge, no existing work on the TER task can be found. Therefore, we build the following baselines for comparison with our approach: (1) the first baseline is a static method using an ad hoc ranking function without considering the given time range t, defined as $Score(e, q) = \sum_{e_q} C(e_q) R_S(e_q, e)$, where $C(e_q)$ represents the commonness of each query entity e_q w.r.t. the corresponding mention in the query q, which has been introduced by [11,21], and $R_S(e_q, e)$ denotes the Wikipedia link based relatedness between each query entity e_q and the candidate entity e [13]; (2) the second baseline is similar to our probabilistic model, but without taking into account the time range t, defined as $P(e, q) = \sum_{e_q} P(e) P(e_q|e) P(s_{e_q}|e_q) P(ce_q|e_q)$, where $P(e)$ and $P(s_{e_q}|e_q)$ are estimated using our popularity and mention models respectively, $P(e_q|e)$ and $P(ce_q|e_q)$ are also estimated using our relatedness and context models, but with $\lambda = 1$ (see Eq. 11), i.e., only the static relatedness between entities is considered in these models. For a comparative analysis, we have conducted the experiments with several methods: the above described two baselines, denoted by *BSL1* and *BSL2*, respectively; our proposed method leaving out each of the popularity, temporality, relatedness, mention and context models, denoted by $-\Phi_e$, $-\Phi_{t|e}$, $-\Phi_{e'|e,t}$, $-\Phi_{s|e}$ and $-\Phi_{c|e,t}$, respectively; and our method with all these five models, denoted by *Full Model*.

The existing work, such as the Spark system from Yahoo! [4] and the similar one published by Microsoft [9,10], could also be used for comparison with our method, even though they are not dedicated to the TER task. However, these systems assume that a query refers to only one entity, so they cannot deal with our more general case, where the query could involve multiple query entities. More importantly, these systems rely on the datasets that only major Web search engines have and are not publicly accessible. Due to these reasons, it is difficult to re-implement such systems and compare them with our method.

4.2   Results of Entity Retrieval

To assess the quality of entities retrieved by our method, we employ Normalized Discounted Cumulative Gain (nDCG) at rank k [22] as quality criteria, which is defined as $nDCG@k = \frac{DCG@k}{IDCG@k}$, where $DCG@k = \sum_{i=0}^{k} \frac{2^{rel_i} - 1}{log_2(i+1)}$ and rel_i is

Table 1. $nDCG@k$ of retrieved entities (with the best results in bold)

	nDCG@k							
	$BSL1$	$BSL2$	$-\Phi_e$	$-\Phi_{t\|e}$	$-\Phi_{e'\|e,t}$	$-\Phi_{s\|e}$	$-\Phi_{c\|e,t}$	Full Model
$k=5$	0.597	0.622	0.805	0.778	0.140	0.800	0.797	**0.824**
$k=10$	0.594	0.621	0.817	0.786	0.176	0.803	0.804	**0.839**
$k=15$	0.596	0.640	0.846	0.810	0.505	0.830	0.823	**0.859**
$k=20$	0.616	0.642	0.865	0.831	0.521	0.853	0.847	**0.879**
$k=30$	0.635	0.658	0.898	0.877	0.552	0.895	0.887	**0.925**

Table 2. $Recall@k$ of temporally related entities (with the best results in bold)

	Recall@k							
	$BSL1$	$BSL2$	$-\Phi_e$	$-\Phi_{t\|e}$	$-\Phi_{e'\|e,t}$	$-\Phi_{s\|e}$	$-\Phi_{c\|e,t}$	Full Model
$k=5$	0.273	0.264	0.464	0.464	0.091	0.491	0.491	**0.518**
$k=10$	0.318	0.309	0.582	0.591	0.146	0.591	0.600	**0.646**
$k=15$	0.318	0.336	0.655	0.655	0.182	0.700	0.700	**0.736**
$k=20$	0.346	0.346	0.709	0.682	0.255	0.746	0.736	**0.755**
$k=30$	0.364	0.364	0.791	0.827	0.318	0.846	0.809	**0.855**

the graded relevance assigned to the result at position i and $IDCG@k$ is the maximum attainable $DCG@k$. This measure captures the goodness of a retrieval model based on the graded relevance of the top-k results. For each information need, all the entities retrieved by different methods are judged on 1–5 relevance scale by the 6 volunteers based on the criteria including both relevance and timeliness w.r.t. the underlying information needs. The final relevance of each candidate entity is determined by the relevance score voted by most judges and ties are resolved by the authors. More details about the description of each graded relevance are available in our datasets.

The experimental results of $nDCG@k$ with varying k for different methods are shown in Table 1. Our method with *Full Model* performs the best for different k. Compared with the static baseline $BSL2$ using a similar probabilistic model, it achieves 32.5 %, 35.1 %, 34.2 %, 36.9 % and 40.6 % improvements when k is 5, 10, 15, 20 and 30, respectively. The baselines only obtain better results compared with our method without the relatedness model, while our method leaving out any other model still greatly outperforms the baselines. By comparing the two static baselines, $BSL2$ clearly outperforms $BSL1$, which also shows the advantage of the method based on our probabilistic model over the ad hoc method.

As we focus on the TER task, the capability of our method to find temporally related entities is of great importance such that we have created an additional dataset consisting of only temporally related entities, which are also determined based on the votes of the 6 judges. Firstly, they are asked to select the entities

Table 3. The gold-standard ranking of 10 entities (with dynamically related ones in bold) for the query *"Germany Brazil"* and the date range *"July 2014"* as well as the rankings by the baseline *BSL2* and our method with *Full Model*

Gold standard	BSL2	Full model
Germany nat'l football team	Latin America	**Brazil nat'l football team**
Brazil nat'l football team	**Brazil nat'l football team**	**Germany nat'l football team**
2014 FIFA World Cup	Brazil nat'l basketball team	**2014 FIFA World Cup**
Joachim Löw	**2014 FIFA World Cup**	**Luiz Felipe Scolari**
Toni Kroos	**Germany nat'l football team**	FIFA World Rankings
Luiz Felipe Scolari	FIFA World Rankings	**Toni Kroos**
Neymar	**Luiz Felipe Scolari**	Neymar
FIFA World Rankings	**Neymar**	**Joachim Löw**
Latin America	Joachim Löw	Latin America
Brazil nat'l basketball team	**Toni Kroos**	Brazil nat'l basketball team

that are temporally related to each information need and such entities are then ranked by the number of times being selected. Only the top-5 ranked candidates are included into the final dataset, where ties are resolved by the authors. This results in 110 entities in total (5 for each of the 22 information needs).

In this experimental setting, we are concerned with whether these temporally related entities can appear on top of the ranked list of the retrieved entities. For this, we consider recall at rank k ($recall@k$) as quality criteria, where recall defines the number of relevant results that are retrieved in relation to the total number of relevant results and $recall@k$ is defined by only taking into account the top-k results. The experimental results of $recall@k$ with varying k for different methods are shown in Table 2. While the two static baselines exhibit only minor differences, our method with *Full Model* achieves a considerable performance improvement over the baselines for different k.

For both measures of $nDCG@k$ and $recall@k$, we observe that our method achieves better results by adding each individual model and the relatedness model that incorporates both static and dynamic entity relatedness contributes the most. For example, when $k = 30$, $nDCG@k$ and $recall@k$ decrease 40.1 % and 62.8 % respectively, by ablating the relatedness model, while the performance reduction without the other models ranges from 5.2 % to 2.9 % for $nDCG@k$ and from 7.5 % to 1.1 % for $recall@k$.

4.3 Results of Entity Ranking

The measures of nDCG@k and recall@k assess the quality of only top-k results, while we would like to evaluate the ranking of entities from highly relevant ones to only remotely relevant or even not relevant ones. Therefore, we have created another dataset, where the authors select 10 candidate entities for each information need in a way that their relevances are clearly distinguishable among each other. Similar to [23], the gold-standard ranking of the 10 candidate entities is then created in the following way: (1) for all possible comparisons of the

Table 4. Spearman rank correlation between the gold-standard ranking and the ranking generated by different methods (with the best results in bold)

Domain (#Query)	$BSL1$	$BSL2$	$-\Phi_e$	$-\Phi_{t\|e}$	$-\Phi_{e'\|e,t}$	$-\Phi_{s\|e}$	$-\Phi_{c\|e,t}$	Full model
Sports (6)	0.149	0.289	0.531	0.572	0.240	0.646	0.529	**0.663**
Entertainment (4)	0.191	0.252	0.594	0.645	0.188	0.667	0.673	**0.688**
Business (3)	0.596	0.596	0.790	0.834	−0.139	0.838	**0.855**	0.838
Emergencies (4)	−0.130	−0.082	0.473	0.421	0.470	0.503	0.467	**0.494**
Others (5)	0.365	0.358	**0.612**	0.522	0.232	0.576	0.527	0.581
Average	0.216	0.272	0.586	0.582	0.219	0.634	0.588	**0.642**

10 candidate entities (45 in total), the 6 judges are asked which of the given two entities is more related to the information need by considering both relevance and timeliness; (2) all comparisons are then aggregated into a single confidence value for each entity and the 10 candidate entities are ranked by these confidence values as described by [24]. The final output is a set of 22 ranked lists consisting of 10 entities for each, against which we compare the automatically generated rankings by different methods using Spearman rank correlation, which measures the strength of association between two ranked variables. Some examples of different rankings are shown in Table 3.

The Spearman rank correlation between the gold-standard ranking and the automatically generated rankings by all these methods is given in Table 4. It shows that the experimental results of entity ranking are consistent with the results obtained in the entity retrieval experiments. The static baseline *BSL2* with a probabilistic model yields slightly better results than the baseline *BSL1* that is based on an ad hoc method. Clearly, our method with *Full Model* achieves the best results and considerably outperforms the baselines. Similarly, all the individual models contribute to the final performance improvement, where the relatedness model contributes the most. By respectively ablating the models Φ_e, $\Phi_{t\|e}$, $\Phi_{e'\|e,t}$, $\Phi_{s\|e}$ and $\Phi_{c\|e,t}$, the performance correspondingly reduces 8.7%, 9.3%, 65.8%, 1.2% and 8.4%.

Our method is sensitive to the parameter λ used in the relatedness model (see Eq. 11). Intuitively, a smaller λ reflects that the dynamic entity relatedness

Table 5. Spearman rank correlation between the gold-standard ranking and the ranking by our *Full Model* for different λ (with the best results in bold)

Domain	$\lambda = 0$	$\lambda = .1$	$\lambda = .2$	$\lambda = .3$	$\lambda = .4$	$\lambda = .5$	$\lambda = .6$	$\lambda = .7$	$\lambda = .8$	$\lambda = .9$	$\lambda = 1$
Sports	0.620	0.653	**0.663**	0.636	0.634	0.610	0.604	0.564	0.541	0.489	0.285
Entertainment	0.573	0.670	**0.688**	0.636	0.612	0.530	0.512	0.473	0.445	0.439	0.348
Business	0.737	0.838	0.838	**0.842**	**0.842**	**0.842**	0.822	0.826	0.834	0.794	0.657
Emergencies	**0.530**	0.518	0.494	0.509	0.467	0.479	0.458	0.412	0.367	0.303	−0.058
Others	0.537	0.576	**0.581**	0.564	0.537	0.503	0.505	0.534	0.537	0.493	0.280
Average	0.592	0.639	**0.642**	0.625	0.606	0.579	0.568	0.549	0.531	0.489	0.284

measure plays a more important role in the model. Table 5 shows the impact of λ on the ranking performance of our method with *Full Model*, where $\lambda = 0.2$ yields the best results on average, which has been used as the default value in our experiments. We observe that only using the dynamic relatedness measure, i.e., $\lambda = 0$, achieves the best results for the *Emergencies* domain. This is because in this domain there are more entities that are only dynamically related to the query. For example, given the information need about the crash of Indonesia AirAsia Flight 8501 into the Java sea in December 2014, where the query is *"Indonesia Java"* and the date range is *"December 2014"*, the related entities AirAsia, Aviation_accidents_and_incidents and Search_and_rescue do not have a static connection with the query. Another tunable parameter is γ (see Eq. 22). We observe that $\gamma = 0.9$ achieves the best results, which has been set as the default value in our experiments. For the sake of space, we omit the results based on different γ because they exhibit only minor differences.

5   Related Work

The TER task can be placed in the context of (1) entity search, (2) related entity recommendation and (3) temporal information retrieval.

Entity search has been defined by [1] as finding entities explicitly named in the query. Recently, entity search becomes more complex and closer to question answering when the query only provides a description of the target entity, where a list of member relationships to a single entity is given in the query. A recent development in evaluating entity search of this type was the introduction of the Related Entity Finding using Linked Open Data (REF-LOD) task at the TREC Entity Track in 2010 and 2011 [25], where the type of relation to the target entity and the type of the target entity are both given as constraints.

For *related entity recommendation*, the Spark system developed at Yahoo! extracts several features from a variety of data sources and uses a machine learning model to produce a recommendation of entities to a Web search query, where neither the relation type nor the type of the target entity are specified [4]. Following Spark, Sundog aims to improve entity recommendation, in particular with respect to freshness, by exploiting Web search log data using a stream processing based implementation [8]. Microsoft has also developed a similar system that performs personalized entity recommendation by analyzing user click logs and entity pane logs [9,10].

In recent years, the time dimension has received a large share of attention in *temporal information retrieval* [5]. The temporal characteristics of queries [26] and dynamics of document content [27] have been leveraged in relevance ranking. The real-time information extracted from Twitter has been used to train learning to rank models [28]. To improve Web search results, the temporal information has also been used for query understanding [29] and auto-completion of queries [30].

6   Conclusions

In this paper, we introduce a novel task of *time-aware entity recommendation* (TER), since we argue that time-awareness should be a crucial factor in entity recommendation, which has not been addressed so far. To tackle this challenge, we propose a probabilistic model that aims to rank related entities according to a time-specific information need presented as a keyword query and a date range. The main contribution of our approach is that we decompose the TER task into several well defined probability distributions, each representing the context of a different component in the model. Through these components, heterogeneous entity knowledge extracted from different data sources that are publicly available on the Web can be incorporated into our model. Experimental results show that our method clearly outperforms approaches that are not context-aware, specifically when being time-agnostic.

Acknowledgments. The research leading to these results has received funding from the European Union Seventh Framework Programme (FP7/2007–2013) under grant agreement no. 611346.

References

1. Pound, J., Mika, P., Zaragoza, H.: Ad-hoc object retrieval in the web of data. In: WWW, pp. 771–780(2010)
2. van Zwol, R., Pueyo, L.G., Muralidharan, M., Sigurbjörnsson, B.: Machine learned ranking of entity facets. In: SIGIR, pp. 879–880 (2010)
3. Kang, C., Vadrevu, S., Zhang, R., van Zwol, R., Pueyo, L.G., Torzec, N., He, J., Chang, Y.: Ranking related entities for web search queries. In: WWW Companion, vol. 67–68 (2011)
4. Blanco, R., Cambazoglu, B.B., Mika, P., Torzec, N.: Entity recommendations in web search. In: Alani, H., et al. (eds.) ISWC 2013, Part II. LNCS, vol. 8219, pp. 33–48. Springer, Heidelberg (2013)
5. Kanhabua, N., Blanco, R., Nørvåg, K.: Temporal information retrieval. Found. Trends Inf. Retr. **9**(2), 91–208 (2015)
6. Nørvåg, K.: Supporting temporal text-containment queries in temporal document databases. Data Knowl. Eng. **49**(1), 105–125 (2004)
7. Berberich, K., Bedathur, S.J., Neumann, T., Weikum, G.: A time machine for text search. In: SIGIR, pp. 519–526 (2007)
8. Fischer, L., Blanco, R., Mika, P., Bernstein, A.: Timely semantics: a study of a stream-based ranking system for entity relationships. In: Arenas, M., et al. (eds.) ISWC 2015. LNCS, vol. 9367, pp. 429–445. Springer, Heidelberg (2015). doi:10.1007/978-3-319-25010-6_28
9. Yu, X., Ma, H., Hsu, B.P., Han, J.: On building entity recommender systems using user click log and freebase knowledge. In: WSDM, pp. 263–272 (2014)
10. Bi, B., Ma, H., Hsu, B.P., Chu, W., Wang, K., Cho, J.: Learning to recommend related entities to search users. In: WSDM, pp. 139–148 (2015)
11. Shen, W., Wang, J., Luo, P., Wang, M.: LINDEN: linking named entities with knowledge base via semantic knowledge. In: WWW, pp. 449–458 (2012)

12. Singh, S., Subramanya, A., Pereira, F., McCallum, A.: Wikilinks: a large-scale cross-document coreference corpus labeled via links to Wikipedia. Technical report, UM-CS-2012-015 (2012)
13. Milne, D., Witten, I.H.: An effective, low-cost measure of semantic relatedness obtained from Wikipedia links. In: AAAI Workshop on Wikipedia and Artificial Intelligence (2008)
14. Ciglan, M., Nørvåg, K.: WikiPop: personalized event detection system based on Wikipedia page view statistics. In: CIKM, pp. 1931–1932 (2010)
15. Trampuš, M., Novak, B.: Internals of an aggregated web news feed. In: SiKDD, pp. 431–434 (2012)
16. Zhang, L., Rettinger, A.: X-LiSA: Cross-lingual semantic annotation. PVLDB 7(13), 1693–1696 (2014)
17. Osborne, M., Petrovic, S., McCreadie, R., Macdonald, C., Ounis, I.: Bieber no more: first story detection using Twitter and Wikipedia. In: SIGIR 2012 Workshop on Time-aware Information Access (2012)
18. Bron, M., Balog, K., de Rijke, M.: Ranking related entities: components and analyses. In: CIKM, pp. 1079–1088 (2010)
19. Li, X., Croft, W.B.: Time-based language models. In: CIKM, pp. 469–475(2003)
20. Kanhabua, N., Nørvåg, K.: Determining time of queries for re-ranking search results. In: Lalmas, M., Jose, J., Rauber, A., Sebastiani, F., Frommholz, I. (eds.) ECDL 2010. LNCS, vol. 6273, pp. 261–272. Springer, Heidelberg (2010)
21. Milne, D.N., Witten, I.H.: Learning to link with Wikipedia. In: CIKM, pp. 509–518 (2008)
22. Järvelin, K., Kekäläinen, J.: IR evaluation methods for retrieving highly relevant documents. In: SIGIR, pp. 41–48 (2000)
23. Hoffart, J., Seufert, S., Nguyen, D.B., Theobald, M., Weikum, G.: KORE: keyphrase overlap relatedness for entity disambiguation. In: CIKM, pp. 545–554 (2012)
24. Coppersmith, D., Fleischer, L., Rudra, A.: Ordering by weighted number of wins gives a good ranking for weighted tournaments. ACM Trans. Algorithms 6(3), Article no. 55 (2010)
25. Balog, K., Serdyukov, P., de Vries, A.P.: Overview of the TREC 2011 entity track. In: TREC (2011)
26. Dai, N., Davison, B.D.: Freshness matters: in flowers, food, and web authority. In: SIGIR, pp. 114–121 (2010)
27. Elsas, J.L., Dumais, S.T.: Leveraging temporal dynamics of document content in relevance ranking. In: WSDM, pp. 1–10 (2010)
28. Dong, A., Zhang, R., Kolari, P., Bai, J., Diaz, F., Chang, Y., Zheng, Z., Zha, H.: Time is of the essence: improving recency ranking using Twitter data. In: WWW, pp. 331–340 (2010)
29. Kulkarni, A., Teevan, J., Svore, K.M., Dumais, S.T.: Understanding temporal query dynamics. In: WSDM, 167–176 (2011)
30. Shokouhi, M., Radinsky, K.: Time-sensitive query auto-completion. In: SIGIR, pp. 601–610 (2012)

A Knowledge Base Approach to Cross-Lingual Keyword Query Interpretation

Lei Zhang$^{(\boxtimes)}$, Achim Rettinger, and Ji Zhang

Institute AIFB, Karlsruhe Institute of Technology (KIT), Karlsruhe, Germany
l.zhang@kit.edu

Abstract. The amount of entities in large knowledge bases available on the Web has been increasing rapidly, making it possible to propose new ways of intelligent information access. In addition, there is an impending need for technologies that can enable cross-lingual information access. As a simple and intuitive way of specifying information needs, keyword queries enjoy widespread usage, but suffer from the challenges including *ambiguity*, *incompleteness* and *cross-linguality*. In this paper, we present a knowledge base approach to cross-lingual keyword query interpretation by transforming keyword queries in different languages to their semantic representation, which can facilitate query disambiguation and expansion, and also bridge language barriers. The experimental results show that our approach achieves both high efficiency and effectiveness and considerably outperforms the baselines.

1 Introduction

The ever-increasing quantities of entities in large knowledge bases (KBs), such as Wikipedia, DBpedia, Freebase and YAGO, pose new challenges but at the same time open up new opportunities of intelligent information access on the Web. In recent years, many research activities involving *entities* have emerged, such as entity tagging/extraction from texts and entity linking/disambiguation with KBs. Furthermore, there is an increasing portion of Web search queries involving entities. For example, through query log analysis, Pound et al. [1] found that more than half of Web queries are related to entities. In this regard, the exploitation of *entities and their relations* in information retrieval (IR) research beyond the term-based paradigm has become an area of particular interest. Recently, almost every major commercial Web search engine has announced their work on incorporating entity information from knowledge bases into its search process, including Google's Knowledge Graph, Yahoo!'s Web of Objects and Microsoft's Satori Graph / Bing Snapshots.

Within the context of globalization, *multilingual* and *cross-lingual* access to information has drawn increasing attention. Nowadays, more and more people from different countries are connecting to the Internet and many Web users are able to understand more than one language, e.g., more than half of the citizens in the European Union can speak at least one other language than their mother tongue. While the diversity of languages on the Web has been growing in recent

© Springer International Publishing AG 2016
P. Groth et al. (Eds.): ISWC 2016, Part I, LNCS 9981, pp. 615–631, 2016.
DOI: 10.1007/978-3-319-46523-4_37

years, for most people there is still very little content in their native language. As a consequence of the ability to understand more than one language, users are also interested in Web content in other languages.

In addition, keyword search has proven to be a simple and intuitive paradigm for expressing information needs of users. However, traditional keyword search systems mainly suffer from the following challenges.

Ambiguity. Keyword queries are naturally ambiguous due to the fact that keywords could refer to different things in different contexts. In the multilingual and cross-lingual settings, this problem is more serious, e.g., "*WM*" could refer to the entity Windows_Mobile in English and FIFA_World_Cup in German[1].

Incompleteness. Keyword queries are often incomplete in the sense that instead of the full entity names, only the aliases, acronyms and misspellings are usually given in the queries. In addition, keyword queries might contain concept names representing a set of entities, e.g., "*Internet companies of China*".

Cross-linguality. Multilingual users probably formulate their information needs using native language. However, they are interested in relevant information in any language that they can understand. In some other cases, multilingual users could issue queries consisting of keywords in multiple languages. For example, Chinese users might represent a foreign company using its original name and a local company using its Chinese name, such as "*Google 百度*" with the aim of finding the relationship between Google and Baidu, the largest search engines for English and Chinese, respectively. In addition, specifying the query language should not be the burden of users, which poses new challenges since existing techniques for language detection, such as the well-known character n-gram probability language model, do not work well for short keyword queries [2].

In order to address these challenges, we present a knowledge base approach to cross-lingual keyword query interpretation. The goal is to find entity graphs in the KB matching the keyword query, called *query entity graphs* (QEG), which reflect different semantic interpretations of the keyword query. More specifically, our approach aims to eliminate the ambiguity of keyword queries by exploiting the semantic graph of the KB to generate the top-k QEGs. It supports keyword queries matching entities in their incomplete forms, such as aliases, acronyms and misspellings instead of the full names. In addition, the matching concepts in keyword queries are automatically expanded into sets of associated entities. To the best of our knowledge, this is the first work that allows users to issue keyword queries in any language, which can even contain keywords in multiple languages, for finding the query interpretations grounded in any other languages.

It is noteworthy that this work has been incorporated into XKnowSearch![2], a novel system to entity-based cross-lingual information retrieval (IR) [3]. With the help of the resulting QEGs, XKnowSearch! allows users to further explore entity relations to refine the queries. For bridging the language barriers between queries and documents, XKnowSearch! leverages the cross-lingual query interpretation

---

[1] *WM* is the abbreviation of *Weltmeisterschaft* in German, which means *World Cup*.

[2] http://km.aifb.kit.edu/sites/XKnowSearch/.

technique in this paper and a cross-lingual semantic annotation system [4] to construct semantic representation of keyword queries and documents in different languages, which are then used for document retrieval.

The main contributions of this paper are: (1) the introduction of a *knowledge base approach to cross-lingual query interpretation* by representing information needs of users as entity graphs to *address the challenges* of traditional keyword search; (2) a *scoring mechanism* for *effective query interpretation ranking* by exploiting various structures in the multilingual KB; (3) a new *top-k query graph exploration algorithm* aimed for *efficient query interpretation generation*; and (4) a *separate evaluation* of the ranking mechanism and the top-k graph exploration algorithm to show that both of them lead to a *considerable improvement* over the baseline methods on *effectiveness and efficiency*, respectively.

The rest of the paper is organized as follows. We firstly introduce the problem in Sect. 2 and provide an overview of our approach in Sect. 3. Details on the scoring mechanism and the top-k query graph exploration algorithm are then presented in Sects. 4 and 5, respectively. Experimental results are presented in Sect. 6. Finally, we survey the related work in Sect. 7 and conclude in Sect. 8.

2   Problem Definition

We deal with the scenarios where queries formulated by users are sets of keywords in any language or even in multiple languages, which are unknown in advance. Given such queries, we first introduce the concepts of *key term* and *key term set* and then define the *query entity graph* (QEG) as the interpretation of a query.

Definition 1 *(Key Term and Key Term Set). Given a query Q consisting of a sequence of keywords $\langle k_1, \cdots, k_n \rangle$, a key term $t = \langle k_i, \cdots, k_j \rangle$ is a subsequence of Q with the start index $start(t) = i$ and the end index $end(t) = j$, for which at least one matching entity or concept can be found in the knowledge base. A key term set $T = \{t_1, \cdots, t_m\}$ is a set of* non-overlapping *key terms resulting from Q such that for any t and t' in T either $start(t) \leq end(t')$ or $end(t) \geq start(t')$.*

For example, the keywords *"online companies of US"* could result in many key terms like *online, companies, online companies, US* and *online companies of US*, which could lead to different key term sets, such as {*online, companies, US*} and {*online companies of US*}. The key terms like *online* and *US* could refer to the entities Online_game and United_States, respectively, while *online companies of US* might refer to the concept Internet_companies_of_the_United_States, which has a list of associated entities belonging to it, such as Google, Yahoo! and EBay.

We consider the KB as a directed graph $G_{KB}(N, E)$, where each node $n \in N$ represents an entity and each edge $e(n_i, n_j) \in E$ denotes the relation between entities n_i and n_j. Given the key term sets resulting from a keyword query Q, the query interpretation of Q, i.e., the query entity graph, is defined as follows:

Definition 2 *(Query Entity Graph). A query entity graph (QEG) to a keyword query Q, denoted by $G_Q = (N_Q, E_Q)$, is a subgraph of $G_{KB}(N, E)$, which*

618   L. Zhang et al.

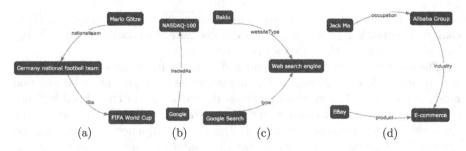

Fig. 1. Example QEGs generated by our system for the queries (a) *"WM Götze"*, (b) *"online companies of US NDX"*, (c) *"Google* 百度 and (d) *"eBay* 马云"

satisfies the following conditions: (1) there exists at least one key term set T and for each key term $t \in T$ there is at least one entity $n_t \in N$ that matches t. The set of matching entities containing one for every $t \in T$ is $N_T \subseteq N_Q$; (2) for every possible pair $n_i, n_j \in N_T$ and $n_i \neq n_j$, there is a path $n_i \leadsto n_j$, i.e., an edge $e(n_i, n_j) \in E$ or a sequence of edges $e(n_i, n_k) \ldots e(n_l, n_j)$ in E, such that every $n_i \in N_T$ is connected to every other $n_j \in N_T$.

Problem. We are concerned with the computation of QEGs from keywords in any language or even in multiple languages. Given a query Q, the goal is to find the top-k ranked QEGs, where the ranking is produced by the application of a scoring function $S : G_Q \to s$. For any given QEG G_Q, S assigns a score s that captures the degree to which G_Q matches the information need of users.

Some examples of the top-ranked QEGs generated by our system for different queries are shown in Fig. 1. To avoid the users' burden of specifying the query languages, our approach does not assume any input language given by users for all the queries. In the query *"WM Götze"*, the keyword *"WM"*, which could refer to 212 entities in German and 11 entities in English, has been disambiguated as FIFA_World_Cup based on the relation to Mario_Götze. Regarding the query *"online companies of US NDX"*, the alias *"online companies of US"* referring to the concept Internet_companies_of_the_United_States has been resolved to the entity Google, which is listed in NASDAQ-100 referred to by the acronym *"NDX"*. For the multilingual queries *"Google* 百度" and*"eBay* 马云", our approach can deal with them by supporting query keywords in multiple languages.

3   Overview of the Approach

In this section, we provide an overview of the off-line preprocessing and online computation required in our approach to cross-lingual query interpretation.

Preprocessing. In this work, we use DBpedia as the knowledge base, which is a crowd-sourced community effort to extract structured information from Wikipedia in different languages. In the following, we briefly introduce the offline cross-lingual grounding extraction, where we construct the cross-lingual lexica[3] by exploiting multilingual Wikipedia to extract the cross-lingual groundings of DBpedia entities and concepts, which correspond to Wikipedia articles and categories, respectively. As Wikipedia provides several useful structures, such as titles of pages, redirect pages, disambiguation pages and link anchors, which associate entities and concepts in DBpedia with terms including words and phrases, also called *labels or surface forms*, all of them can be used to refer to the corresponding resources. In addition, Wikipedia pages in different languages that provide information about the equivalent resources are often connected through the cross-language links. Based on the above sources, for each DBpedia entity or concept grounded in one language we extract its possible surface forms in different languages. More details can be found in our previous work [5,6]. The cross-lingual lexica and the knowledge extracted from DBpedia are indexed for online computation. Based on such indexed data, we are concerned with ranking the query interpretations effectively and propose a scoring mechanism for it, which will be discussed in Sect. 4.

Query Interpretation Computation. In order to compute the QEGs as query interpretations for a keyword query Q, all the key terms are first extracted from Q based on the cross-lingual lexica, which has been also used for finding the matching entities n_t for each key term t, where either t can be used to refer to n_t directly or n_t belongs to a concept that can be referred to by t. Such key terms then result in different key term sets, each of which reflects one possible information need of users. For each key term set T and all the matching entities of its key terms, the exploration of the knowledge graph G_{KB} starts from each matching entity n_t of a key term $t \in T$ to find a connecting element, denoted by n_c, namely an entity that connects at least one starting entity n_t for all $t \in T$. Once a connecting element n_c is found, a QEG can be constructed from a set of paths that start at each n_t and meet at n_c. This process of exploration continues until the top-k QEGs have been achieved. In this paper, we are concerned with performing this query interpretation computation efficiently and propose a new top-k graph exploration algorithm, which will be discussed in Sect. 5.

4   Query Graph Scoring

A keyword query could result in many QEGs all corresponding to possible query interpretations. This section introduces a scoring mechanism that aims to assess the relevance of QEGs for *effective query interpretation ranking*.

4.1   Key Term Set Score

Our approach supports query keywords in multiple languages and we assume that the languages of keywords in a query Q are unknown, such that key terms extracted from Q could be entity/concept names in any language. Therefore,

---

[3] http://km.aifb.kit.edu/sites/xlid-lexica/.

for each language L, we define the probability $P(t_L)$ that the key term t in L, denoted by $t_L{}^4$, is an entity name or a concept name as

$$P(t_L) = \frac{count_{link}(t_L)}{count_{link}(t_L) + count_{text}(t_L)} \tag{1}$$

where $count_{link}(t_L)$ denotes the number of links using t as anchor text and $count_{text}(t_L)$ denotes the frequency of t mentioned in plain text without links in Wikipedia of language L. This estimation is further smoothed by the Laplace smoothing method for the zero probability problem. As the languages of query keywords are not specified, we define the probability $P(t)$ that the key term t refers to an entity or a concept for a set of supported languages $\mathcal{L}$ as

$$P(t) = \max_{L \in \mathcal{L}} P(t_L) \tag{2}$$

All the possible key terms might result in many key term sets that reflect different information needs. Therefore, we define the score of each key term set in the following. Given a keyword query Q, for each resulting key term set T, we take into account both its *importance* and *informativeness*. In general, the more often a key term t is selected as anchor text for the corresponding resources, i.e., t has larger $P(t)$, the more likely that t is important. In addition, the more keywords in Q are covered by all key terms $t \in T$, the more likely that T is informative, since it can reflect more aspects of the initial keyword query. Based on the above observation, we calculate the score of T as

$$S(T) = \frac{\sum_{t \in T} P(t) \cdot \sum_{t \in T} |t|}{|T|} \tag{3}$$

where $|t|$ is the number of keywords in t and $|T|$ is the number of key terms in T. While $\sum_{t \in T} P(t)$ reflects the *importance* of T, $\sum_{t \in T} |t|$ captures its *informativeness*. The denominator $|T|$ is a normalization factor used to reduce the advantage of T with more key terms. For example, {*online, companies, US*} might result in a larger numerator compared with {*online companies of US*}.

4.2   Entity Matching Score

For each key term t, there might be many entities that can be referred to by t. Assuming that t is in language L, denoted by t_L, we define the probability $P(n_{L'}|t_L)$ that t_L refers to the entity $n_{L'}$ grounded in the target language L' as

$$P(n_{L'}|t_L) = \frac{count_{link}(n_L, t_L) \cdot \tau(n_L, n_{L'})}{\sum_{n_L \in N_L} count_{link}(n_L, t_L)} \tag{4}$$

where $count_{link}(n_L, t_L)$ denotes the number of links using t_L as anchor text pointing to n_L in Wikipedia of language L and N_L is the set of entities that

---

[4] We use t for a term whose language is not observed and t_L for the same term t whose language is considered as L.

have name t_L. The language mapping function $\tau(n_L, n_{L'})$ is defined as

$$\tau(n_L, n_{L'}) = \begin{cases} 1 & \text{if } n_L \overset{LL}{\leftrightarrow} n_{L'} \text{ or } n_L = n_{L'}, \\ 0 & \text{otherwise} \end{cases} \tag{5}$$

where n_L and $n_{L'}$ are considered to be an equivalent entity if they are connected by cross-language links in Wikipedia, denoted by $n_L \overset{LL}{\leftrightarrow} n_{L'}$. Given a key term t, for which the language is not specified, we calculate the matching score of entity $n_{L'}$ based on the maximal probability $P(n_{L'}|t_L)$ as

$$S_m(n_{L'}, t) = \max_{L \in \mathcal{L}} P(n_{L'}|t_L) \tag{6}$$

In addition, for each key term t_L in language L that could be a concept name, we first map t_L to the matching concepts C_L in the same language L and then expand each C_L into a set of associated entities in the target language L', denoted by $N_{L'}^{t_L}$, based on the associations between entities and concepts as well as the cross-language links between entities available in the KB (see more details about concept matching and expansion in our TR [7]). Let $|N_{L'}^{t_L}|$ denote the number of entities in $N_{L'}^{t_L}$. For each entity $n_{L'} \in N_{L'}^{t_L}$, we calculate its score based on a uniform distribution over all entities in $N_{L'}^{t_L}$, Similarly, the matching score of entity $n_{L'}$ is calculated based on the maximal score w.r.t. t_L as

$$S_m(n_{L'}, t) = \max_{L \in \mathcal{L}} \frac{1}{|N_{L'}^{t_L}|} \tag{7}$$

4.3   Query Entity Graph Score

Given a key term set T extracted from a keyword query Q and the set of matching entities N_T containing one for each key term $t \in T$, each QEG, denoted by G_Q^T, is constructed from a set of paths that start at each $n_s \in N_T$ matching a key term $t \in T$ and meet at a *connecting element* n_c. Based on that, we introduce a scoring function to assess the relevance of QEGs as follows

$$S(G_Q^T) = \sum_{n_s \in N_T} S(T) \cdot S_m(n_s, t) \cdot S(P_{n_s \leadsto n_c}) \tag{8}$$

where $S(T)$ is the score of key term set T defined in Eq. 3, $S_m(n_s, t)$ is the matching score of entity n_s defined in Eqs. 6 and 7, and $S(P_{n_s \leadsto n_c})$ captures the score of edges $\langle n_i, n_j \rangle$ along the path $P_{n_s \leadsto n_c}$ from n_s to n_c, defined as

$$S(P_{n_s \leadsto n_c}) = \prod_{\langle n_i, n_j \rangle \in P_{n_s \leadsto n_c}} \frac{S_r(n_i, n_j) \cdot (S_p(n_i) + S_p(n_j))}{2} \tag{9}$$

where $S_r(n_i, n_j)$ measures the relatedness between entities n_i and n_j, and $S_p(n)$ reflects the popularity of entity n.

For each pair of entities n_i and n_j, we adopt the Wikipedia link-based measure described in [8] to calculate their relatedness score as follows

$$S_r(n_i, n_j) = 1 - \frac{\log(\max(|N_i|, |N_j|)) - \log(|N_i \cap N_j|)}{\log(|N|) - \log(min(|N_i|, |N_j|))} \qquad (10)$$

where N_i and N_j are the sets of entities that link to n_i and n_j respectively, and N is the set of all entities in the KB.

To measure entity popularity, we exploit both Wikipedia link structure and page view statistics. The second source captures the number of times Wikipedia pages are requested and can be treated as a query log of entities. By leveraging the two sources, we calculate the frequency of entity n as

$$freq(n) = freq_{link}(n) + \beta \cdot freq_{view}(n) \qquad (11)$$

where $freq_{link}(n)$ denotes the number of links pointing to n in Wikipedia and $freq_{view}(n)$ denotes the average number of page view requests on n per day. While $freq_{link}(n)$ represents the prior popularity of n in the KB, $freq_{view}(n)$ captures the popularity of n based on user interests. Due to the different scales between Wikipedia link frequency and page view request frequency, $freq_{view}(n)$ is adjusted by a balance parameter $\beta = \frac{\text{total number of links in Wikipedia}}{\text{average number of page views per day}}$, which accounts for the difference in frequencies of Wikipeida links and per-day page view requests. Then the popularity score of each entity $n \in N$ is calculated as

$$S_p(n) = \frac{freq(n)}{\sum_{n_i \in N} freq(n_i)} \qquad (12)$$

5   Top-K Query Graph Exploration

In this section, we present the top-k query graph exploration for *efficient query interpretation generation*. The goal is to find top-k QEGs that connect at least one entity for each key term in a key term set. For pragmatic reasons, existing solutions [9–11] use a maximal path length d_{max}, such that only paths of length d_{max} or less between entities n_i and n_j, denoted by $n_i \leadsto^{d_{max}} n_j$, will be taken into account. Such restriction has also been applied to graph exploration in this work, where d_{max} is set as 6. The algorithm is shown in Algorithm 1.

Input and Data Structures. The input to the algorithm comprises the list of top-m key term sets $LT = \{T_1, \cdots, T_m\}$ and the list $LN = \{N_{t_1}, \cdots, N_{t_n}\}$, where each N_{t_i} is a set of entities matching key term t_i. And d_{max} is the maximal path length applied to the graph exploration. For each entity n, we keep track of the information of paths from an entity n_{start} matching $t_j^i \in T_i$[5] to n, where $n.S_{t_j^i}$ is used to store each pair of the starting entity n_{start} and the score $s_{n_{start}}$ of the path from n_{start} to n, $n.s_{t_j^i}$ and $n.d_{t_j^i}$ are employed to store the maximal score

---

[5] We use t_j^i to denote a key term t_j belonging to a specific key term set T_i, while t_j represents the same key term without considering the key term sets it belongs to.

Algorithm 1. Top-k Exploration of QEGs

Input: $LT = \{T_1, \cdots, T_m\}$; $LN = \{N_{t_1}, \cdots, N_{t_n}\}$; d_{max}.
Data: $n.S_{t_j^i} = \{\langle n_1, s_{n_1}\rangle, \cdots, \langle n_l, s_{n_l}\rangle\}$; $n.s_{t_j^i}$; $n.d_{t_j^i}$; $LQ_{T_i} = \{NQ_{t_1^i}, \cdots, NQ_{t_{|T_i|}^i}\}$;

$UB_{T_i} = \{ub_{t_1^i}, \cdots, ub_{t_{|T_i|}^i}\}$; $\overline{S(G_Q^{T_i})}$; R; θ.

Result: the top-k optimal QEGs.

1 **foreach** $T_i \in LT$ **do**
2   **foreach** $t_j^i \in T_i$ **do**
3   **foreach** $n_{start} \in N_{t_j}$ **do**
4   **if** $\forall t_{k \neq j}^i \in T_i, \exists n'_{start} \in N_{t_k} : n_{start} \overset{d_{max}}{\rightsquigarrow} n'_{start}$ **then**
5          $s_{n_{start}} \leftarrow S(T_i) \cdot S_m(n_{start})$;
6          $n_{start}.S_{t_j^i}.add(\langle n_{start}, s_{n_{start}}\rangle)$;
7          $n_{start}.s_{t_j^i} \leftarrow s_{n_{start}}$;
8          $n_{start}.d_{t_j^i} \leftarrow 0$;
9          $NQ_{t_j^i}.add(n_{start})$;
10   **end**
11   **end**
12      $ub_{t_j^i} \leftarrow \max_{n \in NQ_{t_j^i}} n.s_{t_j^i}$;
13   **end**
14    $\overline{S(G_Q^{T_i})} \leftarrow \sum_{ub_{t_j^i} \in UB_{T_i}} ub_{t_j^i}$;
15 **end**
16 **while** *not all $NQ \in LQ$ are empty* **do**
17    $T_i \leftarrow \arg\max_{T_i \in LT} \overline{S(G_Q^{T_i})}$;
18    $t_j^i \leftarrow \arg\max_{t_j^i \in T_i} ub_{t_j^i}$;
19    $n \leftarrow NQ_{t_j^i}.pop()$;
20   **foreach** $n' \in n.neighbors()$ **do**
21      $n'.d_{t_j^i} \leftarrow n.d_{t_j^i} + 1$;
22   **if** $n'.d_{t_j^i} < d_{max}$ *and* $\forall t_{k \neq j}^i \in T_i, \exists n'_{start} \in N_{t_k} : n' \overset{d_{max} - n'.d_{t_j^i}}{\rightsquigarrow} n'_{start}$ **then**
23   **foreach** $\langle n_{start}, s_{n_{start}}\rangle \in n.S_{t_j^i}$ **do**
24          $s'_{n_{start}} \leftarrow s_{n_{start}} \cdot \frac{S_r(n,n') \cdot (S_p(n) + S_p(n'))}{2}$;
25          $n'.S_{t_j^i}.add(\langle n_{start}, s'_{n_{start}}\rangle)$;
26   **end**
27        $n'.s_{t_j^i} \leftarrow n'.S_{t_j^i}.maxScore()$;
28        $NQ_{t_j^i}.add(n')$;
29        $ub_{t_j^i} \leftarrow \max_{n \in NQ_{t_j^i}} n.s_{t_j^i}$;
30        $\overline{S(G_Q^{T_i})} \leftarrow \sum_{ub_{t_j^i} \in UB_{T_i}} ub_{t_j^i}$;
31   **if** $\forall t_j^i \in T_i : n'.S_{t_j^i}$ *is not empty* **then**
32          $R.add(newQEGsByMergingPath(n'))$;
33   **if** $R.size() \geq k$ *and* $\max_{T_i \in LT} \overline{S(G_Q^{T_i})} < \theta$ **then**
34   **return** Top-$k(R)$;
35   **end**
36   **end**
37   **end**
38   **end**
39 **end**
40 **return** Top-$k(R)$;

---

extracted from $n.S_{t^i_j}$ and the length of shortest path from entities matching t^i_j to n, respectively. For each T_i, LQ_{T_i} is a list of $NQ_{t^i_j}$, each of which is a priority queue of entities on the paths starting at entities matching t^i_j and UB_{T_i} is a list of upper bound scores $ub_{t^i_j}$ for paths starting at entities matching all $t^i_j \in T_i$. For supporting top-k, R is used to keep track of the obtained candidate QEGs during graph exploration and θ denotes the lowest top-k score of the QEG in R.

Initialization. Instead of starting at entities matching each query keyword as described in [9–12], our exploration starts with each matching entity $n_{start} \in N_{t_j}$ for a *key term* $t^i_j \in T_i$ (Lines 1–3). For each starting entity n_{start}, we first check its connectivity (Line 4) to avoid unproductive exploration, which will be discussed later. When the connectivity condition is satisfied, we initialize the score $s_{n_{start}}$ stored in $n_{start}.S_{t^i_j}$, the maximal score $n_{start}.s_{t^i_j}$ and the distance $n_{start}.d_{t^i_j}$ (Lines 5–8). Such starting entities n_{start} are then added into the respective queue $NQ_{t^i_j} \in LQ_{T_i}$ (Line 9) and the upper bound score $ub_{t^i_j}$ for each t^i_j is initialized as the maximal score for all $n_{start} \in NQ_{t^i_j}$ (Line 12).

Connectivity Checking. The aim of checking the connectivity (Lines 4 and 22) is to predict whether an entity n could participate in any QEGs. Given an entity n with path of length $n.d_{t^i_j}$ from n_{start} matching $t^i_j \in T_i$ to n, if it cannot reach some entities n'_{start} matching $t^i_k \in T_i$ ($k \neq j$) within distance $d_{max} - n.d_{t^i_j}$, it is guaranteed not to be a connecting element and thus the exploration involving n can be avoided. For efficient entity connectivity indexing, we model paths between entities in G_{KB} with length no larger than d as a boolean matrix M^d_{KB}, where each entry m^d_{ij} is 1, if there is a path between entities n_i and n_j of length no larger than d; otherwise, m^d_{ij} is 0. The matrix $M^{d_{max}}_{KB}$ is constructed iteratively using the formula $M^{d_{max}}_{KB} = M^{d_{max}-1}_{KB} \times M^1_{KB}$.

Upper Bound Principle. The upper bound principle captures the goal of exploring only necessary entities for generating the top-k QEGs. The key is to effectively bound the ultimate score of potential QEGs based on the currently explored paths. Since the score of each edge $\langle n_i, n_j \rangle$ defined in Eq. 9 is less than 1, the score of paths satisfy the subset monotonic property, namely $S(P_{n_{start} \rightsquigarrow n}) \geq S(P_{n_{start} \rightsquigarrow n'})$ if $P_{n_{start} \rightsquigarrow n} \subseteq P_{n_{start} \rightsquigarrow n'}$. This implies that the score of a path cannot increase after path expansion during graph exploration and thus the score of all paths starting at entities matching t^i_j can be upper bounded by the maximal score for all $n \in NQ_{t^i_j}$. i.e., $ub_{t^i_j} = \max_{n \in NQ_{t^i_j}} n.s_{t^i_j}$, where $n.s_{t^i_j} = n.S_{t^i_j}.maxScore()$. These upper bound scores indicate the best the potential QEGs resulting from T_i, denoted by $G^{T_i}_Q$, can eventually achieve, such that we define the maximal possible score for all $G^{T_i}_Q$ as $\overline{S(G^{T_i}_Q)} = \sum_{ub_{t^i_j} \in UB_{T_i}} ub_{t^i_j}$, which will guide our graph exploration and help with early termination.

Graph Exploration. The graph exploration starts with entities in $NQ \in LQ$ (Line 16). To avoid the unnecessary exploration, our algorithm prioritizes the entity by the maximal possible score of the potential QEGs. At each iteration, the most promising T_i that could result in the optimal QEG and the key term $t_j^i \in T_i$ with the largest upper bound score $ub_{t_j^i}$ are selected (Lines 17–18). Then the entity n achieving the maximal score of paths from entities matching t_j^i to n is taken from $NQ_{t_j^i}$ (Line 19) and the algorithm continues to explore the neighborhood of n, i.e., all adjacent entities n'. In case that the distance $n'.d_{t_j^i}$ does not exceed d_{max} and the connectivity condition is satisfied (Line 22), we expand the path from each n_{start} to n by adding n', and the score $s'_{n_{start}}$ of each expanded path is calculated and added into $n'.S_{t_j^i}$ (Lines 24–25), where the maximal score $n'.s_{t_j^i}$ is extracted (Line 27). All newly explored entities n' are then added into $NQ_{t_j^i}$ for further exploration (Line 28). Since the maximal score of paths from entities matching t_j^i might change after expansion, the upper bound score $ub_{t_j^i}$ and the maximal possible score $\overline{S(G_Q^{T_i})}$ of potential QEGs are updated accordingly (Lines 29–30). If n' is verified to be an connecting element, i.e., for all $t_j^i \in T_i$, there exists a path from n_{start} matching t_j^i to n' (Line 31), the new QEGs generated by merging paths resulted from n' are added into R (Line 32). Finally, we check whether the exploration can terminate to retrieve the top-k QEGs (Lines 33–35), which will be discussed in the following.

Early Termination. The exploration terminates when one of the following conditions is satisfied: (1) all possible entities have been explored such that there are no further entities in any $NQ \in LQ$ or (2) the top-k QEGs are guaranteed to be obtained. With the goal of retrieving the top-k QEGs, all entities have to be considered as connecting element in order to keep track of all possible QEGs. However, the upper bound principle deals with the requirement of early termination. The maximal possible score $\overline{S(G_Q^{T_i})}$ for all T_i indicates the best the potential QEGs can achieve and the lowest top-k score of the obtained QEGs captures the threshold θ such that only the QEGs with score higher than or equal to θ have a chance to make into the top-k. To conclude that the current k top-ranked QEGs in R are guaranteed to qualify for the final top-k and thus the exploration can terminate, there should be at least k QEGs in R and $\overline{S(G_Q^{T_i})}$ for all T_i must be below θ, i.e., $\max_{T_i \in LT} \overline{S(G_Q^{T_i})} < \theta$ (Lines 33–35).

6 Experimental Results

The experiments were conducted on a virtual machine with 8 Cores at 2.0 GHz and 40GB memory and our system is implemented in Java 8. To assess both effectiveness and efficiency of our approach addressed by Sects. 4 and 5 respectively, we asked volunteers to provide keyword queries along with the underlying information needs. It results in 21 English queries, 10 German queries, 5 Chinese queries and 14 multilingual queries[6], where the query length ranges from 2 to 7

---

[6] It is a realistic phenomenon that queries consist of keywords in different languages, especially for Chinese users, which is also reflected in the 14 multilingual queries in our experiments, where only English and Chinese keywords are contained.

with an average of 3.24. We assume that the language of each keyword query is unknown and the target language of query interpretations is English[7].

6.1   Effectiveness Evaluation

For evaluating the effectiveness of query interpretation ranking, which is mainly addressed by Sect. 4, we consider the normalized Discounted Cumulative Gain at rank k, denoted by nDCG@k, as quality criteria, which measures the goodness of a retrieval model based on the graded relevance of the top-k results. According to our query interpretation problem, the results are judged by the volunteers who provide the keyword queries on 0–5 relevance scale based on the criteria such as relevance, completeness and correctness w.r.t. the underlying information needs.

For a comparative analysis, we conducted the experiments with the following approaches: (1) the baseline using an online *machine translation* service[8] and a *keyword-based scoring* function described in [11], denoted by *MT+KS*; (2) the baseline using our *cross-lingual lexica* for keyword-to-entity mapping and the *keyword-based scoring* same as (1), denoted by *CL+KS*; (3) the baseline using the *machine translation* service same as (1) and an adaption of our query entity *graph scoring* based on *key term* sets, denoted by *MT+GS+KT*; (4) our approach using the *cross-lingual lexica* for entity matching and the query entity *graph scoring* based on *key term* sets as discussed in Sect. 4, denoted by *CL+GS+KT*.

Figure 2(a) illustrates the nDCG@20 of different approaches for the individual queries (Q1-Q50). Our approach CL+GS+KT achieves the best results for 38 queries, while MT+KS, CL+KS and MT+GS+KT perform the best for 9, 16 and

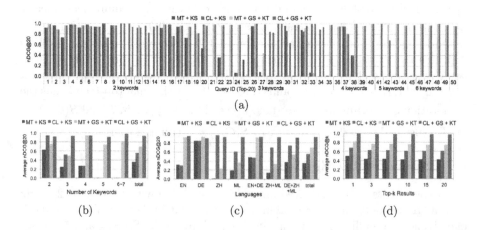

Fig. 2. Experimental results of query interpretation effectiveness

---

[7] In our experiments, we use English as the target language of query interpretations, but it can be easily extended to other languages.

[8] In our experiments, we used GOOGLE TRANSLATE for translating queries in different languages to English by selecting the input language option as "Detect language".

28 queries, respectively. Comparing the two methods with keyword-based scoring function, i.e., MT+KS and CL+KS, it is observed that using our cross-lingual lexica (CL) performs better than the machine translation service (MT) in most cases (e.g., Q10-Q14). There is a similar conclusion for the approaches based on our query entity graph scoring, i.e., MT+GS+KT and CL+GS+KT (e.g., Q27-Q31). Based on the further comparison between MT+KS and MT+GS+KT as well as CL+KS and CL+GS+KT, our query entity graph scoring based on key term sets (GS+KT) considerably outperforms the keyword-based scoring (KS) (e.g., Q38-Q50). By taking advantage of both CL and GS+KT compared with MT and KS, CL+GS+KT apparently achieves the best results in most cases.

Figure 2(b) illustrates the impact of query length l, i.e., the number of keywords, on query interpretation effectiveness. While our approach CL+GS+KT is stable for different l, the results of other approaches change considerably when l varies. More specifically, the performance of the approaches using keyword-based scoring (KS), i.e., MT+KS and CL+KS, decreases rapidly when l increases. This is due to the fact that when l is larger, the query entities are usually expressed by more than one keyword such that the keyword-to-entity mapping doesn't work well.

The impact of languages on query interpretation is shown in Fig. 2(c). For English queries (EN), by comparing MT+KS with CL+KS and MT+GS+KT with CL+GS+KT, MT and CL exhibit only minor differences because no cross-lingual mapping is needed when the input and target languages are both English. However, MT+GS+KT and CL+GS+KT still considerably outperform MT+KS and CL+KS respectively, because GS+KT has a clear advantage over KS. For German queries (DE), all approaches achieve comparable results for two reasons: (1) the entities in German queries are usually expressed by compound keywords or their abbreviations, e.g., *"Fußball-Weltmeisterschaft"* or *"WM"* corresponding to *"FIFA World Cup"*, such that the keyword-based scoring yields a similar performance to ours; (2) the machine translation service works well when translating from German to English. For Chinese queries (ZH), CL+KS and CL+GS+KT considerably outperform MT+KS and MT+GS+KT because the machine translation service (MT) doesn't work well for translating entity names from Chinese to English compared with our cross-lingual lexica (CL). In addition, in Chinese queries each entity is usually split by users as one compound keyword such that CL+KS even yields slightly better results than CL+GS+KT. Obviously, CL+GS+KT achieves the best results for multilingual queries (ML), where MT+KS and MT+GS+KT perform the worst because the machine translation service (MT) cannot deal with the keywords in multiple languages simultaneously. The experimental results for different combinations of the query languages are also shown in Fig. 2(c), where our approach CL+GS+KT achieves the best results (with nDCG@20 > 0.9) for most cases.

Figure 2(d) illustrates the results of nDCG@k for different k. We observe that the performance of all approaches decreases slightly when k becomes larger. Among these approaches, CL+GS+KT achieves the most stable performance, e.g., MT+KS, CL+KS, MT+GS+KT and CL+GS+KT yield 15 %, 10 %, 8 % and 2 % performance degradation respectively, when k varies from 1 to 20.

6.2   Efficiency Study

For assessing the efficiency of query interpretation generation, which is mainly
addressed by Sect. 5, we conducted the experiments with several approaches:
(1) the *keyword-based exploration* from each keyword matching entity [12],
denoted by *KE*; (2) the *top-k algorithm* on top of the *keyword-based exploration* [11], denoted by *KE+Top-k*; (3) our key term *set-based exploration*
starting from the entities matching the extracted key terms, denoted by *SE*;
(4) our graph exploration incorporating only *connectivity checking*, denoted
by *SE+CC*; (5) our graph exploration incorporating only *early termination*,
denoted by *SE+ET*; (6) our approach incorporating both *connectivity checking*
and *early termination* into the graph exploration as discussed in Sect. 5, denoted
by *SE+CC+ET*.

We start with a comparison between different approaches for the individual
queries. The experimental results for computing the top-20 query interpretations
for Q21-Q50 with query length from 3 to 7 are illustrated in Fig. 3(a). For the sake
of space, we omit the results for Q1-Q20 with query length 2, where individual
times do not exhibit significant differences. Clearly, SE outperforms KE for the
long queries (e.g., Q36-Q50), where 42 % performance improvement has been
achieved on average, while the performance of SE for short queries is slightly
better than KE (e.g., Q21-Q35) or similar to KE (e.g., Q1-Q20). Such differences
are primarily due to the number of starting entities for the graph exploration as
shown in Fig. 3(b). While both connectivity checking (CC) and early termination
(ET) contribute to the performance improvement individually, the incorporation
of both of them into SE yields the best results. Compared with the baselines KE
and KE+Top-k, our approach SE+CC+ET achieves a considerable performance
improvement in most cases.

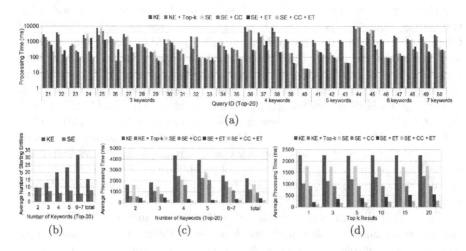

Fig. 3. Experimental results of query interpretation efficiency

We have investigated the impact of query length l on the performance of different approaches. Figure 3(c) shows the average processing time for different l. Compared with KE, the processing time for SE is relatively stable. The reason might be the number of starting entities generated by SE is less sensitive to l as shown in Fig. 3(b). Furthermore, our approaches SE+ET and SE+CC+ET are not sensitive to l due to the application of early termination (ET), while the performance of other approaches changes with varying l.

Figure 3(d) shows the average time for computing top-k query interpretations for different k. The time needed by KE+Top-k, SE+ET and SE+CC+ET decreases rapidly when k becomes smaller. For example, KE+Top-k, SE+ET and SE+CC+ET yield 24 %, 61 % and 62 % time reduction respectively, when k varies from 20 to 1, while the performance of other approaches doesn't change with k since they have to process all results no matter what the value of k is. In total, our approach SE+CC+ET outperforms KE by one order of magnitude and is 5 times faster than KE+Top-k when $k = 20$, and it achieves even more considerable performance improvement for smaller k, e.g., 22 times and 10 times faster than KE and KE+Top-k respectively, when $k = 1$.

7   Related Work

We firstly present the related work to keyword query interpretation and then review some existing work on cross-lingual and concept-based IR.

Keyword Query Interpretation. The main challenges in dealing with keyword queries are their *ambiguity* and *incompleteness*. The use of thesauri to deal with the ambiguity of keywords has a long history. Most commonly, WordNet thesaurus has been found beneficial in disambiguating keywords and in choosing their senses [13]. There are also proposals for mapping keyword queries to elements in an ontology [14], where the resulting semantics provides the basis for identifying the search intents of users. In addition, graph-based approaches [9,11,12] have been widely used to find substructures in structured data, including relational, XML and RDF data. The recent work [15] also aimed to boost the scalability of interactive query construction over large scale data from the perspective of both user interaction cost and performance.

While existing methods only deal with individual keywords in the query, our approach relies on the extracted key terms referring to entities in KBs, which helps to improve both efficiency and effectiveness as shown in our experiments. In addition, most existing methods only focus on the *ambiguity* of keywords. The *cross-linguality* issue has not been studied in the previous work.

Cross-lingual and Concept-based IR. Traditional IR is normally based on the bag-of-words (BOW) models, which have the limitation of retrieving only the syntactically relevant but not the semantically relevant documents. Meanwhile, they suffer from the vocabulary mismatch problem, i.e., queries and documents, which are semantically very related, might contain only few terms in common. This problem is more serious in cross-lingual IR due to the fact that queries and documents in different languages rarely share common terms. In order to address the problem, different concept-based solutions [16–19] and their cross-lingual extensions [20,21] have been proposed. Instead of the BOW models used in the classic IR, the goal is to capture queries and documents as concepts, such that the relevance can be estimated in the concept space even in the presence of vocabulary gap, especially for cross-lingual IR.

Unlike the previous studies, we developed XKnowSearch!, a novel system to entity-based cross-lingual IR by exploiting multilingual knowledge bases. Based on our cross-lingual query interpretation, XKnowSearch!, to the best of our knowledge, is the first entity-centric system to cross-lingual IR, where users can issue keyword queries in any language (even in multiple languages), for retrieving documents related to the query entities in any other languages.

8   Conclusions and Future Work

We present a knowledge base approach to cross-lingual query interpretation by transforming keywords in different languages to their semantic representation. As the main contributions of this work, we propose a scoring mechanism for effective query interpretation ranking and a top-k graph exploration algorithm for efficient query interpretation generation. A separate evaluation on each of these two aspects has been performed and it shows that our approach achieves promising results w.r.t. both effectiveness and efficiency. In addition, this work has been integrated into XKnowSearch!, a novel system for entity-based cross-lingual IR. As future work, we would like to extend our approach by taking into account entity relations expressed in keyword queries to construct the QEGs. And it is essential to perform further evaluation to show the promising results of our query interpretation can carry over to the performance of cross-lingual IR.

Acknowledgments. The research leading to these results has received funding from the European Union Seventh Framework Programme (FP7/2007–2013) under grant agreement no. 611346.

References

1. Pound, J., Mika, P., Zaragoza, H.: Ad-hoc object retrieval in the web of data. In: WWW, pp. 771–780 (2010)
2. Baldwin, T., Lui, M.: Language identification: the long and the short of the matter. In: HLT-NAACL, pp. 229–237 (2010)
3. Zhang, L., Färber, M., Rettinger, A.: XKnowSearch! exploiting knowledge bases for entity-based cross-lingual information retrieval. In: CIKM (2016)
4. Zhang, L., Rettinger, A.: X-LiSA: cross-lingual semantic annotation. PVLDB **7**(13), 1693–1696 (2014)
5. Zhang, L., Färber, M., Rettinger, A.: xLiD-Lexica: cross-lingual linked data lexica. In: LREC, pp. 2101–2105 (2014)
6. Zhang, L., Rettinger, A., Thoma, S.: Bridging the gap between cross-lingual NLP and DBpedia by exploiting Wikipedia. In: NLP&DBpedia Workshop (2014)

7. Zhang, L., Rettinger, A.: Exploiting knowledge bases for entity-based multilingual and cross-lingual information retrieval. Technical report. http://km.aifb.kit.edu/sites/XKnowSearch/TR.pdf

8. Witten, I., Milne, D.: An effective, low-cost measure of semantic relatedness obtained from Wikipedia links. In: WIKIAI, pp. 25–30 (2008)

9. He, H., Wang, H., Yang, J., Yu, P.S.: BLINKS: ranked keyword searches on graphs. In: SIGMOD, pp. 305–316 (2007)

10. Li, G., Ooi, B.C., Feng, J., Wang, J., Zhou, L.: Ease: an effective 3-in-1 keyword search method for unstructured, semi-structured and structured data. In: SIGMOD, pp. 903–914(2008)

11. Tran, T., Wang, H., Rudolph, S., Cimiano, P.: Top-k exploration of query candidates for efficient keyword search on graph-shaped (RDF) data. In: ICDE, pp. 405–416 (2009)

12. Kacholia, V., Pandit, S., Chakrabarti, S., Sudarshan, S., Desai, R., Karambelkar, H.: Bidirectional expansion for keyword search on graph databases. In: VLDB, pp. 505–516 (2005)

13. Voorhees, E.M.: Query expansion using lexical-semantic relations. In: SIGIR, pp. 61–69 (1994)

14. Tran, T., Cimiano, P., Rudolph, S., Studer, R.: Ontology-based interpretation of keywords for semantic search. In: Aberer, K., et al. (eds.) ASWC 2007 and ISWC 2007. LNCS, vol. 4825, pp. 523–536. Springer, Heidelberg (2007)

15. Demidova, E., Zhou, X., Nejdl, W.: Efficient query construction for large scale data. In: SIGIR, pp. 573–582 (2013)

16. Wei, X., Croft, W.B.: LDA-based document models for ad-hoc retrieval. In: SIGIR, pp. 178–185 (2006)

17. Bendersky, M., Croft, W.B.: Discovering key concepts in verbose queries. In: SIGIR, pp. 491–498(2008)

18. Gabrilovich, E., Markovitch, S.: Computing semantic relatedness using Wikipedia-based explicit semantic analysis. In: IJCAI, pp. 1606–1611 (2007)

19. Egozi, O., Markovitch, S., Gabrilovich, E.: Concept-based information retrieval using explicit semantic analysis. ACM Trans. Inf. Syst. **29**(2), 8 (2011)

20. Sorg, P., Cimiano, P.: Cross-language information retrieval with explicit semantic analysis. In: CLEF (Working Notes) (2008)

21. Potthast, M., Stein, B., Anderka, M.: A Wikipedia-based multilingual retrieval model. In: Macdonald, C., Ounis, I., Plachouras, V., Ruthven, I., White, R.W. (eds.) ECIR 2008. LNCS, vol. 4956, pp. 522–530. Springer, Heidelberg (2008)

Context-Free Path Queries on RDF Graphs

Xiaowang Zhang[1,2], Zhiyong Feng[1,2(✉)], Xin Wang[1,2], Guozheng Rao[1,2],
and Wenrui Wu[1,2]

[1] School of Computer Science and Technology, Tianjin University, Tianjin, China
{xiaowangzhang,zyfeng,wangx,rgz,wenruiwu}@tju.edu.cn
[2] Tianjin Key Laboratory of Cognitive Computing and Application, Tianjin, China

Abstract. Navigational graph queries are an important class of queries
that can extract implicit binary relations over the nodes of input graphs.
Most of the navigational query languages used in the RDF community,
e.g. property paths in W3C SPARQL 1.1 and nested regular expressions in nSPARQL, are based on the regular expressions. It is known
that regular expressions have limited expressivity; for instance, some natural queries, like *same generation-queries*, are not expressible with regular expressions. To overcome this limitation, in this paper, we present
cfSPARQL, an extension of SPARQL query language equipped with
context-free grammars. The cfSPARQL language is strictly more expressive than property paths and nested expressions. The additional expressivity can be used for modelling graph similarities, graph summarization
and ontology alignment. Despite the increasing expressivity, we show
that cfSPARQL still enjoys a low computational complexity and can be
evaluated efficiently.

1 Introduction

The Resource Description Framework (RDF) [30] recommended by World Wide
Web Consortium (W3C) is a standard graph-oriented model for data interchange
on the Web [6]. RDF has a broad range of applications in the semantic web,
social network, bio-informatics, geographical data, etc. [1]. Typical access to
graph-structured data is its navigational nature [12,16,21]. Navigational queries
on graph databases return binary relations over the nodes of the graph [9]. Many
existing navigational query languages for graphs are based on binary relational
algebra such as XPath (a standard navigational query language for trees [25])
or regular expressions such as RPQ (regular path queries) [24].

SPARQL [32] recommended by W3C has become the standard language for
querying RDF data since 2008 by inheriting classical relational languages such
as SQL. However, SPARQL only provides limited navigational functionalities
for RDF [28,37]. Recently, there are several proposed languages with navigational capabilities for querying RDF graphs [3–5,7,11,19,26,28,35]. Roughly,
Versa [26] is the first language for RDF with navigational capabilities by using
XPath over the XML serialization of RDF graphs. SPARQLeR proposed by
Kochut et al. [19] extends SPARQL by allowing path variables. SPARQL2L proposed by Anyanwu et al. [7] allows path variables in graph patterns and offers

© Springer International Publishing AG 2016
P. Groth et al. (Eds.): ISWC 2016, Part I, LNCS 9981, pp. 632–648, 2016.
DOI: 10.1007/978-3-319-46523-4_38

good features in nodes and edges such as constraints. PSPARQL proposed by Alkhateeb et al. [5] extends SPARQL by allowing regular expressions in general triple patterns with possibly blank nodes and CASPAR further proposed by Alkhateeb et al. [3,4] allows constraints over regular expressions in PSPARQL where variables are allowed in regular expressions. nSPARQL proposed by Pérez et al. [28] extends SPARQL by allowing nested regular expressions in triple patterns. Indeed, nSPARQL is still expressible in SPARQL if the transitive closure relation is absent [37]. In March 2013, SPARQL 1.1 [33] recommended by W3C allows property paths which strengthen the navigational capabilities of SPARQL1.0 [11,35].

However, those regular expression-based extensions of SPARQL are still limited in representing some more expressive navigational queries which are not expressed in regular expressions. Let us consider a fictional biomedical ontology mentioned in [31] (see Fig. 1). We are interested in a navigational query about those paths that confer similarity (e.g., between *Gene(B)* and *Gene(C)*), which suggests a causal relationship (e.g., between *Gene(S)* and *Phenotype(T)*). This query about similarity arises from the well-known *same generation-query* [2], which is proven to be inexpressible in any regular expression. To express the query, we have to introduce a query embedded with a context-free grammar (CFG) for expressing the language of $\{ww^T | w \text{ is a string}\}$[31] where w^T is the converse of w. For instance, if $w = \text{``}abcdfe\text{''}$ then $w^T = \text{``}e^{-1}f^{-1}d^{-1}c^{-1}b^{-1}a^{-1}\text{''}$. As we know, CFG has more expressive power than any regular expression [18]. Moreover, the context-free grammars can provide a simplified more user-friendly dialect of Datalog [1] which still allows powerful recursion [18]. Besides, the context-free graph queries have also practical query evaluation strategies. For instance, there are some applications in verification [20]. So it is interesting to introduce a navigational query embedded with context-free grammars to express more practical queries like the same generation-query.

A proposal of conjunctive context-free path queries (written by *Helling's CCFPQ*) for edge-labeled directed graphs has been presented by Helling [14] by allowing context-free grammars in path queries. A naive idea to express same generation-queries is transforming this RDF graph to an edge-labeled directed graph via navigation axes [28] and then using Helling's CCFPQ since an RDF graph can be intuitively taken as an edge-labeled directed graph. However, this transformation is difficult to capture the full information of this RDF graph since there exist some slight differences between RDF graphs and edge-labeled directed graphs, particularly regarding the connectivity [13], thus it could not express some regular expression-based path queries on RDF graphs. For instance, a nested regular expression (nre) of the form $axis::[e]$ on RDF graphs in nSPARQL [28], is always evaluated to the empty set over any edge-labeled directed graph. That is to say, an nre of the form "$axis::[e]$" is hardly expressible in Helling's CCFPQ.

To represent more expressive queries with efficient query evaluation is a renewed interest topic in the classical topic of graph databases [2]. Hence, in this paper, we present a context-free extension of path queries and SPARQL

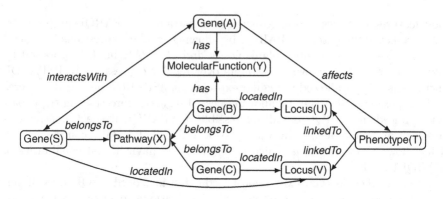

Fig. 1. A biomedical ontology [31]

queries on RDF graphs which can express both nre and nSPARQL [28]. Furthermore, we study several fundamental properties of the proposed context-free path queries and context-free SPARQL queries. The main contributions of this paper can be summarized as follows:

- We present *context-free path queries* (CFPQ) (including *conjunctive context-free path queries* (CCFPQ), *union of simple conjunctive context-free path queries* (UCCFPQ^s), and *union of conjunctive context-free path queries* (UCCFPQ) for RDF graphs and find that CFPQ, CCFPQ, and UCCFPQ have efficient query evaluation where the query evaluation has the polynomial data complexity and the NP-complete combined complexity. Finally, we implement our CFPQs and evaluate experiments on some popular ontologies.
- We discuss the expressiveness of CFPQs by referring to nested regular expressions (nre). We show that CFPQ, CCFPQ, UCCFPQ^s, and UCCFPQ exactly express four fragments of nre, basic nre "nre_0", union-free nre "$nre_0(N)$", nesting-free nre "$nre_0(|)$", and full nre, respectively (see Fig. 2). The query evaluation of cfSPARQL has the same complexity as SPARQL.
- We propose *context-free SPARQL* (cfSPARQL) and *union of conjunctive context-free SPARQL* (uccfSPARQL) based on CFPQ and UCCFPQ, respectively. It shows that cfSPARQL has the same expressiveness as that of uccf-SPARQL. Furthermore, we prove that cfSPARQL can strictly express both SPARQL and nSPARQL (even nSPARQL$^\neg$: a variant of nSPARQL by allowing nre with negation "$nre^\neg$") (see Fig. 3).

Organization of the Paper. Section 2 recalls nSPARQL and context-free grammar. Section 3 defines CFPQ. Section 4 discusses the expressiveness of CFPQ. Section 5 presents cfSPARQL and Sect. 6 discusses the relations on nre with negation. Section 7 evaluates experiments. We conclude in Sect. 8. Due to the space limitation, all proofs and some further preliminaries are omitted but they are available in an extended technical report in arXiv.org [36].

2 Preliminaries

In this section, we introduce the language nSPARQL and context-free grammar.

2.1 The Syntax and Semantics of nSPARQL

In this subsection, we recall the syntax and semantics of nSPARQL, largely following the excellent expositions [27,28].

RDF Graphs. An RDF statement is a *subject-predicate-object* structure, called *RDF triple* which represents resources and the properties of those resources. For the sake of simplicity similar to [28], we assume that RDF data is composed only IRIs[1]. Formally, let U be an infinite set of *IRIs*. A triple $(s, p, o) \in U \times U \times U$ is called an *RDF triple*. An *RDF graph* G is a finite set of RDF triples. We use $adom(G)$ to denote the *active domain* of G, i.e., the set of all elements from U occurring in G.

For instance, a biomedical ontology shown in Fig. 1 can be modeled in an RDF graph named as G_{bio} where each labeled-edge of the form $a \xrightarrow{p} b$ is directly translated into a triple (a, p, b).

Paths and Traces. Let G be an RDF graph. A *path* $\pi = (c_1 c_2 \ldots c_m)$ in G is a non-empty finite sequence of constants from G, where, for every $i \in \{1, \ldots, m-1\}$, c_i and c_{i+1} exactly occur in the same triple of G (i.e., $(c_i, c, c_{i+1}), (c_i, c_{i+1}, c)$, and (c, c_i, c_{i+1}) etc.). Note that the precedence between c_i and c_{i+1} in a path is independent of the positions of c_i, c_{i+1} in a triple.

In nSPARQL, three different *navigation axes*, namely, *next*, *edge*, and *node*, and their inverses, i.e., $next^{-1}, edge^{-1}$, and $node^{-1}$, are introduced to move through an RDF triple (s, o, o) [28].

Let $\Sigma = \{axis, axis::c | c \in U\}$ where $axis \in \{self, next, edge, node, next^{-1}, edge^{-1}, node^{-1}\}$. Let G be an RDF graph. We use $\Sigma(G)$ to denote the set of all symbols $\{axis, axis::c | c \in adom(G)\}$ occurring in G.

Let $\pi = (c_1 \ldots c_m)$ be a path in G. A *trace* of path π is a string over $\Sigma(G)$ written by $\mathcal{T}(\pi) = l_1 \ldots l_{m-1}$ where, for all $i \in \{1, \ldots, m-1\}$, $(c_i c_{i+1})$ is labeled by l_i and l_i is of the form $axis, axis::c, axis^{-1}$, or $axis^{-1}::c$ [28]. We use $Trace(\pi)$ to denote the set of all traces of π.

Note that it is possible that a path has multiple traces since any two nodes possibly occur in the multiple triples. For example, consider an RDF graph $G = \{(a, b, c), (a, c, b)\}$ and given a path $\pi = (abc)$, both $(edge::c)(node::a)$ and $(next::c)(node^{-1}::a)$ are traces of π.

For instance, in the RDF graph G_{bio} (see Fig. 1), a path from *Gene(B)* to *Gene(C)* has a trace: $(next::locatedIn)(next^{-1}::linkedTo)(next::linkedTo)$ $(next^{-1}::locatedIn)$.

---

[1] A standard RDF data is composed of IRIs, blank nodes, and literals. For the purposes of this paper, the distinction between IRIs and literals will not be important.

Nested Regular Expressions. Nested regular expressions (*nre*) are defined by the following formal syntax:

$$e := axis\,|\,axis{::}c\,(c \in U)\,|\,axis{::}[e]\,|\,e/e\,|\,e\,|\,e^*.$$

Here the *nesting nre* is of the form $axis{::}[e]$.

For simplification, we denote some interesting fragments of nre as follows:

- nre_0: *basic nre*, i.e., nre only consisting of "*axis*", "/", and "∗";
- $\text{nre}_0(|)$: basic nre by adding the operator "|";
- $\text{nre}_0(\text{N})$ to basic nre by adding nesting nre $axis{::}[e]$.

Patterns. Assume an infinite set V of *variables*, disjoint from U. A *nested regular expression triple* (or *nre-triple*) is a tuple of the form $(?x, e, ?y)$, where $?x, ?y \in V$ and e is an nre[2].

Formally, nSPARQL (graph) patterns are recursively constructed from nre-triples:

- An nre-triple is an nSPARQL pattern;
- All P_1 UNION P_2, P_1 AND P_2, and P_1 OPT P_2 are nSPARQL patterns if P_1 and P_2 are nSPARQL patterns;
- P FILTER C if P is an nSPARQL pattern and C is a constraint;
- $\text{SELECT}_S(P)$ if P is an nSPARQL pattern and S is a set of variables.

Semantics. Given an RDF graph G and an nre e, the evaluation of e on G, denoted by $[\![e]\!]_G$, is a binary relation. More details can be found in [28]. Here, we recall the semantics of nesting nre of the form $axis{::}[e]$ as follows:

$$[\![axis{::}[e]]\!]_G = \{(a,b)\,|\,\exists c,d \in adom(G),\ (a,b) \in [\![axis{::}c]\!]_G \text{ and } (c,d) \in [\![e]\!]_G\}.$$

The semantics of nSPARQL patterns is defined in terms of sets of so-called *mappings*, which are simply total functions $\mu\colon S \to U$ on some finite set S of variables. We denote the domain S of μ by $\text{dom}(\mu)$.

Basically, the semantics of an nre-triple (u, e, v) is defined as follows:

$$[\![(u,e,v)]\!]_G = \{\mu\colon \{u,v\} \cap V \to U\,|\,(\mu(u),\mu(v)) \in [\![e]\!]_G\}.$$

Here, for any mapping μ and any constant $c \in U$, we agree that $\mu(c)$ equals c itself.

Let P be an nSPARQL pattern, the semantics of P on G, denoted by $[\![P]\!]_G$, is analogously defined as usual following the semantics of SPARQL [27,28].

Query Evaluation. A *SPARQL (SELECT) query* is an nSPARQL pattern. Given a RDF graph G, a pattern P, and a mapping μ, the query evaluation problem of nSPARQL is to decide whether μ is in $[\![P]\!]_G$. The complexity of query evaluation problem is PSpace-complete [27].

---

[2] In nSPARQL [28], nre-triples allow a general form (v, e, u) where $u, v \in U \cup V$. In this paper, we mainly consider the case $u, v \in V$ to simplify our discussion.

2.2   Context-Free Grammar

In this subsection, we recall context-free grammar. For more details, we refer the interested readers to some references about formal languages [18].

A *context-free grammar* (COG) is a 3-tuple $\mathcal{G} = (N, A, R)$ [3] where

- N is a finite set of variables (called *non-terminals*);
- A is a finite set of constants (called *terminals*);
- R is a finite set of production rules r of the form $v \rightarrow S$, where $v \in N$ and $S \in (N \cup A)^*$ (the asterisk $*$ represents the Kleene star operation). We write $v \rightarrow \epsilon$ if ϵ is the empty string.

A string over $N \cup A$ can be written to a new string over $N \cup A$ by applying production rules. Consider a string avb and a production rule $r : v \rightarrow avb$, we can obtain a new string $aavbb$ by applying this rule r one time and another new string $aaavbbb$ by applying the rule r twice. Analogously, strings with increasing length can be obtained in this rule.

Let $S, T \in (N \cup A)^*$. We write $(S \xrightarrow{\mathcal{G}} T)$ if T can be obtained from S by applying production rules of $\mathcal{G}$ within a finite number of times.

The *language* of grammar $\mathcal{G} = (N, A, R)$ w.r.t. start non-terminal $v \in N$ is defined by $\mathcal{L}(\mathcal{G}_v) = \{S \text{ a finite string over } A \mid v \xrightarrow{\mathcal{G}} S\}$.

For example, $\mathcal{G} = (N, A, R)$ where $N = \{v\}, A = \{a, b\}$, and $R = \{v \rightarrow ab, v \rightarrow avb\}$. Thus $\mathcal{L}(\mathcal{G}_v) = \{a^n b^n \mid n \geq 1\}$.

3   Context-Free Path Queries

In this section, we introduce context-free path queries on RDF graphs based on context-free path queries on directed graphs [14] and nested regular expressions [28].

3.1   Context-Free Path Queries and Their Extensions

In this subsection, we firstly define *conjunctive context-free path queries* on RDF graphs and then present some variants (it also can been seen as extensions).

Conjunctive Context-Free Path Queries. In this paper, we assume that $N \cap V = \emptyset$ and $A \subseteq \Sigma$ for all CFG $\mathcal{G} = (N, A, R)$.

Definition 1. *Let $\mathcal{G} = (N, A, R)$ be a CFG and m a positive integer. A conjunctive context-free path query (CCFPQ) is of the form* $\mathbf{q}(?x, ?y)$[4], *where,*

$$\mathbf{q}(?x, ?y) := \bigwedge_{i=1}^{m} \alpha_i, \tag{1}$$

where

---

[3] We deviate from the usual definition of context-free grammar by not including a special start non-terminal following [14].

[4] In this paper, we simply write a conjunctive query as a Datalog rule.

- α_i is a triple pattern either of the form $(?x, ?y, ?z)$ or of the form $v(?x, ?y)$;
- $\{?x, ?y\} \subseteq vars(\mathbf{q})$ where $vars(\mathbf{q})$ denotes a collection of all variables occurring in the body of $\mathbf{q}$;
- $\{v_1, \ldots, v_m\} \subseteq N$.

We regard the name of query $\mathbf{q}(?x, ?y)$ as $\mathbf{q}$ and call the right of Eq. (1) as the body of $\mathbf{q}$.

Remark 1. In our CCFPQ, we allow a triple pattern of the form $(?x, ?y, ?z)$ to characterize those queries w.r.t. ternary relationships such as nre-triple patterns of nSPARQL [28] to be discussed in Sect. 4. The formula $v(?x, ?y)$ is used to capture context-free path queries [14].

We say a *simple conjunctive context-free path query* (written by $CCFPQ^s$) if only the form $v(?x, ?y)$ is allowed in the body of a CCFPQ. We also say a *context-free path query* (written by $CFPQ$) if $m = 1$ in the body of a $CCFPQ^s$.

Semantically, let $\mathcal{G} = (N, A, R)$ be a CFG and G an RDF graph, given a CCFPQ $\mathbf{q}(?x, ?y)$ of the form (1), $[\![\mathbf{q}(?x, ?y)]\!]_G$ is defined as follows:

$$\{\mu|_{\{?x, ?y\}} | \mathrm{dom}(\mu) = vars(\mathbf{q}) \text{ and } \forall i = 1, \ldots, m, \mu|_{vars(\alpha_i)} \in [\![\alpha_i]\!]_G\}, \qquad (2)$$

where the semantics of $v(?x, ?y)$ over G is defined as follows:

$$[\![v(?x, ?y)]\!]_G = \{\mu | \mathrm{dom}(\mu) = \{?x, ?y\} \text{ and }$$
$$\exists \pi = (\mu(?x)c_1 \ldots c_m \mu(?y)) \text{ a path in } G, Trace(\pi) \cap \mathcal{L}(\mathcal{G}_v) \neq \emptyset\}.$$

Intuitively, $[\![v(?x, ?y)]\!]_G$ returns all pairs connected by a path in G which contains a trace belonging to the language generated from this CFG starting at non-terminal v.

Example 1. Let $\mathcal{G} = (N, A, R)$ be a CFG where $N = \{u, v\}, A = \{next::locatedIn, next^{-1}::locatedIn, next::linkedTo, next^{-1}::linkedTo\}$, and $P = \{v \to (next::locatedIn) u (next^{-1}::locatedIn), u \to (next^{-1}::linkedTo) u (next::linkedTo), u \to \epsilon\}$. Consider a CFPQ $\mathbf{q}$ be of the form $v(?x, ?y)$. The query $\mathbf{q}$ represents the relationship of similarity (between two genes) since $\mathcal{L}(\mathcal{G}_v) = \{(next^{-1}::locatedIn)^n (next^{-1}::linkedTo)(next::linkedTo)(next::locatedIn)^n | n \geq 1\}$. Consider the RDF graph G_{bio} in Fig. 1, $[\![\mathbf{q}(?x, ?y)]\!]_{G_{bio}} = \{(?x = Gene(B), ?y = Gene(C))\}$. Clearly, the query $\mathbf{q}$ returns all pairs with similarity.

Query Evaluation. Let $\mathcal{G} = (N, A, R)$ be a CFG and G an RDF graph. Given a CCFPQ $\mathbf{q}(?x, ?y)$ and a tuple $\mu = (?x = a, ?y = b)$, the *query evaluation problem* is to decide whether $\mu \in [\![\mathbf{q}(?x, ?y)]\!]_G$, that is, whether the tuple μ is in the result of the query $\mathbf{q}$ on the RDF graph G. There are two kinds of computational complexity in the query evaluation problem [1,2]:

- the *data complexity* refers to the complexity w.r.t. the size of the RDF graph G, given a fixed query $\mathbf{q}$; and
- the *combined complexity* refers to the complexity w.r.t. the size of query $\mathbf{q}$ and the RDF graph G.

A CFG $\mathcal{G} = (N, A, R)$ is said to be in *norm form* if all of its production rules are of the form $v \rightarrow uw, v \rightarrow a$, or $v \rightarrow \epsilon$ where $v, u, w \in N$ and $a \in A$. Note that this norm form deviates from the usual *Chomsky Normal Form* [22] where the start non-terminals are absent. Indeed, every CFG is equivalent to a CFG in norm form, that is, for every CFG $\mathcal{G}$, there exists some CFG $\mathcal{G}'$ in norm form constructed from $\mathcal{G}$ in polynominal time such that $\mathcal{L}(\mathcal{G}_v) = \mathcal{L}(\mathcal{G}'_v)$ for every $v \in N$ [14].

Let G be an RDF graph and $\mathcal{G} = (N, A, R)$ a CFG. Given a non-terminal $v \in N$, let $\mathcal{R}_v(G)$ be *the context-free relation* of G w.r.t. v can be defined as follows:

$$\mathcal{R}_v(G) := \{(a, b) | \exists \pi = (ac_1 \dots c_m b) \text{ a path in } G, \text{ } Trace(\pi) \cap \mathcal{L}(\mathcal{G}_v) \neq \emptyset\}. \quad (3)$$

Conveniently, the query evaluation of CCFPQ over an RDF graph can be reduced into the conjunctive first-order query over the context-free relations. Based on the conjunctive context-free recognizer for graphs presented in [14], we directly obtain a conjunctive context-free recognizer (see Algorithm 1) for RDF graphs by adding a convertor to transform an RDF graph into an edge-labeled directed graph (see Algorithm 2).

Algorithm 1. Conjunctive context-free recognizer for RDF

Input: G: an RDF graph; $\mathcal{G} = (N, A, R)$: a CFG in norm form; $v \in N$.
Output: $\{(v, a, b) | (a, b) \in \mathcal{R}_v(G)\}$

```
 1:  Θ:={(v, a, a)|(a ∈ adom(G)) ∧ (v → ε ∈ P)}
 2:    ∪{(v, a, b)|((a, l, b) ∈ Convertor(G)) ∧ (v → l) ∈ P}
 3:  Θ_new := Θ
 4:  while Θ_new ≠ ∅ do
 5:    pick and remove a (v, a, b) from Θ_new
 6:    for all (u, a', a) ∈ Θ do
 7:      for all v' → uv ∈ R ∧ ((v', a', b) ∉ Θ) do
 8:        Θ_new := Θ_new ∪ {(v', a', b)}
 9:        Θ := Θ ∪ {(v', a', b)}
10:      end for
11:    end for
12:    for all (u, b, b') ∈ Θ do
13:      for all u' → vu ∈ R ∧ ((u', a, b') ∉ Θ) do
14:        Θ_new := Θ_new ∪ {(u', a, b')}
15:        Θ := Θ ∪ {(u', a, b')}
16:      end for
17:    end for
18:  end while
19:  return Θ
```

Given a path π and a context-free grammar $\mathcal{G}$, Algorithm 1 is sound and complete to decide whether the path π in RDF graphs has a trace generated from the grammar $\mathcal{G}$.

Algorithm 2. RDF convertor

Input: G: an RDF graph
Output: $Convertor(G) = (\mathcal{V}, \mathcal{E})$
 1: $\mathcal{V} := adom(G)$
 2: $\mathcal{E} := \{(c, self, c), (c, self::c, c) | c \in adom(G)\}$
 3: $G_{new} := G$
 4: **while** $G_{new} \neq \emptyset$ **do**
 5:    pick and remove a triple (s, p, o) from G_{new}
 6:    $\mathcal{E} := \mathcal{E} \cup \{(s, next::p, o), (s, next, o), (o, next^{-1}::p, s), (o, next^{-1}, s),$
               $(s, edge::o, p), (s, edge, p), (p, edge^{-1}::o, s), (p, edge^{-1}, s),$
               $(p, node::s, o), (p, node, o), (o, node^{-1}::s, p), (o, node^{-1}, p)\}$
 7: **end while**
 8: **return** $Convertor(G)$

Proposition 1. *Let G be an RDF graph and $\mathcal{G} = (N, A, R)$ a CFG in norm form. For every $v \in N$, let Θ be the result computed in Algorithm 1, $(v, a, b) \in \Theta$ if and only if $(a, b) \in \mathcal{R}_v(G)$.*

Moreover, we can easily observe the worst-case complexity of Algorithm 1 since the complexity of Algorithm 2 is $\mathcal{O}(|G|)$.

Proposition 2. *Let G be an RDF graph and $\mathcal{G} = (N, A, R)$ a CFG. Algorithm 1 applied to G and $\mathcal{G}$ has a worst-case complexity of $\mathcal{O}((|N||G|)^3)$.*

As a result, we can conclude the following proposition.

Proposition 3. *The followings hold:*

1. *The query evaluation of CCFPQ has polynomial data complexity;*
2. *The query evaluation of CCFPQ has NP-complete combined complexity.*

Union of CCFPQ. An extension of CCFPQ capturing more expressive power such as disjunctive capability is introducing the union of CCFPQ. For instance, given a gene (e.g., $Gene(B)$) in the biomedical ontology (see Fig. 1), we wonder to find those genes which are relevant to this gene, that is, those genes either are similar to it (e.g., $Gene(C)$) or belong to the same pathway (e.g., $Gene(S)$).

A *union of conjunctive context-free path query* (UCCFPQ) is of the form

$$\mathbf{q}(?x, ?y) := \bigvee_{i=1}^{m} \mathbf{q}_i(?x, ?y), \tag{4}$$

where $\mathbf{q}_i(?x, ?y)$ is a CCFPQ for all $i = 1, \ldots, m$.

Analogously, we can define *union of simple conjunctive context-free path query* written by $UCCFPQ^s$.

Semantically, let G be an RDF graph, we define

$$[\![\mathbf{q}(?x, ?y)]\!]_G = \bigcup_{i=1}^{m} [\![\mathbf{q}_i(?x, ?y)]\!]_G, \tag{5}$$

where $[\![\mathbf{q}_i(?x, ?y)]\!]_G$ is defined as the semantics of CCFPQ for all $i = 1, \ldots, m$.

In Example 1, based on $\mathcal{G} = (N, A, R)$, we construct a CFG $\mathcal{G}' = (N', A', R')$ where $N' = N \cup \{s\}, A = A \cup \{next::belongsTo, next^{-1}::belongsTo\}$, and $R' = R \cup \{s \rightarrow (next::belongsTo)s(next^{-1}::belongsTo)\}$. Consider a UCCFPQ $\mathbf{q}(?x, ?y) := v(?x, ?y) \vee s(?x, ?y)$, $[\![\mathbf{q}(?x, ?y)]\!]_{G_{\text{bio}}} = \{(?x = Gene(B), ?y = Gene(C)), (?x = Gene(B), ?y = Gene(S))\}$. That is, $[\![\mathbf{q}(?x, ?y)]\!]_{G_{\text{bio}}}$ returns all pairs where the first gene is relevant to the latter.

Note that the query evaluation of UCCFPQ has the same complexity as that of the evaluating of CCFPQ since we can simply evaluate a number (linear in the size of a UCCFPQ) of CCFPQs in isolation [2].

4   Expressivity of (U)(C)CFPQ

In this section, we investigate the expressivity of (U)(C)CFPQ by referring to nested regular expressions [28] and fragments of nre.

We discuss the relations between variants of UCCFPQ and variants of (nested) regular expressions and obtain the following results:

1. nre_0-triples can be expressed in CFPQ;
2. $nre_0(N)$-triples can be expressed in CCFPQ;
3. $nre_0(|)$-triples can be expressed in UCCFPQ^s;
4. nre-triples can be expressed in UCCFPQ.

1. nre_0 in CFPQ. The following proposition shows that CFPQ can express nre_0-triples.

Proposition 4. *For every nre_0-triple $(?x, e, ?y)$, there exist some CFG $\mathcal{G} = (N, A, R)$ and some CFPQ $\mathbf{q}(?x, ?y)$ such that for every RDF graph G, we have $[\![(?x, e, ?y)]\!]_G = [\![\mathbf{q}(?x, ?y)]\!]_G$.*

2. $nre_0(N)$ in CCFPQ. Let $\mathcal{G}$ be a CFG. A *CCFPQ* $\mathbf{q}(?x, ?y)$ is in *nested norm form* if the following holds:

$$\mathbf{q}(?x, ?y) := ((?x', ?y', ?z') \wedge v(?x, ?y)) \wedge \mathbf{q}_1(?u, ?w), \tag{6}$$

where

- $\{?x, ?y\} \cap \{?x', ?y', ?z'\} \neq \emptyset$;
- $\{?x', ?y', ?z'\} \cap \{?u, ?w\} \neq \emptyset$;
- $\mathbf{q}_1(?u, ?w)$ is a CCFPQ.

Note that $(?x', ?y', ?z')$ is used to express a nested nre of the form $axis::[e]$ and $v(?x, ?y)$ is necessary to express a nested nre of the form $self::[e]$.

The following proposition shows that CCFPQ can express $nre_0(N)$-triples.

Proposition 5. *For every $nre_0(N)$-triple $(?x, e, ?y)$, there exist a CFG $\mathcal{G} = (N, A, R)$ and a CCFPQ $\mathbf{q}(?x, ?y)$ in nested norm form (6) such that for every RDF graph G, we have $[\![(?x, e, ?y)]\!]_G = [\![\mathbf{q}(?x, ?y)]\!]_G$.*

3. $nre_0(|)$ in $UCCFPQ^s$. Let e be an nre. We say e is in *union norm form* if e is of the following form $e_1|e_2|\ldots|e_m$ where e_i is an $nre_0(N)$ for all $i = 1,\ldots,m$.

We can conclude that each nre-triple is equivalent to an nre in union norm form.

Proposition 6. *For every nre-triple $(?x, e, ?y)$, there exists some e' in union norm form such that $[\![(?x, e, ?y)]\!]_G = [\![(?x, e', ?y)]\!]_G$ for every RDF graph G.*

The following proposition shows that $UCCFPQ^s$ can express $nre_0(|)$.

Proposition 7. *For every $nre_0(|)$-triple $(?x, e, ?y)$, there exists some CFG $\mathcal{G} = (N, A, R)$ and some $UCCFPQ^s$ $\mathbf{q}(?x, ?y)$ in nested norm form such that for every RDF graph G, we have $[\![(?x, e, ?y)]\!]_G = [\![\mathbf{q}(?x, ?y)]\!]_G$.*

4. nre in UCCFPQ. By Propositions 5 and 7, we can conclude that

Proposition 8. *For every nre-triple $(?x, e, ?y)$, there exists some CFG $\mathcal{G} = (N, A, R)$ and some $UCCFPQ$ $\mathbf{q}(?x, ?y)$ in nested norm form such that for every RDF graph G, we have $[\![(?x, e, ?y)]\!]_G = [\![\mathbf{q}(?x, ?y)]\!]_G$.*

However, those results above in this subsection are not vice versa since the context-free language is not expressible in any nre.

Proposition 9. *CFPQ is not expressible in any nre.*

5   Context-Free SPARQL

In this section, we introduce an extension language *context-free SPARQL* (for short, *cfSPARQL*) of SPARQL by using context-free triple patterns, plus SPARQL basic operators UNION, AND, OPT, FILTER, and SELECT and its expressiveness.

A *context-free triple pattern* (cftp) is of the form $(?x, \mathbf{q}, ?y)$ where $\mathbf{q}(?x, ?y)$ is a CFPQ. Analogously, we can define *union of conjunctive context-free triple pattern* (for short, *uccftp*) by using UCCFPQ.

cfSPARQL and Query Evaluation. Formally, cfSPARQL (graph) patterns are then recursively constructed from context-free triple patterns:

- A cftp is a cfSPARQL pattern;
- A triple pattern of the form $(?x, ?y, ?z)$ is a cfSPARQL pattern;
- All P_1 UNION P_2, P_1 AND P_2, and P_1 OPT P_2 are cfSPARQL patterns if P_1, P_2 are cfSPARQL patterns;
- P FILTER C if P is a cfSPARQL pattern and C is a constraint;
- $SELECT_S(P)$ if P is a cfSPARQL pattern and S is a set of variables.

Remark 2. In cfSPARQL, we allow triple patterns of form $(?x, ?y, ?z)$ (see Item 2), which can express any SPARQL triple pattern together with FILTER [38], to ensure that SPARQL is still expressible in cfSPARQL while SPARQL is not expressible in nSPARQL since any triple pattern $(?x, ?y, ?z)$ is not expressible in nSPARQL [28]. Our generalization of nSPARQL inherits the power of queries without more cost and maintains the coherence between CFPQ and "nested" nre of the form $axis::[e]$. Moreover, this extension in cfSPARQL coincides with our proposed CCFPQ where triple patterns of the form $(?x, ?y, ?z)$ are allowed.

Semantically, let P be a cfSPARQL pattern and G an RDF graph, $[\![(?x, \mathbf{q}, ?y)]\!]_G$ is defined as $[\![\mathbf{q}(?x, ?y)]\!]_G$ and other expressive cfSPARQL patterns are defined as normal [27,28].

Proposition 10. *SPARQL is expressible in cfSPARQL but not vice versa.*

A *cfSPARQL query* is a pattern.

We can define *union of conjunctive context-free SPARQL query* (for short, *uccfSPARQL*) by using uccftp in the analogous way.

At the end of this subsection, we discuss the complexity of evaluation problem of uccfSPARQL queries.

For a given RDF graph G, a uccftp P, and a mapping μ, the query evaluation problem is to decide whether μ is in $[\![P]\!]_G$.

Proposition 11. *The evaluation problem of uccfSPARQL queries has the same complexity as the evaluation problem of SPARQL queries.*

As a direct result of Proposition 8, we can conclude

Corollary 1. *nSPARQL is expressible in uccfSPARQL but not vice versa.*

On the Expressiveness of cfSPARQL. In this subsection, we show that cfSPARQL has the same expressiveness as uccfSPARQL. In other words, cfSPARQL is enough to express UCCFPQ on RDF graphs.

Since every cfSPARQL pattern is a uccfSPARQL pattern, we merely show that uccfSPARQL is expressible in cfSPARQL.

Proposition 12. *For every uccfSPARQL pattern P, there exists some cfSPARQL pattern Q such that $[\![P]\!]_G = [\![Q]\!]_G$ for any RDF graph G.*

6 Relations on (Nested) Regular Expressions with Negation

In this section, we discuss both the relation between UCCFPQ and nested regular expressions with negation and the relation between cfSPARQL and variants of nSPARQL.

Nested Regular Expressions with Negation. *A nested regular expression with negation ($nre^{\neg}$) is an extension of nre by adding two new operators "difference ($e_1 - e_2$)" and "negation (e^c)"* [37].

Semantically, let e, e_1, e_2 be three $nre^{\neg}$s and G an RDF graph,

- $[\![e_1 - e_2]\!]_G = \{(a, b) \in [\![e_1]\!]_G | (a, b) \notin [\![e_2]\!]_G\}$;
- $[\![e^c]\!]_G = \{(a, b) \in adom(G) \times adom(G) | (a, b) \notin [\![e]\!]_G\}$.

Analogously, an $nre^{\neg}$-triple pattern is of the form $(?x, e, ?y)$ where e is an $nre^{\neg}$. Clearly, $nre^{\neg}$-triple pattern is non-monotone.

Since nre is monotone, nre is strictly subsumed in $nre^{\neg}$ [37]. Though property paths in SPARQL 1.1 [29,33] are not expressible in nre since property paths allow the negation of IRIs, property paths can be still expressible in the following subfragment of nre$^\neg$: let $c, c_1, \ldots, c_{n+m} \in U$,

$$e := next{::}c|e/e|self{::}[e]|e^*|e^+|next^{-1}{::}[e]|$$
$$(next{::}c_1|\ldots|next{::}c_n|next^{-1}{::}c_{n+1}|\ldots|next^{-1}{::}c_{n+m})^c.$$

Note that e^+ can be expressible as the expression $e^* - self$.

Proposition 13. *uccftp is not expressible in any nre$^\neg$-triple pattern.*

Due to the non-monotonicity of nre$^\neg$, we have that nre$^\neg$ is beyond the expressiveness of any union of conjunctive context-free triple patterns even the star-free nre$^\neg$ (for short, sf-nre$^\neg$) where the Kleene star ($*$) is not allowed in nre$^\neg$.

Proposition 14. *sf-nre$^\neg$-triple pattern is not expressible in any uccftp.*

In short, nre$^\neg$-triple pattern and uccftp cannot express each other. Indeed, negation could make the evaluation problem hard even allowing a limited form of negation such as property paths [23].

cfSPARQL Can Express nSPARQL$^\neg$. Following nSPARQL, we can analogously construct the language nSPARQL$^\neg$ which is built on nre$^\neg$, by adding SPARQL operators UNION, AND, OPT, FILTER, and SELECT.

Though uccftps cannot express nre$^\neg$-triple patterns by Proposition 13, cfSPARQL can express nSPARQL$^\neg$ since nSPARQL$^\neg$ is still expressible in nSPARQL [37].

Corollary 2. *nSPARQL$^\neg$ is expressible in cfSPARQL.*

6.1 Overview

Finally, Figs. 2 and 3 provide the implication of the results on RDF graphs for the general relations between variants of CFPQ and nre and the general relations between cfSPARQL and nSPARQL where $\mathcal{L}_1 \to \mathcal{L}_2$ denotes that $\mathcal{L}_1$ is expressible in $\mathcal{L}_2$ and $\mathcal{L}_1 \leftrightarrow \mathcal{L}_2$ denotes that $\mathcal{L}_1 \to \mathcal{L}_2$ and $\mathcal{L}_2 \to \mathcal{L}_1$. Analogously, nSPARQL^{sf} is an extension of SPARQL by allowing star-free nre$^\neg$-triple patterns.

7 Implementation and Evaluation

In this section, we have implemented the two algorithms for CFPQs without any optimization. Two context-free path queries over RDF graphs were evaluated and we found some results which cannot be captured by any regular expression-based path queries from RDF graphs.

The experiments were performed under Windows 7 on a Intel i5-760, 2.80GHz CPU system with 6GB memory. The program was written in Java 7 with maximum 2GB heap space allocated for JVM. Ten popular ontologies like *foaf*, *wine*, and *pizza* were used for testing.

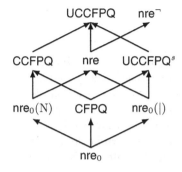

Fig. 2. Known relations between variants of CFPQ and variants of nre

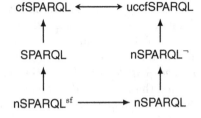

Fig. 3. Known relations between variants of cfSPARQL and variants of nSPARQL

Table 1. The evaluation results of Q_1 and Q_2

Ontology	# triples	Query 1		Query 2	
		Time (ms)	# results	Time (ms)	# results
protege	41	468	509	5	0
funding	144	499	296	125	77
skos	254	1044	810	16	1
foaf	454	5027	1929	1154	324
generation	319	6091	2164	13	0
univ-bench	306	20981	2540	532	228
travel	327	13971	2499	281	151
people+pets	703	82081	9472	247	120
biomedical-measure-primitive	459	420604	15156	1068851	9178
atom-primitive	561	515285	15454	4711499	13940
pizza	1980	3233587	56195	255853	4694
wine	2012	4075319	66572	273	79

Query 1. Consider a CFG $\mathcal{G}_1 = (N, A, R)$ where $N = \{S\}, A = \{next^{-1}{::}subClassOf, next{::}subClassOf, next^{-1}{::}type, next{::}type\}$, and $R = \{S \rightarrow (next^{-1}{::}subClassOf)\ S\ (next{::}subClassOf), S \rightarrow (next^{-1}{::}type)\ S\ (next{::}type), S \rightarrow \varepsilon\}$. The query Q_1 based on the grammar $\mathcal{G}_1$ can return all pairs of concepts or individuals at the same layer of the hierarchy of RDF graphs. Table 1 shows the experimental results of Q_1 over the testing ontologies. Note that #*results* denotes that number of pairs of concepts or individuals corresponding to Q_1.

Taking the ontology *foaf*, for example, the query Q_1 over *foaf* returns pairs of concepts like (*foaf:Document, foaf:Person*), which shows that the two concepts, *Document* and *Person*, are at the same layer of the hierarchy of *foaf*, where the top concept (*owl:Thing*) is at the first layer.

Query 2. Similarly, consider a CFG $\mathcal{G}_2 = (N, A, R)$ where $N = \{S, B\}$, $A = \{next^{-1}::subClassOf, next::subClassOf\}$, and $R = \{S \rightarrow BS, B \rightarrow (next::subClassOf) B (next^{-1}::subClassOf), B \rightarrow B(next^{-1}::subClassOf) B \rightarrow (next::subClassOf)(next^{-1}::subClassOf), S \rightarrow \varepsilon\}$. The query Q_2 based on the grammar $\mathcal{G}_2$ can return all pairs of concepts which are at adjacent two layers of the hierarchy of RDF graphs. We also take the ontology *foaf*, for example, the query Q_2 over *foaf* returns pairs of concepts like (*foaf:Person, foaf:PersonalProfileDocument*), which denotes that *Person* is at higher layer than *PersonalProfileDocument*, since *PersonalProfileDocument* is a subclass of *Document*. Table 1 shows the experimental results of Q_2 over the testing ontologies.

8   Conclusions

In this paper, we have proposed context-free path queries (including some variants) to navigate through an RDF graph and the context-free SPARQL query language for RDF built on context-free path queries by adding the standard SPARQL operators. Some investigation about some fundamental properties of those context-free path queries and their context-free SPARQL query languages has been presented. We proved that CFPQ, CCFPQ, UCCFPQ^s, and UCCFPQ strictly express basic nested regular expression (nre_0), $nre_0(N)$, $nre_0(|)$, and nre, respectively. Moreover, uccfSPARQL has the same expressiveness as cfSPARQL; and both SPARQL and nSPARQL are expressible in cfSPARQL. Furthermore, we looked at the relationship between context-free path queries and nested regular expressions with negation (which can express property paths in SPARQL1.1) and the relationship between cfSPARQL queries and nSPARQL queries with negation (nSPARQL$^{\neg}$). We found that neither CFPQ nor $nre^{\neg}$ can express each other while nSPARQL$^{\neg}$ is still expressible in cfSPARQL. Finally, we discussed the query evaluation problem of CFPQ and cfSPARQL on RDF graphs. The query evaluation of UCCFPQ maintains the polynomial time data complexity and NP-complete combined complexity the same as conjunctive first-order queries and the query evaluation of cfSPARQL maintains the complexity as the same as SPARQL. These results provide a starting point for further research on expressiveness of navigational languages for RDF graphs and the relationships among regular path queries, nested regular path queries, and context-free path queries on RDF graphs.

There are a number of practical open problems. In this paper, we restrict that RDF data does not contain *blank nodes* as the same treatment in nSPARQL. We have to admit that blank nodes do make RDF data more expressive since a blank node in RDF is taken as an existentially quantified variable [17]. An interesting future work is to extend our proposed (U)(C)CFPQ for general RDF data with blank nodes by allowing path variables which are already valid in some extensions of SPARQL such as SPARQ2L [7], SPARQLeR [19], PSPARQL [5], and CPSPARQL [3,4], which are popular in querying general RDF data with blank nodes.

Acknowledgments. The authors thank Jelle Hellings and Guohui Xiao for their help-ful and constructive comments. This work is supported by the program of the National Key Research and Development Program of China (2016YFB1000603) and the National Natural Science Foundation of China (NSFC) (61502336, 61373035). Xiaowang Zhang is supported by Tianjin Thousand Young Talents Program.

References

1. Abiteboul, S., Buneman, P., Suciu, D.: Data on the Web: From Relations to Semi-structured Data and XML. Morgan Kaufmann, Burlington (2000)
2. Abiteboul, S., Hull, R., Vianu, V.: Foundations of Databases. Addison-Wesley, Boston (1995)
3. Alkhateeb, F., Baget, J.-F., Euzenat, J.: Constrained regular expressions in SPARQL. In: Proceedings of SWWS 2008, pp. 91–99 (2008)
4. Alkhateeb, F., Euzenat, J.: Constrained regular expressions for answering RDF-path queries modulo RDFS. Int. J. Web Inf. Syst. **10**(1), 24–50 (2014)
5. Alkhateeb, F., Baget, J.-F., Euzenat, J.: Extending SPARQL with regular expres-sion patterns (for querying RDF). J. Web Semant. **7**(2), 57–73 (2009)
6. Angles, R., Gutierrez, C.: Survey of graph database models. ACM Comput. Surv. **40**(1), 1 (2008)
7. Anyanwu, K., Maduko, A., Sheth, A.P.: SPARQ2L: towards support for subgraph extraction queries in RDF databases. In: Proceedings of WWW 2007, pp. 797–806 (2007)
8. Arenas, M., Pérez, J., Gutierrez, C.: On the semantics of SPARQL. In: De Virgilio, R., Giunchiglia, F., Tanca, L. (eds.) Semantic Web Information Management - A Model-Based Perspective, pp. 281–307. Springer, Berlin (2009)
9. Barceló, P.: Querying graph databases. In: Proceedings of PODS 2013, pp. 175–188 (2013)
10. Bischof, S., Martin, C., Polleres, A., Schneider, P.: Collecting, integrating, enrich-ing and republishing open city data as linked data. In: Arenas, M., et al. (eds.) ISWC 2015. LNCS, vol. 9367, pp. 57–75. Springer, Heidelberg (2015). doi:10.1007/978-3-319-25010-6_4
11. Fionda, V., Pirrò, G., Consens, M.P.: Extended property paths: writing more SPARQL queries in a succinct way. In: Proceedings of AAAI 2015, pp. 102–108 (2015)
12. Fletcher, G.H.L., Gyssens, M., Leinders, D., Surinx, D., den Bussche, J.V., Gucht, D.V., Vansummeren, S., Wu, Y.: Relative expressive power of navigational querying on graphs. Inf. Sci. **298**, 390–406 (2015)
13. Hayes, J., Gutierrez, C.: Bipartite graphs as intermediate model for RDF. In: McIlraith, S.A., Plexousakis, D., van Harmelen, F. (eds.) ISWC 2004. LNCS, vol. 3298, pp. 47–61. Springer, Heidelberg (2004)
14. Hellings, J.: Conjunctive context-free path queries. In: Proceedings of ICDT 2014, pp. 119–130 (2014)
15. Hellings, J., Fletcher, G.H.L., Haverkort, H.J.: Efficient external-memory bisimu-lation on DAGs. In: Proceedings of SIGMOD 2012, pp. 553–564 (2012)
16. Hellings, J., Kuijpers, B., Van den Bussche, J., Zhang, X.: Walk logic as a frame-work for path query languages on graph databases. In: Proceedings of ICDT 2013, pp. 117–128 (2013)
17. Hogan, A., Arenas, M., Mallea, A., Polleres, A.: Everything you always wanted to know about blank nodes. J. Web Semant. **27**, 42–69 (2014)

18. Hopcroft, J., Ullman, J.: Introduction to Automata Theory, Languages, and Computation. Addison-Wesley, Boston (1979)
19. Kochut, K.J., Janik, M.: SPARQLeR: extended SPARQL for semantic association discovery. In: Franconi, E., Kifer, M., May, W. (eds.) ESWC 2007. LNCS, vol. 4519, pp. 145–159. Springer, Heidelberg (2007)
20. Lange, M.: Model checking propositional dynamic logic with all extras. J. Appl. Log. **4**(1), 39–49 (2006)
21. Libkin, L., Reutter, J.L., Vrgoc, D.: Trial for RDF: adapting graph query languages for RDF data. In Proceedings of PODS 2013, pp. 201–212 (2013)
22. Linz, P.: An Introduction to Formal Languages and Automata, 5th edn. Jones & Bartlett Publishers, Burlington (2012)
23. Losemann, K., Martens, W.: The complexity of regular expressions and property paths in SPARQL. ACM Trans. Database Syst. **38**(4), 24 (2013)
24. Reutter, J.L., Romero, M., Vardi, M.Y.: Regular queries on graph databases. In: Proceedings of ICDT 2015, pp. 177–194 (2015)
25. Marx, M., de Rijke, M.: Semantic characterizations of navigational XPath. SIGMOD Rec. **34**(2), 41–46 (2005)
26. Olson, M., Ogbuij, U.: The Versa Specification, October 2001
27. Pérez, J., Arenas, M., Gutierrez, C.: Semantics and complexity of SPARQL. ACM Trans. Database Syst. **34**(3), 16 (2009)
28. Pérez, J., Arenas, M., Gutierrez, C.: nSPARQL: a navigational language for RDF. J. Web Semant. **8**(4), 255–270 (2010)
29. Polleres, A., Wallner, J.P.: On the relation between SPARQL1.1 and answer set programming. J. Appl. Non-Class. Log. **23**(1–2), 159–212 (2013)
30. RDF primer. W3C Recommendation, Febraury 2004
31. Sevon, P., Eronen, L.: Subgraph queries by context-free grammars. J. Integr. Bioinform. **5**(2), 100 (2008)
32. SPARQL query language for RDF. W3C Recommendation, January 2008
33. SPARQL 1.1 query language. W3C Recommendation, March 2013
34. Tian, Y., Hankins, R.A., Patel, J.M.: Efficient aggregation for graph summarization. In: Proceedings of SIGMOD 2008, pp. 567–580 (2008)
35. Kostylev, E.V., Reutter, J.L., Romero, M., Vrgoč, D.: SPARQL with property paths. In: Arenas, M., et al. (eds.) ISWC 2015. LNCS, vol. 9366, pp. 3–18. Springer, Berlin (2015)
36. Zhang, X., Feng, Z., Wang, X., Rao, G., Wu, W.: Context-free path queries on RDF graphs. Revised version (2016). arXiv:1506.00743
37. Zhang, X., den Bussche, J.V.: On the power of SPARQL in expressing navigational queries. Comput. J. **58**(11), 2841–2851 (2015)
38. Zhang, X., den Bussche, J.V., Picalausa, F.: On the satisfiability problem for SPARQL patterns. J. Artif. Intel. Res. **56**, 403–428 (2016)

Unsupervised Entity Resolution
on Multi-type Graphs

Linhong Zhu[✉], Majid Ghasemi-Gol, Pedro Szekely, Aram Galstyan,
and Craig A. Knoblock

Information Sciences Institute, University of Southern California,
Marina Del Rey, USA
{linhong,ghasemig,pszekely,galstyan,knoblock}@isi.edu

Abstract. Entity resolution is the task of identifying all mentions that
represent the same real-world entity within a knowledge base or across
multiple knowledge bases. We address the problem of performing entity
resolution on RDF graphs containing multiple types of nodes, using the
links between instances of different types to improve the accuracy. For
example, in a graph of products and manufacturers the goal is to resolve
all the products and all the manufacturers. We formulate this problem
as a multi-type graph summarization problem, which involves clustering
the nodes in each type that refer to the same entity into one super node
and creating weighted links among super nodes that summarize the inter-
cluster links in the original graph. Experiments show that the proposed
approach outperforms several state-of-the-art generic entity resolution
approaches, especially in data sets with missing values and one-to-many,
many-to-many relations.

1 Introduction

The increasing number of entities created online raises the problem of integrating
and relating entities from different sources. In this work, we focus on the entity
resolution problem. It is a common challenge in various domains including digital libraries, E-commerce, natural language understanding, etc. For example, in
digital libraries, a challenging problem is to automatically group references that
refer to the same publication and disambiguate author names, venues, etc. In
E-commerce, a difficult problem is to match products from one domain (e.g.,
Amazon) to another domain (e.g., eBay).

Consider the example in Fig. 1, where we have five products from different
sellers represented by RDF. The entity resolution task is to group vertices of
the same product entity (e.g., 1 and 2) and vertices of the same manufacturer
entity (e.g., Bose and Bose Electronic) together and relate product entities with
manufacture entities.

There are several challenges in tacking entity resolution tasks. The first challenge is due to the poor quality of data, such as different spellings (cancel and
cancelling), missing values (e.g., missing price for product 1) and ambiguity
(e.g., the title of product 1 "Apple Noise Cancel Headphones" actually means

© Springer International Publishing AG 2016
P. Groth et al. (Eds.): ISWC 2016, Part I, LNCS 9981, pp. 649–667, 2016.
DOI: 10.1007/978-3-319-46523-4_39

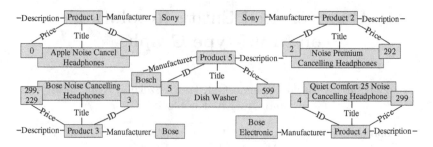

Fig. 1. An example of RDF graph for five products, where values of description field are omitted

that the headphones are suitable for apple products, but not manufactured by Apple). This makes traditional pair-wise distance measures approaches [9, 25, 30] less effective with noisy content and context (see related work). The second challenge is due to the one-to-many and many-to-many relation between entities. For instance, in the product entity resolution example, a product might be associated with many prices (normal or discount), and each manufacturer is associated with many products. The heterogeneous nature of relationships brings in an additional challenge when performing collective entity resolution [5, 10, 12]: to determine which kind of relationship is best suited for resolving a particular type of entity.

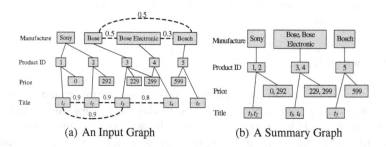

(a) An Input Graph   (b) A Summary Graph

Fig. 2. An example of multi-type graph representation for Fig. 1 and its corresponding summary graph. The description type of vertices is omitted.

To address the aforementioned challenges, we model the observed RDF graph as a multi-type graph and formulate the collective entity resolution as a multi-type graph summarization problem. Particularly, the goal is to transform the original k-type graph into another k-type *summary graph* composed of *super nodes* and *super edges*. Each super node is a cluster of original vertices (of the same type) representing a latent entity, while super edges encode potentially valuable relations between those entities. As shown in Fig. 2(a), we model the observed RDF graph as a multi-type graph, where vertices represent different types of objects, and edges represent either co-occurrence between two-types of vertices (solid edge), or similarity between the same-type vertices (dashed edge). The dashed similarity edge can be computed by any similarity measures such as graph proximity based similarity (e.g., number of common neighbors) or content-based similarity (e.g., string similarity). An example of a summary graph is shown in Fig. 2(b), where each super node is a cluster of original vertices representing a hidden entity and each super edge relates one type entity (e.g., product 1, 2) to another type entity (e.g., Sony). The weights of the super edges also indicate which type of information is more useful when resolving certain types of entities. For instance, for product 1 and 2 disambiguation, manufacturer is more reliable than price, while for disambiguating product 3 and 4, price is a more reliable indicator than manufacturer.

In this work, we thus propose a unified, multi-type graph co-summarization based entity resolution framework (**CoSum**), which (1) jointly condenses a set of similar vertices in the observation into a super node in the summary graph so that each super node (hidden entity) is coherent; (2) reveals how entities of different types are related with each other. Our main contributions are summarized as follows:

1. A novel formulation for the entity resolution problem, where we model the observed relations between different types of mentions as a multi-type graph and reduce the entity resolution to a graph summarization problem.
2. A multi-type graph co-summarization based generic entity resolution framework, which supports determining how many entities are discussed, entity disambiguation and entity relation discovery simultaneously.
3. A generic entity resolution framework that supports different user-supplied similarity measures as inputs. Those similarity measures can be of any general form and are not restricted to simple distance-based metrics.

We validate the proposed approach on real-world networks from both an E-commerce and a citation domain. The results show that the proposed approach outperforms other state-of-the-art approaches.

2   Related Work

Entity resolution has been extensively studied under different names such as record linkage [2,7,30], reference reconciliation [12], coreference resolution [23,29]. In the following, we review a set of representative traditional entity resolution approaches and collective entity resolution approaches; while we refer to tutorials [13] and surveys [6,8,36] for more throughout reviews.

Distance-based entity resolution approaches focus on learning a pairwise distance metric between entities, and then either set a distance threshold or build a pairwise classifier to determine which entities are merged. The entire process can be unsupervised [9,25,30], or supervised [29], or a hybrid of these two [7,15]. Limes [30] and Silk [15] are two representative entity resolution systems that focus on a pair of records at a time, and decide whether they are the same or not according to acceptance metrics and thresholds. Unfortunately, pairwise-based decision is very sensitive to noise and cannot capture the dependency between two pair-wise decisions.

To address the limitation of pairwise distance-based resolution decision, recently collective entity resolution has been extensively studied. This work can be categorized into three types. First, traditional collective entity resolution focuses on capturing the dependence among the same-type entities. For example, Pasula et al. [31] proposed the Relational Probabilistic Model for capturing the dependence among multiple coreference decisions. Conditional random fields (CRFs) [21] have been successfully applied to the entity resolution domain [26] and is one of the most popular approaches in generic entity resolution. On another hand, Singla and Domingos [34] proposed a well-founded, integrated solution to the entity-resolution problem based on Markov logic. Bhattacharya and Getoor [4] proposed a novel relational clustering algorithm that uses both attribute and relational information between the same-type entities for determining the underlying entities.

With heterogeneous data becoming more widespread, two additional types of collective entity resolution have emerged: (1) Collective resolution for entities with different types [5]. For instance, an extended LDA model was used in [5] to perform entity resolution for authors and publications simultaneously; (2) Collective resolution for entities with the same type from different domains. For example, Dong et al. [12] models a pair of mentions or attributes from two different domains as a node and then applies a label propagation algorithm to perform collective entity resolution. Cudré-Mauroux et al. [10] adopt the factor-graph model to perform collective entity resolution for personal profiles. In this work, we propose a multi-type graph model for collective entity resolution, which supports the three aforementioned different types of collective entity resolutions in the same generic framework.

There is another direction of work that focused on methods to scale up entity resolution algorithms, such as using indexing [8] or blocking techniques [17,18,35] to facilitate pairwise similarity computation. A representative example is the Serf system [3], which developed strategies that minimize the number of invocations to these potentially expensive black-box entity resolution algorithms. Our framework is very generic, and any indexing/blocking technique can be seamlessly embedded into it.

Our work is related but less relevant to named-entity relation extraction, tagging [28,32] and entity linking [16]. This work aims to extract named entities from a corpus and find the relation between entities deploying a fixed or universal schema, and implicitly do entity resolution along with extraction and tagging. However, our work focuses on resolving the same entities in a structured or semi-structured dataset, possibly extracted from different sources. In the outputted summary graph, our approach relates one type of entities to another type with weighted edges, but it does not support tagging the edge with a schema type such as "is-produced-by" between a product entity and a manufacturer entity.

3   Problem Definition

3.1   Notations

Let $G = (\cup_{t=1}^{k} V_t, \cup_{t=1}^{k} \cup_{t'=t+1}^{k} E_{tt'})$ be a k-type graph where each V_t denote a set of vertices of type t, and each $E_{tt'}$ denote the set of edges connecting two different types of vertices. Note that $E_{tt'}$ can be empty if none of the t-type vertices is connected to t'-type vertices. In addition, we also allow connections between vertices with the same type by introducing the similarity function sim. Each $\text{sim}_t(x, y) \geq 0$ evaluates the similarity between two t-type vertices x and y.

Given an input k-type graph G, a summary graph $\mathcal{S}(G) = (\cup_{t=1}^{k} S_t, \cup_{t=1}^{k} C_t, \cup_{t'>t}^{k} L_{tt'})$ is another k-type graph that consists of:

- k sets of super nodes $\{S_1, \cdots, S_k\}$, where each super node $s \in S_t$ ($t = 1$ to k) denotes a cluster of t-type vertices in the original graph,
- $\binom{k}{2}$ sets of super links $L_{tt'}$ where each weighted edge $(s_t, s_{t'})$ denotes the expected probability that a t-type super node s_t is connected with a t'-type super node $s_{t'}$,
- k sets of "zoom-in" mappings $\{C_1, \cdots, C_k\}$ such that each C_t denotes probabilistic mapping between t-type vertices V_t and super nodes S_t.

Table 1. Notations and explanations

Notations	Explanations
n, m, p, q, k	Number of vertices, edges, super nodes, super links, types
V, E, S, L	The set of vertices, edges, super nodes, super links
$E_{tt'}$	Coreference links between t-type and t'-type vertices in the original graph
$L_{tt'}$	Super links between t-type and t'-type super nodes
$C \in R^{n \times p}$	The mapping between vertices and super nodes
sim	The similarity function between the same-type vertices
$C(x)$	The xth row of a matrix C
$d(x), CN(x, y)$	The degree of vertex x, the common neighbors of vertex x and y
$J(x), \mathcal{J}(x)$	Objective function, Lagrangian function of x
$\circ$	Element-wise multiplicative operator

Note that we use terms vertex and edge to refer to node and edge in original graph and terms super node and super link to refer to node and edge in summary graph. For simplicity, we use V to denote the set of vertices, E to denote the set of edges in original graph G, S to denote the set of super nodes and L to denote the set of super links in summary graph $\mathcal{S}(G)$. The total number of vertices and edges in G are denoted as n and m, and the total number of super nodes and super links in $\mathcal{S}(G)$ are denoted as p and q. We use symbols with subscript t to denote notations that are related to type t. A summary of all the notations and explanations are presented in Table 1.

3.2   Problem Formulation

As explained in Sect. 1, our goal is to reduce the entity resolution problem to a graph summarization problem, where the nodes representing different mentions of the same hidden entity are summarized, or condensed, into the same super node. There are numerous ways to summarize a graph depending on specific objectives. We now provide some intuition about what constitutes a good summarization in the context of the entity resolution task. In particular, we postulate that the super nodes in our summary graph need to be coherent, in the sense that the nodes comprising a given super node should be similar to each other. The rationale behind this assumption is that different mentions of the same hidden entity needs to share some similarities, otherwise the problem becomes infeasible. Furthermore, we differentiate between *inherent similarity*, as described by the content of those nodes themselves (e.g., string similarity between their labels), and *structural similarity*, which reflects similar connectivity patterns in the multi-type graph.

To accommodate for the first type of similarity, we define the following optimization problem:

$$\arg\min_{\mathcal{S}(G)} \sum_t \sum_{x,y \in V_t} \text{sim}_t(x,y) \|C_t(x) - C_t(y)\|_F^2 \tag{1}$$

This objective function ensures that any summary graph in which two highly similar vertices (x, y) are not mapped to the same super node, incurs a penalty. The intuition behind this term is illustrated in the example in Fig. 2. If the titles t_1 and t_2 are very similar, then it is very likely that t_1 and t_2 will be condensed into the same super node.

To accommodate for structural similarity, we note that if two t-type vertices are connected to the same t'-type vertex (or a set of t'-type vertices representing the same entity), it is likely that those two vertices are referring to the same entity as well. For instance, as shown in Fig. 2, since both record 1 and record 2 are connected to the manufacturer "Sony" (and their connected titles/descriptions are very similar), it is likely that the records 1 and 2 are about the same product. Based on this intuition, we define the following optimization criterion [1]:

$$\arg\min_{\mathcal{S}(G)} \sum_t \sum_{t'>t} \|G_{tt'} - C_t L_{tt'} C_{t'}{}^T\|_F^2 \tag{2}$$

Next, we combine Eq. (1) with Eq. (2) and formulate the following optimization problem:

Problem 1. Given an input k-type graph G and the similarity function sim_t for each vertex type t, find a summary graph $\mathcal{S}(G)$ for G that minimizes the following objective:

---

[1] Note that since the input graphs we focused are undirected, we save the half computation by assuming that types of vertices are ordered and restricting edges from a precedent type t to t'.

$$J(\mathcal{S}(G)) = \sum_t \sum_{x,y \in V_t} \mathtt{sim}_t(x,y) \|C_t(x) - C_t(y)\|_F^2 + \sum_t \sum_{t'>t} \|G_{tt'} - C_t L_{tt'} {C_{t'}}^T\|_F^2$$

(3)

It is worthwhile to note that while both terms in Eq. 3 tend to produce more coherent super nodes, there are also certain important differences. Namely, the first term becomes irrelevant if two nodes are very dissimilar ($\mathtt{sim}_t(x,y) \approx 0$), whereas the second term will tend to assign structurally dissimilar nodes to different super nodes. Furthermore, the second term favors a larger number of super nodes, whereas the first term tends to condense similar nodes as much as possible. These differences introduce some non-trivial tradeoffs in the optimization process, which allow us to arrive at good summary graphs.

4   Methodology

4.1   Solution Overview

In this section, we introduce our solution to Problem 1. The overview of our solution is as follows (as well as outlined in Algorithm 1): Start with a random summary graph (Line 1), we first search for an improved summary graph with fewer super nodes, by crossing out one or many super nodes (Sect. 4.3). The second step is to fix the number of super nodes $[p_1, \cdots, p_k]$, and compute the vertex-to-clustering mapping C and super links L (Lines 4–10). These two procedures are performed alternately, until they reach a locally optimal summary graph (Lines 2–11).

Algorithm 1. The graph summarization framework for k-partite graphs

---

Input: A k-type Graph G
Output: A k-type summary graph $\mathcal{S}(G)$
01: Initialize a random k-type summary graph, with number of super nodes $[n_1, \cdots, n_k]$
02: **repeat**
          /∗ vertex allocation optimization (Section 4.3)∗/
03:     $\mathcal{S}(G)$=Search(G, $\mathcal{S}(G)$) (see Alg. 2)
          /∗ fix the number of super nodes, and optimize super nodes assignment (Section 4.2)∗/
04:     **do**
05:         **for** each t-type vertices
06:             update C_t with Eq. (4)
07:         **for** each non-empty edge set between t- and t'-type vertices
08:             update $L_{tt'}$ with Eq. (5)
09:     **while** C and L converge
10:     construct the new summary graph $\mathcal{S}(G)$
11: **until** $J(\mathcal{S}(G))$ converges
12: **return** $\mathcal{S}(G)$

4.2   Graph Summarization with Given Super Nodes

We first study the summarization algorithms with a simplified condition, in which we assume that the number of super nodes in the summary graph

$([p_1, \cdots, p_t])$ is given. With this assumption, we show that the vertex to super nodes mapping C and the connections among super nodes L can be computed with a standard multiplicative update rule [22]. The intuition of the multiplicative rule is that whenever the solution is smaller than the local optimum, it multiplies with a larger value; otherwise, it multiplies with a smaller value.

Lemma 1. *With a non-negative initialization of $C_t \in R^{n_t \times p_t}$, C_t can be iteratively improved via the following update rule:*

$$C_t = C_t \circ \sqrt{\frac{\sum_{t'>t} G_{tt'} C_{t'} L_{tt'}{}^T + sim_t C_t}{\sum_{t'>t} C_t E_{tt'} C_{t'}{}^T C_{t'} E_{tt'}{}^T + D_t C_t}} \tag{4}$$

where D_t is the diagonal weighted degree matrix of the similarity matrix sim_t, and $\circ$ (/) is the element-wise multiplicative (division) operator.

Proof (sketch). The update rule can be derived following the similar proof procedure proposed by Ding et al. [11] and Zhu et al. [37]. For each C_t, we introduce the Lagrangian multiplier Λ for non-negative constraint (i.e., $C_t \geq 0$) in Eq. (3), which leads to the following Lagrangian function $\mathcal{J}(C_t)$:

$$\mathcal{J}(C_t) = \sum_{t'>t} \|G_{tt'} - C_t L_{tt'} C_{t'}{}^T\|_F^2 + \sum_{x,y \in V_t} sim_t(x,y) \|C_t(x) - C_t(y)\|_F^2) + \mathbf{tr}(\Lambda_{C_t} C_t^T)$$

The next step is to optimize the above terms w.r.t. C_t. We set the deviation of $\mathcal{J}(C_t)$ to zero ($\nabla_{C_t} \mathcal{J}(C_t) = 0$), and obtain:

$$\Lambda_{C_t} = -2 \left(\sum_{t'>t} G_{tt'} C_{t'} L_{tt'}{}^T + sim_t C_t \right) + 2 \left(\sum_{t'>t} C_t E_{tt'} C_{t'}{}^T C_{t'} E_{tt'}{}^T + D_t C_t \right)$$

Using the KKT condition $\Lambda_{C_t} \circ C_t = 0$ [20], we obtain:

$$\left[-2 \left(\sum_{t'>t} G_{tt'} C_{t'} L_{tt'}{}^T + sim_t C_t \right) + 2 \left(\sum_{t'>t} C_t E_{tt'} C_{t'}{}^T C_{t'} E_{tt'}{}^T + D_t C_t \right) \right] \circ C_t = 0$$

Since C_t is non-negative, we show that when the solution converges, the above equation is identical to the fixed point condition of following term:

$$\left[-2 \left(\sum_{t'>t} G_{tt'} C_{t'} L_{tt'}{}^T + sim_t C_t \right) + 2 \left(\sum_{t'>t} C_t E_{tt'} C_{t'}{}^T C_{t'} E_{tt'}{}^T + D_t C_t \right) \right] \circ C_t^2 = 0$$

That is, either an entry of C_t or the corresponding entry of the left factor is zero. We thus have:

$$C_t = C_t \circ \sqrt{\frac{\sum_{t'>t} G_{tt'} C_{t'} L_{tt'}{}^T + \texttt{sim}_t C_t}{\sum_{t'>t} C_t E_{tt'} C_{t'}{}^T C_{t'} E_{tt'}{}^T + D_t C_t}}$$

This completes the proof.   □

Note that in Eq. (4), when we compute the vertex to super nodes mapping C_t for t-type vertices, we utilize their connections to all the other t'-type vertices (i.e., $G_{tt'}$) and the vertex to super nodes mapping for all the other t'-type vertices (i.e., C_t').

Similarly, the connections among super nodes $L_{tt'} \in R^{p_t \times p_{t'}}$ can be computed via the following Lemma:

Lemma 2. *The solution of $L_{tt'}$ can be approximated via the following multiplicative update rule:*

$$L_{tt'} = L_{tt'} \circ \sqrt{\frac{C_t{}^T G_{tt'} C_{t'}}{C_t^T C_t L_{tt'} C_{t'}{}^T C_{t'}}} \tag{5}$$

Proof (sketch). The proof is omitted since it is similar to that of Lemma 1.

To develop some intuition about the above solution, let us again consider the example in Fig. 2. Assume that the product IDs 3 and 4 share many discriminative words in their respective descriptions. After the first iteration of the algorithm, this evidence will be captured by Lemma 1 and those nodes will be grouped together in $C_{Product}$ mapping. After this step, using Lemma 2, the links between the new super-node and other-type nodes will be updated. The updated links show that "*Bose*" and "*Bose Electronic*" nodes in the manufacturer type have a common neighbor in the product type (share the same product). This evidence, along with the similarity link between those two nodes, will be captured by Lemma 1, so that those two nodes will be clustered together.   □

4.3   Searching for the Optimal Number of Super Nodes

We have discussed the proposed algorithm that computes the "best-effort" summary graph and mapping between the original graph and the summary graph with the assumption that the number of super nodes in the summary graph is known in advance. However, a remaining challenge is to determine the actual number of entities (super nodes). A possible approach is to enumerate all the combinations of numbers of super nodes for each type of vertices and then pick the "best" one with an exhaustive search. Unfortunately, such trial-and-error procedures can be inefficient in practice. In the following, we propose a greedy local search algorithm that can automatically determine the number of super nodes for each type of vertices. The intuition of our approach is to utilize a backward search procedure: starting with an initialization of a summary graph, where each type of vertices is assigned to a maximum number of clusters, it repeatedly removes one or many super nodes from a summary graph with the lowest information. The details of the above procedure are presented in Algorithm 2, where $\texttt{Info}(s)$ denotes the information of a super node s.

Algorithm 2. $\texttt{Search}(G,\ \mathcal{S}(G),\ p)$

Input: A k-type Graph G, a summary graph $\mathcal{S}(G)$
Output: A refined summary graph $\mathcal{S}_{new}(G)$
01: **for** each t-type super nodes and vertices
02:    $\theta = \min_{s \in S_t} \sum_{v \in V_t} C_t(v,s)$
03:    **for** each $s \in S_t$
04:       $\texttt{Info}(s) = \sum_{v \in V_t} C_t(v,s)$
05:       **if** $\texttt{Info}(s) == \theta$ and $(J(\mathcal{S}(G)) - J(\mathcal{S} \setminus \{s\}(G))) > 0$
06:          delete s from $\mathcal{S}(G)$
07: **return** $\mathcal{S}(G)$

Note that our algorithm differs from the traditional bottom-up merging or top-down split algorithm. Bottom-up merging iteratively picks two clusters such that merging of these two cluster leads to improved performance. Therefore, at each iteration, it requires to search over all cluster pairs, which is computationally very expensive (p^2 in a naïve implementation and $p \log p$ with a heap implementation). In contrast, in our search algorithm, we only have to decide whether a super node will be removed (lines 3–6), which results in a time-complexity that is linear in the number of super nodes p. For the top-down split algorithm, although the computational cost of searching for the best cluster to be split is linear, the algorithm requires sophisticated heuristics to perform a split, which entails reassigning each vertex from one cluster to one of two smaller clusters. In our algorithm, on the other hand, the vertices within a removed super node can be merged into the remaining super nodes through the procedure presented in Sect. 4.2.

4.4   Complexity Analysis

In this section, we analyze the time complexity of our proposed graph summarization algorithm. The time complexity for each basic operation is summarized in Table 2. In addition, for the Algorithm 2, the computational cost is dominated by the computation of the objective function in Line 7. Fortunately, instead of computing the objective function, we are only required to compute the change in the objective function after removing a super node. The differences (i.e., $J(\mathcal{S}(G)) - J(\mathcal{S} \setminus \{s\}(G)))$) can be computed in time that is linear in the number of nodes. Therefore, the time complexity of $\texttt{Search}(G,\ \mathcal{S}(G),\ p)$ is $O(\sum_t p_t n_t)$.

With the above analysis, the overall time complexity of Algorithm 1 is $O(r_o r_i [\sum_t \sum_{t' > t} n_t n_{t'} (p_t + p_{t'}) + \sum_t (n_t)^2 p_t])$ for dense matrices and

Table 2. The time complexity for each basic operator with both dense and sparse matrices representation. Here $(nz)_t$ is number of non-zero entries in the matrix sim_t

	Dense	Sparse
C_t	$O(n_t^2 p_t + n_t \sum_{t'>t}(n_t p_t + p_t p_t))$	$O((nz)_t p_t + \sum_{t'>t}(m_{tt'} + n_t q_{tt'}))$
$L_{tt'}$	$O(n_t p_t^2 + p_t(n_t n_{t'} + p_{t'} n_{t'} + p_t p_{t'}))$	$O(n_t p_t^2 + p_t(m_{tt'} + q_{tt'} + p_t n_{t'}))$

$O(r_o r_i[\sum_t \sum_{t'>t} m_{tt'} + q_{tt'}(n_t + n_{t'})])$ for sparse matrices, where r_i/r_o is number of iterations within inner loops (Line 4–10)/outer loops (Line 2–11). Both r_i and r_o are small in practice, which are around 20–200.

5 Experiments

5.1 Dataset and Comparable Methods

Data. We use two datasets from different domains: Product [19], and Citeseer [4]. Product consists of product entities from two online retailers *Amazon.com* and *Google Products*. Each record has attributes ID, title, description, manufacturer and price. The RDF schema of Product data is shown in Fig. 1. Note that we only use a flat schema to model the Product data because we cannot retrieve many-to-many relations (e.g., many-to-many relations between products and manufacturers) due to the fact that only the product field has a unique identifier. Based on the schema, we create an input multi-type graph that consists of two types of vertices: product and word. Each product is connected to a word that appears in the title, manufacturer, and description. In addition, we also provide product to product similarity and word to word similarity. The available ground truth presents product equivalences but not manufacturer equivalences.

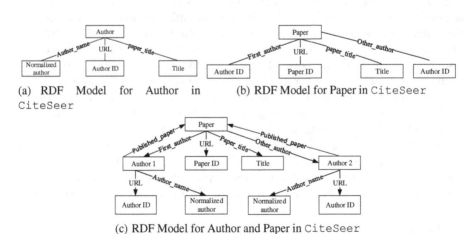

(a) RDF Model for Author in CiteSeer

(b) RDF Model for Paper in CiteSeer

(c) RDF Model for Author and Paper in CiteSeer

Fig. 3. RDF Model Schema of Citeseer Data.

In the `Citeseer` data set, each publication has a title and multiple authors. We modeled the `Citeseer` data in two ways, with a multi-object schema preserving author-to-paper, paper-to-author, and author-to-author relations (see Fig. 3(c)), and with a flat schema only preserving author-to-paper and paper-to-author relations (see Fig. 3(a) and (b)). Compared to the flat schema, the former model is more informative in terms that it supports accessing the list of co-authors in a paper for each author. However, some entity resolution systems need the data to be in CSV/XML format, and the latter flat model is more suitable for such systems. Based on the flat schema, we create a multi-type graph that consists of four types of vertices: normalized name, author ID, paper ID, and word. Each author ID is connected to its normalized name and its related paper ID; while each paper ID is connected to words from its title and authors. In addition, we also provide the author to author similarity and paper to paper similarity. The ground truths are whether two paper IDs refer to the same publication and whether two author IDs refer to the same author entity.

The statistics of two multi-type graphs are summarized in Table 3.

Table 3. The statistics of graphs

Data	#types	# records	# nodes	# edges	# entities	Full input mapping
Citeseer	4	2892	8591	17521	author:1165, paper:899	8.4 Million
Product	2	4589	12397	41165	product:1104	4.4 Million

Comparable approaches. We compare our approach (**CoSum**) with representative state-of-the-art unsupervised entity resolution systems **Limes** [30], **Silk** [15], and **Serf** [3]. For `Product` data, we also report the best performance achieved by all the entity resolution approaches and unsupervised entity resolution approaches reported in the original paper that provide the data [19]. For `Citeseer` data, we report the best performance achieved by the collective entity resolution method [4]. Moreover, Limes and Silk support reading data from a RDF store, which takes advantage of the graph representation and therefore more complicated data models. Thus, for Limes/Silk, we use **Limes-F/Silk-F** to denote running Limes/Silk using flat models (e.g., Fig. 3(a) and (b)), and **Limes-MO/Silk-MO** using multi-object models (e.g., Fig. 3(c)).

Note that various graph summarization techniques have been proposed in terms of other purposes such as compressing minimum description length [24, 27, 33]. We also compare our approach to one representative minimum-description-length-based graph summarization approach **GSum** [33] in terms of entity resolution task.

Evaluation metrics. We evaluate the entity resolution quality using the usual measures: precision, recall, and F-measure. We also report the running time comparison of different approaches, although the comparison is unfair since they are implemented in different languages C++ (GSum), Java (Serf, Limes and Silk), and Matlab (CoSum). All the experiments are conducted on a single machine, with a 4-core 2.7GHZ CPU and 16 GB memory.

5.2 Configuration

Limes and Silk require a configuration file, describing the input/output format, as well as the acceptance metric and thresholds which determine whether a pair of records are the same or not. In Serf, the user is required to implement a decision-maker function that receives a pair of records and returns a true/false decision. For all the three systems, we need to determine which attribute/field to choose, their best similarity metrics, and how important their roles are in the acceptance decision. In our experiments, we tried our best to choose the best fitted metric functions based on what each system offers and the characteristic of data. Tables 4 and 5 illustrate a summary of the acceptance metrics for different systems on Product and Citeseer domains respectively. The details are described as follows.

Table 4. Configurations of different systems on Product data

	Name (N)	Price (P)	Description (D)	Manufacturer (M)	Acceptance Metric
Limes	Trigrams	Normalized difference	Cosine	-	N > 0.6 AND P > 0.5 AND D > 0.5
Silk			Trigrams	Jaro	$20N + 10M + 5P + D$ [2]
Serf			Jaccard+4-grams		N > 0.6 AND P > 0.5 AND D > 0.5

[2] In Silk, when choosing the weighted average score aggregation, the user just introduces rejection thresholds and weights for each attribute

We first select the set of attributes according to different systems. For instance, in Limes, the user first introduces all the attributes he wants to use for record comparison, and Limes requires all the specified attributes to be available in both records in order to compare them. As a result, we had to ignore the manufacturer name in Product domain, since more than 90 % of the records in Google product dataset do not have the manufacturer name. The configuration in Silk is very similar to Limes, except that Silk allows the user to specify which attributes are not required for record-pair comparison and can be ignored if their value is missing. Therefore, we still use the manufacturer attribute in Silk. With the selected attributes and the details reported in the original work [4,19] that provide these two benchmark datasets, we have tried combinations of various string and set similarity measures that are available in the systems (including Levenshtein, Jaro, N-grams, and Jaccard) as metrics. Finally, we perform multi-level grid search for optimal weights of attributes and the threshold. For instance, we search for the best acceptance threshold for Limes by a top-level grid search between [0, 1] with step size 0.2, following the bottom-level grid search with step size 0.01. Therefore, our manual configuration performs better than the active learning method within Silk because the learning method is based on a genetic algorithm (ActiveGenLink [14]).

Table 5. Configurations of different systems on `Citeseer` data

	Papers			Author		Acceptance Metric	
	Title (T)	First-author (F)	Authors (A)	Name (N)	Co-authors (C)	Papers	Authors
Limes-F	Trigrams	Jaro-Winkler	Jaccard	Jaro-Winkler	-	0.5T+0.4F+0.1A >0.75	N>0.85
Limes-MO					Jaccard		0.8N+0.2C>0.85
Silk-F			soft Jaccard		-	20T+20F+A	N>0.85
Silk-MO					soft Jaccard		6N+C
Serf					-	0.5T+0.4F+0.1A >0.75	N>0.85

For graph summarization approaches, the configuration is much easier. We do not need any acceptance metric since the decision is automatically given by the summary graph. In addition, if no domain knowledge is available, we could simply compute the similarity between the same-type vertices using graph proximity measures. In the experiments, CoSum-B denotes that the similarity between t-type vertices are computed using the weighted common neighbor approach proposed by [1]. That is, for each $x, y \in V_t$,

$$\mathrm{sim}_t(x, y) = \sum_{z \in CN(x,y)} \frac{1}{\log d(z)} \qquad (6)$$

where $CN(x, y)$ is the set of common neighbors shared by vertices x and y in the given k-type graph, and $d(z)$ denotes the weighted degree of vertex z. CoSum-P denotes that we use the string similarity metrics between the same-type vertices configured in Tables 4 and 5.

5.3   Quality Comparisons

In this section, we evaluate the performance of proposed approach in terms of precision, recall and F-measure for entity resolution tasks.

Question 1. *F-measure: How does CoSum perform compared to the state-of-the-art entity resolution systems?*

> **Result 1** *CoSum outperforms several existing state-of-the-art generic entity resolution systems (including Limes, Silk, Serf) in terms of F-measure. In addition, the quality is comparable to the best performance reported in the literature on both* `Citeseer` *and* `Product`*.*

Question 2. *Algorithm: How does the proposed CoSum perform compared to other graph summarization algorithms?*

> **Result 2** *The poor quality achieved by GSum shows that minimum-description-based graph summarization algorithms may not work well for the entity-resolution task. On the contrary, the significant improvement achieved by CoSum compared to Gsum verified the advantage of the proposed graph summarization algorithm.*

Table 6. Quality comparisons of different approaches

	Precision			Recall			F-measure		
	Author	Paper	Product	Author	Paper	Product	Author	Paper	Product
Limes-F	0.958	0.827	0.446	0.864	0.761	0.160	0.909	0.792	0.236
Silk-F	0.846	0.877	0.459	0.986	0.756	0.348	0.910	0.812	0.395
Gsum	0.727	0.668	0.01	0.569	0.624	0.587	0.638	0.645	0.02
CoSum-B	0.993	0.871	0.58	0.940	0.611	0.477	*0.966*	*0.718*	*0.524*
Limes-MO	0.912	0.827	0.446	0.944	0.761	0.160	0.928	0.792	0.236
Silk-MO	0.932	**0.877**	0.459	0.958	0.756	0.348	0.945	0.812	0.395
Serf	0.985	0.837	0.436	0.687	0.808	0.186	0.809	0.822	0.261
CoSum-P	**0.999**	0.771	**0.639**	**0.997**	**0.997**	**0.695**	*0.998*	*0.87*	*0.666*
Best in Literature	NA	NA	0.615 [19]	NA	NA	0.63 [19]	0.995 [4]	NA	0.622 [19]

Question 3. *Modeling: What's the effect of modeling on state-of-the-art entity resolution systems?*

> **Result 3** *As shown in Table 6, both Limes and Silk are very sensitive to modeling. For publication entity resolution performance, Limes-MO and Silk-MO perform much better than Limes-F and Silk-F by using the multi-object modeling that captures the co-authorship information.*

Question 4. *Similarity: What's the effect of similarity measures to the proposed CoSum approach?*

> **Result 4** *Compared to CoSum-B, CoSum-P achieves much better performance in disambiguating publication and product entities. This is because for publication, title information is a dominant feature and thus calculating paper to paper similarity using additional title trigram similarity improves the performance. Similarly, in* `Product` *data, using additional word-to-word similarity calculated with Jaro-Winkler captures correlation between noisy texts (e.g., Apple and Apple®)*

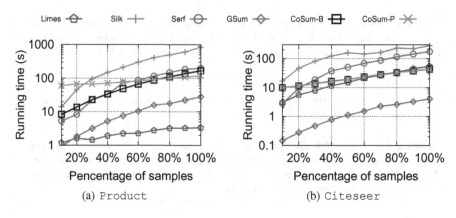

Fig. 4. Running time comparisons for different approaches.

5.4   Efficiency Comparisons

In this section, we evaluate the scalability of the proposed approach. Unfortunately, examining the total running time only is unfair since the compared approaches are implemented in different languages: C++ (GSum), Java (Limes, Silk and Serf) and Matlab (CoSum). Therefore, we report how running time varies with the size of data to evaluate the scalability.

Question 5. *Scalability with Sample Size: How does the proposed CoSum scale compared to other approaches?*

> **Result 5** *Among all the approaches, Limes is the most efficient and highly scalable with sample size. In terms of total running time, GSum is the most efficient because it is implemented in C++. However, we observed that the run-time for GSum, Serf and Silk scales super-linearly with the sample size, while for CoSum and Limes it scales almost linearly.*

Question 6. *CoSum-B Versus CoSum-P: How does the proposed CoSum scale compared to other approaches?*

> **Result 6** *Compared to CoSum-B, CoSum-P requires more I/O time on* Product *data because the word to word string similarity is much denser than weighted common neighbor similarity. However, the word to word string similarity is very informative and thus it helps convergence. Therefore, on product data, when the sample size is small, the I/O time is dominant for CoSum-P and thus CoSum-P is slower than CoSum-B. When the sample size becomes larger, the CPU time is dominant and thus CoSum-P is more efficient than CoSum-B (faster convergence). On* Citeseer *data, both paper-to-paper string similarity and paper-to-paper weighted common neighbor similarity are sparse. Therefore, the running time of CoSum-B and CoSum-P are very close.*

6   Conclusion

In this work, we proposed a multi-graph co-summarization-based method that simultaneously identifies entities and their connections. This framework is very generic, and does not require any domain-specific knowledge such as RDF modeling or tuning the pairwise similarity threshold. We applied the proposed approach to real multi-type graphs from different domains and obtained good results in terms of F-measure for entity-resolution tasks. The proposed method has some limitations. First, the quality of entity-resolution solution depends on the quality of the user-supplied same-type vertex similarity. We plan to extend the current method by adaptively refining the same-type vertex similarity with a small number of training samples. Second, if the same-type vertex similarity matrices and the observed graphs are very dense, the proposed algorithm is not scalable. In the future, we will improve the efficiency bottleneck by embedding the blocking techniques with the graph summarization algorithm. Finally, we plan to apply the graph summarization algorithm to the entity linking tasks, to evaluate the quality of super links in summary graphs.

Acknowledgments. This research is supported by the Defense Advanced Research Projects Agency (DARPA) and the Air Force Research Laboratory (AFRL) under contract number FA8750-14-C-0240. The views and conclusions contained herein are those of the authors and should not be interpreted as necessarily representing the official policies or endorsements, either expressed or implied, of DARPA, AFRL, or the U.S. Government.

References

1. Adamic, L.A., Adar, E.: Friends and neighbors on the web. Soc. Netw. **25**(3), 211–230 (2003)
2. Arasu, A., Götz, M., Kaushik, R.: On active learning of record matching packages. In: Proceedings of the ACM SIGMOD International Conference on Management of Data, pp. 783–794 (2010)
3. Benjelloun, O., Garcia-Molina, H., Menestrina, D., Su, Q., Whang, S.E., Widom, J.: Swoosh: a generic approach to entity resolution. VLDB J. **18**(1), 255–276 (2009)
4. Bhattacharya, I., Getoor, L.: Collective entity resolution in relational data. ACM Trans. Knowl. Disc. Data **1**(1), Article no. 5 (2007)
5. Bhattacharya, I., Getoor, L.: A latent dirichlet model for unsupervised entity resolution. In: SIAM International Conference on Data Mining (2007)
6. Brizan, D.G., Tansel, A.U.: A survey of entity resolution and record linkage methodologies. Commun. IIMA **6**(3), 5 (2015)
7. Christen, P.: Automatic record linkage using seeded nearest neighbour and support vector machine classification. In: Proceedings of the ACM SIGKDD International Conference on Knowledge Discovery and Data Mining, pp. 151–159 (2008)
8. Christen, P.: A survey of indexing techniques for scalable record linkage and deduplication. IEEE Trans. Knowl. Data Eng. **24**(9), 1537–1555 (2012)
9. Cohen, W.W., Richman, J.: Learning to match and cluster large high-dimensional data sets for data integration. In: Proceedings of the ACM SIGKDD International Conference on Knowledge Discovery and Data Mining. pp. 475–480 (2002)
10. Cudré-Mauroux, P., Haghani, P., Jost, M., Aberer, K., De Meer, H.: idMesh: graph-based disambiguation of linked data. In: Proceedings of the 18th International Conference on World Wide Web, pp. 591–600 (2009)

11. Ding, C., Li, T., Peng, W., Park, H.: Orthogonal nonnegative matrix t-factorizations for clustering. In: Proceedings of the ACM SIGKDD International Conference on Knowledge Discovery and Data Mining, pp. 126–135 (2006)
12. Dong, X., Halevy, A., Madhavan, J.: Reference reconciliation in complex information spaces. In: Proceedings of the ACM SIGMOD International Conference on Management of Data, pp. 85–96 (2005)
13. Getoor, L., Machanavajjhala, A.: Entity resolution: theory, practice & open challenges. Proc. VLDB Endow. **5**(12), 2018–2019 (2012)
14. Isele, R., Bizer, C.: Active learning of expressive linkage rules using genetic programming. Web Semant. Sci., Serv. Agents World Wide Web **23**, 2–15 (2013)
15. Isele, R., Jentzsch, A., Bizer, C.: Silk server-adding missing links while consuming linked data. In: Proceedings of the First International Conference on Consuming Linked Data, vol. 665, pp. 85–96 (2010)
16. Ji, H., Nothman, J., Hachey, B.: Overview of TAC-KBP2014 entity discovery and linking tasks. In: Text Analysis Conference (2014)
17. Kejriwal, M., Miranker, D.P.: An unsupervised instance matcher for schema-free RDF data. Web Semant. Sci. Serv. Agents World Wide Web **35**, 102–123 (2015)
18. Kolb, L., Thor, A., Rahm, E.: Load balancing for map-reduce based entity resolution. In: Proceedings of the IEEE International Conference on Data Engineering, pp. 618–629 (2012)
19. Köpcke, H., Thor, A., Rahm, E.: Evaluation of entity resolution approaches on real-world match problems. Proc. VLDB Endow. **3**(1–2), 484–493 (2010)
20. Kuhn, H.W., Tucker, A.W.: Nonlinear programming. In: Proceedings of the 2nd Berkeley Symposium on Mathematical Statistics and Probability, pp. 481–492 (1950)
21. Lafferty, J.D., McCallum, A., Pereira, F.C.N.: Conditional random fields: probabilistic models for segmenting and labeling sequence data. In: Proceedings of the Eighteenth International Conference on Machine Learning, pp. 282–289 (2001)
22. Lee, D.D., Seung, H.S.: Algorithms for non-negative matrix factorization. In: Proceedings of the Annual Conference on Neural Information Processing Systems, pp. 556–562 (2000)
23. Lee, H., Chang, A., Peirsman, Y., Chambers, N., Surdeanu, M., Jurafsky, D.: Deterministic coreference resolution based on entity-centric, precision-ranked rules. Comput. Linguist. **39**(4), 885–916 (2013)
24. LeFevre, K., Terzi, E.: Grass: graph structure summarization. In: SIAM International Conference on Data Mining, pp. 454–465 (2010)
25. McCallum, A., Nigam, K., Ungar, L.H.: Efficient clustering of high-dimensional data sets with application to reference matching. In: Proceedings of the ACM SIGKDD International Conference on Knowledge Discovery and Data Mining, pp. 169–178 (2000)
26. McCallum, A., Wellner, B.: Conditional models of identity uncertainty with application to proper noun coreference. In: Proceedings of the Annual Conference on Neural Information Processing Systems (2005)
27. Navlakha, S., Rastogi, R., Shrivastava, N.: Graph summarization with bounded error. In: Proceedings of the ACM SIGMOD International Conference on Management of Data, pp. 419–432 (2008)
28. Neelakantan, A., Roth, B., McCallum, A.: Compositional Vector Space Models for Knowledge Base Completion, April 2015. http://arxiv.org/abs/org/abs/1504.06662

29. Ng, V., Cardie, C.: Improving machine learning approaches to coreference resolution. In: Proceedings of the 40th Annual Meeting on Association for Computational Linguistics, pp. 104–111 (2002)
30. Ngomo, A.C.N., Auer, S.: Limes: a time-efficient approach for large-scale link discovery on the web of data. In: Proceedings of the Twenty-Second International Joint Conference on Artificial Intelligence, pp. 2312–2317. AAAI Press (2011)
31. Pasula, H., Marthi, B., Milch, B., Russell, S.J., Shpitser, I.: Identity uncertainty and citation matching. In: Proceedings of the Annual Conference on Neural Information Processing Systems (2002)
32. Riedel, S., Yao, L., McCallum, A., Marlin, B.M.: Relation extraction with matrix factorization and universal schemas. In: HLT-NAACL (2013)
33. Riondato, M., Garcia-Soriano, D., Bonchi, F.: Graph summarization with quality guarantees. In: Proceedings of the IEEE International Conference on Data Mining, pp. 947–952 (2014)
34. Singla, P., Domingos, P.: Entity resolution with Markov logic. In: Proceedings of the IEEE International Conference on Data Mining, pp. 572–582 (2006)
35. Whang, S.E., Menestrina, D., Koutrika, G., Theobald, M., Garcia-Molina, H.: Entity resolution with iterative blocking. In: Proceedings of the ACM SIGMOD International Conference on Management of Data, pp. 219–232 (2009)
36. Winkler, W.E.: Matching and record linkage. Wiley Interdisc. Rev.: Comput. Stat. **6**(5), 313–325 (2014)
37. Zhu, L., Galstyan, A., Cheng, J., Lerman, K.: Tripartite graph clustering for dynamic sentiment analysis on social media. In: Proceedings of the ACM SIGMOD International Conference on Management of Data, pp. 1531–1542 (2014)

Author Index

Printed in the United States
By Bookmasters